KEVIN P. BUTT
520-A Brookhaven Cr.
Waterloo, Ont.
N2L-4R7
PH: 519-746-1851

OR

86 CARTIER CRESCENT
RICHMOND HILL, ONT.
L4C-2N5
PH: 416: 884-5328

GROUNDWATER

Department of Geological Sciences
University of British Columbia
Vancouver, British Columbia

John A. Cherry

Department of Earth Sciences
University of Waterloo
Waterloo, Ontario

GROUNDWATER

Prentice-Hall, Inc.
Englewood Cliffs, New Jersey 07632

Library of Congress Cataloging in Publication Data

FREEZE, R ALLAN.
Groundwater.

Bibliography: p.
Includes index.
1. Water, Underground. I. Cherry, John A., joint author. II. Title.
GB1003.2.F73 551.4′98 78-25796
ISBN 0-13-365312-9

Editorial/production supervision by Cathy Brenn/Kim McNeily
Interior design by Chris Gadekar
Manufacturing buyer: Harry Baisley
Chapter logos: Peter Russell

Printed in the United States of America

10 9 8 7 6

PRENTICE-HALL INTERNATIONAL, INC., *London*
PRENTICE-HALL OF AUSTRALIA PTY. LIMITED, *Sydney*
PRENTICE-HALL OF CANADA, LTD., *Toronto*
PRENTICE-HALL OF INDIA PRIVATE LIMITED, *New Delhi*
PRENTICE-HALL OF JAPAN, INC., *Tokyo*
PRENTICE-HALL OF SOUTHEAST ASIA PTE. LTD., *Singapore*
WHITEHALL BOOKS LIMITED, *Wellington, New Zealand*

This may well be the only book that either of us will ever write. We cannot save our dedications, as novelists do, to let them forth one by one. We recognize and appreciate the life-long influences of our parents, our wives, our families, our teachers, and our students. This book is dedicated to all of them.

This book is also dedicated to the taxpayers of Canada and the United States, few of whom will ever read it, but all of whom have contributed to its birth through scholarships in our student days and through research support and sabbatical periods in more recent years.

CONTENTS

5 Flow Nets 167

6 Groundwater and the Hydrologic Cycle 192

7 Chemical Evolution of Natural Groundwater 237

8 Groundwater Resource Evaluation 303

Appendices 525

PREFACE

We perceive a trend in the study and practice of groundwater hydrology. We see a science that is emerging from its geological roots and its early hydraulic applications into a full-fledged environmental science. We see a science that is becoming more interdisciplinary in nature and of greater importance in the affairs of man.

This book is our response to these perceived trends. We have tried to provide a text that is suited to the study of groundwater during this period of emergence. We have made a conscious attempt to integrate geology and hydrology, physics and chemistry, and science and engineering to a greater degree than has been done in the past.

This book is designed for use as a text in introductory groundwater courses of the type normally taught in the junior or senior year of undergraduate geology, geological engineering, or civil engineering curricula. It has considerably more material than can be covered in a course of one-semester duration. Our intention is to provide a broad coverage of groundwater topics in a manner that will enable course instructors to use selected chapters or chapter segments as a framework for a semester-length treatment. The remaining material can serve as a basis for a follow-up undergraduate course with more specialization or as source material for an introductory course at the graduate level. We recognize that the interdisciplinary approach may create some difficulties for students grounded only in the earth sciences, but we are convinced that the benefits of the approach far outweigh the cost of the additional effort that is required.

The study of groundwater at the introductory level requires an understanding of many of the basic principles of geology, physics, chemistry, and mathematics. This text is designed for students who have a knowledge of these subjects at the level normally covered in freshman university courses. Additional background in these subjects is, of course, desirable. Elementary calculus is used frequently in several of the chapters. Although knowledge of topics of more advanced calculus is definitely an asset to students wishing to pursue specialized groundwater topics, we hope that for students without this background this text will serve as a pathway to the understanding of the basic physical principles of groundwater flow. Differential equations have been used very sparingly, but are included where we view their use as essential. The physical meaning of the equations and their boundary conditions is held paramount. To avoid mathematical disruptions in continuity of presentation of physical concepts, detailed derivations and solution methods are restricted to the appendices.

Until recently, groundwater courses at the university level were normally viewed in terms of only the geologic and hydraulic aspects of the topic. In response to the increasing importance of natural groundwater quality and groundwater contamination by man, we have included three major chapters primarily chemical in emphasis. We assume that the reader is conversant with the usual chemical symbols and can write and balance equations for inorganic chemical reactions. On this basis, we describe the main principles of physical chemistry that are necessary for an introductory coverage of the geochemical aspects of the groundwater environment. Students wishing for a more advanced treatment of these topics would require training in thermodynamics at a level beyond the scope of this text.

Although we have attempted to provide a broad interdisciplinary coverage of groundwater principles, we have not been able to include detailed information on the technical aspects of such topics as well design and installation, operation of well pumps, groundwater sampling methods, procedures for chemical analysis of groundwater, and permeameter and consolidation tests. The principles of these practical and important techniques are discussed in the text but the operational aspects must be gleaned from the many manuals and technical papers cited throughout the text.

Acknowledgments

The manuscript for this text was reviewed in its entirety by Pat Domenico, Eugene Simpson, and Dave Stephenson. Their comments and suggestions aided us immeasurably in arriving at the final presentation. We are also indebted to Bill Back, Lee Clayton, Shlomo Neuman, Eric Reardon, and Leslie Smith, who provided valuable reviews of portions of the book. In addition, we requested and received help on individual sections from Bob Gillham, Gerry Grisak, Bill Mathews, Dave McCoy, Steve Moran, Nari Narasimhan, Frank Patton, John Pickens, Doug Piteau, Joe Poland, Dan Reynolds, and Warren Wood. In addition, we would be remiss not to recognize the vital influence of our long-time associations with Paul Witherspoon and Bob Farvolden.

We also owe a debt to the many graduate and undergraduate students in groundwater hydrology at U.B.C. and Waterloo who identified flaws in the presentation and who acted as guinea pigs on the problem sets.

R. ALLAN FREEZE
Vancouver, British Columbia

JOHN A. CHERRY
Waterloo, Ontario

GROUNDWATER

CHAPTER 1

Introduction

1.1 Groundwater, the Earth, and Man

This book is about groundwater. It is about the geological environments that control the occurrence of groundwater. It is about the physical laws that describe the flow of groundwater. It is about the chemical evolution that accompanies flow. It is also about the influence of man on the natural groundwater regime; and the influence of the natural groundwater regime on man.

The term *groundwater* is usually reserved for the subsurface water that occurs beneath the water table in soils and geologic formations that are fully saturated. We shall retain this classical definition, but we do so in full recognition that the study of groundwater must rest on an understanding of the subsurface water regime in a broader sense. Our approach will be compatible with the traditional emphasis on shallow, saturated, groundwater flow; but it will also encompass the near-surface, unsaturated, soil-moisture regime that plays such an important role in the hydrologic cycle, and it will include the much deeper, saturated regimes that have an important influence on many geologic processes.

We view the study of groundwater as interdisciplinary in nature. There is a conscious attempt in this text to integrate chemistry and physics, geology and hydrology, and science and engineering to a greater degree than has been done in the past. The study of groundwater is germane to geologists, hydrologists, soil scientists, agricultural engineers, foresters, geographers, ecologists, geotechnical engineers, mining engineers, sanitary engineers, petroleum reservoir analysts, and probably others. We hope that our introductory treatment is in tune with these broad interdisciplinary needs.

If this book had been written a decade ago, it would have dealt almost entirely with groundwater as a resource. The needs of the time would have dictated that approach, and books written in that era reflected those needs. They emphasize the development of water supplies through wells and the calculation of aquifer yields. The groundwater problems viewed as such are those that threaten that yield. The water supply aspects of groundwater are still important and they will be treated in this text with the deference they deserve. But groundwater is more than a resource. It is an important feature of the natural environment; it leads to environmental problems, and may in some cases offer a medium for environmental solutions. It is part of the hydrologic cycle, and an understanding of its role in this cycle is mandatory if integrated analyses are to be promoted in the consideration of watershed resources, and in the regional assessment of environmental contamination. In an engineering context, groundwater contributes to such geotechnical problems as slope stability and land subsidence. Groundwater is also a key to understanding a wide variety of geological processes, among them the generation of earthquakes, the migration and accumulation of petroleum, and the genesis of certain types of ore deposits, soil types, and landforms.

The first five chapters of this book lay the physical, chemical, and geologic foundations for the study of groundwater. The final six chapters apply these principles in the several spheres of interaction between groundwater, the earth, and man. The following paragraphs can be viewed as an introduction to each of the later chapters.

Groundwater and the Hydrologic Cycle

The endless circulation of water between ocean, atmosphere, and land is called the *hydrologic cycle*. Our interest centers on the land-based portion of the cycle as it might be operative on an individual watershed. Figures 1.1 and 1.2 provide two schematic diagrams of the hydrologic cycle on a watershed. They are introduced here primarily to provide the reader with a diagrammatic introduction to hydro-

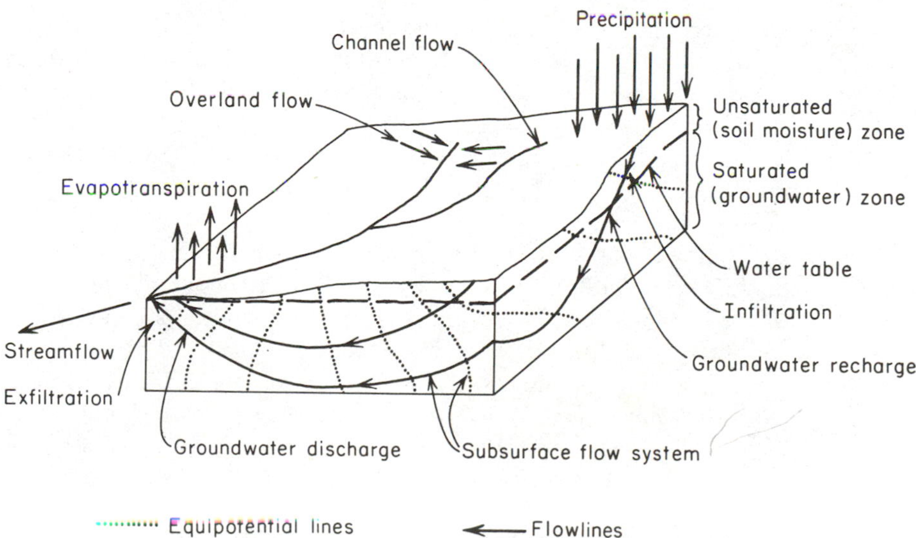

Figure 1.1 Schematic representation of the hydrologic cycle.

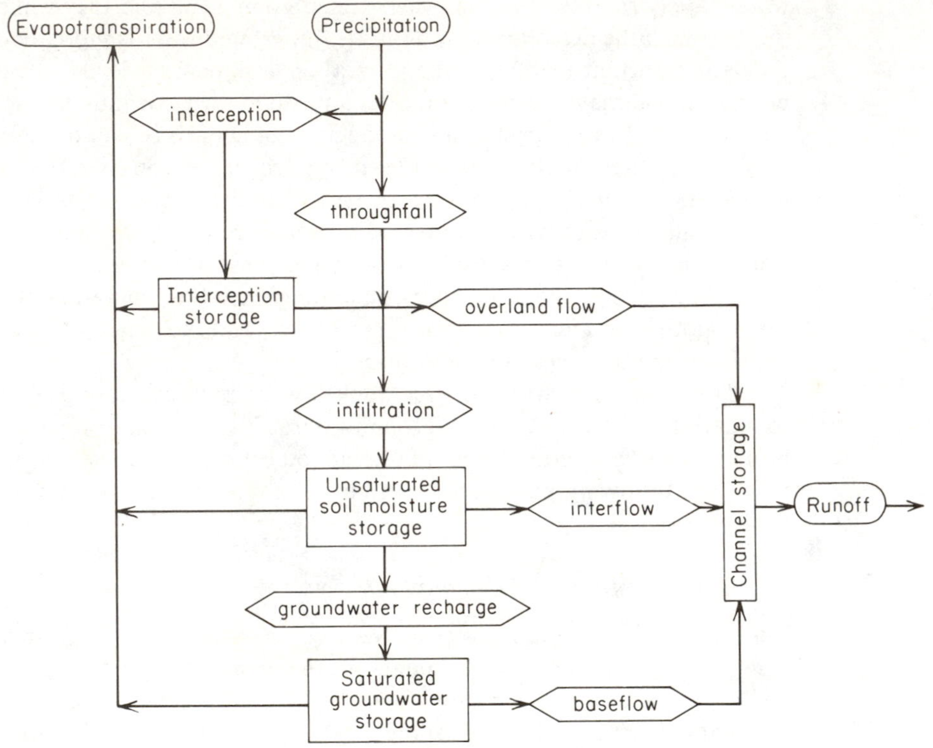

Figure 1.2 Systems representation of the hydrologic cycle.

logic terminology. Figure 1.1 is conceptually the better in that it emphasizes processes and illustrates the flow-system concept of the hydrologic cycle. The pot-and-pipeline representation of Figure 1.2 is often utilized in the systems approach to hydrologic modeling. It fails to reflect the dynamics of the situation, but it does differentiate clearly between those terms that involve rates of movement (in the hexagonal boxes) and those that involve storage (in the rectangular boxes).

Inflow to the hydrologic system arrives as *precipitation*, in the form of rainfall or snowmelt. Outflow takes place as *streamflow* (or runoff) and as *evapotranspiration*, a combination of evaporation from open bodies of water, evaporation from soil surfaces, and transpiration from the soil by plants. Precipitation is delivered to streams both on the land surface, as *overland flow* to tributary channels; and by subsurface flow routes, as *interflow* and *baseflow* following *infiltration* into the soil. Figure 1.1 makes it clear that a watershed must be envisaged as a combination of both the surface drainage area and the parcel of subsurface soils and geologic formations that underlie it. The subsurface hydrologic processes are just as important as the surface processes. In fact, one could argue that they are more important, for it is the nature of the subsurface materials that controls infiltration rates, and the infiltration rates influence the timing and spatial distribution of surface runoff. In Chapter 6, we will examine the nature of regional groundwater flow patterns in some detail, and we will investigate the relations among infiltration, groundwater recharge, groundwater discharge, baseflow, and streamflow generation. In Chapter 7, we will look at the chemical evolution of groundwater that accompanies its passage through the subsurface portion of the hydrologic cycle.

Before closing this section, it is worth looking at some data that reflect the quantitative importance of groundwater relative to the other components of the hydrologic cycle. In recent years there has been considerable attention paid to the concept of the *world water balance* (Nace, 1971; Lvovitch, 1970; Sutcliffe, 1970), and the most recent estimates of these data emphasize the ubiquitous nature of groundwater in the hydrosphere. With reference to Table 1.1, if we remove from consideration the 94% of the earth's water that rests in the oceans and seas at high levels of salinity, then groundwater accounts for about two-thirds of the freshwater resources of the world. If we limit consideration to the utilizable freshwater resources (minus the icecaps and glaciers), groundwater accounts for almost the total volume. Even if we consider only the most "active" groundwater regimes, which Lvovitch (1970) estimates at 4×10^6 km³ (rather than the 60×10^6 km³ of Table 1.1), the freshwater breakdown comes to: groundwater, 95%; lakes, swamps, reservoirs, and river channels, 3.5%; and soil moisture, 1.5%.

Table 1.1 Estimate of the Water Balance of the World

Parameter	Surface area (km²) ×10⁶	Volume (km³) ×10⁶	Volume (%)	Equivalent depth (m)*	Residence time
Oceans and seas	361	1370	94	2500	~4000 years
Lakes and reservoirs	1.55	0.13	<0.01	0.25	~10 years
Swamps	<0.1	<0.01	<0.01	0.007	1–10 years
River channels	<0.1	<0.01	<0.01	0.003	~2 weeks
Soil moisture	130	0.07	<0.01	0.13	2 weeks–1 year
Groundwater	130	60	4	120	2 weeks–10,000 years
Icecaps and glaciers	17.8	30	2	60	10–1000 years
Atmospheric water	504	0.01	<0.01	0.025	~10 days
Biospheric water	<0.1	<0.01	<0.01	0.001	~1 week

SOURCE: Nace, 1971.

*Computed as though storage were uniformly distributed over the entire surface of the earth.

This volumetric superiority, however, is tempered by the average residence times. River water has a turnover time on the order of 2 weeks. Groundwater, on the other hand, moves slowly, and residence times in the 10's, 100's, and even 1000's of years are not uncommon. The principles laid out in Chapter 2 and the regional flow considerations of Chapter 6 will clarify the hydrogeologic controls on the large-scale movement of groundwater.

Most hydrology texts contain detailed discussions of the hydrologic cycle and of the global water balance. Wisler and Brater (1959) and Linsley, Kohler, and

Paulhus (1975) are widely used introductory hydrology texts. A recent text by Eagleson (1970) updates the science at a more advanced level. The massive *Handbook of Applied Hydrology*, edited by Chow (1964a), is a valuable reference.

The history of development of hydrological thought is an interesting study. Chow (1964b) provides a concise discussion; Biswas' (1970) booklength study provides a wealth of detail, from the contributions of the early Egyptians and the Greek and Roman philosophers, right up to and through the birth of scientific hydrology in western Europe in the eighteenth and nineteenth centuries.

Groundwater as a Resource

The primary motivation for the study of groundwater has traditionally been its importance as a resource. For the United States, the significance of the role of groundwater as a component of national water use can be gleaned from the statistical studies of the U.S. Geological Survey as reported most recently for the year 1970 by Murray and Reeves (1972) and summarized by Murray (1973).

Table 1.2 documents the growth in water utilization in the United States during the period 1950–1970. In 1970 the nation used 1400×10^6 m³/day. Of this, 57% went for industrial use and 35% for irrigation. Surface water provided 81% of the total, groundwater 19%. Figure 1.3 graphically illustrates the role of groundwater relative to surface water in the four major areas of use for the 1950–1970 period. Groundwater is less important in industrial usage, but it provides a significant percentage of the supply for domestic use, both rural and urban, and for irrigation.

The data of Table 1.2 and Figure 1.3 obscure some striking regional variations. About 80% of the total irrigation use occurs in the 17 western states, whereas 84% of the industrial use is in the 31 eastern states. Groundwater is more widely used in the west, where it accounts for 46% of the public supply and 44% of the industrial use (as opposed to 29% and 16%, respectively, in the east).

Table 1.2 Water Use in the United States, 1950–1970

	Cubic meters/day × 10^6*					Percent of 1970 use
	1950	1955	1960	1965	1970	
Total water withdrawals	758	910	1023	1175	1400	100
Use						
Public supplies	53	64	80	91	102	7
Rural supplies	14	14	14	15	17	1
Irrigation	420	420	420	455	495	35
Industrial	292	420	560	667	822	57
Source						
Groundwater	130	182	190	227	262	19
Surface water	644	750	838	960	1150	81

SOURCE: Murray, 1973.

*$1 \text{ m}^3 = 10^3\ \ell = 264$ U.S. gal.

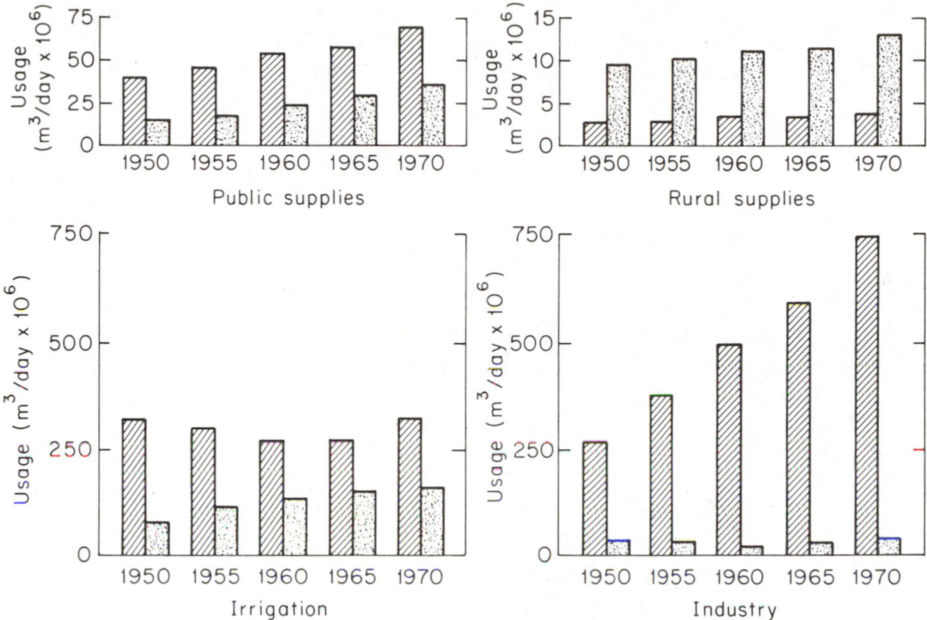

Figure 1.3 Surface water (hatched) and groundwater (stippled) use in the United States, 1950–1970 (after Murray, 1973).

In Canada, rural and municipal groundwater use was estimated by Meyboom (1968) at 1.71×10^6 m³/day, or 20% of the total rural and municipal water consumption. This level of groundwater use is considerably lower than that of the United States, even when one considers the population ratio between the two countries. A more detailed look at the figures shows that rural groundwater development in Canada is relatively on a par with rural development in the United States, but municipal groundwater use is significantly smaller. The most striking differences lie in irrigation and industrial use, where the relative total water consumption in Canada is much less than in the United States and the groundwater component of this use is extremely small.

McGuinness (1963), quoting a U.S. Senate committee study, has provided predictions of future U.S. national water requirements. It is suggested that water needs will reach 1700×10^6 m³/day by 1980 and 3360×10^6 m³/day by the year 2000. The attainment of these levels of production would represent a significant acceleration in the rate of increase in water use outlined in Table 1.2. The figure for the year 2000 begins to approach the total water resource potential of the nation, which is estimated to be about 4550×10^6 m³/day. If the requirements are to be met, it is widely accepted that groundwater resources will have to provide a greater proportion of the total supply. McGuinness notes that for the predictions above, if the percent groundwater contribution is to increase from 19% to 33%, groundwater usage would have to increase from its current 262×10^6 m³/day to 705 ×

10^6 m^3/day in 1980 and 1120×10^6 m^3/day in the year 2000. He notes that the desirable properties of groundwater, such as its clarity, bacterial purity, consistent temperature, and chemical quality, may encourage the needed large-scale development, but he warns that groundwater, especially when large quantities are sought, is inherently more difficult and expensive to locate, to evaluate, to develop, and to manage than surface water. He notes, as we have, that groundwater is an integral phase of the hydrologic cycle. The days when groundwater and surface water could be regarded as two separate resources are past. Resource planning must be carried out with the realization that groundwater and surface water have the same origin.

In Chapter 8, we will discuss the techniques of groundwater resource evaluation: from the geologic problems of aquifer exploration, through the field and laboratory methods of parameter measurement and estimation, to the simulation of well performance, aquifer yield, and basin-wide groundwater exploitation.

Groundwater Contamination

If groundwater is to continue to play an important role in the development of the world's water-resource potential, then it will have to be protected from the increasing threat of subsurface contamination. The growth of population and of industrial and agricultural production since the second world war, coupled with the resulting increased requirements for energy development, has for the first time in man's history begun to produce quantities of waste that are greater than that which the environment can easily absorb. The choice of a waste-disposal method has become a case of choosing the least objectionable course from a set of objectionable alternatives. As shown schematically on Figure 1.4, there are no currently-feasible, large-scale waste disposal methods that do not have the potential for serious pollution of some part of our natural environment. While there has been a growing concern over air- and surface-water pollution, this activism has not yet encompassed the subsurface environment. In fact, the pressures to reduce surface pollution are in part responsible for the fact that those in the waste management field are beginning to covet the subsurface environment for waste disposal. Two of the disposal techniques that are now being used and that are viewed most optimistically for the future are deep-well injection of liquid wastes and sanitary landfill for solid wastes. Both these techniques can lead to subsurface pollution. In addition, subsurface pollution can be caused by leakage from ponds and lagoons which are widely used as components of larger waste-disposal systems, and by leaching of animal wastes, fertilizers, and pesticides from agricultural soils.

In Chapter 9 we will consider the analysis of groundwater contamination. We will treat the principles and processes that allow us to analyze the general problems of municipal and industrial waste disposal, as well as some more specialized problems associated with agricultural activities, petroleum spills, mining activities, and radioactive waste. We will also discuss contamination of coastal groundwater supplies by salt-water intrusion. In all of these problems, physical considerations of groundwater flow must be coupled with the chemical properties and principles

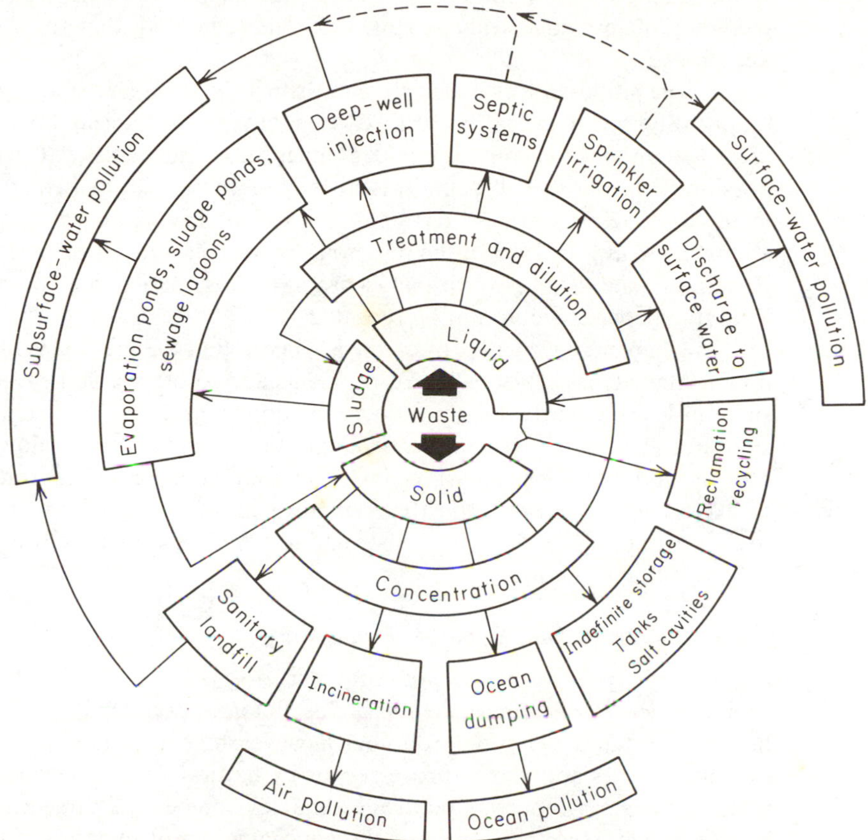

Figure 1.4 Spectrum of waste disposal alternatives.

introduced in Chapter 3; and the coupling must be carried out in light of the concepts of natural geochemical evolution discussed in Chapter 7.

Groundwater as a Geotechnical Problem

Groundwater is not always a blessing. During the construction of the San Jacinto tunnel in California, tunnel driving on this multi-million-dollar water aqueduct was held up for many months as a result of massive unexpected inflows of groundwater from a system of highly fractured fault zones.

In Mexico City during the period 1938–1970, parts of the city subsided as much as 8.5 m. Differential settlements still provide severe problems for engineering design. The primary cause of the subsidence is now recognized to be excessive groundwater withdrawals from subsurface aquifers.

At the Jerome dam in Idaho, the dam "failed," not through any structural weakness in the dam itself, but for the simple reason that the dam would not hold water. The groundwater flow systems set up in the rock formations adjacent to the reservoir provided leakage routes of such efficiency that the dam had to be abandoned.

At the proposed Revelstoke dam in British Columbia, several years of exploratory geological investigation were carried out on an ancient landslide that was identified on the reservoir bank several miles above the damsite. The fear lay in the possibility that increased groundwater pressures in the slide caused by the impoundment of the reservoir could retrigger slope instability. An event of this type took almost 2500 lives in 1963 in the infamous Vaiont reservoir disaster in Italy. At the Revelstoke site, a massive drainage program was carried out to ensure that the Vaiont experience would not be repeated.

In Chapter 10 we will explore the application of the principles of groundwater flow to these types of geotechnical problems and to others. Some of the problems, such as leakage at dams and inflows to tunnels and open pit mines, arise as a consequence of excessive rates and quantities of groundwater flow. For others, such as land subsidence and slope instability, the influence arises from the presence of excessive fluid pressures in the groundwater rather than from the rate of flow itself. In both cases, flow-net construction, which is introduced in Chapter 5, is a powerful analytic tool.

Groundwater and Geologic Processes

There are very few geologic processes that do not take place in the presence of groundwater. For example, there is a close interrelationship between groundwater flow systems and the geomorphological development of landforms, whether by fluvial processes and glacial processes, or by natural slope development. Groundwater is the most important control on the development of karst environments.

Groundwater plays a role in the concentration of certain economic mineral deposits, and in the migration and accumulation of petroleum.

Perhaps the most spectacular geologic role played by groundwater lies in the control that fluid pressures exert on the mechanisms of faulting and thrusting. One exciting recent outgrowth of this interaction is the suggestion that it may be possible to control earthquakes on active faults by manipulating the natural fluid pressures in the fault zones.

In Chapter 11, we will delve more deeply into the role of groundwater as an agent in various geologic processes.

1.2 The Scientific Foundations for the Study of Groundwater

The study of groundwater requires knowledge of many of the basic principles of geology, physics, chemistry and mathematics. For example, the flow of groundwater in the natural environment is stongly dependent on the three-dimensional configuration of geologic deposits through which flow takes place. The groundwater hydrologist or geologist must therefore have some background in the interpretation of geologic evidence, and some flair for the visualization of geologic environments. He should have training in sedimentation and stratigraphy, and an understanding of the processes that lead to the emplacement of volcanic and intrusive igneous rocks. He should be familiar with the basic concepts of structural geology and be able to recognize and predict the influence of faulting and folding on geologic systems. Of particular importance to the student of groundwater is an understanding of the nature of surficial deposits and landforms. A large proportion of groundwater flow and a significant percentage of groundwater resource development takes place in the unconsolidated surficial deposits created by fluvial, lacustrine, glacial, deltaic, and aeolian geologic processes. In the northern two-thirds of North America an understanding of the occurrence and flow of groundwater rests almost entirely on an understanding of the glacial geology of the Pleistocene deposits.

Geology provides us with a qualitative knowledge of the framework of flow, but it is physics and chemistry that provide the tools for quantitative analysis. Groundwater flow exists as a *field* just as heat and electricity do, and previous exposure to these more classic fields provides good experience for the analysis of groundwater flow. The body of laws that controls the flow of groundwater is a special case of that branch of physics known as fluid mechanics. Some understanding of the basic mechanical properties of fluids and solids, and a dexterity with their dimensions and units, will aid the student in grasping the laws of groundwater flow. Appendix I provides a review of the elements of fluid mechanics. Any reader who does not feel facile with such concepts as density, pressure, energy, work, stress, and head would be well advised to peruse the appendix before attacking Chapter 2. If a more detailed treatment of fluid mechanics is desired, Streeter (1962) and Vennard (1961) are standard texts; Albertson and Simons (1964) provide a useful short review. For the specific topic of flow through porous media, a more advanced treatment of the physics than is attempted in this text can be found in Scheidegger (1960) and Collins (1961), and especially in Bear (1972).

The analysis of the natural chemical evolution of groundwater and of the behavior of contaminants in groundwater requires use of some of the principles of inorganic and physical chemistry. These principles have long been part of the methodology of geochemists and have in recent decades come into common use in groundwater studies. Principles and techniques from the field of nuclear chemistry are now contributing to our increased understanding of the groundwater environment. Naturally-occurring stable and radioactive isotopes, for example, are being used to determine the age of water in subsurface systems.

Groundwater hydrology is a quantitative science, so it should come as no surprise to find that mathematics is its language, or at the very least one of its principal dialects. It would be almost impossible, and quite foolish, to ignore the powerful tools of the groundwater trade that rest on an understanding of mathematics. The mathematical methods upon which classical studies of groundwater flow are based were borrowed by the early researchers in the field from areas of applied mathematics originally developed for the treatment of problems of heat flow, electricity, and magnetism. With the advent of the digital computer and its widespread availability, many of the important recent advances in the analysis of groundwater systems have been based on much different mathematical approaches generally known as numerical methods. Although in this text neither the classical analytical methods nor numerical methods are pursued in any detail, our intention has been to include sufficient introductory material to illustrate some of the more important concepts.

Our text is certainly not the first to be written on groundwater. There is much material of interest in several earlier texts. Todd (1959) has for many years been the standard introductory engineering text in groundwater hydrology. Davis and De Wiest (1966) place a much heavier emphasis on the geology. For a text totally committed to the resource evaluation aspects of groundwater, there are none better than Walton (1970), and Kruseman and De Ridder (1970). A more recent text by Domenico (1972) differs from its predecessors in that it presents the basic theory in the context of hydrologic systems modeling. Among the best texts from abroad are those of Schoeller (1962), Bear, Zaslavsky, and Irmay (1968), Custodio and Llamas (1974), and the advanced Russian treatise of Polubarinova-Kochina (1962).

There are several other applied earth sciences that involve the flow of fluids through porous media. There is a close kinship between groundwater hydrology, soil physics, soil mechanics, rock mechanics, and petroleum reservoir engineering. Students of groundwater will find much of interest in textbooks from these fields such as Baver, Gardner, and Gardner (1972), Kirkham and Powers (1972), Scott (1963), Jaeger and Cook (1969), and Pirson (1958).

1.3 The Technical Foundations for the Development of Groundwater Resources

The first two sections of this chapter provide an introduction to the topics we plan to cover in this text. It is equally important that we set down what we do not intend to cover. Like most applied sciences, the study of groundwater can be broken into three broad aspects: science, engineering, and technology. This textbook places heavy emphasis on the scientific principles; it includes much in the way of engineering analysis; it is *not* in any sense a handbook on the technology.

Among the technical subjects that are not discussed in any detail are: methods of drilling; the design, construction, and maintenance of wells; and geophysical logging and sampling. All are required knowledge for the complete groundwater specialist, but all are treated well elsewhere, and all are learned best by experience rather than rote.

There are several books (Briggs and Fiedler, 1966; Gibson and Singer, 1971; Campbell and Lehr, 1973; U.S. Environmental Protection Agency, 1973a, 1976) that provide technical descriptions of the various types of water well drilling equipment. They also contain information on the design and setting of well screens, the selection and installation of pumps, and the construction and maintenance of wells.

On the subject of geophysical logging of boreholes, the standard reference in the petroleum industry, where most of the techniques arose, is Pirson (1963). Patten and Bennett (1963) discuss the various techniques with specific reference to groundwater exploration. We will give brief mention to subsurface drilling and borehole logging in Section 8.2, but the reader who wants to see examples in greater number in the context of case histories of groundwater resource evaluation is directed to Walton (1970).

There is one other aspect of groundwater that is technical, but in a different sense, that is not considered in this text. We refer to the subject of groundwater law. The development and management of groundwater resources must take place within the framework of water rights set down by existing legislation. Such legislation is generally established at the state or provincial level, and the result in North America is a patchwork quilt of varying traditions, rights, and statutes. Piper (1960) and Dewsnut et al. (1973) have assessed the situation in the United States. Thomas (1958) has drawn attention to some of the paradoxes that arise out of conflicts between hydrology and the law.

Suggested Readings

CHOW, V. T. 1964. Hydrology and its development. *Handbook of Applied Hydrology*, ed. V. T. Chow. McGraw-Hill, New York, pp. 1.1–1.22.

MCGUINNESS, C. L. 1963. The role of groundwater in the national water situation. *U.S. Geol. Surv. Water-Supply Paper 1800*.

MURRAY, C. R. 1973. Water use, consumption, and outlook in the U.S. in 1970. *J. Amer. Water Works Assoc.*, 65, pp. 302–308.

NACE, R. L., ed. 1971. Scientific framework of world water balance. *UNESCO Tech. Papers Hydrol.*, 7, 27 pp.

CHAPTER 2

Physical Properties and Principles

2.1 Darcy's Law

The birth of groundwater hydrology as a quantitative science can be traced to the year 1856. It was in that year that a French hydraulic engineer named Henry Darcy published his report on the water supply of the city of Dijon, France. In the report Darcy described a laboratory experiment that he had carried out to analyze the flow of water through sands. The results of his experiment can be generalized into the empirical law that now bears his name.

Consider an experimental apparatus like that shown in Figure 2.1. A circular cylinder of cross section A is filled with sand, stoppered at each end, and outfitted with inflow and outflow tubes and a pair of manometers. Water is introduced into the cylinder and allowed to flow through it until such time as all the pores are filled with water and the inflow rate Q is equal to the outflow rate. If we set an arbitrary datum at elevation $z = 0$, the elevations of the manometer intakes are

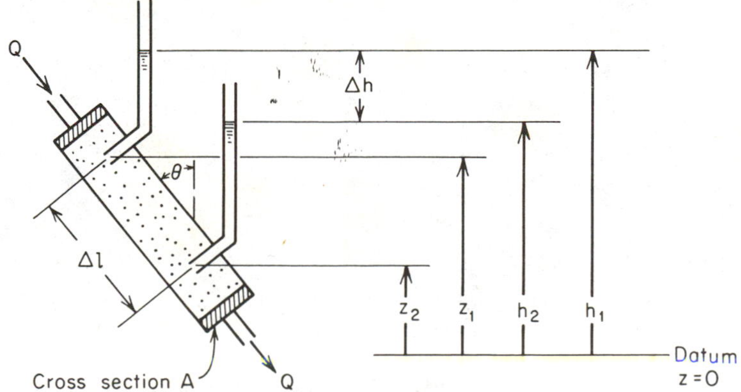

Figure 2.1 Experimental apparatus for the illustration of Darcy's law.

z_1 and z_2 and the elevations of the fluid levels are h_1 and h_2. The distance between the manometer intakes is Δl.

We will define v, the *specific discharge* through the cylinder, as

$$v = \frac{Q}{A} \tag{2.1}$$

If the dimensions of Q are $[L^3/T]$ and those of A are $[L^2]$, v has the dimensions of a velocity $[L/T]$.

The experiments carried out by Darcy showed that v is directly proportional to $h_1 - h_2$ when Δl is held constant, and inversely proportional to Δl when $h_1 - h_2$ is held constant. If we define $\Delta h = h_2 - h_1$ (an arbitrary sign convention that will stand us in good stead in later developments), we have $v \propto -\Delta h$ and $v \propto 1/\Delta l$. *Darcy's law* can now be written as

$$v = -K\frac{\Delta h}{\Delta l} \tag{2.2}$$

or, in differential form,

$$v = -K\frac{dh}{dl} \tag{2.3}$$

In Eq. (2.3), h is called the *hydraulic head* and dh/dl is the *hydraulic gradient*. K is a constant of proportionality. It must be a property of the soil in the cylinder, for were we to hold the hydraulic gradient constant, the specific discharge would surely be larger for some soils than for others. In other words, if dh/dl is held constant, $v \propto K$. The parameter K is known as the *hydraulic conductivity*. It has high values for sand and gravel and low values for clay and most rocks. Since Δh and Δl both have units of length $[L]$, a quick dimensional analysis of Eq. (2.2) shows that K has the dimensions of a velocity $[L/T]$. In Section 2.3, we will show that K is a function not only of the media, but also of the fluid flowing through it.

An alternative form of Darcy's law can be obtained by substituting Eq. (2.1) in Eq. (2.3) to yield

$$Q = -K\frac{dh}{dl}A \tag{2.4}$$

This is sometimes compacted even further into the form

$$Q = -KiA \tag{2.5}$$

where i is the hydraulic gradient.

Darcy's law is valid for groundwater flow in any direction in space. With regard to Figure 2.1 and Eq. (2.3), if the hydraulic gradient dh/dl and the hydraulic conductivity K are held constant, v is independent of the angle θ. This is true even for θ values greater than 90° when the flow is being forced up through the cylinder against gravity.

We have noted that the specific discharge v has the dimensions of a velocity, or flux. For this reason it is sometimes known as the *Darcy velocity* or *Darcy flux*. The specific discharge is a macroscopic concept and it is easily measured. It must be clearly differentiated from the microscopic velocities associated with the actual paths of individual particles of water as they wind their way through the grains of sand (Figure 2.2). The microscopic velocities are real, but they are probably impossible to measure. In the remainder of the chapter we will work exclusively with concepts of flow on a macroscopic scale. Despite its dimensions we will not refer to v as a velocity; rather we will utilize the more correct term, *specific discharge*.

Figure 2.2 Macroscopic and microscopic concepts of groundwater flow.

This last paragraph may appear innocuous, but it announces a decision of fundamental importance. When we decide to analyze groundwater flow with the Darcian approach, it means, in effect, that we are going to replace the actual ensemble of sand grains (or clay particles or rock fragments) that make up the porous medium by a representative continuum for which we can define macroscopic parameters, such as the hydraulic conductivity, and utilize macroscopic laws, such as Darcy's law, to provide macroscopically averaged descriptions of the microscopic behavior. This is a conceptually simple and logical step to take, but it rests on some knotty theoretical foundations. Bear (1972), in his advanced text on porous-media flow, discusses these foundations in detail. In Section 2.12, we will further explore the interrelationships between the microscopic and macroscopic descriptions of groundwater flow.

Darcy's law is an empirical law. It rests only on experimental evidence. Many attempts have been made to derive Darcy's law from more fundamental physical laws, and Bear (1972) also reviews these studies in some detail. The most successful approaches attempt to apply the Navier-Stokes equations, which are widely known in the study of fluid mechanics, to the flow of water through the pore channels of idealized conceptual models of porous media. Hubbert (1956) and Irmay (1958) were apparently the earliest to attempt this exercise.

This text will provide ample evidence of the fundamental importance of Darcy's law in the analysis of groundwater flow, but it is worth noting here that it is equally important in many other applications of porous-media flow. It describes the flow of soil moisture and is used by soil physicists, agricultural engineers, and soil mechanics specialists. It describes the flow of oil and gas in deep geological formations and is used by petroleum reservoir analysts. It is used in the design of filters by chemical engineers and in the design of porous ceramics by materials scientists. It has even been used by bioscientists to describe the flow of bodily fluids across porous membranes in the body.

Darcy's law is a powerful empirical law and its components deserve our more careful attention. The next two sections provide a more detailed look at the physical significance of the hydraulic head h and the hydraulic conductivity K.

2.2 Hydraulic Head and Fluid Potential

The analysis of a physical process that involves flow usually requires the recognition of a potential gradient. For example, it is known that heat flows through solids from higher temperatures toward lower and that electrical current flows through electrical circuits from higher voltages toward lower. For these processes, the temperature and the voltage are potential quantities, and the rates of flow of heat and electricity are proportional to these potential gradients. Our task is to determine the potential gradient that controls the flow of water through porous media.

Fortunately, this question has been carefully considered by Hubbert in his classical treatise on groundwater flow (Hubbert, 1940). In the first part of this section we will review his concepts and derivations.

Hubbert's Analysis of the Fluid Potential

Hubbert (1940) defines *potential* as "a physical quantity, capable of measurement at every point in a flow system, whose properties are such that flow always occurs from regions in which the quantity has higher values to those in which it has lower, regardless of the direction in space" (p. 794). In the Darcy experiment (Figure 2.1) the hydraulic head h, indicated by the water levels in the manometers, would appear to satisfy the definition, but as Hubbert points out, "to adopt it empirically without further investigation would be like reading the length of the mercury column of a thermometer without knowing that temperature was the physical quantity being indicated" (p. 795).

Two obvious possibilities for the potential quantity are elevation and fluid pressure. If the Darcy apparatus (Figure 2.1) were set up with the cylinder vertical ($\theta = 0$), flow would certainly occur down through the cylinder (from high elevation to low) in response to gravity. On the other hand, if the cylinder were placed in a horizontal position ($\theta = 90°$) so that gravity played no role, flow could presumably be induced by increasing the pressure at one end and decreasing it at the other. Individually, neither elevation nor pressure are adequate potentials, but we certainly have reason to expect them to be components of the total potential quantity.

It will come as no surprise to those who have been exposed to potential concepts in elementary physics or fluid mechanics that the best way to search out our quarry is to examine the energy relationships during the flow process. In fact, the classical definition of potential as it is usually presented by mathematicians and physicists is in terms of the work done during the flow process; and the work done in moving a unit mass of fluid between any two points in a flow system is a measure of the energy loss of the unit mass.

Fluid flow through porous media is a mechanical process. The forces driving the fluid forward must overcome the frictional forces set up between the moving fluid and the grains of the porous medium. The flow is therefore accompanied by an irreversible transformation of mechanical energy to thermal energy through the mechanism of frictional resistance. The direction of flow in space must therefore be away from regions in which the mechanical energy per unit mass of fluid is higher and toward regions in which it is lower. In that the mechanical energy per unit mass at any point in a flow system can be defined as the work required to move a unit mass of fluid from an arbitrarily chosen standard state to the point in question, it is clear that we have uncovered a physical quantity that satisfies both Hubbert's definition of a potential (in terms of the direction of flow) and the classical definition (in terms of the work done). The *fluid potential* for flow through porous media is therefore the *mechanical energy per unit mass of fluid.*

It now remains to relate this quantity to the elevation and pressure terms that we anticipated earlier. Consider an arbitrary standard state (Figure 2.3) at elevation $z = 0$ and pressure $p = p_0$, where p_0 is atmospheric. A unit mass of fluid of density ρ_0 will occupy a volume V_0, where $V_0 = 1/\rho_0$. We wish to calculate the work required to lift the unit mass of fluid from the standard state to some point P in the flow system which is at elevation z and where the fluid pressure is p. Here, a unit mass of the fluid may have density ρ and will occupy a volume $V = 1/\rho$. In addition, we will consider the fluid to have velocity $v = 0$ at the standard state and velocity v at the point P.

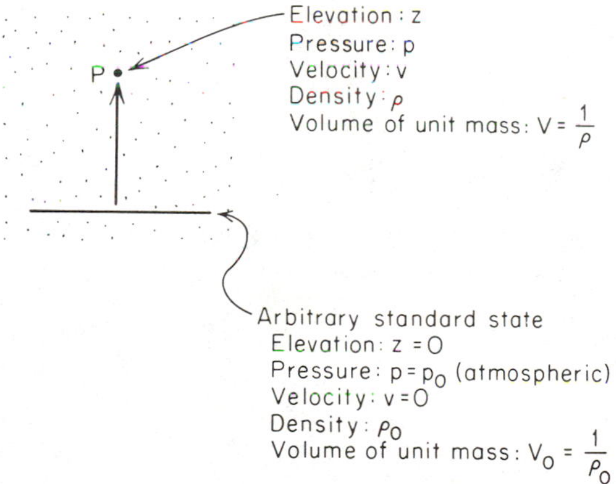

Figure 2.3 Data for calculation of mechanical energy of unit mass of fluid.

There are three components to the work calculation. First, there is the work required to lift the mass from elevation $z = 0$ to elevation z:

$$w_1 = mgz \tag{2.6}$$

Second, there is the work required to accelerate the fluid from velocity $v = 0$ to velocity v:

$$w_2 = \frac{mv^2}{2} \tag{2.7}$$

Third, there is the work done on the fluid in raising the fluid pressure from $p = p_0$ to p:

$$w_3 = m\int_{p_0}^{p} \frac{V}{m}\,dp = m\int_{p_0}^{p} \frac{dp}{\rho} \tag{2.8}$$

If the fluid were to flow from point P to a point at the standard state, Eq. (2.6) represents the loss in potential energy, Eq. (2.7) is the loss in kinetic energy, and Eq. (2.8) is the loss in elastic energy, or p–V work.

The fluid potential Φ (the mechanical energy per unit mass) is the sum of w_1, w_2, and w_3. For a unit mass of fluid, $m = 1$ in Eqs. (2.6), (2.7), and (2.8), and we have

$$\Phi = gz + \frac{v^2}{2} + \int_{p_0}^{p} \frac{dp}{\rho} \tag{2.9}$$

For porous-media flow, velocities are extremely low, so the second term can almost always be neglected. For incompressible fluids (fluids with a constant density, so that ρ is not a function of p), Eq. (2.9) can be simplified further to give

$$\Phi = gz + \frac{p - p_0}{\rho} \tag{2.10}$$

Our earlier premonitions as to the likely components of the fluid potential are now seen to be correct. The first term of Eq. (2.10) involves the elevation z and the second term involves the fluid pressure p.

But how do these terms relate to the hydraulic head h? Let us return to the Darcy manometer (Figure 2.4). At P, the fluid pressure p is given by

$$p = \rho g \psi + p_0 \tag{2.11}$$

where ψ is the height of the liquid column above P and p_0 is atmospheric pressure or pressure at the standard state. It is clear from Figure 2.4 and Eq. (2.11) that

$$p = \rho g(h - z) + p_0 \tag{2.12}$$

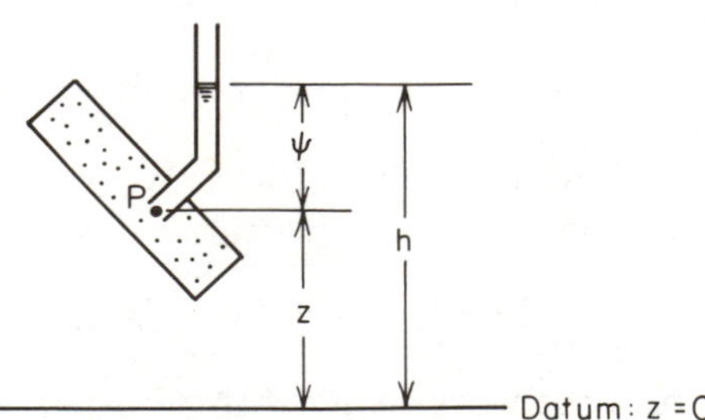

Figure 2.4 Hydraulic head h, pressure head ψ, and elevation head z for a laboratory manometer.

Substituting Eq. (2.12) in Eq. (2.10) yields

$$\Phi = gz + \frac{[\rho g(h - z) + p_0] - p_0}{\rho} \tag{2.13}$$

or, canceling terms,

$$\Phi = gh \tag{2.14}$$

Our long exercise has led us to the simplest of conclusions. The fluid potential Φ at any point P in a porous medium is simply the hydraulic head at the point multiplied by the acceleration due to gravity. Since g is very nearly constant in the vicinity of the earth's surface, Φ and h are almost perfectly correlated. To know one is to know the other. The hydraulic head h is therefore just as suitable a potential as is Φ. To recall Hubbert's definition: it is a physical quantity, it is capable of measurement, and flow always occurs from regions where h has higher values to regions where it has lower. In fact reference to Eq. (2.14) shows that, if Φ is energy per unit mass, then h is energy per unit weight.

It is common in groundwater hydrology to set the atmospheric pressure p_0 equal to zero and work in gage pressures (i.e., pressures above atmospheric). In this case Eqs. (2.10) and (2.14) become

$$\Phi = gz + \frac{p}{\rho} = gh \tag{2.15}$$

Dividing through by g, we obtain

$$h = z + \frac{p}{\rho g} \tag{2.16}$$

Putting Eq. (2.11) in terms of gage pressures yields

$$p = \rho g \psi \tag{2.17}$$

and Eq. (2.16) becomes

$$h = z + \psi \tag{2.18}$$

The hydraulic head h is thus seen to be the sum of two components: the elevation of the point of measurement, or *elevation head*, z, and the *pressure head* ψ. This fundamental head relationship is basic to an understanding of groundwater flow. Figure 2.4 displays the relationship for the Darcy manometer, Figure 2.5 for a field measurement site.

Those who are familiar with elementary fluid mechanics may already have recognized Eq. (2.9) as the *Bernoulli equation*, the classical formulation of energy loss during fluid flow. Some authors (Todd, 1959; Domenico, 1972) use the Bernoulli equation as the starting point for their development of the concepts of fluid

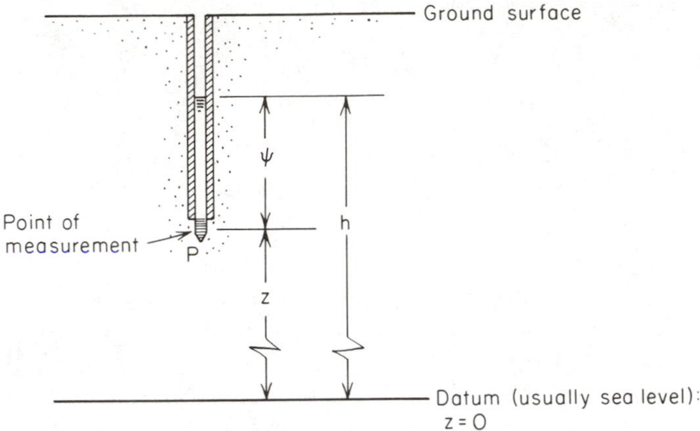

Figure 2.5 Hydraulic head *h*, pressure head *ψ*, and elevation head *z* for a field piezometer.

potential and hydraulic head. If we put Eq. (2.9) in terms of head and use a simplified notation, it becomes

$$h_T = h_z + h_p + h_v \tag{2.19}$$

where h_z is the elevation head, h_p the pressure head, and h_v the velocity head. In our earlier notation: $h_z = z$, $h_p = \psi$, and $h_v = v^2/2g$. The term h_T is called the *total head*, and for the special case where $h_v = 0$, it is equal to the hydrualic head h, and Eq. (2.18) holds.

Dimensions and Units

The dimensions of the head terms h, ψ, and z are those of length $[L]$. They are usually expressed as "meters of water" or "feet of water." The specification "of water" emphasizes that head measurements are dependent on fluid density through the relationship of Eq. (2.17). Given the same fluid pressure p at point P in Figure 2.5, the hydraulic head h and pressure head ψ would have different values if the fluid in the pores of the geological formation were oil rather than water. In this text, where we will always be dealing with water, the adjectival phrase will usually be dropped and we will record heads in meters.

As for the other terms introduced in this section; in the SI system, with its $[M][L][T]$ base, pressure has dimensions $[M/LT^2]$, mass density has dimensions $[M/L^3]$, and the fluid potential, from its definition, is energy per unit mass with dimensions $[L^2/T^2]$. Table 2.1 clarifies the dimensions and common units for all the important parameters introduced thus far. Reference to Appendix I should resolve any confusion. In this text, we will use SI metric units as our basic system of units, but Table 2.1 includes the FPS equivalents. Table A1.3 in Appendix I provides conversion factors.

Note in Table 2.1 that the weight density of water, γ, defined by

$$\gamma = \rho g \tag{2.20}$$

is a more suitable parameter than the mass density ρ for the FPS system of units, which has force as one of its fundamental dimensions.

Table 2.1 Dimensions and Common Units for Basic Groundwater Parameters*

		Système International† SI		Foot-pound-second system,‡ FPS	
Parameter	Symbol	Dimension	Units	Dimension	Units
Hydraulic head	h	$[L]$	m	$[L]$	ft
Pressure head	ψ	$[L]$	m	$[L]$	ft
Elevation head	z	$[L]$	m	$[L]$	ft
Fluid pressure	p	$[M/LT^2]$	N/m^2 or Pa	$[F/L^2]$	lb/ft^2
Fluid potential	Φ	$[L^2/T^2]$	m^2/s^2	$[L^2/T^2]$	ft^2/s^2
Mass density	ρ	$[M/L^3]$	kg/m^3	—	—
Weight density	γ	—	—	$[F/L^3]$	lb/ft^3
Specific discharge	v	$[L/T]$	m/s	$[L/T]$	ft/s
Hydraulic conductivity	K	$[L/T]$	m/s	$[L/T]$	ft/s

*See also Tables A1.1, A1.2, and A1.3, Appendix I.

†Basic dimensions are length $[L]$, mass $[M]$, and time $[T]$.

‡Basic dimensions are length $[L]$, force $[F]$, and time $[T]$.

Piezometers and Piezometer Nests

The basic device for the measurement of hydraulic head is a tube or pipe in which the elevation of a water level can be determined. In the laboratory (Figure 2.4) the tube is a *manometer;* in the field (Figure 2.5) the pipe is called a *piezometer*. A piezometer must be sealed along its length. It must be open to water flow at the bottom and be open to the atmosphere at the top. The intake is usually a section of slotted pipe or a commercially available *well point*. In either case the intake must be designed to allow the inflow of water but not of the sand grains or clay particles that make up the geologic formation. It must be emphasized that the point of measurement in a piezometer is at its base, not at the level of the fluid surface. One can view the functioning of a piezometer much like that of a thermometer. It is simply the instrument, if such it can be called, used to determine the value of h at some point P in a groundwater reservoir. In recent years, the simple standpipe piezometer has been replaced in some applications by more complex designs utilizing pressure transducers, pneumatic devices, and electronic components.

Piezometers are usually installed in groups so that they can be used to determine directions of groundwater flow. In Figure 2.6(a) three piezometers tap a water-bearing geological formation. It is a worthwhile exercise to remove the instruments of measurement from the diagram [Figure 2.6(b)] and consider only

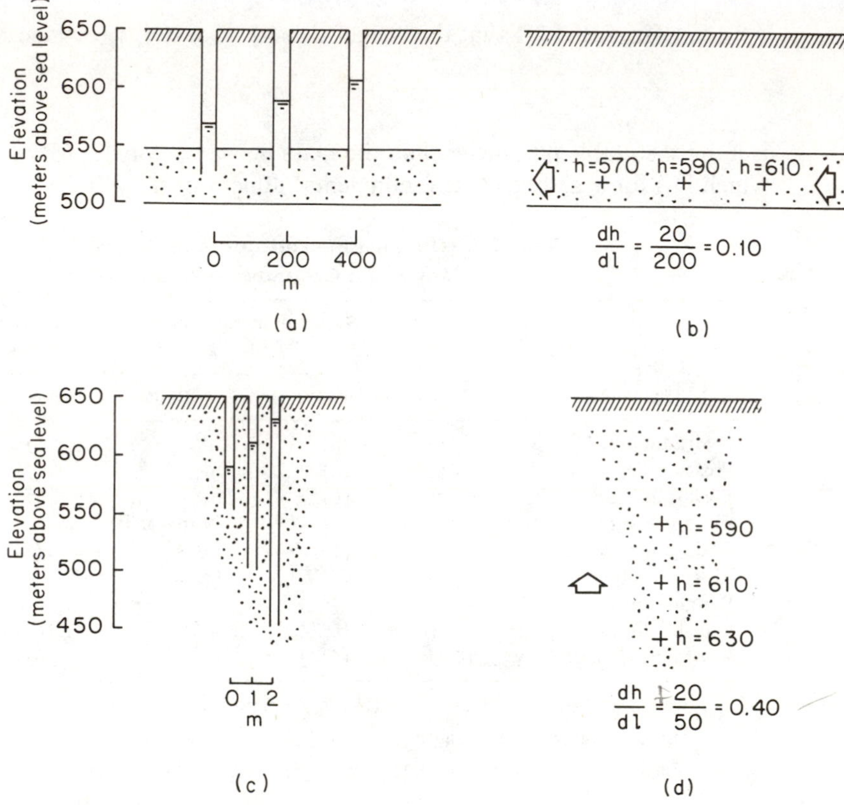

Figure 2.6 Determination of hydraulic gradients from piezometer installations.

the measured values. Flow is from higher h to lower, in this case from right to left. If the distance between the piezometers were known, the hydraulic gradient dh/dl could be calculated; and if the hydraulic conductivity K of the geological formation were known, Darcy's law could be used to calculate the specific discharge (or volume rate of flow through any cross-sectional area perpendicular to the flow direction).

Sometimes it is the vertical potential gradient that is of interest. In such cases a *piezometer nest* is utilized, with two or more piezometers installed side by side at the same location (or possibly in the same hole), each bottoming at a different depth and possibly in a different geological formation. Figure 2.6(c) and (d) shows a piezometer nest in a region of upward groundwater flow.

The distribution of hydraulic heads in a groundwater system is three-dimensional through space. The piezometer groupings shown in Figure 2.6 only prove the existence of *components* of flow in the directions indicated. If a large number of piezometers could be distributed throughout the three-dimensional hydrogeologic system, it would be possible to contour the positions of equal hydraulic head.

In three dimensions the locus of such points forms an *equipotential surface*. In any two-dimensional cross section, be it horizontal, vertical or otherwise, the traces of the equipotential surfaces on the section are called *equipotential lines*. If the pattern of hydraulic heads is known in a cross section, *flowlines* can be constructed perpendicular to the equipotential lines (in the direction of the maximum potential gradient). The resulting set of intersecting equipotential lines and flowlines is known as a *flow net*. Chapter 5 will provide detailed instructions on the construction of flow nets, and Chapter 6 will prove their usefulness in the interpretation of regional groundwater flow.

Coupled Flow

There is now a large body of experimental and theoretical evidence to show that water can be induced to flow through porous media under the influence of gradients other than that of hydraulic head. For example, the presence of a *temperature gradient* can cause groundwater flow (as well as heat flow) even when hydraulic gradients do not exist (Gurr et al., 1952; Philip and de Vries, 1957). This component becomes important in the formation of frost wedges in soil (Hoekstra, 1966; Harlan, 1973).

An *electrical gradient* can create a flow of water from high voltage to low when earth currents are set up in a soil. The mechanism of flow involves an interaction between charged ions in the water and the electrical charge associated with clay minerals in the soil (Casagrande, 1952). The principle is used in soil mechanics in the *electrokinetic* approach to soil drainage (Terzaghi and Peck, 1967).

Chemical gradients can cause the flow of water (as well as the movement of chemical constituents *through* the water) from regions where water has higher salinity to regions where it has lower salinity, even in the absence of other gradients. The role of chemical gradients in the production of water flow is relatively unimportant, but their direct influence on the movement of chemical constituents is of major importance in the analysis of groundwater contamination. These concepts will come to the fore in Chapters 3, 7, and 9.

If each of these gradients plays a role in producing flow, it follows that a more general flow law than Eq. (2.3) can be written in the form

$$v = -L_1 \frac{dh}{dl} - L_2 \frac{dT}{dl} - L_3 \frac{dc}{dl} \tag{2.21}$$

where h is hydraulic head, T is temperature, and c is chemical concentration; L_1, L_2, and L_3 are constants of proportionality. For the purposes of discussion, let us set $dc/dl = 0$. We are left with a situation where fluid flow is occurring in response to both a hydraulic head gradient and a temperature gradient:

$$v = -L_1 \frac{dh}{dl} - L_2 \frac{dT}{dl} \tag{2.22}$$

In general, $L_1\, dh/dl \gg L_2\, dT/dl$.

If a temperature gradient can cause fluid flow as well as heat flow in a porous medium, it should come as no surprise to find that a hydraulic gradient can cause heat flow as well as fluid flow. This mutual interdependency is a reflection of the well-known thermodynamic concept of *coupled flow*. If we set $dh/dl = i_1$ and $dT/dl = i_2$, we can write a pair of equations patterned after Eq. (2.22):

$$v_1 = -L_{11}i_1 - L_{12}i_2 \tag{2.23}$$

$$v_2 = -L_{21}i_1 - L_{22}i_2 \tag{2.24}$$

where v_1 is the specific discharge of *fluid* through the medium and v_2 is the specific discharge of *heat* through the medium. The L's are known as *phenomenological coefficients*. If $L_{12} = 0$ in Eq. (2.23), we are left with Darcy's law of groundwater flow and L_{11} is the hydraulic conductivity. If $L_{21} = 0$ in Eq. (2.24), we are left with Fourier's law of heat flow and L_{22} is the thermal conductivity.

It is possible to write a complete set of coupled equations. The set of equations would have the form of Eq. (2.23) but would involve all the gradients of Eq. (2.21) and perhaps others. The development of the theory of coupled flows in porous media was pioneered by Taylor and Cary (1964). Olsen (1969) has carried out significant experimental research. Bear (1972) provides a more detailed development of the concepts than can be attempted here. The thermodynamic description of the physics of porous media flow is conceptually powerful, but in practice there are very few data on the nature of the off-diagonal coefficients in the matrix of phenomenological coefficients L_{ij}. In this text we will assume that groundwater flow is fully described by Darcy's law [Eq. (2.3)]; that the hydraulic head [Eq. (2.18)], with its elevation and pressure components, is a suitable representation of the total head; and that the hydraulic conductivity is the only important phenomenological coefficient in Eq. (2.21).

2.3 Hydraulic Conductivity and Permeability

As Hubbert (1956) has pointed out, the constant of proportionality in Darcy's law, which has been christened the hydraulic conductivity, is a function not only of the porous medium but also of the fluid. Consider once again the experimental apparatus of Figure 2.1. If Δh and Δl are held constant for two runs using the same sand, but water is the fluid in the first run and molasses in the second, it would come as no surprise to find the specific discharge v much lower in the second run than in the first. In light of such an observation, it would be instructive to search for a parameter that can describe the conductive properties of a porous medium independently from the fluid flowing through it.

To this end experiments have been carried out with ideal porous media consisting of uniform glass beads of diameter d. When various fluids of density ρ and dynamic viscosity μ are run through the apparatus under a constant hydraulic gradient dh/dl, the following proportionality relationships are observed:

$$v \propto d^2$$

$$v \propto \rho g$$

$$v \propto \frac{1}{\mu}$$

Together with Darcy's original observation that $v \propto -dh/dl$, these three relationships lead to a new version of Darcy's law:

$$v = -\frac{Cd^2\rho g}{\mu}\frac{dh}{dl} \tag{2.25}$$

The parameter C is yet another constant of proportionality. For real soils it must include the influence of other media properties that affect flow, apart from the mean grain diameter: for example, the distribution of grain sizes, the sphericity and roundness of the grains, and the nature of their packing.

Comparison of Eq. (2.25) with the original Darcy equation [Eq. (2.3)] shows that

$$K = \frac{Cd^2\rho g}{\mu} \tag{2.26}$$

In this equation, ρ and μ are functions of the fluid alone and Cd^2 is a function of the medium alone. If we define

$$k = Cd^2 \tag{2.27}$$

then

$$K = \frac{k\rho g}{\mu} \tag{2.28}$$

The parameter k is known as the *specific* or *intrinsic permeability*. If K is always called hydraulic conductivity, it is safe to drop the adjectives and refer to k as simply the permeability. That is the convention that will be followed in this text, but it can lead to some confusion, especially when dealing with older texts and reports where the hydraulic conductivity K is sometimes called the *coefficient of permeability*.

Hubbert (1940) developed Eqs. (2.25) through (2.28) from fundamental principles by considering the relationships between driving and resisting forces on a microscopic scale during flow through porous media. The dimensional considerations inherent in his analysis provided us with the foresight to include the constant g in the proportionality relationship leading to Eq. (2.25). In this way C emerges as a dimensionless constant.

The permeability k is a function only of the medium and has dimensions $[L^2]$. The term is widely used in the petroleum industry, where the existence of gas, oil, and water in multiphase flow systems makes the use of a fluid-free conductance parameter attractive. When measured in m^2 or cm^2, k is very small, so petroleum engineers have defined the *darcy* as a unit of permeability. If Eq. (2.28) is substituted in Eq. (2.3), Darcy's law becomes

$$v = \frac{-k\rho g}{\mu}\frac{dh}{dl} \tag{2.29}$$

Referring to this equation, 1 darcy is defined as the permeability that will lead to a specific discharge of 1 cm/s for a fluid with a viscosity of 1 cp under a hydraulic gradient that makes the term $\rho g\, dh/dl$ equal to 1 atm/cm. One darcy is approximately equal to 10^{-8} cm^2.

In the water well industry, the unit gal/day/ft^2 is widely used for hydraulic conductivity. Its relevance is clearest when Darcy's law is couched in terms of Eq. (2.4):

$$Q = -K\frac{dh}{dl}A$$

The early definitions provided by the U.S. Geological Survey with regard to this unit differentiate between a laboratory coefficient and a field coefficient. However, a recent updating of these definitions (Lohman, 1972) has discarded this formal differentiation. It is sufficient to note that differences in the temperature of measurement between the field environment and the laboratory environment can influence hydraulic conductivity values through the viscosity term in Eq. (2.28). The effect is usually small, so correction factors are seldom introduced. It still makes good sense to report whether hydraulic conductivity measurements have been carried out in the laboratory or in the field, because the methods of measurement are very different and the interpretations placed on the values may be dependent on the type of measurement. However, this information is of practical rather than conceptual importance.

Table 2.2 indicates the range of values of hydraulic conductivity and permeability in five different systems of units for a wide range of geological materials. The table is based in part on the data summarized in Davis' (1969) review. The primary conclusion that can be drawn from the data is that hydraulic conductivity varies over a very wide range. There are very few physical parameters that take on values over 13 orders of magnitude. In practical terms, this property implies that an order-of-magnitude knowledge of hydraulic conductivity can be very useful. Conversely, the third decimal place in a reported conductivity value probably has little significance.

Table 2.3 provides a set of conversion factors for the various common units of k and K. As an example of its use, note that a k value in cm^2 can be converted to one in ft^2 by multiplying by 1.08×10^{-3}. For the reverse conversion from ft^2 to cm^2, multiply by 9.29×10^2.

Table 2.2 Range of Values of Hydraulic Conductivity and Permeability

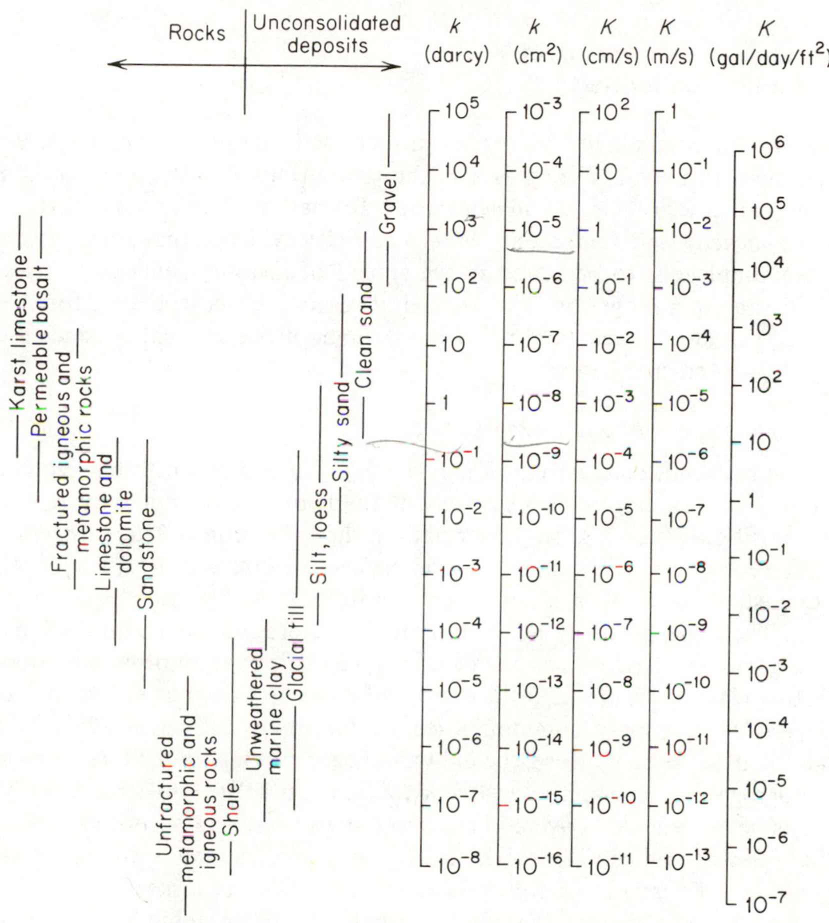

Table 2.3 Conversion Factors for Permeability and Hydraulic Conductivity Units

	Permeability, *k**			Hydraulic conductivity, *K*		
	cm^2	ft^2	darcy	m/s	ft/s	U.S. gal/day/ft^2
cm^2	1	1.08×10^{-3}	1.01×10^{8}	9.80×10^{2}	3.22×10^{3}	1.85×10^{9}
ft^2	9.29×10^{2}	1	9.42×10^{10}	9.11×10^{5}	2.99×10^{6}	1.71×10^{12}
darcy	9.87×10^{-9}	1.06×10^{-11}	1	9.66×10^{-6}	3.17×10^{-5}	1.82×10^{1}
m/s	1.02×10^{-3}	1.10×10^{-6}	1.04×10^{5}	1	3.28	2.12×10^{6}
ft/s	3.11×10^{-4}	3.35×10^{-7}	3.15×10^{4}	3.05×10^{-1}	1	6.46×10^{5}
U.S. gal/day/ft^2	5.42×10^{-10}	5.83×10^{-13}	5.49×10^{-2}	4.72×10^{-7}	1.55×10^{-6}	1

*To obtain k in ft^2, multiply k in cm^2 by 1.08×10^{-3}.

The various approaches to the measurement of hydraulic conductivity in the laboratory and in the field are described in Sections 8.4 through 8.6.

2.4 Heterogeneity and Anisotropy of Hydraulic Conductivity

Hydraulic conductivity values usually show variations through space within a geologic formation. They may also show variations with the direction of measurement at any given point in a geologic formation. The first property is termed *heterogeneity* and the second *anisotropy*. The evidence that these properties are commonplace is to be found in the spread of measurements that arises in most field sampling programs. The geological reasoning that accounts for their prevalence lies in an understanding of the geologic processes that produce the various geological environments.

Homogeneity and Heterogeneity

If the hydraulic conductivity K is *independent* of position within a geologic formation, the formation is *homogeneous*. If the hydraulic conductivity K is *dependent* on position within a geologic formation, the formation is *heterogeneous*. If we set up an xyz coordinate system in a homogeneous formation, $K(x, y, z) = C$, C being a constant; whereas in a heterogeneous formation, $K(x, y, z) \neq C$.

There are probably as many types of heterogeneous configurations as there are geological environments, but it may be instructive to draw attention to three broad classes. Figure 2.7(a) is a vertical cross section that shows an example of *layered heterogeneity*, common in sedimentary rocks and unconsolidated lacustrine and marine deposits. Here, the individual beds making up the formation each have a homogeneous conductivity value $K_1, K_2, \ldots$, but the entire system can be thought of as heterogeneous. Layered heterogeneity can result in K contrasts of almost the full 13-order range (Table 2.2), as, for example, in interlayered deposits of clay and sand. Equally large contrasts can arise in cases of *discontinuous heterogeneity* caused by the presence of faults or large-scale stratigraphic features. Perhaps the most ubiquitous discontinuous feature is the overburden-bedrock contact. Figure 2.7(b) is a map that shows a case of *trending heterogeneity*. Trends are possible in any type of geological formation, but they are particularly common in response to the sedimentation processes that create deltas, alluvial fans, and glacial outwash plains. The A, B, and C soil horizons often show vertical trends in hydraulic conductivity, as do rock types whose conductivity is primarily dependent on joint and fracture concentration. Trending heterogeneity in large consolidated or unconsolidated sedimentary formations can attain gradients of 2–3 orders of magnitude in a few miles.

Many hydrogeologists and petroleum geologists have used statistical distributions to provide a quantitative description of the degree of heterogeneity in a geological formation. There is now a large body of direct evidence to support the

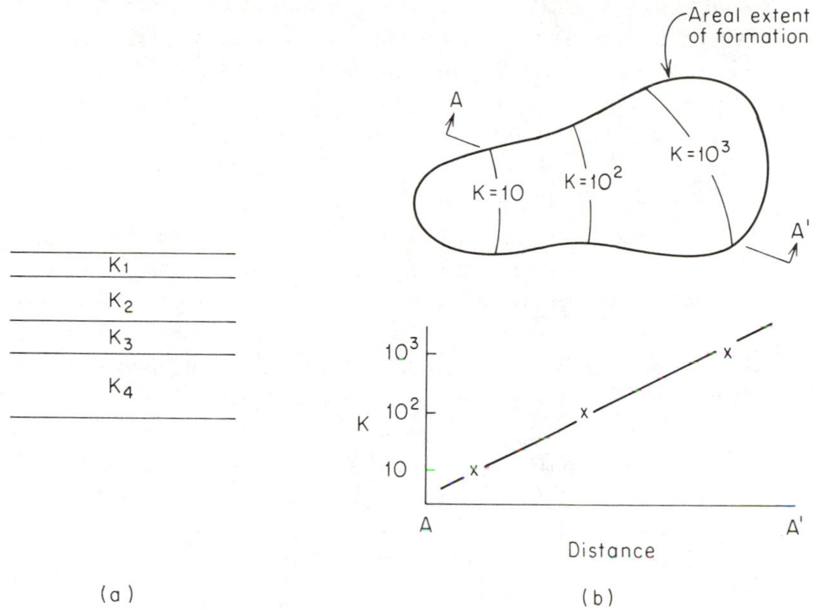

Figure 2.7 Layered heterogeneity and trending heterogeneity.

statement that the probability density function for hydraulic conductivity is log-normal. Warren and Price (1961) and Bennion and Griffiths (1966) found this to be the case in oilfield reservoir rocks, and Willardson and Hurst (1965) and Davis (1969) support the conclusion for unconsolidated water-bearing formations. A log-normal distribution for K is one for which a parameter Y, defined as $Y = \log K$, shows a normal distribution. Freeze (1975) provides a table, based on the references above, that shows the standard deviation on Y (which is independent of the units of measurement) is usually in the range 0.5–1.5. This means that K values in most geological formations show internal heterogeneous variations of 1–2 orders of magnitude. Trending heterogeneity within a geological formation can be thought of as a trend in the *mean value* of the probability distribution. The same standard deviation may be evident in measurements at different positions in the formation, but the trending means lead to an increase in the overall observed range for the formation.

Greenkorn and Kessler (1969) have provided a set of definitions of heterogeneity that are consistent with the statistical observations. In effect, they argue that if all geologic formations display spatial variations in K, then under the classical definitions, there is no such thing as a homogeneous formation. They redefine a homogeneous formation as one in which the probability density function of hydraulic conductivity is monomodal. That is, it shows variations in K, but maintains a constant mean K through space. A heterogeneous formation is defined as one in which the probability density function is multimodal. To describe a porous medium that satisfies the classical definition of homogeneity (K constant everywhere, such as in experimental glass beads of diameter d)they use the term *uniform*. If we wish to adapt the classical definitions given at the start of this section to this more rational set of concepts, we can do so by adding the adjective "mean" and couching the original definitions in terms of *mean hydraulic conductivity*.

Isotropy and Anisotropy

If the hydraulic conductivity K is *independent* of the direction of measurement at a point in a geologic formation, the formation is *isotropic* at that point. If the hydraulic conductivity K *varies* with the direction of measurement at a point in a geologic formation, the formation is *anisotropic* at that point.

Consider a two-dimensional vertical section through an anisotropic formation. If we let θ be the angle between the horizontal and the direction of measurement of a K value at some point in the formation, then $K = K(\theta)$. The directions in space corresponding to the θ angle at which K attains its maximum and minimum values are known as the *principal directions of anisotropy*. They are always perpendicular to one another. In three dimensions, if a plane is taken perpendicular to one of the principal directions, the other two principal directions are the directions of maximum and minimum K in that plane.

If an xyz coordinate system is set up in such a way that the coordinate directions coincide with the principal directions of anisotropy, the hydraulic conductivity values in the principal directions can be specified as K_x, K_y, and K_z. At any point (x, y, z), an isotropic formation will have $K_x = K_y = K_z$, whereas an anisotropic formation will have $K_x \neq K_y \neq K_z$. If $K_x = K_y \neq K_z$, as is common in horizontally bedded sedimentary deposits, the formation is said to be *transversely isotropic*.

To fully describe the nature of the hydraulic conductivity in a geologic formation, it is necessary to use two adjectives, one dealing with heterogeneity and one with anisotropy. For example, for a homogeneous, isotropic system in two dimensions: $K_x(x, z) = K_z(x, z) = C$ for all (x, z), where C is a constant. For a homogeneous, anisotropic system, $K_x(x, z) = C_1$ for all (x, z) and $K_z(x, z) = C_2$ for all (x, z) but $C_1 \neq C_2$. Figure 2.8 attempts to further clarify the four possible combinations. The length of the arrow vectors is proportional to the K_x and K_z values at the two points (x_1, z_1) and (x_2, z_2).

The primary cause of anisotropy on a small scale is the orientation of clay minerals in sedimentary rocks and unconsolidated sediments. Core samples of clays and shales seldom show horizontal to vertical anisotropy greater than 10 : 1, and it is usually less than 3:1.

On a larger scale, it can be shown (Maasland, 1957; Marcus and Evenson, 1961) that there is a relation between layered heterogeneity and anisotropy. Consider the layered formation shown in Figure 2.9. Each layer is homogeneous and isotropic with hydraulic conductivity values $K_1, K_2, \ldots, K_n$. We will show that the system as a whole acts like a single homogeneous, anistropic layer. First, consider flow perpendicular to the layering. The specific discharge v must be the

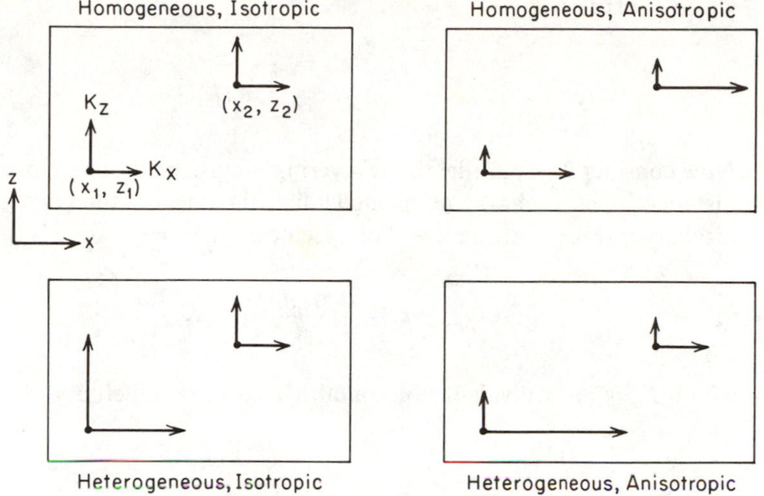

Figure 2.8 Four possible combinations of heterogeneity and anisotropy.

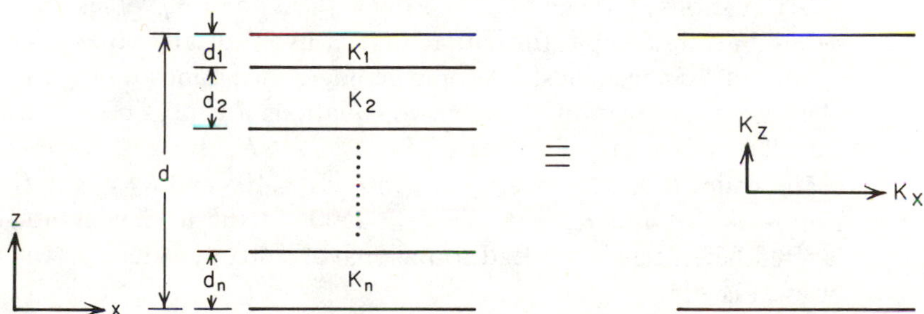

Figure 2.9 Relation between layered heterogeneity and anisotropy.

same entering the system as it is leaving; in fact, it must be constant throughout the system. Let Δh_1 be the head loss across the first layer, Δh_2 across the second layer, and so on. The total head loss is then $\Delta h = \Delta h_1 + \Delta h_2 + \ldots + \Delta h_n$, and from Darcy's law,

$$v = \frac{K_1\,\Delta h_1}{d_1} = \frac{K_2\,\Delta h_2}{d_2} = \ldots = \frac{K_n\,\Delta h_n}{d_n} = \frac{K_z\,\Delta h}{d} \tag{2.30}$$

where K_z is an equivalent vertical hydraulic conductivity for the system of layers. Solving the outside relationship of Eq. (2.30) for K_z and using the inside relationships for $\Delta h_1, \Delta h_2, \ldots$, we obtain

$$K_z = \frac{vd}{\Delta h} = \frac{vd}{\Delta h_1 + \Delta h_2 + \ldots + \Delta h_n}$$

$$= \frac{vd}{vd_1/K_1 + vd_2/K_2 + \ldots + vd_n/K_n}$$

which leads to

$$K_z = \frac{d}{\sum_{i=1}^{n} d_i/K_i} \tag{2.31}$$

Now consider flow parallel to the layering. Let Δh be the head loss over a horizontal distance l. The discharge Q through a unit thickness of the system is the sum of the discharges through the layers. The specific discharge $v = Q/d$ is therefore given by

$$v = \sum_{i=1}^{n} \frac{K_i d_i}{d} \frac{\Delta h}{l} = K_x \frac{\Delta h}{l}$$

where K_x is an equivalent horizontal hydraulic conductivity. Simplification yields

$$K_x = \sum_{i=1}^{n} \frac{K_i d_i}{d} \tag{2.32}$$

Equations (2.31) and (2.32) provide the K_x and K_z values for a single homogeneous but anisotropic formation that is hydraulically equivalent to the layered system of homogeneous, isotropic geologic formations of Figure 2.9. With some algebraic manipulation of these two equations it is possible to show that $K_x > K_z$ for all possible sets of values of $K_1, K_2, \ldots, K_n$. In fact, if we consider a set of cyclic couplets $K_1, K_2, K_1, K_2, \ldots$ with $K_1 = 10^4$ and $K_2 = 10^2$, then $K_x/K_z = 25$. For $K_1 = 10^4$ and $K_2 = 1$, $K_x/K_z = 2500$. In the field, it is not uncommon for layered heterogeneity to lead to regional anisotropy values on the order of 100:1 or even larger.

Snow (1969) showed that fractured rocks also behave anisotropically because of the directional variations in joint aperture and spacing. In this case, it is quite common for $K_z > K_x$.

Darcy's Law in Three Dimensions

For three-dimensional flow, in a medium that may be anisotropic, it is necessary to generalize the one-dimensional form of Darcy's law [Eq. (2.3)] presented earlier. In three dimensions the velocity v is a vector with components v_x, v_y, and v_z, and the simplest generalization would be

$$\begin{aligned} v_x &= -K_x \frac{\partial h}{\partial x} \\ v_y &= -K_y \frac{\partial h}{\partial y} \\ v_z &= -K_z \frac{\partial h}{\partial z} \end{aligned} \tag{2.33}$$

where K_x, K_y, and K_z are the hydraulic conductivity values in the x, y, and z direction. Since h is now a function of x, y, and z, the derivatives must be partial.

In this text we will assume this simple generalization to be an adequate description of three-dimensional flow, but it is worth noting that a more generalized set of equations could be written in the form

$$v_x = -K_{xx}\frac{\partial h}{\partial x} - K_{xy}\frac{\partial h}{\partial y} - K_{xz}\frac{\partial h}{\partial z}$$

$$v_y = -K_{yx}\frac{\partial h}{\partial x} - K_{yy}\frac{\partial h}{\partial y} - K_{yz}\frac{\partial h}{\partial z} \qquad (2.34)$$

$$v_z = -K_{zx}\frac{\partial h}{\partial x} - K_{zy}\frac{\partial h}{\partial y} - K_{zz}\frac{\partial h}{\partial z}$$

This set of equations exposes the fact that there are actually nine components of hydraulic conductivity in the most general case. If these components are put in matrix form, they form a second-rank symmetric tensor known as the *hydraulic conductivity tensor* (Bear, 1972). For the special case $K_{xy} = K_{xz} = K_{yx} = K_{yz} = K_{zx} = K_{zy} = 0$, the nine components reduce to three and Eq. (2.33) is a suitable generalization of Darcy's law. The necessary and sufficient condition that allows use of Eq. (2.33) rather than Eq. (2.34) is that the principal directions of anisotropy coincide with the x, y, and z coordinate axes. In most cases it is possible to choose a coordinate system that satisfies this requirement, but one can conceive of heterogeneous anisotropic systems in which the principal directions of anisotropy vary from one formation to another, and in such systems the choice of suitable axes would be impossible.

Hydraulic Conductivity Ellipsoid

Consider an arbitrary flowline in the xz plane in a homogeneous, anisotropic medium with principal hydraulic conductivities K_x and K_z [Figure 2.10(a)]. Along

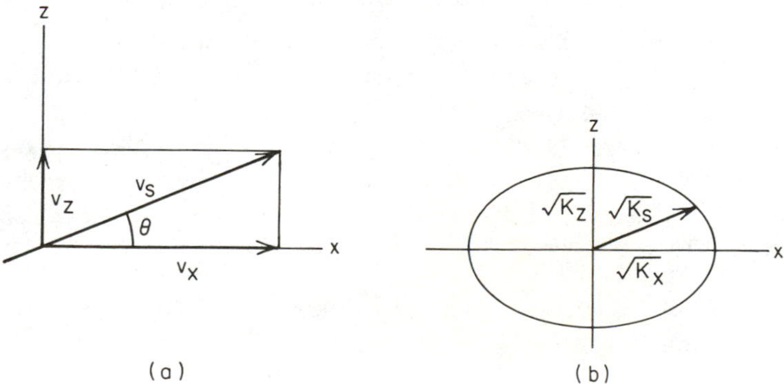

Figure 2.10 (a) Specific discharge v_s in an arbitrary direction of flow. (b) Hydraulic conductivity ellipse.

the flowline

$$v_s = -K_s \frac{\partial h}{\partial s} \tag{2.35}$$

where K_s is unknown, although it presumably lies in the range $K_x - K_z$. We can separate v_s into its components v_x and v_z, where

$$\begin{aligned} v_x &= -K_x \frac{\partial h}{\partial x} = v_s \cos\theta \\ v_z &= -K_z \frac{\partial h}{\partial z} = v_s \sin\theta \end{aligned} \tag{2.36}$$

Now, since $h = h(x, z)$,

$$\frac{\partial h}{\partial s} = \frac{\partial h}{\partial x} \cdot \frac{\partial x}{\partial s} + \frac{\partial h}{\partial z} \cdot \frac{\partial z}{\partial s} \tag{2.37}$$

Geometrically, $\partial x/\partial s = \cos\theta$ and $\partial z/\partial s = \sin\theta$. Substituting these relationships together with Eqs. (2.35) and (2.36) in Eq. (2.37) and simplifying yields

$$\frac{1}{K_s} = \frac{\cos^2\theta}{K_x} + \frac{\sin^2\theta}{K_z} \tag{2.38}$$

This equation relates the principal conductivity components K_x and K_z to the resultant K_s in any angular direction θ. If we put Eq. (2.38) into rectangular coordinates by setting $x = r\cos\theta$ and $z = r\sin\theta$, we get

$$\frac{r^2}{K_s} = \frac{x^2}{K_x} + \frac{z^2}{K_z} \tag{2.39}$$

which is the equation of an ellipse with major axes $\sqrt{K_x}$ and $\sqrt{K_z}$ [Figure 2.10(b)]. In three dimensions, it becomes an ellipsoid with major axes $\sqrt{K_x}$, $\sqrt{K_y}$, and $\sqrt{K_z}$, and it is known as the *hydraulic conductivity ellipsoid*. In Figure 2.10(b), the conductivity value K_s for any direction of flow in an anisotropic medium can be determined graphically if K_x and K_z are known.

In Section 5.1, the construction of flow nets in anisotropic media will be discussed, and it will be shown there that, in contrast to isotropic media, flowlines are not perpendicular to equipotential lines in anisotropic media.

2.5 Porosity and Void Ratio

If the total unit volume V_T of a soil or rock is divided into the volume of the solid portion V_s and the volume of the voids V_v, the *porosity* n is defined as $n = V_v/V_T$. It is usually reported as a decimal fraction or a percent.

Figure 2.11 shows the relation between various rock and soil textures and porosity. It is worth distinguishing between *primary porosity*, which is due to the

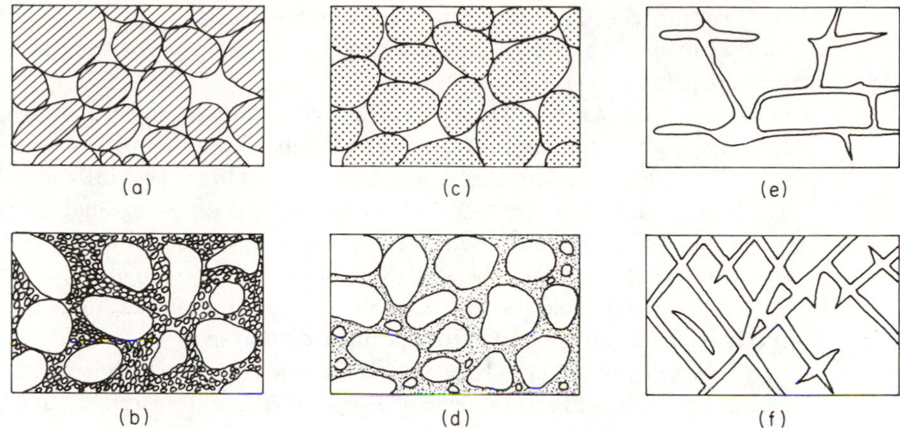

Figure 2.11 Relation between texture and porosity. (a) Well-sorted sedimentary deposit having high porosity; (b) poorly sorted sedimentary deposit having low porosity; (c) well-sorted sedimentary deposit consisting of pebbles that are themselves porous, so that the deposit as a whole has a very high porosity; (d) well-sorted sedimentary deposit whose porosity has been diminished by the deposition of mineral matter in the interstices; (e) rock rendered porous by solution; (f) rock rendered porous by fracturing (after Meinzer, 1923).

soil or rock matrix [Figure 2.11(a), (b), (c), and (d)], and *secondary porosity*, which may be due to such phenomena as secondary solution [Figure 2.11(e)] or structurally controlled regional fracturing [Figure 2.11(f)].

Table 2.4, based in part on data summarized by Davis (1969), lists representative porosity ranges for various geologic materials. In general, rocks have lower porosities than soils; gravels, sands, and silts, which are made up of angular and

Table 2.4 Range of Values of Porosity

	n(%)
Unconsolidated deposits	
Gravel	25–40
Sand	25–50
Silt	35–50
Clay	40–70
Rocks	
Fractured basalt	5–50
Karst limestone	5–50
Sandstone	5–30
Limestone, dolomite	0–20
Shale	0–10
Fractured crystalline rock	0–10
Dense crystalline rock	0–5

rounded particles, have lower porosities than soils rich in platy clay minerals; and poorly sorted deposits [Figure 2.11(b)] have lower porosities than well-sorted deposits [Figure 2.11(a)].

The porosity n can be an important controlling influence on hydraulic conductivity K. In sampling programs carried out within deposits of well-sorted sand or in fractured rock formations, samples with higher n generally also have higher K. Unfortunately, the relationship does not hold on a regional basis across the spectrum of possible rock and soil types. Clay-rich soils, for example, usually have higher porosities than sandy or gravelly soils but lower hydraulic conductivities. In Section 8.7 techniques will be presented for the estimation of hydraulic conductivity from porosity and from grain-size analyses.

The porosity n is closely related to the *void ratio* e, which is widely used in soil mechanics. The void ratio is defined as $e = V_v/V_s$, and e is related to n by

$$e = \frac{n}{1 - n} \quad \text{or} \quad n = \frac{e}{1 + e} \tag{2.40}$$

Values of e usually fall in the range 0–3.

The measurement of porosity on soil samples in the laboratory will be treated in Section 8.4.

2.6 Unsaturated Flow and the Water Table

Up until this point, Darcy's law and the concepts of hydraulic head and hydraulic conductivity have been developed with respect to a *saturated* porous medium, that is, one in which all the voids are filled with water. It is clear that some soils, especially those near the ground surface, are seldom saturated. Their voids are usually only partially filled with water, the remainder of the pore space being taken up by air. The flow of water under such conditions is termed *unsaturated* or *partially saturated*. Historically, the study of unsaturated flow has been the domain of soil physicists and agricultural engineers, but recently both soil scientists and groundwater hydrologists have recognized the need to pool their talents in the development of an integrated approach to the study of subsurface flow, both saturated and unsaturated.

Our emphasis in this section will be on the hydraulics of *liquid-phase* transport of water in the unsaturated zone. We will not discuss *vapor-phase* transport, nor will we consider *soil water-plant interactions*. These latter topics are of particular interest in the agricultural sciences and they play an important role in the interpretation of soil geochemistry. More detailed consideration of the physics and chemistry of moisture transfer in unsaturated soils can be found at an introductory level in Baver et al. (1972) and at a more advanced level in Kirkham and Powers (1972) and Childs (1969).

Moisture Content

If the total unit volume V_T of a soil or rock is divided into the volume of the solid portion V_s, the volume of the water V_w, and the volume of the air V_a, the *volumetric moisture content* θ is defined as $\theta = V_w/V_T$. Like the porosity n, it is usually reported as a decimal fraction or a percent. For saturated flow, $\theta = n$; for unsaturated flow, $\theta < n$.

Water Table

The simplest hydrologic configuration of saturated and unsaturated conditions is that of an *unsaturated zone* at the surface and a *saturated zone* at depth [Figure 2.12(a)]. We commonly think of the water table as being the boundary between the two, yet we are aware that a saturated capillary fringe often exists above the water table. With this type of complication lurking in the background, we must take care to set up a consistent set of definitions for the various saturated-unsaturated concepts.

The *water table* is best defined as the surface on which the fluid pressure p in the pores of a porous medium is exactly atmospheric. The location of this surface is revealed by the level at which water stands in a shallow well open along its length and penetrating the surficial deposits just deeply enough to encounter standing water in the bottom. If p is measured in gage pressure, then on the water table, $p = 0$. This implies $\psi = 0$, and since $h = \psi + z$, the hydraulic head at any point on the water table must be equal to the elevation z of the water table at that point. On figures we will often indicate the position of the water table by means of a small inverted triangle, as in Figure 2.12(a).

Negative Pressure Heads and Tensiometers

We have seen that $\psi > 0$ (as indicated by piezometer measurements) in the saturated zone and that $\psi = 0$ on the water table. It follows that $\psi < 0$ in the unsaturated zone. This reflects the fact that water in the unsaturated zone is held in the soil pores under surface-tension forces. A microscopic inspection would reveal a concave meniscus extending from grain to grain across each pore channel [as shown in the upper circular inset on Figure 2.12(c)]. The radius of curvature on each meniscus reflects the surface tension on that individual, microscopic air-water interface. In reference to this physical mechanism of water retention, soil physicists often call the pressure head ψ, when $\psi < 0$, the *tension head* or *suction head*. In this text, on the grounds that one concept deserves only one name, we will use the term *pressure head* to refer to both positive and negative ψ.

Regardless of the sign of ψ, the hydraulic head h is still equal to the algebraic sum of ψ and z. However, above the water table, where $\psi < 0$, piezometers are no longer a suitable instrument for the measurement of h. Instead, h must be obtained indirectly from measurements of ψ determined with *tensiometers*. Kirkham (1964) and S. J. Richards (1965) provide detailed descriptions of the design

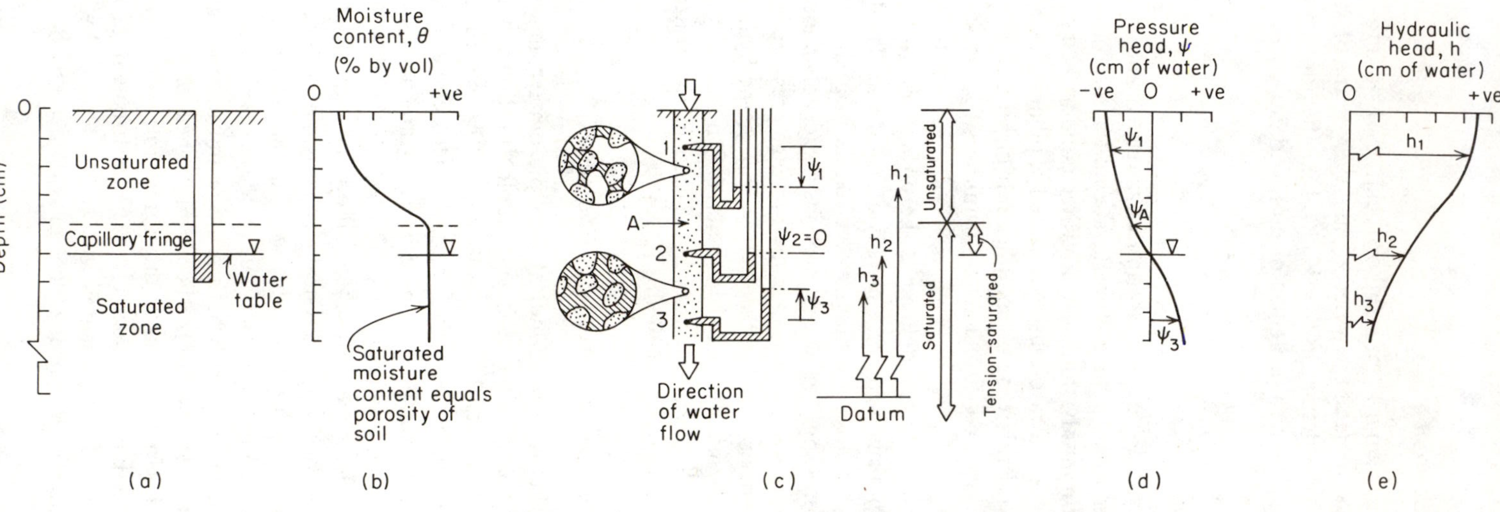

Figure 2.12 Groundwater conditions near the ground surface. (a) Saturated and unsaturated zones; (b) profile of moisture content versus depth; (c) pressure-head and hydraulic-head relationships; insets: water retention under pressure heads less than (top) and greater than (bottom) atmospheric; (d) profile of pressure head versus depth; (e) profile of hydraulic head versus depth.

and use of these instruments. Very briefly, a tensiometer consists of a porous cup attached to an airtight, water-filled tube. The porous cup is inserted into the soil at the desired depth, where it comes into contact with the soil water and reaches hydraulic equilibrium. The equilibration process involves the passage of water through the porous cup from the tube into the soil. The vacuum created at the top of the airtight tube is a measure of the pressure head in the soil. It is usually measured by a vacuum gage attached to the tube above the ground surface, but it can be thought of as acting like the inverted manometer shown for point 1 in the soil profile of Figure 2.12(c). To obtain the hydraulic head h, the negative ψ value indicated by the vacuum gage on a tensiometer must be added algebraically to the elevation z of the point of measurement. In Figure 2.12(c) the instrument at point 1 is a tensiometer; the one at point 3 is a piezometer. The diagram is, of course, schematic. In practice, the tensiometer would be a tube with a gage and a porous cup at the base; the piezometer would be an open pipe with a well point at the base.

Characteristic Curves of the Unsaturated Hydraulic Parameters

There is a further complication to the analysis of flow in the unsaturated zone. Both the moisture content θ and the hydraulic conductivity K are functions of the pressure head ψ. On reflection, the first of these conditions should come as no great surprise. In that soil moisture is held between the soil grains under surface-tension forces that are reflected in the radius of curvature of each meniscus, we might expect that higher moisture contents would lead to larger radii of curvature, lower surface-tension forces, and lower tension heads (i.e., less-negative pressure heads). Further, it has been observed experimentally that the θ–ψ relationship is hysteretic; it has a different shape when soils are wetting than when they are drying. Figure 2.13(a) shows the hysteretic functional relationship between θ and ψ for a naturally occurring sand soil (after Liakopoulos, 1965a). If a sample of this soil were saturated at a pressure head greater than zero and the pressure was then lowered step by step until it reached levels much less than atmospheric ($\psi \ll 0$), the moisture contents at each step would follow the *drying curve* (or *drainage curve*) on Figure 2.13(a). If water were then added to the dry soil in small steps, the pressure heads would take the return route along the *wetting curve* (or *imbibition curve*). The internal lines are called *scanning curves*. They show the course that θ and ψ would follow if the soil were only partially wetted, then dried, or vice versa.

One would expect, on the basis of what has been presented thus far, that the moisture content θ would equal the porosity n for all $\psi > 0$. For coarse-grained soils this is the case, but for fine-grained soils this relationship holds over a slightly larger range $\psi > \psi_a$, where ψ_a is a small negative pressure head [Figure 2.13(a)] known as the *air entry pressure head*. The corresponding pressure p_a is called the *air entry pressure* or the *bubbling pressure*.

Figure 2.13(b) displays the hysteretic curves relating the hydraulic conductivity K to the pressure head ψ for the same soil. For $\psi > \psi_a$, $K = K_0$, where K_0

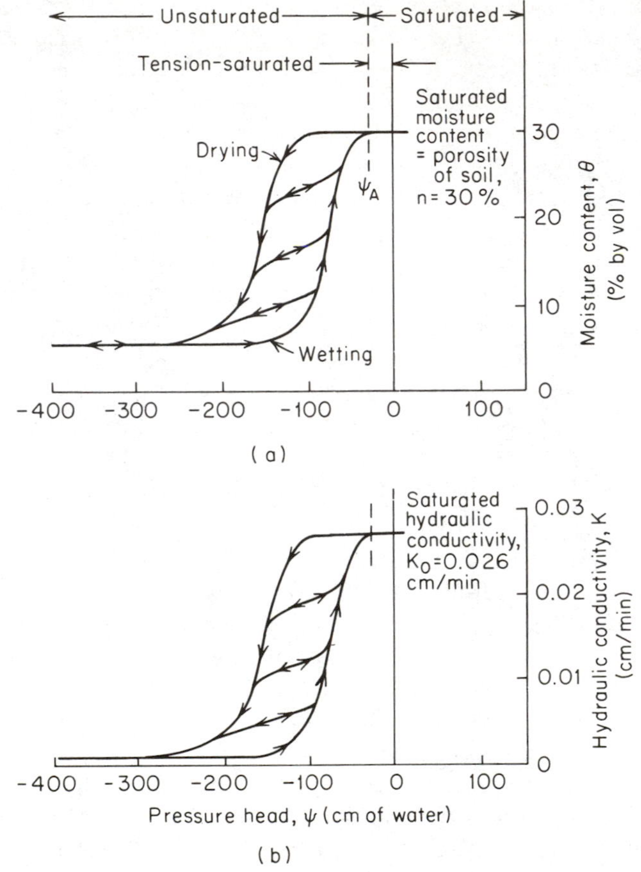

Figure 2.13 Characteristic curves relating hydraulic conductivity and moisture content to pressure head for a naturally occurring sand soil (after Liakopoulos, 1965a).

is now known as the *saturated hydraulic conductivity*. Since $K = K(\psi)$ and $\theta = \theta(\psi)$, it is also true that $K = K(\theta)$. The curves of Figure 2.13(b) reflect the fact that the hydraulic conductivity of an unsaturated soil increases with increasing moisture content. If we write Darcy's law for unsaturated flow in the x direction in an isotropic soil as

$$v_x = -K(\psi)\frac{\partial h}{\partial x} \tag{2.41}$$

we see that the existence of the $K(\psi)$ relationship implies that, given a constant hydraulic gradient, the specific discharge v increases with increasing moisture content.

In actual fact, it would be impossible to hold the hydraulic gradient constant while increasing the moisture content. Since $h = \psi + z$ and $\theta = \theta(\psi)$, the hydraulic head h is also affected by the moisture content. In other words, a hydraulic-head gradient infers a pressure-head gradient (except in pure gravity flow), and this in turn infers a moisture-content gradient. In Figure 2.12, the vertical profiles for these three variables are shown schematically for a hypothetical case of downward infiltration from the surface. Flow must be downward because the hydraulic heads displayed in Figure 2.12(e) decrease in that direction. The large positive values of h infer that $|z| \gg |\psi|$. In other words, the $z = 0$ datum lies at some depth. For a real case, these three profiles would be quantitatively interlinked through the $\theta(\psi)$ and $K(\psi)$ curves for the soil at the site. For example, if the $\theta(\psi)$ curve were known for the soil and the $\theta(z)$ profile measured in the field, then the $\psi(z)$ profile, and hence the $h(z)$ profile, could be calculated.

The pair of curves $\theta(\psi)$ and $K(\psi)$ shown in Figure 2.13 are characteristic for any given soil. Sets of measurements carried out on separate samples from the same homogeneous soil would show only the usual statistical variations associated with spatially separated sampling points. The curves are often called the *characteristic curves*. In the saturated zone we have the two fundamental hydraulic parameters K_0 and n; in the unsaturated zone these become the functional relationships $K(\psi)$ and $\theta(\psi)$. More succinctly,

$$\begin{aligned} \theta &= \theta(\psi) \qquad \psi < \psi_a \\ \theta &= n \qquad\quad \psi \geq \psi_a \end{aligned} \tag{2.42}$$

$$\begin{aligned} K &= K(\psi) \qquad \psi < \psi_a \\ K &= K_0 \qquad\quad \psi \geq \psi_a \end{aligned} \tag{2.43}$$

Figure 2.14 shows some hypothetical single-valued characteristic curves (i.e., without hysteresis) that are designed to show the effect of soil texture on the shape of the curves. For a more complete description of the physics of moisture retention in unsaturated soils, the reader is directed to White et al. (1971).

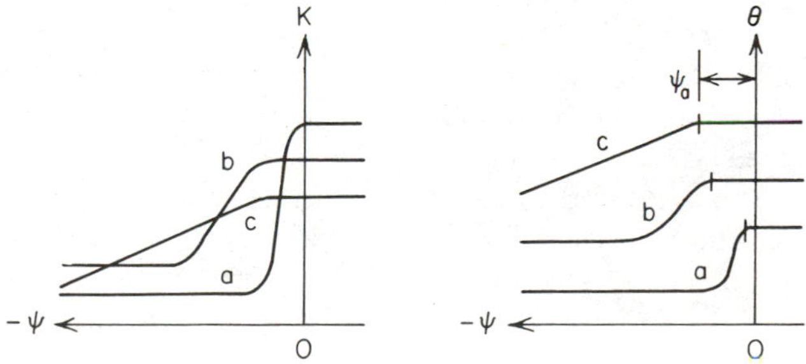

Figure 2.14 Single-valued characteristic curves for three hypothetical soils. (a) Uniform sand; (b) silty sand; (c) silty clay.

Saturated, Unsaturated, and Tension-Saturated Zones

It is worthwhile at this point to summarize the properties of the saturated and unsaturated zones as they have been unveiled thus far. For the *saturated zone*, we can state that:

1. It occurs below the water table.
2. The soil pores are filled with water, and the moisture content θ equals the porosity n.
3. The fluid pressure p is greater than atmospheric, so the pressure head ψ (measured as gage pressure) is greater than zero.
4. The hydraulic head h must be measured with a piezometer.
5. The hydraulic conductivity K is a constant; it is not a function of the pressure head ψ.

For the *unsaturated zone* (or, as it is sometimes called, the *zone of aeration* or the *vadose zone*):

1. It occurs above the water table and above the capillary fringe.
2. The soil pores are only partially filled with water; the moisture content θ is less than the porosity n.
3. The fluid pressure p is less than atmospheric; the pressure head ψ is less than zero.
4. The hydraulic head h must be measured with a tensiometer.
5. The hydraulic conductivity K and the moisture content θ are both functions of the pressure head ψ.

In short, for saturated flow, $\psi > 0$, $\theta = n$, and $K = K_0$; for unsaturated flow, $\psi < 0$, $\theta = \theta(\psi)$, and $K = K(\psi)$.

The capillary fringe fits into neither of the groupings above. The pores there are saturated, but the pressure heads are less than atmospheric. A more descriptive name that is now gaining acceptance is the *tension-saturated zone*. An explanation of its seemingly anomalous properties can be discovered in Figure 2.13. It is the existence of the air entry pressure head $\psi_a < 0$ on the characteristic curves that is responsible for the existence of a capillary fringe. ψ_a is the value of ψ that will exist at the top of the tension-saturated zone, as shown by ψ_A for point A in Figure 2.12(d). Since ψ_a has greater negative values in clay soils than it does in sands, these fine-grained soils develop thicker tension-saturated zones than do coarse-grained soils.

Some authors consider the tension-saturated zone as part of the saturated zone, but in that case the water table is no longer the boundary between the two zones. From a physical standpoint it is probably best to retain all three zones—saturated, tension-saturated, and unsaturated—in one's conception of the complete hydrologic system.

A point that follows directly from the foregoing discussion in this section may warrant a specific statement. In that fluid pressures are less than atmospheric, there can be no natural outflow to the atmosphere from an unsaturated or tension-saturated face. Water can be transferred from the unsaturated zone to the atmosphere by evaporation and transpiration, but natural outflows, such as springs on streambanks or inflows to well bores, must come from the saturated zone. The concept of a saturated seepage face is introduced in Section 5.5 and its importance in relation to hillslope hydrology is emphasized in Section 6.5.

Perched and Inverted Water Tables

The simple hydrologic configuration that we have considered thus far, with a single unsaturated zone overlying the main saturated groundwater body, is a common one. It is the rule where homogeneous geologic deposits extend to some depth. Complex geological environments, on the other hand, can lead to more complex saturated-unsaturated conditions. The existence of a low-permeability clay layer in a high-permeability sand formation, for example, can lead to the formation of a discontinuous saturated lense, with unsaturated conditions existing both above and below. If we consider the line *ABCDA* in Figure 2.15 to be the $\psi = 0$ isobar, we can refer to the *ABC* portion as a *perched water table* and *ADC* as an *inverted water table*. *EF* is the true water table.

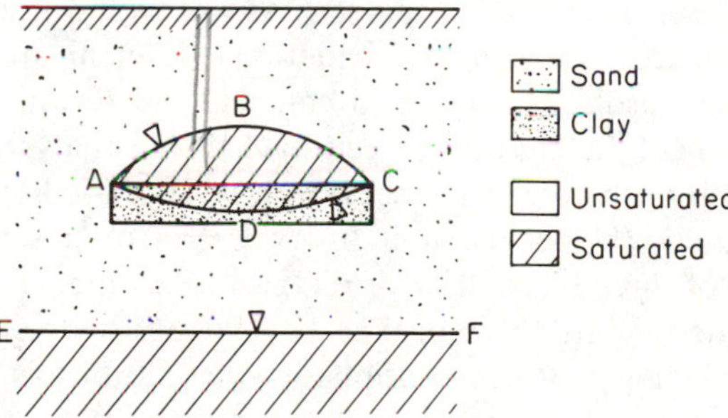

Figure 2.15 Perched water table *ABC*, inverted water table *ADC*, and true water table *EF*.

Saturated conditions can be discontinuous in time as well as space. Heavy rainfall can lead to the formation of a temporary saturated zone at the ground surface, its lower boundary being an inverted water table underlain by unsaturated conditions. Saturated zones of this type dissipate with time under the influence of downward percolation and evaporation from the surface. In Chapter 6 we will examine the interactions of rainfall and infiltration in saturated-unsaturated systems in greater detail.

Multiphase Flow

The approach to unsaturated flow outlined in this section is the one used almost universally by soil physicists, but it is, at root, an approximate method. Unsaturated flow is actually a special case of *multiphase flow* through porous media, with two phases, air and water, coexisting in the pore channels. Let θ_w be the volumetric moisture content (previously denoted by θ) and θ_a be the volumetric air content, defined analogously to θ_w. There are now two fluid pressures to consider: p_w for the water phase and p_a for the air phase; and two pressure heads, ψ_w and ψ_a. Each soil now possesses two characteristic curves of fluid content versus pressure head, one for the water, $\theta_w(\psi_w)$, and one for the air, $\theta_a(\psi_a)$.

When it comes to conductivity relationships, it makes sense to work with the permeability k [Eq. (2.28)] rather than the hydraulic conductivity K, since k is independent of the fluid and K is not. The flow parameters k_w and k_a are called the *effective permeabilities* of the medium to water and air. Each soil has two characteristic curves of effective permeability versus pressure head, one for water, $k_w(\psi_w)$, and one for air, $k_a(\psi_a)$.

The single-phase approach to unsaturated flow leads to techniques of analysis that are accurate enough for almost all practical purposes, but there are some unsaturated flow problems where the multiphase flow of air and water must be considered. These commonly involve cases where a buildup in air pressure in the entrapped air ahead of a wetting front influences the rate of propagation of the front through a soil. Wilson and Luthin (1963) encountered the effects experimentally, Youngs and Peck (1964) provide a theoretical discussion, and McWhorter (1971) presents a complete analysis. As will be shown in Section 6.8, air entrapment also influences water-table fluctuations. Bianchi and Haskell (1966) discuss air entrapment problems in a field context, and Green et al. (1970) describe a field application of the multiphase approach to the analysis of a subsurface flow system.

Much of the original research on multiphase flow through porous media was carried out in the petroleum industry. Petroleum reservoir engineering involves the analysis of three-phase flow of oil, gas, and water. Pirson (1958) and Amyx et al. (1960) are standard references in the field. Stallman (1964) provides an interpretive review of the petroleum multiphase contributions as they pertain to groundwater hydrology.

The two-phase analysis of unsaturated flow is an example of *immiscible displacement*; that is, the fluids displace each other without mixing, and there is a distinct fluid-fluid interface within each pore. The simultaneous flow of two fluids that are soluble in each other is termed *miscible displacement*, and in such cases a distinct fluid-fluid interface does not exist. Bear (1972) provides an advanced theoretical treatment of both miscible and immiscible displacement in porous media. In this text, the only examples of immiscible displacement are those that have been discussed in this subsection. In the rest of the text, unsaturated flow will be treated as a single-phase problem using the concepts and approach of the first part of this section. The most common occurrences of miscible displacement in groundwater hydrology involve the mixing of two waters with different chemistry (such as seawater and fresh-water, or pure water and contaminated water). The transport processes associated with miscible displacement and the techniques of analysis of groundwater contamination will be discussed in Chapter 9.

2.7 Aquifers and Aquitards

Of all the words in the hydrologic vocabulary, there are probably none with more shades of meaning than the term *aquifer*. It means different things to different people, and perhaps different things to the same person at different times. It is used to refer to individual geologic layers, to complete geologic formations, and even to groups of geologic formations. The term must always be viewed in terms of the scale and context of its usage.

Aquifers, Aquitards, and Aquicludes

An *aquifer* is best defined as a saturated permeable geologic unit that can transmit significant quantities of water under ordinary hydraulic gradients. An *aquiclude* is defined as a saturated geologic unit that is incapable of transmitting significant quantities of water under ordinary hydraulic gradients.

An alternative pair of definitions that are widely used in the water-well industry states that an aquifer is permeable enough to yield economic quanitities of water to wells, whereas aquicludes are not.

In recent years the term *aquitard* has been coined to describe the less-permeable beds in a stratigraphic sequence. These beds may be permeable enough to transmit water in quantities that are significant in the study of regional groundwater flow, but their permeability is not sufficient to allow the completion of production wells within them. Most geologic strata are classified as either aquifers or aquitards; very few formations fit the classical definition of an aquiclude. As a result, there is a trend toward the use of the first two of these terms at the expense of the third.

The most common aquifers are those geologic formations that have hydraulic conductivity values in the upper half of the observed range (Table 2.2): unconsolidated sands and gravels, permeable sedimentary rocks such as sandstones and limestones, and heavily fractured volcanic and crystalline rocks. The most common aquitards are clays, shales, and dense crystalline rocks. In Chapter 4, the principal aquifer and aquitard types will be examined more fully within the context of a discussion on geological controls on groundwater occurrence.

The definitions of aquifer and aquitard are purposely imprecise with respect to hydraulic conductivity. This leaves open the possibility of using the terms in a relative sense. For example, in an interlayered sand-silt sequence, the silts may be considered aquitards, whereas in a silt-clay system, they are aquifers.

Aquifers are often called by their stratigraphic names. The Dakota Sandstone, for example, owes its geological fame largely to Meinzer's (1923) assessment of its properties as an aquifer. Two other well-known North American aquifers are the St. Peter Sandstone in Illinois and the Ocala Limestone in Florida. A summary of the principal aquifer systems in the United States can be found in McGuinness (1963) and Maxey (1964), who build on the earlier compilations of Meinzer (1923), Tolman (1937), and Thomas (1951). Brown (1967) provides information on Canada's major aquifers.

In the ideal world of analysis where many of the expositary sections of this book must reside, aquifers tend to appear as homogeneous, isotropic formations of constant thickness and simple geometry. We hope the reader will bear in mind that the real world is somewhat different. The hydrogeologist constantly faces complex aquifer-aquitard systems of heterogeneous and anisotropic formations rather than the idealized cases pictured in texts. It will often seem that the geological processes have maliciously conspired to maximize the interpretive and analytical difficulties.

Confined and Unconfined Aquifers

A *confined aquifer* is an aquifer that is confined between two aquitards. An *unconfined aquifer*, or *water-table aquifer*, is an aquifer in which the water table forms the upper boundary. Confined aquifers occur at depth, unconfined aquifers near the ground surface (Figure 2.16). A saturated lense that is bounded by a perched water table (Figure 2.15) is a special case of an unconfined aquifer. Such lenses are sometimes called *perched aquifers*.

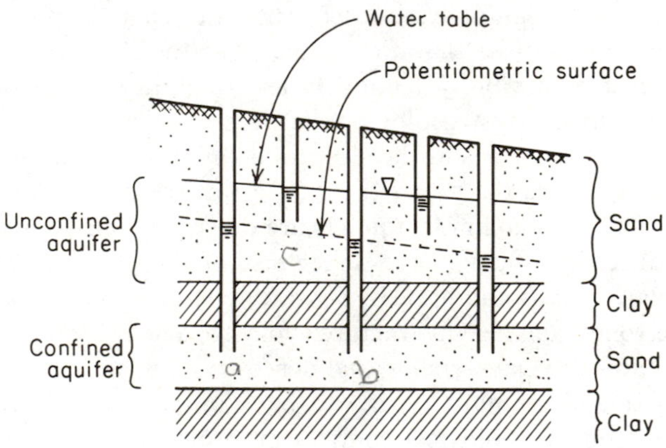

Figure 2.16 Unconfined aquifer and its water table; confined aquifer and its potentiometric surface.

In a confined aquifer, the water level in a well usually rises above the top of the aquifer. If it does, the well is called an *artesian well* and the aquifer is said to exist under *artesian conditions*. In some cases the water level may rise above the ground surface, in which case the well is known as a *flowing artesian well* and the aquifer is said to exist under *flowing artesian conditions*. In Section 6.1, we will examine the topographic and geologic circumstances that lead to flowing artesian conditions. The water level in a well in an unconfined aquifer rests at the water table.

Potentiometric Surface

For confined aquifers, which are extensively tapped by wells for water supply, there has grown up a traditional concept that is not particularly sound but which is firmly entrenched in usage. If the water-level elevations in wells tapping a confined aquifer are plotted on a map and contoured, the resulting surface, which is actually a map of the hydraulic head in the aquifer, is called a *potentiometric surface*. A potentiometric map of an aquifer provides an indication of the directions of groundwater flow in the aquifer.

The concept of a potentiometric surface is only rigorously valid for horizontal flow in horizontal aquifers. The condition of horizontal flow is met only in aquifers with hydraulic conductivities that are much higher than those in the associated confining beds. Some hydrogeological reports contain potentiometric surface maps based on water-level data from sets of wells that bottom near the same elevation but that are not associated with a specific well-defined confined aquifer. This type of potentiometric surface is essentially a map of hydraulic head contours on a two-dimensional horizontal cross section taken through the three-dimensional hydraulic head pattern that exists in the subsurface in that area. If there are vertical components of flow, as there usually are, calculations and interpretations based on this type of potentiometric surface can be grossly misleading.

It is also possible to confuse a potentiometric surface with the water table in areas where both confined and unconfined aquifers exist. Figure 2.16 schematically distinguishes between the two. In general, as we shall see from the flow nets in Chapter 6, the two do not coincide.

2.8 Steady-State Flow and Transient Flow

Steady-state flow occurs when at any point in a flow field the magnitude and direction of the flow velocity are constant with time. *Transient flow* (or *unsteady flow*, or *nonsteady flow*) occurs when at any point in a flow field the magnitude or direction of the flow velocity changes with time.

Figure 2.17(a) shows a steady-state groundwater flow pattern (dashed equipotentials, solid flowline) through a permeable alluvial deposit beneath a concrete dam. Along the line AB, the hydraulic head $h_{AB} = 1000$ m. It is equal to the elevation of the surface of the reservoir above AB. Similarly, $h_{CD} = 900$ m (the elevation of the tailrace pond above CD). The hydraulic head drop Δh across the system is 100 m. If the water level in the reservoir above AB and the water level in the tailrace pond above CD do not change with time, the flow net beneath the dam will not change with time. The hydraulic head at point E, for example, will be $h_E = 950$ m and will remain constant. Under such circumstances the velocity $v = -K\, \partial h/\partial l$ will also remain constant through time. In a steady-state flow system the velocity may vary from point to point, but it will not vary with time at any given point.

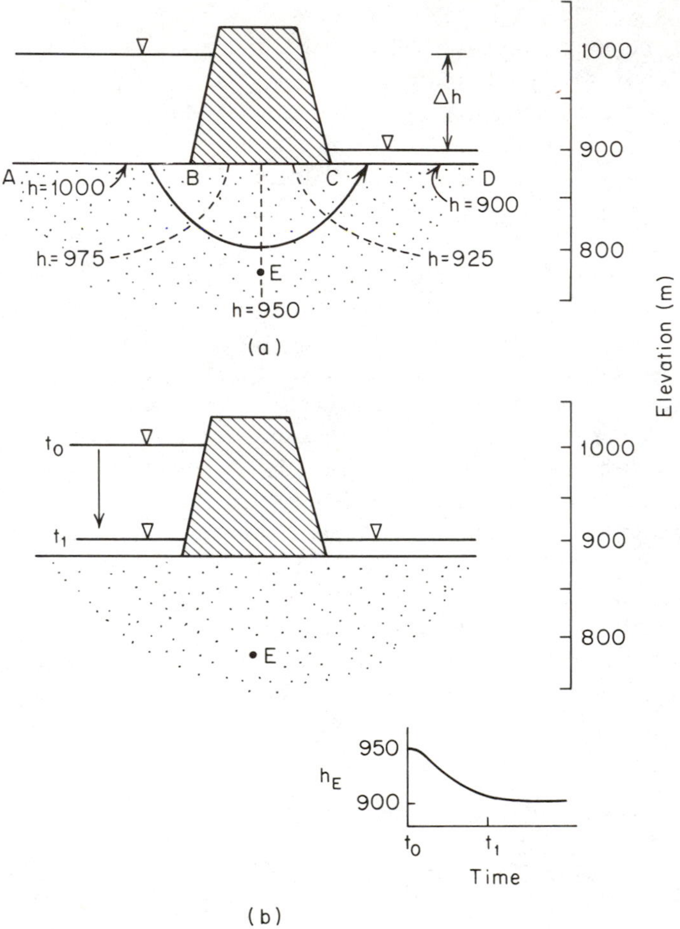

Figure 2.17 Steady-state and transient groundwater flow beneath a dam.

Let us now consider the transient flow problem schematically shown in Figure 2.17(b). At time t_0 the flow net beneath the dam will be identical to that of Figure 2.17(a) and h_E will be 950 m. If the reservoir level is allowed to drop over the period t_0 to t_1, until the water levels above and below the dam are identical at time t_1, the ultimate conditions under the dam will be static with no flow of water from the upstream to the downstream side. At point E the hydraulic head h_E will undergo a time-dependent decline from $h_E = 950$ m at time t_0 to its ultimate value of $h_E = 900$ m. There may well be time lag in such a system so that h_E will not necessarily reach the value $h_E = 900$ m until some time after $t = t_1$.

One important difference between steady and transient systems lies in the relation between their flowlines and pathlines. *Flowlines* indicate the instantaneous directions of flow throughout a system (at all times in a steady system, or at a given instant in time in a transient system). They must be orthogonal to the equipotential lines throughout the region of flow at all times. *Pathlines* map the route that an individual particle of water follows through a region of flow during a steady or transient event. In a steady flow system a particle of water that enters the system at an inflow boundary will flow toward the outflow boundary along a pathline that coincides with a flowline such as that shown in Figure 2.17(a). In a transient flow system, on the other hand, pathlines and flowlines do not coincide. Although a flow net can be constructed to describe the flow conditions at any given instant in time in a transient system, the flowlines shown in such a snapshot represent only the directions of movement at that instant in time. In that the configuration of flowlines changes with time, the flowlines cannot describe, in themselves, the complete path of a particle of water as it traverses the system. The delineation of transient pathlines has obvious importance in the study of groundwater contamination.

A groundwater hydrologist must understand the techniques of analysis for both steady-state flow and transient flow. In the final sections of this chapter the equations of flow will be developed for each type of flow, under both saturated and unsaturated conditions. The practical methodology that is presented in later chapters is often based on the theoretical equations, but it is not usually necessary for the practicing hydrogeologist to have the mathematics at his or her fingertips. The primary application of steady-state techniques in groundwater hydrology is in the analysis of regional groundwater flow. An understanding of transient flow is required for the analysis of well hydraulics, groundwater recharge, and many of the geochemical and geotechnical applications.

2.9 Compressibility and Effective Stress

The analysis of transient groundwater flow requires the introduction of the concept of *compressibility*, a material property that describes the change in volume, or strain, induced in a material under an applied stress. In the classical approach to the strength of elastic materials, the *modulus of elasticity* is a more familiar material property. It is defined as the ratio of the change in stress $d\sigma$ to the resulting change in strain $d\epsilon$. Compressibility is simply the inverse of the modulus of elasticity. It is defined as strain/stress, $d\epsilon/d\sigma$, rather than stress/strain, $d\sigma/d\epsilon$. The term is utilized for both elastic materials and nonelastic materials. For the flow of water through porous media, it is necessary to define two compressibility terms, one for the water and one for the porous media.

Compressibility of Water

Stress is imparted to a fluid through the fluid pressure p. An increase in pressure dp leads to a decrease in the volume V_w of a given mass of water. The *compressibility of water* β is therefore defined as

$$\beta = \frac{-dV_w/V_w}{dp} \tag{2.44}$$

The negative sign is necessary if we wish β to be a positive number.

Equation (2.44) implies a linear elastic relationship between the volumetric strain dV_w/V_w and the stress induced in the fluid by the change in fluid pressure dp. The compressibility β is thus the slope of the line relating strain to stress for water, and this slope does not change over the range of fluid pressures encountered in groundwater hydrology (including those less than atmospheric that are encountered in the unsaturated zone). For the range of groundwater temperatures that are usually encountered, temperature has a small influence on β, so that for most practical purposes we can consider β a constant. The dimensions of β are the inverse of those for pressure or stress. Its value can be taken as 4.4×10^{-10} m^2/N (or Pa^{-1}).

For a given mass of water it is possible to rewrite Eq. (2.44) in the form

$$\beta = \frac{d\rho/\rho}{dp} \tag{2.45}$$

where ρ is the fluid density. Integration of Eq. (2.45) yields the *equation of state* for water:

$$\rho = \rho_0 \exp\,[\beta(p - p_0)] \tag{2.46}$$

where ρ_0 is the fluid density at the datum pressure p_0. For p_0 atmospheric, Eq. (2.46) can be written in terms of gage pressures as

$$\rho = \rho_0 e^{\beta p} \tag{2.47}$$

An incompressible fluid is one for which $\beta = 0$ and $\rho = \rho_0 =$ constant.

Effective Stress

Let us now consider the compressibility of the porous medium. Assume that a stress is applied to a unit mass of saturated sand. There are three mechanisms by which a reduction in volume can be achieved: (1) by compression of the water in the pores, (2) by compression of the individual sand grains, and (3) by a rearrangement of the sand grains into a more closely packed configuration. The first of these mechanisms is controlled by the fluid compressibility β. Let us assume that the second mechanism is negligible, that is, that the individual soil grains are incompressible. Our task is to define a compressibility term that will reflect the third mechanism.

To do so, we will have to invoke the principle of effective stress. This concept was first proposed by Terzaghi (1925), and has been analyzed in detail by Skempton

(1961). Most soil mechanics texts, such as those by Terzaghi and Peck (1967) and Scott (1963), provide a full discussion.

For our purposes, consider the stress equilibrium on an arbitrary plane through a saturated geological formation at depth (Figure 2.18). σ_T is the total stress acting downward on the plane. It is due to the weight of overlying rock and water. This stress is borne in part by the granular skeleton of the porous medium and in part by the fluid pressure p of the water in the pores. The portion of the total stress that is not borne by the fluid is called the *effective stress* σ_e. It is this stress that is actually applied to the grains of the porous medium. Rearrangement of the soil grains and the resulting compression of the granular skeleton is caused by changes in the effective stress, not by changes in the total stress. The two are related by the simple equation

$$\sigma_T = \sigma_e + p \tag{2.48}$$

or, in terms of the changes,

$$d\sigma_T = d\sigma_e + dp \tag{2.49}$$

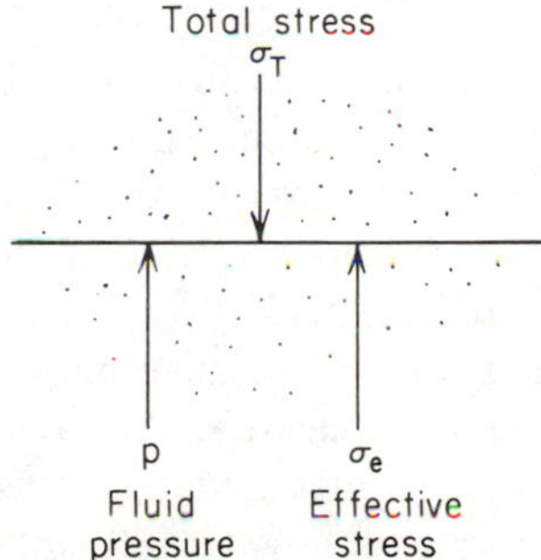

Figure 2.18 Total stress, effective stress, and fluid pressure on an arbitrary plane through a saturated porous medium.

Many of the transient subsurface flow problems that must be analyzed do not involve changes in the total stress. The weight of rock and water overlying each point in the system often remains essentially constant through time. In such cases, $d\sigma_T = 0$ and

$$d\sigma_e = -dp \tag{2.50}$$

Under these circumstances, if the fluid pressure increases, the effective stress decreases by an equal amount; and if the fluid pressure decreases, the effective stress increases by an equal amount. For cases where the total stress does not change with time, the effective stress at any point in the system, and the resulting volumetric deformations there, are controlled by the fluid pressures at that point. Since $p = \rho g \psi$ and $\psi = h - z$ (z being constant at the point in question), changes in the effective stress at a point are in effect governed by changes in the hydraulic head at that point:

$$d\sigma_e = -\rho g\, d\psi = -\rho g\, dh \tag{2.51}$$

Compressibility of a Porous Medium

The compressibility of a porous medium is defined as

$$\alpha = \frac{-dV_T/V_T}{d\sigma_e} \tag{2.52}$$

where V_T is the total volume of a soil mass and $d\sigma_e$ the change in effective stress.

Recall that $V_T = V_S + V_v$, where V_S is the volume of the solids and V_v is the volume of the water-saturated voids. An increase in effective stress $d\sigma_e$ produces a reduction dV_T in the total volume of the soil mass. In granular materials this reduction occurs almost entirely as a result of grain rearrangements. It is true that individual grains may themselves be compressible, but the effect is usually considered to be negligible. In general, $dV_T = dV_S + dV_v$; but for our purposes we will assume that $dV_S = 0$ and $dV_T = dV_v$.

Consider a sample of saturated soil that has been placed in a laboratory loading cell such as the one shown in Figure 2.19(a). A total stress $\sigma_T = L/A$ can be applied to the sample through the pistons. The sample is laterally confined by the cell walls, and entrapped water is allowed to escape through vents in the pistons to an external pool held at a constant known fluid pressure. The volumetric reduction in the size of the soil sample is measured at several values of L as L is increased in a stepwise fashion. At each step, the increased total stress $d\sigma_T$ is initially borne by the water under increased fluid pressures, but drainage of the water from the sample to the external pool slowly transfers the stress from the water to the granular skeleton. This transient process is known as *consolidation*, and the time required for the consolidation process to reach hydraulic equilibrium at each L can be considerable. Once attained, however, it is known that $dp = 0$ within the sample, and from Eq. (2.49), $d\sigma_e = d\sigma_T = dL/A$. If the soil sample has an original void ratio e_0 (where $e = V_v/V_s$) and an original height b [Figure 2.19(a)], and assuming that $dV_T = dV_v$, Eq. (2.52) can be written as

$$\alpha = \frac{-db/b}{d\sigma_e} = \frac{-de/(1+e_0)}{d\sigma_e} \tag{2.53}$$

The compressibility α is usually determined from the slope of a strain-stress plot in the form of e versus σ_e. The curve AB in Figure 2.19(b) is for loading (increasing σ_e), BC is for unloading (decreasing σ_e). In general, the strain-stress relation is neither linear nor elastic. In fact, for repeated loadings and unloadings, many fine-grained soils show hysteretic properties [Figure 2.19(c)]. The soil compressibility α, unlike the fluid compressibility β, is not a constant; it is a function of the applied stress and it is dependent on the previous loading history.

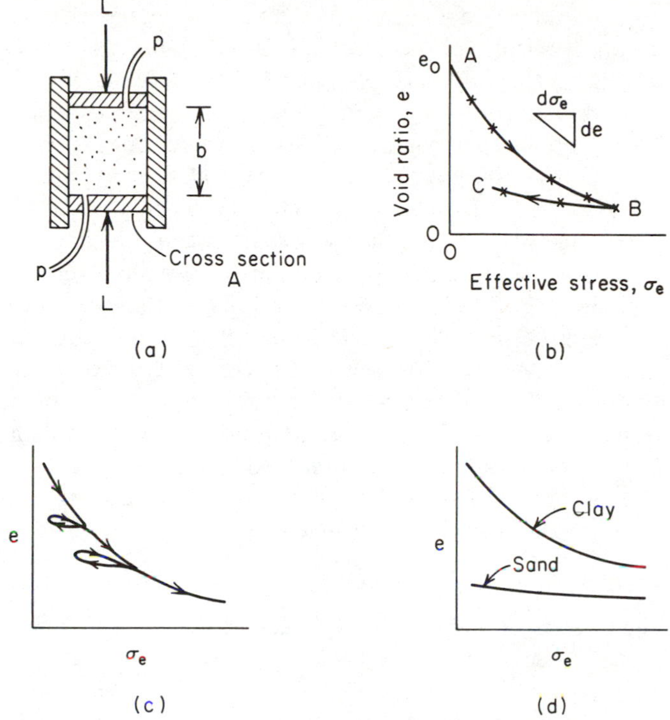

Figure 2.19 (a) Laboratory loading cell for the determination of soil compressibility; (b), (c), and (d) schematic curves of void ratio versus effective stress.

Figure 2.19(d) provides a schematic comparison of the e–σ_e curves for clay and sand. The lesser slope for the sand curve implies a smaller α, and its linearity implies an α value that stays constant over a wide range of σ_e. In groundwater systems, the time-dependent fluctuations in σ_e are often quite small, so that even for clays, a constant α can have some meaning. Table 2.5 is a table of compres-

Table 2.5 Range of Values of Compressibility*

	Compressibility, α (m^2/N or Pa^{-1})
Clay	10^{-6}–10^{-8}
Sand	10^{-7}–10^{-9}
Gravel	10^{-8}–10^{-10}
Jointed rock	10^{-8}–10^{-10}
Sound rock	10^{-9}–10^{-11}
Water (β)	4.4×10^{-10}

*See Table A1.3, Appendix I, for conversion factors.

sibility values that indicates the ranges of values that have been measured for various types of geologic materials. Original sources of compressibility data include Domenico and Mifflin (1965) and Johnson et al. (1968). The dimensions of α, like β, are the inverse of those for stress. Values are expressed in SI units of m^2/N or Pa^{-1}. Note that the compressibility of water is of the same order of magnitude as the compressibility of the less-compressible geologic materials.

As noted in Figure 2.19(b) and (c), the compressibility of some soils in expansion (expansibility?) is much less than in compression. For clays, the ratio of the two α's is usually on the order of 10 : 1; for uniform sands, it approaches 1 : 1. For soils that have compressibility values that are significantly less in expansion than compression, volumetric deformations that occur in response to increasing effective stress [perhaps due to decreasing hydraulic heads as suggested by Eq. (2.51)] are largely irreversible. They are not recovered when effective stresses subsequently decrease. In a clay-sand aquifer-aquitard system, the large compactions that can occur in the clay aquitards (due to the large α values) are largely irrecoverable; whereas the small deformations that occur in the sand aquifers (due to the small α values) are largely elastic.

Aquifer Compressibility

The concept of compressibility inherent in Eq. (2.53) and in Figures 2.18 and 2.19 is one-dimensional. In the field, at depth, a one-dimensional concept has meaning if it is assumed that the soils and rocks are stressed only in the vertical direction. The total vertical stress σ_T at any point is due to the weight of overlying rock and water; the neighboring materials provide the horizontal confinement. The effective vertical stress σ_e is equal to $\sigma_T - p$. Under these conditions the *aquifer compressibility* α is defined by the first equality of Eq. (2.53), where b is now the aquifer thickness rather than a sample height. The parameter α is a vertical compressibility. If it is to be determined with a laboratory apparatus like that of Figure 2.19(a), the soil cores must be oriented vertically and loading must be applied at right angles to any horizontal bedding. Within an aquifer, α may vary with horizontal position; that is, α may be heterogeneous with $\alpha = \alpha(x, y)$.

In the most general analysis, it must be recognized that the stress field existing at depth is not one-dimensional but three-dimensional. In that case, aquifer compressibility must be considered as an anisotropic parameter. The vertical compressibility α is then invoked by changes in the vertical component of effective stress, and the horizontal compressibilities are invoked by changes in the horizontal components of effective stress. Application of the concepts of three-dimensional stress analysis in the consideration of fluid flow through porous media is an advanced topic that cannot be pursued here. Fortunately, for many practical cases, changes in the horizontal stress field are very small, and in most analyses it can be assumed that they are negligible. It is sufficient for our purposes to think of the aquifer compressibility α as a single isotropic parameter, but it should be kept in mind that it is actually the compressibility in the vertical direction, and that this is the only direction in which large changes in effective stress are anticipated.

To illustrate the nature of the deformations that can occur in compressible aquifers, consider the aquifer of thickness b shown in Figure 2.20. If the weight of overlying material remains constant and the hydraulic head in the aquifer is decreased by an amount $-dh$, the increase in effective stress $d\sigma_e$ is given by Eq. (2.51) as $\rho g\, dh$, and the aquifer compaction, from Eq. (2.53) is

$$db = -\alpha b\, d\sigma_e = -\alpha b \rho g\, dh \tag{2.54}$$

The minus sign indicates that the decrease in head produces a reduction in the thickness b.

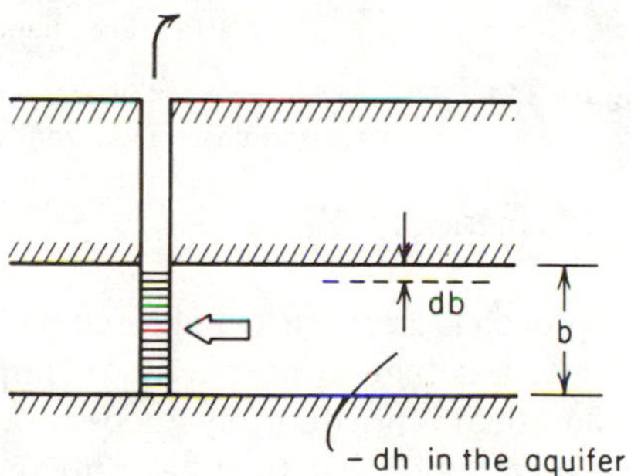

Figure 2.20 Aquifer compaction caused by groundwater pumping.

One way that the hydraulic head might be lowered in an aquifer is by pumping from a well. Pumping induces horizontal hydraulic gradients toward the well in the aquifer, and as a result the hydraulic head is decreased at each point near the well. In response, effective stresses are increased at these points, and aquifer compaction results. Conversely, pumping water into an aquifer increases hydraulic heads, decreases effective stresses, and causes aquifer expansion. If the compaction of an aquifer-aquitard system due to groundwater pumping is propagated to the ground surface, the result is *land subsidence*. In Section 8.12 this phenomenon is considered in detail.

Effective Stress in the Unsaturated Zone

The first equality in Eq. (2.51) indicates that the relationship between the effective stress σ_e and the presure head ψ should be linear. This relation, and the concept of Figure 2.18 on which it is based, holds in the saturated zone, but there is now abundant evidence to suggest that it does not hold in the unsaturated zone (Narasimhan, 1975). For unsaturated flow, Bishop and Blight (1963) suggest that Eq. (2.51) should be modified to read

$$d\sigma_e = -\rho g \chi\, d\psi \tag{2.55}$$

where the parameter χ depends on the degree of saturation, the soil structure, and the wetting-drying history of the soil. Curve ABC in Figure 2.21 shows such a

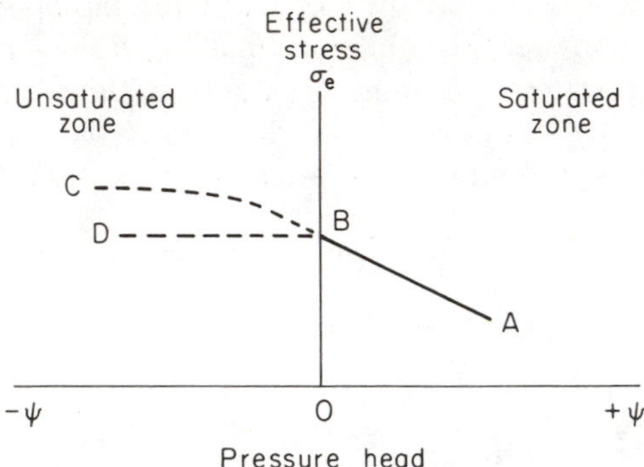

Figure 2.21 Relationship between effective stress and pressure head in the saturated and unsaturated zones (after Narasimhan, 1975).

relationship schematically. For $\psi > 0$, $\chi = 1$; for $\psi < 0$, $\chi \leq 1$; and for $\psi \ll 0$, $\chi = 0$.

The χ approach is an empirical one and its use reflects the fact that the capacity of fluid pressures less than atmospheric to support a part of the total stress in an unsaturated flow field is not yet fully understood. As a first approximation, it is not unreasonable to suppose that they have no such capacity, as suggested by curve *ABD* in Figure 2.21. Under this assumption, for $\psi < 0$, $\chi = 0$, $d\sigma_e = d\sigma_T$, and changes in pressure head (or moisture content) in the unsaturated zone do not lead to changes in effective stress.

The definition of the compressibility of a porous medium in the unsaturated zone is still given by Eq. (2.52) just as it is in the saturated zone, but the influence of the fluid pressure on the effective stress is now considered to be muted or non-existent.

2.10 Transmissivity and Storativity

There are six basic physical properties of fluid and porous media that must be known in order to describe the hydraulic aspects of saturated groundwater flow. These six have all now been introduced. They are, for the water, *density* ρ, *viscosity* μ, and *compressibility* β; and for the media, *porosity* n (or void ratio e), *permeability* k, and *compressibility* α. All the other parameters that are used to describe the hydrogeologic properties of geologic formations can be derived from these six. For example, we have seen from Eq. (2.28) that the saturated hydraulic conductivity K is a combination of k, ρ, and μ. In this section, we will consider the concepts of specific storage S_s, storativity S, and transmissivity T.

Specific Storage

The *specific storage* S_s of a saturated aquifer is defined as the volume of water that a unit volume of aquifer releases from storage under a unit decline in hydraulic head. From Section 2.9 we now know that a decrease in hydraulic head h infers a decrease in fluid pressure p and an increase in effective stress σ_e. The water that is released from storage under conditions of decreasing h is produced by two mechanisms: (1) the compactions of the aquifer caused by increasing σ_e, and (2) the expansion of the water caused by decreasing p. The first of these mechanisms is controlled by the aquifer compressibility α and the second by the fluid compressibility β.

Let us first consider the water produced by the compaction of the aquifer. The volume of water expelled from the unit volume of aquifer during compaction will be equal to the reduction in volume of the unit volume of aquifer. The volumetric reduction dV_T will be negative, but the amount of water produced dV_w will be positive, so that, from Eq. (2.52),

$$dV_w = -dV_T = \alpha V_T \, d\sigma_e \tag{2.56}$$

For a unit volume, $V_T = 1$, and from Eq. (2.51), $d\sigma_e = -\rho g \, dh$. For a unit decline in hydraulic head, $dh = -1$, and we have

$$dV_w = \alpha \rho g \tag{2.57}$$

Now consider the volume of water produced by the expansion of the water. From Eq. (2.44),

$$dV_w = -\beta V_w \, dp \tag{2.58}$$

The volume of water V_w in the total unit volume V_T is nV_T, where n is the porosity. With $V_T = 1$ and $dp = \rho g \, d\psi = \rho g \, d(h - z) = \rho g \, dh$, Eq. (2.58) becomes, for $dh = -1$,

$$dV_w = \beta n \rho g \tag{2.59}$$

The specific storage S_s is the sum of the two terms given by Eqs. (2.57) and (2.59):

$$S_s = \rho g(\alpha + n\beta) \tag{2.60}$$

A dimensional inspection of this equation shows S_s to have the peculiar dimensions of $[L]^{-1}$. This also follows from the definition of S_s as a volume per volume per unit decline in head.

Transmissivity and Storativity of a Confined Aquifer

For a confined aquifer of thickness b, the *transmissivity* (or *transmissibility*) T is defined as

$$T = Kb \tag{2.61}$$

and the *storativity* (or *storage coefficient*) S is defined as

$$S = S_s b \tag{2.62}$$

If we substitute Eq. (2.60) in Eq. (2.62), the expanded definition of S is seen to be

$$S = \rho g b(\alpha + n\beta) \tag{2.63}$$

The storativity of a saturated confined aquifer of thickness b can be defined in words as the volume of water that an aquifer releases from storage per unit surface area of aquifer per unit decline in the component of hydraulic head normal to that surface. The hydraulic head for a confined aquifer is usually displayed in the form of a potentiometric surface, and Figure 2.22(a) illustrates the concept of storativity in this light.

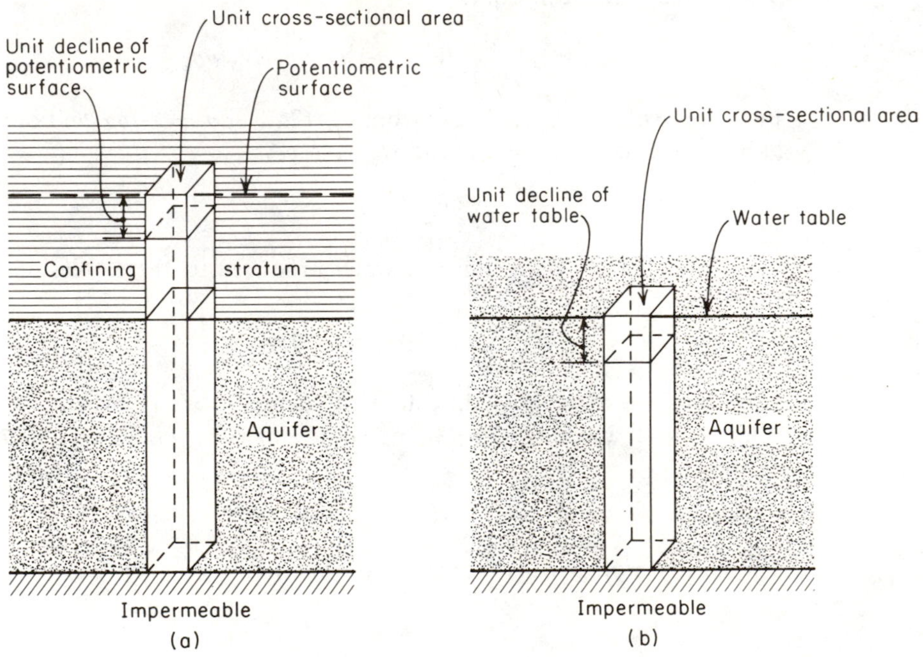

Figure 2.22 Schematic representation of the storativity in (a) confined and (b) unconfined aquifers (after Ferris et al., 1962).

In that the hydraulic conductivity K has dimensions $[L/T]$, it is clear from Eq. (2.61) that the transmissivity T has dimensions $[L^2/T]$. The SI metric unit is m^2/s. T and S are widely used terms in the North American water well industry and are often expressed in FPS engineering units. If K is expressed in gal/day/ft^2, then T has units of gal/day/ft. The range of values of T can be calculated by multiplying the pertinent K values from Table 2.2 by the range of reasonable aquifer thicknesses, say 5–100 m. Transmissivities greater than 0.015 m^2/s (or 0.16 ft^2/s or 100,000 gal/day/ft) represent good aquifers for water well exploitation. Storativities are dimensionless. In confined aquifers, they range in value from 0.005 to 0.00005. Reference to the definition of S, coupled with a realization of its range of values, makes it clear that large head changes over extensive areas are required to produce substantial water yields from confined aquifers.

Transmissivities and storativities can be specified for aquitards as well as aquifers. However, in most applications, the vertical hydraulic conductivity of an aquitard has more significance than its transmissivity. It might also be noted that in clay aquitards, $\alpha \gg \beta$, and the $n\beta$ term in the definition of storativity [Eq. (2.63)] and specific storage [Eq. (2.60)] becomes negligible.

It is possible to define a single formation parameter that couples the transmission properties T or K, and the storage properties S or S_s. The *hydraulic diffusivity* D is defined as

$$D = \frac{T}{S} = \frac{K}{S_s} \tag{2.64}$$

The term is not widely used in practice.

The concepts of *transmissivity* T and *storativity* S were developed primarily for the analysis of well hydraulics in confined aquifers. For two-dimensional, horizontal flow toward a well in a confined aquifer of thickness b, the terms are well defined; but they lose their meaning in many other groundwater applications. If a groundwater problem has three-dimensional overtones, it is best to revert to the use of *hydraulic conductivity* K and *specific storage* S_s; or perhaps even better, to the fundamental parameters *permeability* k, *porosity* n, and *compressibility* α.

Transmissivity and Specific Yield in Unconfined Aquifers

In an unconfined aquifer, the transmissivity is not as well defined as in a confined aquifer, but it can be used. It is defined by the same equation [Eq. (2.61)], but b is now the saturated thickness of the aquifer or the height of the water table above the top of the underlying aquitard that bounds the aquifer.

The storage term for unconfined aquifers is known as the *specific yield* S_y. It is defined as the volume of water that an unconfined aquifer releases from storage per unit surface area of aquifer per unit decline in the water table. It is sometimes called the unconfined storativity. Figure 2.22(b) illustrates the concept schematically.

The idea of specific yield is best visualized with reference to the saturated-unsaturated interaction it represents. Figure 2.23 shows the water-table position and the vertical profile of moisture content vs. depth in the unsaturated zone at two times, t_1 and t_2. The crosshatched area represents the volume of water released from storage in a column of unit cross section. If the water-table drop represents a unit decline, the crosshatched area represents the specific yield.

The specific yields of unconfined aquifers are much higher than the storativities of confined aquifers. The usual range of S_y is 0.01–0.30. The higher values reflect the fact that releases from storage in unconfined aquifers represent an actual dewatering of the soil pores, whereas releases from storage in confined aquifers represent only the secondary effects of water expansion and aquifer compaction

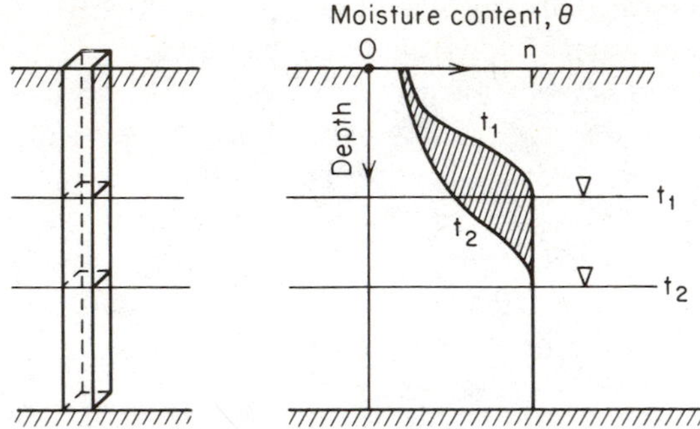

Figure 2.23 Concept of specific yield viewed in terms of the unsaturated moisture profiles above the water table.

caused by changes in the fluid pressure. The favorable storage properties of unconfined aquifers make them more efficient for exploitation by wells. When compared to confined aquifers, the same yield can be realized with smaller head changes over less extensive areas.

Storage in the Unsaturated Zone

In an unsaturated soil, changes in moisture content θ, such as those shown in Figure 2.23, are accompanied by changes in the pressure head ψ, through the $\theta(\psi)$ relationship displayed on the characteristic curve of Figure 2.13(a). The slope of this characteristic curve represents the unsaturated storage property of a soil. It is called the *specific moisture capacity* C and is defined as

$$C = \frac{d\theta}{d\psi} \tag{2.65a}$$

An increase of $d\psi$ in the pressure head (say, from -200 cm to -100 cm on Figure 2.13) must be accompanied by an increase of $d\theta$ in the moisture stored in the unsaturated soil. Since $\theta(\psi)$ is nonlinear and hysteretic, so too, is C. It is not a constant; it is a function of the pressure head $\psi : C = C(\psi)$. In the saturated zone, in fact for all $\psi > \psi_a$, the moisture content θ is equal to the porosity n, a constant, so that $C = 0$. A parallel formulation to Eq. (2.42) for C is

$$\begin{aligned} C &= C(\psi) \qquad \psi < \psi_a \\ C &= 0 \qquad\qquad \psi \geq \psi_a \end{aligned} \tag{2.65b}$$

The transmission and storage properties of an unsaturated soil are fully specified by the characteristic curve $K(\psi)$ and one of the two curves $\theta(\psi)$ or $C(\psi)$.

In an analogous manner to Eq. (2.64), the *soil-water diffusivity* can be defined as

$$D(\psi) = \frac{K(\psi)}{C(\psi)} \tag{2.66}$$

2.11 Equations of Groundwater Flow

In almost every field of science and engineering the techniques of analysis are based on an understanding of the physical processes, and in most cases it is possible to describe these processes mathematically. Groundwater flow is no exception. The basic law of flow is Darcy's law, and when it is put together with an equation of continuity that describes the conservation of fluid mass during flow through a porous medium, a partial differential equation of flow is the result. In this section, we will present brief developments of the equations of flow for (1) steady-state saturated flow, (2) transient saturated flow, and (3) transient unsaturated flow. All three of the equations of flow are well known to mathematicians, and mathematical techniques for their manipulation are widely available and in common use in science and engineering. Generally, the equation of flow appears as one component of a *boundary-value problem*, so in the last part of this section we will explore this concept.

In that so many of the standard techniques of analysis in groundwater hydrology are based on boundary-value problems that involve partial differential equations, it is useful to have a basic understanding of these equations as one proceeds to learn the various techniques. Fortunately, it is not an absolute requirement. In most cases, the techniques can be explained and understood without returning at every step to the fundamental mathematics. The research hydrogeologist must work with the equations of flow on a daily basis; the practicing hydrogeologist can usually avoid the advanced mathematics if he or she so desires.

Steady-State Saturated Flow

Consider a unit volume of porous media such as that shown in Figure 2.24. Such an element is usually called an *elemental control volume*. The law of conservation of mass for steady-state flow through a saturated porous medium requires that the rate of fluid mass flow *into* any elemental control volume be equal to the rate of fluid mass flow *out* of any elemental control volume. The *equation of continuity* that

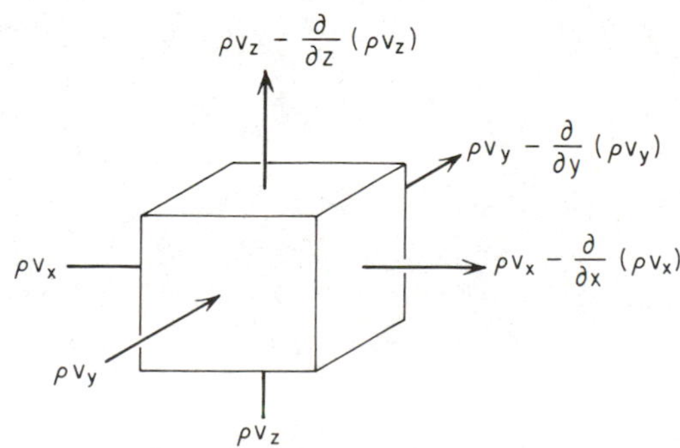

Figure 2.24 Elemental control volume for flow through porous media.

translates this law into mathematical form can be written, with reference to Figure 2.24, as

$$-\frac{\partial(\rho v_x)}{\partial x}-\frac{\partial(\rho v_y)}{\partial y}-\frac{\partial(\rho v_z)}{\partial z}=0 \tag{2.67}$$

A quick dimensional analysis on the ρv terms will show them to have the dimensions of a mass rate of flow across a unit cross-sectional area of the elemental control volume. If the fluid is incompressible, $\rho(x, y, z)$ = constant and the ρ's can be removed from Eq. (2.67). Even if the fluid is compressible and $\rho(x, y, z) \neq$ constant, it can be shown that terms of the form $\rho\, \partial v_x/\partial x$ are much greater than terms of the form $v_x\, \partial \rho/\partial x$, both of which arise when the chain rule is used to expand Eq. (2.67). In either case, Eq. (2.67) simplifies to

$$-\frac{\partial v_x}{\partial x}-\frac{\partial v_y}{\partial y}-\frac{\partial v_z}{\partial z}=0 \tag{2.68}$$

Substitution of Darcy's law for v_x, v_y, and v_z in Eq. (2.68) yields the equation of flow for steady-state flow through an anisotropic saturated porous medium:

$$\frac{\partial}{\partial x}\left(K_x\frac{\partial h}{\partial x}\right)+\frac{\partial}{\partial y}\left(K_y\frac{\partial h}{\partial y}\right)+\frac{\partial}{\partial z}\left(K_z\frac{\partial h}{\partial z}\right)=0 \tag{2.69}$$

For an isotropic medium, $K_x = K_y = K_z$, and if the medium is also homogeneous, then $K(x, y, z)$ = constant. Equation (2.69) then reduces to the equation of flow for steady-state flow through a homogeneous, isotropic medium:

$$\frac{\partial^2 h}{\partial x^2}+\frac{\partial^2 h}{\partial y^2}+\frac{\partial^2 h}{\partial z^2}=0 \tag{2.70}$$

Equation (2.70) is one of the most basic partial differential equations known to mathematicians. It is called *Laplace's equation*. The solution of the equation is a function $h(x, y, z)$ that describes the value of the hydraulic head h at any point in a three-dimensional flow field. A solution to Eq. (2.70) allows us to produce a contoured equipotential map of h, and with the addition of flowlines, a flow net.

For steady-state, saturated flow in a two-dimensional flow field, say in the xz plane, the central term of Eq. (2.70) would drop out and the solution would be a function $h(x, z)$.

Transient Saturated Flow

The law of conservation of mass for transient flow in a saturated porous medium requires that the *net* rate of fluid mass flow into any elemental control volume be equal to the time rate of change of fluid mass storage within the element. With reference to Figure 2.24, the equation of continuity takes the form

$$-\frac{\partial(\rho v_x)}{\partial x}-\frac{\partial(\rho v_y)}{\partial y}-\frac{\partial(\rho v_z)}{\partial z}=\frac{\partial(\rho n)}{\partial t} \tag{2.71}$$

or, expanding the right-hand side,

$$-\frac{\partial(\rho v_x)}{\partial x}-\frac{\partial(\rho v_y)}{\partial y}-\frac{\partial(\rho v_z)}{\partial z}=n\frac{\partial \rho}{\partial t}+\rho\frac{\partial n}{\partial t} \tag{2.72}$$

The first term on the right-hand side of Eq. (2.72) is the mass rate of water produced by an expansion of the water under a change in its density ρ. The second term is the mass rate of water produced by the compaction of the porous medium as reflected by the change in its porosity n. The first term is controlled by the compressibility of the fluid β and the second term by the compressibility of the aquifer α. We have already carried out the analysis (in Section 2.10) that is necessary to simplify the two terms on the right of Eq. (2.72). We know that the change in ρ *and* the change in n are both produced by a change in hydraulic head h, and that the *volume* of water produced by the two mechanisms for a unit decline in head is S_s, where S_s is the specific storage given by $S_s = \rho g(\alpha + n\beta)$. The mass rate of water produced (time rate of change of fluid mass storage) is $\rho S_s\, \partial h/\partial t$, and Eq. (2.72) becomes

$$-\frac{\partial(\rho v_x)}{\partial x}-\frac{\partial(\rho v_y)}{\partial y}-\frac{\partial(\rho v_z)}{\partial z}=\rho S_s\frac{\partial h}{\partial t} \tag{2.73}$$

Expanding the terms on the left-hand side by the chain rule and recognizing that terms of the form $\rho\, \partial v_x/\partial_x$ are much greater than terms of the form $v_x\, \partial\rho/\partial x$ allows us to eliminate ρ from both sides of Eq. (2.73). Inserting Darcy's law, we obtain

$$\frac{\partial}{\partial x}\left(K_x\frac{\partial h}{\partial x}\right)+\frac{\partial}{\partial y}\left(K_y\frac{\partial h}{\partial y}\right)+\frac{\partial}{\partial z}\left(K_z\frac{\partial h}{\partial z}\right)=S_s\frac{\partial h}{\partial t} \tag{2.74}$$

This is the equation of flow for transient flow through a saturated anisotropic porous medium. If the medium is homogeneous and isotropic, Eq. (2.74) reduces to

$$\frac{\partial^2 h}{\partial x^2}+\frac{\partial^2 h}{\partial y^2}+\frac{\partial^2 h}{\partial z^2}=\frac{S_s}{K}\frac{\partial h}{\partial t} \tag{2.75}$$

or expanding S_s,

$$\frac{\partial^2 h}{\partial x^2}+\frac{\partial^2 h}{\partial y^2}+\frac{\partial^2 h}{\partial z^2}=\frac{\rho g(\alpha+n\beta)}{K}\frac{\partial h}{\partial t} \tag{2.76}$$

Equation (2.76) is known as the *diffusion equation*. The solution $h(x, y, z, t)$ describes the value of the hydraulic head at any point in a flow field at any time. A solution requires knowledge of the three basic hydrogeological parameters, K, α, and n, and the fluid parameters, ρ and β.

For the special case of a horizontal confined aquifer of thickness b, $S = S_s b$ and $T = Kb$, and the two-dimensional form of Eq. (2.75) becomes

$$\frac{\partial^2 h}{\partial x^2}+\frac{\partial^2 h}{\partial y^2}=\frac{S}{T}\frac{\partial h}{\partial t} \tag{2.77}$$

The solution $h(x, y, t)$ describes the hydraulic head field at any point on a horizontal plane through the horizontal aquifer at any time. Solution requires knowledge of the aquifer parameters S and T.

The equation of flow for transient, saturated flow [in any of the forms given by Eqs. (2.74) through (2.77)] rests on the law of flow established by Darcy (1856), on the clarification of the hydraulic potential by Hubbert (1940), and on the recognition of the concepts of aquifer elasticity by Meinzer (1923), and effective stress by Terzaghi (1925). The classical development was first put forward by Jacob (1940) and can be found in its most complete form in Jacob (1950). The development presented in this section, together with the storage concepts of the earlier sections, is essentially that of Jacob.

In recent years there has been considerable reassessment of the classical development. Biot (1955) recognized that in compacting aquifers it is necessary to cast Darcy's law in terms of a relative velocity of fluid to grains, and Cooper (1966) pointed out the inconsistency of taking a fixed elemental control volume in a deforming medium. Cooper showed that Jacob's classical development is correct if one views the velocity as relative and the coordinate system as deforming. He also showed that De Wiest's (1966) attempt to address this problem (which also appears in Davis and De Wiest, 1966) is incorrect. Appendix II contains a more rigorous presentation of the Jacob-Cooper development than has been attempted here.

The classical development, through its use of the concept of vertical aquifer compressibility, assumes that stresses and deformations in a compacting aquifer occur only in the vertical direction. The approach couples a three-dimensional flow field and a one-dimensional stress field. The more general approach, which couples a three-dimensional flow field and a three-dimensional stress field, was first considered by Biot (1941, 1955). Verruijt (1969) provides an elegant summary of the approach.

For almost all practical purposes it is not necessary to consider relative velocities, deforming coordinates, or three-dimensional stress fields. The classical equations of flow presented in this section are sufficient.

Transient Unsaturated Flow

Let us define the degree of saturation θ' as $\theta' = \theta/n$, where θ is the moisture content and n is the porosity. For flow in an elemental control volume that may be only partially saturated, the equation of continuity must reveal the time rate of change of moisture content as well as the time rate of change of storage due to water expansion and aquifer compaction. The ρn term in Eq. (2.71) must become $\rho n\theta'$, and Eq. (2.72) becomes

$$-\frac{\partial(\rho v_x)}{\partial x} - \frac{\partial(\rho v_y)}{\partial y} - \frac{\partial(\rho v_z)}{\partial z} = n\theta'\frac{\partial \rho}{\partial t} + \rho\theta'\frac{\partial n}{\partial t} + n\rho\frac{\partial \theta'}{\partial t} \qquad (2.78)$$

For unsaturated flow, the first two terms on the right-hand side of Eq. (2.78) are much smaller than the third term. Discarding these two terms, canceling the ρ's from both sides in the usual way, inserting the unsaturated form of Darcy's law [Eq. (2.41)], and recognizing that $n\,d\theta' = d\theta$, leads to

$$\frac{\partial}{\partial x}\left[K(\psi)\frac{\partial h}{\partial x}\right] + \frac{\partial}{\partial y}\left[K(\psi)\frac{\partial h}{\partial y}\right] + \frac{\partial}{\partial z}\left[K(\psi)\frac{\partial h}{\partial z}\right] = \frac{\partial \theta}{\partial t} \tag{2.79}$$

It is usual to put Eq. (2.79) into a form where the independent variable is either θ or ψ. For the latter case it is necessary to multiply the top and bottom of the right-hand side by $\partial\psi$. Then, recalling the definition of the specific moisture capacity C [Eq. (2.65)], and noting that $h = \psi + z$, we obtain

$$\frac{\partial}{\partial x}\left[K(\psi)\frac{\partial \psi}{\partial x}\right] + \frac{\partial}{\partial y}\left[K(\psi)\frac{\partial \psi}{\partial y}\right] + \frac{\partial}{\partial z}\left[K(\psi)\left(\frac{\partial \psi}{\partial z} + 1\right)\right] = C(\psi)\frac{\partial \psi}{\partial t} \tag{2.80}$$

Equation (2.80) is the ψ-based equation of flow for transient flow through an unsaturated porous medium. It is often called *Richards equation*, in honor of the soil physicist who first developed it (Richards, 1931). The solution $\psi(x, y, z, t)$ describes the pressure head field at any point in a flow field at any time. It can easily be converted into an hydraulic head solution $h(x, y, z, t)$ through the relation $h = \psi + z$. Solution requires knowledge of the characteristic curves $K(\psi)$ and $C(\psi)$ or $\theta(\psi)$.

The coupling of the unsaturated flow equation [Eq. (2.80)] and the saturated flow equation [Eq. (2.74)] has been attempted by Freeze (1971a) and by Narasimhan (1975). Improvements in the theory underlying saturated-unsaturated systems must await a better understanding of the principle of effective stress in the unsaturated zone.

Boundary-Value Problems

A boundary-value problem is a *mathematical model.* The technique of analysis inferred by this latter term is a four-step process, involving (1) examination of the physical problem, (2) replacement of the physical problem by an equivelant mathematical problem, (3) solution of the mathematical problem with the accepted techniques of mathematics, and (4) interpretation of the mathematical results in terms of the physical problem. Mathematical models based on the physics of flow usually take the form of boundary-value problems of the type pioneered by the developers of potential field theory and as applied in physics to such problems as the conduction of heat in solids (Carslaw and Jaeger, 1959).

To fully define a transient boundary-value problem for subsurface flow, one needs to know (1) the size and shape of the region of flow, (2) the equation of flow within the region, (3) the boundary conditions around the boundaries of the region, (4) the initial conditions in the region, (5) the spatial distribution of the hydrogeologic parameters that control the flow, and (6) a mathematical method of solution. If the boundary-value problem is for a steady-state system, requirement (4) is removed.

Consider the simple groundwater flow problem illustrated in Figure 2.25(a). The region $ABCD$ contains a homogeneous, isotropic porous medium of hydraulic

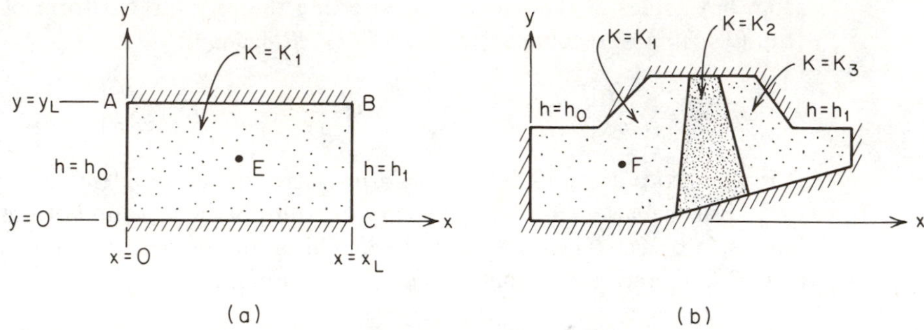

Figure 2.25 Two steady-state boundary-value problems in the xy plane.

conductivity K_1. The boundaries AB and CD are impermeable; the hydraulic heads on AD and BC are h_0 and h_1, respectively. Assuming steady flow and setting $h_0 = 100$ m and $h_1 = 0$ m, we can see by inspection that the hydraulic head at point E will be 50 m. Apparently, we made implicit use of properties (1), (3), and (5) from the list above; our method of solution (6) was one of inspection. It is not clear that we needed to know the equation of flow within the region. If we move to a more difficult problem such as that shown in Figure 2.25(b) (an earthfill dam resting on a sloping base), the value of the hydraulic head at point F does not come so easily. Here we would have to invoke a mathematical method of solution, and it would require that we know the equation of flow.

The methods of solution can be categorized roughly into five approaches: (1) solution by inspection, (2) solution by graphical techniques, (3) solution by analog model, (4) solution by analytical mathematical techniques, and (5) solution by numerical mathematical techniques. We have just seen an example of solution by inspection. The methods of flow-net construction presented in Chapter 5 can be viewed as graphical solutions to boundary-value problems. Electrical analog models are discussed in Sections 5.2 and 8.9. Numerical solutions are the basis of modern computer simulation techniques as described in Sections 5.3 and 8.8.

The most straightforward approach to the solution of boundary-value problems is that of analytical solutions. Many of the standard groundwater techniques presented later in the text are based on analytical solutions, so it is pertinent to examine a simple example. Consider, once again, the boundary-value problem of Figure 2.25(a). The analytical solution is

$$h(x, y) = h_0 - (h_0 - h_1)\frac{x}{x_L} \tag{2.81}$$

This is the equation of a set of equipotential lines traversing the field $ABCD$ parallel to the boundaries AD and BC. Since the equipotentials are parallel to the y axis, h is not a function of y and y does not appear on the right-hand side of Eq. (2.81). At point E, $x/x_L = 0.5$, and if $h_0 = 100$ m and $h_1 = 0$ m as before, then

h_E from Eq. (2.81) is 50 m, as expected. In Appendix III, the *separation-of-variables* technique is used to obtain the analytical solution Eq. (2.81), and it is shown that the solution satisfies the equation of flow and the boundary conditions.

2.12 Limitations of the Darcian Approach

Darcy's law provides an accurate description of the flow of groundwater in almost all hydrogeological environments. In general, Darcy's law holds (1) for saturated flow and for unsaturated flow, (2) for steady-state flow and for transient flow, (3) for flow in aquifers and for flow in aquitards, (4) for flow in homogeneous systems and for flow in heterogeneous systems, (5) for flow in isotropic media and for flow in anisotropic media, and (6) for flow in both rocks and granular media. In this text, we will assume that Darcy's law is a valid basis for our quantitative analyses.

Despite this soothing statement, or perhaps because of it, it is important that we examine the theoretical and practical limitations of the Darcian approach. It is necessary to look at the assumptions that underlie our definition of a continuum; examine the concepts of microscopic and macroscopic flow; investigate the upper and lower limits of Darcy's law; and consider the particular problems associated with flow in fractured rock.

Darcian Continuum and Representative Elementary Volume

In Section 2.1, it was noted that the definition of Darcy's law requires the replacement of the actual ensemble of grains that make up a porous medium by a representative continuum. It was further stated that this continuum approach is carried out at a macroscopic rather than a microscopic scale. If Darcy's law is a macroscopic law, there must be a lower limit to the size of an element of porous media for which the law is valid. Hubbert (1940) has addressed this problem. He defined the term *macroscopic* with the aid of Figure 2.26. This diagram is a hypothetical plot of the porosity of a porous medium as it might be measured on samples of

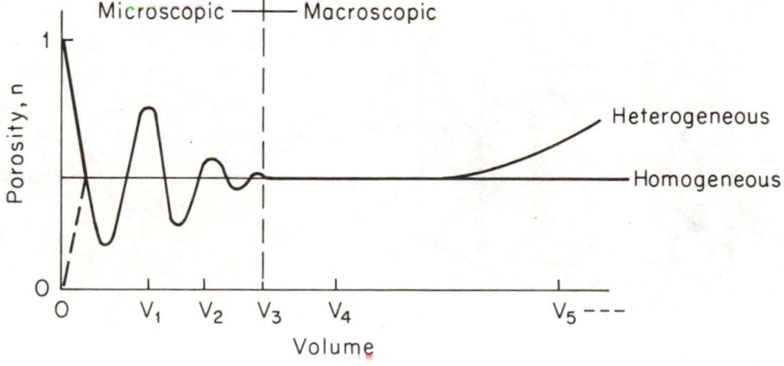

Figure 2.26 Microscopic and macroscopic domains and the representative elementary volume V_3 (after Hubbert, 1956; Bear, 1972).

increasing volume $V_1, V_2, \ldots$, taken at a point P within a porous medium. Bear (1972) defines the volume V_3 in Figure 2.26 as the *representative elementary volume*. He notes that it is a volume that must be larger than a single pore. In fact, it must include a sufficient number of pores to permit the meaningful statistical average required in the continuum approach. Below this volume there is no single value that can represent the porosity at P. Throughout this text the values of porosity, hydraulic conductivity, and compressibility refer to measurements that could be carried out on a sample larger than the representative elementary volume. In a more practical sense, they refer to values that can be measured on the usual sizes of cored soil samples. Where the scale of analysis involves volumes, such as V_5 in Figure 2.26, that may encompass more than one stratum in heterogeneous media, the scale is sometimes called *megascopic*.

The development of each of the equations of flow presented in Section 2.11 included the invocation of Darcy's law. It must be recognized, then, that the methods of analysis that are based on boundary-value problems involving these equations apply on a macroscopic scale, at the level of the Darcian continuum. There are some groundwater phenomena, such as the movement of a tracer through a porous medium, that cannot be analyzed on this scale. It is therefore necessary to examine the interrelationship that exists between the Darcy velocity (or specific discharge) defined for the macroscopic Darcian continuum and the microscopic velocities that exist in the liquid phase of the porous medium.

Specific Discharge, Macroscopic Velocity, and Microscopic Velocity

Our development will be more rigorous if we first differentiate, as Bear (1972) has done, between the *volumetric porosity*, n, which was defined in Section 2.5, and the *areal porosity*, n_A, which can be defined for any areal cross section through a unit volume, as $n_A = A_v/A_T$, where A_v is the area occupied by the voids and A_T is the total area. As suggested by Figure 2.27(a), various cross sections within a given unit volume may exhibit differing areal porosities $n_{A_1}, n_{A_2}, \ldots$. The volumetric porosity, n, is an average of the various possible areal porosities, n_{A_i}.

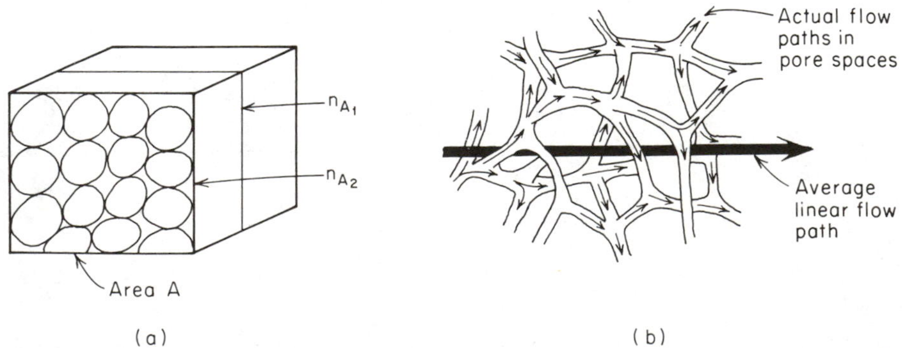

Figure 2.27 Concepts of (a) areal porosity and (b) average linear velocity.

For any cross section A, the *specific discharge*, v, is defined from Eq. (2.1) as

$$v = \frac{Q}{A}$$

In that the volumetric flux Q is divided by the full cross-sectional area (voids and solids alike), this velocity is identified as being pertinent to the macroscopic continuum approach. In actual fact, the flow passes through only that portion of the cross-sectional area occupied by voids. For cross section A_1 we can define a velocity $\bar{v}_1 = Q/n_{A_1}A$ that represents the volumetric flux divided by the actual cross-sectional area through which flow occurs. For the various sections $A_1, A_2, \ldots$ we can define $\bar{v}_1, \bar{v}_2, \ldots$. If we denote their average by $\bar{v}$, then

$$\bar{v} = \frac{Q}{nA} = \frac{v}{n} = \frac{-K}{n}\frac{\partial h}{\partial l} \tag{2.82}$$

The velocity $\bar{v}$ is known under a variety of names. We will refer to it as the *average linear velocity*. In that Q, n, and A are measurable macroscopic terms, so is $\bar{v}$. It should be emphasized that $\bar{v}$ does not represent the average velocity of the water particles traveling through the pore spaces. These true, microscopic velocities are generally larger than $\bar{v}$, because the water particles must travel along irregular paths that are longer than the linearized path represented by $\bar{v}$. This is shown schematically in Figure 2.27(b). The true, microscopic velocities that exist in the pore channels are seldom of interest, which is indeed fortunate, for they are largely indeterminate. For all the situations that will be considered in this text, the Darcy velocity v and the average linear velocity $\bar{v}$ will suffice.

As a basis for further explanation of $\bar{v}$, consider an experiment where a tracer is used to determine how much time is required for the bulk mass of groundwater to move a short but significant distance AB along a flow path. $\bar{v}$ is then defined as the ratio of travel distance to travel time, where the travel distance is defined as the linear distance from A to B and the travel time is the time required for the tracer to travel from A to B. In light of this conceptualization of $\bar{v}$, Nelson (1968) has suggested a slightly different form of Eq. (2.82):

$$\bar{v} = \frac{Q}{\epsilon nA} = \frac{v}{\epsilon n} \tag{2.83}$$

where ϵ is an empirical constant dependent on the characteristics of the porous medium. Data obtained in laboratory experiments by Ellis et al. (1968) using relatively uniform sands indicate values of ϵ in the range 0.98–1.18. Values of ϵ for nonuniform sands and for other materials do not exist at present. In studies of groundwater tracers and groundwater contamination the almost universal unstated assumption is that $\epsilon = 1$. For granular media this probably introduces little error. In fractured media the assumption may have less validity.

Upper and Lower Limits of Darcy's Law

Even if we limit ourselves to the consideration of specific discharge on a macroscopic scale through the Darcian continuum, there may be limitations on the applicability of Darcy's law. Darcy's law is a linear law. If it were universally valid, a plot of the specific discharge v versus the hydraulic gradient dh/dl would reveal a straight-line relationship for all gradients between 0 and ∞. For flow through granular materials there are at least two situations where the validity of this linear relationship is in question. The first concerns flow through low-permeability sediments under very low gradients and the second concerns large flows through very high permeability sediments. In other words, there may be both a lower limit and an upper limit to the range of validity of Darcy's law. It has been suggested that a more general form of the porous media flow law might be

$$v = -K\left(\frac{dh}{dl}\right)^m \tag{2.84}$$

If $m = 1$, as it does in all the common situations, the flow law is linear and is called Darcy's law; if $m \neq 1$, the flow law is not linear and should not be called Darcy's law.

For fine-grained materials of low permeability, it has been suggested on the basis of laboratory evidence that there may be a threshold hydraulic gradient below which flow does not take place. Swartzendruber (1962) and Bolt and Groenevelt (1969) review the evidence and summarize the various hypotheses that have been put forward to explain the phenomenon. As yet, there is no agreement on the mechanism, and the experimental evidence is still open to some doubt. In any event, the phenomenon is of very little practical importance; at the gradients being considered as possible threshold gradients, flow rates will be exceedingly small in any case.

Of greater practical importance is the upper limit on the range of validity of Darcy's law. It has been recognized and accepted for many years (Rose, 1945; Hubbert, 1956) that at very high rates of flow, Darcy's law breaks down. The evidence is reviewed in detail by both Todd (1959) and Bear (1972). The upper limit is usually identified with the aid of the *Reynolds number* R_e, a dimensionless number that expresses the ratio of inertial to viscous forces during flow. It is widely used in fluid mechanics to distinguish between *laminar flow* at low velocities and *turbulent flow* at high velocities. The Reynolds number for flow through porous media is defined as

$$R_e = \frac{\rho v d}{\mu} \tag{2.85}$$

where ρ and μ are the fluid density and viscosity, v the specific discharge, and d a representative length dimension for the porous medium, variously taken as a mean pore dimension, a mean grain diameter, or some function of the square root of the permeability k. Bear (1972) summarizes the experimental evidence with the statement that "Darcy's law is valid as long as the Reynolds number based on average grain diameter does not exceed some value between 1 and 10" (p. 126). For this range of Reynolds numbers, all flow through granular media is laminar.

Flow rates that exceed the upper limit of Darcy's law are common in such important rock formations as karstic limestones and dolomites, and cavernous volcanics. Darcian flow rates are almost never exceeded in nonindurated rocks and granular materials. Fractured rocks (and we will use this term to refer to rocks rendered more permeable by joints, fissures, cracks, or partings of any genetic origin) constitute a special case that deserves separate attention.

Flow in Fractured Rocks

The analysis of flow in fractured rocks can be carried out either with the *continuum* approach that has been emphasized thus far in this text or with a *noncontinuum* approach based on the hydraulics of flow in individual fractures. As with granular porous media, the continuum approach involves the replacement of the fractured media by a representative continuum in which spatially defined values of hydraulic conductivity, porosity, and compressibility can be assigned. This approach is valid as long as the fracture spacing is sufficiently dense that the fractured media acts in a hydraulically similar fashion to granular porous media. The conceptualization is the same, although the representative elementary volume is considerably larger for fractured media than for granular media. If the fracture spacings are irregular in a given direction, the media will exhibit trending heterogeneity. If the fracture spacings are different in one direction than they are in another, the media will exhibit anisotropy. Snow (1968, 1969) has shown that many fracture-flow problems can be solved using standard porous-media techniques utilizing Darcy's law and an anisotropic conductivity tensor.

If the fracture density is extremely low, it may be necessary to analyze flow in individual fissures. This approach has been used in geotechnical applications where rock-mechanics analyses indicate that slopes or openings in rock may fail on the basis of fluid pressures that build up on individual critical fractures. The methods of analysis are based on the usual fluid mechanics principles embodied in the Navier-Stokes equations. These methods will not be discussed here. Wittke (1973) provides an introductory review.

Even if we limit ourselves to the continuum approach there are two further problems that must be addressed in the analysis of flow through fractured rock. The first is the question of non-Darcy flow in rock fractures of wide aperture. Sharp and Maini (1972) present laboratory data that support a nonlinear flow law for fractured rock. Wittke (1973) suggests that separate flow laws be specified for the linear-laminar range (Darcy range), a nonlinear laminar range, and a turbulent range. Figure 2.28 puts these concepts into the context of a schematic curve of specific discharge vs. hydraulic gradient. In wide rock fractures, the specific discharges and Reynolds numbers are high, the hydraulic gradients are usually less

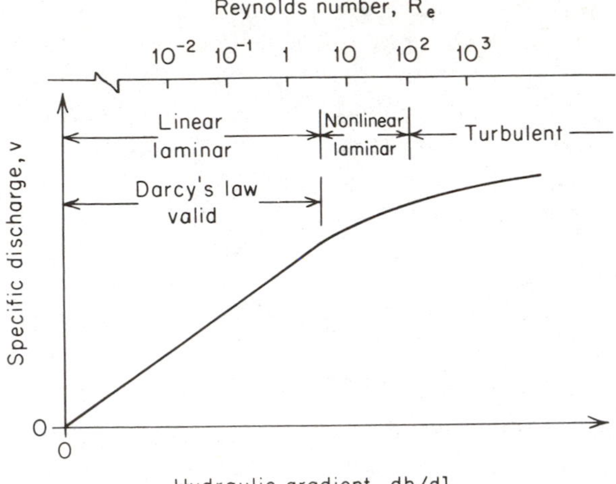

Figure 2.28 Range of validity of Darcy's law.

than 1, and the exponent m in Eq. (2.84) is greater than 1. These conditions lead to a downward deflection in the curve in Figure 2.28.

The second problem concerns the interaction of the three-dimensional stress field and the three-dimensional fluid flow field in rock. The general theoretical requirement for the coupling of these two fields was briefly discussed in Section 2.11, and reference was made there to the classic work of Biot (1941, 1955) for flow through porous media. For fractured rock, however, there is a further complication. Because the porosity of fractured rock is so low, the expansions and contractions of the fracture apertures that occur under the influence of changes in stress affect the values of hydraulic conductivity, K. The interaction between the fluid pressure $p(x, y, z, t)$, or the hydraulic head $h(x, y, z, t)$, and the effective stress $\sigma_e(x, y, z, t)$ is thus complicated by the fact that K must be represented by a function, $K(\sigma_e)$. The analysis of such systems, and the experimental determination of the nature of the $K(\sigma_e)$ function, is a continuing subject of research in the fields of rock mechanics and groundwater hydrology.

Many researchers involved in the application of groundwater theory in rock mechanics have proposed formulas that relate the fracture porosity n_f and the hydraulic conductivity K of jointed rocks to the joint geometry. Snow (1968) notes that for a parallel array of planar joints of aperture, b, with N joints per unit distance across the rock face, $n_f = Nb$, and

$$K = \left(\frac{\rho g}{\mu}\right)\left(\frac{Nb^3}{12}\right) \tag{2.86}$$

or

$$k = \frac{Nb^3}{12} \tag{2.87}$$

where k is the permeability of the rock. N and b have dimensions $1/L$ and L, respectively, so that k comes out in units of L^2, as it should. Equation (2.86) is based on the hydrodynamics of flow in a set of planar joints. It holds in the linear-laminar range where Darcy's law is valid. It must be applied to a block of rock of sufficient size that the block acts as a Darcian continuum. A permeability k, calculated with Eq. (2.87), can be considered as the permeability of an equivalent porous medium; one that acts hydraulically like the fractured rock.

Snow (1968) states that a cubic system of like fractures creates an isotropic system with a porosity $n_f = 3Nb$ and a permeability twice the permeability that any one of its sets would contribute; that is, $k = Nb^3/6$. Snow (1969) also provides predictive interrelationships between porosity and the anisotropic permeability tensor for three-dimensional joint geometries in which fracture spacings or apertures differ with direction. Sharp and Maini (1972) provide further discussion of the hydraulic properties of anisotropic jointed rock.

2.13 Hydrodynamic Dispersion

It is becoming increasingly common in the investigation of groundwater flow systems to view the flow regime in terms of its ability to transport dissolved substances known as solutes. These solutes may be natural constituents, artificial tracers, or contaminants. The process by which solutes are transported by the bulk motion of the flowing groundwater is known as *advection*. Owing to advection, nonreactive solutes are carried at an average rate equal to the average linear velocity, $\bar{v}$, of the water. There is a tendency, however, for the solute to spread out from the path that it would be expected to follow according to the advective hydraulics of the flow system. This spreading phenomenon is called *hydrodynamic dispersion*. It causes dilution of the solute. It occurs because of mechanical mixing during fluid advection and because of molecular diffusion due to the thermal-kinetic energy of the solute particles. Diffusion, which is a dispersion process of importance only at low velocities, is described in Section 3.4. Emphasis in the present discussion is on dispersion that is caused entirely by the motion of the fluid. This is known as *mechanical dispersion* (or *hydraulic dispersion*). Figure 2.29 shows a schematic example of the results of this dispersive process in a homogeneous, granular medium.

Mechanical dispersion is most easily viewed as a microscopic process. On the microscopic scale, dispersion is caused by three mechanisms (Figure 2.30). The first occurs in individual pore channels because molecules travel at different velocities at different points across the channel due to the drag exerted on the fluid by the roughness of the pore surfaces. The second process is caused by the difference in pore sizes along the flow paths followed by the water molecules. Because of differences in surface area and roughness relative to the volume of water in individual pore channels, different pore channels have different bulk fluid velocities.

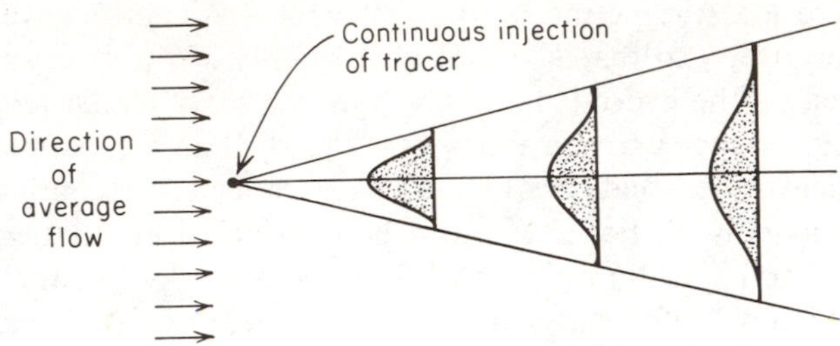

Figure 2.29 Schematic representation of the dilution process caused by mechanical dispersion in granular porous media.

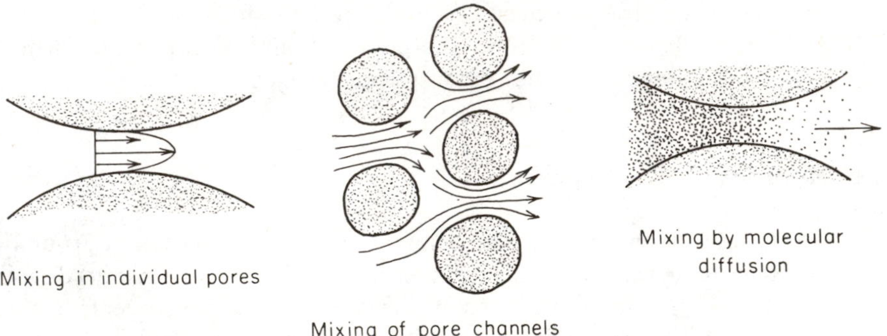

Figure 2.30 Processes of dispersion on a microscopic scale.

The third dispersive process is related to the tortuosity, branching, and interfingering of pore channels. The spreading of the solute in the direction of bulk flow is known as *longitudinal dispersion.* Spreading in directions perpendicular to the flow is called *transverse dispersion.* Longitudinal dispersion is normally much stronger than lateral dispersion.

Dispersion is a mixing process. Qualitatively, it has a similar effect to turbulence in surface-water regimes. For porous media, the concepts of average linear velocity and longitudinal dispersion are closely related. Longitudinal dispersion is the process whereby some of the water molecules and solute molecules travel more rapidly than the average linear velocity and some travel more slowly. The solute therefore spreads out in the direction of flow and declines in concentration.

When a tracer experiment is set up in the laboratory, the only dispersion that can be measured is that which is observable at the macroscopic scale. It is assumed that this macroscopic result has been produced by the microscopic processes described above. Some investigators believe that heterogeneities on the macroscopic scale can cause additional dispersion to that caused by the microscopic processes. The concept of macroscopic dispersion is still not well understood. Dispersive processes are pursued further in Chapter 9.

Suggested Readings

BEAR, J. 1972. *Dynamics of Fluids in Porous Media.* American Elsevier, New York, pp. 15–24, 52–56, 85–90, 122–129, 136–148.

HUBBERT, M. K. 1940. The theory of groundwater motion. *J. Geol.*, 48, pp. 785–822.

JACOB, C. E. 1940. On the flow of water in an elastic artesian aquifer. *Trans. Amer. Geophys. Union*, 2, pp. 574–586.

MAASLAND, M. 1957. Soil anisotropy and land drainage. *Drainage of Agricultural Lands* ed. J. N. Luthin. American Society of Agronomy, Madison, Wisc., pp. 216–246.

SKEMPTON, A. W. 1961. Effective stress in soils, concrete and rocks. *Conference on Pore Pressures and Suction in Soils.* Butterworth, London, pp. 4–16.

STALLMAN, R. W. 1964. Multiphase fluids in porous media—a review of theories pertinent to hydrologic studies. *U.S. Geol. Surv. Prof. Paper 411E.*

VERRUIJT, A. 1969. Elastic storage of aquifers. *Flow Through Porous Media*, ed. R. J. M. De Wiest. Academic Press, New York, pp. 331–376.

Problems

1. The following field notes were taken at a nest of piezometers installed side by side at a single site:

Piezometer	*a*	*b*	*c*
Elevation at surface (m a.s.l.)	450	450	450
Depth of piezometer (m)	150	100	50
Depth to water (m)	27	47	36

Let *A*, *B*, and *C* refer to the points of measurement of piezometers *a*, *b*, and *c*. Calculate:

(a) The hydraulic head at *A*, *B*, and *C* (m).

(b) The pressure head at *A*, *B*, and *C* (m).

(c) The elevation head at *A*, *B*, and *C* (m).

(d) The fluid pressure at *B* (N/m^2).

(e) The hydraulic gradients between *A* and *B* and between *B* and *C*. Can you conceive of a hydrogeological situation that would lead to the directions of flow indicated by these data?

2. Draw diagrams of two realistic field situations in which three piezometers installed side by side, but bottoming at different depths, would have the same water-level elevation.

3. Three piezometers located 1000 m apart bottom in the same horizontal aquifer. Piezometer *A* is due south of piezometer *B* and piezometer *C* is to the east of

the line AB. The surface elevations of A, B, and C are 95, 110, and 135 m, respectively. The depth to water in A is 5 m, in B is 30 m, and in C is 35 m. Determine the direction of groundwater flow through the triangle ABC and calculate the hydraulic gradient.

4. Show that the fluid potential Φ is an energy term, by carrying out a dimensional analysis on the equation $\Phi = gz + p/\rho$. Do so for both the SI system of units and the FPS system of units.

5. Three formations, each 25 m thick, overlie one another. If a constant-velocity vertical flow field is set up across the set of formations with $h = 120$ m at the top and $h = 100$ m at the bottom, calculate h at the two internal boundaries. The hydraulic conductivity of the top formation is 0.0001 m/s, the middle formation 0.0005 m/s, and the bottom formation 0.0010 m/s.

6. A geologic formation has a permeability of 0.1 darcy (as determined by a petroleum company for the flow of oil). What is the hydraulic conductivity of the formation for the flow of water? Give your answer in m/s and in gal/day/ft^2. What kind of rock would this likely be?

7. (a) Four horizontal, homogeneous, isotropic geologic formations, each 5 m thick, overlie one another. If the hydraulic conductivities are 10^{-4}, 10^{-6}, 10^{-4}, and 10^{-6} m/s, respectively, calculate the horizontal and vertical components of hydraulic conductivity for the equivalent homogeneous-but-anisotropic formation.

(b) Repeat for hydraulic conductivities of 10^{-4}, 10^{-8}, 10^{-4}, and 10^{-8} m/s, and for hydraulic conductivities of 10^{-4}, 10^{-10}, 10^{-4}, and 10^{-10} m/s. Put the results of the three sets of calculations in a table relating orders of magnitude of layered heterogeneity to resulting equivalent anisotropy.

8. (a) From the volumetric definitions of porosity and void ratio, develop the relationships given in Eq. (2.40).

(b) Is the porosity ever greater than the void ratio when both are measured on the same soil sample?

9. The elevation of the ground surface at a soil-moisture measurement site is 300 cm. The soil is a sand and its unsaturated properties are represented by the drying curves of Figure 2.13. Draw a quantitatively accurate set of vertical profiles of moisture content, pressure head, and hydraulic head versus depth (as in Figure 2.12) for a 200-cm depth under the following conditions:

(a) The moisture content is 20% throughout the profile.

(b) The pressure head is -50 cm throughout the profile.

(c) The hydraulic head is 150 cm throughout the profile (static case).

For cases (a) and (b), calculate the hydraulic gradients and the rates of flow through the profile. For case (c), determine the depth to the water table.

10. Given a potentiometric surface with a regional slope of 7 m/km, calculate the natural rate of groundwater discharge through a confined aquifer with transmissivity, $T = 0.002$ m^2/s.

11. Show by dimensional analysis on the equation $S = \rho g b(\alpha + n\beta)$ that the storativity is dimensionless.

12. (a) A horizontal aquifer is overlain by 50 ft of saturated clay. The specific weight (or unit dry weight) of the clay is 120 lb/ft^3. The specific weight of water is 62.4 lb/ft^3. Calculate the total stress acting on the top of the aquifer.
    (b) If the pressure head in the aquifer is 100 ft, calculate the effective stress in the aquifer.
    (c) If the aquifer is pumped and the hydraulic head at some point is reduced 10 ft, what will be the resulting changes in the pressure head, the fluid pressure, the effective stress, and the total stress?
    (d) If the compressibility of the aquifer is 10^{-6} ft^2/lb and its thickness is 25 ft, how much compaction will the aquifer undergo during the head reduction in part (c)?
    (e) If the porosity and hydraulic conductivity of the aquifer are 0.30 and 10 gal/day/ft^2, calculate the transmissivity and storativity for the aquifer. The compressibility of water is 2.1×10^{-8} ft^2/lb.

13. Review the problems that arise in the definition or use of the following classical groundwater terms: potentiometric surface, permeability, and groundwater flow velocity.

CHAPTER 3

Chemical Properties and Principles

The chemical and biochemical constituents in groundwater determine its usefulness for industry, agriculture, and the home. Dissolved constituents in the water provide clues on its geologic history, its influence on the soil or rock masses through which it has passed, the presence of hidden ore deposits, and its mode of origin within the hydrologic cycle. Chemical processes in the groundwater zone can influence the strength of geologic materials, and in situations where they are not recognized, can cause failure of artificial slopes, dams, mining excavations, and other features of importance to man. It is becoming increasingly common for industrial, agricultural, and domestic wastes to be stored or disposed on or beneath the land surface. This can be a safe or hazardous practice, the consequences of which depend greatly on the chemical and microbiological processes in the groundwater zone. In the study of landscape evolution the assumption is commonly made that the physical processes of mechanical erosion, thermal expansion and contraction, frost action, and slope movements are the dominant influences, but on closer examination it is often found that chemical processes in the groundwater zone are the controlling influences.

The purpose of this chapter is to describe the geochemical properties and principles that control the behavior of dissolved constituents in the groundwater environment. A more comprehensive coverage of the study and interpretation of the chemical characteristics of natural water is provided by Hem (1970) and by Stumm and Morgan (1970). Most of the geochemical principles described in this chapter are based on equilibrium concepts. Examples described in Chapter 7 indicate that many hydrochemical processes in the groundwater zone proceed slowly toward chemical equilibrium and some rarely achieve equilibrium. At times, the reader may doubt the usefulness of equilibrium approaches. Equilibrium concepts or models have great value, however, because of their capability for establishing boundary conditions on chemical processes. Differences between observed hydrochemical conditions and computed equilibrium conditions can provide insight into the behavior of the system and at a minimum can provide a quantitative framework within which appropriate questions can be posed.

3.1 Groundwater and Its Chemical Constituents

Water and Electrolytes

Water is formed by the union of two hydrogen atoms with one oxygen atom. The oxygen atom is bonded to the hydrogen atoms unsymmetrically, with a bond angle of 105°. This unsymmetrical arrangement gives rise to an unbalanced electrical charge that imparts a polar characteristic to the molecule. Water in the liquid state, although given the formula H_2O or HOH, is composed of molecular groups with the HOH molecules in each group held together by hydrogen bonding. Each group or molecular cluster is estimated to have an average of 130 molecules at 0°C, 90 molecules at 20°C, and 60 molecules at 72°C (Choppin, 1965). $H_{180}O_{90}$ is an approximate formula for the cluster at 20°C.

Water is unusual in that the density of the solid phase, ice, is substantially lower than the density of the liquid phase, water. In the liquid phase the maximum density is achieved at 4°C. With further cooling below this temperature there is a significant density decrease.

All chemical elements have two or more isotopes. In this book, however, we will be concerned only with the isotopes that provide useful hydrological or geochemical information. The formula H_2O is a gross simplification from the structural viewpoint and is also a simplification from the atomic viewpoint. Natural water can be a mixture of the six nuclides listed in Table 3.1. The atomic nature of the hydrogen isotopes is illustrated in Figure 3.1. Eighteen combinations of H—O—H are possible using these nuclides. $^2H_2{}^{16}O$, $^1H_2{}^{18}O$, $^3H_2{}^{17}O$ are some examples of the molecules that comprise water, which in its most common form is $^1H_2{}^{16}O$. Of the six isotopes of hydrogen and oxygen in Table 3.1, five are stable and one, ^3H, known as tritium, is radioactive, with a half-life of 12.3 years.

Table 3.1 Natural Isotopes of Hydrogen, Oxygen, and Radioactive Carbon and Their Relative Abundance in Water of the Hydrologic Cycle

Isotope		Relative abundance (%)	Type
^1H	proteum	99.984	Stable
^2H	deuterium	0.016	Stable
^3H	tritium	$0\text{–}10^{-15}$	Radioactive half-life 12.3 years
^{16}O	oxygen	99.76	Stable
^{17}O	oxygen	0.04	Stable
^{18}O	oxygen	0.20	Stable
^{14}C	carbon	<0.001	Radioactive half-life 5730 years

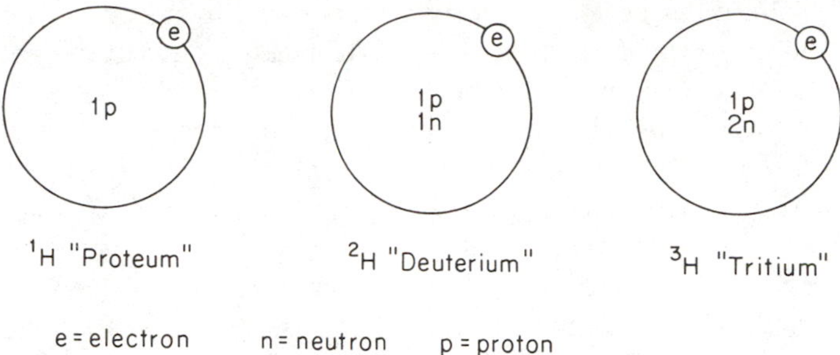

Figure 3.1 Isotopes of hydrogen.

Pure water contains hydrogen and oxygen in ionic form as well as in the combined molecular form. The ions are formed by the dissociation of water,

$$H_2O \rightleftharpoons H^+ + OH^- \tag{3.1}$$

where the plus and minus signs indicate the charge on the ionic species. Hydrogen can occur in vastly different forms, as illustrated in Figure 3.2. Although the ionic form of hydrogen in water is usually expressed in chemical equations as H^+, it is normally in the form H_3O^+, which denotes a hydrogen core surrounded by oxygen with four electron-cloud pairs. In discussions of groundwater mineral interactions, a process known as *proton transfer* denotes the transfer of an H^+ between components or phases.

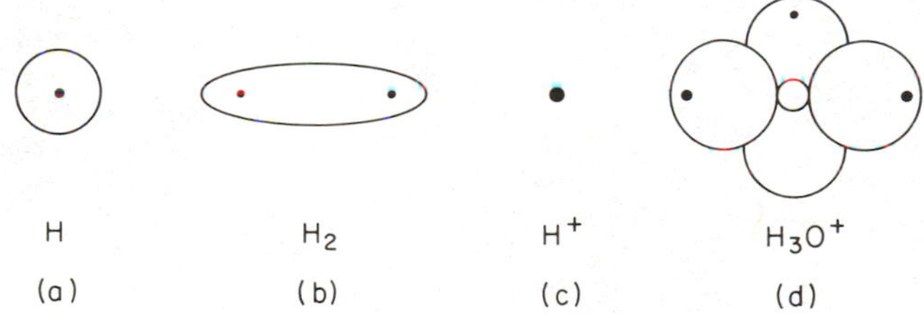

Figure 3.2 Four forms of hydrogen drawn to relative scale. (a) The hydrogen atom, a proton with one electron. (b) The hydrogen molecule, two separated protons in a cloud of two electrons. (c) The hydrogen core, or H^+, a proton. (d) The hydronium ion, oxygen with four electron cloud pairs, three of which are protonated in H_3O^+.

Water is a solvent for many salts and some types of organic matter. Water is effective in dissolving salts because it has a very high dielectric constant and because its molecules tend to combine with ions to form hydrated ions. The thermal agitation of ions in many materials is great enough to overcome the relatively weak charge attraction that exists when surrounded by water, thus allowing large numbers of ions to dissociate into aqueous solution. Stability of the ions in the aqueous solution is promoted by the formation of hydrated ions. Each positively charged ion, known as a *cation*, attracts the negative ends of the polar water molecules and binds several molecules in a relatively stable arrangement. The number of water molecules attached to a cation is determined by the size of the cation. For example, the small cation Be^{2+} forms the hydrated ion $Be(H_2O)_4{}^{2+}$. Larger ions, such as Mg^{2+} or Al^{3+}, have hydrated forms such as $Mg(H_2O)_6{}^{2+}$ and $Al(H_2O)_6{}^{3+}$. Negatively charged species, known as *anions*, exhibit a much weaker tendency for hydration. In this case the anions attract the positive ends of the polar water molecules. The sizes of the ions in their hydrated form are important with respect to many processes that occur in the groundwater environment.

As a result of chemical and biochemical interactions between groundwater and the geological materials through which it flows, and to a lesser extent because of contributions from the atmosphere and surface-water bodies, groundwater contains a wide variety of dissolved inorganic chemical constituents in various concentrations. The concentration of total dissolved solids (TDS) in groundwater is determined by weighing the solid residue obtained by evaporating a measured volume of filtered sample to dryness. The solid residue almost invariably consists of inorganic constituents and very small amounts of organic matter. The TDS concentrations in groundwater vary over many orders of magnitude. A simple but widely used scheme for categorizing groundwater based on TDS is presented in Table 3.2. To put the concentration ranges in perspective, it may be useful to note that water containing more than 2000–3000 mg/ℓ TDS is generally too salty to drink. The TDS of seawater is approximately 35,000 mg/ℓ.

Table 3.2 Simple Groundwater Classification Based on Total Dissolved Solids

Category	Total dissolved solids (mg/ℓ or g/m^3)
Fresh water	0–1000
Brackish water	1000–10,000
Saline water	10,000–100,000
Brine water	More than 100,000

Groundwater can be viewed as an electrolyte solution because nearly all its major and minor dissolved constituents are present in ionic form. A general indication of the total dissolved ionic constituents can be obtained by determining the capability of the water to conduct an applied electrical current. This property is usually reported as electrical conductance and is expressed in terms of the conductance of a cube of water 1 cm^2 on a side. It is the reciprocal of electrical resistance and has units known as siemens (S) or microsiemens (μS) in the SI system. In the past these units have been known as millimhos and micromhos. The values are the same; only the designations have changed. The conductance of groundwater ranges from several tens of microsiemens for water nearly as nonsaline as rainwater to hundreds of thousands of microsiemens for brines in deep sedimentary basins.

A classification of the inorganic species that occur in groundwater is shown in Table 3.3. The concentration categories are only a general guide. In some groundwaters, the concentration ranges are exceeded. The major constituents in Table 3.3 occur mainly in ionic form and are commonly referred to as the *major ions* (Na^+, Mg^{2+}, Ca^{2+}, Cl^-, HCO_3^-, SO_4^{2-}). The total concentration of these six major ions normally comprises more than 90% of the total dissolved solids in the water, regardless of whether the water is dilute or has salinity greater than seawater. The

Table 3.3 Classification of Dissolved Inorganic Constituents in Groundwater

Major constituents (greater than 5 mg/ℓ)	
Bicarbonate	Silicon
Calcium	Sodium
Chloride	Sulfate
Magnesium	Carbonic acid
Minor constituents (0.01–10.0 mg/ℓ)	
Boron	Nitrate
Carbonate	Potassium
Fluoride	Strontium
Iron	
Trace constituents (less than 0.1 mg/ℓ)	
Aluminum	Molybdenum
Antimony	Nickel
Arsenic	Niobium
Barium	Phosphate
Beryllium	Platinum
Bismuth	Radium
Bromide	Rubidium
Cadmium	Ruthenium
Cerium	Scandium
Cesium	Selenium
Chromium	Silver
Cobalt	Thallium
Copper	Thorium
Gallium	Tin
Germanium	Titanium
Gold	Tungsten
Indium	Uranium
Iodide	Vanadium
Lanthanum	Ytterbium
Lead	Yttrium
Lithium	Zinc
Manganese	Zirconium

SOURCE: Davis and De Wiest, 1966.

concentrations of the major, minor, and trace inorganic constituents in groundwater are controlled by the availability of the elements in the soil and rock through which the water has passed, by geochemical constraints such as solubility and adsorption, by the rates (kinetics) of geochemical processes, and by the sequence in which the water has come into contact with the various minerals occurring in the geologic materials along the flow paths. It is becoming increasingly common for the concentrations of the dissolved inorganic constituents to be influenced by man's activities. In some cases contributions from man-made sources can cause some of the elements listed as minor or trace constituents in Table 3.3 to occur as contaminants at concentration levels that are orders of magnitude above the normal ranges indicated in this table.

Organic Constituents

Organic compounds are those that have carbon and usually hydrogen and oxygen as the main elemental components in their structural framework. By definition, carbon is the key element. The species H_2CO_3, CO_2, HCO_3^-, and CO_3^{2-}, which are important constituents in all groundwater, however, are not classified as organic compounds.

Dissolved organic matter is ubiquitous in natural groundwater, although the concentrations are generally low compared to the inorganic constituents. Little is known about the chemical nature of organic matter in groundwater. Investigations of soil water suggest that most dissolved organic matter in subsurface flow systems is fulvic and humic acid. These terms refer to particular types of organic compounds that persist in subsurface waters because they are resistant to degradation by microorganisms. The molecular weights of these compounds range from a few thousand to many thousand grams. Carbon is commonly about half of the formula weight. Although little is known about the origin and composition of organic matter in groundwater, analyses of the total concentrations of dissolved organic carbon (DOC) are becoming a common part of groundwater investigations. Concentrations in the range 0.1–10 mg/ℓ are most common, but in some areas values are as high as several tens of milligrams per liter.

Dissolved Gases

The most abundant dissolved gases in groundwater are N_2, O_2, CO_2, CH_4 (methane), H_2S, and N_2O. The first three make up the earth's atmosphere and it is, therefore, not surprising that they occur in subsurface water. CH_4, H_2S, and N_2O can often exist in groundwater in significant concentrations because they are the product of biogeochemical processes that occur in nonaerated subsurface zones. As will be shown later in this chapter and in Chapter 7, the concentrations of these gases can serve as indicators of geochemical conditions in groundwater.

Dissolved gases can have a significant influence on the subsurface hydrochemical environment. They can limit the usefulness of groundwater and, in some cases, can even cause major problems or even hazards. For example, because of its odor, H_2S at concentrations greater than about 1 mg/ℓ renders water unfit for human consumption. CH_4 bubbling out of solution can accumulate in wells or buildings and cause explosion hazards. Gases coming out of solution can form bubbles in wells, screens, or pumps, causing a reduction in well productivity or efficiency. Radon 222 (^{222}Rn), which is a common constituent of groundwater because it is a decay product of radioactive uranium and thorium, which are common in rock or soil, can accumulate to undesirable concentrations in unventilated basements. Decay products of radon 222 can be hazardous to human health.

Other species of dissolved gases, which occur in groundwater in minute amounts, can provide information on water sources, ages, or other factors of hydrologic or geochemical interest. Noteworthy in this regard are Ar, He, Kr, Ne, and Xe, for which uses in groundwater studies have been described by Sugisaki (1959, 1961) and Mazor (1972).

Concentration Units

To have a meaningful discussion of the chemical aspects of groundwater, the relative amounts of solute (the dissolved inorganic or organic constituents) and the solvent (the water) must be specified. This is accomplished by means of concentrations units. Various types of concentration units are in use.

Molality is defined as the number of moles of solute dissolved in a 1-kg mass of solution. This is an SI unit with the symbol mol/kg. The derived SI symbol for this quantity is m_B, where B denotes the solute. A is normally used to designate the solvent. One mole of a compound is the equivalent of one molecular weight.

Molarity is the number of moles of solute in 1 m^3 of solution. The SI unit for molarity is designated as mol/m^3. It is useful to note that 1 mol/m^3 equals 1 mmol/ℓ. Moles per liter, with the symbol mol/ℓ, is a permitted unit for molarity in the SI system and is commonly used in groundwater studies.

Mass concentration is the mass of solute dissolved in a specified unit volume of solution. The SI unit for this quantity is kilograms per cubic meter, with the symbol kg/m^3. Grams per liter (g/ℓ) is a permitted SI unit. The most common mass concentration unit reported in the groundwater literature is milligrams per liter (mg/ℓ). Since 1 mg/ℓ equals 1 g/m^3, there is no difference in the magnitude of this unit (mg/ℓ) and the permitted SI concentration unit (g/m^3).

There are many other non-SI concentration units that commonly appear in the groundwater literature. *Equivalents per liter* (epℓ) is the number of moles of solute, multiplied by the valence of the solute species, in 1 liter of solution:

$$\text{ep}\ell = \frac{\text{moles of solute} \times \text{valence}}{\text{liter of solution}}$$

Equivalents per million (epm) is the number of moles of solute multiplied by the valence of the solute species, in 10^6 g of solution, or this can be stated as the number of milligram equivalents of solute per kilogram of solution:

$$\text{epm} = \frac{\text{moles of solute} \times \text{valence}}{10^6 \text{ g of solution}}$$

Parts per million (ppm) is the number of grams of solute per million grams of solution

$$\text{ppm} = \frac{\text{grams of solute}}{10^6 \text{ g of solution}}$$

For nonsaline waters, 1 ppm equals 1 g/m^3 or 1 mg/ℓ. *Mole fraction* (X_B) is the ratio of the number of moles of a given solute species to the total number of moles of all components in the solution. If n_B is moles of solute, n_A is moles of solvent, and $n_C, n_D, \ldots$ denote moles of other solutes, the mole fraction of solute B is

$$X_B = \frac{n_B}{n_A + n_B + n_C + n_D + \cdots}$$

or X_B for aqueous solutions can be expressed as

$$X_B = \frac{m_B}{55.5 + \sum m_{B,C,D} \cdots}$$

where m denotes molality.

In procedures of chemical analysis, quantities are most conveniently obtained by use of volumetric glassware. Concentrations are therefore usually expressed in the laboratory in terms of solute mass in a given volume of water. Most chemical laboratories report analytical results in milligrams per liter or, in SI units, as kilograms per cubic meter. When the results of chemical analyses are used in a geochemical context, it is usually necessary to use data expressed in molality or molarity, since elements combine to form compounds according to relations between moles rather than mass or weight amounts. To convert between molarity and kilograms per cubic meter or milligrams per liter, the following relation is used:

$$\text{molarity} = \frac{\text{milligrams per liter or grams per cubic meter}}{1000 \times \text{formula weight}}$$

If the water does not have large concentrations of total dissolved solids and if the temperature is close to 4°C, 1 ℓ of solution weighs 1 kg, in which case molality and molarity are equivalent and 1 mg/ℓ = 1 ppm. For most practical purposes, water with less than about 10,000 mg/ℓ total dissolved solids and at temperatures below about 100°C can be considered to have a density close enough to 1 kg/ℓ for the unit equivalents above to be used. If the water has higher salinity or temperature, density corrections should be used when converting between units with mass and volume denominators.

3.2 Chemical Equilibrium

The Law of Mass Action

One of the most useful relations in the analysis of chemical processes in groundwater is the *law of mass action*. It has been known for more than a century that the driving force of a chemical reaction is related to the concentrations of the constituents that are reacting and the concentrations of the products of the reaction. Consider the constituents B and C reacting to produce the products D and E,

$$b\text{B} + c\text{C} \rightleftharpoons d\text{D} + e\text{E} \tag{3.2}$$

where b, c, d, and e are the number of moles of the chemical constituents B, C, D, E, respectively.

The law of mass action expresses the relation between the reactants and the products when the reaction is at equilibrium,

$$K = \frac{[D]^d[E]^e}{[B]^b[C]^c} \tag{3.3}$$

where K is a coefficient known as the *thermodynamic equilibrium constant* or the *stability constant*. The brackets specify that the concentration of the constituent is the thermodynamically effective concentration, usually referred to as the *activity*. Equation (3.3) indicates that for any initial condition, the reaction expressed in Eq. (3.2) will proceed until the reactants and products attain their equilibrium activities. Depending on the initial activities, the reaction may have to proceed to the left or to the right to achieve this equilibrium condition.

The law of mass action contains no parameters that express the rate at which the reaction proceeds, and therefore tells us nothing about the kinetics of the chemical process. It is strictly an equilibrium statement. For example, consider the reaction that occurs when groundwater flows through a limestone aquifer composed of the mineral calcite ($CaCO_3$). The reaction that describes the thermodynamic equilibrium of calcite is

$$CaCO_3 \rightleftharpoons Ca^{2+} + CO_3{}^{2-} \tag{3.4}$$

This reaction will proceed to the right (mineral dissolution) or to the left (mineral precipitation) until the mass-action equilibrium is achieved. It may take years or even thousands of years for equilibrium to be achieved. After a disturbance in the system, such as an addition of reactants or removal of products, the system will continue to proceed toward the equilibrium condition. If the temperature or pressure changes, the system will proceed toward a new equilibrium because the magnitude of K changes. If disturbances are frequent compared to the reaction rate, equilibrium will never be achieved. As we will see in Chapter 7, some chemical interactions between groundwater and its host materials never do attain equilibrium.

Activity Coefficients

In the law of mass action, solute concentrations are expressed as activities. Activity and molality are related by

$$a_i = m_i \gamma_i \tag{3.5}$$

where a_i is the activity of solute species i, m_i the molality, and γ_i the activity coefficient. γ carries the dimensions of reciprocal molality (kg/mol), and a_i is therefore dimensionless. Except for waters with extremely high salt concentrations, γ_i is less than 1 for ionic species. In the previous section, *activity* was referred to as the thermodynamically effective concentration, because it is conceptually convenient to consider it to be that portion of m_i that actually participates in the reaction. The activity coefficient is therefore just an adjustment factor that can be used to convert concentrations into the form suitable for use in most thermodynamically based equations.

The activity coefficient of a given solute is the same in all solutions of the same ionic strength. Ionic strength is defined by the relation

$$I = \tfrac{1}{2} \sum m_i z_i^2 \tag{3.6}$$

where m_i is the molality of species i, and z_i is the valence, or charge, that the ion carries. For groundwater, in which the six common major ions are the only ionic constituents that exist in significant concentration,

$$I = \tfrac{1}{2}[(Na^+) + 4(Mg^{2+}) + 4(Ca^{2+}) + (HCO_3^-) + (Cl^-) + 4(SO_4{}^{2-})] \tag{3.7}$$

where the quantities in parentheses are molalities. To obtain values for γ_i the graphical relations of γ versus I shown in Figure 3.3 can be used for the common inorganic constituents, or at dilute concentrations a relation known as the Debye-Hückel equation can be used (Appendix IV). At ionic strengths below about 0.1, the activity coefficients for many of the less common ions can be estimated from the Kielland table, which is also included in Appendix IV. For a discussion of the theoretical basis for activity coefficient relations, the reader is referred to Babcock (1963). Comparison between experimental and calculated values of activity coefficients are made by Guenther (1968).

Equilibrium and Free Energy

From a thermodynamic viewpoint the equilibrium state is a state of maximum stability toward which a closed physicochemical system proceeds by irreversible processes (Stumm and Morgan, 1970). The concepts of stability and instability for a simple mechanical system serve as an illustrative step toward development of the thermodynamic concept of equilibrium. Similar examples have been used by Guggenheim (1949) and others. Consider three different "equilibrium" positions of a rectangular box on a horizontal surface [Figure 3.4(a)]. Position 3 is the most *stable* position that the box can achieve. In this position the gravitational potential energy is at a minimum, and if the position is slightly disturbed, it will return to

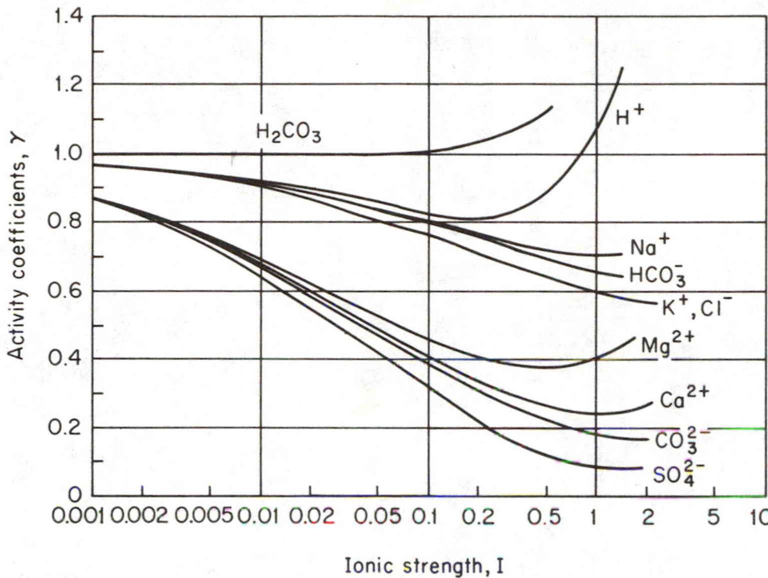

Figure 3.3 Activity coefficient versus ionic strength relations for common ionic constituents in groundwater.

the condition of *stable equilibrium*. In position 1, the box is also in an equilibrium position to which it will return if only *slightly* disturbed. But in this position the potential energy is not a minimum, so it is referred to as a condition of *metastable equilibrium*. If the box in position 2 is disturbed only slightly, the box will move to a new position. Position 2 is therefore a condition of *unstable equilibrium*.

An analogy between the mechanical system and the thermodynamic system is illustrated in Figure 3.4(b). Following the development by Stumm and Morgan (1970), a hypothetical, generalized energy or entropy profile is shown as a function of the state of the system. The conditions of stable, metastable, and unstable equilibrium are represented by troughs and peaks on the energy or entropy function. If the chemical system exists in closed conditions under constant temperature and pressure, its response to change can be described in terms of a particular energy function known as the Gibbs free energy, named after Willard Gibbs, the founder of classical thermodynamics. This direction of possible change in response to change in a composition variable is that accompanied by a decrease in Gibbs free energy. *State C is the most stable state because it has an absolute minimum Gibbs free energy under closed-system conditions at constant temperature and pressure.* State A is stable with respect to infinitesimally near states of the system, but is unstable with respect to a finite change toward state C. Natural processes proceed toward equilibrium states and never away from them. Therefore, thermodynamic equilibrium is found in metastable and stable conditions of equilibrium but not in unstable equilibrium.

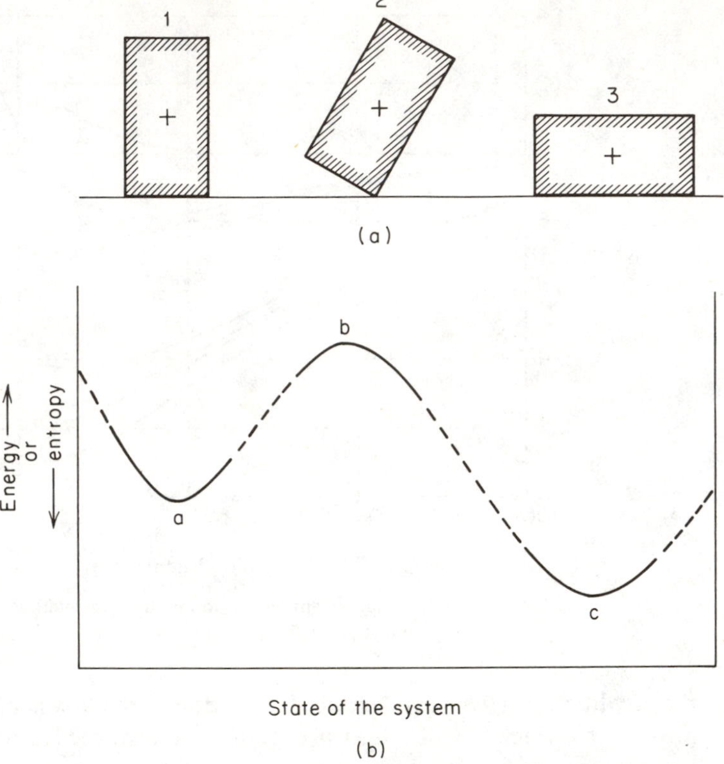

Figure 3.4 Concepts in mechanical and chemical equilibrium. (a) Metastable, unstable, and stable equilibrium in a mechanical system. (b) Metastability, instability, and stability for different energetic states of a thermodynamic system (after Stumm and Morgan, 1970).

The driving force in a chemical reaction is commonly represented by the Gibbs free energy of reaction denoted as ΔG_r. For systems at constant temperature and pressure, ΔG_r represents the change in internal energy per unit mass and is a measure of the reaction's capability to do nonmechanical work. Since in this text our objective in the use of thermodynamic data focuses on determining the directions in which reactions will proceed and on obtaining numerical values for equilibrium constants, there is little need to directly consider the thermodynamic components that make up ΔG_r. For a development of the theory of chemical thermodynamics, the reader is referred to the text by Denbigh (1966), and the comprehensive discussion of the thermodynamics of soil systems by Babcock (1963).

The condition of chemical equilibrium can be defined as

$$\sum \text{free-energy: products} - \sum \text{free-energy: reactants} = 0 \tag{3.8}$$

The next step in this development is to relate free-energy changes of reactions to their equilibrium constants. To do this a convenient free-energy accounting system is needed. The *standard free energy of formation*, ΔG_f^0, is defined as the free energy of the reaction to produce 1 mol of a substance from the stable elements under conditions that are specified as *standard-state* conditions. The standard free energy of elements in their most stable pure chemical state is assigned a value of zero by convention. Similarly, it is convenient to take as zero the ΔG_f^0 of hydrogen ion. For example, carbon as graphite and oxygen as O_2 have ΔG_f^0 values of zero, but 1 mol of gaseous carbon dioxide has a ΔG_f^0 value of −386.41 kJ (−92.31 kcal), which is the energy released when CO_2 forms from the stable elements in their standard state. The standard state of pure water is defined as unity at the temperature and pressure of the reaction, and for solutes the standard state is a unimolal concentration in a hypothetical condition where the activity coefficient is unity, or, in other words, in a condition where the activity equals the molality. For gas, the standard state is pure (ideal) gas at 1 bar total pressure at the temperature of the reaction. This system of arbitrarily defined standard states may at first seem unnecessarily complex, but in practice it leads to a tidy consistent system of bookkeeping. A more detailed discussion of standard states is provided by Berner (1971).

The *standard free-energy change of reaction*, ΔG_r^0, is the sum of the free energies of formation of the products in their standard states minus the free energies of formation of the reactants in their standard states:

$$\Delta G_r^0 = \sum \Delta G_f^0 \text{ products} - \sum \Delta G_f^0 \text{ reactants} \tag{3.9}$$

For the general reaction in Eq. (3.2), the change in free energy of the reaction is related to the standard free-energy change and to the activities of each of the reactants and products, measured at the same temperature, by the expression

$$\Delta G_r = \Delta G_r^0 + RT \ln \frac{[D]^d[E]^e}{[B]^b[C]^c} \tag{3.10}$$

where R is the universal gas constant and T is temperature in degrees Kelvin. At 25°C, $R = 8.314$ J/K·mol or 0.001987 kcal/K·mol. Conversion of temperatures on the Celsius scale to those on the Kelvin scale is made through the relation K = °C + 273.15. For a chemical reaction to proceed spontaneously as written, ΔG_r must be less than zero, or, in other words, there must be a net decrease in free energy. If $\Delta G_r > O$, the reaction can only proceed from right to left. If $\Delta G_r = O$, the reaction will not proceed in either direction, in which case the equilibrium condition has been achieved. In accordance with our definition of the standard state for solutes (unimolal conditions where $\gamma = 1$), $\Delta G_r^0 = \Delta G_r$ in the standard state because $[D]^d[E]^e/[B]^b[C]^c = 1$, and hence the natural logarithm of this term is zero. Substitution of the equilibrium constant relation [Eq. (3.3)] into Eq. (3.10) yields, for equilibrium conditions,

$$\Delta G_r^0 = -RT \ln K \tag{3.11}$$

For standard-state conditions, the equilibrium constant can be obtained from free-energy data by means of the relations

$$\log K = -0.175\Delta G_r^0 \text{ (for } \Delta G_r^0 \text{ in kJ/mol)}$$
$$\log K = -0.733\Delta G_r^0 \quad \text{(for } \Delta G_r^0 \text{ in kcal/mol)} \tag{3.12}$$

where ΔG_r^0 can be obtained from Eq. (3.9) using ΔG_f^0 data. Values for ΔG_f^0 at 25°C and 1 bar have been tabulated for thousands of minerals, gases, and aqueous species that occur in geologic systems (Rossini et al., 1952; Sillen and Martell, 1964, 1971). Less comprehensive tables that are convenient for student use are included in the texts by Garrels and Christ (1965), Krauskopf (1967), and Berner (1971).

Compared to the abundance of ΔG_f^0 data for conditions of 25°C and 1 bar total pressure, there is a paucity of data for other temperatures and pressures. Pressure has only a slight effect on ΔG_f^0 values and consequently has little influence on the equilibrium constant. For practical purposes, the variation in K over the fluid pressures normally encountered in the upper few hundred meters of the earth's crust are negligible. Changes of several degrees, however, can cause significant changes in the equilibrium constant. To obtain estimates of K values at other temperatures, an expression known as the van't Hoff relation, named after a Dutch physical chemist who made important contributions in the late 1800's and early 1900's to the understanding of solution behavior, can be used:

$$\log K_T = \log K_{T^*} - \frac{\Delta H_{T^*}}{2.3R}\left(\frac{1}{T} - \frac{1}{T^*}\right) \tag{3.13}$$

where T^* is the reference temperature, usually 298.15 K (25°C), T the temperature of the solution, and ΔH_{T^*} the enthalpy. Enthalpy data for many of the minerals, gases, and dissolved species of interest are tabulated in the tables referred to above. Since the van't Hoff equation considers only two temperatures and assumes a linear relationship between them, it yields only approximate values. The best approach is to develop specific interpolation relations from free-energy data over a wide range of temperatures, if such data are available.

To illustrate the use of ΔG_f^0 data to obtain equilibrium constants, consider the calcite dissolution reaction as expressed in Eq. (3.4). ΔG_f^0 values for pure $CaCO_3$, Ca^{2+}, and $CO_3{}^{2-}$ are −1129.10, −553.04, and −528.10 kJ, respectively, at 25°C and 1 bar. The standard free energy of the reaction is therefore

$$\Delta G_r^0 = (-553.04 - 528.10) - (-1129.10)$$

From Eq. (3.12) we obtain, for 25°C and 1 bar,

$$\log K_{\text{calcite}} = -8.40 \quad \text{or} \quad K_c = 10^{-8.40}$$

Dissolved Gases

When water is exposed to a gas phase, an equilibrium is established between the gas and the liquid through the exchange of molecules across the liquid-gas interface. If the gaseous phase is a mixture of more than one gas, an equilibrium will be established for each gas. The pressure that each gas in the mixture exerts is its *partial pressure*, which is defined as the pressure that the specific component of the gas would exert if it occupied the same volume alone. *Dalton's law of partial pressures* states that in a mixture of gases, the total pressure equals the sum of the partial pressures. The partial pressure of a vapor is also referred to as the *vapor pressure*.

Groundwater contains dissolved gases as a result of (1) exposure to the earth's atmosphere prior to infiltration into the subsurface environment, (2) contact with soil gases during infiltration through the unsaturated zone, or (3) gas production below the water table by chemical or biochemical reactions involving the groundwater, minerals, organic matter, and bacterial activity.

Probably the most important of the dissolved gases in groundwater is CO_2. Two reactions that describe the interaction between gaseous CO_2 and its dissolved species are,

$$CO_2(g) + H_2O \rightleftharpoons CO_2(aq) + H_2O \tag{3.14}$$

$$CO_2(g) + H_2O \rightleftharpoons H_2CO_3(aq) \tag{3.15}$$

where the suffixes (g) and (aq) denote gaseous and dissolved species, respectively. The ratio of $CO_2(aq)/H_2CO_3$ is much greater than unity in aqueous solutions; *however, it is customary to denote all dissolved CO_2 in water as H_2CO_3 (carbonic acid)*. This usage results in no loss of generality as long as consistency is maintained elsewhere in the treatment of this dissolved molecular species. These matters are discussed in detail by Kern (1960).

The partial pressure of a dissolved gas is the partial pressure with which the dissolved gas would be in equilibrium if the solution were in contact with a gaseous phase. It is common practice to refer to the partial pressure of a solute such as H_2CO_3 or dissolved O_2 even though the water may be isolated from the gas phase. For example, we can refer to the partial pressure of dissolved CO_2 in groundwater even though the water is isolated from the earth's atmosphere and from the gases in the open pore spaces above the water table.

In dilute solutions the partial pressure of a solute, expressed in bars (1 bar $= 10^5$ N/m^2), is proportional to its molality. This is a statement of *Henry's law*. It is applicable to gases that are not very soluble, such as CO_2, O_2, N_2, CH_4, and H_2S. From application of the law of mass action to Eq. (3.15),

$$K_{CO_2} = \frac{[H_2CO_3]}{[H_2O][CO_2(g)]} \tag{3.16}$$

Because the activity of H_2O is unity except for very saline solutions and because the partial pressure of CO_2 in bars is equal to its molality, Eq. (3.16) can be expressed as

$$K_{CO_2} = \frac{[H_2CO_3]}{\gamma_{CO_2} \cdot P_{CO_2}} \tag{3.17}$$

where γ_{CO_2} is the activity coefficient for dissolved CO_2 and P_{CO_2} is the partial pressure in bars. With this expression the partial pressure of CO_2 that would exist at equilibrium with a solution of specified H_2CO_3 activity can be computed. The activity coefficients for uncharged solute species such as dissolved gases (CO_2, O_2, H_2S, N_2, etc.) are greater than unity. The solubility of these gases in water therefore decreases with increasing ionic strength. This effect is known as the *salting-out effect*.

In addition to its dependence on ionic strength, the activity coefficient can be influenced by the type of electrolyte present in the water. For example, at a given ionic strength, CO_2 is less soluble (i.e., has a larger activity coefficient) in a NaCl solution than in a KCl solution. Most geochemical problems of interest in groundwater hydrology involve solutions at ionic strengths of less than 0.1 or 0.2. It is common practice, therefore, for the activity coefficient of the dissolved gas to be approximated as unity. Consequently, under these conditions, Eq. (3.17) reduces to the relation

$$K_{CO_2} = \frac{[H_2CO_3]}{P_{CO_2}} \tag{3.18}$$

3.3 Association and Dissociation of Dissolved Species

The Electroneutrality Condition

Before proceeding with a discussion of the processes and consequences of the chemical interactions between groundwater and the geologic materials through which it flows, the behavior of dissolved constituents in the liquid phase without interactions with solid phases will be considered.

A fundamental condition of electrolyte solutions is that on a macroscopic scale, rather than the molecular scale, a condition of electroneutrality exists. The sum of the positive ionic charges equals the sum of the negative ionic charges, or

$$\sum zm_c = \sum zm_a \tag{3.19}$$

where z is the ionic valence, m_c the molality of cation species, and m_a the molality of anion species. This is known as the *electroneutrality equation*, or the *charge-balance equation*, and it is used in nearly all calculations involving equilibrium interactions between water and geologic materials.

An indication of the accuracy of water analysis data can be obtained using the charge-balance equation. For example, if a water sample is analyzed for the major constituents listed in Table 3.3, and if the concentration values are substituted into Eq. (3.19) as

$$(Na^+) + 2(Mg^{2+}) + 2(Ca^{2+}) = (Cl^-) + (HCO_3^-) + 2(SO_4^{2-}) \qquad (3.20)$$

the quantities obtained on the left- and right-hand sides of the equation should be approximately equal. Silicon is not included in this relation because it occurs in a neutral rather than in a charged form. If significant deviation from equality occurs, there must be (1) analytical errors in the concentration determinations or (2) ionic species at significant concentration levels that were not included in the analysis. It is common practice to express the deviation from equality in the form

$$E = \frac{\sum zm_c - \sum zm_a}{\sum zm_c + \sum zm_a} \times 100 \qquad (3.21)$$

where E is the *charge-balance error* expressed in percent and the other terms are as defined above.

Water analysis laboratories normally consider a charge-balance error of less than about 5% to be acceptable, although for some types of groundwater many laboratories consistently achieve results with errors that are much smaller than this. It should be kept in mind that an acceptable charge-balance error may occur in situations where large errors in the individual ion analyses balance one another. Appraisal of the charge-balance error therefore cannot be used as the only means of detecting analytical errors.

For the purpose of computing the charge-balance error, the results of chemical analyses are sometimes expressed as millequivalents per liter. When these units are used, the valence terms are omitted from Eq. (3.20).

Dissociation and Activity of Water

In the liquid state water undergoes the equilibrium dissociation,

$$H_2O \rightleftharpoons H^+ + OH^- \qquad (3.22)$$

which, from the law of mass action, can be expressed as

$$K_w = \frac{[H^+][OH^-]}{[H_2O]} \qquad (3.23)$$

where brackets denote activities. It will be recalled that the activity of *pure* water is defined as unity at standard-state conditions. The reference condition of 25°C at 1 bar will be used. Since water vapor at low or moderate pressures behaves as an ideal gas, the activity of water in aqueous solution can be expressed as

$$[H_2O] = \frac{P_{H_2O}}{P^*_{H_2O}} \qquad (3.24)$$

where $P^*_{H_2O}$ is the partial pressure of the vapor for pure water and P_{H_2O} is the partial pressure of the vapor for the aqueous solution. At 25°C and 1 bar, the activity of water in a solution of NaCl at a concentration similar to that in seawater, which is approximately 3%, is 0.98, and in a 20% NaCl solution is 0.84. Thus, except for highly concentrated waters such as brines, the activity of water can, for practical purposes, be taken as unity. In this case

$$K_w = [H^+][OH^-] \tag{3.25}$$

Values for K_w at temperatures between 0 and 50°C are listed in Table 3.4. Because the effect of fluid pressure is very slight, this expression is also acceptable for pressures as high as about 100 bars. At 1000 bars and 25°C, the activity of water is 2.062 (Garrels and Christ, 1965).

Table 3.4 Equilibrium Constants for Dissociation of Water, 0–60°C

t(°C)	$K_w \times 10^{-14}$
0	0.1139
5	0.1846
10	0.2920
15	0.4505
20	0.6809
25	1.008
30	1.469
35	2.089
40	2.919
45	4.018
50	5.474
55	7.297
60	9.614

SOURCE: Garrels and Christ, 1965.

Since pH is defined as the negative logarithm of the hydrogen-ion activity, water at 25°C and pH 7 has equal H^+ and OH^- activities ($[H^+] = [OH^-] = 1.00 \times 10^{-7}$). At lower temperatures the equality of H^+ and OH^- activities occurs at higher pH values and vice versa for higher temperatures. For example, at 0°C the equality occurs at a pH 7.53 and at 50°C at pH 6.63.

Polyprotic Acids

The most important acid in natural groundwater and in many contaminated groundwaters is carbonic acid (H_2CO_3), which forms when carbon dioxide (CO_2) combines with water [Eq. (3.15)]. Carbonic acid can dissociate in more than one step by transferring hydrogen ions (protons) through the reactions

$$H_2CO_3 \rightleftharpoons H^+ + HCO_3^- \tag{3.26}$$

$$HCO_3^- \rightleftharpoons H^+ + CO_3{}^{2-} \tag{3.27}$$

Because hydrogen ions are commonly referred to as protons by chemists and because more than one hydrogen-ion dissociation is involved, carbonic acid is known as a *polyprotic acid*. Another polyprotic acid that occurs in groundwater, although in much smaller concentrations than carbonic acid, is phosphoric acid, which dissociates in three steps;

$$H_3PO_4 \rightleftharpoons H_2PO_4^- + H^+ \tag{3.28}$$

$$H_2PO_4^- \rightleftharpoons HPO_4^{2-} + H^+ \tag{3.29}$$

$$HPO_4^{2-} \rightleftharpoons PO_4^{3-} + H^+ \tag{3.30}$$

Since the dissociation equations for the polyprotic acids all involve H^+, it is possible to calculate the fraction of the acid in its molecular form or in any one of its anionic forms as a function of pH. For example, for carbonic acid, dissociation constants for Eqs. (3.26) and (3.27) can be expressed according to the law of mass action as

$$K_{H_2CO_3} = \frac{[H^+][HCO_3^-]}{[H_2CO_3]} \tag{3.31}$$

$$K_{HCO_3^-} = \frac{[H^+][CO_3^{2-}]}{[HCO_3^-]} \tag{3.32}$$

A mass-balance expression for the carbon in the acid and its dissociated anionic species, expressed in molality, is

$$\text{DIC} = (H_2CO_3) + (HCO_3^-) + (CO_3^{2-}) \tag{3.33}$$

where DIC is the concentration of total dissolved inorganic carbon in these species. If we select an arbitrary value of 1 for DIC, and reexpress Eq. (3.33) in terms of pH, HCO_3^-, $K_{H_2CO_3}$, and $K_{HCO_3^-}$, and then in terms of pH, CO_3^{2-}, and the dissociation constants, equations for the relative concentration of H_2CO_3, HCO_3^-, and CO_3^{2-} as a function of pH are obtained. They are expressed graphically in Figure 3.5(a).

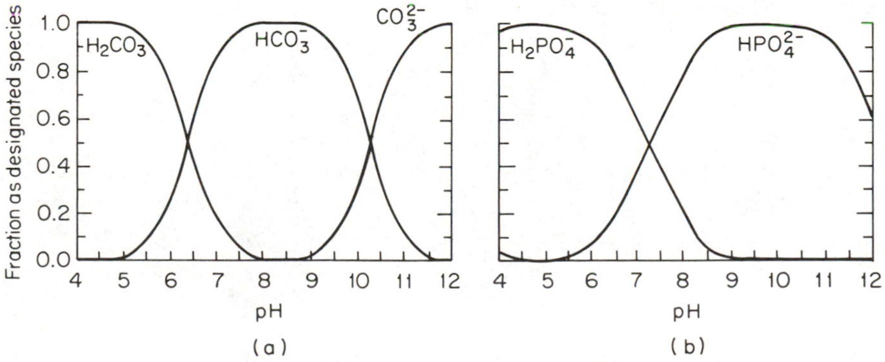

Figure 3.5 Distribution of major species of (a) dissolved inorganic carbon and (b) inorganic phosphorus in water at 25°C.

At lower pH, H_2CO_3 is the dominant species and at high pH, $CO_3{}^{2-}$ is the dominant species. Over most of the normal pH range of groundwater (6–9), HCO_3^- is the dominant carbonate species. This is why HCO_3^-, rather than $CO_3{}^{2-}$ or H_2CO_3, is listed in Table 3.3 as one of the major dissolved inorganic constituents in groundwater. Following a similar analysis, the relative concentrations of the dissolved phosphate species shown in Figure 3.5(b) are obtained. In the normal pH range of groundwater, $H_2PO_4^-$ and $HPO_4{}^{2-}$ are the dominant species.

Ion Complexes

Chemical analyses of dissolved constituents in groundwater indicate the total concentrations of the constituents, but not the form in which the constituents occur in the water. Some constituents are present almost entirely in the simple ionic form. For instance, chlorine is present as the chloride ion, Cl^-. Calcium and magnesium, however, are present in the free ionic form, Ca^{2+} and Mg^{2+}, in inorganic ion associations such as the uncharged (zero-valence) species, $CaSO_4^\circ$, $CaCO_3^\circ$, $MgSO_4^\circ$, and $MgCO_3^\circ$, and the charged associations, $CaHCO_3^+$ and $MgHCO_3^+$. These charged and uncharged associations are known as complexes or in some cases, as ion pairs. The complexes form because of the forces of electrical attraction between the ions of opposite charge. Some inorganic species such as aluminum occur in dissolved form as Al^{3+}, as the positively charged complex or ion pair, $[Al(OH)]^{2+}$, and as complexes with covalent bonds such as $[Al_2(OH)_2]^{4+}$, $[Al_6(OH_{15})]^{3+}$, and $[Al(OH)_4]^-$. The total dissolved concentration of an inorganic species C_i can be expressed as

$$C_i = \sum C_{\text{free ion}} + \sum C_{\text{inorganic complexes}} + \sum C_{\text{organic complexes}} \tag{3.34}$$

The occurrence of ion complexes can be treated using the law of mass action. For example, the formation of $CaSO_4^\circ$ can be expressed as

$$Ca^{2+} + SO_4{}^{2-} \rightleftharpoons CaSO_4^\circ \tag{3.35}$$

with the equilibrium relation

$$K_{CaSO_4{}^\circ} = \frac{[Ca^{2+}][SO_4{}^{2-}]}{[CaSO_4^\circ]} \tag{3.36}$$

where $K_{CaSO_4{}^\circ}$ is the thermodynamic equilibrium constant, sometimes referred to as the dissociation constant, and the terms in brackets are activities. Concentration values for the free ions are related to activities through the ionic strength versus activity coefficient relations described in Section 3.2. The activity coefficient for the neutral complex, $CaSO_4^\circ$, is taken as unity. Values for $K_{CaSO_4{}^\circ}$ and equilibrium constants for other inorganic pairs and complexes can be computed using Eq. (3.12).

Table 3.5 shows the results of a chemical analysis of groundwater expressed in both milligrams per liter (or grams per cubic meter) and molality. The concentrations of free ions and inorganic ion complexes were calculated from the total

Table 3.5 Chemical Analysis of Groundwater Expressed as Analytical Results and as Computed Dissolved Species

Dissolved constituent	Analytical results from laboratory		Computed dissolved species			
	mg/ℓ or g/m³	molality × 10^{-3}	Free-ion concentration (molality × 10^{-3})	SO_4^{2-} ion pairs* (molality × 10^{-3})	HCO_3^- ion pairs† (molality × 10^{-3})	CO_3^{2-} ion pairs‡ (molality × 10^{-3})
Ca	136	3.40	2.61	0.69	0.09	0.007
Mg	63	2.59	2.00	0.47	0.12	0.004
Na	325	14.13	14.0	0.07	0.06	0.001
K	9.0	0.23	0.23	0.003	<0.0001	<0.0001
Cl	40	1.0	1.0			
SO_4	640	6.67	5.43			
HCO_3	651	10.67	10.4			
CO_3	0.12	0.020	0.0086			
DIC	147.5	12.29	CO_3^{2-}, DIC calculated from HCO_3 and pH data			

Temp. = 10°C, pH = 7.20, partial pressure CO_2 (calculated) = 3.04×10^{-2} bar

$\sum$ cations (analytical) = 26.39 meq/ℓ $\sum$ anions (analytical) = 25.17

$\sum$ cations (computed) = 23.64 meq/ℓ $\sum$ anions (computed) = 22.44

Error in cation-anion total (charge-balance error) (analytical) = 2.9%

Error in cation-anion totals (charge-balance error) (computed) = 2.7%

*SO_4^{2-} complexes = $CaSO_4^0$, $MgSO_4^0$, $NaSO_4^-$, $Na_2SO_4^0$, KSO_4^-.

†HCO_3^- complexes = $CaHCO_3^-$, $MgHCO_3^-$, $NaHCO_3^0$.

‡CO_3^{2-} complexes = $CaCO_3^0$, $MgCO_3^0$, $NaCO_3^-$.

analytical concentrations in the manner described below. In this sample (Table 3.5), the only complexes that occur in appreciable concentrations are those of sulfate; 18% of the total sulfate is complexed. When groundwater has large sulfate concentrations, sulfate complexes are normally quite important. The procedure by which the concentrations of the free-ion complexes in Table 3.5 were calculated is described by Garrels and Christ (1965) and Truesdell and Jones (1974).

Inorganic constituents in groundwater can also form dissolved complexes with organic compounds such as fulvic and humic acids. In natural groundwater, which rarely has dissolved organic carbon at concentrations of more than 10 mg/ℓ, complexing of major ions with the dissolved organic matter is probably insignificant. In contaminated groundwaters, however, the movement of hazardous inorganic compounds as organic complexes can be very important.

Calculation of Dissolved Species

Depending on the methods of analysis used in the laboratory, results of analysis of inorganic carbon may be expressed as total dissolved inorganic carbon (DIC) or as HCO_3^-. Each of these types of data can be used, in conjunction with pH values, to compute the concentrations of H_2CO_3, $CO_3{}^{2-}$, HCO_3^- or DIC, and the partial pressure of CO_2. Equations (3.18), (3.20), (3.31), (3.32), and (3.33) serve as a basis for the calculations. If the water is nonsaline, the activity of H_2O and the activity coefficients for CO_2 and H_2CO_3 are taken as unity. It must be kept in mind that Eqs. (3.18), (3.31), and (3.32) are expressed in activities, whereas Eqs. (3.20) and (3.33) require molalities. If in the chemical analysis of the water HCO_3^- concentration and pH are determined, Eq. (3.31) can be used, along with Eq. (3.5) for conversion between concentrations and activities, to obtain the activity and concentration of H_2CO_3. Substitution of the H_2CO_3 activity in Eq. (3.18) yields the partial pressure of CO_2 in bars. The activity of $CO_3{}^{2-}$ can be computed from Eq. (3.32) and then converted to concentration by Eq. (3.5). Substitution of the concentration values in Eq. (3.33) then yields the concentration of DIC. The accuracy of the calculated result is strongly dependent on the accuracy of pH measurement. To obtain reliable pH data, it is necessary to make the pH measurements in the field. This is discussed further in Section 3.9.

In the following illustration of the method for calculating free-ion and complex concentrations, it will be assumed that only the cation-sulfate complexes occur in significant concentrations. The equilibrium relations of interest are, therefore,

$$K_{CaSO_4{}^\circ} = \frac{[Ca^{2+}][SO_4{}^{2-}]}{[CaSO_4^\circ]} \tag{3.37}$$

$$K_{MgSO_4{}^\circ} = \frac{[Mg^{2+}][SO_4{}^{2-}]}{[MgSO_4^\circ]} \tag{3.38}$$

$$K_{NaSO_4{}^-} = \frac{[Na^+][SO_4{}^{2-}]}{[NaSO_4^-]} \tag{3.39}$$

From the conservation of mass principle, we can write

$$\text{Ca(total)} = (Ca^{2+}) + (CaSO_4^\circ) \tag{3.40}$$

$$\text{Mg(total)} = (Mg^{2+}) + (MgSO_4^\circ) \tag{3.41}$$

$$\text{Na(total)} = (Na^+) + (NaSO_4^-) \tag{3.42}$$

$$SO_4\text{(total)} = (SO_4{}^{2-}) + (CaSO_4^\circ) + (MgSO_4^\circ) + (NaSO_4^-) \tag{3.43}$$

The concentrations of Ca(total), Mg(total), Na(total), and SO_4(total) are those obtained from the laboratory analysis. We therefore have seven equations and seven unknowns (Na^+, Mg^{2+}, Ca^{2+}, $SO_4{}^{2-}$, $NaSO_4^-$, $MgSO_4^\circ$, and $CaSO_4^\circ$). The equations can be solved manually using the method of successive approximations described by Garrels and Christ (1965). Conversion between activities and molalities is accomplished using the ionic strength versus activity coefficient relations indicated in the discussions of Eqs. (3.5) and (3.6). In many cases the ionic strength calculated from the total concentration values has acceptable accuracy. In saline solutions, however, the ionic strength should be adjusted for the effect of complexes.

The process of computing the concentrations of free ions and complexes can be quite tedious and time-consuming, particularly when the sulfate, bicarbonate, and carbonate complexes are all included in the calculations. In recent years it has become common for the computations to be done by digital computer. There are several well-documented and widely used computer programs available for this purpose. Two of the most readily available programs are those by Truesdell and Jones (1974), which were used to obtain the results listed in Table 3.5, and Kharaka and Barnes (1973). Processing of chemical data on groundwater using programs of this type is becoming a relatively standard procedure in situations where one wishes to interpret chemical analyses in a geochemical framework.

3.4 Effects of Concentration Gradients

Diffusion in solutions is the process whereby ionic or molecular constituents move under the influence of their kinetic activity in the direction of their concentration gradient. Diffusion occurs in the absence of any bulk hydraulic movement of the solution. If the solution is flowing, diffusion is a mechanism, along with mechanical dispersion, that causes mixing of ionic or molecular constituents. Diffusion ceases only when concentration gradients become nonexistent. The process of diffusion is often referred to as *self-diffusion*, *molecular diffusion*, or *ionic diffusion*.

The mass of diffusing substance passing through a given cross section per unit time is proportional to the concentration gradient. This is known as *Fick's first law*. It can be expressed as

$$F = -D\frac{dC}{dx} \tag{3.44}$$

where F, which is the mass flux, is the mass of solute per unit area per unit time $[M/L^2T]$; D is the diffusion coefficient $[L^2/T]$; C is the solute concentration $[M/L^3]$; and dC/dx is the concentration gradient, which is a negative quantity in the direction of diffusion. The diffusion coefficients for electrolytes in aqueous solutions are well known. The major ions in groundwater (Na^+, K^+, Mg^{2+}, Ca^{2+}, Cl^-, HCO_3^-, $SO_4{}^{2-}$) have diffusion coefficients in the range 1×10^{-9} to 2×10^{-9} m^2/s at 25°C (Robinson and Stokes, 1965). The coefficients are temperature-dependent. At 5°C, for example, the coefficients are about 50% smaller. The effect of ionic strength is very small.

In porous media the apparent diffusion coefficients for these ions are much smaller than in water because the ions follow longer paths of diffusion caused by the presence of the particles in the solid matrix and because of adsorption on the solids. The apparent diffusion coefficient for nonadsorbed species in porous media, D^*, is represented by the relation

$$D^* = \omega D \tag{3.45}$$

where ω, which is less than 1, is an empirical coefficient that takes into account the effect of the solid phase of the porous medium on the diffusion. In laboratory studies of diffusion of nonadsorbed ions in porous geologic materials, ω values between about 0.5 and 0.01 are commonly observed.

From Fick's first law and the equation of continuity, it is possible to derive a differential equation that relates the concentration of a diffusing substance to space and time. In one dimension, this expression, known as *Fick's second law*, is

$$\frac{\partial C}{\partial t} = D^* \frac{\partial^2 C}{\partial x^2} \tag{3.46}$$

To obtain an indication of the rates at which solutes can diffuse in porous geological materials, we will consider a hypothetical situation where two strata containing different solute concentrations are in contact. It will be assumed that the strata are saturated with water and that the hydraulic gradients in these strata are negligible. At some initial time, one of the strata has solute species i at concentration C_0. In the other bed the initial concentration of C is small enough to be approximated as zero. Because of the concentration gradient across the interface, the solute will diffuse from the higher concentration layer to the lower concentration layer. It will also be assumed that the solute concentration in the higher concentration layer remains constant with time, as would be the case if the solute concentration were maintained at an equilibrium by mineral dissolution. Values of C in the x direction over time t can be calculated from the relation (Crank, 1956)

$$C_i(x, t) = C_0 \operatorname{erfc}(x/2\sqrt{D^*t}) \tag{3.47}$$

where erfc is the complementary error function (Appendix V). Assuming a value of 5×10^{-10} m^2/s for D^*, the solute concentration profile at specified time intervals can be computed. For instance, if we choose a relative concentration C/C_0 of 0.1 and a distance x of 10 m, Eq. (3.47) indicates that the diffusion time would be approximately 500 years. It is evident, therefore, that diffusion is a relatively slow process. In zones of active groundwater flow its effects are usually masked by the effects of the bulk water movement. In low-permeability deposits such as clay or shale, in which the groundwater velocities are small, diffusion over periods of geologic time can, however, have a strong influence on the spatial distribution of dissolved constituents. This is discussed further in Sections 7.8 and 9.2.

Laboratory investigations have shown that compacted clays can act as semipermeable membranes (Hanshaw, 1962). Semipermeable membranes restrict the passage of ions while allowing relatively unrestricted passage of neutral species. If the pore waters in strata on either side of a compacted clay layer have different ionic concentrations, the concentration of the water in these strata must also be different. Because water molecules as uncharged species can move through semipermeable clay membranes, it follows that under conditions of negligible hydraulic gradients across the membrane, movement from the higher water-concentration zone (lower salinity zone) to the lower water-concentration zone (higher salinity zone) would occur by diffusion. If the higher salinity zone is a closed system, movement of water into the zone by diffusion across the clay will cause the fluid pressure in it to rise. If the lower-salinity zone is a closed system, its fluid pressure will decline. This process of development of a pressure differential across the clay is known as *osmosis*. The *equilibrium osmotic pressure* across the clay is the pressure differential that would exist when the effect of water diffusion is balanced by the pressure differential. When this occurs, migration of water across the clay ceases. In laboratory experiments the osmotic pressure across a semipermeable membrane separating solutions of different concentrations is measured by applying a pressure differential just sufficient to prevent water diffusion. In sedimentary basins osmosis may cause significant pressure differentials across clayey strata even if the equilibrium osmotic pressure differential is not achieved.

Many equations have been used to express the relation between the osmotic pressure differential and the difference in solution concentration across semipermeable membranes. One of these, which can be derived from thermodynamic arguments (Babcock, 1963), is

$$P_0 = \frac{RT}{\bar{V}_{H_2O}} \ln\left(\frac{[H_2O]^{I}}{[H_2O]^{II}}\right) \tag{3.48}$$

where P_0 is the hydrostatic pressure differential caused by osmosis, R is the gas constant (0.0821 liter·bar/K·mol), T is degrees Kelvin, $\bar{V}_{H_2O}$ is the molal volume of pure water, (0.018 ℓ/mol at 25°C) and $[H_2O]^{I}$ and $[H_2O]^{II}$ are the activities of water in the more saline solution and the less saline solution, respectively. Values for the activity of water in various salt solutions are listed in Robinson and Stokes (1965). Using Eq. (3.48), it can be shown that salinity differences that are not uncommon in groundwater of sedimentary basins can cause large osmotic pressures, provided of course that there is a compacted, unfractured clay or shale separating the salinity zones. For example, consider two sandstone aquifers, I and II, separated by a layer of compacted clay. If the water in both of the aquifers has high NaCl concentrations, one with 6% NaCl and the other with 12% NaCl, the H_2O activity ratio will be 0.95, which upon substitution in Eq. (3.48) yields an osmotic pressure difference between the two aquifers of 68 bars. This is the equivalent of 694 m of hydrostatic head (expressed in terms of pure water). This would indeed be a striking head differential in any sedimentary basin. For large osmotic pressure differentials to actually occur, however, it is necessary for the hydrostratigraphic conditions to be such that osmotic pressure develops much more quickly than the pressure that is dissipated by fluid flow from the high-pressure zone and by flow into the low-pressure zone.

3.5 Mineral Dissolution and Solubility

Solubility and the Equilibrium Constant

When water comes into contact with minerals, dissolution of the minerals begins and continues until equilibrium concentrations are attained in the water or until all the minerals are consumed. The *solubility* of a mineral is defined as the mass of the mineral that will dissolve in a unit volume of solution under specified conditions. The solubilities of minerals that are encountered by groundwater as it moves along its flow paths vary over many orders of magnitude. Thus, depending on the minerals that the water has come into contact with during its flow history, groundwater may be only slightly higher in dissolved solids than rainwater, or it many become many times more salty than seawater.

Table 3.6 indicates the solubilities of several sedimentary minerals in pure water at 25°C and 1 bar total pressure. This table also lists the dissolution reactions for these minerals and the equilibrium constants for the reactions at 25°C and 1 bar. The solubility of carbonate minerals is dependent on the partial pressure of CO_2. The solubilities of calcite and dolomite at two partial pressures (10^{-3} bar and 10^{-1} bar) are listed in Table 3.6 as an indication of the range of values that are relevant for natural groundwater.

Table 3.6 Dissociation Reactions, Equilibrium Constants, and Solubilities of Some Minerals That Dissolve Congruently in Water at 25°C and 1 Bar Total Pressure

Mineral	Dissociation reaction	Equilibrium constant, K_{eq}	Solubility at pH 7 (mg/ℓ or g/m³)
Gibbsite	$Al_2O_3 \cdot 2H_2O + H_2O = 2Al^{3+} + 6OH^-$	10^{-34}	0.001
Quartz	$SiO_2 + 2H_2O = Si(OH)_4$	$10^{-3.7}$	12
Hydroxylapatite	$Ca_5OH(PO_4)_3 = 5Ca^{2+} + 3PO_4{}^{3-} + OH^-$	$10^{-55.6}$	30
Amorphous silica	$SiO_2 + 2H_2O = Si(OH)_4$	$10^{-2.7}$	120
Fluorite	$CaF_2 = Ca^{2+} + 2F^-$	$10^{-9.8}$	160
Dolomite	$CaMg(CO_3)_2 = Ca^{2+} + Mg^{2+} + 2CO_3{}^{2-}$	$10^{-17.0}$	90,* 480†
Calcite	$CaCO_3 = Ca^{2+} + CO_3{}^{2-}$	$10^{-8.4}$	100,* 500†
Gypsum	$CaSO_4 \cdot 2H_2O = Ca^{2+} + SO_4{}^{2-} + 2H_2O$	$10^{-4.5}$	2100
Sylvite	$KCl = K^+ + Cl^-$	$10^{+0.9}$	264,000
Epsomite	$MgSO_4 \cdot 7H_2O = Mg^{2+} + SO_4{}^{2-} + 7H_2O$	—	267,000
Mirabillite	$Na_2SO_4 \cdot 10H_2O = 2Na^+ + SO_4{}^{2-} + 10H_2O$	$10^{-1.6}$	280,000
Halite	$NaCl = Na^+ + Cl^-$	$10^{+1.6}$	360,000

*Partial pressure of $CO_2 = 10^{-3}$ bar.

†Partial pressure of $CO_2 = 10^{-1}$ bar.

SOURCE: Solubility data from Seidell, 1958.

Comparison of the mineral solubilities and equilibrium constants indicates that the relative magnitudes of the equilibrium constant are a poor indication of the relative solubilities of the minerals because in the equilibrium relations, the activities of the ions or molecules are raised to the power of the number of moles in the balanced dissociation expression. For example, the solubility of calcite in pure water at $P_{CO_2} = 10^{-1}$ bar is 500 mg/ℓ, and the solubility of dolomite under the same conditions is nearly the same (480 mg/ℓ), but the equilibrium constants differ by eight orders of magnitude because the term $[CO_3{}^{2-}]$ is raised to the second power in the K_{dol} expression. Another example is hydroxylapatite, which has a solubility of 30 mg/ℓ at pH 7 and yet has an equilibrium constant of $10^{-55.6}$, a value that might give the erroneous impression that this mineral has no significant solubility.

All the minerals listed in Table 3.6 normally dissolve *congruently*. This statement means that the products of the mineral dissolution reaction are all dissolved species. Many minerals that affect the chemical evolution of groundwater dissolve incongruently; that is, one or more of the dissolution products occur as minerals or as amorphous solid substances. Most aluminum silicate minerals dissolve incongruently. The feldspar, albite, is a good example

$$\underbrace{NaAlSi_3O_8(s)}_{\text{albite}} + H_2CO_3 + \tfrac{9}{2}H_2O \rightleftharpoons$$

$$Na^+ + HCO_3^- + 2H_4SiO_4 + \tfrac{1}{2}\underbrace{Al_2Si_2O_5(OH)_4(s)}_{\text{kaolinite}} \qquad (3.49)$$

In this reaction albite dissolves under the leaching action of carbonic acid (H_2CO_3) to produce dissolved products and the clay mineral kaolinite. This is a common reaction in groundwater zones in granitic terrain. From the law of mass action,

$$K_{\text{alb-kaol}} = \frac{[Na^+][HCO_3^-][H_4SiO_4]^2}{[H_2CO_3]} \qquad (3.50)$$

where the equilibrium constant K depends on temperature and pressure. If the partial pressure of CO_2 is specified, it is evident from Eqs. (3.18), (3.31), and (3.32) that $[H_2CO_3]$ and $[HCO_3^-]$ are also specified. The solubility of albite and other cation aluminosilicates increases with increasing partial pressure of CO_2.

Effect of Ionic Strength

Comparison of the solubilities of minerals in pure water versus water with a high salt content indicates that the salinity increases the solubilities. This is known as the *ionic strength effect* because the increased solubility is caused by decreases in activity coefficients as a result of increased ionic strength. For example, the expression for the equilibrium constant for gypsum can be written

$$K_{gyp} = [\gamma_{Ca^{2+}} \cdot \gamma_{SO_4{}^{2-}}][(Ca^{2+})(SO_4{}^{2-})] \qquad (3.51)$$

where γ is the activity coefficient and the species in parentheses are expressed in molality. Figure 3.3 indicates that as ionic strength increases, $\gamma_{Ca^{2+}}$ and the $\gamma_{SO_4^{2-}}$ value decrease. To compensate, in Eq. (3.51), the concentrations of Ca^{2+} and SO_4^{2-} must increase. This results in greater solubility of the mineral under the specified conditions of temperature and pressure. This effect is illustrated in Figure 3.6, which shows that the solubility of gypsum more than triples as a result of the ionic strength effect. Other examples, described in Chapter 7, indicate that the ionic strength effect can play an important role in the chemical evolution of natural and contaminated groundwater.

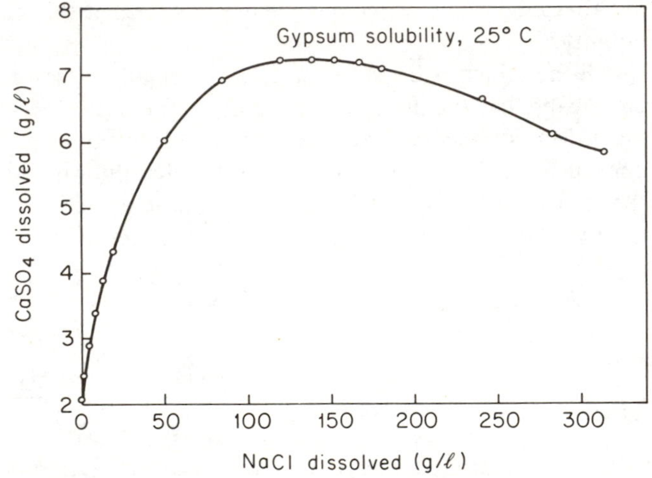

Figure 3.6 Solubility of gypsum in aqueous solutions of different NaCl concentrations, 25°C, and 1 bar (after Shternina, 1960).

The Carbonate System

It is estimated that over 99% of the earth's carbon exists in carbonate minerals, the most important of which are calcite, $CaCO_3$, and dolomite, $CaMg(CO_3)_2$. In nearly all sedimentary terrain and in many areas of metamorphic and igneous rocks, groundwater is in contact with carbonate minerals during at least part of its flow history. The ability of the groundwater zone to minimize adverse effects of many types of pollutants can be dependent on interactions that involve water and carbonate minerals. Interpretation of carbon 14 age dates of groundwater requires an understanding of how the water has interacted with these minerals.

At equilibrium, the reactions between water and the carbonate minerals calcite and dolomite can be expressed as

$$K_{cal} = [Ca^{2+}][CO_3^{2-}] \tag{3.52}$$

$$K_{dol} = [Ca^{2+}][Mg^{2+}][CO_3^{2-}]^2 \tag{3.53}$$

where the equilibrium constants depend on temperature and pressure.

If the minerals dissolve in water that has an abundant supply of $CO_2(g)$ at a constant partial pressure, the concentration of dissolved CO_2 (expressed as carbonic acid, H_2CO_3) remains constant, as indicated by Eq. (3.18). It is instructive to represent the calcite dissolution process as

$$CaCO_3 + H_2CO_3 \longrightarrow Ca^{2+} + 2HCO_3^- \tag{3.54}$$

which indicates that the dissolution is accompanied by consumption of carbonic acid. The higher the P_{CO_2}, the greater is the amount of H_2CO_3 available for consumption, and hence the reaction proceeds farther to the right to achieve equilibrium.

An aqueous system in which the dissolved CO_2 is constant because of relatively unobstructed interaction with an abundant gaseous environment of constant P_{CO_2}, such as the earth's atmosphere, is commonly referred to in the context of mineral dissolution as an *open system*. If the H_2CO_3 consumed by mineral-water reactions is not replenished from a gaseous reservoir, the system is denoted as a *closed system*.

Substitutions of Eqs. (3.18), (3.31), and (3.32) in Eq. (3.52) and rearranging yields

$$[H^+] = \left\{\frac{K_{H_2CO_3} \cdot K_{HCO_3^-} \cdot K_{CO_2}}{K_{cal}} \cdot P_{CO_2}[Ca^{2+}]\right\}^{1/2} \tag{3.55}$$

The bracketed terms are activities and P_{CO_2} is expressed in bars. Values for the equilibrium constants in the range 0–30°C are listed in Table 3.7. At 25°C, Eq. (3.55) simplifies to

$$[H^+] = 10^{-4.9}\{[Ca^{2+}]P_{CO_2}\}^{1/2} \tag{3.56}$$

To calculate the solubility of calcite under the specified CO_2 partial pressure, another equation is required. At this stage in this type of mineral-water equilibrium

Table 3.7 Equilibrium Constants for Calcite, Dolomite, and Major Aqueous Carbonate Species in Pure Water, 0–30°C, and 1 Bar Total Pressure

Temperature (°C)	$pK^*_{CO_2}$	$pK_{H_2CO_3}$	$pK_{HCO_3^-}$	pK_{cal}	pK_{dol}
0	1.12	6.58	10.62	8.340	16.56
5	1.20	6.52	10.56	8.345	16.63
10	1.27	6.47	10.49	8.355	16.71
15	1.34	6.42	10.43	8.370	16.79
20	1.41	6.38	10.38	8.385	16.89
25	1.47	6.35	10.33	8.400	17.0
30	1.67	6.33	10.29	8.51	17.9

*$pK = -\log K$.

SOURCES: Garrels and Christ, 1965; Langmuir, 1971.

problem, it is appropriate to make use of the electroneutrality equation. For the case of calcite dissolution in pure water, the electroneutrality expression is

$$2(Ca^{2+}) + (H^+) = (HCO_3^-) + 2(CO_3{}^{2-}) + (OH^-) \tag{3.57}$$

The terms in this equation are expressed in molality. For the P_{CO_2} range of interest in groundwater studies, the (H^+) and (OH^-) terms are negligible compared to the other terms in this equation. Equations (3.56) and (3.57) can be combined and with substitution of Eqs. (3.18), (3.31), and (3.32) can be expressed as a polynomial in terms of two of the variables and the activity coefficients. For a specific P_{CO_2}, iterative solutions by computer can be obtained. Manual solutions can also be obtained with little difficulty using the method of successive approximations outlined by Garrels and Christ (1965) and Guenther (1975). As a first approximation a convenient approach is to assume that (HCO_3^-) is large compared to $(CO_3{}^{2-})$. Figure 3.5(a) indicates that this assumption is valid for solutions with pH values less than about 9, which includes nearly all natural waters, provided that concentrations of cations that complex $CO_3{}^{2-}$ are low. Equation (3.57) therefore reduces to

$$(Ca^{2+}) = \frac{(HCO_3^-)}{2} \tag{3.58}$$

After calcite has dissolved to equilibrium at a specified P_{CO_2}, the dissolved species in the water can now be obtained through the following steps: (1) assign an arbitrary value of $[H^+]$ to Eq. (3.55) and then calculate a value for $[Ca^{2+}]$; (2) estimate an ionic strength value using the $[Ca^{2+}]$ obtained from step (1) and an HCO_3^- value obtained from Eq. (3.58); (3) obtain an estimate for $\gamma_{Ca^{2+}}$ and $\gamma_{HCO_3^-}$ from Figure 3.3 and then calculate (Ca^{2+}) from the relation $(Ca^{2+}) = [Ca^{2+}]/\gamma_{Ca^{2+}}$; (4) using the specified P_{CO_2} and the $[H^+]$ chosen in step (1), calculate $[HCO_3^-]$ from Eqs. (3.18) and (3.31); (5) convert $[HCO_3^-]$ to (HCO_3^-) through the activity coefficient relation; and (6) compare the (Ca^{2+}) value obtained from step (1) with the calculated value of $(HCO_3^-)/2$ from step (5). If the two computed values are equal or nearly so, Eq. (3.57) has been satisfied and a solution to the problem has been obtained. If they are unequal, the sequence of computational steps must be repeated using a new selection for $[H^+]$. In these types of problems an acceptable solution can usually be obtained after two or three iterations. The results of these types of calculations for equilibrium calcite dissolution in pure water under various fixed P_{CO_2} and temperature conditions are illustrated in Figure 3.7, which indicates that the solubility is strongly dependent on the P_{CO_2} and that the equilibrium values of $[H^+]$ or pH vary strongly with P_{CO_2}. The calculation procedure did not include ion pairs such as $CaCO_3^\circ$ and $CaHCO_3^+$, which occur in small concentrations in dilute aqueous solutions saturated with $CaCO_3$. From Figure 3.7 it is apparent that the Ca^{2+} and HCO_3^- concentration lines are parallel (just 0.30 unit apart). This indicates that the reaction in Eq. (3.54) accurately represents the dissolution process under the range of P_{CO_2} conditions that are characteristic of the groundwater environment, where P_{CO_2} is almost invariably greater than 10^{-4} bar.

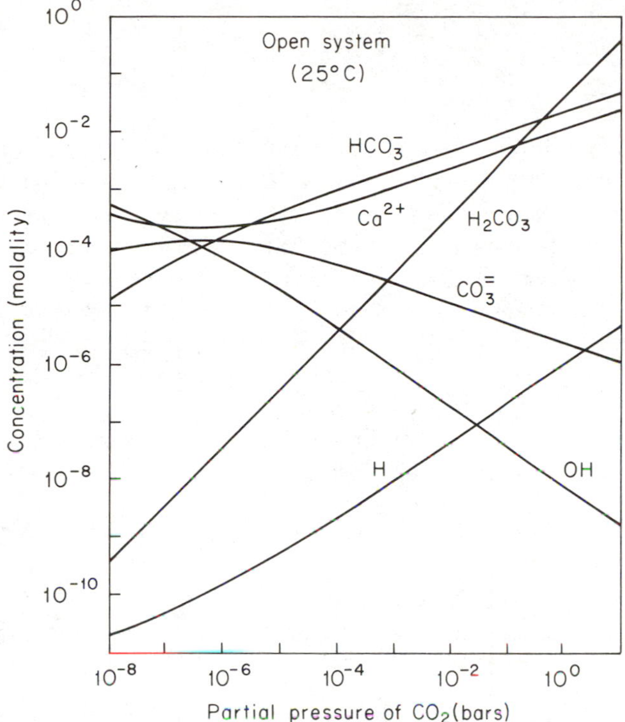

Figure 3.7 Dissolved species in water in equilibrium with calcite as a function of P_{CO_2} open-system dissolution (after Guenther, 1975).

This explains why HCO_3^- rather than $CO_3{}^{2-}$ is the dominant ionic species of dissolved inorganic carbon in groundwater.

If water becomes charged with CO_2, which may occur because of contact with the atmosphere of the earth or with the soil-zone atmosphere, and then comes into contact with calcite or dolomite in a zone isolated from the gaseous CO_2 source, such as the groundwater zone, dissolution will occur but the concentration of dissolved species at equilibrium will be different. In this process of closed-system dissolution, carbonic acid is consumed and not replenished from outside the system as dissolution proceeds. For this condition Eq. (3.18) indicates that the P_{CO_2} must also decline as the reaction proceeds toward equilibrium.

The carbonate minerals are less soluble under closed-system conditions, and have higher equilibrium pH values. In the closed-system case, the dissolved inorganic carbon is derived from the dissolved CO_2 present as dissolution begins and from the calcite and/or dolomite which dissolves. In the open-system case, CO_2 continues to enter the solution from the atmosphere as dissolution proceeds. In this case the total dissolved inorganic carbon consists of carbon from the initial CO_2 and from the replenished CO_2 and also from the minerals. As indicated in Chapter 7, these differences can be crucial to the interpretation of the chemical evolution of groundwater in carbonate terrain and in the evaluation of carbon 14 age dates.

The Common-Ion Effect

In some situations the addition of ions by dissolution of one mineral can influence the solubility of another mineral to a greater degree than the effect exerted by the change in activity coefficients. If an electrolyte that does not contain Ca^{2+} or $CO_3{}^{2-}$ is added to an aqueous solution saturated with calcite, the solubility of calcite will increase because of the ionic strength effect. However, if an electrolyte is added which contains either Ca^{2+} or $CO_3{}^{2-}$, calcite will eventually precipitate because the product $[Ca^{2+}][CO_3{}^{2-}]$ must adjust to attain a value equal to the equilibrium constant K_{cal}. This process is known as the *common-ion effect.*

Water moving through a groundwater zone that contains sufficient Ca^{2+} and $CO_3{}^{2-}$ for their activity product to equal K_{cal} may encounter strata that contain gypsum. Dissolution of gypsum,

$$CaSO_4 \cdot 2H_2O \longrightarrow Ca^{2+} + SO_4{}^{2-} + 2H_2O \tag{3.59}$$

causes the ionic strength to increase and the concentration of Ca^{2+} to rise. Expressed in terms of molality and activity coefficients, the equilibrium expression for calcite is

$$K_{cal} = \gamma_{Ca^{2+}} \cdot \gamma_{CO_3{}^{2-}} \cdot (Ca^{2+})(CO_3{}^{2-})$$

Gypsum dissolution causes the activity coefficient product $\gamma_{Ca^{2+}} \cdot \gamma_{CO_3{}^{2-}}$ to decrease. But because of the contribution of (Ca^{2+}) from dissolved gypsum, the product $(Ca^{2+})(CO_3{}^{2-})$ increases by a much greater amount. Therefore, for the solution to remain in equilibrium with respect to calcite, precipitation of calcite must occur.

The solubilities of calcite and gypsum in water at various NaCl concentrations are shown in Figure 3.8. For a given NaCl content, the presence of each mineral, calcite or gypsum, causes a decrease in the solubility of the other. Because of the ionic strength effect, both minerals increase in solubility at higher NaCl concentrations.

Disequilibrium and the Saturation Index

Considering Eq. (3.2) in a condition of disequilibrium, the relation between the reactants and the products can be expressed as

$$Q = \frac{[B]^b[C]^c}{[D]^d[E]^e} \tag{3.60}$$

where Q is the reaction quotient and the other parameters are as expressed in Eq. (3.3). The following ratio is a useful comparison between the status of a mineral dissolution-precipitation reaction at a particular point in time or space and the

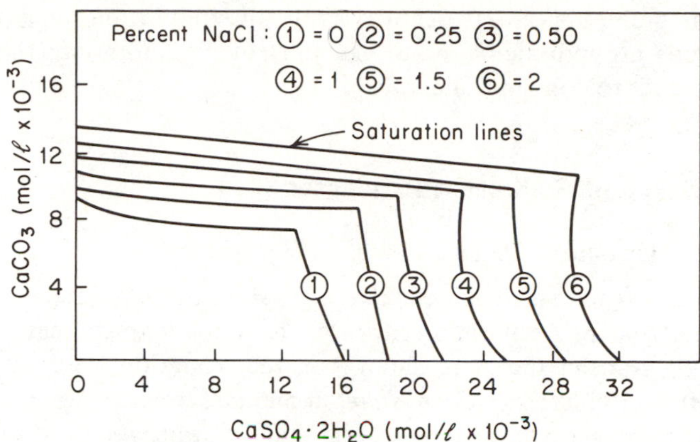

Figure 3.8 Solubility of gypsum and calcite in water with various concentrations of dissolved NaCl, 25°C, $P_{CO_2} = 1$ bar (after Shternina and Frolova, 1945).

thermodynamic equilibrium condition:

$$S_i = \frac{Q}{K_{eq}} \tag{3.61}$$

where S_i is called the *saturation index*. For calcite in contact with groundwater (see Section 3.2), the saturation index is

$$S_i = \frac{[Ca^{2+}][CO_3{}^{2-}]}{K_{cal}} \tag{3.62}$$

The ion activities in the numerator can be obtained from analysis of groundwater samples and Eq. (3.32) and the equilibrium constant K_{cal} can be obtained from the free-energy data, or directly from equilibrium-constant tabulations, such as Table 3.7.

If $S_i > 1$, the water contains an excess of the ionic constituents. The reaction [Eq. (3.4)] must therefore proceed to the left, which requires that mineral precipitation occur. If $S_i < 1$, the reaction proceeds to the right as the mineral dissolves. If $S_i = 1$, the reaction is at equilibrium, which means that it is saturated with respect to the mineral in question. With the saturation index relation, it is possible for specified mineral-water reactions to compare the status of actual water samples to computed equilibrium conditions. For the saturation index to be of interest, the mineral need not actually be present in the groundwater zone. Knowledge of the mineralogical composition is necessary, however, if one desires to obtain a detailed understanding of the geochemical behavior and controls on the water.

Some authors express the saturation index in logarithmic form, in which case an index value of zero denotes the equilibrium condition. The saturation index is in some publications denoted as the disequilibrium index because in some situations groundwater is more generally in disequilibrium than in equilibrium with respect to common minerals.

3.6 Oxidation and Reduction Processes

Oxidation States and Redox Reactions

Many reactions that occur in the groundwater environment involve the transfer of electrons between dissolved, gaseous, or solid constituents. As a result of the electron transfer there are changes in the oxidation states of the reactants and the products. The *oxidation state*, sometimes referred to as the *oxidation number*, represents a hypothetical charge that an atom would have if the ion or molecule were to dissociate. The oxidation states that can be achieved by the most important multioxidation state elements that occur in groundwater are listed in Table 3.8, which also contains some rules that can be used to deduce the oxidation state from the formula of a substance. Sometimes there are uncertainties in the assignment of electron loss or electron gain to a particular atom, especially when the reactions involve covalent bonds. In this book Roman numerals are used to represent odixation states and Arabic numbers represent actual valence.

Table 3.8 Rules for Assigning Oxidation States and Some Examples

Rules for assigning oxidation states:

1. The oxidation state of free elements, whether in atomic or molecular form, is zero.
2. The oxidation state of an element in simple ionic form is equal to the charge on the ion.
3. The oxidation state of oxygen in oxygen compounds is -2. The only exceptions are O_2, O_3 (see rule 1), OF_2 (where it is $+2$), and peroxides such as H_2O_2 and other compounds with –0–0– bonds, where it is -1.
4. The oxidation state of hydrogen is $+1$ except in H_2 and in compounds where H is combined with a less electronegative element.
5. The sum of oxidation states is zero for molecules, and for ion pairs or complexes it is equal to the formal charge on the species.

Examples:

Carbon compounds		Sulfur compounds		Nitrogen compounds		Iron compounds	
Substance	C state	Substance	S state	Substance	N state	Substance	Fe state
HCO_3^-	+IV	S	0	N_2	0	Fe	0
CO_3^{2-}	+IV	H_2S	−II	SCN^-	+II	FeO	+II
CO_2	+IV	HS^-	−II	N_2O	−III	$Fe(OH)_2$	+II
CH_2O	0	FeS_2	−I	NH_4	+III	$FeCO_3$	+II
$C_6H_{12}O_6$	0	FeS	−II	NO_2^-	+V	Fe_2O_3	+III
CH_4	−IV	SO_3^{2-}	+IV	NO_3^-	−III	$Fe(OH)_3$	+III
CH_3OH	−II	SO_4^{2-}	+VI	HCN	−I	FeOOH	+III

SOURCE: Gymer, 1973; Stumm and Morgan, 1970.

In oxidation-reduction reactions, which will be referred to as redox reactions, there are no free electrons. Every oxidation is accompanied by a reduction and vice versa, so that an electron balance is always maintained. By definition, oxidation is the loss of electrons and reduction is the gain in electrons. This is illustrated by expressing the redox reaction for the oxidation of iron:

$$O_2 + 4Fe^{2+} + 4H^+ = 4Fe^{3+} + 2H_2O \tag{3.63}$$

For every redox system *half-reactions* in the following form can be written:

$$\text{oxidized state} + ne = \text{reduced state} \tag{3.64}$$

The redox reaction for iron can therefore be expressed in half-reactions,

$$O_2 + 4H^+ + 4e = 2H_2O \qquad \text{(reduction)} \tag{3.65}$$

$$4Fe^{2+} = 4Fe^{3+} + 4e \qquad \text{(oxidation)} \tag{3.66}$$

In the reduction half-reaction the oxidation state of oxygen goes from zero (oxygen as O_2) to $-$II (oxygen in H_2O). There is, therefore, a release of four electrons because 2 mol of H_2O forms from 1 mol of O_2 and 4 mol of H^+. In the oxidation half-reaction, 4 mol of Fe(+II) goes to 4 mol of Fe (+III), with a gain of four electrons. The complete redox reaction [Eq. (3.63)] expresses the net effect of the electron transfer and therefore contains no free electrons. When writing half-reactions, care must be taken to ensure that the electrons on each side of the equation are balanced. These reactions need not involve oxygen or hydrogen, although most redox reactions that occur in the groundwater zone do involve one or both of these elements. The concept of oxidation and reduction in terms of changes in oxidation states is illustrated in Figure 3.9.

A list of half-reactions that represent most of the redox processes occurring in groundwater is presented in Table 3.9.

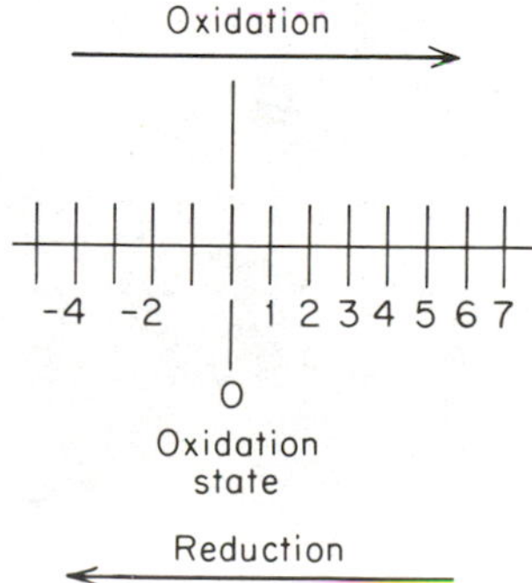

Figure 3.9 Oxidation and reduction in relation to oxidation states.

Table 3.9 Redox Half-Reactions for Many Constituents That Occur in the Groundwater Environment

(1)	$\frac{1}{4}O_2 + H^+ + e = \frac{1}{2}H_2O$	(18)	$\frac{1}{8}SO_4{}^{2-} + \frac{9}{8}H^+ + e = \frac{1}{8}HS^- + \frac{1}{2}H_2O$
(2)	$H^+ + e = \frac{1}{2}H_2(g)$	(19)	$\frac{1}{2}S(s) + H^+ + e = \frac{1}{2}H_2S(g)$
(3)	$H_2O + e = \frac{1}{2}H_2(g) + OH^-$	(20)	$Fe^{3+} + e = Fe^{2+}$
(4)	$\frac{1}{5}NO_3^- + \frac{6}{5}H^+ + e = \frac{1}{10}N_2(g) + \frac{3}{5}H_2O$	(21)	$Fe(OH)_3(s) + HCO_3^- + 2H^+ + e = Fe(CO_3)(s) + 3H_2O$
(5)	$\frac{1}{2}NO_3^- + H^+ + e = \frac{1}{2}NO_2^- + \frac{1}{2}H_2O$	(22)	$Fe(OH)_3(s) + 3H^+ + e = Fe^{2+} + 3H_2O$
(6)	$\frac{1}{8}NO_3^- + \frac{5}{4}H^+ + e = \frac{1}{8}NH_4^+ + \frac{3}{8}H_2O$	(23)	$Fe(OH)_3(s) + H^+ + e = Fe(OH)_2(s) + H_2O$
(7)	$\frac{1}{6}NO_2^- + \frac{4}{3}H^+ + e = \frac{1}{6}NH_4^+ + \frac{1}{3}H_2O$	(24)	$\frac{1}{2}FeS_2(s) + 2H^+ + e = \frac{1}{2}Fe^{2+} + H_2S(g)$
(8)	$\frac{1}{4}NO_3^- + \frac{5}{4}H^+ + e = \frac{1}{8}N_2O(g) + \frac{5}{8}H_2O$	(25)	$\frac{1}{2}Fe^{2+} + S(s) + e = \frac{1}{2}FeS_2(s)$
(9)	$\frac{1}{2}NO_2^- + \frac{3}{2}H^+ + e = \frac{1}{4}N_2O(g) + \frac{3}{4}H_2O$	(26)	$\frac{1}{16}Fe^{2+} + \frac{1}{8}SO_4{}^{2-} + e = \frac{1}{16}FeS_2(s) + \frac{1}{2}H_2O$
(10)	$\frac{1}{6}N_2(g) + \frac{4}{3}H^+ + e = \frac{1}{3}NH_4^+$	(27)	$\frac{1}{14}Fe(OH)_2(s) + \frac{1}{7}SO_4{}^{2-} + \frac{9}{7}H^+ + e = \frac{1}{14}FeS_2(s) + \frac{5}{7}H_2O$
(11)	$\frac{1}{4}CH_2O + H^+ + e = \frac{1}{4}CH_4(g) + \frac{1}{4}H_2O$	(28)	$\frac{1}{14}Fe(CO_3)(s) + \frac{1}{7}SO_4{}^{2-} + \frac{17}{14}H^+ + e = \frac{1}{14}FeS_2(s) + \frac{4}{7}H_2O + \frac{1}{14}HCO_3^-$
(12)	$\frac{1}{4}CO_2(g) + H^+ + e = \frac{1}{4}CH_2O + \frac{1}{4}H_2O$	(29)	$\frac{1}{2}MnO_2(s) + \frac{1}{2}HCO_3^- + \frac{3}{2}H^+ + e = \frac{1}{2}MnCO_3(s) + \frac{3}{8}H_2O$
(13)	$\frac{1}{2}CH_2O + H^+ + e = \frac{1}{2}CH_3OH$	(30)	$Mn^{2+} + 2e = Mn(s)$
(14)	$\frac{1}{8}CO_2(g) + H^+ + e = \frac{1}{8}CH_4(g) + \frac{1}{4}H_2O$	(31)	$\frac{1}{2}MnCO_3 + \frac{1}{2}H^+ + e = \frac{1}{2}Mn(s) + \frac{1}{2}HCO_3^-$
(15)	$\frac{1}{2}CH_3OH + H^+ + e = \frac{1}{2}CH_4(g) + \frac{1}{2}H_2O$	(32)	$MnOOH(s) + HCO_3^- + 2H^+ + e = MnCO_3 + 2H_2O$
(16)	$\frac{1}{6}SO_4{}^{2-} + \frac{4}{3}H^+ + e = \frac{1}{6}S(s) + \frac{2}{3}H_2O$	(33)	$MnO_2 + H^+ + e = MnOOH$
(17)	$\frac{1}{8}SO_4{}^{2-} + \frac{5}{4}H^+ + e = \frac{1}{8}H_2S(g) + \frac{1}{2}H_2O$		

Consumption of Oxygen and Organic Matter

Unpolluted rivers and lakes generally have oxidizing conditions because of mixing with oxygen from the earth's atmosphere. The tendency in groundwater systems, however, is toward oxygen depletion and reducing conditions. Because the water that circulates through the groundwater zone is generally isolated from the earth's atmosphere, oxygen that is consumed by hydrochemical and biochemical reactions is not replenished. In order for reduction of inorganic constituents to occur, some other constituents must be oxidized. The oxidized compound is generally organic matter. The reactions are catalyzed by bacteria or isolated enzymes that derive energy by facilitating the process of electron transfer. In the present discussion we will assume that reactions proceed in an appropriate thermodynamic direction, without clarification of the associated biochemical processes. To illustrate the process of organic-matter oxidation, a simple carbohydrate, CH_2O, is used as the electron donor, even though numerous other organic compounds, such as polysaccharides, saccharides, fatty acids, amino acids, and phenols, may be the actual organic compound involved in the redox process

$$\tfrac{1}{4}CH_2O + \tfrac{1}{4}H_2O = \tfrac{1}{4}CO_2(g) + H^+ + e \qquad (3.67)$$

To obtain full reactions for redox processes, the half-reaction for the oxidation of organic matter, represented by Eq. (3.67), can be combined with many of the half-reactions for reduction of inorganic compounds given in Table 3.9. Combinations of Eq. (3.67) and reaction (1) in Table 3.9 yields the redox relation

$$O_2(g) + CH_2O = CO_2(g) + H_2O \qquad (3.68)$$

which represents the process of organic-matter oxidation in the presence of bacteria and free molecular oxygen. This redox process is the main source of dissolved CO_2. CO_2 combines with H_2O to produce H_2CO_3 [Eq. (3.15)], which is an acid of considerable strength when viewed in a geochemical context.

Because the solubility of O_2 in water is low (9 mg/ℓ at 25°C and 11 mg/ℓ at 5°C), and because O_2 replenishment in subsurface environments is limited, oxidation of only a small amount of organic matter can result in consumption of all the dissolved O_2. For example, from the mass conservation relations inherent in Eq. (3.68), oxidation of only 8.4 mg/ℓ (0.28 mmol/ℓ of CH_2O would consume 9 mg/ℓ (0.28 mmol/ℓ) of O_2. This would result in the water having no dissolved O_2. Water that infiltrates through the soil zone is normally in contact with soil organic matter. O_2 consumption and CO_2 production is therefore a widespread process in the very shallow part of the subsurface environment.

Table 3.10 lists some redox reactions in which dissolved oxygen is consumed. In all these reactions, H^+ ions are produced. In many groundwater systems the H^+ ions are consumed by reactions with minerals. The pH therefore does not decrease

Table 3.10 Some Inorganic Oxidation Processes That Consume Dissolved Oxygen in Groundwater

Process	Reaction*	
Sulfide oxidation	$O_2 + \frac{1}{2}HS^- = \frac{1}{2}SO_4{}^{2-} + \frac{1}{2}H^+$	(1)
Iron oxidation	$\frac{1}{4}O_2 + Fe^{2+} + H^+ = Fe^{3+} + \frac{1}{2}H_2O$	(2)
Nitrification	$O_2 + \frac{1}{2}NH_4^+ = \frac{1}{2}NO_3^- + H^+ + \frac{1}{2}H_2O$	(3)
Manganese oxidation	$O_2 + 2Mn^{2+} + 2H_2O = 2MnO_2(s) + 4H^+$	(4)
Iron sulfide oxidation†	$\frac{15}{4}O_2 + FeS_2(s) + \frac{7}{2}H_2O = Fe(OH)_3(s) + 2SO_4{}^{2-} + 4H^+$	(5)

*(s), solid.

†Expressed as a combined reaction.

appreciably. In some systems, however, minerals that react in this manner are not present, in which case the oxidation processes cause the water to become acidic.

When all the dissolved O_2 in groundwater is consumed, oxidation of organic matter can still occur, but the oxidizing agents (i.e., constituents that undergo reduction) are NO_3^-, MnO_2, $Fe(OH)_3$, $SO_4{}^{2-}$, and others, as indicated in Table 3.11. As these oxidizing agents are consumed, the groundwater environment becomes more and more reduced. If the processes proceed far enough, the environment may become so strongly reducing that organic compounds may undergo anaerobic degradation. An equation for this process, which represents the conversion of organic matter to methane and carbon dioxide, is shown by reaction (5) in Table 3.11. The sequence of redox processes represented by reactions (1) to (5) in Table 3.11 proceed from aerobic oxidation through to methane fermentation provided that (1) organic matter in a consumable form continues to be available in the water, (2) the bacteria that mediate the reactions have sufficient nutrients to sustain their existence, and (3) the temperature variations are not large enough to disrupt the biochemical processes. In many groundwater systems one or more of these factors is limiting, so the groundwater does not proceed through all the redox stages. The evolution of groundwater through various stages of oxidation and reduction is described in more detail in Chapter 7.

Table 3.11 Some Redox Processes That Consume Organic Matter and Reduce Inorganic Compounds in Groundwater

Process	Equation*	
Denitrification†	$CH_2O + \frac{4}{5}NO_3^- = \frac{2}{5}N_2(g) + HCO_3^- + \frac{1}{5}H^+ + \frac{2}{5}H_2O$	(1)
Manganese(IV) reduction	$CH_2O + 2MnO_2(s) + 3H^+ = 2Mn^{2+} + HCO_3^- + 2H_2O$	(2)
Iron(III) reduction	$CH_2O + 4Fe(OH)_3(s) + 7H^+ = 4Fe^{2+} + HCO_3^- + 10H_2O$	(3)
Sulfate reduction‡	$CH_2O + \frac{1}{2}SO_4{}^{2-} = \frac{1}{2}HS^- + HCO_3^- + \frac{1}{2}H^+$	(4)
Methane fermentation	$CH_2O + \frac{1}{2}H_2O = \frac{1}{2}CH_4 + \frac{1}{2}HCO_3^- + \frac{1}{2}H^+$	(5)

*(g), gaseous or dissolved form; (s), solid.

†CH_2O represents organic matter; other organic compounds can also be oxidized.

‡H_2S exists as a dissolved species in the water: $HS^- + H^+ = H_2S$. H_2S is the dominant species at pH < 7.

Equilibrium Redox Conditions

Aqueous solutions do not contain free electrons, but it is nevertheless convenient to express redox processes as half-reactions and then manipulate the half-reactions as if they occur as separate processes. Within this framework a parameter known as the pE is used to describe the relative electron activity. By definition,

$$pE = -\log [e] \tag{3.69}$$

pE, which is a dimensionless quantity, is analogous to the pH expression for proton (hydrogen-ion) activity. The pE of a solution is a measure of the oxidizing or reducing tendency of the solution. In parallel to the convention of arbitrarily assigning $\Delta G^\circ = 0$ for the hydration of H^+ (i.e., $K_{H^+} = 0$ for the reaction $H^+ + H_2O = H_3O^+$) the free-energy change for the reduction of H^+ to $H_2(g)$ [$H^+ + e = \frac{1}{2}H_2(g)$] is zero. p$E$ and pH are functions of the free energy involved in the transfer of 1 mol of protons or electrons, respectively.

For the general half-reaction

$$b\text{B} + c\text{C} + ne = d\text{D} + e\text{E} \tag{3.70}$$

the law of mass action can be written as

$$K = \frac{[\text{D}]^d[\text{E}]^e}{[\text{e}]^n[\text{B}]^b[\text{C}]^c} \tag{3.71}$$

For example, consider the oxidation of Fe(II) to Fe(III) by free oxygen:

$$\tfrac{1}{2}O_2 + 2H^+ + 2e = H_2O \qquad \text{(reduction)} \tag{3.72}$$

$$2Fe^{2+} = 2Fe^{3+} + 2e \qquad \text{(oxidation)} \tag{3.73}$$

$$\tfrac{1}{2}O_2 + 2Fe^{2+} + 2H^+ = 2Fe^{3+} + H_2O \qquad \text{(redox reaction)} \tag{3.74}$$

In this book the equilibrium constants for half-reactions are always expressed in the reduction form. The oxidized forms and electrons are written on the left and the reduced products on the right. This is known as the *Stockholm* or *IUPAC* (International Union of Physical and Analytical Chemistry) *convention*. Expressing the half-reactions [Eq. (3.72) and (3.73)] in terms of equilibrium constants [Eq. (3.71)] for conditions at 25°C and 1 bar yields

$$K = \frac{1}{P_{O_2}^{1/2}[H^+]^2[e]^2} = 10^{41.55} \tag{3.75}$$

$$K = \frac{[Fe^{2+}]}{[Fe^{3+}][e]} = 10^{12.53} \tag{3.76}$$

The numerical values for the equilibrium constants were computed from Eq. (3.12) using Gibbs' free-energy data for 25°C and 1 bar. To obtain expressions for redox conditions expressed as pE, Eqs. (3.75) and (3.76) can be rearranged to yield

$$\text{p}E = 20.78 + \tfrac{1}{4}\log(P_{O_2}) - \text{pH} \tag{3.77}$$

$$\text{p}E = 12.53 + \log\left(\frac{[\text{Fe}^{3+}]}{[\text{Fe}^{2+}]}\right) \tag{3.78}$$

If the redox reaction [Eq. (3.74)] is at equilibrium, and if the concentrations of iron, the P_{O_2}, and the pH are known, the pE obtained from both these relations is the same. Even though there may be many dissolved species in the solution involved in reactions with electron and hydrogen ion transfer, at equilibrium there is only one pE condition, just as there is only one pH condition. In groundwater systems there is an interdependency of pH and pE. Nearly all the reactions listed in Table 3.9 involve both electron and proton transfers. If equilibrium is assumed, the reactions that include pH can be written as pE expressions. Graphical representations of pH–pE relations are described below.

Although the discussion above was based entirely on the assumption that the redox processes are at equilibrium, in field situations the concentrations of oxidizable and reducible species may be far from those predicted using equilibrium models. Many redox reactions proceed at a slow rate and many are irreversible. It is possible, therefore, to have several different redox levels existing in the same locale. There is also the possibility that the bacteria required to catalyze many of the redox reactions exist in microenvironments in the porous media that are not representative of the overall macroenvironment in which the bulk flow of groundwater occurs. Equilibrium considerations can, however, greatly aid in our efforts to understand in a general way the redox conditions observed in subsurface waters. Stumm and Morgan (1970), in their comprehensive text on aquatic chemistry, state: "In all circumstances equilbrium calculations provide boundary conditions towards which the systems must be proceeding. Moreover, partial equilibria (those involving some but not all redox couples) are approximated frequently, even though total equilibrium is not reached. . . . Valuable insight is gained even when differences are observed between computations and observations. The lack of equilibrium and the need for additional information or more sophisticated theory are then made clear" (p. 300).

The redox condition for equilibrium processes can be expressed in terms of pE (dimensionless), Eh (volts), or ΔG (joules or calories). Although in recent years pE has become a commonly used parameter in redox studies, Eh has been used in many investigations, particularly prior to the 1970's. Eh is commonly referred to as the *redox potential* and is defined as the energy gained in the transfer of 1 mol of electrons from an oxidant to H_2. The h in Eh indicates that the potential is on the hydrogen scale and E symbolizes the electromotive force. pE and Eh are related by

$$\text{p}E = \frac{nF}{2.3RT}Eh \tag{3.79}$$

where F is the faraday constant (9.65×10^4 C·mol^{-1}), R the gas constant, T the absolute temperature, and n the number of electrons in the half-reaction. For reactions at 25°C in which the half-reactions are expressed in terms of transfer of a single electron, Eq. (3.79) becomes

$$\text{p}E = 16.9Eh \tag{3.80}$$

Eh is defined by a relation known as the *Nernst equation*,

$$Eh(\text{volts}) = Eh^\circ + \frac{2.3RT}{nF} \log \left(\frac{[\text{oxidant}]}{[\text{reductant}]}\right) \tag{3.81}$$

where Eh° is a standard or reference condition at which all substances involved are at unit activity and n is the number of transferred electrons. This is a thermodynamic convenience. Unit activities could only exist in solutions of infinite dilution; this condition is therefore only hypothetical. The equation relating Eh° directly to the equilibrium constant is

$$Eh^\circ = \frac{RT}{nF} \ln K \tag{3.82}$$

In the study of aqueous systems the same objectives can be met using either pE or Eh to represent redox conditions. pE is often the preferred parameter because its formulation follows so simply from half-cell representations of redox reactions in combination with the law of mass action. Facility in making computations interchangeably between pE and Eh is desirable because tabulations of thermodynamic data for redox reactions are commonly expressed as Eh° values and because in some aqueous systems a convenient method of obtaining an indication of the redox conditions involves measurements of electrode potentials as voltage.

Microbiological Factors

Microorganisms catalyze nearly all the important redox reactions that occur in groundwater. This means that although the reactions are spontaneous thermodynamically, they require the catalyzing effect of microorganisms in order to proceed at a significant rate. Although it is not customary for microorganisms to be regarded as important components of the groundwater environment, their influence cannot be dismissed if we wish to understand the causes and effects of redox processes.

The microorganisms that are most important in redox processes in the groundwater zone are bacteria. Other types of microorganisms, such as algae, fungi, yeasts, and protozoans, can be important in other aqueous environments. Bacteria generally range in size from about 0.5 to 3 μm. They are small compared to the pore sizes in most nonindurated geological materials and are large in relation to the size of hydrated inorganic ions and molecules. The catalytic capability of bacteria is produced by the activity of enzymes that normally occur within the bacteria. Enzymes are protein substances formed by living organisms that have the power to increase the rate of redox reactions by decreasing the activation energies of the reactions. They accomplish this by strongly interacting with complex molecules representing molecular structures halfway between the reactant and the product (Pauling and Pauling, 1975). The local molecular environment of many enzyme reactions is very different from the bulk environment of the aqueous system.

Bacteria and their enzymes are involved in redox processes in order to acquire energy for synthesis of new cells and maintenance of old cells. An important step in the process of bacterial cell growth is the construction of molecules forming an energy-storage substance known as adenosine triphosphate (ATP). After its formation, molecules of this high-energy material can by hydrolyzed through a sequence of energy-releasing reactions that provide for synthesis of new cell material. The growth of bacteria is therefore directly related to the number of moles of ATP formed from the available nutrients. Some of the energy obtained from the redox reactions is maintenance energy required by bacterial cells for such things as mobility, to prevent an undesirable flow of solutes either into or out of the cell, or for resynthesis of proteins that are constantly degrading (McCarty, 1965).

For bacteria to be able to make use of an energy yield from a redox reaction, a minimum free-energy change of approximately 60 kJ/mol between the reactants and the products is required (Delwiche, 1967). The main source of energy for bacteria in the groundwater zone is the oxidation of organic matter.

Bacteria that can thrive only in the presence of dissolved oxygen are known as *aerobic bacteria. Anaerobic bacteria* require an absence of dissolved oxygen. *Facultative bacteria* can thrive with or without oxygen. The lower limit of dissolved O_2 for the existence of most aerobic bacteria is considered to be about 0.05 mg/ℓ, but some aerobic species can persist at lower levels. Since most methods commonly used for measuring dissolved O_2 have a lower detection limit of about 0.1 mg/ℓ, it is possible that aerobic bacteria can mediate redox reactions in situations that might appear to be anaerobic based on the lack of detectable oxygen.

Bacteria of different varieties can withstand fluid pressures of many hundreds of bars, pH conditions from 1 to 10, temperatures from near 0 to greater than 75°C, and salinities much higher than that of seawater. They can migrate through porous geological materials and in unfavorable environments can evolve into resistant bodies that may be activated at a later time (Oppenheimer, 1963). In spite of these apparent characteristics of hardiness, there are many groundwater environments in which organic matter is not being oxidized at an appreciable rate. As a result, the redox conditions have not declined to low levels even though hundreds or thousands of years or more have been available for the reactions to proceed. If the redox reactions that require bacterial catalysis are not occurring at significant rates, a lack of one or more of the essential nutrients for bacterial growth is likely the cause. There are various types of nutrients. Some are required for incorporation into the cellular mass of the bacteria. Carbon, nitrogen, sulfur, and phosphorous compounds and many metals are in this category. Other nutrients are substances that function as electron donors or energy sources, such as water, ammonia, glucose, and H_2S, and substances that function as electron acceptors, such as oxygen, nitrate, and sulfate. *Macronutrients* are those substances that are required in large amounts as direct building blocks in cell construction. *Micronutrients* are required in amounts so small as to be difficult to detect. The macronutrient requirements of many bacteria are similar or identical. The micronutrient requirements are more likely to differ from species to species (Brock, 1966).

Although bacteria play an important role in the geochemical environment of groundwater, the study of bacteria at depths below the soil zone is in its infancy. The next decade or two should yield interesting developments in this area of research.

pE–pH Diagrams

Graphs that show the equilibrium occurrence of ions or minerals as domains relative to p*E* (or *Eh*) and pH are known as p*E*–pH or *Eh*–pH diagrams. During the 1950's diagrams of this type were developed by M. J. N. Pourbaix and coworkers at the Belgian Center for Study of Corrosion as a practical tool in applied chemistry. The results of this work are summarized by Pourbaix et al. (1963). Following the methods developed by the Belgian group, R. M. Garrels and coworkers pioneered applications in the analysis of geological systems. The use of p*E*–pH diagrams has become widespread in geology, limnology, oceanography, and pedology. In groundwater quality investigations, considerable emphasis is now being placed on developing an understanding of the processes that control the occurrence and mobility of minor and trace elements. p*E*–pH diagrams are an important aid in this endeavor. The following discussion of these diagrams is only a brief introduction. The redox condition will be represented by p*E* rather than *Eh*, but this is just a matter of convenience. Comprehensive treatments of the subject are presented in the texts by Garrels and Christ (1965), Stumm and Morgan (1970), and Guenther (1975). A concise outline of methods for construction of *Eh*-pH diagrams is provided by Cloke (1966).

Since we are interested in the equilibrium occurrence (i.e., stability) of dissolved species and minerals in aqueous environments, an appropriate first step in the consideration of p*E*–pH relations is to determine conditions under which H_2O is stable. From the redox half-reactions

$$O_2(g) + 4H^+ + 4e = 2H_2O \tag{3.83}$$

$$2H^+ + 2e = H_2(g) \tag{3.84}$$

we obtain for conditions at 25°C,

$$pE = 20.8 - pH + \tfrac{1}{4}\log P_{O_2} \tag{3.85}$$

$$pE = -pH - \tfrac{1}{2}\log P_{H_2} \tag{3.86}$$

These relations plot as straight lines (1 and 2) on the pE–pH diagram shown in Figure 3.10(a).

As an example explanation of the stability domains of ions and minerals, the Fe–H_2O system represented in Figure 3.10 will be considered. In groundwater, iron in solution is normally present mainly as Fe^{2+} and Fe^{3+}. These are the only species that are accounted for in our analysis. In a more detailed treatment, complexes such as $Fe(OH)^{2+}$, $Fe(OH)_2^+$, and $HFeO_2^-$ would be included. The solid compounds that can occur in the Fe–H_2O system are listed in Table 3.12. A series of reduction reactions involving a solid material (iron compound) and H^+ and e as reactants and a more reduced solid compound and water as products can be written for the compounds in this table. For example,

$$Fe(OH)_3 + H^+ + e = Fe(OH)_2 + H_2O \tag{3.87}$$

Expressing this reaction in mass-action form, with activities of water and the solid phases taken as unity (for reasons indicated in Section 3.2), yields

$$K = \frac{1}{[H][e]} \tag{3.88}$$

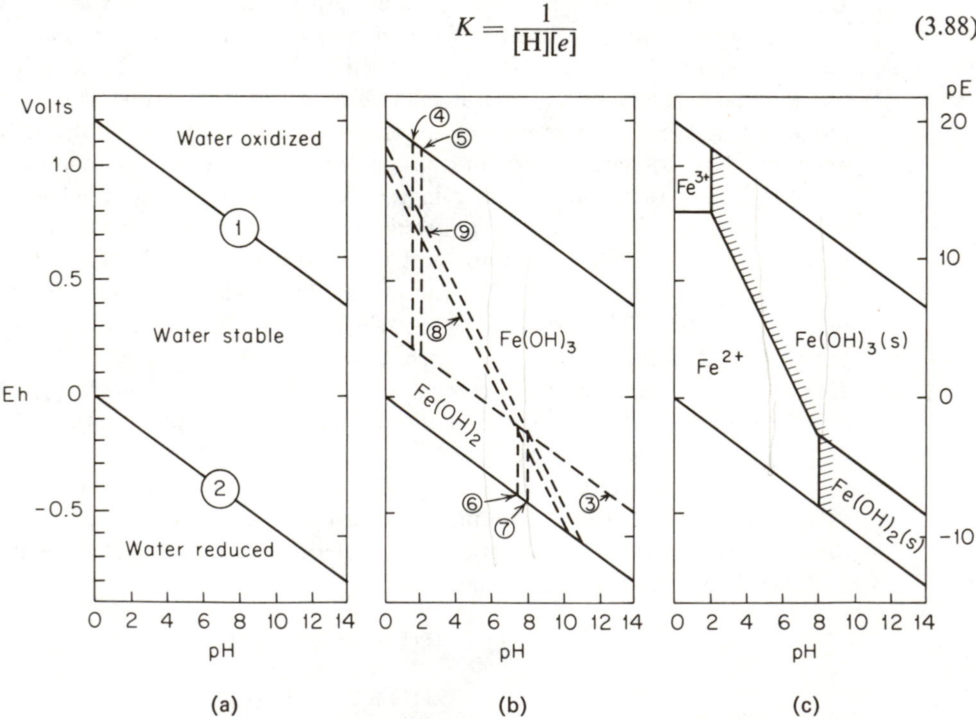

Figure 3.10 pE–pH diagrams, 25°C and 1 bar. (a) Stability field for water; (b) construction lines for the Fe–H_2O system (see the text for equations representing number-designated lines); (c) completed diagram showing stability fields for major dissolved species and solid phases.

Table 3.12 Oxides and Hydroxides in the Fe–H_2O System

Oxidation state	Solid substances
0	Fe
II	FeO, $Fe(OH)_2$
II and III	Fe_3O_4
III	Fe_2O_3, $Fe(OH)_3$, FeOOH

and in logarithmic form,

$$\log K - \text{pH} - \text{p}E = 0 \tag{3.89}$$

The equilibrium constant in this equation can be obtained using Eq. (3.12) and tables of values of Gibbs' free energy of formation (ΔG_f°), as indicated in Section 3.2. Equation (3.89) is represented as a line on a p*E*–pH diagram as shown in Figure 3.10(b) (line 3). In the p*E*–pH domain above this line, $Fe(OH)_3$ is stable; below the line it is reduced to $Fe(OH)_2$. These are known as the *stability fields* for these two solid compounds of iron. Lines representing the many other reduction equations obtained by reacting the solids in Table 3.12 with H^+ and *e* to form more reduced compounds and H_2O can be constructed on the p*E*–pH diagram. However, these lines are located outside the stability field for H_2O [i.e., above and below lines (1) and (2)], consequently are of no interest in groundwater studies.

In most studies of natural waters, interest is focused on the dissolved species as well as on the mineral phases. Therefore, information on the equilibrium concentrations of dissolved species is commonly included on p*E*–pH diagrams. For illustration, the Fe–H_2O system will be considered further. The oxidation state of iron in $Fe(OH)_3$ is +III. The dissociation of moderately crystalline $Fe(OH)_3$ in water is

$$Fe(OH)_3 + 3H^+ = Fe^{3+} + 3H_2O \qquad \Delta G_r^0 = -1.84 \text{ kJ} \tag{3.90}$$

The law of mass action yields

$$K_{Fe(OH)_3} = \frac{[Fe^{3+}]}{[H^+]^3} \tag{3.91}$$

From Eq. (3.12), a value of +0.32 is obtained for log *K*. The mass-action relation can be expressed as

$$\log [Fe^{3+}] = 0.32 - 3\text{ pH} \tag{3.92}$$

which plots as a vertical line on the p*E*–pH diagram. If the pH is specified, the line obtained from this expression represents the equilibrium activity of Fe^{3+} that will exist in an aqueous solution in contact with the solid phase, $Fe(OH)_3$. Equation

(3.92) indicates that Fe^{3+} activity increases at lower pH values. In the construction of p*E*–pH diagrams, a common procedure is to choose a pH condition at which the activity of the dissolved species is at a level considered to be negligible. The choice of this level depends on the nature of the problem. For illustration purposes, two lines are shown on Figure 3.10(b) [lines (4) and (5), which represent Fe^{3+} activities of 10^{-5} and 10^{-6}]. Although in theory these lines represent activities, and therefore are dimensionless, they can be valid as representing molality, because in low-salinity solutions activity coefficients are nearly equal to unity.

Under lower p*E* conditions, Fe^{2+} is the important species of dissolved iron. The reaction of interest is

$$Fe(OH)_2 + 3H^+ = Fe^{2+} + 2H_2O \qquad \Delta G_r^0 = +26.33 \text{ kJ} \tag{3.93}$$

From the mass-action relation, the following expression is derived:

$$\log [Fe^{2+}] = 10.23 - 2 \text{ pH} \tag{3.94}$$

For $[Fe^{2+}]$ values of 10^{-5} and 10^{-6}, this equation is represented in Figure 3.10 by lines (6) and (7), respectively. The lines have been superimposed only on the part of the diagram in which $Fe(OH)_2$ is the stable solid phase. But Fe^{2+} also exists at some equilibrium activity in the part of the diagram in which $Fe(OH)_3$ is the stable solid phase. Fe^{2+} and $Fe(OH)_3$ are releated by the redox half-reaction

$$Fe(OH)_3 + 3H^+ + e = Fe^{2+} + 3H_2O \qquad \Delta G_r^0 = -76.26 \text{ kJ} \tag{3.95}$$

From the mass-action relation

$$\log [Fe^{2+}] = \log K_{Fe(OH)_3} - 3\text{pH} - \text{p}E \tag{3.96}$$

where $\log K_{Fe(OH)_3} = 13.30$. On Figure 3.10(b) this expression is represented as lines (8) and (9) for $[Fe^{2+}]$ values of 10^{-5} and 10^{-6}, respectively.

Figure 3.10(c) is a "cleaned-up" version of the p*E*–pH diagram. It illustrates the general form in which p*E*–pH diagrams are normally presented in the literature. It is important to keep in mind that the boundary lines between solid phases and dissolved species are based on specified activity values, and that the validity of all lines as thermodynamically defined equilibrium conditions is dependent on the reliability of the free-energy data used in construction of the diagram. In the example above, there is considerable uncertainty in the position of some of the boundaries because the solid phase, $Fe(OH)_3$, is a substance of variable crystallinity which has different ΔG_f^0 values depending on its crystallinity.

In Chapter 9, p*E*–pH diagrams are used in the consideration of other dissolved constituents in groundwater. Although some p*E*–pH diagrams appear complex, their construction can be accomplished by procedures not much more elaborate than those described above.

3.7 Ion Exchange and Adsorption

Mechanisms

Porous geological materials that are composed of an appreciable percentage of colloidal-sized particles have the capability to exchange ionic constituents adsorbed on the particle surfaces. Colloidal particles have diameters in the range 10^{-3}–10^{-6} mm. They are large compared to the size of small molecules, but are sufficiently small so that interfacial forces are significant in controlling their behavior. Most clay minerals are of colloidal size. The geochemical weathering products of rocks are often inorganic, amorphous (uncrystallized or poorly crystallized) colloids in a persistent metastable state. These colloidal weathering products may occur as coatings on the surfaces of much larger particles. Even a deposit that appears to be composed of clean sand or gravel can have a significant colloid content.

Ion-exchange processes are almost exclusively limited to colloidal particles because these particles have a large electrical charge relative to their surface areas. The surface charge is a result of (1) imperfections or ionic substitutions within the crystal lattice or (2) chemical dissociation reactions at the particle surface. Ionic substitutions cause a net positive or negative charge on the crystal lattice. This charge imbalance is compensated for by a surface accumulation of ions of opposite charge, known as *counterions*. The counterions comprise an adsorbed layer of changeable composition. Ions in this layer can be exchanged for other ions providing that the electrical charge imbalance in the crystal lattice continues to be balanced off.

In geologic materials the colloids that characteristically exhibit surface charge caused primarily by ionic substitution are clay minerals. The common clay minerals can be subdivided into five groups: the kaolinite group, the montmorillonite group (often referred to as the smectite group), the illite group, the chlorite group, and the vermiculite group. Each group may include a few or many compositional and structural varieties with separate mineral names. The five groups, however, are all layer-type aluminosilicates. The structure and composition of these groups is described in detail in the monographs on clay mineralogy by Grim (1968) and on ion exchange by van Olphen (1963).

Silica, which is the most common oxide in the earth's crust and one of the simpler oxides, is characterized by electrically charged surfaces. The surfaces contain ions that are not fully coordinated and hence have unbalanced charge. In a vacuum the net charge is extremely small. On exposure to water, the charged sites are converted to surface hydroxide groups that control the charge on the mineral surface. Surface charge is developed because of the dissociation of the adsorbed OH^- groups on the particular surface. To neutralize this charge, an adsorbed layer of cations and anions forms in a zone adjacent to the hydroxylated layer. Parks (1967) states that hydroxylated surface conditions should be expected on all oxide materials that have had a chance to come to equilibrium with the aqueous environment. Depending on whether the hydroxyl-group dissociation is predominantly acidic or basic, the net charge on the hydroxylated layer can be negative or positive. Surface charge may also be produced by adsorption of charged ionic complexes.

The nature of the surface charge is a function of pH. At low pH a positively charged surface prevails; at a high pH a negatively charged surface develops. At some intermediate pH, the charge will be zero, a condition known as the *zero point of charge* (pH_{zpc}). The tendency for adsorption of cations or anions therefore depends on the pH of the solution.

Cation Exchange Capacity

The *cation exchange capacity* (CEC) of a colloidal material is defined by van Olphen (1963) as the excess of counter ions in the zone adjacent to the charged surface or layer which can be exchanged for other cations. The cation exchange capacity of geological materials is normally expressed as the number of milliequivalents of cations that can be exchanged in a sample with a dry mass of 100 g. The standard test for determing the CEC of these materials involves (1) adjustment of the pore water pH to 7.0, (2) saturation of the exchange sites with NH_4^+ by mixing the soil sample with a solution of ammonium acetate, (3) removal of the absorbed NH_4^+ by leaching with a strong solution of NaCl (Na^+ replaces NH_4^+ on the exchange sites), and (4) determination of the NH_4^+ content of the leaching solution after equilibrium has been attained. CEC values obtained from standard laboratory tests are a measure of the exchange capacity under the specified conditions of the test. For minerals that owe their exchange capacity to chemical dissociation reactions on their surfaces, the actual exchange capacity can be strongly dependent on pH.

The concept of cation exchange capacity and its relation to clay minerals and isomorphous substitution is illustrated by the following example adapted from van Olphen (1963). Consider a montmorillonitic clay in which 0.67 mol of Mg occurs in isomorphous substitution for Al in the alumina octahedra of the crystal lattice. The unit cell formula for the montmorillonite crystal lattice can be expressed as

$$Ex\ (Si_8)(Al_{3.33}Mg_{0.67})O_{20}(OH)_4$$

where *Ex* denotes exchangeable cations. It will be assumed that the exchangeable cations are entirely Na^+. From the atomic weights of the elements, the formula weight of this montmorillonite is 734. Hence, from Avogadro's number, 734 g of this clay contains 6.02×10^{23} unit cells. The unit cell is the smallest structural unit from which clay particles are assembled. Typical unit cell dimensions for montmorillonite determined from X-ray diffraction analyses are 5.15 Å and 8.9 Å (angstroms) in the plane of the octahedral-tetrahedral sheets. The spacing between sheets varies from 9 to 15 Å depending on the nature of the adsorbed cations and water molecules. The total surface area of 1 g of clay is

$$\frac{1}{734} \times 6.02 \times 10^{23} \times 2 \times 5.15 \times 8.9\ \text{Å}^2/\text{g} = 750\ \text{m}^2/\text{g}$$

To balance the negative charge caused by Mg substitution, 0.67 mol of monovalent cations, in this case Na^+, is required per 734 g of clay. Expressed in the units normally used, the cation exchange capacity is therefore

$$\text{CEC} = \frac{0.67}{734} \times 10^3 \times 100 = 91.5 \text{ meq/100 g}$$

which is equivalent to $0.915 \times 6.02 \times 10^{20}$ monovalent cations per gram.

Since the number of cations that are required to balance the surface charge per unit mass of clay and the surface area per unit mass of clay are now known, the surface area available for each monovalent exchangeable cation can be calculated:

$$\frac{750 \times 10^{20}}{0.915 \times 6.02 \times 10^{20}} = 136 \text{ Å}^2/\text{ion}$$

The hydrated radius of Na^+ is estimated to be in the range 5.6–7.9 Å, which corresponds to areas of 98.5–196.1 Å^2. Comparison of these areas to the surface area available per monovalent cation indicates that little more than a monolayer of adsorbed cations is required to balance the surface charge caused by isomorphous substitution.

A similar calculation for kaolinite indicates that for this clay the surface area is 1075 m^2/g (Wayman, 1967). The cation exchange capacity for kaolinite is typically in the range 1–10 meq/100 g, and therefore a monolayer of adsorbed cations would satisfy the charge-balance requirements.

Mass-Action Equations

Following the methodology used in consideration of many of the other topics covered in the chapter, we will develop quantitative relations for cation exchange processes by applying the law of mass action. To proceed on this basis it is assumed that the exchange system consists of two discrete phases, the solution phase and the exchange phase. The exchange phase consists of all or part of the porous medium. The process of ion exchange is then represented simply as an exchange of ions between these two phases,

$$a\text{A} + b\text{B(ad)} = a\text{A(ad)} + b\text{B} \tag{3.97}$$

where A and B are the exchangeable ions, a and b are the number of moles, and the suffix (ad) represents an adsorbed ion. The absence of this suffix denotes an ion in solution. From the law of mass action,

$$K_{\text{A-B}} = \frac{[\text{A}_{(\text{ad})}]^a[\text{B}]^b}{[\text{A}]^a[\text{B}_{(\text{ad})}]^b} \tag{3.98}$$

where the quantities in brackets represent activities. For the exchange between

Na^+ and Ca^{2+}, which is very important in many natural groundwater systems, the exchange equation is

$$2Na^+ + Ca(ad) = Ca^{2+} + 2Na(ad) \tag{3.99}$$

$$K_{Na\text{-}Ca} = \frac{[Ca^{2+}][Na_{(ad)}]^2}{[Na^+]^2[Ca_{(ad)}]} \tag{3.100}$$

The activity ratio of ions in solution can be expressed in terms of molality and activity coefficients as

$$\frac{[B]^b}{[A]^a} = \frac{\gamma_B^b(B)^b}{\gamma_A^a(A)^a} \tag{3.101}$$

where activity coefficient values (γ_A, γ_B) can be obtained in the usual manner (Section 3.2). For Eq. (3.98) to be useful it is necessary to obtain values for the activities of the ions adsorbed on the exchange phase. Vanselow (1932) proposed that the activities of the adsorbed ions be set equal to their mole fractions (Section 3.2 includes a definition of this quantity). The mole fractions of A and B are

$$N_A = \frac{(A)}{(A) + (B)} \quad \text{and} \quad N_B = \frac{(B)}{(A) + (B)}$$

where (A) and (B), expressed in moles, are adsorbed constituents. The equilibrium expression becomes

$$\bar{K}_{(A\text{-}B)} = \frac{\gamma_B^b(B)^b}{\gamma_A^a(A)^a}\frac{N_{A(ad)}^a}{N_{B(ad)}^b} \tag{3.102}$$

Vanselow and others have found experimentally that for some exchange systems involving electrolytes and clays, $\bar{K}$ is a constant. Consequently, $\bar{K}$ has become known as the *selectivity coefficient*. In cases where it is not a constant, it is more appropriately called a *selectivity function* (Babcock, 1963). In many investigations the activity coefficient terms in Eq. (3.101) are not included. Babcock and Schulz (1963) have shown, however, that the activity coefficient effect can be particularly important in the case of monovalent-divalent cation exchange.

Argersinger et al. (1950) extended Vanselow's theory to more fully account for the effects of the adsorbed ions. Activity coefficients for adsorbed ions were introduced in a form analogous to solute activity coefficients.

$$\gamma_{A(ad)} = \frac{[A_{(ad)}]}{N_{A(ad)}} \quad \text{and} \quad \gamma_{B(ad)} = \frac{[B_{(ad)}]}{N_{B(ad)}} \tag{3.103}$$

The mass-action equilibrium constant, $K_{A\text{-}B}$, is therefore related to the selectivity function by

$$K_{A\text{-}B} = \frac{\gamma_{A(ad)}^a}{\gamma_{B(ad)}^b}\bar{K}_{(A\text{-}B)} \tag{3.104}$$

Although in theory this equation should provide a valid method for predicting the effects of ion exchange on cation concentrations in groundwater, with the notable exceptions of the investigations by Jensen and Babcock (1973) and El-Prince and Babcock (1975), cation exchange studies generally do not include determination of K and $\gamma_{(ad)}$ values. Information on selectivity coefficients is much more common in the literature. For the Mg^{2+}–Ca^{2+} exchange pair, Jensen and Babcock and others have observed that the selectivity coefficient is constant over large ranges of ratios of $(Mg^{2+})_{ad}/(Ca^{2+})_{ad}$ and ionic strength. $\bar{K}_{Mg\text{-}Ca}$ values are typically in the range 0.6–0.9. This indicates that Ca^{2+} is adsorbed preferentially to Mg^{2+}.

Interest in cation exchange processes in the groundwater zone commonly focuses on the question of what will happen to the cation concentrations in groundwater as water moves into a zone in which there is significant cation exchange capacity. Strata that can alter the chemistry of groundwater by cation exchange may possess other important geochemical properties. For simplicity these are excluded from this discussion. When groundwater of a particular composition moves into a cation exchange zone, the cation concentrations will adjust to a condition of exchange equilibrium. The equilibrium cation concentrations depend on initial conditions, such as: (1) cation concentrations of the water entering the pore space in which the exchange occurs and (2) the mole fractions of adsorbed cations on the pore surfaces immediately prior to entry of the new pore water. As each new volume of water moves through the pore space, a new equilibrium is established in response to the new set of initial conditions. Continual movement of groundwater through the cation exchange zone can be accompanied by a gradually changing pore chemistry, even though exchange equilibrium in the pore water is maintained at all times. This condition of changing equilibrium is particularly characteristic of cation exchange processes in the groundwater zone, and is also associated with other hydrochemical processes where hydrodynamic flow causes continual pore water replacement as rapid mineral-water reactions occur.

The following example illustrates how exchange reactions can influence groundwater chemistry. Consider the reaction

$$(Mg^{2+}) + (Ca^{2+})_{ad} \rightleftharpoons (Ca^{2+}) + (Mg^{2+})_{ad} \tag{3.105}$$

which leads to

$$\bar{K}_{Mg\text{-}Ca} = \frac{\gamma_{Ca}(Ca^{2+})}{\gamma_{Mg}(Mg^{2+})}\frac{N_{Mg}}{N_{Ca}} \tag{3.106}$$

where $\bar{K}_{Mg\text{-}Ca}$ is the selectivity coefficient, γ denotes activity coefficient, (Ca^{2+}) and (Mg^{2+}) are molalities, and N_{Mg} and N_{Ca} are the mole fractions of adsorbed Mg^{2+} and Ca^{2+}. At low and moderate ionic strengths, the activity coefficients of Ca^{2+} and Mg^{2+} are similar (Figure 3.3), and Eq. (3.106) can be simplified to

$$\bar{K}_{Mg\text{-}Ca} = \frac{(Ca^{2+})}{(Mg^{2+})}\frac{N_{Mg}}{N_{Ca}} \tag{3.107}$$

In this example, exchange occurs when groundwater of low ionic strength with Mg^{2+} and Ca^{2+} molalities of 1×10^{-3} flows through a clayey stratum with a cation exchange capacity of 100 meq/100 g. Concentrations of other cations in the water are insignificant. It is assumed that prior to entry of the groundwater into the clay stratum, the exchange positions on the clay are shared equally by Mg^{2+} and Ca^{2+}. The initial adsorption condition is therefore $N_{Mg} = N_{Ca}$. To compute the equilibrium cation concentrations, information on the porosity or bulk dry mass density of the clay is required. It is assumed that the porosity is 0.33 and that the mass density of the solids is 2.65 g/cm³. A reasonable estimate for the bulk dry mass density is therefore 1.75 g/cm³. It is convenient in this context to express the cation concentrations in solution as moles per liter, which at low concentrations is the same as molality. Since the porosity is 0.33, expressed as a fraction, each liter of water in the clayey stratum is in contact with 2×10^3 cm³ of solids that have a mass 5.3×10^3 g. Because the CEC is 1 meq/g and because 1 mol of Ca^{2+} or $Mg^{2+} = 2$ equivalents, 5.3×10^3 g of clay will have a total of 5.3 equivalents, which equals 1.33 mol of adsorbed Mg^{2+} and 1.33 mol of Ca^{2+}. It is assumed that the groundwater flows into the water-saturated clay and totally displaces the original pore water. The Ca^{2+} and Mg^{2+} concentrations in the groundwater as it enters the clayey stratum can now be calculated. A $\bar{K}_{Mg\text{-}Ca}$ value of 0.6 will be used, and it will be assumed that pore-water displacement occurs instantaneously with negligible hydrodynamic dispersion. Because the initial conditions are specified as $N_{Mg} = N_{Ca}$, a liter of water is in contact with clay that has 1.33 mol of Mg^{2+} and 1.33 mol of Ca^{2+} on the exchange sites. Compared to the concentrations of Mg^{2+} and Ca^{2+} in the groundwater, the adsorbed layer on the clay particles is a large reservoir of exchangeable cations.

Substitution of the initial values into the right-hand side of Eq. (3.107) yields a value for the reaction quotient [Eq. (3.60)]:

$$Q_{Mg\text{-}Ca} = \frac{1 \times 10^{-3}}{1 \times 10^{-3}} \times \frac{0.5}{0.5} = 1$$

For the reaction to proceed to equilibrium with respect to the new pore water, $Q_{Mg\text{-}Ca}$ must decrease to a value of 0.6 to attain the condition of $Q = K$. This occurs by adsorption of Ca^{2+} and release of Mg^{2+} to the solution. The equilibrium is achieved when $(Ca^{2+}) = 0.743 \times 10^{-3}$, $(Mg^{2+}) = 1.257 \times 10^{-3}$, $N_{Ca} = 0.500743$, and $N_{Mg} = 0.499257$. The ratio of adsorbed cations is not changed significantly, but the $(Mg^{2+})/(Ca^{2+})$ ratio for the dissolved species has increased from 1 to 1.7. If the groundwater continues to flow through the clayey stratum, the equilibrium cation concentrations will remain as indicated above until a sufficient number of pore volumes pass through to cause the ratio of adsorbed cations to gradually change. Eventually, the N_{Mg}/N_{Ca} ratio decreases to a value of 0.6, at which time the clay will no longer be capable of changing the Mg^{2+} and Ca^{2+} concentrations of the incoming groundwater. If the chemistry of the input water changes, the steady-state equilibrium will not be achieved.

This example illustrates the dynamic nature of cation exchange equilibria. Because exchange reactions between cations and clays are normally fast, the cation concentrations in groundwater can be expected to be in exchange equilibrium, but many thousands or millions of pore volumes may have to pass through the porous medium before the ratio of adsorbed cations completely adjusts to the input water. Depending on the geochemical and hydrologic conditions, time periods of millions of years may be necessary for this to occur.

Exchange involving cations of the same valence is characterized by preference for one of the ions if the selectivity coefficient is greater or less than unity. The normal order of preference for some monovalent and divalent cations for most clays is

Affinity for adsorption

$$Cs^+ > Rb^+ > K^+ > Na^+ > Li^+$$

$$\text{stronger} \longrightarrow \text{weaker}$$

$$Ba^{2+} > Sr^{2+} > Ca^{2+} > Mg^{2+}$$

The divalent ions normally have stronger adsorption affinity than the monovalent ions, although this depends to some extent on the nature of the exchanger and the concentration of the solutions (Wiklander, 1964). Both affinity sequences proceed in the direction of increasing hydrated ionic radii, with strongest adsorption for the smaller hydrated ions and weakest adsorption for the largest ions. It must be kept in mind, however, that the direction in which a cation exchange reaction proceeds also depends on the ratio of the adsorbed mole fractions at the initial condition and on the concentration ratio of the two ions in solution. For example, if we consider the Mg–Ca exchange condition used in the equilibrium calculations presented above but alter the initial condition of adsorbed ions to $N_{Mg} = 0.375$ and $N_{Ca} = 0.625$, there would be no change in the Mg^{2+} and Ca^{2+} concentrations as the groundwater passes through the clay. If the initial adsorbed ion conditions were such that the N_{Mg}/N_{Ca} ratio was less than 0.6, the exchange reaction would proceed in the reverse direction [to the right in Eq. (3.105)], thereby causing the ratio $(Mg^{2+})/(Ca^{2+})$ to decrease. This indicates that to determine the direction in which an ion exchange reaction will proceed, more information than the simple adsorption affinity series presented above is required.

The most important cation exchange reactions in groundwater systems are those involving monovalent and divalent cations such as Na^+–Ca^{2+}, Na^+–Mg^{2+}, K^+–Ca^{2+}, and K^+–Mg^{2+}. For these reactions,

$$2A^+ + B(ad) = B^{2+} + 2A(ad) \tag{3.108}$$

$$\bar{K}_{A\text{-}B} = \frac{[B^{2+}]N_A^2}{[A^+]^2 N_B} \tag{3.109}$$

The Na^+–Ca^{2+} exchange reaction is of special importance when it occurs in montmorillonitic clays (smectite) because it can cause large changes in permeability.

Clays of the montmorillonite group can expand and contract in response to changes in the composition of the adsorbed cation between the clay platelets. The hydrated radii of Na^+ and Ca^{2+} are such that two hydrated Na^+ require more space than one Ca^{2+}. Hence, replacement of Ca^{2+} by Na^+ on the exchange sites causes an increase in the dimension of the crystal lattice. This results in a decrease in permeability. This can cause a degradation in the agricultural productivity of soils.

3.8 Environmental Isotopes

Since the early 1950's naturally occurring isotopes that exist in water in the hydrologic cycle have been used in investigations of groundwater and surface water systems. Of primary importance in these studies are tritium (^3H) and carbon 14 (^{14}C), which are radioactive, and oxygen 18 (^{18}O) and deuterium (^2H), which are nonradioactive (Table 3.1). The latter are known as *stable isotopes*. Tritium and deuterium are often represented as T and D, respectively. ^3H and ^{14}C are used as a guide to the age of groundwater. ^{18}O and ^2H serve mainly as indicators of groundwater source areas and as evaporation indicators in surface-water bodies.

In this text these four isotopes are the only environmental isotopes for which hydrogeologic applications are described. For discussions of the theory and hydrologic or hydrochemical use of other naturally occurring isotopes, such as carbon 13, nitrogen 15, and sulfur 34, the reader is referred to Back and Hanshaw (1965), Kreitler and Jones (1975), and Wigley (1975). There are many situations where isotopic data can provide valuable hydrologic information that could not otherwise be obtained. Sophisticated techniques for the measurement of the abovementioned isotopes in water have been available for several decades, during which time the use of these isotopes in groundwater studies has gradually increased.

Carbon-14

Prior to the advent of large aboveground thermonuclear tests in 1953, ^{14}C in the global atmosphere was derived entirely from the natural process of nitrogen transmutation caused by bombardment of cosmic rays. This ^{14}C production has been estimated to be about 2.5 atoms/s · cm^2 (Lal and Suess, 1968). Oxidation to CO_2 occurs quickly, followed by mixing with the atmospheric CO_2 reservoir. The steady-state concentration of ^{14}C in the atmosphere is about one ^{14}C atom in 10^{12} atoms of ordinary carbon (^{12}C). Studies of the ^{14}C content of tree rings indicate that this concentration of ^{14}C has varied only slightly during the past 7000 years. Other evidence suggests that there have been no major shifts in the atmospheric ^{14}C concentrations during the past several tens of thousands of years.

The law of radioactive decay describes the rate at which the activity of ^{14}C and all other radioactive substances decreases with time. This is expressed as

$$A = A_0 2^{-t/T} \tag{3.110}$$

where A_0 is the radioactivity level at some initial time, A the level of radioactivity after time t, and T the half-life of the isotope. This law, in conjunction with measurements of the ^{14}C content of groundwater, can be used as a guide to *groundwater age*. In this context the term *age* refers to the period of time that has elapsed since the water moved deep enough into the groundwater zone to be isolated from the earth's atmosphere.

Use of ^{14}C for dating of groundwater was first proposed by Münnich (1957), following the development of techniques for ^{14}C dating of solid carbonaceous materials pioneered by the Nobel laureate W. F. Libby in 1950. When water moves below the water table and becomes isolated from the earth's CO_2 reservoir, radioactive decay causes the ^{14}C content in the dissolved carbon to gradually decline. The expression for radioactive decay [Eq. (3.110)] can be rearranged, and upon substitution of $T = 5730$ years yields

$$t = -8270 \ln \left(\frac{A}{A_0}\right) \tag{3.111}$$

where A_0 is the specific activity (disintegrations per unit time per unit mass of sample) of carbon 14 in the earth's atmosphere, A the activity per unit mass of sample, and t the decay age *of the carbon*. In groundwater investigations ^{14}C determinations are made on samples of inorganic carbon that are extracted from samples of groundwater that generally range in volume from 20 to 200 ℓ. The mass of carbon needed for accurate analysis by normal methods is about 3 g. The ^{14}C values obtained in this manner are a measure of the ^{14}C content of the $CO_2(aq)$, H_2CO_3, $CO_3{}^{2-}$, and HCO_3^- in the water at the time of sampling. ^{14}C may also be present in dissolved organic carbon such as fulvic and humic acids, but this ^{14}C source is small and is normally not included in studies of groundwater age.

The specific activity of ^{14}C in carbon that was in equilibrium with the atmosphere of the earth prior to atmospheric testing of thermonuclear devices is approximately 10 disintegrations per minute per gram (dpm/g). Modern equipment for ^{14}C measurement can detect ^{14}C activity levels as low as approximately 0.02 dpm/g. Substitution of these specific activities in Eq. (3.111) yields a maximum age of 50,000 years. It must be emphasized that this is an apparent age of the dissolved inorganic carbon. To gain some useful hydrologic information from this type of data, it is necessary to determine the source of the inorganic carbon. Calcite or dolomite occur in many groundwater environments. Carbon that enters the groundwater by dissolution of these minerals can cause dilution of the ^{14}C content of the total inorganic carbon in the water. This is the case because in most groundwater systems the calcite and dolomite are much older than 50,000 years. Their carbon is therefore devoid of significant amounts of ^{14}C and is often referred to as "dead" carbon. To obtain ^{14}C estimates of the actual groundwater age it is necessary to determine the extent to which this dead carbon has reduced the relative ^{14}C content of the groundwater. An indication of how this can be done is described in Chapter 7.

Tritium

The occurrence of tritium in waters of the hydrological cycle arises from both natural and man-made sources. In a manner similar to ^{14}C production, ^{3}H is produced naturally in the earth's atmosphere by interaction of cosmic-ray-produced neutrons with nitrogen. Lal and Peters (1962) estimated that the atmospheric production rate is 0.25 atoms/s · cm^2. In 1951, Van Grosse and coworkers discovered that ^{3}H occurred naturally in precipitation. Two years later large quantities of man-made tritium entered the hydrological cycle as a result of large-scale atmospheric testing of thermonuclear bombs. Unfortunately, few measurements of natural tritium in precipitation were made before atmospheric contamination occurred. It has been estimated that prior to initiation of atmospheric testing in 1952, the natural tritium content of precipitation was in the range of about 5–20 tritium units (Payne, 1972). A tritium unit is the equivalent of 1 tritium atom in 10^{18} atoms of hydrogen. Since the half-life of ^{3}H is 12.3 years, groundwater that was recharged prior to 1953 is therefore expected to have ^{3}H concentrations below about 2–4 TU. The first major source of man-made ^{3}H entered the atmosphere during the initial tests of large thermonuclear devices in 1952. These tests were followed by additional tests in 1954, 1958, 1961, and 1962 before the moratorium on atmospheric testing agreed upon by the United States and the USSR.

Since the onset of thermonuclear testing, the tritium content in precipitation has been monitored at numerous locations in the northern hemisphere and at a smaller but significant number of locations in the southern hemisphere. Considering the data separately by hemispheres, there is a strong parallelism in ^{3}H concentration with time, although absolute values vary from place to place (Payne, 1972). In the southern hemisphere, ^{3}H values are much lower because of the higher ratio of oceanic area to land area. The longest continuous record of ^{3}H concentrations in precipitation is from Ottawa, Canada, where sampling was begun in 1952. The ^{3}H versus time record for this location is shown in Figure 3.11. The trends displayed in this graph are representative of the ^{3}H trends recorded elsewhere in the northern hemisphere. Tritium data obtained by the International Atomic Energy Agency (IAEA) from a global sampling network enable the estimation of ^{3}H versus time trends for areas in which there are no sampling stations or only short-term records. At a given latitude the concentrations of tritium in precipitation at sampling stations near the coast are lower than those inland because of dilution from oceanic water vapor, which is low in tritium.

Measurements of tritium concentrations can be a valuable aid in many types of groundwater investigations. If a sample of groundwater from a location in the northern hemisphere contains tritium at concentration levels of hundreds or thousands of TU, it is evident that the water, or at least a large fraction of the water, originally entered the groundwater zone sometime after 1953. If the water has less than 5–10 TU, it must have originated prior to 1953. Using routine methods for measurement of low-level tritium in water samples, concentrations as low as about 5–10 TU can be detected. Using special methods for concentrating ^{3}H

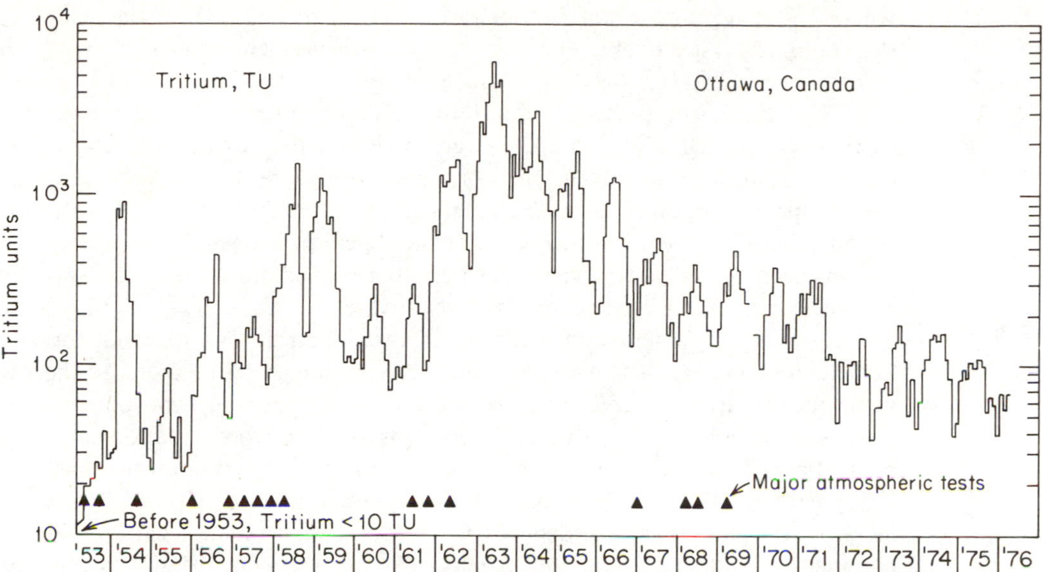

Figure 3.11 Variations of tritium in precipitation (mean monthly concentrations, TU) at Ottawa, Canada.

from water samples, values as low as about 0.1 TU can be measured. If samples contain no detectable ^3H in routine measurements, it is usually reasonable to conclude that significant amounts of post-1953 water are not present. Post-1953 water is often referred to as *modern water* or *bomb tritium water*.

Tritium data from detailed sampling patterns can sometimes be used to distinguish different age zones within the modern-water part of groundwater flow systems. For this type of tritium use, the stratigraphic setting should be simple so that complex flow patterns do not hinder the identification of tritium trends. In situations where the ^3H concentrations of two adjacent flow zones are well defined, tritium data can be useful to distinguish zones of mixing. The usefulness of tritium in groundwater studies is enhanced by the fact that it is not significantly affected by reactions other than radioactive decay.

Oxygen and Deuterium

With the advent of the mass spectrometer, it became possible in the early 1950's to make rapid accurate measurements of isotope ratios. Of special interest to hydrologists are the ratios of the main isotopes that comprise the water molecule, $^{18}O/^{16}O$ and $^2H/^1H$. The isotope ratios are expressed in delta units (δ) as per mille (parts per thousand or ‰) differences relative to an arbitrary standard known as *standard mean ocean water* (SMOW):

$$\delta‰ = [(R - R_{\text{standard}})/R_{\text{standard}}] \times 1000 \tag{3.112}$$

where R and R_{standard} are the isotope ratios, $^2H/^1H$ or $^{18}O/^{16}O$, of the sample and the standard, respectively. The accuracy of measurement is usually better than $\pm 0.2‰$ and $\pm 2‰$ for $\delta^{18}O$ and δ^2H, respectively.

The various isotopic forms of water have slightly different vapor pressures and freezing points. These two properties give rise to differences in ^{18}O and ^2H concentrations in water in various parts of the hydrologic cycle. The process whereby the isotope content of a substance changes as a result of evaporation, condensation, freezing, melting, chemical reactions, or biological processes is known as *isotopic fractionation.* When water evaporates from the oceans, the water vapor produced is depleted in ^{18}O and ^2H relative to ocean water, by about 12–15‰ in ^{18}O and 80–120‰ in ^2H. When water vapor condenses, the rain or snow that forms has higher ^{18}O and ^2H concentrations than the remaining water vapor. As the water vapor moves farther inland as part of regional or continential atmospheric circulation systems, and as the process of condensation and precipitation is repeated many times, rain or snow becomes characterized by low concentrations of the heavy isotopes ^{18}O and ^2H. The ^{18}O and ^2H content of precipitation at a given locality at a particular time depends in a general way on the location within the continental land mass, and more specifically on the condensation-precipitation history of the atmospheric water vapor. Since both condensation and isotope fractionation are temperature-dependent, the isotopic composition of precipitation is also temperature-dependent. The combined effect of these factors is that (1) there are strong continental trends in the average annual isotopic composition of precipitation, (2) there is a strong seasonal variation in the time-averaged isotopic composition of precipitation at a given location, and (3) the isotopic composition of rain or snow during an individual precipitation event is very variable and unpredictable. In continental areas, rain values can vary between about 0 and $-25‰$ for ^{18}O and 0 to $-150‰$ for ^2H, even though the average annual values have little variation. Because of temperature changes in the zone of atmospheric condensation or isotopic depletion effects, large variations can even occur during individual rainfall events. Changes can also occur in the raindrop during its fall, especially at the beginning of a rainstorm and in arid or semiarid regions.

In deep subsurface zones where temperatures are above 50–100°C, the ^{18}O and ^2H content of groundwater can be significantly altered as a result of chemical interactions with the host rock. In shallower groundwater systems with normal temperatures, the concentrations of these isotopes are little, if at all, affected by chemical processes. In these flow regimes, ^{18}O and ^2H are nonreactive, naturally occurring tracers that have concentrations determined by the isotopic composition of the precipitation that falls on the ground surface and on the amount of evaporation that occurs before the water penetrates below the upper part of the soil zone. Once the water moves below the upper part of the soil zone, the ^{18}O and ^2H concentrations become a characteristic property of the subsurface water mass, which in many hydrogeologic settings enables the source areas and mixing patterns to be determined by sampling and analysis for these isotopes.

^2H and ^{18}O concentrations obtained from global precipitation surveys correlate according to the relation (Dansgaard, 1964)

$$\delta^2H‰ = 8\delta^{18}O‰ + 10 \tag{3.113}$$

which is known as the *meteoric water line*. Linear correlations with coefficients only slightly different than this are obtained from studies of local precipitation. When water evaporates from soil- or surface-water bodies under natural conditions, it becomes enriched in ^{18}O and ^2H. The relative degree of enrichment is different than the enrichment that occurs during condensation. The ratio of $\delta^{18}O/\delta^2H$ for precipitation that has partially evaporated is greater than the ratio for normal precipitation obtained from Eq. (3.113). The departure of ^{18}O and ^2H concentrations from the meteoric water line is a feature of the isotopes that can be used in a variety of hydrologic investigations, including studies of the influence of groundwater on the hydrologic balance of lakes and reservoirs and the effects of evaporation on infiltration.

3.9 Field Measurement of Index Parameters

Description of the laboratory techniques that are used in the chemical or isotopic analysis of water samples is beyond the scope of this text. For this type of information the reader is referred to Rainwater and Thatcher (1960) and U.S. Environmental Protection Agency (1974b). Our purpose here is to briefly describe methods by which several important index parameters are measured in field investigations. These parameters are specific electrical conductance, pH, redox potential, and dissolved oxygen. In groundwater studies each of these parameters can be measured in the field by immersing probes in samples of water or by lowering probes down wells or piezometers.

Electrical conductivity is the ability of a substance to conduct an electrical current. It has units of reciprocal ohm-meters, denoted in the SI system as siemens per meter (S/m). Electrical conductance is the conductivity of a body or mass of fluid of unit length and unit cross section at a specified temperature. In the groundwater literature electrical conductance has normally been reported as reciprocal milliohms or reciprocal microohms, known as millimhos and micromhos. In the SI system, 1 millimho is denoted as 1 millisiemen (mS) and 1 micromho as 1 microsiemen (μS).

Pure liquid water has a very low electrical conductance, less than a tenth of a microsiemen at 25°C (Hem, 1970). The presence of charged ionic species in solution makes the solution conductive. Since natural waters contain a variety of both ionic and uncharged species in various amounts and proportions, conductance determinations cannot be used to obtain accurate estimates of ion concentrations or total dissolved solids. As a general indication of total dissolved solids (TDS), however, specific conductance values are often useful in a practical manner. For conversion between conductance values and TDS, the following relation is used (Hem, 1970):

$$\text{TDS} = AC \tag{3.114}$$

where C is the conductance in microsiemens or micromhos, TDS is expressed in g/m^3 or mg/ℓ, and A is a conversion factor. For most groundwater, A is between 0.55 and 0.75, depending on the ionic composition of the solution.

Measurements of electrical conductance can be made in the field simply by immersing a conductance cell in water samples or lowering it down wells and then recording the conductance on a galvanometer. Rugged equipment that is well suited for field use is available from numerous commercial sources. In groundwater studies conductance measurements are commonly made in the field so that variations in dissolved solids can be determined without the delay associated with transportation of samples to the laboratory. As distributions of groundwater conductance values are mapped in the field, sampling programs can be adjusted to take into account anomalies or trends that can be identified as the field work proceeds.

To avoid changes caused by escape of CO_2 from the water, measurements of the pH of groundwater are normally made in the field immediately after sample collection. Carbon dioxide in groundwater normally occurs at a much higher partial pressure than in the earth's atmosphere. When groundwater is exposed to the atmosphere, CO_2 will escape and the pH will rise. The amount of pH rise for a given decrease in P_{CO_2} can be calculated using the methods described in Section 3.3. For field measurements of pH, portable pH meters and electrodes are generally used. Samples are usually brought to ground surface by pumping or by means of down-hole samplers rather than by lowering electrodes down wells. A detailed description of the theory and methods of pH measurement in water are presented by Langmuir (1970).

Dissolved oxygen is another important hydrochemical parameter that is commonly measured in the field by immersing a small probe in water samples or down wells. In a dissolved oxygen probe, oxygen gas molecules diffuse through a membrane into a measuring cell at a rate proportional to the partial pressure of oxygen in the water. Inside the sensor the oxygen reacts with an electrolyte and is reduced by an applied voltage. The current that is generated is directly proportional to the partial pressure of oxygen in the water outside the sensor (Back and Hanshaw, 1965). Rugged dissolved oxygen probes that connect to portable meters are commercially available. These probes can be lowered down wells or piezometers to obtain measurements that are representative of *in situ* conditions. Dissolved oxygen can also be measured in the field by a titration technique known as the Winkler method (U.S. Environmental Protection Agency, 1974b).

Dissolved oxygen probes of the type that are generally used have a detection limit of about 0.1 mg/ℓ. High-precision probes can measure dissolved oxygen at levels as low as 0.01 mg/ℓ. Even at dissolved oxygen contents near these detection limits, groundwater can have enough oxygen to provide considerable capability for oxidation of many types of reduced constituents. *Eh* or p*E* values can be computed from measured values of dissolved oxygen by means of Eq. (3.77). The concentration of dissolved oxygen is converted to P_{O_2} using Henry's law ($P_{O_2} = O_2$ dissolved/K_{O_2}), where K_{O_2} at 25°C is 1.28×10^{-3} mol/bar). At pH 7, p*E* values obtained in this manner using dissolved oxygen values at the detection limits indicated above are 13.1 and 12.9, or expressed as *Eh*, 0.78 and 0.76 V, respectively. Figure 3.10 indicates that these values are near the upper limit of the p*E*–pH domain for water. If the water is saturated with dissolved oxygen (i.e., in equilibrium with oxygen in the earth's atmosphere), the calculated p*E* is 13.6. For p*E* values calculated from dissolved oxygen concentrations to serve as a true indication of the redox condition of the water, dissolved oxygen must be the controlling oxidative species in the water with redox conditions at or near equilibrium. Measurements values of other dissolved multivalent constituents can also be used to obtain estimates of redox conditions of groundwater. Additional discussion of this topic is included in Chapter 7.

Another approach to obtaining estimates of the redox condition of groundwater is to measure electrical potential in the water using an electrode system that includes an inert metallic electrode (platinum is commonly used). Electrode systems known as *Eh probes* are commercially available. To record the electrical potential, they can be attached to the same meters used for pH. For these readings to have significance, the probes must be lowered into wells or piezometers or be placed in sample containers that prevent the invasion of air. For some groundwater zones the potentials measured in this manner are an indication of the redox conditions, but in many cases they are not. Detailed discussions of the theory and significance of the electrode approach to redox measurements are provided by Stumm and Morgan (1970) and Whitfield (1974).

Suggested Readings

BLACKBURN, T. R. 1969. *Equilibrium, A Chemistry of Solutions.* Holt, Rinehart and Winston, New York, pp. 93–111.

GARRELS, R. M., and C. L. CHRIST. 1965. *Solutions, Minerals, and Equilibria.* Harper & Row, New York, pp. 1–18, 50–71.

KRAUSKOPF, K. 1967. *Introduction to Geochemistry.* McGraw-Hill, New York, pp. 3–23, 29–54, 206–226, 237–255.

STUMM, W., and J. J. MORGAN. 1970. *Aquatic Chemistry.* Wiley-Interscience, New York, pp. 300–377.

Problems

In the problems listed below for which calculations are required, neglect the occurrence of ion associations or complexes such as $CaSO_4^\circ$, $MgSO_4^\circ$, $NaSO_4^-$, $CaHCO_3^+$, and $CaCO_3^\circ$. Information that can be obtained from some of the figures in this text should serve as a guide in many of the problems.

1. A laboratory analysis indicates that the total dissolved inorganic carbon in a sample of aquifer water is 100 mg/ℓ (expressed as *C*). The temperature in the aquifer is 15°C, the pH is 7.5, and the ionic strength is 0.05. What are the concentrations of H_2CO_3, CO_3^{2-}, and HCO_3^- and the partial pressure of CO_2? Is the P_{CO_2} within the range that is common for groundwater?

2. Saline water is injected into an aquifer that is confined below by impervious rock and above by a layer of dense unfractured clay that is 10 m thick. A freshwater aquifer occurs above this aquitard. The Cl^- content of the injected water is 100,000 mg/ℓ. Estimate the length of time that would be required for Cl^- to move by molecular diffusion through the clayey aquitard into the freshwater aquifer. Express your answer in terms of a range of time that would be reasonable in light of the available information. Assume that the velocity of hydraulic flow through the clay is insignificant relative to the diffusion rate.

3. Two permeable horizontal sandstone strata in a deep sedimentary basin are separated by a 100-m-thick bed of unfractured montmorillontic shale. One of the sandstone strata has a total dissolved solids of 10,000 mg/ℓ; the other has 100,000 mg/ℓ. Estimate the largest potential difference that could develop, given favorable hydrodynamic conditions, across the shale as a result of the effect of osmosis (for water activity in salt solutions, see Robinson and Stokes, 1965). The system has a temperature of 25°C. What factors would govern the actual potential difference that would develop?

4. Rainwater infiltrates into a deposit of sand composed of quartz and feldspar. In the soil zone, the water is in contact with soil air that has a partial pressure of $10^{-1.5}$ bar. The system has a temperature of 10°C. Estimate the pH of the soil water. Assume that reactions between the water and the sand are so slow that they do not significantly influence the chemistry of the water.

5. The results of a chemical analysis of groundwater are as follows (expressed in mg/ℓ): $K^+ = 5.0$, $Na^+ = 19$, $Ca^{2+} = 94$, $Mg^{2+} = 23$, $HCO_3^- = 334$, $Cl^- = 9$, and $SO_4^{2-} = 85$; pH 7.21; temperature 25°C. Determine the saturation indices with respect to calcite, dolomite, and gypsum. The water sample is from an aquifer composed of calcite and dolomite. Is the water capable of dissolving the aquifer? Explain.

6. Is there any evidence indicating that the chemical analysis listed in Problem 5 has errors that would render the analysis unacceptable with regard to the accuracy of the analysis?

7. Groundwater at a temperature of 25°C and a P_{CO_2} of 10^{-2} bar flows through strata rich in gypsum and becomes gypsum saturated. The water then moves into a limestone aquifer and dissolves calcite to saturation. Estimate the composition of the water in the limestone after calcite dissolves to equilibrium. Assume that gypsum does not precipitate as calcite dissolves.

8. A sample of water from an aquifer at a temperature of 5°C has the following composition (expressed in mg/ℓ): $K^+ = 9$, $Na^+ = 56$, $Ca^{2+} = 51$, $Mg^{2+} = 104$, $HCO_3^- = 700$, $Cl^- = 26$, and $SO_4^{2-} = 104$; pH 7.54. The pH was obtained from a measurement made in the field immediately after sampling. If the sample is allowed to equilibrate with the atmosphere, estimate what the pH will be. (*Hint*: The P_{CO_2} of the earth's atmosphere is $10^{-3.5}$ bar; assume that calcite and other minerals do not precipitate at a significant rate as equilibration occurs with respect to the earth's atmosphere.)

9. Field measurements indicate that water in an unconfined aquifer has a pH of 7.0 and a dissolved oxygen concentration of 4 mg/ℓ. Estimate the p*E* and *Eh* of the water. Assume that the redox system is at equilibrium and that the water is at 25°C and 1 bar.

10. The water described in Problem 9 has a HCO_3^- content of 150 mg/ℓ. If the total concentration of iron is governed by equilibria involving $FeCO_3(s)$ and $Fe(OH)_3(s)$, estimate the concentrations of Fe^{3+} and Fe^{2+} in the water. What are the potential sources of error in your estimates?

11. A water sample has a specific conductance of 2000 μS at a temperature of 25°C. Estimate the total dissolved solids and ionic strength of the water. Present your answer as a range in which you would expect the TDS and *I* values to occur.

12. Groundwater has the following composition (expressed in mg/ℓ): $K^+ = 4$, $Na^+ = 460$, $Ca^{2+} = 40$, $Mg^{2+} = 23$, $HCO_3^- = 1200$, $Cl^- = 8$, and $SO_4^{2-} = 20$; pH 6.7. How much water would have to be collected to obtain sufficient carbon for a ^{14}C determination by normal methods?

13. Compute the P_{CO_2} for the water described in Problem 12. The P_{CO_2} is far above the P_{CO_2} for the earth's atmosphere and is above the normal range for most groundwaters. Suggest a reason for the elevated P_{CO_2}.

14. In the normal pH range of groundwater (6–9), what are the dominant dissolved species of phosphorus? Explain why.

15. Prepare a percent occurrence versus pH graph similar in general form to Figure 3.5 for dissolved sulfide species (HS^-, S^{2-}, H_2S) in water at 25°C.

16. Radiometric measurements on a sample of inorganic carbon from well water indicate a ^{14}C activity of 12 disintegrations per minute (dpm). The background activity is 10 dpm. What is the apparent age of the sample?

17. Groundwater at 5°C has a pH of 7.1. Is the water acidic or alkaline?

18. Does precipitation of calcite in zones below the water table (i.e., closed-system conditions) cause the pH of the water to rise or fall? Explain.

CHAPTER 4

Groundwater Geology

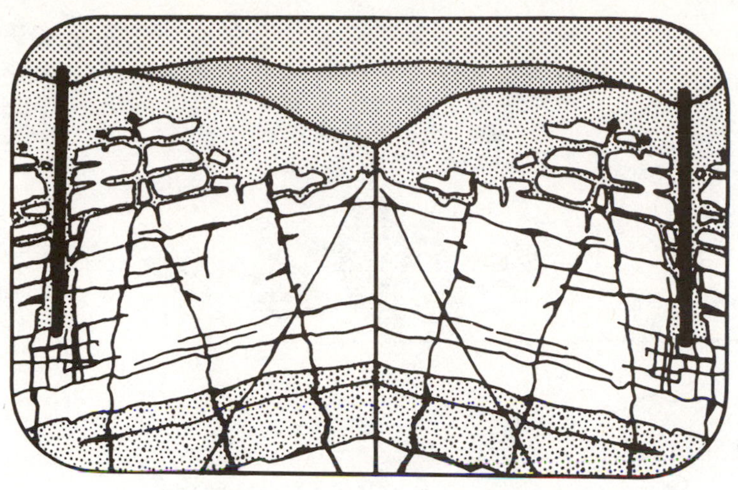

4.1 Lithology, Stratigraphy, and Structure

The nature and distribution of aquifers and aquitards in a geologic system are controlled by the lithology, stratigraphy, and structure of the geologic deposits and formations. The *lithology* is the physical makeup, including the mineral composition, grain size, and grain packing, of the sediments or rocks that make up the geological systems. The *stratigraphy* describes the geometrical and age relations between the various lenses, beds, and formations in geologic systems of sedimentary origin. *Structural features*, such as cleavages, fractures, folds, and faults are the geometrical properties of the geologic systems produced by deformation after deposition or crystallization. In unconsolidated deposits, the lithology and stratigraphy constitute the most important controls. In most regions knowledge of the lithology, stratigraphy, and structure leads directly to an understanding of the distribution of aquifers and aquitards.

Situations in which the stratigraphy and structure control the occurrence of aquifers and aquitards are illustrated in Figure 4.1. In the Great Plains states and in western Canada, there are many occurrences of Cretaceous or Paleozoic sandstones warped up along the Rocky Mountains or along igneous intrusions such as the Black Hills. The permeable sandstones are regional artesian aquifers [Figure 4.1(a)] fed by recharge in the outcrop areas and by leakage through clayey confining beds. In the intermountain basins in the western United States, aquifers of permeable sand and gravel formed in alluvial fans interfinger with layers of clay and silt deposited in playa lakes [Figure 4.1(b)]. Water recharges the aquifers along the mountains. Confined conditions develop as the aquifers lense out toward the basin flats. In the Sahara region of Africa, gently warped permeable beds form regional aquifers that receive water along distant mountain fronts and by vertical leakage. Occurrences of surface water are controlled by faults or folds or where the desert floor is eroded close to the top of the aquifers [Figure 4.1(c)].

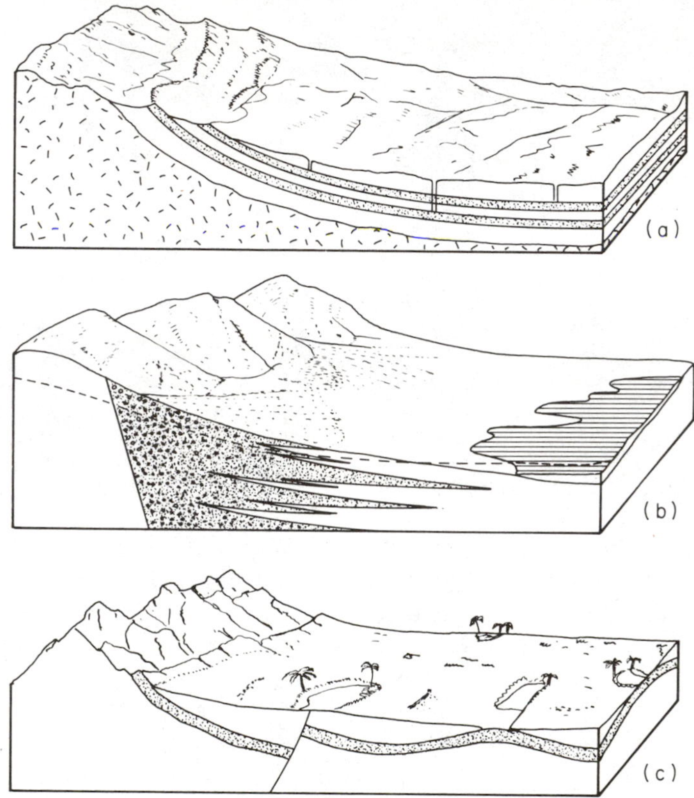

Figure 4.1 Influence of stratigraphy and structure on regional aquifer occurrence. (a) Gently dipping sandstone aquifers with outcrop area along mountain front; (b) interfingering sand and gravel aquifers extending from uplands in intermountain region; (c) faulted and folded aquifer in desert region. Surface water bodies reflect structural features (after Hamblin, 1976).

Unconformities are stratigraphic features of particular importance in hydrogeology. An *unconformity* is a surface that represents an interval of time during which deposition was negligible or nonexistent, or more commonly during which the surface of the existing rocks was weathered, eroded, or fractured. Often the underlying rocks were warped or tilted prior to the deposition of new materials over the unconformity. Aquifers are commonly associated with unconformities, either in the weathered or fractured zone immediately below the surface of the buried landscape or in permeable zones in coarse-grained sediments laid down on top of this surface when the system entered a new era of accretion. In many of the tectonically stable parts of the interior of North America, where near-horizontal sedimentary rocks occur beneath the overburden, the occurrence of unconformities are the key to the distribution of aquifers and aquitards and the quality of water within them.

In terrain that has been deformed by folding and faulting, aquifers can be difficult to discern because of the geologic complexity. In these situations the main ingredient in groundwater investigations is often large-scale structural analysis of the geologic setting.

4.2 Fluvial Deposits

Nonindurated deposits are composed of particles of gravel, sand, silt, or clay size that are not bound or hardened by mineral cement, by pressure, or by thermal alteration of the grains. *Fluvial deposits* are the materials laid down by physical processes in river channels or on floodplains. The materials are also known as *alluvial deposits*. In this section emphasis is on fluvial materials deposited in non-glacial environments. Deposits formed by meltwater rivers are discussed in Section 4.4.

Fluvial materials occur in nearly all regions. In many areas aquifers of fluvial origin are important sources of water supply. Figure 4.2 illustrates the morphology and variations in deposits formed by braided rivers and by meandering rivers. Because of the shifting position of river channels and the ever-changing depositional velocities, river deposits have characteristic textural variability that causes much heterogeneity in the distribution of hydraulic properties. Braided rivers generally occur in settings where the sediment available for transport has considerable coarse-grained sand or gravel and where velocities are large because of steep regional topographic slopes. Shifting positions of channels and bars and changing velocity can result in extensive deposits of bedded sand and gravel with minor zones of silty or clayey sediments filling in abandoned channels. Meandering rivers and their associated floodplain environments also have coarse and fine-grained deposits. The relative abundances and stratigraphic relations of the sediments, however, are generally much different than in braided-river deposits. Silty or clayey channel-fill deposits are more abundant than in braided-river deposits. Cross-bedded sand, which is commonly fine- or medium-grained with variable contents of silt and clay, is deposited on the levees and floodplains. Coarse sand and gravel commonly form along point bars. Gravel deposits develop as channel lag. The relative abundance of the various deposits laid down in meandering rivers and their floodplains are greatly influenced by the nature of the sediments supplied to the river from the watershed. Because of the variability of sediment sources and flow, delineation of aquifer zones in these deposits using borehole data is a difficult task that often involves much speculation.

Large numbers of hydraulic conductivity tests, both in the field and in the laboratory, have been made on fluival deposits. Results of permeameter tests on core samples characteristically indicate variations within the permeable zones of

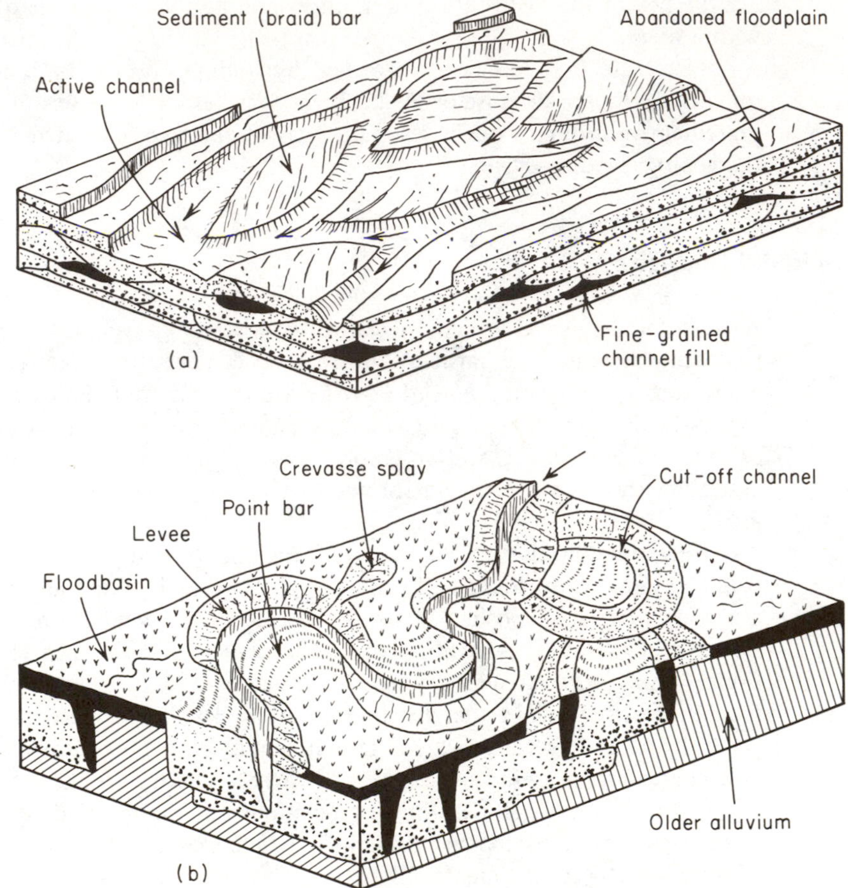

Figure 4.2 Surface features and nature of deposits formed in (a) the braided-river environment and (b) the floodplain environment of meandering rivers (after Allen, 1970).

more than 2 or 3 orders of magnitude. These variations reflect the difference in grain-size distributions in individual layers within the bulk deposit.

When the average properties of large volumes are considered, the bedded character of fluvial deposits imparts a strong anisotropy to the system. On a smaller scale represented by core samples tested in the laboratory, anisotropy of permeability is present but not as marked. Johnson and Morris (1962) report both vertical and horizontal hydraulic conductivities of 61 laboratory samples of fluvial and lacustrine sediments from the San Joaquim Valley, California. Forty-six of the samples had a greater horizontal than vertical hydraulic conductivity, 11 of the samples were isotropic and only 4 had greater vertical conductivities. The horizontal conductivities were between 2 and 10 times larger than the vertical values.

4.3 Aeolian Deposits

Materials that are transported and deposited by wind are known as *aeolian deposits*. Aeolian deposits consist of sand or silt. Sand dunes form along coasts and in inland areas where rainfall is sparse and surface sand is available for transportation and deposition. Nonindurated aeolian sand is characterized by lack of silt and clay fractions, by uniform texture with particles in the fine- or medium-grain-size range, and by rounded grains. These sands are moderately permeable (10^{-4}–10^{-6} m/s) and form aquifers in areas where appreciable saturated thicknesses occur. Porosities are between 30 and 45%. In comparison with alluvial deposits, aeolian sands are quite homogeneous and are about as isotropic as any deposits occurring in nature. The sorting action of wind tends to produce deposits that are uniform on a local scale and in some cases quite uniform over large areas.

The most extensive nonindurated aeolian deposits in North America are blanket deposits of silt, which are known as *loess*. Loess occurs at the surface or in the shallow subsurface in large areas in the Midwest and Great Plains regions of North America. Loess was deposited during Pleistocene and post-Pleistocene time as a result of wind activity that caused clouds of silt to be swept across the landscape. Because of small amounts of clay and calcium carbonate cement that are almost always present, loess is slightly to moderately cohesive. The porosity of loess is normally in the range 40–50%. Hydraulic conductivity varies from about 10^{-5} m/s for coarse, clean loess to 10^{-7} m/s or lower in fine or slightly clayey loess that has no secondary permeability.

Fractures, root channels, and animal burrows commonly cause secondary permeability in the vertical direction that may greatly exceed the primary permeability. As a result of repeated episodes of atmospheric silt movement, buried soils are common in loess. Zones of secondary permeability are often associated with these soils. In some loess areas sufficient permeability occurs at depth to provide farm or household water supplies. Major aquifers, however, do not occur in loess. In some situations blankets of loess act as aquitards overlying major aquifers. For further information on the occurrence and hydraulic properties of loess, the reader is referred to Gibbs and Holland (1960) and McGary and Lambert (1962).

4.4 Glacial Deposits

Of particular hydrogeologic importance in the northern part of the United States and in Canada and Europe are deposits formed by or in association with continental glaciers. The deposits include glacial till, glaciofluvial sediments, and glaciolacustrine sediments. In meltwater lakes that existed during Pleistocene time, deposits of glaciolacustrine silt and clay were laid down offshore. These deposits form some of the most extensive shallow aquitards in North America. Sand and gravel deposits laid down near shore and on beaches are aquifers in some areas. In comparison to aquifers of glaciofluvial origin, these aquifers of glaciolacustrine origin are generally of minor importance.

Glacial till is the most abundant material that was deposited on the land surface during Pleistocene time. In the Precambrian Shield region, till is generally sandy, with variable amounts of silt and little clay. Sandy till forms local aquifers in some areas. In the regions of sedimentary bedrock in North America, glacial erosion produced till that generally has considerable silt and clay and therefore has low permeability. Till layers of this type are aquitards.

Figure 4.3 is a schematic diagram of the occurrence of aquifers and aquitards in the Midwest and Great Plains regions of North America. Most aquifers in these regions are composed of glaciofluvial sand and gravel confined by deposits of till or glaciolacustrine silt or clay. The aquifers occur as extensive blanket bodies or as channel deposits in surface or buried valleys. The deposits of sand and gravel in buried valleys form aquifers that are generally many tens of kilometers long and several kilometers wide. The largest buried valleys are many tens of kilometers wide. In many cases there are no surface indications of the presence of buried valley aquifers. The overlying till is usually some tens of meters thick or less, but occasionally may be on the order of a hundred meters thick.

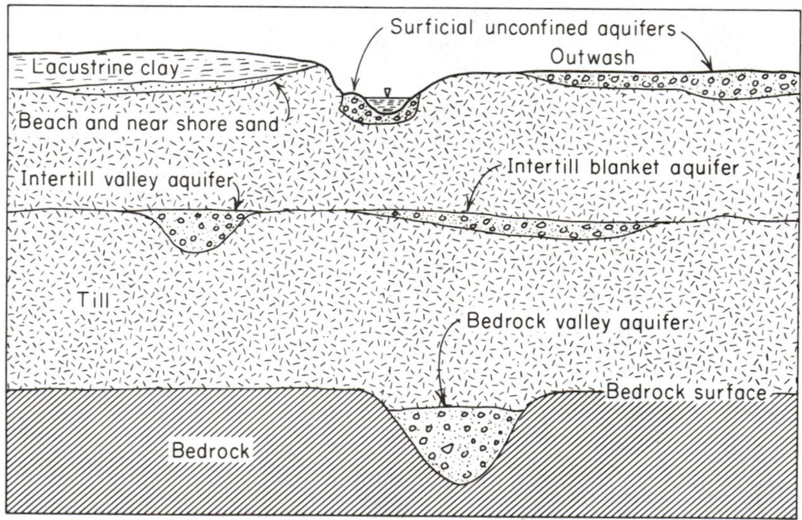

Figure 4.3 Schematic diagram of aquifer occurrence in the glaciated regions of the Midwest and Great Plains physiographic provinces.

Many of the meltwater rivers that formed aquifer deposits were braided in the manner shown in Figure 4.2(a); others flowed in deep channels or valleys eroded in glacial terrain or in sedimentary bedrock. Examples of aquifers that originated in valleys and as blanket deposits are included in Figure 4.3.

In addition to the classical types of meltwater deposits laid down by rivers of meltwater flowing beyond the margin of glaciers, many glaciated areas have deposits of sand and gravel that formed on top of masses of stagnant ice during episodes of glacial retreat. These deposits are known as collapsed outwash, stagnant-ice outwash, or ice-contact deposits. An example of this type of deposit is illustrated in Figure 4.4. Sand and gravel aquifers of this origin occur at the surface or buried beneath till deposited during periods of ice readvance.

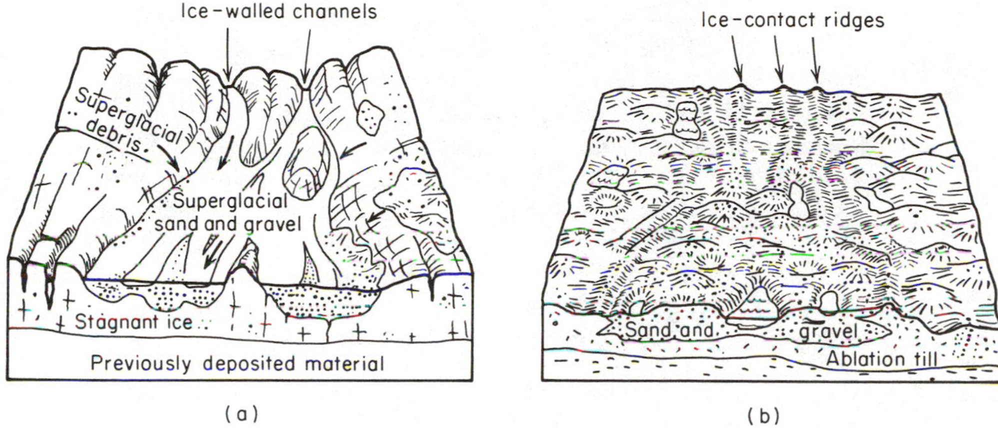

Figure 4.4 Formation of collapsed outwash in an environment of continental glacier stagnation (after Parizek, 1969).

Dense, fine-grained glacial till, and deposits of glaciolacustrine silt and clay are the most common aquitards in most of the northern part of the United States and in the southern part of Canada. These deposits have intergranular hydraulic conductivities that are very low, with values typically in the range 10^{-10}–10^{-12} m/s. With a hydraulic gradient of 0.5, which is near the upper limit of gradients observed in these aquitards, and a hydraulic conductivity of 10^{-11} m/s, nearly 10,000 years would be required for water to flow through a 10-m-thick unfractured layer of this material. Extensive deposits of clayey till or glaciolacustrine clay can cause isolation of buried aquifers from zones of near-surface groundwater flow.

In the Great Plains region, in parts of the American Midwest, and in southern Ontario, it has been observed that in some locations deposits of clayey or silty till and glaciolacustrine clay have networks of hairline fractures. These features are sometimes referred to as *fissures* or *joints*. The fractures are predominantly vertical or nearly vertical. The distance between the fractures varies from several centimeters to many meters. Infillings of calcite or gypsum are common. The soil matrix adjacent to the fractures is commonly distinguished by a color change caused by different degrees of oxidation or reduction. In some areas thin rootlets are observed along the fractures to depths of 5–10 m below ground surface. In some cases the fractures pass through successive layers of till and clay. In other cases they are restricted to individual layers.

In many areas the fractures impart an enhanced capability for groundwater flow. The bulk hydraulic conductivity of the fractured till and clay determined by field tests is commonly between 1 and 3 orders of magnitude larger than values of intergranular hydraulic conductivity determined by laboratory tests on unfractured samples. As a result of increased lateral stresses caused by overburden loading, the hydraulic conductivity of fractured till and clay decreases with depth, but because of the stiffness of many of these materials the fractures can provide significant secondary permeability to depths of hundreds of meters.

In areas of glacial till and glaciolacustrine clay, highly fractured zones are common within several meters of the ground surface. Shallow fractures are caused primarily by stress changes resulting from cycles of wetting and drying and freezing and thawing. Openings caused by roots also cause secondary permeability. The origin of fracture networks at greater depths is more problematic. Mechanisms such as stress release related to glacial unloading and crustal rebound, and volume changes due to geochemical processes such as cation exchange, have been suggested by various investigators. For more detailed discussions of the nature and hydrogeologic significance of fractures in till and glaciolacustrine clay, the reader is referred to Rowe (1972), Williams and Farvolden (1969), Grisak and Cherry (1975) and Grisak et al. (1976).

4.5 Sedimentary Rocks

Sandstone

About 25% of the sedimentary rock of the world is sandstone. In many countries sandstone strata form regional aquifers that have vast quantities of potable water. Sandstone bodies of major hydrologic significance owe their origin to various depositional environments, including floodplain, marine shoreline, deltaic, aeolian, and turbidity-current environments. Knowledge of the distribution of permeability in sandstones can best be acquired within an interpretive framework that is based on an understanding of depositional environments in which the sand bodies were formed. In this endeavor a knowledge of sedimentology is necessary. The monograph by Blatt et al. (1972) provides a comprehensive discussion of the origin and character of sandstone.

Nonindurated sands have porosities in the range 30–50%. Sandstones, however, commonly have lower porosities because of compaction and because of cementing material between the grains. In extreme cases porosities are less than 1% and hydraulic conductivities approach those of unfractured siltstone and shale (i.e., less than about 10^{-10} m/s). The most common cementing materials are quartz, calcite, and clay minerals. These minerals form as a result of precipitation or mineral alteration during groundwater circulation through the sand. Compaction is important at great depth, where temperatures and pressures are high. Studies by Chilingar (1963), Maxwell (1964), and Atwater (1966) show that the porosity of sandstone decreases systematically with depth. In Louisiana petroleum reservoirs, Atwater found that the decrease averages about 1.3% for every 300-m increase in depth of burial. Chilingar (1963) showed that when sand and sandstone are grouped according to grain-size categories, there are well-defined trends of increasing permeability with increasing porosity (Figure 4.5). An increase in porosity of several percent corresponds to a large increase in permeability.

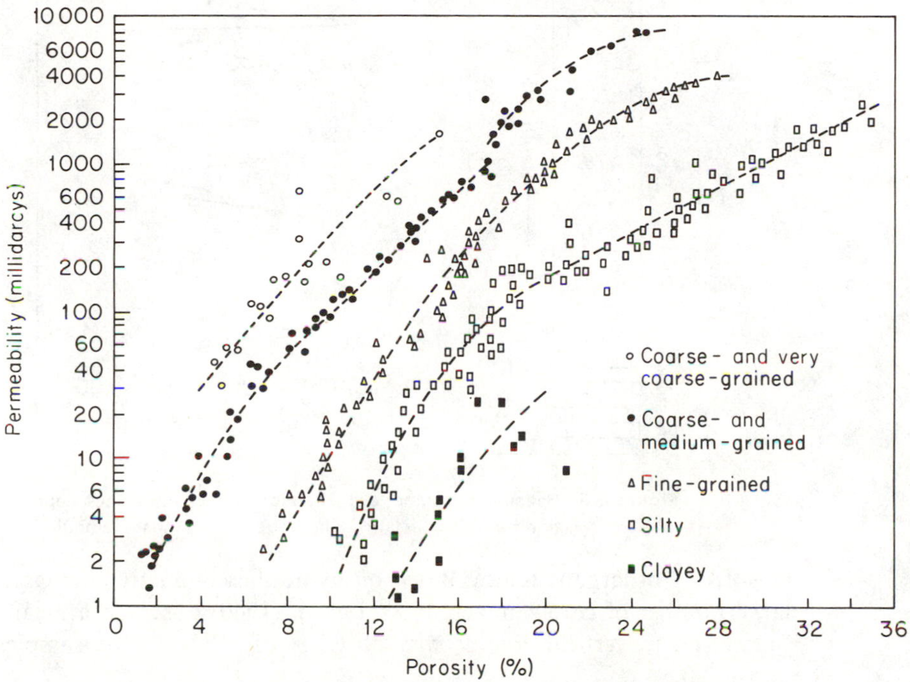

Figure 4.5 Relationship between porosity and permeability for sandstone in various grain-size categories (after Chilingar, 1963).

Permeameter tests on core samples from sandstone strata indicate that the conductivity can vary locally by a factor of as much as 10–100 in zones that appear, on the basis of visual inspection, to be relatively homogeneous. Figure 4.6 is a schematic illustration of a vertical hydraulic conductivity profile through a thick relatively homogeneous sandstone. Conductivity variations reflect minor changes in the depositional conditions that existed as the sand was deposited.

Davis (1969) suggests that the presence of small-scale stratification in sandstone enables the permeability of very large samples to be considered to be uniformly anisotropic. He indicates that the gross effect of permeability stratification is that the effective vertical permeability of large masses of sandstone can be low even in zones where the horizontal permeability is quite high. Davis states that knowledge concerning small-scale anisotropy of sandstone is rather incomplete, but is nevertheless much better founded than our understanding of the gross

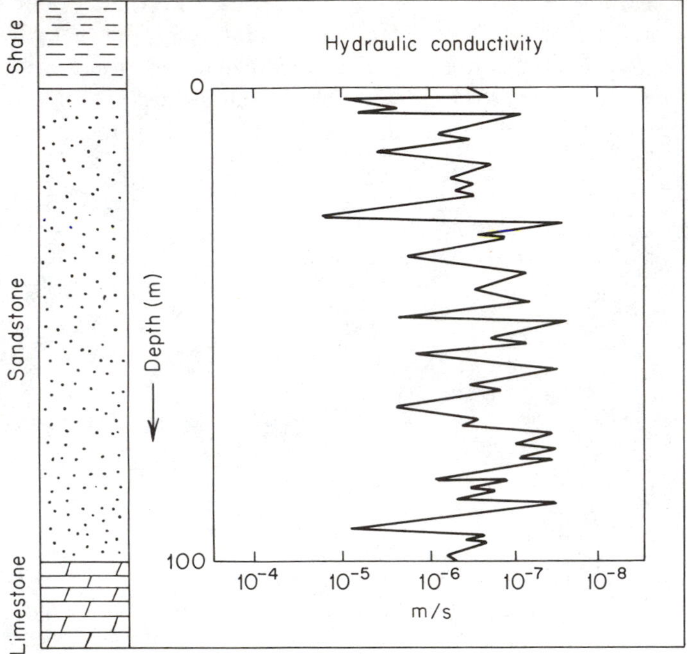

Figure 4.6 Schematic diagram of hydraulic conductivity versus depth relation for a thick, relatively homogeneous sandstone aquifer.

anisotropy of large volumes. Based on hydraulic conductivity measurements of a large number of core samples, Piersol et al. (1940) observed a mean ratio of the horizontal to vertical conductivity of 1.5. Only 12% of the samples had ratios above 3.0.

As sands become more cemented and compacted (i.e., more lithified) the contribution of fractures to the bulk permeability of the material increases. The tendency of large permeability values to occur in the horizontal direction is replaced by a preference for higher fracture permeability in the vertical direction. The nature of the anistropy in the fractured medium can reflect a complex geological history involving many stress cycles.

Carbonate Rock

Carbonate rocks, in the form of limestone and dolomite, consist mostly of the minerals calcite and dolomite, with very minor amounts of clay. Some authors refer to dolomitic rock as dolostone. In this text, dolomite is used to denote both the mineral and the rock. Nearly all dolomite is secondary in origin, formed by geochemical alteration of calcite. This mineralogical transformation causes an increase in porosity and permeability because the crystal lattice of dolomite occupies about 13% less space than that of calcite. Geologically young carbonate rocks commonly have porosities that range from 20% for coarse, blocky limestone to more than 50% for poorly indurated chalk (Davis, 1969). With increasing depth of burial, the matrix of soft carbonate minerals is normally compressed and recrystallized into a more dense, less porous rock mass. The primary permeability of old unfractured limestone and dolomite is commonly less than 10^{-7} m/s at near-surface temperature. Carbonate rocks with primary permeability of this magnitude can be important in the production of petroleum but are not significant sources of groundwater supply.

Many carbonate strata have appreciable secondary permeability as a result of fractures or openings along bedding planes. Secondary openings in carbonate rock caused by changes in the stress conditions may be enlarged as a result of calcite or dolomite dissolution by circulating groundwater. For the water to cause enlargement of the permeability network, it must be undersaturated with respect to these minerals. The origin of solution openings in carbonate rock is described in Chapter 11.

Observations in quarries and other excavations in flat-lying carbonate rocks indicate that solution openings along vertical joints generally are widely spaced. Openings along bedding planes are more important from the point of view of water yield from wells (Walker, 1956; Johnston, 1962). In nearly horizontal carbonate rocks with regular vertical fractures and horizontal bedding planes, there is usually a much higher probability of wells encountering horizontal openings than vertical fractures. This is illustrated in Figure 4.7. In fractured carbonate rocks, successful and unsuccessful wells can exist in close proximity, depending on the frequency of encounter of fractures by the well bore. Seasonally, the water

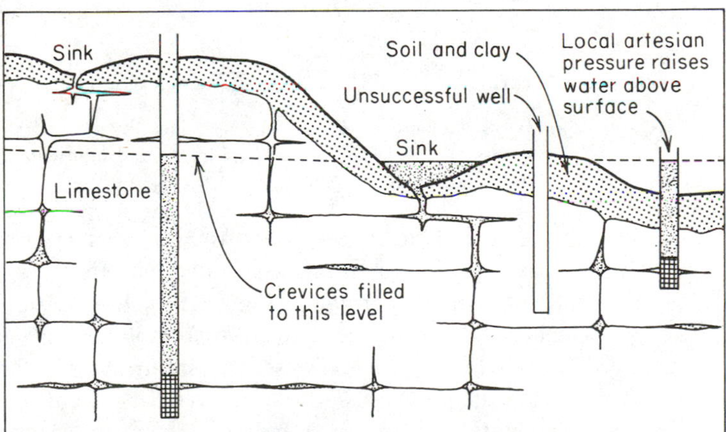

Figure 4.7 Schematic illustration of the occurrence of groundwater in carbonate rock in which secondary permeability occurs along enlarged fractures and bedding plane openings (after Walker, 1956; Davis and De Wiest, 1966).

levels in shallow wells can vary greatly because the bulk fracture porosity is generally a few percent or less.

In some carbonate rocks lineations of concentrated vertical fractures provide zones of high permeability. Figure 4.8 illustrates a situation where the fracture intersections and lineaments are reflected in the morphology of the land surface. Zones in which fractures are concentrated are the zones of most rapid groundwater flow. Dissolution may cause the permeability of such zones to increase. Intensive studies of lineaments in carbonate rock by Parizek and coworkers have shown that the probability of obtaining successful wells is greatly enhanced if drilling sites are located along lineaments or at their intersections (Lattman and Parizek, 1964; Parizek and Drew, 1966). In some areas, however, excessive thicknesses of overburden prevent recognition of bedrock lineaments, and the search for favorable drill sites in this manner is not feasible.

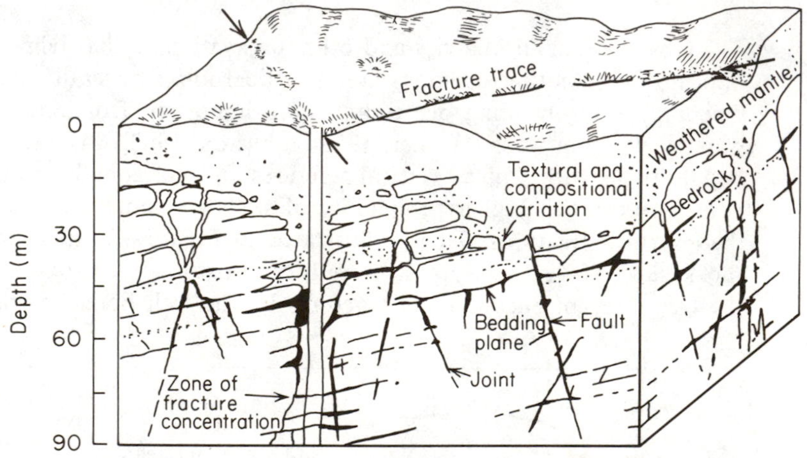

Figure 4.8 Occurrence of permeability zones in fractured carbonate rock. Highest well yields occur in fracture intersection zones (after Lattman and Parizek, 1964).

In areas of folded carbonate rocks, the zones of fracture concentration and solution enlargement are commonly associated with the crest of anticlines and to a lesser extent with synclinal troughs (Figure 4.9). In situations where rapid direct recharge can occur, fracture enlargement by dissolution has great influence. In the situation illustrated in Figure 4.9, water that infiltrates into the fractured carbonate rock beneath the alluvium will cause solution enlargement if the alluvium is devoid of carbonate minerals. If the alluvium has a significant carbonate-mineral content, groundwater would normally become saturated with respect to calcite and dolomite prior to entry into the fracture zones in the carbonate rock. In fractured carbonate rock in which solution channeling has been active in the geologic past,

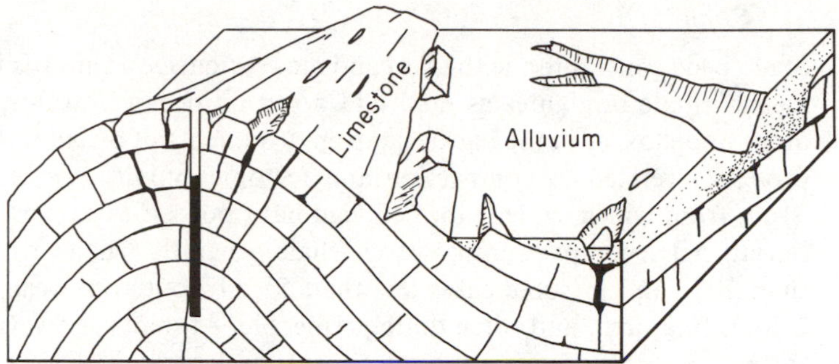

Figure 4.9 Occurrence of high-permeability zone in solution-enlarged fractures along the exposed crest of an anticline in carbonate rock (after Davis and De Wiest, 1966).

caverns or large tunnels can form, causing local permeability to be almost infinite compared to other parts of the same formation.

Coal

Beds of coal are common occurrences within sequences of sedimentary rocks formed in floodplain or deltaic environments. In a large part of the interior of North America, particularly in parts of North Dakota, Montana, Wyoming, Saskatchewan, and Alberta, beds of soft lignite coal form significant aquifers. The coal strata, which are of Tertiary or Cretaceous age, are generally less than 10–20 m thick, and many are only a meter or two thick. These aquifers are a common source of water supply for farms and small towns in this region.

Despite their importance, little is known about the hydrogeologic properties of coal aquifers. Investigations of the hydraulic conductivity of shallow lignite coal strata by Van Voast and Hedges (1975) and Moran et al. (1976) indicate values generally in the range 10^{-6}–10^{-4} m/s, with decreasing values at depths greater than 50–100 m. Below about 100 m the coal strata are rarely capable of supplying water at rates adequate for water supply. The bulk hydraulic conductivity of coal seams can be attributed to joints and to openings along bedding planes. The bulk fracture porosity is generally a small fraction of 1%.

The hydrogeologic role of coal in the Great Plains region has recently become a focus of interest as a result of the rapid increase in strip mining in this region. The near-surface coal aquifers are being drained as mining proceeds in some areas. Deeper coal seams may serve as alternative water supplies. Most of the coal layers are overlain and underlain by deposits of silt or clay that act as regional aquitards. Less commonly, the coal seams occur above or below sandstone of floodplain origin. Where coal and sandstone occur together, they often act as a single aquifer system.

Shale

Shale beds constitute the thickest and most extensive aquitards in most sedimentary basins. Shale originates as mud laid down on ocean bottoms, in the gentle-water areas of deltas, or in the backswamp environments of broad floodplains. Diagenetic processes related to compaction and tectonic activity convert the clay to shale. Mud, from which shale is formed, can have porosities as high as 70–80% prior to burial. After compaction, however, shale generally has a primary porosity of less than 20% and in some cases less than 5%. In outcrop areas, shale is commonly brittle, fractured, and often quite permeable. At depth, however, shale is generally softer, fractures are much less frequent, and permeability is generally very low. Some shale beds are quite plastic and fractures are insignificant.

Values of the hydraulic conductivity of intact samples of shale tested in the laboratory (Peterson, 1954; Young et al., 1964; Davis, 1969; Moran et al., 1976) are rarely larger than 10^{-9} m/s and are commonly in the range 10^{-12}–10^{-10} m/s. It is evident from the Darcy relation that even under strong hydraulic gradients, groundwater in unfractured shale cannot move at rates greater than a few centimeters per century. These rates are hardly significant on a human time scale, but on a geological time scale the flow of groundwater through intact shale can be a significant component in the water budget of regional aquifers confined by shale. Within a few hundreds of meters of ground surface, fractures in shale can impart a significant component of secondary porosity and permeability. Even in situations where hairline fractures exist in relatively wide spacing, the very small secondary porosity they create (perhaps as low as 10^{-4}–10^{-5}) can produce secondary permeability of magnitudes that exceed the primary permeability.

4.6 Igneous and Metamorphic Rocks

Solid samples of unfractured metamorphic rock and plutonic igneous rock have porosities that are rarely larger than 2%. The intercrystalline voids that make up the porosity are minute and many are not interconnected. Because of the small pore sizes and low degree of pore interconnectivity, the primary permeabilities of these rocks are extremely small. Measurements on intact specimens of metamorphic rocks (metasediments) from the Marquette Mining district in Michigan indicate primary permeability values in the range of 0.00019 millidarcy (10^{-11}–10^{-13} m/s) expressed as hydraulic conductivity at room temperature for quartzite, mica schist, chert, slate, and graywacke (Stuart et al., 1954). Measurements of the permeability of granite in boreholes in which fractures are absent generally yield values on the order of 10^{-3} millidarcy (10^{-11} m/s). Permeabilities of this magnitude indicate that these rocks are impermeable within the context of most groundwater problems.

In terrain composed of plutonic igneous rocks and crystalline metamorphic rocks, appreciable fracture permeability generally occurs within tens of meters and in some cases within a few hundred meters of ground surface. The fractures are caused by changes in the stress conditions that have occurred during various episodes in the geologic history of the rocks. The widths of fracture openings are generally less than 1 mm. Since the discharge of groundwater is proportional to the fracture width raised to a power of about 3 [Eq. (2.86)], the difference in permeability between rock masses with fracture widths of tenths of a millimeter and those with fracture widths on the order of millimeters or more is enormous.

Tolman (1937) and Davis (1969) draw attention to the fact that in some cases dissolution of siliceous rocks may cause significant increases in the widths of fracture openings. Davis presented a hypothetical example whereby recharge water passing through the upper 10 m of a quartzite removes sufficient silica to widen fractures by 0.38 mm in 10^5 years. This widening could be very significant in terms of fluid flow. Davis indicates that several factors reduce or negate the tendency toward rapid opening by solution of cracks in crystalline rocks. As groundwater passes through overburden prior to entering the fractured rock, it normally acquires appreciable dissolved silica. It is therefore relatively unaggressive with respect to silicate minerals along the fracture faces. Unlike most carbonate rocks, silica-rich rocks have an insoluble residue in the form of iron and aluminum oxides that will tend to clog the small fractures after weathering is initiated.

One of the most characteristic features of the permeability of crystalline rocks is the general trend of permeability decrease with depth. The results (Figure 4.10)

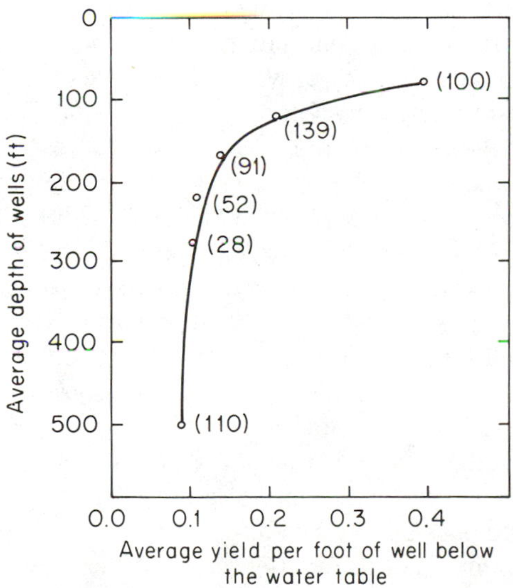

Figure 4.10 Decrease in well yields (gpm/ft of well below the water table) with depth in crystalline rocks of the Statesville area, North Carolina. Numbers near points indicate the number of wells used to obtain the average values that define the curve (after Legrand, 1954; Davis and De Wiest, 1966).

of a study of a crystalline rock area (granite, gabbro, Gneiss, and schist) in North Carolina by LeGrand (1954) are a quantitative expression of the trend that well drillers observe in a more qualitative manner in many crystalline rock regions. Quantitative relations between depth and well yield have also been established by Summers (1972) for a Precambrian-rock area in Wisconsin. Fractured crystalline rocks are less permeable at greater depth because stress variations that cause fractures are larger and, over geologic time, occur more frequently near the ground surface. Fractures tend to close at depth because of vertical and lateral stresses imposed by overburden loads and "locked-in" horizontal stresses of tectonic origin. Rocks maintain much of their brittle character to depths of several kilometers. Fracture permeability can therefore exist to great depth. Striking evidence of this comes from tunnels and from mines at depths of 1 km and more where water flows actively into shafts and adits. In crystalline rock, dry mines are the exception rather than the rule.

In granite, the occurrence of near-horizontal fractures parallel to the ground surface has been attributed by LeGrand (1949) to the removal of overburden load caused by erosion. In an area in Georgia studied by LeGrand, these sheet fractures are an important source of water supply from shallow depths. With depth, fractures of this type decrease rapidly in frequency and aperture width. They are probably unimportant contributors to permeability at depths greater than about 100 m (Davis and De Wiest, 1966).

Because many fractures owe their origin to near-surface stresses related directly or indirectly to topographic conditions, it is not surprising that in many crystalline-rock areas the frequency of wells and well yields is related to the topography. The results of the study by LeGrand (1954) can again be used as a quantitative illustration of well-yield relations, this time with respect to topography. Figure 4.11 indicates that well yields in the crystalline rocks in the North Carolina study area are highest in valleys and broad ravines and lowest at or near the crests of hills. Yields in flat uplands and beneath slopes are between these extremes. In many places, valleys and ravines develop along fault zones. The tendency for fault zones to have greater permeability is the primary factor in the well-yield relationship.

Volcanic rocks form as a result of solidification of magma at or near the ground surface. In a hydrogeologic sense, these rocks generally differ from most other crystalline rocks in that they have primary features that cause permeability within the otherwise-solid rock mass. Davis (1969), in an excellent description of the permeability and porosity of volcanic rocks, notes that these features are related to the history of the rocks.

When magma extrudes to ground surface and flows out as lava, the rocks that form on cooling are generally very permeable. At the surface, rapid cooling and escape of gases causes cooling joints and bubblelike pore spaces. While the lava is in motion, a crust forms on the upper surface as cooling takes place. Flow of the lava beneath the crust causes it to become fractured, producing a blocky mass of rock that is commonly pulled under the leading edge of the lava flow. The final

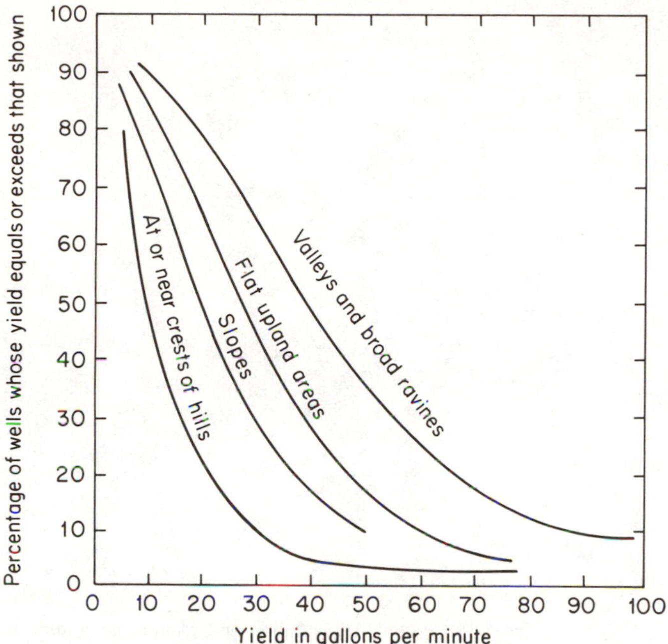

Figure 4.11 Cumulative frequency distribution of well yields with respect to topographic position, Statesville area, North Carolina (after Legrand, 1954; Davis and De Wiest, 1966).

result is a solid mass which in many places has coarse rubble zones above and below more dense rock (Davis, 1969). Gravels deposited by streams on lava landscapes are later covered by new flows. The blocky rock masses and associated gravel interbeds produce a bulk permeability that is very high in most young basalts. Other causes of high permeability in young basalts are gas vents, lava tubes, and tree molds. Alteration by deep burial or by the influx of cementing fluids during geologic time causes the permeability to decrease.

On a large scale the permeability of basalt is very anisotropic. The centers of lava flows are generally impervious. Buried soils that produce high permeability develop in the top of cooled lava flows. Stream deposits occur between the flows. The zones of blocky rubble generally run parallel to the flow trend. The direction of highest permeability is therefore generally parallel to the flows. Davis indicates that within the flow the permeability is normally greatest in the direction of the steepest original dip of the flows. This is illustrated schematically in Figure 4.12, which indicates the orientation and relative magnitude of the overall permeability of young basaltic rocks. In some situations, however, the orientation of the major axes may not be elliptic.

One of the largest accumulations of basaltic rock in the world is located in the northwestern part of the United States in the region known as the Columbia

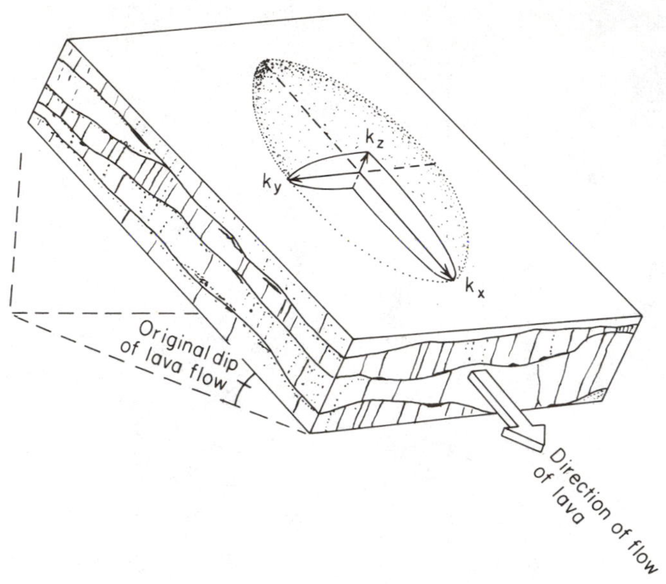

Figure 4.12 Probable orientation and relative magnitude of the bulk permeability of young basaltic rocks (after Davis, 1969).

River Plateau. During Miocene and Pliocene time, enormous volumes of magma welled up through fissures and spread out in broad sheets over areas estimated to be as large as several million square kilometers. Consequently, much of the magma had a low gas content. The basalt in this region is generally quite dense, with only limited zones of vesicular basalt. Extensive river-deposited sediments occur between many of the basalt flows. The average total thickness of the basalt sequence over the Columbia River Plateau is about 550 m.

Studies in boreholes in the lower part of the basalt sequence at a site in the southeastern part of the State of Washington yielded hydraulic conductivity, transmissivity, and porosity data (Atlantic-Richfield Hanford Company, 1976), summarized in Table 4.1. The river-deposited interbeds and the zones of basalts

Table 4.1 Range of Hydrologic Properties of Lower Yakima Basalt Flows and Interbeds

	Hydraulic conductivity (m/s)	Porosity (%)
Dense basalt	10^{-11}–10^{-8}	0.1–1
Vesicular basalt	10^{-9}–10^{-8}	5
Fractured, weathered, or brecciated basalt	10^{-9}–10^{-5}	10
Interbeds	10^{-8}–10^{-5}	20

that are vesicular, fractured, weathered, or brecciated are aquifers in which predominantly horizontal regional flow occurs. The zones of dense basalt have lower hydraulic conductivity and effective porosity but are, nevertheless, generally capable of transmitting considerable water. Some zones of dense, unfractured basalt have very low hydraulic conductivity and probably act as regional aquitards.

4.7 Permafrost

Within the Arctic circle, perennially frozen ground known as *permafrost* is present almost everywhere. In the most northerly regions of Canada, Alaska, Greenland, Scandinavia, and the USSR, permafrost is continuous, but in much of the inhabited or resource frontier northland, the permafrost zones are discontinuous. Except in the high Andes and in Antarctica, permafrost is absent in the southern hemisphere.

Contrary to what one might intuitively expect, permafrost does not necessarily form at all locations where the ground temperature declines to 0°C. Temperatures significantly below 0°C are often required to initiate the change of pore water into ice (Anderson and Morgenstern, 1973; Banin and Anderson, 1974). The occurrence and magnitude of the depression in the initial freezing point depend on a number of factors, including the fluid pressure, the salt content of the pore water, the grain-size distribution of the soil, the soil mineralogy, and the soil structure (van Everdingen, 1976). The relations between liquid water content in the pore water and temperatures of the bulk medium are illustrated in Figure 4.13. When the soil is partially frozen, the material contains both liquid water and ice. The term "permafrost" should be reserved for material in which water persists in the frozen or partially frozen state throughout the year. The 0°C temperature condition indicates little about the exact physical state of the pore water.

The hydrogeologic importance of permafrost lies in the large differences in hydraulic conductivity that exist for most geologic materials between their frozen and unfrozen states. Figure 4.14(a) shows the relation between the content of unfrozen pore water and temperature for several soils, and Figure 4.14(b) shows the effect of this relationship on the hydraulic conductivity. The content of unfrozen water decreases and the pore ice content increases when the bulk temperature of the material is lowered from 0°C toward −1°C. The hydraulic conductivities decline by several orders of magnitude as the temperature declines a few tenths of a degree below 0°C. Fine sand, for example, which could be an aquifer in an unfrozen state under the appropriate stratigraphic conditions, becomes a low-permeability aquitard at a temperature slightly below 0°C. Silt that could have leaky aquitard characteristics in an unfrozen state becomes an impervious aquitard when fully frozen.

The importance of the permafrost configuration on the distribution of aquifers can be shown with reference to cross sections across two alluvial valleys in the

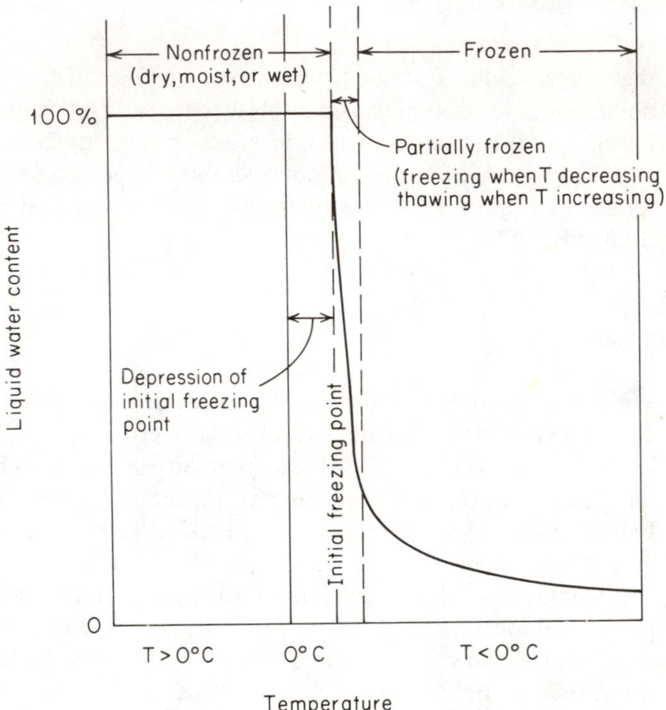

Figure 4.13 Hypothetical graph of liquid water content versus ground temperature showing conditions for the nonfrozen, partially frozen, and frozen soil states (after van Everdingen, 1976).

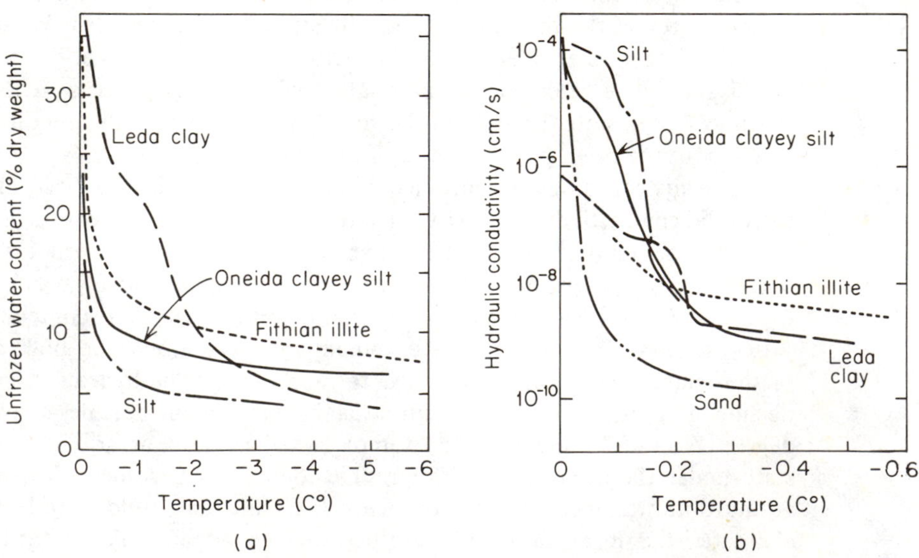

Figure 4.14 Effect of temperature on the hydraulic properties of several saturated soils. (a) Unfrozen water content versus temperature; (b) hydraulic conductivity versus temperature (after Burt and Williams, 1976).

Fairbanks area of northeastern Alaska (Figure 4.15). The gravel and sand deposit beneath the silt aquitard in the Happy Creek valley is an aquifer that yields abundant water. The water is recharged through unfrozen zones on the upper slopes and in the fluvial deposits in the upper reaches of the creeks. Beneath Dome Creek, on the other hand, the base of the permafrost extends into the bedrock beneath the sand and gravel. As a consequence, no water can be obtained from these coarse-grained materials. Because of the confining effect of the permafrost, groundwater in the bedrock zones below the base of the permfrost exhibits hydraulic heads that rise above ground surface, and flowing wells are encountered at depth.

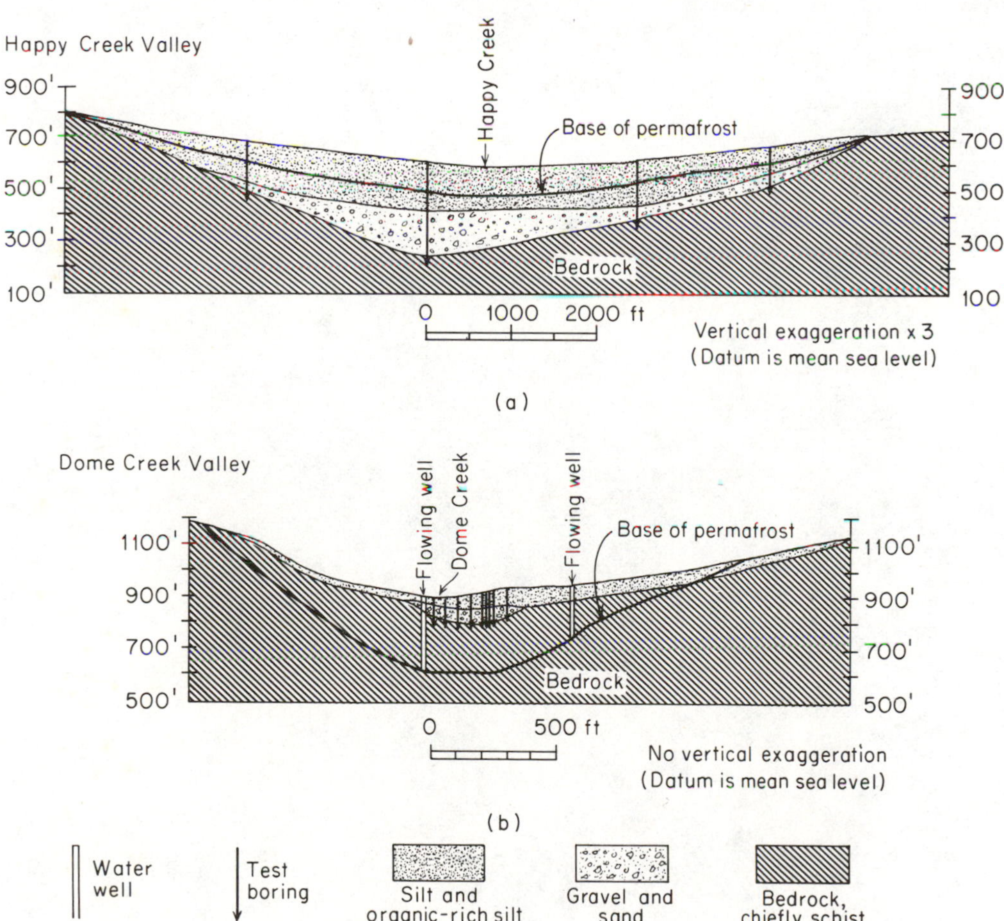

Figure 4.15 Hydrogeologic sections across two valleys in the Fairbanks mining district, Alaska. (a) Occurrence of sand and gravel aquifer below the base of permafrost; (b) sand and gravel deposit that is frozen in the permafrost zone. Flowing wells occur below the base of the permafrost (after Williams, 1970).

Suggested Readings

Brown, I. C., ed. 1967. Groundwater in Canada. *Geol. Surv. Can., Econ. Geol. Rept. No. 24*, pp. 65–171.

Davis, S. N. 1969. Porosity and permeability of natural materials. *Flow Through Porous Media*, ed. R. J. M. De Wiest. Academic Press, New York, pp. 53–89.

Davis, S. N., and R. J. M. De Wiest. 1966. *Hydrogeology*. John Wiley & Sons, New York, pp. 318–417.

McGuinness, C. L. 1963. The role of groundwater in the National Water Situation. *U.S. Geol. Surv. Water-Supply Paper 1800*.

5

CHAPTER

Flow Nets

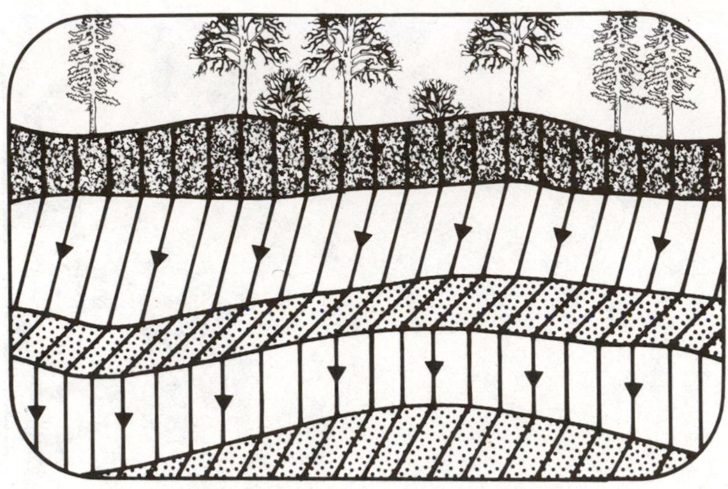

5.1 Flow Nets by Graphical Construction

We have seen in Chapter 2 that a groundwater flow system can be represented by a three-dimensional set of equipotential surfaces and a corresponding set of orthogonal flowlines. If a meaningful two-dimensional cross section can be chosen through the three-dimensional system, the set of equipotential lines and flowlines so exposed constitutes a *flow net*. The construction of flow nets is one of the most powerful analytical tools for the analysis of groundwater flow.

In Section 2.11 and Figure 2.25, we saw that a flow net can be viewed as the solution of a two-dimensional, steady-state, boundary-value problem. The solution requires knowledge of the region of flow, the boundary conditions on the boundaries of the region, and the spatial distribution of hydraulic conductivity within the region. In Appendix III, an analytical mathematical method of solution is presented. In this section, we will learn that flow nets can also be constructed graphically, without recourse to the sophisticated mathematics.

Homogeneous, Isotropic Systems

Let us first consider a region of flow that is homogeneous, isotropic, and fully saturated. For steady-state flow in such a region, three types of boundaries can exist: (1) impermeable boundaries, (2) constant-head boundaries, and (3) water-table boundaries. First, let us consider flow in the vicinity of an impermeable boundary [Figure 5.1(a)]. Since there can be no flow across the boundary, the flowlines adjacent to the boundary must be parallel to it, and the equipotential lines must meet the boundary at right angles. By invoking Darcy's law and setting the specific discharge across the boundary equal to zero, we are led to the mathematical statement of the boundary condition. For boundaries that parallel the axes in an xz plane:

$$\frac{\partial h}{\partial x} = 0 \quad \text{or} \quad \frac{\partial h}{\partial z} = 0 \tag{5.1}$$

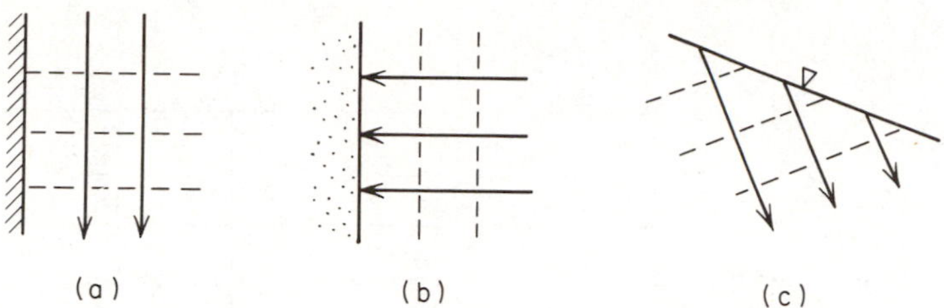

Figure 5.1 Groundwater flow in the vicinity of (a) an impermeable boundary, (b) a constant-head boundary, and (c) a water-table boundary.

In effect, any flowline in a flow net constitutes an imaginary impermeable boundary, in that there is no flow *across* a flowline. In flow-net construction, it is often desirable to reduce the size of the region of flow by considering only those portions of the region on one side or the other of some line of symmetry. If it is clear that the line of symmetry is also a flowline, the boundary condition to be imposed on the symmetry boundary is that of Eq. (5.1).

A boundary on which the hydraulic head is constant [Figure 5.1(b)] is an equipotential line. Flowlines must meet the boundary at right angles, and adjacent equipotential lines must be parallel to the boundary. The mathematical condition is

$$h = c \tag{5.2}$$

On the water table, the pressure head, ψ, equals zero, and the simple head relationship, $h = \psi + z$, yields

$$h = z \tag{5.3}$$

for the boundary condition. As shown in Figure 5.1(c), for a recharge case the water table is neither a flowline nor an equipotential line. It is simply a line of variable but known h.

If we know the hydraulic conductivity K for the material in a homogenous, isotropic region of flow, it is possible to calculate the discharge through the system from a flow net. Figure 5.2 is a completed flow net for the simple case first presented in Figure 2.25(a). The area between two adjacent flowlines is known as a *streamtube* or *flowtube*. If the flowlines are equally spaced, the discharge through each streamtube is the same. Consider the flow through the region $ABCD$ in Figure 5.2. If the distances AB and BC are ds and dm, respectively, and if the hydraulic-head drop between AD and BC is dh, the discharge across this region through a cross-sectional area of unit depth perpendicular to the page is

$$dQ = K\frac{dh}{ds}\,dm \tag{5.4}$$

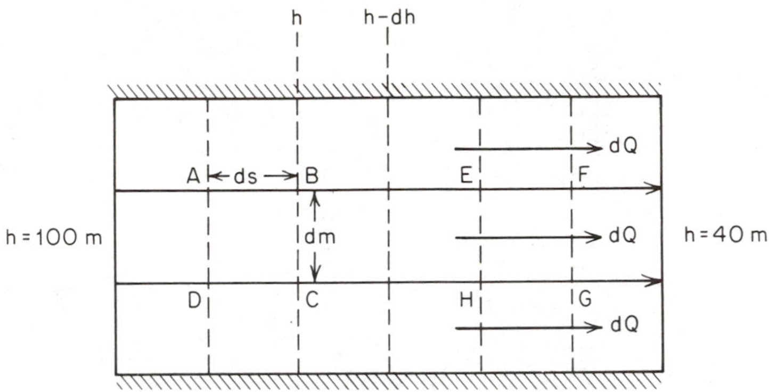

Figure 5.2 Quantitative flow net for a very simple flow system.

Under steady-state conditions, the discharge across any plane of unit depth (say, at AD, EH, or FG) within the streamtube must also be dQ. In other words, the discharge through any part of a streamtube can be calculated from a consideration of the flow in just one element of it.

If we arbitrarily decide to construct the flow net in squares, with $ds = dm$, then Eq. (5.4) becomes

$$dQ = K\,dh \tag{5.5}$$

For a system with m streamtubes, the total discharge is

$$Q = mK\,dh \tag{5.6}$$

If the total head drop across the region of flow is H and there are n divisions of head in the flow net ($H = n\,dh$), then

$$Q = \frac{mKH}{n} \tag{5.7}$$

For Figure 5.2, $m = 3$, $n = 6$, $H = 60$ m, and from Eq. (5.7), $Q = 30K$. For $K = 10^{-4}$ m/s, $Q = 3 \times 10^{-3}$ m³/s (per meter of section perpendicular to the flow net).

Equation (5.7) must be used with care. It is applicable only to simple flow systems with one recharge boundary and one discharge boundary. For more complicated systems, it is best to simply calculate dQ for one streamtube and multiply by the number of streamtubes to get Q.

Figure 5.3 is a flow net that displays the seepage beneath a dam through a foundation rock bounded at depth by an impermeable boundary. It can be used to make three additional points about flow-net construction.

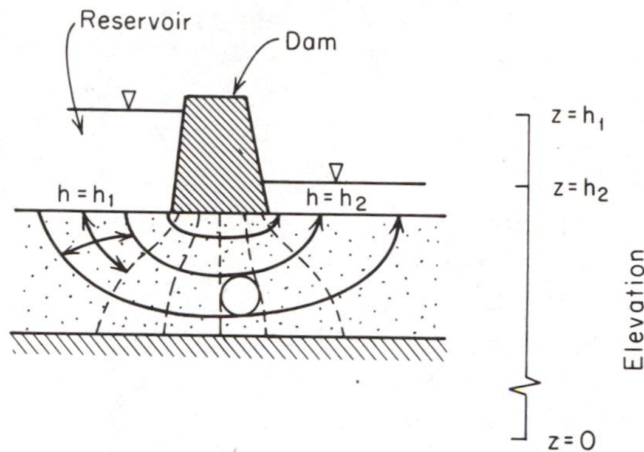

Figure 5.3 Seepage beneath a dam through homogeneous, isotropic foundation rocks.

1. The "squares" in all but the simplest flow nets are actually "curvilinear" squares; that is, they have equal central dimensions; or viewed another way, they enclose a circle that is tangent to all four bounding lines.
2. It is not necessary that flow nets have finite boundaries on all sides; regions of flow that extend to infinity in one or more directions, like the horizontally infinite layer in Figure 5.3, are tractable.
3. A flow net can be constructed with a "partial" streamtube on the edge.

For the flow net shown in Figure 5.3, $m = 3\frac{1}{2}$. If $H = 100$ m and $K = 10^{-4}$ m/s, then, since $n = 6$, we have $Q = 5.8 \times 10^{-3}$ m^3/s (per meter section perpendicular to the flow net).

In homogeneous, isotropic media, the distribution of hydraulic head depends only on the configuration of the boundary conditions. The qualitative nature of the flow net is independent of the hydraulic conductivity of the media. The hydraulic conductivity comes into play only when quantitative discharge calculations are made. It is also worth noting that flow nets are dimensionless. The flow nets of Figures 5.2 and 5.3 are equally valid whether the regions of flow are considered to be a few meters square or thousands of meters square.

The sketching of flow nets is something of an art. One usually pursues the task on a trial-and-error basis. Some hydrologists become extremely talented at arriving at acceptable flow nets quickly. For others, it is a source of continuing frustration. For a flow net in homogeneous, isotropic media, the rules of graphical construction are deceptively simple. We can summarize them as follows: (1) flowlines and equipotential lines must intersect at right angles throughout the system; (2) equipotential lines must meet impermeable boundaries at right angles; (3) equipotential lines must parallel constant-head boundaries; and (4) if the flow net is drawn such that squares are created in one portion of the field, then, with the possible exception of partial flow tubes at the edge, squares must exist throughout the entire field.

Heterogeneous Systems and the Tangent Law

When groundwater flowlines cross a geologic boundary between two formations with different values of hydraulic conductivity, they refract, much as light does when it passes from one medium to another. However, in contradistinction to Snell's law, which is a sine law, groundwater refraction obeys a tangent law.

Consider the streamtube shown in Figure 5.4. Flow proceeds from a medium

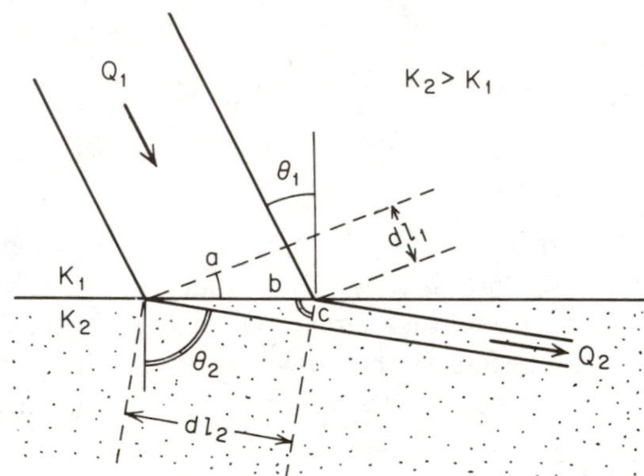

Figure 5.4 Refraction of flowlines at a geologic boundary.

with hydraulic conductivity K_1 to a medium with hydraulic conductivity K_2, where $K_2 > K_1$. The streamtube has a unit depth perpendicular to the page, and the angles and distances are as indicated on the figure. For steady flow, the inflow Q_1 must equal the outflow Q_2; or, from Darcy's law,

$$K_1 a \frac{dh_1}{dl_1} = K_2 c \frac{dh_2}{dl_2} \tag{5.8}$$

where dh_1 is the head drop across the distance dl_1 and dh_2 is the head drop across the distance dl_2. In that dl_1 and dl_2 bound the same two equipotential lines, it is clear that $dh_1 = dh_2$; and from geometrical considerations, $a = b \cos \theta_1$ and $c = b \cos \theta_2$. Noting that $b/dl_1 = 1/\sin \theta_1$ and $b/dl_2 = 1/\sin \theta_2$, Eq. (5.8) becomes

$$K_1 \frac{\cos \theta_1}{\sin \theta_1} = K_2 \frac{\cos \theta_2}{\sin \theta_2} \tag{5.9}$$

or

$$\frac{K_1}{K_2} = \frac{\tan \theta_1}{\tan \theta_2} \tag{5.10}$$

Equation (5.10) constitutes the tangent law for the refraction of groundwater flowlines at a geologic boundary in heterogeneous media. Knowing K_1, K_2, and θ_1, one can solve Eq. (5.10) for θ_2. Figure 5.5 shows the flowline refractions for two cases with $K_1/K_2 = 10$. Flowlines, as if they had a mind of their own, prefer to use high-permeability formations as conduits, and they try to traverse low-permeability formations by the shortest route. In aquifer-aquitard systems with permeability contrasts of 2 orders of magnitude or more, flowlines tend to become almost horizontal in the aquifers and almost vertical in the aquitards. When one considers the wide range in hydraulic conductivity values exhibited in Table 2.2, it is clear that contrasts of 2 orders of magnitude and more are not at all uncommon.

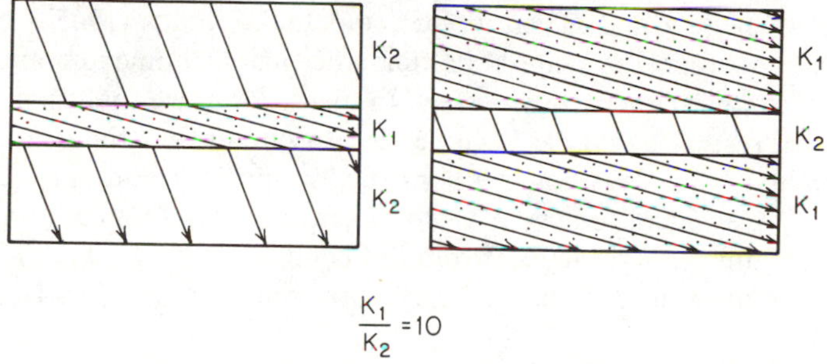

Figure 5.5 Refraction of flowlines in layered systems (after Hubbert, 1940).

If one attempts to draw the equipotential lines to complete the flow systems on the diagrams of Figure 5.5, it will soon become clear that it is not possible to construct squares in all formations. In heterogeneous systems, squares in one formation become rectangles in another.

We can summarize the rules for graphical flow net construction in heterogeneous, isotropic systems as follows: (1) flowlines and equipotential lines must intersect at right angles throughout the system; (2) equipotential lines must meet impermeable boundaries at right angles; (3) equipotential lines must parallel constant-head boundaries; (4) the tangent law must be satisfied at geologic boundaries; and (5) if the flow net is drawn such that squares are created in one portion of one formation, squares must exist throughout that formation and throughout all formations with the same hydraulic conductivity. Rectangles will be created in formations of different conductivity.

The last two rules make it extremely difficult to draw accurate quantitative flow nets in complex heterogeneous systems. However, qualitative flow nets, in which the orthogonality is preserved but no attempt is made to create squares, can be of great help in understanding a groundwater flow system. Figure 5.6 is a qualitatively sketched flow net for the dam seepage problem first introduced in Figure 5.3, but with a foundation rock that is now layered.

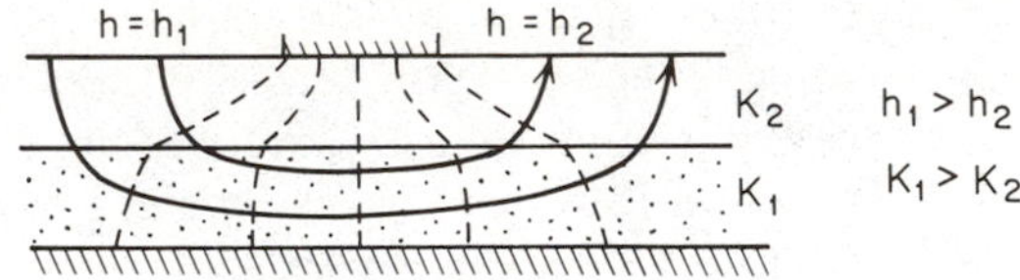

Figure 5.6 Seepage beneath a dam through heterogeneous, isotropic foundation rocks.

Anisotropic Systems and the Transformed Section

In homogeneous but anisotropic media, flow-net construction is complicated by the fact that flowlines and equipotential lines are not orthogonal. Maasland (1957), Bear and Dagan (1965), and Liakopoulos (1965b) provide discussions of the theoretical principles that underlie this phenomenon, and Bear (1972) presents an extensive theoretical review. In this section, we shall look primarily at the practical response that has been devised to circumvent the conditions of nonorthogonality. It involves flow-net construction in the *transformed section.*

Consider the flow in a two-dimensional region in a homogeneous, anisotropic medium with principal hydraulic conductivities K_x and K_z. The hydraulic-conductivity ellipse (Figure 5.7) will have semiaxes $\sqrt{K_x}$ and $\sqrt{K_z}$. Let us transform

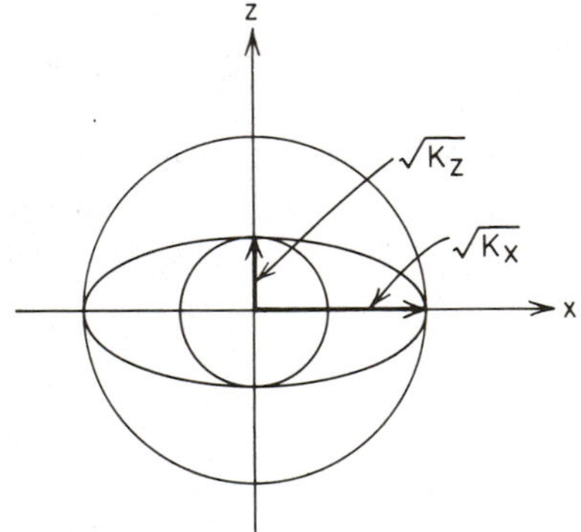

Figure 5.7 Hydraulic conductivity ellipse for an anisotropic medium with $K_x/K_z = 5$. The circles represent two possible isotropic transformations.

the scale of the region of flow such that coordinates in the transformed region with coordinate directions X and Z will be related to those in the original xz system by

$$X = x$$
$$Z = \frac{z\sqrt{K_x}}{\sqrt{K_z}} \qquad (5.11)$$

For $K_x > K_z$, this transformation will expand the vertical scale of the region of flow. It will also expand the hydraulic conductivity ellipse into a circle of radius $\sqrt{K_x}$ (the outer circle in Figure 5.7); and the fictitious, expanded region of flow will then act as if it were homogeneous with conductivity K_x.

The validity of this transformation can be defended on the basis of the steady-state equation of flow. In the original xz coordinate system, for an anisotropic media, we have, from Eq. (2.69),

$$\frac{\partial}{\partial x}\left(K_x \frac{\partial h}{\partial x}\right) + \frac{\partial}{\partial z}\left(K_z \frac{\partial h}{\partial z}\right) = 0 \tag{5.12}$$

Dividing through by K_x yields

$$\frac{\partial^2 h}{\partial x^2} + \frac{\partial}{\partial z}\left(\frac{K_z}{K_x}\frac{\partial h}{\partial z}\right) = 0 \tag{5.13}$$

For the transformed section, we have, from the second expression of Eq. (5.11),

$$\frac{\partial}{\partial z} = \frac{\sqrt{K_x}}{\sqrt{K_z}}\frac{\partial}{\partial Z} \tag{5.14}$$

Noting the first expression of Eq. (5.11), and applying the operation of Eq. (5.14) to the two differentiations of Eq. (5.13), yields

$$\frac{\partial^2 h}{\partial X^2} + \frac{\partial^2 h}{\partial Z^2} = 0 \tag{5.15}$$

which is the equation of flow for a homogeneous, isotropic medium in the transformed section.

An equally valid transformation could be accomplished by contracting the region in the x direction according to the relations

$$X = \frac{x\sqrt{K_z}}{\sqrt{K_x}} \qquad Z = z \tag{5.16}$$

In this case, the conductivity ellipse will be transformed into the small circle in Figure 5.7, and the fictitious, transformed medium will act as if it were homogeneous with hydraulic conductivity K_z.

With the concept of the transformed section in hand, the steps in the graphical construction of a flow net in a homogeneous, anisotropic medium become self-evident: (1) carry out a transformation of coordinates using either Eqs. (5.11) or Eqs. (5.16); (2) construct a flow net in the fictitious, transformed section, according to the rules for a homogeneous, isotropic media; and (3) invert the scaling ratio.

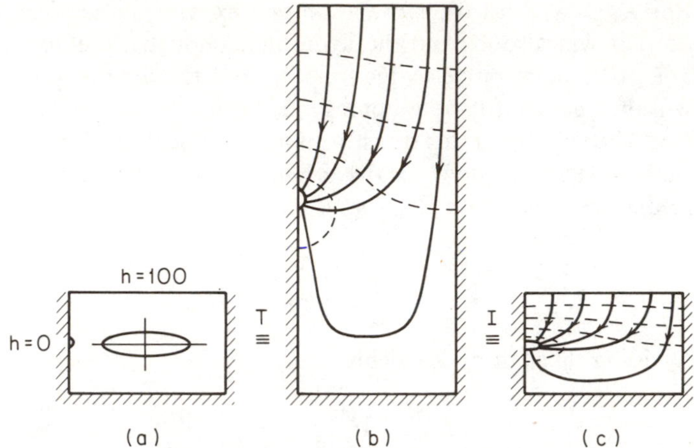

Figure 5.8 (a) Flow problem in a homogeneous anisotropic region with $\sqrt{K_x}/\sqrt{K_z} = 4$. (b) Flow net in the transformed isotropic section. (c) Flow net in the actual anisotropic section. *T*, transformation; *I*, inversion.

Figure 5.8 is an example of the technique. The boundary-value problem illustrated in Figure 5.8(a) is a vertical section that represents flow from a surface pond at $h = 100$ toward a drain at $h = 0$. The drain is considered to be one of many parallel drains set at a similar depth oriented perpendicular to the page. The vertical impermeable boundaries are "imaginary"; they are created by the symmetry of the overall flow system. The lower boundary is a real boundary; it represents the base of the surficial soil, which is underlain by a soil or rock formation with a conductivity several orders of magnitude lower. If the vertical axis is arbitrarily set with $z = 0$ at the drain and $z = 100$ at the surface, then from $h = \psi + z$, and the h values given, we have $\psi = 0$ at both boundaries. At the surface, this condition implies that the soil is just saturated. The "pond" is incipient; it has zero depth. At the drain, $\psi = 0$ implies free-flowing conditions. The soil in the flow field has an anisotropic conductivity of $K_x/K_z = 16$. The transformed section of Figure 5.8(b) therefore has a vertical expansion of $\sqrt{K_x}/\sqrt{K_z} = 4$. Figure 5.8(c) shows the result of the inverse transformation, wherein the homogeneous, isotropic flow net from the transformed section is brought back into the true-scale region of flow. Under the inversion, the hydraulic head at any point (X, Z) in Figure 5.8(b) becomes the hydraulic head at point (x, z) in Figure 5.8(c).

The size of the transformed section is obviously dependent on whether Eqs. (5.11) or Eqs. (5.16) are used for the transformation, but the shape of the region and the resulting flow net are the same in either case.

If discharge quantities or flow velocities are required, it is often easiest to make these calculations in the transformed section. The question then arises as to what hydraulic conductivity value ought to be used for such calculations. It is clear that it would be incorrect to use K_x for a vertically expanded section and K_z for a horizontally contracted one, as might be inferred from Figure 5.7, for this would produce two different sets of quantitative calculations for the two equivalent representations of the same problem. In fact, the correct value to use is

$$K' = \sqrt{K_x \cdot K_z} \tag{5.17}$$

The validity of Eq. (5.17) rests on the condition that flows in each of the two equivalent transformed representations of the flow region must be equal. The proof requires an application of Darcy's law to a single flowtube in each of the two transformations.

The influence of anisotropy on the nature of groundwater flow nets is illustrated in Figure 5.9 for the same boundary-value problem that was brought into play in Figure 5.8. The most important feature of the anisotropic flow nets [Figure 5.9(a) and 5.9(c)] is their lack of orthogonality. It seems to us that the transformation techniques introduced in this section provide an indirect but satisfying explanation of this phenomenon.

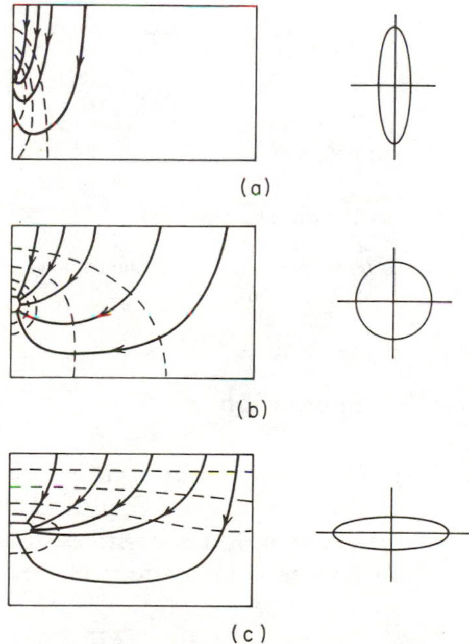

Figure 5.9 Flow nets for the flow problem of Figure 5.8(a) for $\sqrt{K_x}/\sqrt{K_z}$ = (a) 1/4, (b) 1, (c) 4 (after Maasland, 1957).

There are many situations where one may wish to construct a flow net on the basis of piezometric data from the field. If the geologic formations are known to be anisotropic, great care must be exercised in the inference of flow directions from the equipotential data. If the complete flow net is desired, a transformed section is in order, but if flow directions at specific points are all that is required, there is a graphical construction that can be useful. In Figure 5.10 the dashed line represents the directional trend of an equipotential line at some point of interest within an xz field. An *inverse* hydraulic-conductivity ellipse is then constructed about the point. This ellipse has principal semiaxes $1/\sqrt{K_x}$ and $1/\sqrt{K_z}$ (rather than $\sqrt{K_x}$ and $\sqrt{K_z}$, as in Figure 5.7). A line drawn in the direction of the hydraulic gradient intersects the ellipse at the point A. If a tangent is drawn to the ellipse at A, the direction of flow is perpendicular to this tangent line. As an example of the application of this construction, one might compare the results of Figure 5.10 with the flowline/equipotential line intersections in the right-central portion of Figure 5.9(c).

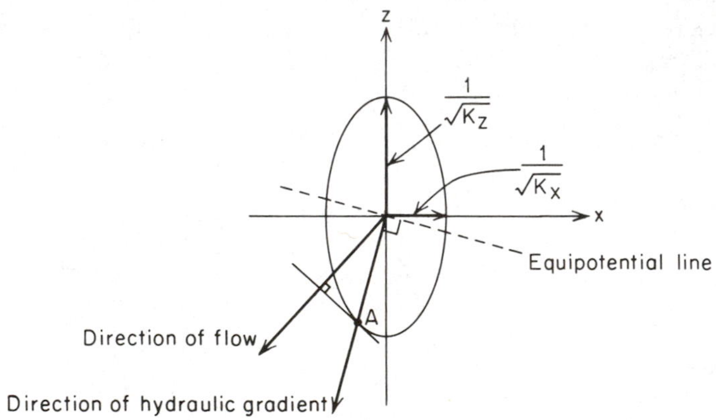

Figure 5.10 Determination of direction of flow in an anisotropic region with $K_x/K_z = 5$.

5.2 Flow Nets by Analog Simulation

For flow in a homogeneous, isotropic medium in an xz coordinate system, the equipotential lines in a flow net are a contoured reflection of the solution, $h(x, z)$, of the boundary-value problem that describes steady-state flow in the region. The construction of the flow net is an indirect solution to *Laplace's equation*:

$$\frac{\partial^2 h}{\partial x^2} + \frac{\partial^2 h}{\partial z^2} = 0 \tag{5.18}$$

This equation is one of the most commonly occurring partial differential equations in mathematical physics. Among the other physical phenomena that it describes are the flow of heat through solids and the flow of electrical current through conductive media. For the latter case, Laplace's equation takes the form

$$\frac{\partial^2 V}{\partial x^2} + \frac{\partial^2 V}{\partial z^2} = 0 \tag{5.19}$$

where V is the electrical potential or voltage.

The similarity of Eqs. (5.18) and (5.19) reveals a mathematical and physical analogy between electrical flow and groundwater flow. The two equations are both developed on the basis of a linear flow law, Darcy's law in one case and Ohm's law in the other; and a continuity relationship, the conservation of fluid mass in one case and the conservation of electrical charge in the other. A comparison of *Ohm's law*,

$$I_x = -\sigma \frac{\partial V}{\partial x} \tag{5.20}$$

and *Darcy's law*,

$$v_x = -K \frac{\partial h}{\partial x} \tag{5.21}$$

clarifies the analogy at once. The specific discharge, v_x (discharge per unit area), is analogous to the current density, I_x (electrical current per unit area); the hydraulic conductivity, K, is analogous to the specific electrical conductivity, σ; and the hydraulic head, h, is analogous to the electrical potential, V.

The analogy between electrical flow and groundwater flow is the basis for two types of analog model that have proven useful for the generation of quantitative flow nets. The first type involves the use of conductive paper and the second type utilizes resistance networks.

Conductive-Paper Analogs

Let us consider once again the hydraulic problem first shown in Figure 5.8 and now reproduced in Figure 5.11(a). The electric analog [Figure 5.11(b)] consists of a sheet of conductive paper cut in the same geometrical shape as the groundwater flow field. A power supply is used to set up a voltage differential across the boundaries, and a sensing probe connected to the circuit through a voltmeter is used to measure the potential distribution throughout the conductive sheet. Constant-head boundaries, such as the $V = 100$ boundary on Figure 5.11(b), are created with highly conductive silver paint; impermeable boundaries are simulated by the unconnected edges of the paper model. It is usually possible to search out the equipotential lines rather efficiently, so that a full equipotential net can be quickly generated.

The method is limited to homogeneous, isotropic systems in two dimensions, but it is capable of handling complex region shapes and boundary conditions. Variations in the conductivity of commercially available paper may lead to random errors that limit the quantitative accuracy of the method. Two of the most detailed applications of the method are Childs' (1943) theoretical analysis of near-surface

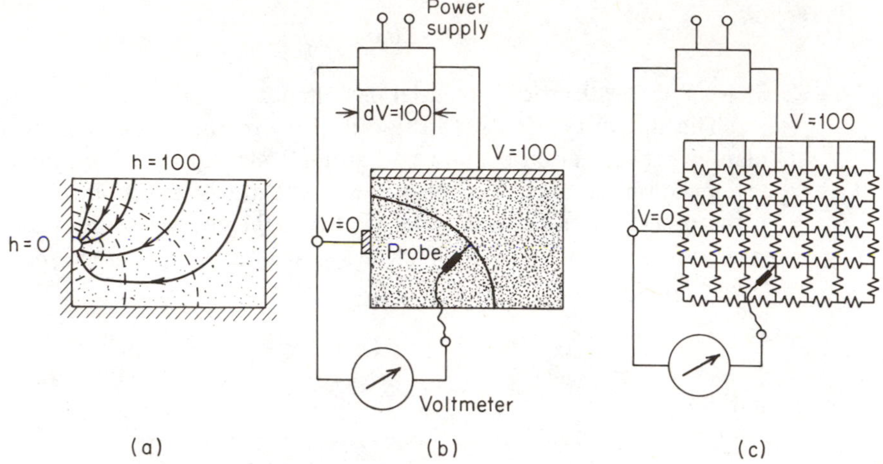

Figure 5.11 Flow nets by electric-analog simulation. (a) Steady-state hydrogeologic boundary-value problem; (b) conducting-paper analog; (c) resistance-network analog.

flow systems in drained land, and Tóth's (1968) consideration of regional groundwater flow nets for a field area in Alberta.

Resistance Network Analogs

The use of a resistance network as an electrical analog is based on the same principles as the conductive-paper analog. In this approach [Figure 5.11(c)] the flow field is replaced by a network of resistors connected to one another at the nodal points of a grid. The flow of electricity through each resistor is analogous to the flow of groundwater through a flow tube parallel to the resistor and having a cross-sectional area reflected by the resistor spacing times a unit depth. For electrical flow through an individual resistor, the I in Eq. (5.20) must now be viewed as the current, and the σ is equal to $1/R$, where R is the resistance of the resistor. As in the paper analog, a potential difference is set up across the constant-head boundaries of the model. A sensing probe is used to determine the voltage at each of the nodal points in the network, and these values, when recorded and contoured, create the equipotential net.

By varying the resistances in the network it is possible to analyze heterogeneous and anisotropic systems with resistance-network analogs. They have an accuracy and versatility that is superior to the paper models, but they are not as flexible as the numerical methods introduced in the next section.

Karplus (1958) provides a detailed handbook for analog simulation. Resistance-network analogs have been used to generate groundwater flow nets by Luthin (1953) in a drainage application, and by Bouwer and Little (1959) for a saturated-unsaturated system. Bouwer (1962) utilized the approach to analyze the configuration of groundwater mounds that develop beneath recharge ponds.

The most widespread use of electrical analog methods in groundwater hydrology is in the form of resistance-capacitance networks for the analysis of *transient* flow in aquifers. This application will be discussed in Section 8.9.

5.3 Flow Nets by Numerical Simulation

The hydraulic-head field, $h(x, z)$, that allows construction of a flow net can be generated *mathematically* from the pertinent steady-state boundary-value problem in two ways. The first approach utilizes analytical solutions as discussed in Section 2.11 and Appendix III; the second approach uses numerical methods of solution. Analytical methods are limited to flow problems in which the region of flow, boundary conditions, and geologic configuration are simple and regular. As we shall see in this section, numerical methods are much more versatile, but their application is bound unequivocally to the use of a digital computer.

Numerical methods are approximate. They are based on a discretization of the continuum that makes up the region of flow. In the discretization, the region is divided into a finite number of blocks, each with its own hydrogeologic properties, and each having a node at the center at which the hydraulic head is defined for the entire block. Figure 5.12(a) shows a 7×5 block-centered nodal grid ($i = 1$ to $i = 7$ in the x direction, and $j = 1$ to $j = 5$ in the z direction) for a rectangular flow region.

Let us now examine the flow regime in the vicinity of one of the interior nodes—say, in the nodal block, $i = 4, j = 3$, and its four surrounding neighbors. To simplify the notation, we will renumber the nodes as indicated in Figure 5.12(b). If flow occurs from node 1 to node 5, we can calculate the discharge, Q_{15}, from Darcy's law:

$$Q_{15} = K_{15} \frac{h_1 - h_5}{\Delta z} \Delta x \tag{5.22}$$

for flow through a cross section of unit depth perpendicular to the page. On the assumption that flow is directed toward the central node in each case, we can write down similar expressions for Q_{25}, Q_{35}, and Q_{45}. For steady-state flow, consideration of the conservation of fluid mass decrees that the sum of these four flows must be zero. If the medium is homogeneous and isotropic, $K_{15} = K_{25} = K_{35} = K_{45}$, and if we arbitrarily select a nodal grid that is square so that $\Delta x = \Delta z$, summation of the four terms leads to

$$h_5 = \tfrac{1}{4}(h_1 + h_2 + h_3 + h_4) \tag{5.23}$$

This equation is known as a *finite-difference equation*. If we revert to the ij notation of Figure 5.12(a), it becomes

$$h_{i,j} = \tfrac{1}{4}(h_{i,j-1} + h_{i+1,j} + h_{i,j+1} + h_{i-1,j}) \tag{5.24}$$

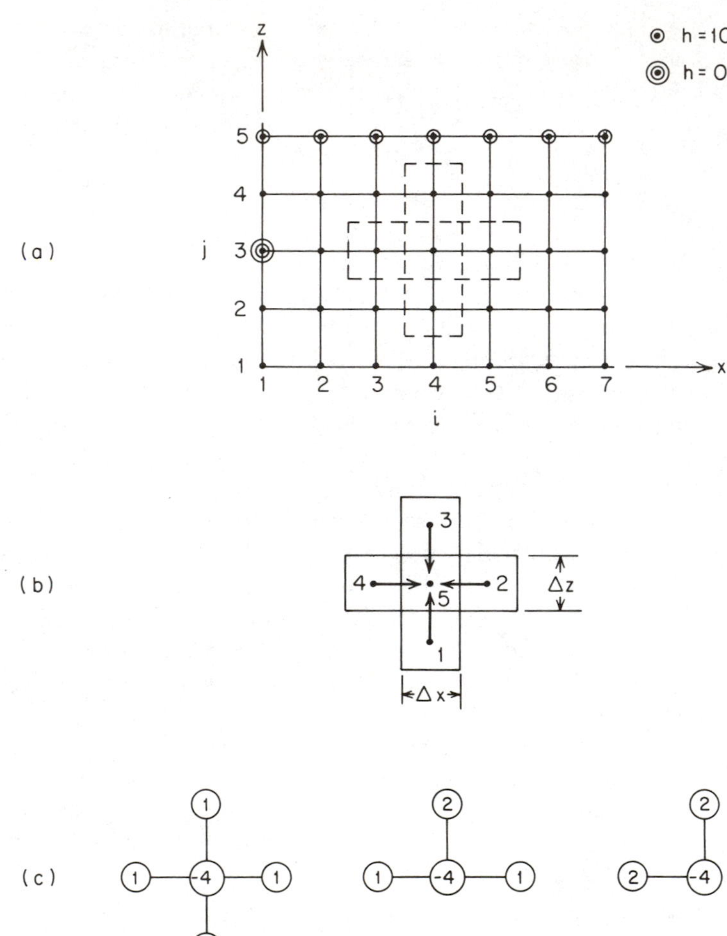

Figure 5.12 (a) Block-centered nodal grid for numerical flow-net simulation. (b) Interior node and its neighboring nodal blocks. (c) Finite-difference stars for an interior node and for nodes on a basal impermeable boundary and an impermeable corner.

In this form, Eq. (5.24) holds for all internal nodes in the nodal grid. It states an elegantly simple truth: in steady flow, in a homogeneous, isotropic medium, the hydraulic head at any node is the average of the four surrounding values.

A similar exercise will reveal that the finite-difference equation for a node along the basal boundary, assuming that boundary to be impermeable, takes the form

$$h_{i,j} = \tfrac{1}{4}(h_{i-1,j} + h_{i+1,j} + 2h_{i,j+1}) \tag{5.25}$$

and, at a corner node,

$$h_{i,j} = \tfrac{1}{4}(2h_{i-1,j} + 2h_{i,j+1}) \tag{5.26}$$

Equations (5.24) through (5.26) are schematically represented, in a self-explanatory way, by the three *finite-difference stars* of Figure 5.12(c).

In short, it is possible to develop a finite-difference equation for every node in the nodal grid. If there are N nodes, there are N finite-difference equations. There are also N unknowns—the N values of h at the N nodes. We have therefore produced N linear, algebraic equations in N unknowns. If N were very small, we could solve the equations directly, using a technique such as Cramer's rule, but where N is large, as it invariably is in numerical-flow-net simulation, we must introduce a more efficient method, known as *relaxation*.

Let us remain faithful to the flow problem of Figure 5.11(a) and assume that we wish to develop its flow net by numerical means. In the nodal grid of Figure 5.12(a), the nodal points at which the head is known are circled: $h = 0$ at $i = 1$, $j = 3$, and $h = 100$ at all the nodes in the $j = 5$ row. Relaxation involves repeated sweeps through the nodal net, top to bottom and left to right (or in any consistent fashion), applying the pertinent finite-difference equation at each node where the head is unknown. One must assume some starting h value at each node. For the problem at hand, $h = 50$ could be assigned as the starting value at all the uncircled nodes. In the application of the finite-difference equations, the most recently calculated h value is always used at every node. Each pass through the system is called an *iteration*, and after each iteration the calculated h values will approach more closely their final answers. The difference in h at any node between two successive iterations is called the *residual*. The maximum residual in the system will decrease as the iterations proceed. A solution has been reached when the maximum residual is reduced below a predetermined *tolerance*.

To test one's understanding of the relaxation process, the reader might carry out a couple of iterations in the upper left-hand portion of the net. If the initial assigned value at the node, $i = 2, j = 4$, for example, is 50, then the value after the first iteration is 62.5 and after the second is 64.0. The final value, attained after several iterations, lies between 65 and 66.

Numerical simulation is capable of handling any shape of flow region and any distribution of boundary conditions. It is easy to redevelop the finite-difference equations for rectangular meshes where $\Delta x \neq \Delta z$, and for heterogeneous, anisotropic conductivity distributions where the K_x and K_z values vary from node to node. In Eq. (5.22), the usual averaging technique would set $K_{15} = (K_1 + K_5)/2$, where K_1 and K_5 in this case refer to the vertical conductivities at nodes 1 and 5, and these might differ from each other and from the horizontal conductivities at these nodes. Numerical simulation allows flow-net construction in cases that are too complex for graphical construction or analytical solution. Numerical simulation is almost always programmed for the digital computer, and computer programs are usually written in a generalized form so that only new data cards are required to handle vastly differing flow problems. This is a distinct advantage over resistance-network analogs, which require a complete breakdown of the assembled hardware to effect a new simulation.

The development of the finite-difference equations presented in this section was rather informal. It *is* possible to begin with Laplace's equation and proceed more mathematically toward the same result. In Appendix VI, we present a brief development along these lines. Perhaps it is worth noting in passing that the informal development utilizes Darcy's law and the continuity relationship to reach the finite-difference expressions. These are the same two steps that led to the development of Laplace's equation in Section 2.11.

The method we have called *relaxation* (after Shaw and Southwell, 1941) has several aliases. It is variously known as the *Gauss-Seidel method*, the *Liebmann method*, and the *method of successive displacements*. It is the simplest, but far from the most efficient, of many available methods for solving the set of finite-difference equations. For example, if the computed heads during relaxation are corrected according to

$$h^k_{\text{corr}} = \omega h^k + (1 - \omega)h^{k-1}_{\text{corr}} \tag{5.27}$$

where h^k is the computed head at the kth iteration and h^{k-1}_{corr} is the corrected head from the previous iteration, then the method is known as *successive overrelaxation* and the number of iterations required to reach a converged solution is significantly reduced. The parameter ω is known as the *overrelaxation parameter*, and it must lie in the range $1 \leq \omega \leq 2$.

There are many texts that will serve the numerical modeler well. McCracken and Dorn (1964) provide an elementary introduction to computer simulation techniques in their Fortran manual. Forsythe and Wasow (1960) deliver their message at a more advanced mathematical level. Remson et al. (1971) discuss a broad spectrum of numerical techniques with particular reference to groundwater flow. Pinder and Gray (1977) treat the subject at a more advanced level.

Numerical methods were introduced to the groundwater hydrology literature by Stallman (1956) in an analysis of regional water levels. Fayers and Sheldon (1962) were among the first to advocate steady-state numerical simulation in the study of regional hydrogeology. Remson et al. (1965) used the method to predict the effect of a proposed reservoir on groundwater levels in a sandstone aquifer. Freeze and Witherspoon (1966) generated many numerical flow nets in their theoretical study of regional groundwater flow. The method was in wide use much earlier in the agricultural drainage field (see Luthin and Gaskell, 1950) and in the derivation of seepage patterns in earth dams (Shaw and Southwell, 1941).

In recent years, the finite-difference method has been equaled in popularity by another numerical method of solution, known as the *finite-element method*. It, too, leads to a set of N equations in N unknowns that can be solved by relaxation, but the nodes in the finite-element method are the corner points of an irregular triangular or quadrilateral mesh that is designed by the modeler for each specific application, rather than the regular rectangular mesh of the finite-difference method. In many cases, a smaller nodal grid suffices and there are resulting economies in computer effort. The finite-element method is also capable of handling one situation that the finite-difference method cannot. The finite-difference method requires that the principal directions of anisotropy in an anisotropic formation parallel the coordinate directions. If there are two anisotropic formations in a flow field, each with different principal directions, the finite-difference method is stymied, whereas the finite-element method can provide a solution. The development of the finite-element equations requires a mathematical sophistication that is out of place in this introductory text. The interested reader is referred to Pinder and Gray (1977).

Numerical methods, both finite-difference and finite-element, are widely used as the basis for digital computer simulation of *transient* flow in groundwater aquifers. This application is discussed in Section 8.8.

5.4 Saturated-Unsaturated Flow Nets

There is another type of flow net that is extremely difficult to construct by graphical means. For flow problems that involve both saturated and unsaturated flow, steady-state flow nets are usually derived by numerical simulation. Consider the flow net illustrated in Figure 5.13. It is similar to the problem that we have repeatedly analyzed in the past sections in that it involves flow to a drain in a system with impermeable boundaries on three sides, but it differs in that the vertical scale has been set up in such a way that the hydraulic head on the upper boundary now infers a pressure-head value that is less than atmospheric. This means that the soil is unsaturated at the surface, although if outflow to the drain is to take place, it must be saturated at depth. The qualitative flow net in Figure 5.13 has been developed for a soil whose unsaturated characteristic curves are those shown on the

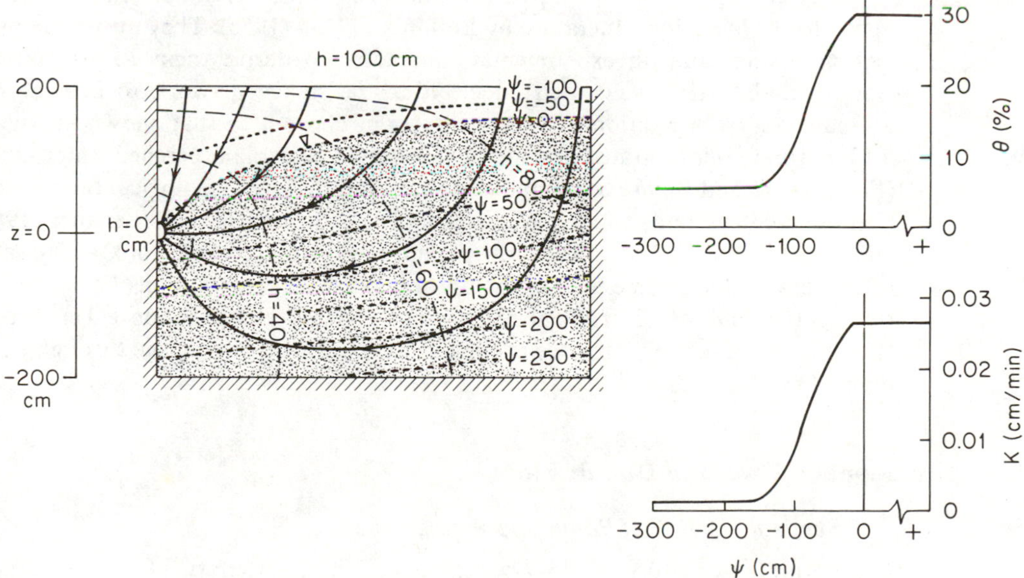

Figure 5.13 Saturated-unsaturated flow net in a homogeneous, isotropic soil. The insets show the unsaturated characteristic curves for the soil.

inset graphs. These curves of hydraulic conductivity, K, and moisture content, θ as a function of ψ, are the wetting curves taken from Figure 2.13.

As in the one-dimensional saturated-unsaturated case that was schematically illustrated in Figure 2.12, there are three types of output from a numerically simulated, steady-state, saturated-unsaturated flow net in two dimensions. First, there is the hydraulic-head pattern, $h(x, z)$, that allows construction of the equipotential net (the dashed lines on Figure 5.13). Second, there is the pressure-head pattern, $\psi(x, z)$ (the dotted lines in Figure 5.13), which is of particular value in defining the position of the water table (the $\psi = 0$ isobar). Third, there is the moisture content pattern, $\theta(x, z)$, which can be determined from the $\psi(x, z)$ pattern with the aid of the $\theta(\psi)$ curve for the soil. For example, along the $\psi = -50$ cm dotted line in Figure 5.13, the moisture content, θ, is 27%.

The flowlines and equipotential lines form a continuous net over the full saturated-unsaturated region. They intersect at right angles throughout the system. A quantitative flow net could be drawn with curvilinear squares in the homogeneous, isotropic, saturated portion, but such flow tubes would not exhibit squares as they traverse the unsaturated zone, even in homogeneous, isotropic soil. As the pressure head (and moisture content) decrease, so does the hydraulic conductivity, and increased hydraulic gradients are required to deliver the same discharge through a given flow tube. This phenomenon can be observed in the flow tubes in the upper left-hand corner of the flow net in Figure 5.13, where the gradients increase toward the surface.

The concept of an integrated saturated-unsaturated flow system was introduced to the hydrologic literature by Luthin and Day (1955). They utilized numerical simulation and an experimental sand tank to derive their $h(x, z)$ pattern. Bouwer and Little (1959) used an electrical resistance network to analyze tile drainage and subirrigation problems similar in concept to that shown in Figure 5.13. Saturated-unsaturated flow nets are required to explain perched water tables (Figures 2.15 and Figure 6.11), and to understand the hydrogeological regime on a hillslope as it pertains to streamflow generation (Section 6.5). Reisenauer (1963) and Jeppson and Nelson (1970) utilized numerical simulation to look at the saturated-unsaturated regime beneath ponds and canals. Their solutions have application to the analysis of artificial recharge of groundwater (Section 8.11). Freeze (1971b) considered the influence of the unsaturated zone on seepage through earth dams (Section 10.2).

5.5 The Seepage Face and Dupuit Flow

Seepage Face, Exit Point, and Free Surface

If a saturated-unsaturated flow system exists in the vicinity of a free-outflow boundary, such as a streambank or the downstream face of an earth dam, a *seepage face* will develop on the outflow boundary. In Figure 5.14(a), BC is a constant-head boundary and DC is impermeable. If there is no source of water at the

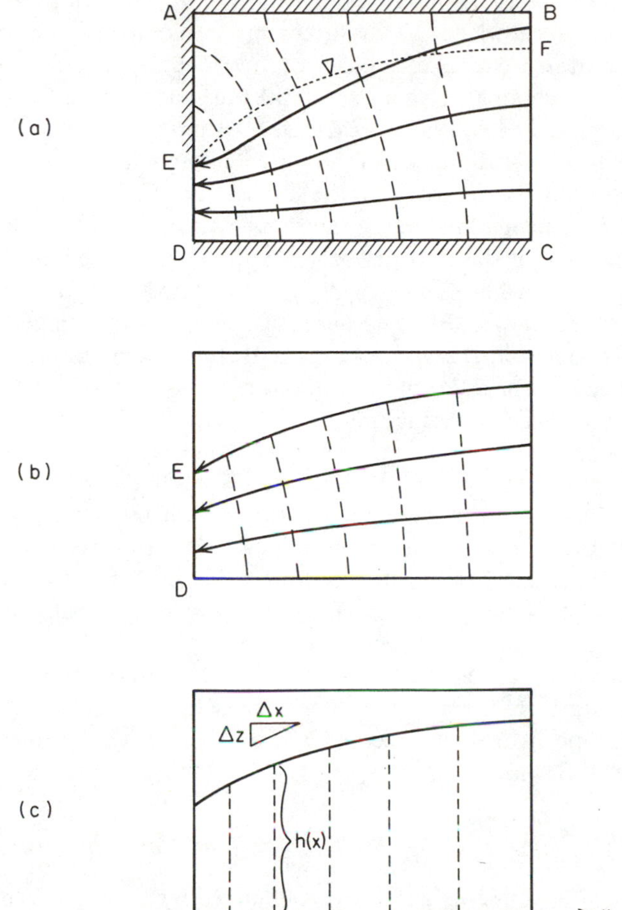

Figure 5.14 Development of a seepage face on a free-outflow boundary. (a) Saturated-unsaturated flow net; (b) free-surface flow net; (c) Dupuit-Forchheimer flow net.

surface, AB will also act like an impermeable boundary. The water table EF intersects the outflow boundary AD at the *exit point E*. All the flow must leave the system across the seepage face ED below the exit point E. Above E, along the line AE, the unsaturated pressure heads, ψ, are less than atmospheric, so outflow to the atmosphere is impossible. In effect, AE acts as an impermeable boundary. The condition on ED is $h = z$, the same as that on the water table. The problem in preparing a flow net for such cases lies in the fact that the position of the exit point that separates the two boundary conditions on the outflow boundary is not known a priori. In numerical simulation, it is necessary to provide an initial prediction for the position of the exit point. The correct exit point is then determined by a series of trial-and-error steady-state solutions.

The construction of a quantitative flow net in a saturated-unsaturated regime requires knowledge of both the saturated hydraulic conductivity, K, and the unsaturated characteristic curve, $K(\psi)$, for the soil. In many engineering applications, including the analysis of seepage through earth dams, the latter data are seldom available. In such cases, the assumption is usually made that flow through the unsaturated portion of the system is negligible, or viewed another way, that the hydraulic conductivity of the soil at moisture contents less than saturation is negligible compared to the saturated hydraulic conductivity. In this case, the upper boundary of the flow net becomes the water table, and the water table itself becomes a flowline. Under these special circumstances, this upper boundary is known as a *free surface*. Flow nets in saturated systems bounded by a free surface can be constructed in the usual way, but there is one complication. The position of the entire free surface (not just the exit point) is unknown a priori. The boundary conditions on a free surface must satisfy both those of a water table ($h = z$), and those of a flowline (equipotential lines must meet it at right angles). Its position is usually determined by graphical trial and error. Texts on engineering seepage, such as Harr (1962) or Cedergren (1967), provide hints on graphical construction and include many examples of steady-state, free-surface flow nets.

Figure 5.14(b) is the equivalent free-surface flow net to the saturated-unsaturated flow net shown in Figure 5.14(a). A glance at the two diagrams confirms that our decision to specify the water table as a flowline is quite a good approximation for this particular flow system. The outflow boundary ED is still known as a seepage face. We will encounter seepage faces in a practical sense when we examine hillslope hydrology (Section 6.5) and when we consider seepage through earth dams (Section 10.2).

Dupuit-Forchheimer Theory of Free-Surface Flow

For flow in unconfined systems bounded by a free surface, an approach pioneered by Dupuit (1863) and advanced by Forchheimer (1930) is often invoked. It is based on two assumptions: (1) flowlines are assumed to be horizontal and equipotentials vertical and (2) the hydraulic gradient is assumed to be equal to the slope of the free surface and to be invariant with depth. Figure 5.14(c) shows the equipotential net for the same problem as in Figure 5.14(a) but with the Dupuit assumptions in effect. The construction of rigorous flowlines is no longer possible. This paradoxical situation identifies Dupuit-Forchheimer theory for what it is, an empirical approximation to the actual flow field. In effect, the theory neglects the vertical flow components. In practice, its value lies in reducing the two-dimensional system to one dimension for the purposes of analysis. Calculations based on the Dupuit assumptions compare favorably with those based on more rigorous methods when the slope of the free surface is small and when the depth of the unconfined flow field is shallow. The discharge Q through a cross section of unit width perpendicular to the page in Figure 5.14(c) is given by

$$Q = Kh(x)\frac{dh}{dx} \tag{5.28}$$

where $h(x)$ is the elevation of the free surface above the base of the flow system at x, and the gradient dh/dx is given by the slope of the free surface $\Delta h/\Delta x$ at x. For steady-state flow, Q must be constant through the system and this can only be true if the free surface is a parabola.

The equation of flow for Dupuit-Forchheimer theory in a homogeneous, isotropic medium can be developed from the continuity relationship, $dQ/dx = 0$. From Eq. (5.28), this leads to

$$\frac{d^2(h^2)}{dx^2} = 0 \tag{5.29}$$

If a three-dimensional unconfined flow field is reduced to a two-dimensional xy horizontal flow field by invocation of Dupuit-Forchheimer theory, the equation of flow in a homogeneous, isotropic medium becomes

$$\frac{\partial^2(h^2)}{\partial x^2} + \frac{\partial^2(h^2)}{\partial y^2} = 0 \tag{5.30}$$

In other words, h^2 rather than h must satisfy Laplace's equation. It is possible to set up steady-state boundary-value problems based on Eq. (5.30) and to solve for $h(x, y)$ in shallow, horizontal flow fields by analog or numerical simulation. It is also possible to develop a transient equation of flow for Dupuit free-surface flow in unconfined aquifers whereby h^2 replaces h in the left-hand side of Eq. (2.77).

Harr (1962) discusses the practical aspects of Dupuit-Forchheimer theory in some detail. Bear (1972) includes a very lengthy theoretical treatment. Kirkham (1967) examines the paradoxes in the theory and provides some revealing explanations. The approach is widely used in engineering applications.

Suggested Readings

CEDERGREN, H. R. 1967. *Seepage, Drainage and Flow Nets.* Chapter 4: Flow Net Construction, John Wiley & Sons, New York, pp. 148–169.

HARR, M. E. 1962. *Groundwater and Seepage.* Chapter 2: Application of the Dupuit Theory of Unconfined Flow, McGraw-Hill, New York, pp. 40–61.

KIRKHAM, D. 1967. Explanation of paradoxes in Dupuit-Forchheimer seepage theory. *Water Resources Res.*, 3, pp. 609–622.

PRICKETT, T. A. 1975. Modeling techniques for groundwater evaluation. *Adv. Hydrosci.*, 10, pp. 42–45, 66–75.

REMSON, I., G. M. HORNBERGER, and F. J. MOLZ. 1971. *Numerical Methods in Subsurface Hydrology.* Chapter 4: Finite-Difference Methods Applied to Steady-Flow Problems, Wiley Interscience, New York, pp. 123–156.

Problems

1. Consider a saturated, homogeneous, isotropic, rectangular, vertical cross section $ABCDA$ with upper boundary AB, basal boundary DC, left-hand boundary AD, and right-hand boundary BC. Make the distance DC twice that of AD. Draw a quantitatively accurate flow net for each of the following cases.
   (a) BC and DC are impermeable. AB is a constant-head boundary with $h =$ 100 m. AD is divided into two equal lengths with the upper portion impermeable and the lower portion a constant-head boundary with $h = 40$ m.
   (b) AD and BC are impermeable, AB is a constant-head boundary with $h =$ 100 m. DC is divided into three equal lengths with the left and right portions impermeable and the central portion a constant-head boundary with $h =$ 40 m.

2. Let the vertical cross section $ABCDA$ from Problem 1 become trapezoidal by raising B vertically in such a way that the elevations of points D and C are 0 m, A is 100 m, and B is 130 m. Let AD, DC, and BC be impermeable and let AB represent a water table of constant slope (on which the hydraulic head equals the elevation).
   (a) Draw a quantitatively accurate flow net for this case. Label the equipotential lines with their correct h value.
   (b) If the hydraulic conductivity in the region is 10^{-4} m/s, calculate the total flow through the system in m^3/s (per meter of thickness perpendicular to the section).
   (c) Use Darcy's law to calculate the entrance or exit velocity of flow at each point where a flowline intersects the upper boundary.

3. (a) Repeat Problems 2(a) and 2(b) for the homogeneous, anisotropic case where the horizontal hydraulic conductivity is 10^{-4} m/s and the vertical hydraulic conductivity is 10^{-5} m/s.
   (b) Draw the hydraulic conductivity ellipse for the homogeneous, anisotropic formation in part (a). Show by suitable constructions on the ellipse that the relation between the direction of flow and the direction of the hydraulic gradient indicated by your flow net is correct at two points on the flow net.

4. Repeat Problem 2(a) for the case where a free-flowing drain (i.e., at atmospheric pressure) is located at the midpoint of BC. The drain is oriented at right angles to the plane of the flow net.

5. (a) Repeat Problems 1(a), 1(b), and 2(a) for the two-layer case where the lower half of the field has a hydraulic conductivity value 5 times greater than that of the upper half.
   (b) Repeat Problem 1(b) for the two-layer case where the upper half of the field has a hydraulic conductivity value 5 times greater than that of the lower half.

6. Sketch a piezometer that bottoms near the center of the field of flow in each of the flow nets constructed in Problems 2, 3, 4, and 5, and show the water levels that would exist in these piezometers according to the flow nets as you have drawn them.

7. (a) Redraw the flow net of Figure 5.3 for a dam that is 150 m wide at its base, overlying a surface layer 120 m thick. Set $h_1 = 150$ m and $h_2 = 125$ m.
   (b) Repeat Problem 7(a) for a two-layer case in which the upper 60-m layer is 10 times less permeable than the lower 60-m layer.

8. Two piezometers, 500 m apart, bottom at depths of 100 m and 120 m in an unconfined aquifer. The elevation of the water level is 170 m above the horizontal impermeable, basal boundary in the shallow piezometer, and 150 m in the deeper piezometer. Utilize the Dupuit-Forchheimer assumptions to calculate the height of the water table midway between the piezometers, and to calculate the quantity of seepage through a 10-m section in which $K = 10^{-3}$ m/s.

9. Sketch flow nets on a horizontal plane through a horizontal confined aquifer:
   (a) For flow toward a single steady-state pumping well (i.e., a well in which the water level remains constant).
   (b) For two steady-state pumping wells pumping at equal rates (i.e., producing equal heads at the well).
   (c) For a well near a linear, constant-head boundary.

Groundwater and the Hydrologic Cycle

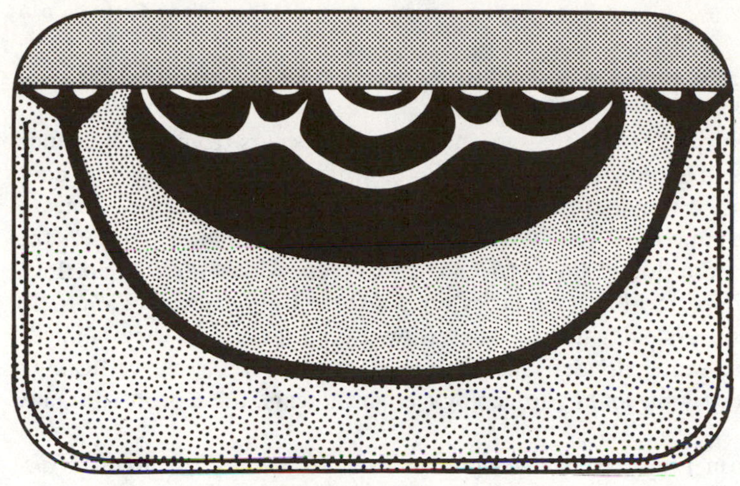

6.1 Steady-State Regional Groundwater Flow

With the methods of construction and simulation of steady-state flow nets in hand, we are now in a position to examine the natural flow of groundwater in hydrogeologic basins.

Recharge Areas, Discharge Areas, and Groundwater Divides

Let us consider the two-dimensional, vertical cross section of Figure 6.1. The section is taken in a direction perpendicular to the strike of a set of long, parallel ridges and valleys in a humid region. The geologic materials are homogeneous and isotropic, and the system is bounded at the base by an impermeable boundary. The water table is coincident with the ground surface in the valleys, and forms a subdued replica of the topography on the hills. The value of the hydraulic head on any one of the dashed equipotential lines is equal to the elevation of the water table at its point of intersection with the equipotential line. The flowlines and equi-

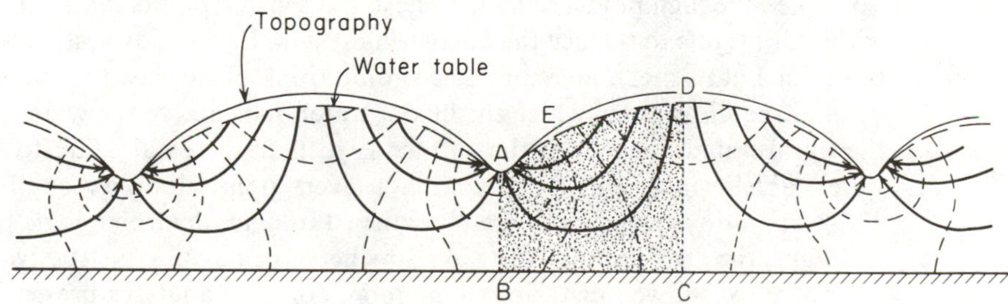

Figure 6.1 Groundwater flow net in a two-dimensional vertical cross section through a homogeneous, isotropic system bounded on the bottom by an impermeable boundary (after Hubbert, 1940).

potential lines were sketched according to the usual rules for graphical flow-net construction in homogeneous, isotropic media.

It is clear from the flow net that groundwater flow occurs from the highlands toward the valleys. The flow net must fill the entire field of flow, and one consequence of this fact is the occurrence of upward-rising groundwater flow beneath the valleys. The symmetry of the system creates vertical boundaries beneath the valleys and ridges (the dotted lines *AB* and *CD*) across which there is no flow. These imaginary impermeable boundaries are known as *groundwater divides*. In the most symmetric systems, such as that shown on Figure 6.1, they coincide exactly with surface-water divides, and their orientation is precisely vertical. In more complex topographic and hydrogeologic environments, these properties may be lost.

The flowlines in Figure 6.1 deliver groundwater from *recharge areas* to *discharge areas*. In a recharge area there is a component to the direction of groundwater flow near the surface that is downward. A recharge area can be defined as that portion of the drainage basin in which the net saturated flow of groundwater is directed away from the water table. In a discharge area there is a component to the direction of groundwater flow near the surface that is upward. A discharge area can be defined as that portion of the drainage basin in which the net saturated flow of groundwater is directed toward the water table. In a recharge area, the water table usually lies at some depth; in a discharge area, it is usually at or very near the surface. For the shaded cell in Figure 6.1, region *ED* is the recharge area and region *AE* is the discharge area. The line that separates recharge areas from discharge areas is called the *hinge line*. For the shaded cell, its intersection with the plane of the section is at point *E*.

The utilization of *steady-state* flow nets for the interpretation of regional flow deserves some discussion. The approach is technically valid only in the somewhat unrealistic case where the water table maintains the same position throughout the entire year. In most actual cases, fluctuations in the water table introduce transient effects in the flow systems. However, if the fluctuations in the water table are small in comparison with the total vertical thickness of the system, and if the relative configuration of the water table remains the same throughout the cycle of fluctuations (i.e., the high points remain highest and the low points remain lowest), we are within our rights to replace the fluctuating system by a steady system with the water table fixed at its mean position. One should think of the steady system as a case of dynamic equilibrium in which the flux of water delivered to the water table through the unsaturated zone from the surface is just the necessary flux to maintain the water table in its equilibrium position at every point along its length at all times. These conditions are approximately satisfied in many hydrogeologic basins, and in this light, the examination of steady flow nets can be quite instructive. Where they are not satisfied, we must turn to the more complex analyses presented in Section 6.3 for transient regional groundwater flow.

Hubbert (1940) was the first to present a flow net of the type shown in Figure 6.1 in the context of regional flow. He presumably arrived at the flow net by graphical construction. Tóth (1962, 1963) was the first to carry this work forward mathematically. He recognized that the flow system in the shaded cell *ABCDA* of Figure 6.1 could be determined from the solution to a boundary-value problem. The equation of flow is Laplace's equation [Eq. (2.70)] and the boundary conditions invoke the water-table condition on *AD* and impermeable conditions on *AB*, *BC*, and *CD*. He used the separation-of-variables technique, similar to that outlined in Appendix III for a simpler case, to arrive at an analytical expression for the hydraulic head in the flow field. The analytical solutions when plotted and contoured provide the equipotential net, and flowlines can easily be added. Appendix VII summarizes Tóth's solutions.

The analytical approach has three serious limitations:

1. It is limited to homogeneous, isotropic systems, or very simple layered systems.
2. It is limited to regions of flow that can be accurately approximated by a rectangle, that is, to water-table slopes, *AD*, that are very small.
3. It is limited to water-table configurations that can be represented by simple algebraic functions. Tóth considered cases with an inclined water table of constant slope, and cases in which a sine curve was superimposed on the incline.

As pointed out by Freeze and Witherspoon (1966, 1967, 1968), all three of these limitations can be removed if numerical simulation, as described in Section 5.3, is used to generate the flow nets. In the following subsections we will look at several flow nets taken from Freeze and Witherspoon's (1967) numerical results in order to examine the effect of topography and geology on the nature of steady-state regional flow patterns.

Effect of Topography on Regional Flow Systems

Figure 6.2 shows the flow nets for two vertical cross sections that are identical in depth and lateral extent. In both cases there is a major valley running perpendicular to the page at the left-hand side of the system, and an upland plateau to the right. In Figure 6.2(a) the upland water-table configuration, which is assumed to closely follow the topography, has a uniform gentle incline such as one might find on a lacustrine plain. Figure 6.2(b), on the other hand, has a hilly upland water-table configuration such as one might find in glacial terrain.

The uniform water table produces a single flow system. The hinge line lies on the valley wall of the major valley; the entire upland plateau is a recharge area. The hilly topography produces numerous subsystems within the major flow system. Water that enters the flow system in a given recharge area may be discharged in the nearest topographic low or it may be transmitted to the regional discharge area in the bottom of the major valley. Tóth (1963) has shown that as the depth to lateral extent of the entire system becomes smaller and as the amplitude of the hummocks becomes larger, the local systems are more likely to reach the basal

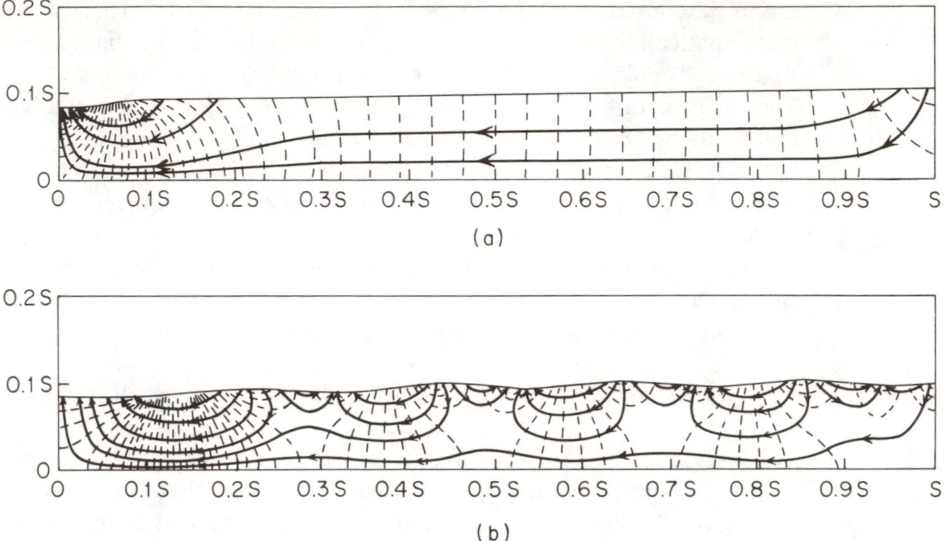

Figure 6.2 Effect of topography on regional groundwater flow patterns (after Freeze and Witherspoon, 1967).

boundary, creating a series of small independent cells such as those shown in Figure 6.1. Tóth (1963) suggests that on most flow nets and in most field areas, one can differentiate between *local* systems of groundwater flow, *intermediate* systems of groundwater flow, and *regional* systems of groundwater flow, as schematically illustrated in Figure 6.3. Where local relief is negligible, only regional systems develop. Where there is pronounced local relief, only local systems develop. These terms are not specific, but they provide a useful qualitative framework for discussion.

Figures 6.2 and 6.3 make it clear that even in basins underlain by homo-

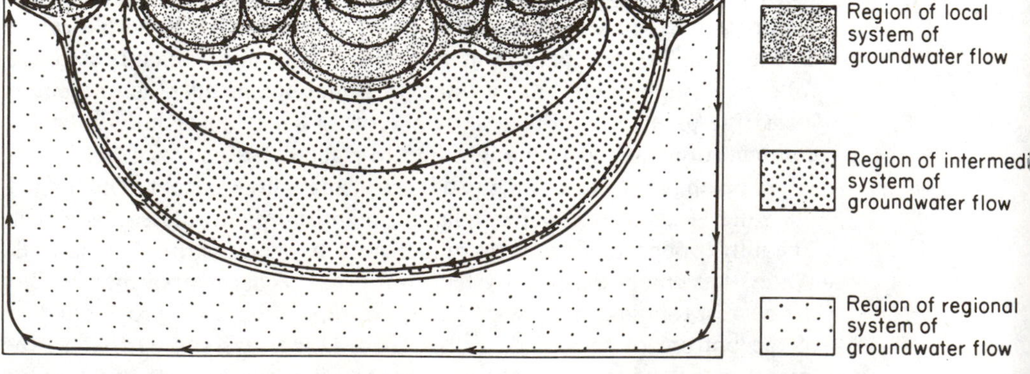

Figure 6.3 Local, intermediate, and regional systems of groundwater flow (after Tóth, 1963).

geneous, isotropic geologic materials, topography can create complex systems of groundwater flow. The only immutable law is that highlands are recharge areas and lowlands are discharge areas. For most common topographic configurations, hinge lines lie closer to valley bottoms than to ridge lines. On an areal map, discharge areas commonly constitute only 5–30% of the surface area of a watershed.

Effect of Geology on Regional Flow Systems

Figure 6.4 shows a sampling of numerically simulated flow nets for heterogeneous systems. Comparison of Figures 6.4(a) and 6.2(a) shows the effect of the introduction of a layer at depth with a permeability 10 times that of the overlying beds. The lower formation is an aquifer with essentially horizontal flow that is being recharged from above. Note the effect of the tangent law at the geologic boundary.

If the hydraulic conductivity contrast is increased [Figure 6.4(b)], the vertical gradients in the overlying aquitard are increased and the horizontal gradients in the aquifer are decreased. The quantity of flow, which can be calculated from the flow net using the methods of Section 5.1, is increased. One result of the increased flow is a larger discharge area, made necessary by the need for the large flows in the aquifer to escape to the surface as the influence of the left-hand boundary is felt.

In hummocky terrain [Figure 6.4(c)] the presence of a basal aquifer creates a highway for flow that passes under the overlying local systems. The existence of a high-permeability conduit thus promotes the possibility of regional systems even in areas of pronounced local relief.

There is a particular importance to the position within the basin of buried lenticular bodies of high conductivity. The presence of a partial basal aquifer in the upstream half of the basin [Figure 6.4(d)] results in a discharge area that occurs in the middle of the uniform upland slope above the stratigraphic pinchout. Such a discharge area cannot occur under purely topographic control. If the partial basal aquifer occurs in the downstream half of the system, the central discharge area will not exist; in fact, recharge in that area will be concentrated.

In the complex topographic and geologic system shown in Figure 6.4(e), the two flowlines illustrate how the difference of just a few meters in the point of recharge can make the difference between recharge water entering a minor local system or a major regional system. Such situations have disturbing implications for the siting of waste disposal projects that may introduce contaminants into the subsurface flow regime.

Subsurface stratigraphy and the resulting subsurface variations in hydraulic conductivity can exist in an infinite variety. It should be clear from these few examples that geological heterogeneity can have a profound effect on regional groundwater flow. It can affect the interrelationship between local and regional systems, it can affect the surficial pattern of recharge and discharge areas, and it can affect the quantities of flow that are discharged through the systems. The dramatic effects shown on Figure 6.4 are the result of conductivity contrasts of 2 orders of magnitude or less. In aquifer-aquitard systems with greater contrasts,

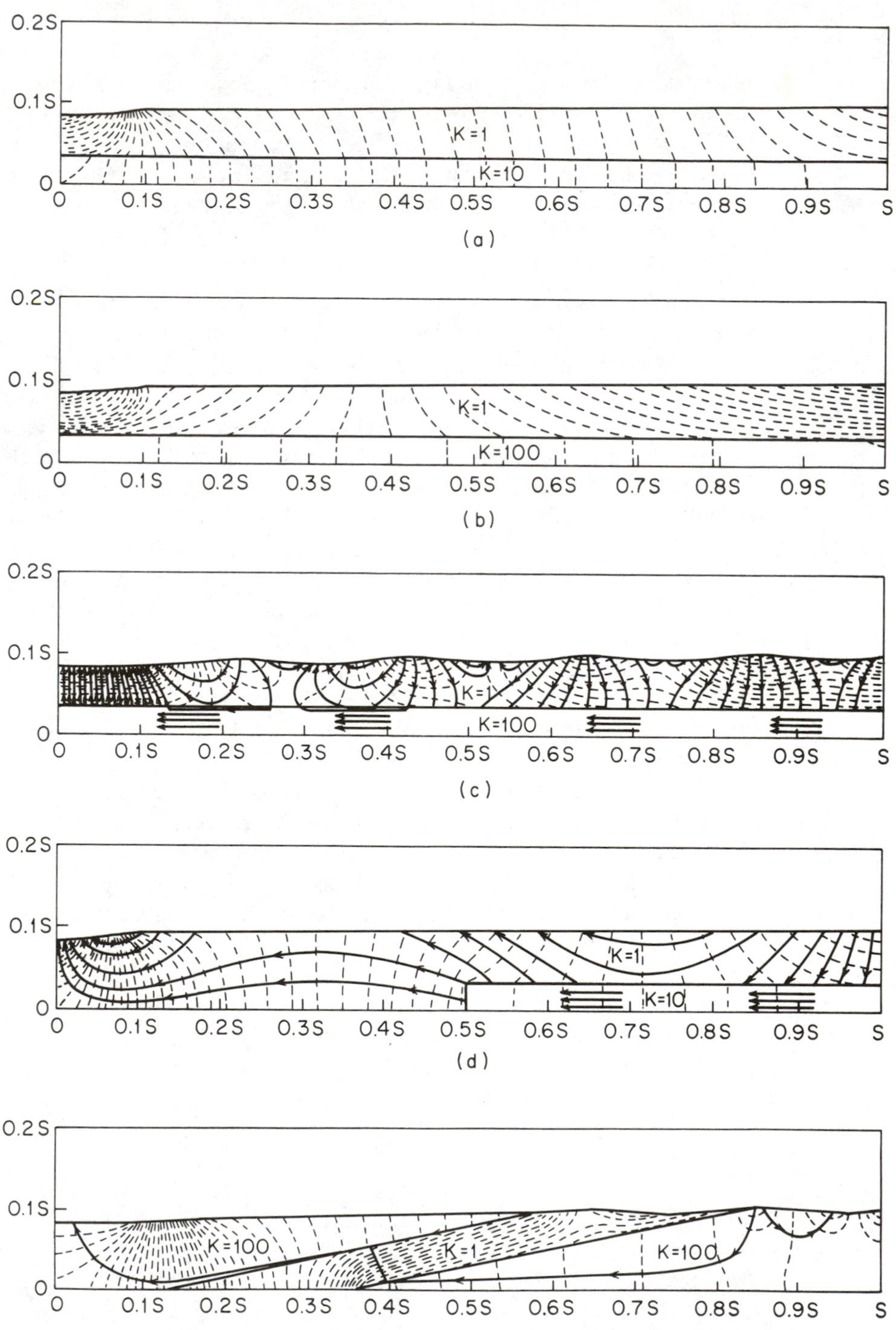

Figure 6.4 Effect of geology on regional groundwater flow patterns (after Freeze and Witherspoon, 1967).

flow patterns become almost rectilinear, with horizontal flow in the aquifers and vertical flow across the aquitards.

Flowing Artesian Wells

Flowing wells (along with springs and geysers) symbolize the presence and mystery of subsurface water, and as such they have always evoked considerable public interest.

The classic explanation of flowing wells, first presented by Chamberlain (1885) and popularized by Meinzer (1923) in connection with the Dakota sandstone, proposed an outcrop-related geologic control. If, as shown in Figure 6.5(a), an aquifer outcrops in an upland and is recharged there, an equipotential net can develop whereby the hydraulic head in the aquifer downdip from the recharge area is higher than the surface elevation. A well tapping the aquifer at such a location, and open at the surface, will flow.

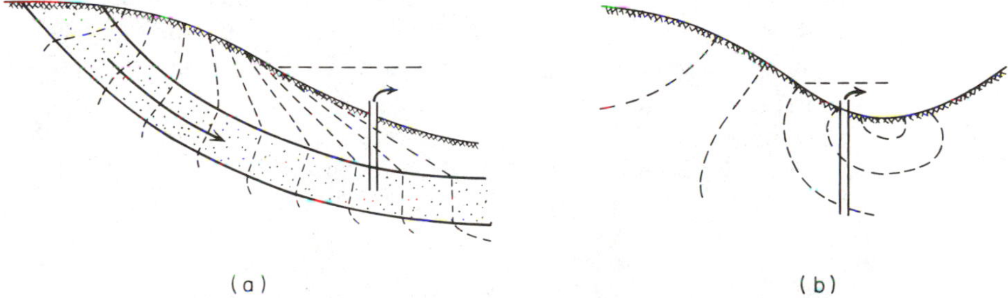

Figure 6.5 Flowing artesian wells: (a) geologically controlled; (b) topographically controlled.

However, it is not necessary to have this geologic environment to get flowing wells, nor is it a particularly common control. The primary control on flowing wells is topography. As shown in Figure 6.5(b), a well in a discharge area that has an intake at some depth below the water table will tap a hydraulic head contour with a head value that lies above the land surface, even in homogeneous, isotropic terrain. If there were a horizontal aquifer at depth beneath the valley in Figure 6.5(b), it need not outcrop to give rise to flowing wells. A well tapping the aquifer in Figure 6.4(b) beneath the valley at the left of the diagram would flow.

Any hydrogeologic system that leads to hydraulic-head values in an aquifer that exceed the surface elevation will breed flowing wells. The importance of topographic control is reflected in the large numbers of flowing wells that occur in valleys of rather marked relief. The specific location of areas of flowing wells within topographically low basins and valleys is controlled by the subsurface stratigraphy.

The Dakota sandstone configuration of Figure 6.5(a) has also been overused as a model of the regional groundwater-recharge process. Aquifers that outcrop in uplands are not particularly ubiquitous. Recharge regimes such as those shown in Figures 6.4(c), 6.4(d), and 6.7(b) are much more common.

Flow-System Mapping

Meyboom (1966a) and Tóth (1966) have shown by means of their work in the Canadian prairies that it is possible to map recharge areas and discharge areas on the basis of field observation. There are five basic types of indicators: (1) topography, (2) piezometric patterns, (3) hydrochemical trends, (4) environmental isotopes, and (5) soil and land surface features.

The simplest indicator is the topography. Discharge areas are topographically low and recharge areas are topographically high. The most direct indicator is piezometric measurement. If it were possible to install piezometer nests at every point in question, mapping would be automatic. The nests would show an upward-flow component in discharge areas and a downward-flow component in recharge areas. Such a course is clearly uneconomical, and in any case comparable information can often be gleaned from the available water-level data on existing wells. A well is not a true piezometer because it is usually open all along its length rather than at one point, but in many geologic environments, especially those where a single aquifer is being tapped, static water-level data from wells can be used as an indicator of potentiometric conditions. If there are many wells of various depths in a single topographic region, a plot of well depth versus depth to static water level can be instructive. Figure 6.6 defines the fields on such a plot where the scatter of points would be expected to fall in recharge areas and discharge areas.

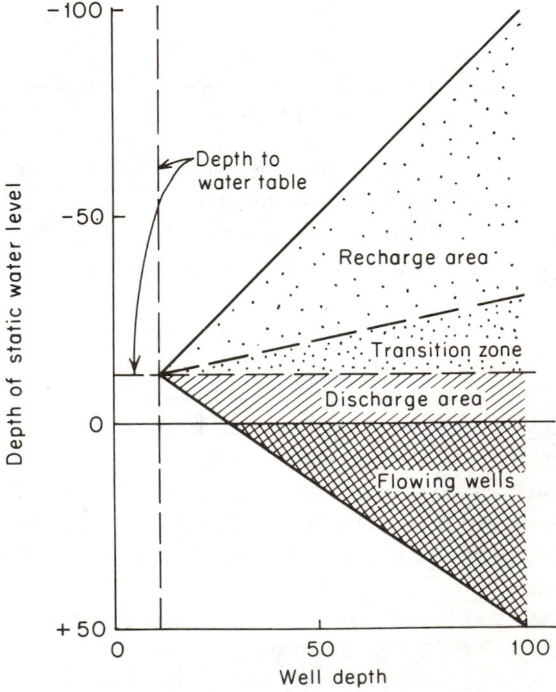

Figure 6.6 Generalized plot of well depth versus depth to static water level.

Geochemical interpretation requires a large number of chemical analyses carried out on water samples taken from a representative set of wells and piezometers in an area. Groundwater as it moves through a flow system undergoes a geochemical evolution that will be discussed in Chapter 7. It is sufficient here to note the general observation that salinity (as measured by total dissolved solids) generally increases along the flow path. Water from recharge areas is usually relatively fresh; water from discharge areas is often relatively saline.

Information on groundwater flow systems is also obtained from analyses of well or piezometer samples for the environmental isotopes ^{2}H, ^{3}H, ^{18}O, and ^{14}C. The nature of these isotopes is described in Section 3.8. Tritium (^{3}H) is used to identify water that has entered the groundwater zone more recently than 1953, when weapons testing in the atmosphere was initiated (Figure 3.11). The distribution of ^{3}H in the groundwater flow system can be used to outline the subsurface zone that is occupied by post-1953 water. In that the tritiated zone extends into the flow system from the recharge area, this approach provides a basis for estimating regional values of the average linear velocity of groundwater flow near the recharge area. Peak concentrations of ^{3}H in the groundwater can sometimes be related to peaks in the long-term record of ^{3}H concentration in rain and snow.

The distribution of ^{14}C can be used to distinguish zones in which old water occurs (Section 3.8). This approach is commonly used in studies of regional flow in large aquifers. ^{14}C is used in favorable circumstances to identify zones of water in the age range several thousand years to a few tens of thousands of years. Case histories of ^{14}C studies of regional flow in aquifers are described by Pearson and White (1967) and Fritz et al. (1974). Hydrochemical methods of interpretation of ^{14}C data are described in Section 7.6.

Particularly in arid and semiarid climates, it is often possible to map discharge areas by direct field observation of springs and seeps and other discharge phenomena, collectively labeled *groundwater outcrops* by Meyboom (1966a). If the groundwater is highly saline, "outcrops" may take the form of saline soils, playas, salinas, or salt precipitates. In many cases, vegetation can provide a significant clue. In discharge areas, the vegetative suite often includes salt-tolerant, water-loving plants such as willow, cottonwood, mesquite, saltgrass, and greasewood. Most of these plants are *phreatophytes*. They can live with their roots below the water table and they extract their moisture requirements directly from the saturated zone. Phreatophytes have been studied in the southwestern United States by Meinzer (1927) and Robinson (1958, 1964) and in the Canadian prairies by Meyboom (1964, 1967). In humid climates, saline and vegetative groundwater outcrops are less evident, and field mapping must rely on springs and piezometric evidence.

As an example of an actual system, consider the flow system near Assiniboia, Saskatchewan (Freeze, 1969a). Figure 6.7(a) shows the topography of the region and the field evidence of groundwater discharge, together with a contoured plot of the hydraulic head values in the Eastend sand member, based on available well records. The stratigraphic position of the Eastend sand is shown along section *A-A′* in Figure 6.7(b). Meyboom (1966a) refers to this hydrogeologic environment,

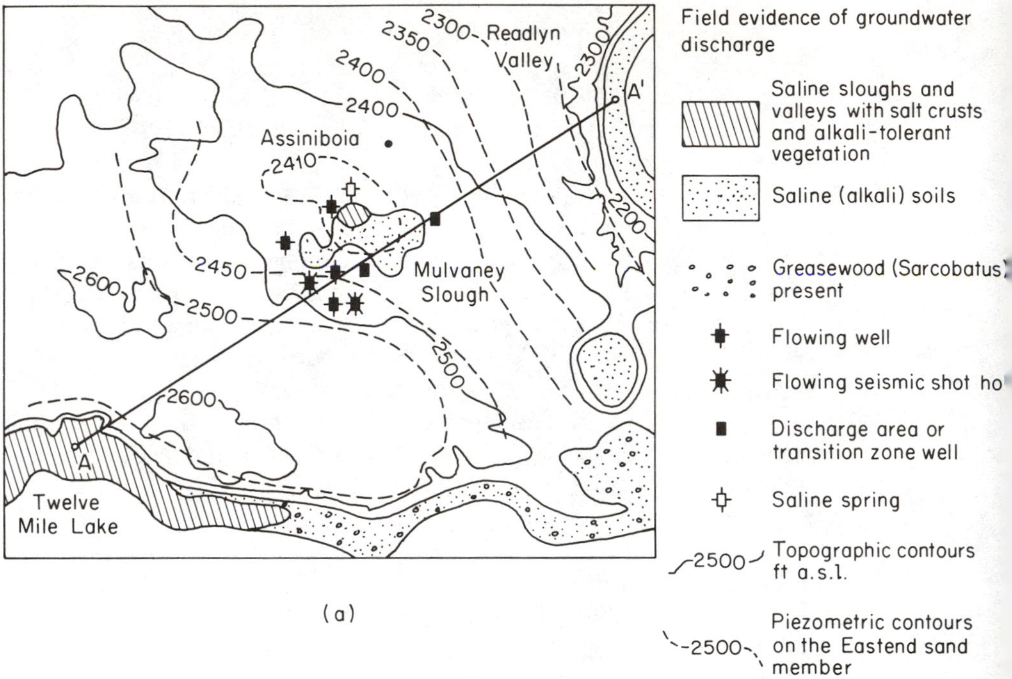

2300
2350
Readlyn Valley
2400
2400
Assiniboia
2410
2450
2600
2500
2500
2600
A
A'
Twelve Mile Lake
Mulvaney Slough
2200
2300
Field evidence of groundwater discharge
Saline sloughs and valleys with salt crusts and alkali-tolerant vegetation
Saline (alkali) soils
Greasewood (Sarcobatus) present
Flowing well
Flowing seismic shot ho
Discharge area or transition zone well
Saline spring
2500 Topographic contours ft a.s.l.
2500 Piezometric contours on the Eastend sand member

(a)

Twelve Mile Lake
Mulvaney Slough
Readlyn Valley
2800
2600
2400
2200
2000
1800
A
A'
Glacial deposits
Sand member of Eastend formation
Clay and silt members of Ravenscrag, Whitemud, and Eastend formations
Bearpaw shale

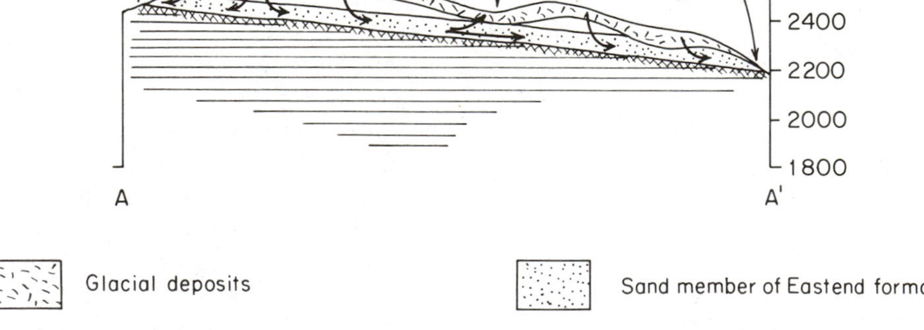

(b)

Figure 6.7 Regional groundwater flow near Assiniboia, Saskatchewan (after Freeze, 1969a). (Reproduced by permission of the Minister of Supply and Services, Canada.)

which is rather common in the Great Plains region of North America, as the *prairie profile*.

The steady-state flow-net approach to the analysis of regional groundwater flow has now been applied in many parts of the world in a wide variety of hydrogeological environments. The approach has generally been applied in drainage basins of small to moderate size, but it has also been utilized on a much larger scale by Hitchon (1969a, b). His analysis of fluid flow in the western Canadian sedimentary basin considered systems that extend from the Rockies to the Canadian Shield. The analysis was carried out to shed new light on the nature of petroleum migration and accumulation. It is discussed more fully in Chapter 11.

6.2 Steady-State Hydrologic Budgets

Steady-state flow nets of regional groundwater flow, whether they are developed on the basis of piezometric measurements and field observations or by mathematical or analog simulation, can be interpreted quantitatively to provide information that is of value in the determination of a hydrologic budget for a watershed.

Quantitative Interpretation of Regional Flow Systems

Figure 6.8 shows a quantitative flow net for a two-dimensional, vertical cross section through a heterogeneous groundwater basin. This particular water-table configuration and set of geologic conditions gives rise to two separate flow systems: a local flow system that is shallow but of large lateral extent (subsystem B), and a larger regional system (subsystem A). The local system is superimposed on the regional system in a way that could hardly have been anticipated by means other than a carefully constructed flow net. With the methods of Section 5.1, we can easily calculate the discharge through each flow system. For $s = 6000$ m, the total relief is 100 m, and since there are 50 increments of potential, $\Delta h = 2$ m. Assuming hydraulic conductivities of 10^{-4} and 10^{-5} m/s, the discharge through each streamtube is 2.0×10^{-4} m^3/s (per meter of thickness of the flow system perpendicular to the diagram). Counting the flow channels in the two subsystems leads to the values: $Q_A = 2.8 \times 10^{-3}$ m^3/s, $Q_B = 2.0 \times 10^{-4}$ m^3/s Quantities calculated in this way represent the regional discharge through an undeveloped basin under natural conditions. As we shall see in Section 8.10, the development of groundwater resources by means of wells leads to new regional systems that may allow total basin yields much greater than the virgin flow rates.

It is also possible to calculate the rate of recharge or discharge at the water table at any point along its length. If the hydraulic conductivities at each point are known and the hydraulic gradient is read off the flow net, Darcy's law can be invoked directly. If the recharge and discharge rates are plotted above the flow net as in Figure 6.8, the smooth line that joins the points is known as a *recharge-discharge profile*. It identifies concentrations of recharge and discharge that would

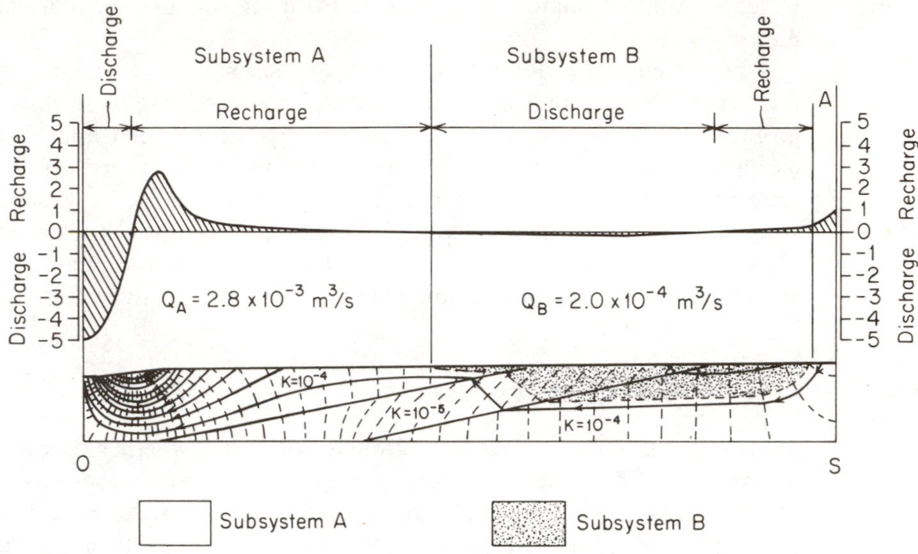

Figure 6.8 Quantitative flow net and recharge-discharge profile in a two-dimensional section through a heterogeneous groundwater basin (after Freeze and Witherspoon, 1968).

be difficult to predict without the use of a quantitative flow net. The crosshatched area above the horizontal zero line in the recharge-discharge profile represents the total groundwater recharge; the crosshatched area below the line represents the total groundwater discharge. For steady flow the two must equal each other.

The three-dimensional equivalent of a recharge-discharge profile is a contoured map of a drainage basin showing the areal distribution of the rates of recharge and discharge. The preparation of such a map in the field would require measurements of the saturated hydraulic conductivity of the near-surface geologic formations, and measurements or estimates of the hydraulic gradient at the water table.

There is one aspect of the arguments presented in this section that leads into a vicious circle. We have noted that the existing water-table configurations, which control the nature of the groundwater flow patterns, will influence rates of groundwater recharge. But it is also true that the patterns and amounts of recharge will control to a certain degree the configuration of the water table. Thus far we have assumed a fixed position of the water table and developed the recharge and discharge patterns. In reality, both the water-table configurations and the recharge patterns are largely controlled by the spatial and temporal patterns of precipitation and evapotranspiration at the ground surface. In the analyses of Sections 6.3 through 6.5, we will look at the saturated-unsaturated interactions that control the response of the water table under various climatic conditions.

Groundwater Recharge and Discharge as Components of a Hydrologic Budget

The recharge-discharge regime has important interrelationships with the other components of the hydrologic cycle. For example, in Figure 6.8 the entire regional flow from subsystem A discharges into the major valley at the left of the diagram. For any given set of topographic and hydrogeologic parameters we can calculate the average rate of discharge over the discharge area in, say, cm/day. In humid areas, this rate of upward-rising groundwater would be sufficient to keep water tables high while satisfying the needs of evapotranspiration, and still provide a base-flow component to a stream flowing perpendicular to the cross section. If such a stream had a tributary flowing across basin A from right to left, parallel to the cross section of Figure 6.8, one would expect the stream to be *influent* (losing water to the subsurface system) as it traverses the recharge area, and *effluent* (gaining water from the subsurface system) as it crosses the discharge area.

Quantification of these concepts requires the introduction of a *hydrologic budget* equation, or *water balance*, that describes the hydrologic regime in a watershed. If we limit ourselves to watersheds in which the surface-water divides and groundwater divides coincide, and for which there are no external inflows or outflows of groundwater, the water-balance equation for an annual period would take the form

$$P = Q + E + \Delta S_S + \Delta S_G \qquad (6.1)$$

where P is the precipitation, Q the runoff, E the evapotranspiration, ΔS_S the change in storage of the surface-water reservoir, and ΔS_G the change in storage of the groundwater reservoir (both saturated and unsaturated) during the annual period.

If we average over many years of record, it can be assumed that $\Delta S_S = \Delta S_G = 0$, and Eq. (6.1) becomes

$$P = Q + E \qquad (6.2)$$

where P is the average annual precipitation, Q the average annual runoff, and E the average annual evapotranspiration. The values of Q and E are usually reported in centimeters over the *drainage basin* so that their units in Eq. (6.2) are consistent with those for P. For example, in Figure 6.9(a), if the average annual precipitation, P, over the drainage basin is 70 cm/yr and the average annual evapotranspiration, E, is 45 cm/yr, the average annual runoff, Q, as measured in the stream at the outlet of the watershed but expressed as the equivalent number of centimeters of water over the drainage basin, would be 25 cm/yr.

Let us consider an idealization of the watershed shown in Figure 6.9(a), wherein most of the watershed comprises a recharge area, and the discharge area is limited to a very small area adjacent to the main stream. The groundwater flow net shown in Figure 6.8 might well be along the section $X - X'$. We can now write two hydrologic-budget equations, one for the recharge area and one for the discharge area.

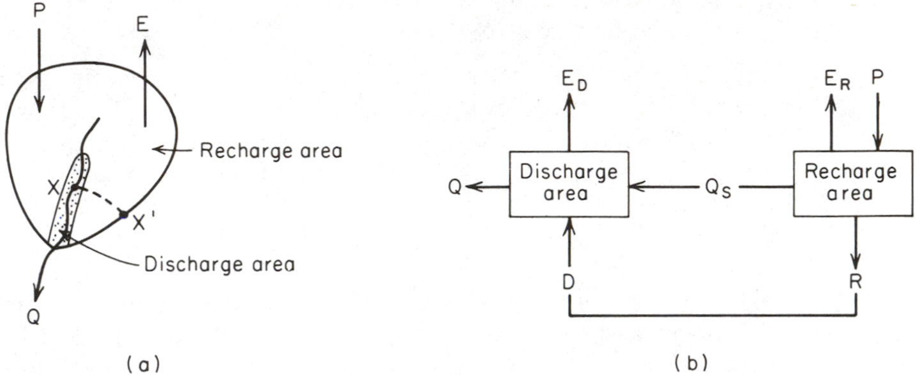

Figure 6.9 Steady-state hydrologic budget in a small watershed.

In the recharge area [Figure 6.9(b)],

$$P = Q_S + R + E_R \tag{6.3}$$

where Q_S is the surface-water component of average annual runoff, R the average annual groundwater recharge, and E_R the average annual evapotranspiration from the recharge area.

In the discharge area [Figure 6.9(b)],

$$Q = Q_S + D - E_D \tag{6.4}$$

where D is the average annual groundwater discharge (and equal to R) and E_D the average annual evapotranspiration from the discharge area. For a discharge area that constitutes a very small percentage of the basin area, P need not appear in Eq. (6.4).

If we set

$$Q_G = D - E_D \tag{6.5}$$

Eq. (6.4) becomes

$$Q = Q_S + Q_G \tag{6.6}$$

where Q_G is the groundwater component of average annual runoff (or average annual baseflow). Equation (6.5) reflects the earlier statement that groundwater discharge in a valley goes to satisfy both evapotranspiration demands and the subsurface component of streamflow. Equation (6.6) suggests that it might be possible to separate streamflow hydrographs into their surface-water and groundwater components; further consideration of this point is deferred until Section 6.6.

The application of the steady-state hydrologic-budget equations provides only a crude approximation of the hydrologic regime in a watershed. In the first place it is a *lumped-parameter* approach (rather than a *distributed-parameter* approach), which does not take into account the areal variations in P, E, R, and D. On an average annual basis, in a small watershed, areal variations in P and E may not be large, but we are aware, on the basis of Figure 6.8, that areal variations in R and D can be significant. Second, the *average annual* approach hides the importance of time-dependent effects. In many cases, the groundwater regime is approximated quite closely by a steady-state regime, but P, E, and Q are strongly time-dependent.

The foregoing discussion of steady-state hydrologic budgets is instructive in that it clarifies many of the interactions between groundwater flow and the other components of the hydrologic cycle. The application of Eqs. (6.2), (6.3), and (6.4) in practice, however, is fraught with problems. One needs several years of records of precipitation, P, and stream runoff, Q, at several sites. In principle, the groundwater components, R and D, can be determined by flow-net analysis, but in practice, the uncertainty surrounding hydraulic conductivity values in heterogeneous groundwater basins leads to a wide range of feasible R and D values. The evapotranspiration parameters, E_R and E_D, must be estimated on the basis of methods of questionable accuracy.

Of all these questions, it is the evapotranspiration estimates that pose the greatest problem. The most widely used methods of calculation utilize the concept of *potential evapotranspiration* (PE), which is defined as the amount of water that would be removed from the land surface by evaporation and transpiration processes if sufficient water were available in the soil to meet the demand. In a discharge area where upward-rising groundwater provides a continuous moisture supply, *actual evapotranspiration* (AE) may closely approach potential evapotranspiration. In a recharge area, actual evapotranspiration is always considerably less than potential. Potential evapotranspiration is dependent on the evaporative capacity of the atmosphere. It is a theoretical calculation based on meterological data. AE is the proportion of PE that is actually evapotranspired under the existing soil moisture supply. It is dependent on the unsaturated moisture storage properties of the soil. It is also affected by vegetative factors such as plant type and stage of growth. The most common methods of calculating potential evapotranspiration are those of Blaney and Criddle (1950), Thornthwaite (1948), Penman (1948), and Van Bavel (1966). The first two of these are based on empirical correlations between evapotranspiration and climatic factors. The last two are energy-budget approaches that have better physical foundations but require more meterological data. Pelton et al. (1960) and Gray et al. (1970) discuss the relative merits of the various techniques. The conversion of PE rates to AE rates in a recharge area is usually carried out with a soil-moisture budget approach. The Holmes and Robertson (1959) technique has been widely applied in the prairie environment.

For the specific case of phreatophytic evapotranspiration from a discharge area with a shallow water table, direct measurements of water-level fluctuations, as outlined in Section 6.8, can be used to calculate the actual evapotranspiration.

For examples of hydrologic-budget studies on small watersheds, in which special attention is paid to the groundwater component, the reader is directed to the reports of Schicht and Walton (1961), Rasmussen and Andreasen (1959), and Freeze (1967).

6.3 Transient Regional Groundwater Flow

Transient effects in groundwater flow systems are the result of time-dependent changes in the inflows and outflows at the ground surface. Precipitation rates, evapotranspiration rates, and snowmelt events are strongly time-dependent. Their transient influence is felt most strongly near the surface in the unsaturated zone, so any analysis of the transient behavior of natural groundwater flow must include both saturated and unsaturated zones.

As with steady-state regional flow, the main features of transient regional flow are most easily illustrated with the aid of numerical simulations carried out in hypothetical groundwater basins. Freeze (1971a), building on the earlier work of Rubin (1968), Hornberger et al. (1969), and Verma and Brutsaert (1970), described a mathematical model for three-dimensional, transient, saturated-unsaturated flow in a groundwater basin. His equation of flow couples the unsaturated flow equation [Eq. (2.80)] and the saturated flow equation [Eq. (2.74)] into an integrated form that allows treatment of the complete subsurface regime. The numerical solutions were obtained with a finite-difference technique known as *successive overrelaxation*. The model allows any generalized region shape and any configuration of time-variant boundary conditions. Here, we will look at the transient response in a two-dimensional cross section to a snowmelt-type infiltration event.

The region of flow is shown in Figure 6.10(a) (at a 2: 1 vertical exaggeration). The boundaries comprise a stream AB at constant hydraulic head, an impermeable basement $AFED$, and the ground surface BCD. The region contains a homogeneous and isotropic soil whose unsaturated characteristic curves are those of Figure 2.13.

As we have seen in Sections 2.6 and 5.4, saturated-unsaturated flow conditions can be presented in three ways: as a pressure-head field, as a moisture-content field, and as a total hydraulic-head field. From the first we can locate the position of the water table, and from the last we can make quantitative flow calculations. Figure 6.10(a), (b), and (c) shows these three fields at time $t = 0$ for initial conditions of steady-state flow resulting from the imposition of a constant hydraulic head along CD. The initial conditions feature a deep, nearly flat water table and very dry surface moisture conditions. At all times $t > 0$, a surface flux equivalent to $0.09K_0$ (where K_0 is the saturated hydraulic conductivity of the soil) is allowed to enter the flow system on the upper boundary. As shown on Figure 6.10(d), this rate of inflow creates a water-table rise that begins after 100 h and approaches the surface after 400 h. Figure 6.10(e) and (f) shows the moisture-content and total hydraulic-head fields at $t = 410$ h.

Figure 6.11 shows the effect on the flow system of the introduction of a heterogeneous geological configuration. The unstippled zone has the same soil properties as those for the homogeneous case of Figure 6.10, but a low-permeability clay layer has been inserted near the surface and a high-permeability basal aquifer at depth. The permeability and porosity relationships are noted in Figure 6.11(a). Figure 6.11(b) illustrates the transient response of the water table to the

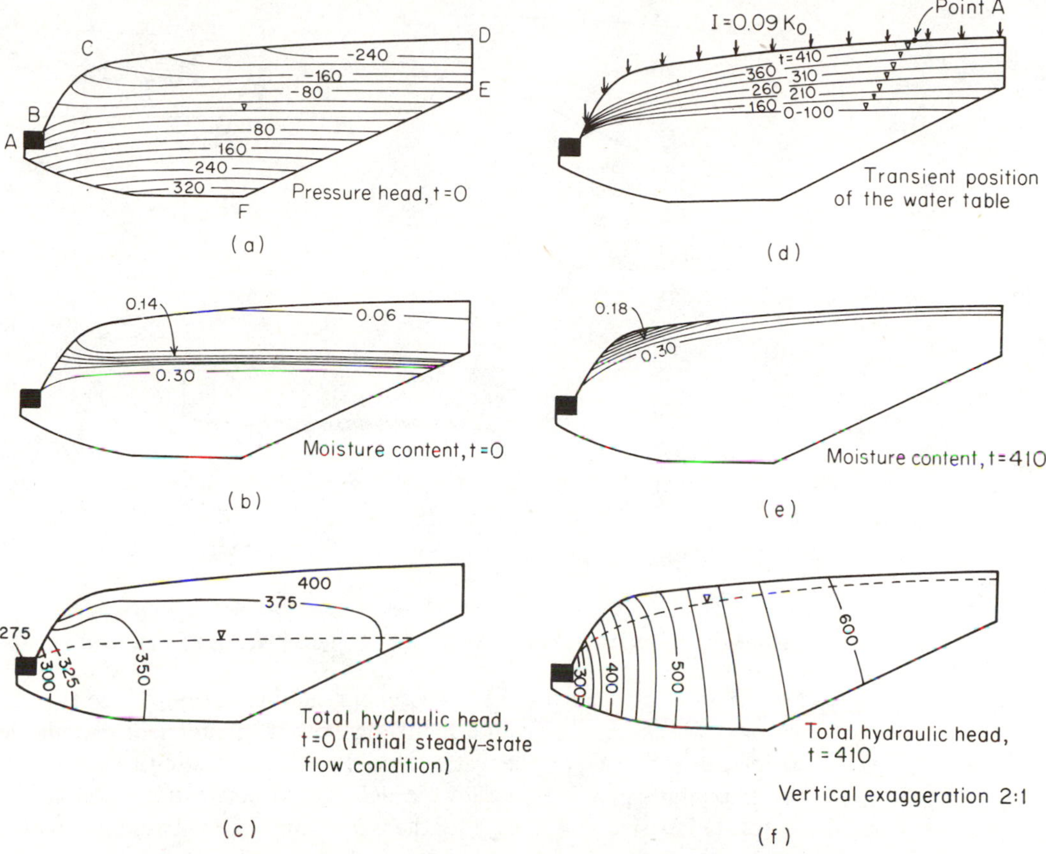

Figure 6.10 Transient response of a saturated-unsaturated flow system to a snowmelt-type infiltration (after Freeze, 1971a).

same surface inflow conditions as those of Figure 6.10. Figure 6.11(c) shows the total hydraulic head pattern at $t = 460$ h. This set of diagrams serves to clarify the saturated-unsaturated mechanisms that are operative in the formation of a perched water table.

If the hydraulic-head field in a watershed can be determined at various times by field measurement or mathematical simulation, it becomes possible to make a direct calculation of the amount of water discharging from the system as a function of time. If the discharge area is limited to a stream valley, the transient rate of groundwater discharge provides a measure of the baseflow hydrograph for the stream. Increased baseflow is the result of increased hydraulic gradients in the saturated zone near the stream, and, as the theoretical models show, this is itself a consequence of increased up-basin gradients created by a water-table rise. The time lag between a surface-infiltration event and an increase in stream baseflow is therefore directly related to the time required for an infiltration event to induce a

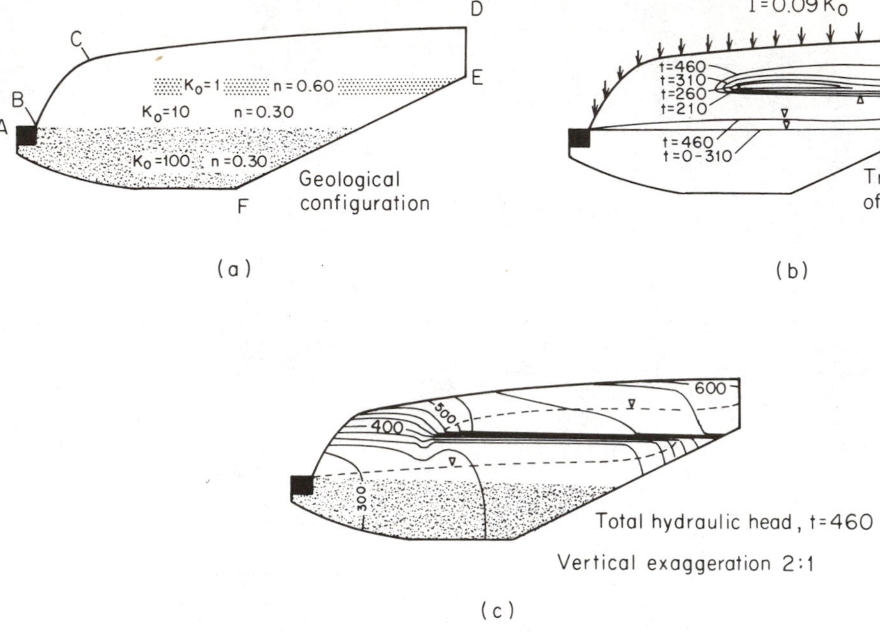

Figure 6.11 Formation of a perched water table (after Freeze, 1971a).

widespread water-table rise. Figure 6.12 is a schematic illustration of the type of baseflow hydrograph that results from a hydrologic event of sufficient magnitude to exert a basin-wide influence on the water table. Baseflow rates must lie between D_{maximum}, the maximum possible baseflow, which would occur under conditions of a fully saturated basin, and D_{minimum}, the minimum likely baseflow, which would occur under conditions of the lowest recorded water-table configuration.

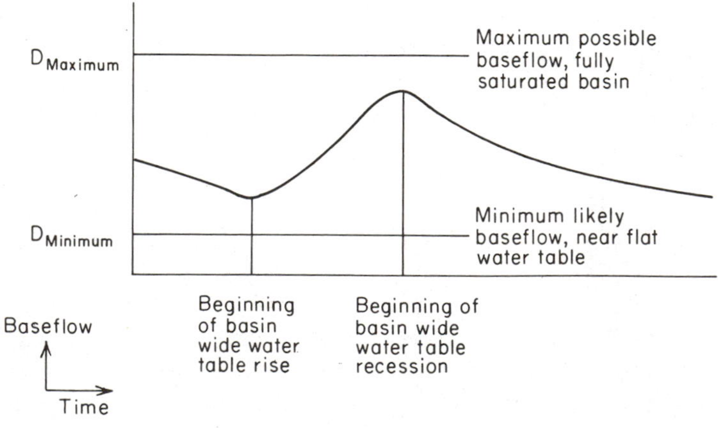

Figure 6.12 Schematic diagram of baseflow hydrograph (after Freeze, 1971a).

Quantitative calculations can also be carried out at the inflow end of the system to examine the interrelationship between infiltration and groundwater recharge. The concepts are clearest, however, when one works with the one-dimensional system outlined in the following section.

6.4 Infiltration and Groundwater Recharge

In Section 6.1, we defined the terms *recharge area* and *discharge area*; in Section 6.2, we first calculated recharge and discharge rates. Let us formalize these concepts with the following definitions for the processes of recharge and discharge.

Groundwater *recharge* can be defined as the entry into the saturated zone of water made available at the water-table surface, together with the associated flow away from the water table within the saturated zone.

Groundwater *discharge* can be defined as the removal of water from the saturated zone across the water-table surface, together with the associated flow toward the water table within the saturated zone.

It should be clear from the previous section that these two saturated processes are intimately interrelated to a pair of parallel processes in the unsaturated zone. Let us define the process of *infiltration* as the entry into the soil of water made available at the ground surface, together with the associated flow away from the ground surface within the unsaturated zone.

In a similar fashion, we will define *exfiltration* as the removal of water from the soil across the ground surface, together with the associated flow toward the ground surface within the unsaturated zone. This term was coined by Philip (1957f), but it is not yet widely used. The process is often referred to as *evaporation*, but this leads to confusion as to whether the meteorological processes in the atmosphere are included.

The Theory of Infiltration

The process of infiltration has been widely studied by both hydrologists and soil physicists. In hydrology, Horton (1933) showed that rainfall, when it reaches the ground surface, infiltrates the surface soils at a rate that decreases with time. He pointed out that for any given soil there is a limiting curve that defines the maximum possible rates of infiltration versus time. For heavy rains, the actual infiltration will follow this limiting curve, which he called the curve of *infiltration capacity* of the soil. The capacity decreases with time after the onset of rainfall and ultimately reaches an approximately constant rate. The decline is caused mainly by the filling of the soil pores by water. Controlled tests carried out on various soil types by many hydrologists over the years have shown that the decline is more rapid and the final constant rate is lower for clay soils with fine pores than for open-textured sandy soils. If at any time during a rainfall event the rate of rainfall exceeds the infiltration capacity, excess water will pond on the soil surface. It is this ponded water that is available for *overland flow* to surface streams.

The hydrologic concept of infiltration capacity is an empirical concept based on observations at the ground surface. A more physically based approach can be found in the soil physics literature, where infiltration is studied as an unsaturated subsurface flow process. Most analyses have considered a one-dimensional vertical flow system with an inflow boundary at the top. Bodman and Colman (1943) provided the early experimental analyses, and Philip (1957a, 1957b, 1957c, 1957d, 1957e, 1958a, 1958b), in his classic seven-part paper, utilized analytical solutions to the one-dimensional boundary-value problem to expose the basic physical principles on which later analyses rest. Almost all of the more recent theoretical treatments have employed a numerical approach to solve the one-dimensional system. This approach is the only one capable of adequately representing the complexities of real systems. Freeze (1969b) provides a review of the numerical infiltration literature in tabular form.

From a hydrologic point of view, the most important contributions are those of Rubin et al. (1963, 1964). Their work showed that Horton's observed curves of infiltration versus time can be theoretically predicted, given the rainfall intensity, the initial soil-moisture conditions, and the set of unsaturated characteristic curves for the soil. If rainfall rates, infiltration rates, and hydraulic conductivities are all expressed in units of $[L/T]$, Rubin and his coworkers showed that the final constant infiltration rate in the Horton curves is numerically equivalent to the saturated hydraulic conductivity of the soil. They also identified the necessary conditions for ponding as twofold: (1) the rainfall intensity must be greater than the saturated hydraulic conductivity, and (2) the rainfall duration must be greater than the time required for the soil to become saturated at the surface.

These concepts become clearer if we look at an actual example. Consider a one-dimensional vertical system (say, beneath point A in Figure 6.10) with its upper boundary at the ground surface and its lower boundary just below the water table. The equation of flow in this saturated-unsaturated system will be the one-dimensional form of Eq. (2.80):

$$\frac{\partial}{\partial z}\left[K(\psi)\left(\frac{\partial \psi}{\partial z} + 1\right)\right] = C(\psi)\frac{\partial \psi}{\partial t} \tag{6.7}$$

where ψ $(= h - z)$ is the pressure head, and $K(\psi)$ and $C(\psi)$ are the unsaturated functional relationships for hydraulic conductivity K and specific moisture capacity C. In the saturated zone below the water table (or more accurately, below the point where $\psi = \psi_a$, ψ_a being the air-entry pressure head), $K(\psi) = K_0$ and $C(\psi) = 0$, where K_0 is the saturated hydraulic conductivity of the soil.

Let us specify a rainfall rate R at the upper boundary. From Darcy's law,

$$R = K(\psi)\frac{\partial h}{\partial z} = K(\psi)\left(\frac{\partial \psi}{\partial z} + 1\right) \tag{6.8}$$

or

$$\frac{\partial \psi}{\partial z} = \frac{R}{K(\psi)} - 1 \tag{6.9}$$

If the rate of groundwater recharge to the regional flow system is Q, then, by analogy to Eq. (6.9), the condition at the saturated base of the system is

$$\frac{\partial \psi}{\partial z} = \frac{Q}{K_0} - 1 \tag{6.10}$$

The boundary-value problem defined by Eqs. (6.7), (6.9), and (6.10) was solved by Freeze (1969b) with a numerical finite-difference method that is briefly outlined in Appendix VIII. Figure 6.13 shows the results of a representative simulation of a hypothetical infiltration event. The three profiles show the time-dependent response of the moisture content, pressure head, and hydraulic head in the upper 100 cm of a soil with unsaturated hydrologic properties identical to those shown in Figure 2.13. The transient behavior occurs in response to a constant-intensity rainfall that feeds the soil surface at the rate $R = 0.13$ cm/min. This rate is 5 times the saturated hydraulic conductivity of the soil, $K_0 = 0.026$ cm/min. The initial conditions are shown by the $t = 0$ curves, and subsequent curves are labeled with the time in minutes.

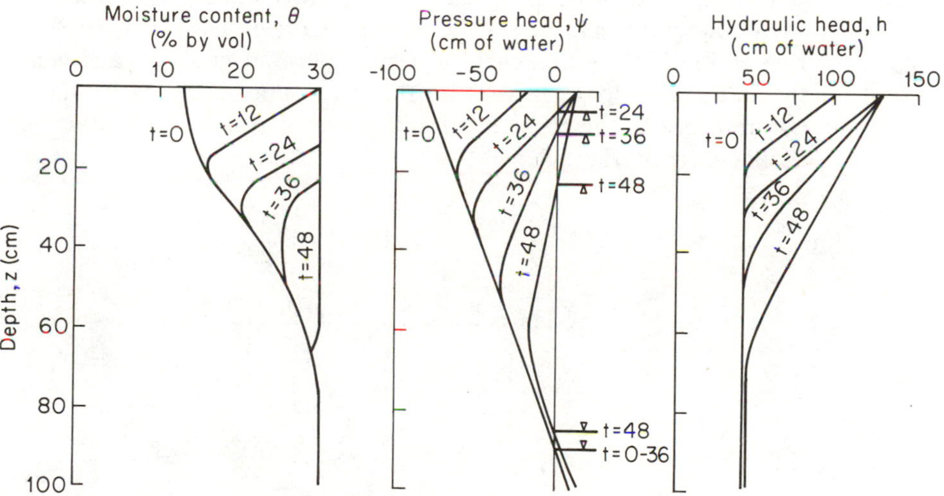

Figure 6.13 Numerical simulation of a hypothetical infiltration event (after Freeze, 1974).

The left-hand diagram shows how the moisture content increases down the profile with time. The surface becomes saturated after 12 min, and the soil pores in the entire profile are almost filled with water after 48 min.

The central diagram shows the pressure-head changes. The pressure-head curve for $t = 12$ min does not reach the $\psi = 0$ point, so the upper few centimeters of surface saturation, indicated by the moisture-content profile, must be "tension-saturated." By the 24-min mark the pressure head at the ground surface has reached $+10$ cm, the indication being that a 10-cm-deep layer of water is ponded on the surface at this point in time. (In this simulation, the maximum allowable ponding depth had been preset at 10 cm.) There is also an inverted water table 5 cm below the ground surface which propagates down the profile with time. The true water table, which is initially at 95 cm depth, remains stationary through the first 36 min but then begins to rise in response to the infiltrating moisture from above.

The hydraulic head profiles near the surface on the right-hand diagram provide the hydraulic-gradient values that can be inserted in Darcy's law to calculate the rate of infiltration at various times. The datum for the values appearing on the horizontal scale at the top was arbitrarily chosen as 125 cm below the ground surface.

Figure 6.14 shows the time-dependent infiltration rate at the ground surface for the constant-rainfall case shown in Figure 6.13. As predicted by Rubin and Steinhardt (1963), the infiltration rate is equal to the rainfall rate until the soil becomes saturated at the surface (and the 10-cm deep pond has been filled); then it decreases asymptotically toward a value equal to K_0. During the early period, as the soil pores are filling up with water, the moisture contents, pressure heads, and hydraulic heads are increasing with time and the downward hydraulic gradient is decreasing. This decrease is balanced by an increase in the hydraulic conductivity values under the influence of the rising pressure heads. The decrease in infiltration rate occurs at the point when the combination of gradients and conductivities in the soil can no longer accept all the water supplied by the rainfall. The rainfall not absorbed by the ground as infiltration nor stored in the 10-cm-deep pond is available for overland flow.

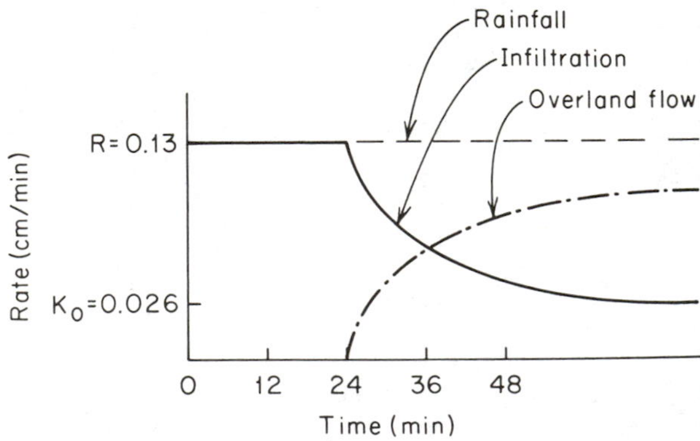

Figure 6.14 Time-dependent rates of infiltration and overland flow for the case shown in Figure 6.13 (after Freeze, 1974).

A similar approach can be used to simulate cases with evaporation at the surface (R negative) or discharge at depth (Q negative), or to analyze redistribution patterns that occur between rainfall events.

The question of whether a given input and a given set of initial conditions and soil type will give rise to groundwater recharge is actually a question of whether this set of conditions will result in a water-table rise. The rise provides the source of replenishment that allows the prevailing rate of recharge to continue. The possibility of a water-table rise is greater for (1) low-intensity rainfalls of long duration rather than high-intensity rainfalls of short duration, (2) shallow water tables rather than deep, (3) low groundwater recharge rates rather than high, (4) wet antecedent moisture conditions rather than dry, and (5) soils whose characteristic curves show high conductivity, low specific moisture capacity, or high moisture content over a considerable range of pressure-head values.

Measurements of Field Sites

In some hydrogeological environments, cases of recharge-sustaining infiltration to the water table are isolated in time and space. In such cases, the types of hydrologic events that lead to recharge are best identified on the basis of field measurement. In the past this was often done on the basis of observation-well hydrographs of water-table fluctuations. However, as indicated in Section 6.8, there are a variety of phenomena that can lead to water-table fluctuations, and not all represent true groundwater recharge. The safest course is to supplement the observation-well records with measurements of hydraulic head both above and below the water table. Figure 6.15 shows a set of field instrumentation designed to this end. Figure 6.16 displays the soil-moisture and water-table response recorded at an instrumented site in east-central Saskatchewan during a dry period punctuated by a single heavy rainfall. The water-table rise is the result of direct infiltration from above.

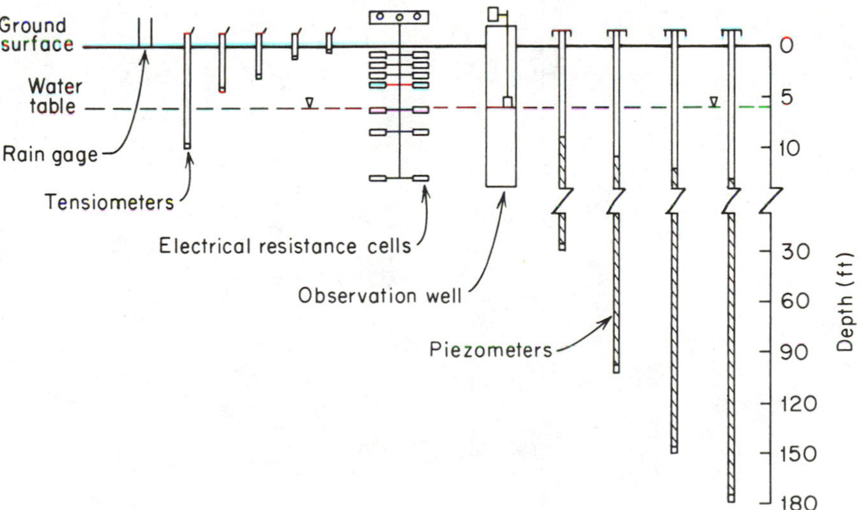

Figure 6.15 Field instrumentation for the investigation of groundwater recharge processes (after Freeze and Banner, 1970).

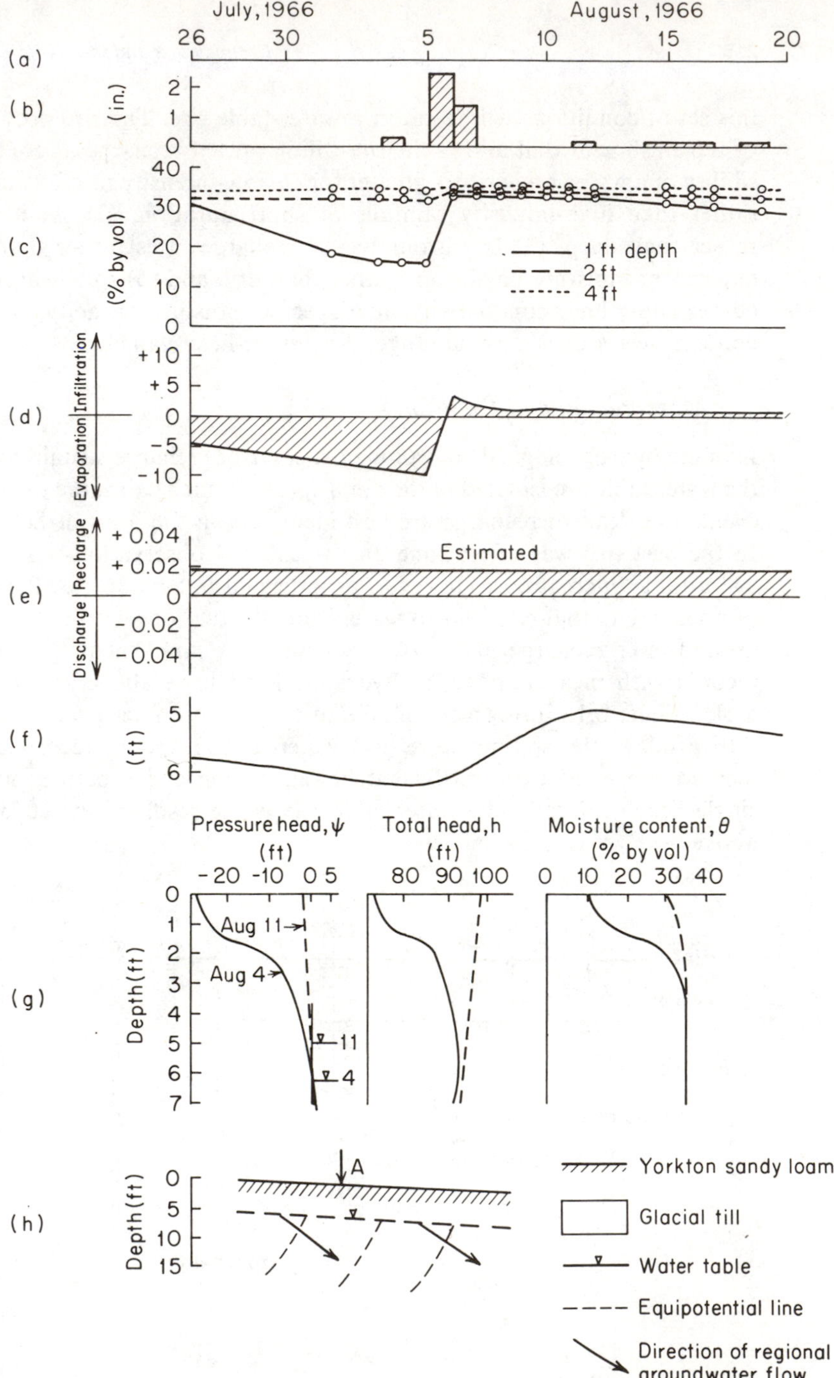

Figure 6.16 Water-table rise created by infiltration from a heavy summer rainfall. (a) Date; (b) rainfall; (c) soil moisture content; (d) vertical hydraulic gradient in the unsaturated zone; (e) vertical hydraulic gradient in the saturated zone; (f) depth to water table; (g) pressure head, total head, and moisture

At another site nearby, the same rainfall did not result in infiltration to the water table, despite the fact that the saturated hydraulic conductivity there was much higher than at the site shown in Figure 6.16. The characteristic curves of the sandy soil at the second site gave rise to a deep water table and very dry near-surface soil-moisture conditions. As noted by Freeze and Banner (1970), estimates of the infiltration and recharge properties of a soil based only on knowledge of the saturated hydraulic conductivity of the soil and its textural classification can often be misleading. One should not map a sand or gravel plain as an effective recharge area without first investigating the water-table depth and the nature of the unsaturated functional relationships for the soil. Small differences in the hydrologic properties of similar field soils can account for large differences in their reaction to the same hydrologic event.

The mechanisms of infiltration and groundwater recharge are not always one-dimensional. In hilly areas certain portions of a groundwater recharge area may never receive direct infiltration to the water table. Rather, recharge may be concentrated in depressions where temporary ponds develop during storms or snowmelt periods. Lissey (1968) has referred to this type of recharge as *depression-focused.* Under such conditions, the water table still undergoes a basinwide rise. The rise is due to vertical infiltration beneath the points of recharge and subsequent horizontal flow toward the water-table depressions created between these points. Further discussion on the interactions between groundwater and ponds is withheld until Section 6.7.

6.5 Hillslope Hydrology and Streamflow Generation

The relationship between rainfall and runoff is at the very core of hydrology. In a scientific sense, there is a need to understand the *mechanisms* of watershed response. In an engineering sense, there is a need for better techniques for the *prediction* of runoff from rainfall. We know, of course, that the larger rivers are fed by smaller tributaries, and it is this network of small tributary streams that drains by far the largest percentage of the land surface. We will therefore focus on the ways in which water moves into small stream channels in upstream tributary drainage basins during and between rainfall events.

The path by which water reaches a stream depends upon such controls as climate, geology, topography, soils, vegetation, and land use. In various parts of the world, and even in various parts of the same watershed, different processes may generate streamflow, or the relative importance of the various processes may differ. Nevertheless, it has been recognized that there are essentially three processes that feed streams. As illustrated in Figure 6.17, these are *overland flow*, *subsurface stormflow* (or interflow), and *groundwater flow*. An insight into the nature of the

---

content profiles; (h) regional hydrogeological setting (after Freeze and Banner, 1970).

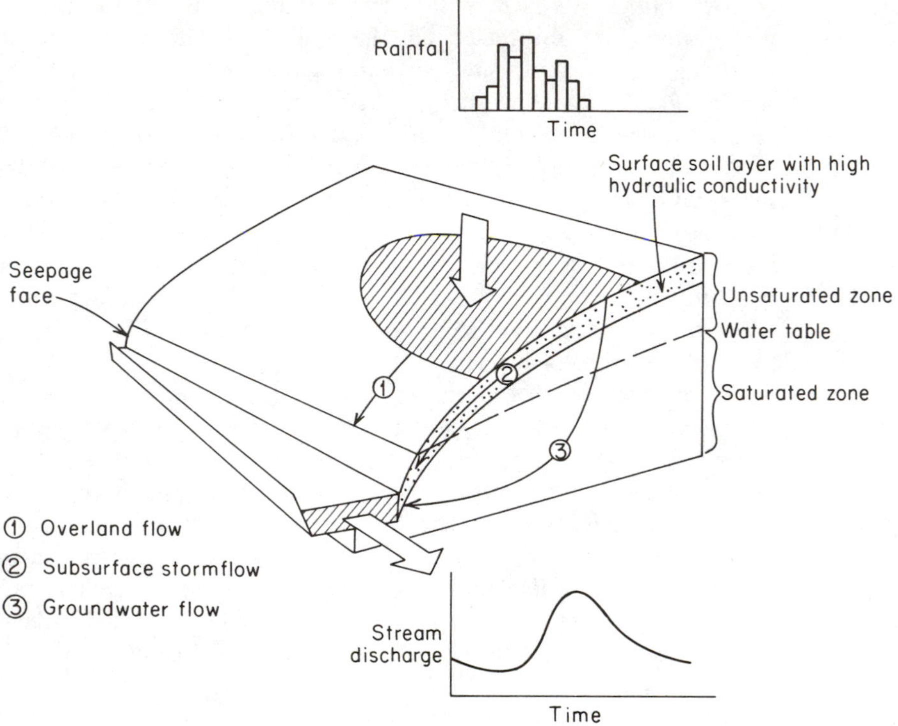

Figure 6.17 Mechanisms of delivery of rainfall to a stream channel from a hillslope in a small tributary watershed (after Freeze, 1974).

subsurface flow regime is necessary for understanding the production of runoff by any of these three mechanisms.

The role of the regional groundwater flow system in delivering baseflow to a stream was covered in Sections 6.2 and 6.3. Although it may sometimes contribute to runoff during storms, its primary role is in sustaining streams during low-flow periods between rainfall and snowmelt events. We will focus our interest in this section on overland flow and subsurface stormflow.

Overland Flow

The classic concept of streamflow generation by overland flow is due to Horton (1933). The dependence of overland flow on the infiltration regime in the unsaturated surface soils of a watershed has been discussed in the previous section. The concepts are summarized in Figure 6.14.

As originally presented, Horton's theory implied that most rainfall events exceed infiltration capacities and that overland flow is common and areally widespread. Later workers recognized that the great heterogeneity in soil types at the ground surface over a watershed and the very irregular patterns of precipitation in both time and space create a very complex hydrologic response on the land surface. This led to the development of the *partial-area-contribution concept* (Betson, 1964;

Ragan, 1968), wherein it is recognized that certain portions of the watershed regularly contribute overland flow to streams, whereas others seldom or never do. The conclusion of most recent field studies is that overland flow is a relatively rare occurrence in time and space, especially in humid, vegetated basins. Most overland-flow hydrographs originate from small portions of the watershed that constitute no more than 10%, and often as little as 1–3%, of the basin area, and even on these restricted areas only 10–30% of the rainfalls cause overland flow.

Freeze (1972b) provided a heuristic argument based on the theory of infiltration and the ponding criteria of Rubin and Steinhardt (1963) to explain the paucity of overland-flow occurrences.

Subsurface Stormflow

The second widely held concept of surface-runoff generation promotes subsurface stormflow as a primary source of runoff. Hewlett and Hibbert (1963) showed the feasibility of such flow experimentally, and Whipkey (1965) and Hewlett and Hibbert (1967) measured lateral inflows to streams from subsurface sources in the field. The prime requirement is a shallow soil horizon of high permeability at the surface. There is reason to suppose that such surface layers are quite common in the form of the A soil horizon, or as agriculturally tilled soils or forest litter.

On the basis of simulations with a mathematical model of transient, saturated-unsaturated, subsurface flow in a two-dimensional, hillslope cross section, Freeze (1972b) concluded that subsurface stormflow can become a quantitatively significant runoff component only on convex hillslopes that feed deeply incised channels, and then only when the permeabilities of the soils on the hillslope are in the very highest bracket of the feasible range. Figure 6.18 shows three simulated

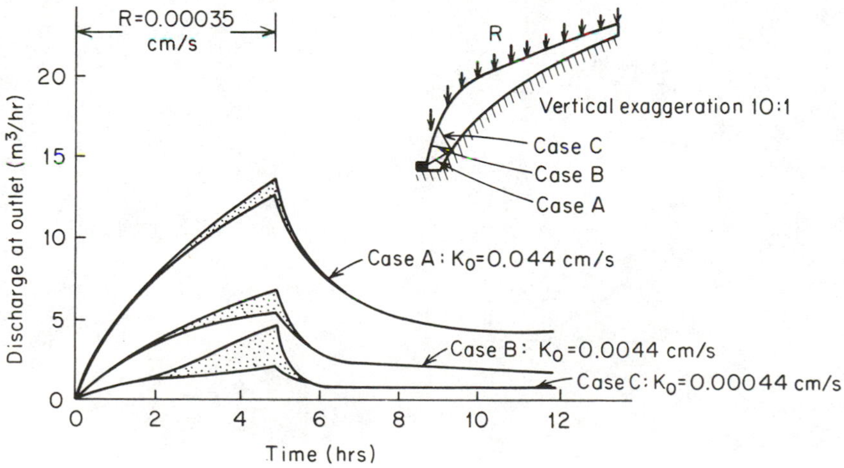

Figure 6.18 Numerically simulated streamflow hydrographs at the outlet of a hypothetical upstream source area in which the stream is fed by a convex hillslope having a shallow surface soil of high conductivity (after Freeze, 1974).

hydrographs for the hillslope cross section shown in the inset. The three cases each differ by an order of magnitude in the saturated hydraulic conductivity, K_0, of the hillslope soil. The line below the stippled regions represents the subsurface stormflow contribution. In each case, one result of the saturated-unsaturated process in the hillslope is a rising water table near the valley (as indicated for $t = 5$ h on the inset). Overland flow from direct precipitation on the saturated wetland created on the streambank by the rising water tables is shown by the stippled portions of the hydrographs. Only curves A and B show a dominance of the storm hydrograph by subsurface stormflow, and the K_0 values for these curves are in the uppermost range of reported field measurement. On concave slopes the saturated valley wetlands become larger more quickly and overland flow from direct precipitation on these areas usually exceeds subsurface stormflow, even where hillslope soils are highly permeable.

In the Sleepers River experimental watershed in Vermont (Figure 6.19), Dunne

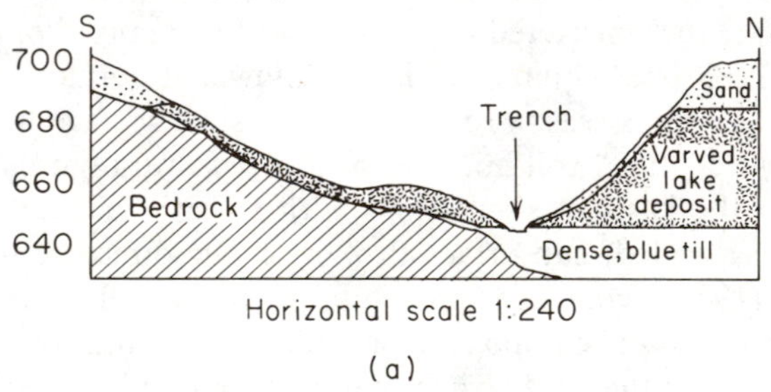

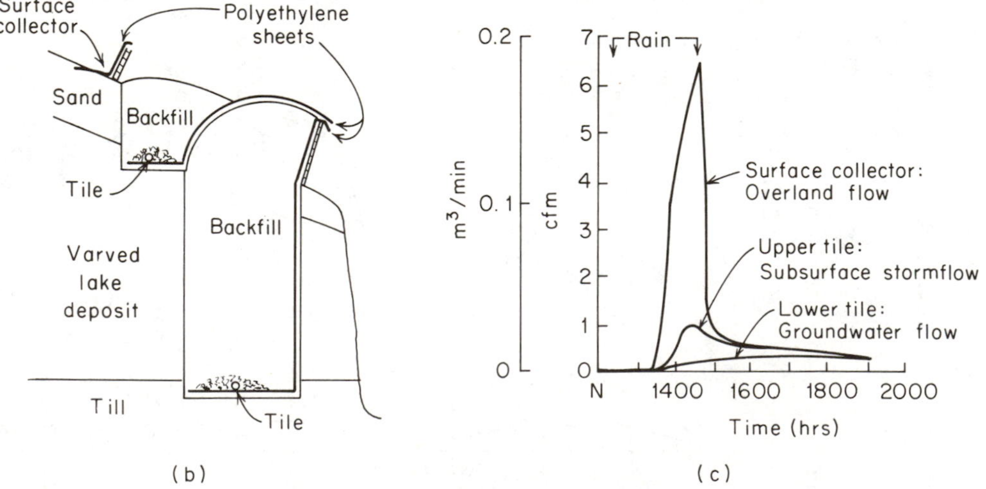

Figure 6.19 Sleepers River experimental watershed, Vermont. (a) Geologic cross section through the hillslope; (b) cross section of the interceptor trench; (c) combined hydrograph of flows during a sample storm (after Dunne and Black, 1970a).

and Black (1970a, b), working with an integrated set of surface and subsurface instrumentation, including an interceptor trench [Figure 6.19(b)], were able to measure simultaneous hydrographs of each of the three component outflows from the hillslope to the stream. The example shown in Figure 6.19(c) displays the preponderance of overland flow that was a recurring feature of measurements in the Sleepers River watershed. Auxiliary instrumentation showed that the contributing areas, as in case *C* in Figure 6.18, were limited to topographically low wetlands created by rising water tables adjacent to the stream channel.

One feature of the streamflow-generating mechanism uncovered at the Sleepers River watershed has been widely reported (Hewlett and Nutter, 1970) in many other watersheds in humid climates. We refer to the expansion and contraction of wetlands during and following storms under the influence of the subsurface flow system. The resulting variation with time in the size of contributing areas is often referred to as the *variable-source-area concept*. It differs from the *partial-area concept* in two ways. First, partial areas are thought of as being more-or-less fixed in location, whereas variable areas expand and contract. Second, partial areas feed water to streams by means of Hortonian overland flow, that is, by water that ponds on the surface due to saturation of the soils at the surface from *above*, whereas variable areas are created when surface saturation occurs from *below*. In the Sleepers River watershed, the majority of the overland flow that arrived at the stream from the variable source areas was created by direct precipitation on the wetlands. In many forested watersheds (Hewlett and Nutter, 1970), a significant proportion of the water arising from variable source areas arrives there by means of subsurface stormflow. Table 6.1 provides a summary of the various storm runoff processes in relation to their major controls.

In recent years there has been rapid growth in the development of physically based hydrologic prediction models that couple surface and subsurface flow. Smith and Woolhiser (1971) have produced a model for the simulation of overland flow on an infiltrating hillslope, and Freeze (1972a) has produced a model that couples saturated-unsaturated flow and streamflow. Stephenson and Freeze (1974) report on the use of the latter model to complement a field study of snowmelt runoff in a small upstream source area in the Reynolds Creek experimental watershed in Idaho.

Chemical and Isotopic Indicators

There are three main approaches that can be used in studies of the processes of streamflow generation during storm runoff: (1) hydrometric monitoring using instruments such as current meters, rain gages, observation wells, and tensiometers; (2) mathematical simulations; and (3) monitoring of dissolved constituents and environmental isotopes, such as ^{2}H, ^{3}H, and ^{18}O. Information obtained from the first two methods served as the basis for the discussion presented above. We will now focus on the hydrochemical and isotope approach.

The chemical mass-balance equation of dissolved constituents in streamflow at a particular sampling location at the stream at a specified time can be expressed

Table 6.1 Schematic Illustration of the Occurrence of Various Storm Runoff Processes in Relation to Their Major Controls

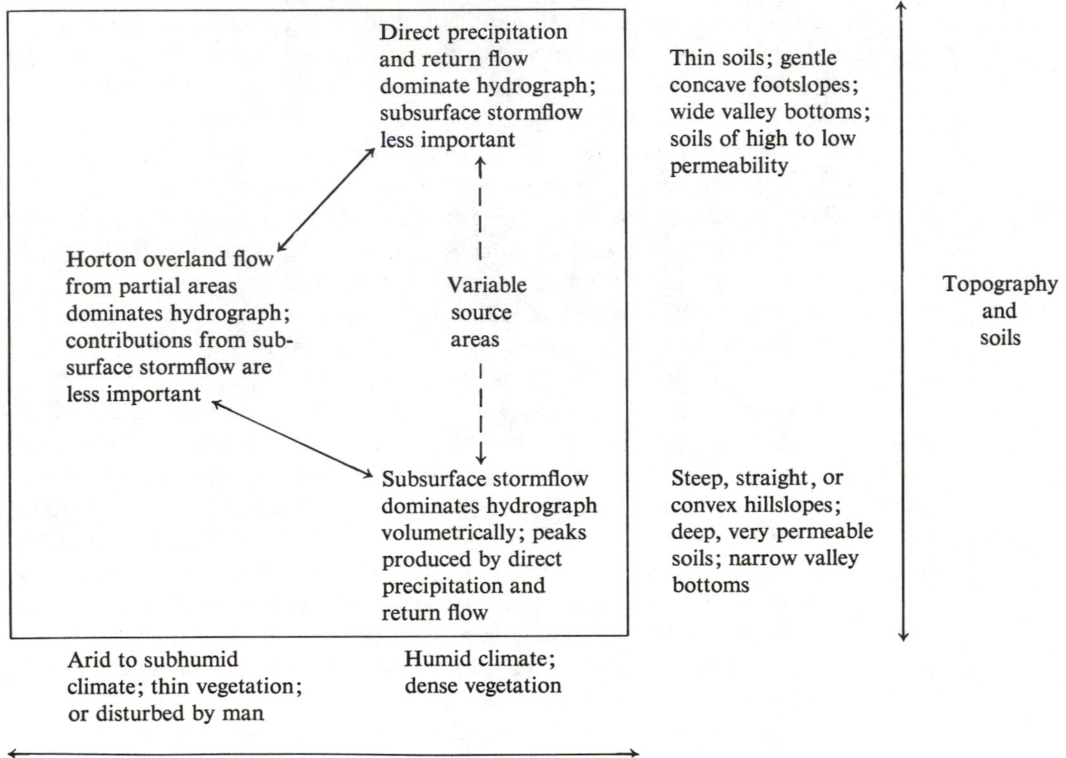

SOURCE: Dunne, 1978.

as

$$CQ = C_pQ_p + C_oQ_o + C_sQ_s + C_gQ_g \tag{6.11}$$

where C is the concentration in the stream water of the constituent under consideration, such as Cl^-, SO_4^{2-}, or HCO_3^-, and Q is the stream discharge $[L^3/T]$. Q_p, Q_o, Q_s, and Q_g represent the contributions to the streamflow from direct rainfall on the stream, overland flow, subsurface stormflow, and groundwater flow, respectively. C_p, C_o, C_s, and C_g are the concentrations of the chemical constituent in these streamflow components. The mass-balance equation for streamflow at the same location is

$$Q = Q_p + Q_o + Q_s + Q_g \tag{6.12}$$

Values for Q are obtained by measuring the streamflow. C is obtained by chemical analysis of samples from the stream at the location where Q is measured. For narrow streams in headwater basins, Q_p is often negligible relative to Q. This leaves two equations with six unknown quantities, C_o, C_s, C_g, Q_o, Q_s, and Q_g. A pragmatic approach at this point is to lump Q_o and Q_s together as a component referred to as direct runoff (Q_d), which represents the component of rainfall that moves rapidly across or through the ground into the stream. C_d is defined as the representative concentration in this runoff water. Substitution of these terms in Eqs. (6.11) and (6.12) and combining these equations yields

$$Q_g = Q\left(\frac{C - C_d}{C_g - C_d}\right) \tag{6.13}$$

Values of C_g are normally obtained by sampling shallow wells or piezometers near the stream or by sampling the stream baseflow prior to, or after, the storm. The second method is appropriate if the stream is fed only by shallow groundwater during baseflow periods. Values of C_d are obtained by sampling surface drainage or soil-zone seepage near the stream during the storm-runoff period. If analyses of these samples yield no excessive variation in space and time, the choice of a representative or average concentration is not unduly subjective. In sedimentary terrain, C_d is generally small relative to C_g because the groundwater has traveled deeper and has a much longer residence time. Substitution of the values of C_d and C_g along with the stream-water parameters, C and Q into Eq. (6.13) yields a value of Q_g, the groundwater component of the streamflow. If C and Q are measured at various times during the storm-runoff period, the variation of Q_g can be computed, as shown schematically in Figure 6.20.

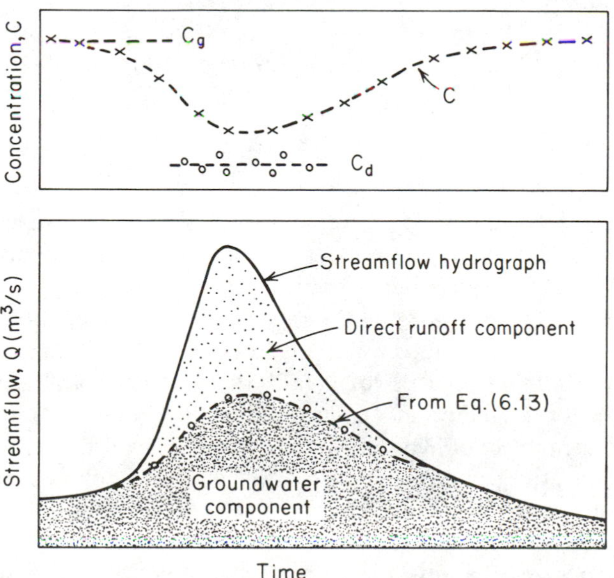

Figure 6.20 Streamflow hydrograph separation by the hydrochemical method.

Pinder and Jones (1969) used variations in Na^{2+}, Ca^{2+}, Mg^{2+}, Cl^{-}, $SO_4{}^{2-}$, and HCO_3^- in their study of storm-runoff components in small headwater basins in sedimentary terrain in Nova Scotia. In a similar investigation in Manitoba, Newbury et al. (1969) found $SO_4{}^{2-}$ and electrical conductance to be the best indicators for identification of the groundwater component in that area. In these and many other investigations using the hydrochemical method, it is commonly concluded that the groundwater-derived component of streamflow during peak runoff is appreciable. Pinder and Jones, for example, reported values in the range 32–42%.

One of the main limitations in the hydrochemical method is that the chemical concentrations used for the shallow groundwater and to represent the direct runoff are lumped parameters that may not adequately represent the water that actually contributes to the stream during the storm. The chemistry of shallow groundwater obtained from wells near streams is commonly quite variable spatially. Direct runoff is a very ephemeral entity that may vary considerably in concentration in time and space.

To avoid some of the main uncertainties inherent in the hydrochemical method, the naturally occurring isotopes ^{18}O, ^{2}H, and ^{3}H can be used as indicators of the groundwater component of streamflow during periods of storm runoff. Fritz et al. (1976) utilized ^{18}O, noting that its concentration is generally very uniform in shallow groundwater and baseflow. Although the mean annual values of ^{18}O in rain at a given location have little variation, the ^{18}O content of rain varies considerably from storm to storm and even during individual rainfall events. The ^{18}O method is suited for the type of rainfall event in which the ^{18}O content of the rain is relatively constant and is much different from the shallow groundwater or baseflow. In this situation the ^{18}O of the rain is a diagnostic tracer of the rainwater that falls on the basin during the storm. From the mass-balance considerations used for Eq. (6.13), the following relation is obtained:

$$Q_g = Q_w\left(\frac{\delta^{18}O_w - \delta^{18}O_R}{\delta^{18}O_g - \delta^{18}O_R}\right) \qquad (6.14)$$

where $\delta^{18}O$ denotes the ^{18}O content in per mille relative to the SMOW standard (Section 3.8) and the subscripts w, g, and R indicate stream water, shallow groundwater, and runoff water derived from the rainfall ($Q_w = Q_g + Q_R$). This relation provides for separation of the rain-derived component from the component of the streamflow represented by water that was in storage in the groundwater zone prior to the rainfall event. Fritz et al. (1976), Sklash et al. (1976), and Sklash (1978) applied this method in studies of streamflow generation in small headwater basins in several types of hydrogeologic settings. They found that even during peak runoff periods, the groundwater component of streamflow is considerable, often as much as half to two-thirds of the total streamflow. The resolution of the apparent contradictions between the mechanisms of streamflow generation suggested by the hydrochemical and isotopic approaches and those suggested by hydrometric measurements remains a subject of active research.

6.6 Baseflow Recession and Bank Storage

It should now be clear that streamflow hydrographs reflect two very different types of contributions from the watershed. The peaks, which are delivered to the stream by overland flow and subsurface stormflow, and sometimes by groundwater flow, are the result of a fast response to short-term changes in the subsurface flow systems in hillslopes adjacent to channels. The baseflow, which is delivered to the stream by deeper groundwater flow, is the result of a slow response to long-term changes in the regional groundwater flow systems.

It is natural to inquire whether these two components can be separated on the basis of a direct examination of the hydrograph alone without recourse to chemical data. Surface-water hydrologists have put considerable effort into the development of such techniques of *hydrograph separation* as a means of improving streamflow-prediction models. Groundwater hydrologists are interested in the indirect evidence that a separation might provide about the nature of the groundwater regime in a watershed. The approach has not led to unqualified success, but the success that has been achieved has been based on the concept of the *baseflow recession curve*.

Consider the stream hydrograph shown in Figure 6.21. Flow varies through

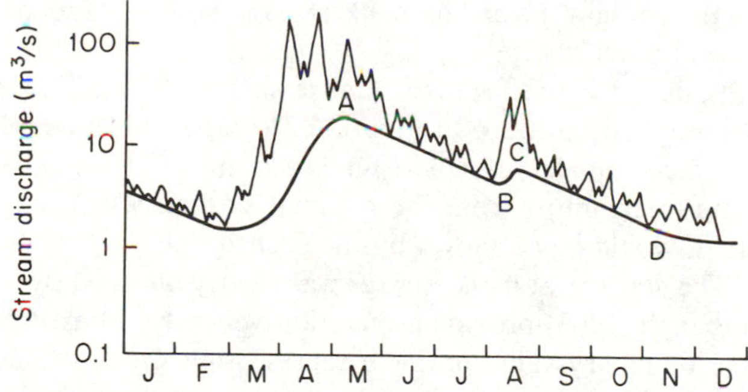

Figure 6.21 Baseflow recession curve for a hypothetical stream hydrograph.

the year from 1 m³/s to over 100 m³/s. The smooth line is the baseflow curve. It reflects the seasonally transient groundwater contributions. The flashy flows above the line represent the fast-response, storm-runoff contributions. If the stream discharge is plotted on a logarithmic scale, as it is in Figure 6.21, the recession portion of the baseflow curve very often takes the form of a straight line or a series of straight lines, such as AB and CD. The equation that describes a straight-line recession on a semilogarithmic plot is

$$Q = Q_0 e^{-at} \tag{6.15}$$

where Q_0 is the baseflow at time $t = 0$ and Q is the baseflow at a later time, t.

The general validity of this equation can be confirmed on theoretical grounds. As first shown by Boussinesq (1904), if one solves the boundary-value problem that represents free-surface flow to a stream in an unconfined aquifer under Dupuit-Forchheimer assumptions (Section 5.5), the analytical expression for the outflow from the system takes the form of Eq. (6.15). Singh (1969) has produced sets of theoretical baseflow curves based on analytical solutions to this type of boundary-value problem. Hall (1968) provides a complete historical review of baseflow recession.

In Figure 6.21 the rising portions of the baseflow hydrograph must fit within the conceptual framework outlined in connection with Figure 6.12. Many authors, among them Farvolden (1963), Meyboom (1961), and Ineson and Downing (1964), have utilized baseflow-recession curves to reach interpretive conclusions regarding the hydrogeology of watersheds.

In the upper reaches of a watershed, subsurface contributions to streamflow aid in the buildup of the flood wave in a stream. In the lower reaches, a different type of groundwater-streamflow interaction, known as *bank storage*, often moderates the flood wave. As shown in Figure 6.22(a), if a large permanent stream undergoes an increase in river stage under the influence of an arriving flood wave, flow may be induced into the stream banks. As the stage declines, the flow is reversed. Figure 6.22(b), (c) and (d) shows the effect of such bank storage on the stream hydrograph, on the bank storage volume, and on the associated rates of inflow and outflow.

Bank-storage effects can cause interpretive difficulties in connection with hydrograph separation. In Figure 6.22(e) the solid line might represent the actual subsurface transfer at a stream bank, including the bank-storage effects. The groundwater inflow from the regional system, which might well be the quantity desired, would be as shown by the dashed line.

The concept of bank storage was clearly outlined by Todd (1955). Cooper and Rorabaugh (1963) provide a quantitative analysis based on an analytical solution to the boundary-value problem representing groundwater flow in an unconfined aquifer adjacent to a fluctuating stream. The numerical solutions of Pinder and Sauer (1971) carry the quantitative analysis a step further by considering the *pair* of boundary-value problems representing both groundwater flow in the bank *and* open-channel flow in the stream. The two systems are coupled through the inflow and outflow terms that represent the passage of water into and out of bank storage.

6.7 Groundwater-Lake Interactions

Stephenson (1971) has shown that the hydrologic regime of a lake is strongly influenced by the regional groundwater flow system in which it sits. Large, permanent lakes are almost always discharge areas for regional groundwater systems. The rates of groundwater inflow are controlled by watershed topography and the hydrogeologic environment as outlined in Section 6.1. Small, permanent lakes in the upland portions of watersheds are usually discharge areas for local flow sys-

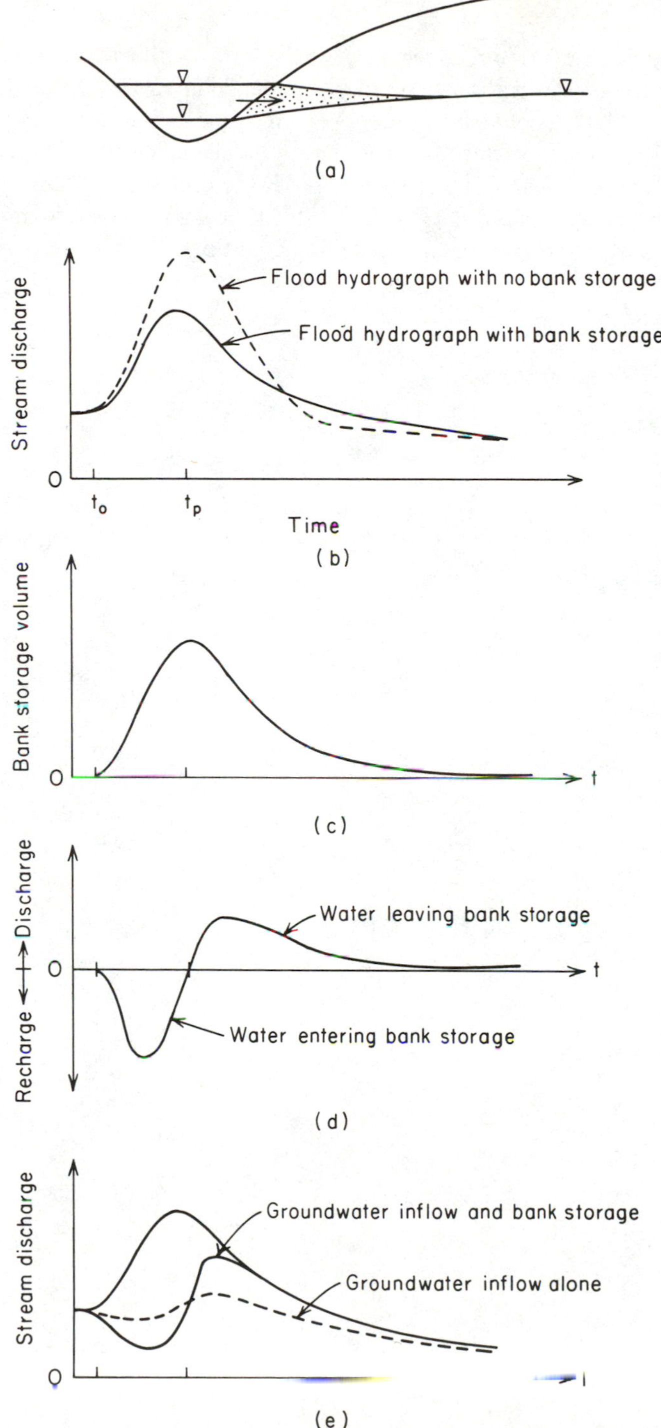

Figure 6.22 Flood-wave modification due to bank-storage effects.

tems, but there are geologic configurations that can cause such lakes to become sites of depression-focused recharge. Winter (1976), on the basis of numerical simulations of steady-state lake and groundwater flow systems, showed that where water-table elevations are higher than lake levels on all sides, a necessary condition for the creation of a recharge lake is the presence of a high-permeability formation at depth. His simulations also show that if a water-table mound exists between two lakes, there are very few geologic settings that lead to groundwater movement from one lake to the other.

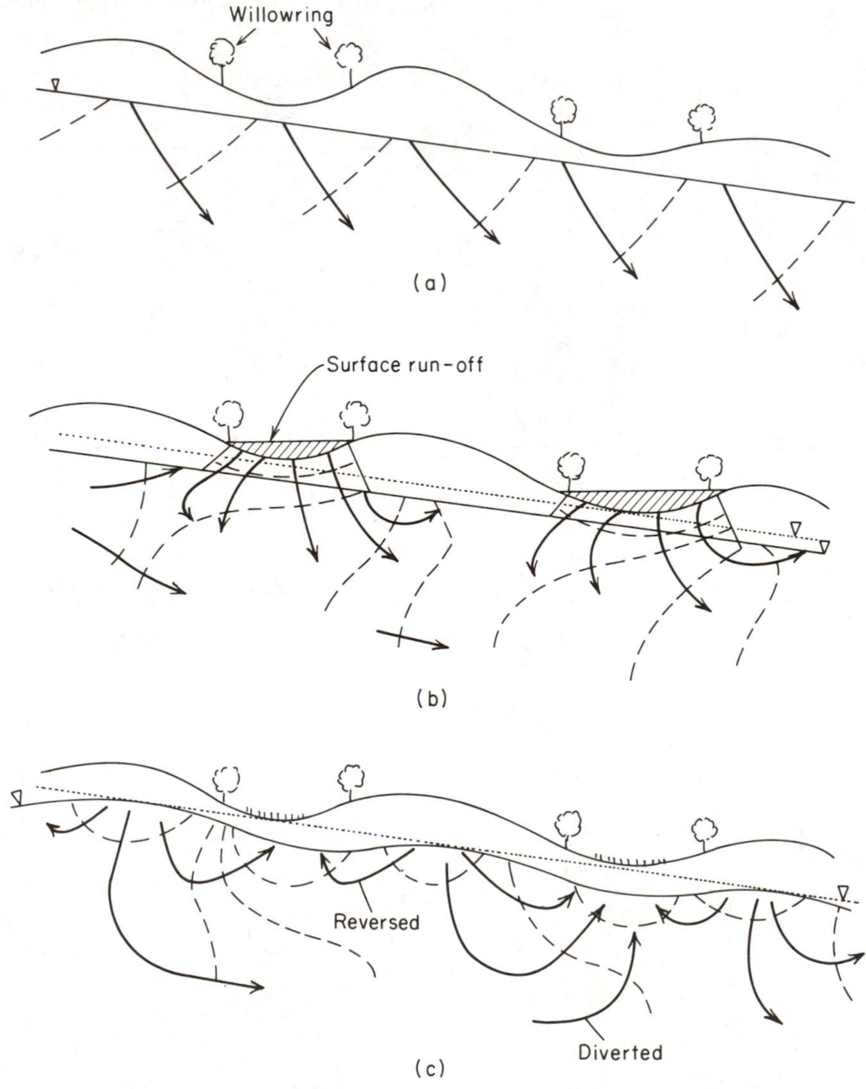

Figure 6.23 Depression-focused groundwater recharge in hummocky terrain (after Meyboom, 1966b).

A recharging lake can leak through part or all of its bed. McBride and Pfannkuch (1975) show, on the basis of theoretical simulation, that for cases where the width of a lake is greater than the thickness of associated high-permeability surficial deposits on which it sits, groundwater seepage into or out of a lake tends to be concentrated near the shore. Lee (1977) has documented this situation by a field study using seepage meters installed in a lake bottom. The design and use of simple, easy-to-use devices for monitoring seepage through lake beds in nearshore zones is described by Lee and Cherry (1978).

In many cases, a steady-state analysis of groundwater-lake interaction is not sufficient. In the hummocky, glaciated terrain of central-western North America, for example, temporary ponds created by runoff from spring snowmelt lead to transient interactions. Meyboom (1966b) made field measurements of transient groundwater flow in the vicinity of a prairie pothole. Figure 6.23 shows the generalized sequence of flow conditions he uncovered in such an environment. The upper diagram shows the normal fall and winter conditions of uniform recharge to a regional system. The middle diagram illustrated the buildup of groundwater mounds beneath the temporary ponds. The third diagram shows the water-table relief during the summer under the influence of phreatophytic groundwater consumption by willows that ring the pond. Meyboom's careful water balance on the willow ring showed that the overall effect of the transient seasonal behavior was a net recharge to the regional groundwater system.

6.8 Fluctuations in Groundwater Levels

The measurement of water-level fluctuations in piezometers and observation wells is an important facet of many groundwater studies. We have seen in Section 6.4, for example, how a water-table hydrograph measured during an infiltration event can be used to analyze the occurrence of groundwater recharge. We will discover in Chapter 8 the importance of detecting long-term regional declines in water levels due to aquifer exploitation. Water-level monitoring is an essential component of field studies associated with the analysis of artificial recharge (Section 8.11), bank storage (Section 6.6), and geotechnical drainage (Chapter 10).

Water-level fluctuations can result from a wide variety of hydrologic phenomena, some natural and some induced by man. In many cases, there may be more than one mechanism operating simultaneously and if measurements are to be correctly interpreted, it is important that we understand the various phenomena. Table 6.2 provides a summary of these mechanisms, classified according to whether they are natural or man-induced, whether they produce fluctuations in confined or unconfined aquifers, and whether they are short-lived, diurnal, seasonal, or long-term in their time frame. It is also noted that some of the mechanisms operate under climatic influence, while others do not. Those checked in the "confined" column produce fluctuations in *hydraulic head* at depth, and it should be recognized that such fluctuations must be measured with a true piezometer, open only at its intake. Those checked in the "unconfined" column produce fluctuations in *water-*

Table 6.2 Summary of Mechanisms That Lead to Fluctuations in Groundwater Levels

	Unconfined	Confined	Natural	Man-induced	Short-lived	Diurnal	Seasonal	Long-term	Climatic influence
Groundwater recharge (infiltration to the water table)	✓		✓				✓		✓
Air entrapment during groundwater recharge	✓		✓		✓				✓
Evapotranspiration and phreatophytic consumption	✓		✓			✓			✓
Bank-storage effects near streams	✓		✓				✓		✓
Tidal effects near oceans	✓	✓	✓			✓			
Atmospheric pressure effects	✓	✓	✓			✓			✓
External loading of confined aquifers		✓		✓	✓				
Earthquakes		✓	✓		✓				
Groundwater pumpage	✓	✓		✓				✓	
Deep-well injection		✓		✓				✓	
Artificial recharge; leakage from ponds, lagoons, and landfills	✓			✓				✓	
Agricultural irrigation and drainage	✓			✓				✓	✓
Geotechnical drainage of open pit mines, slopes, tunnels, etc.	✓			✓				✓	

table elevation near the surface. This type of fluctuation can be measured either with a true piezometer or with a shallow observation well open along its length.

Several of the natural phenomena listed in Table 6.2 have been discussed in some detail in earlier sections. Many of the man-induced phenomena will come into focus in later chapters. In the following paragraphs we will zero in on four types of fluctuations: those caused by phreatophytic consumption in a discharge area, those caused by air entrapment during groundwater recharge, those caused by changes in atmospheric pressure, and those caused by external loading of elastic confined aquifers.

Evapotranspiration and Phreatophytic Consumption

In a discharge area it is often possible to make direct measurements of evapotranspiration on the basis of water-table fluctuations in shallow observation wells. Figure 6.24 (after Meyboom, 1967) displays the diurnal fluctuations observed in the water-table record in a river valley in western Canada. The drawdowns take

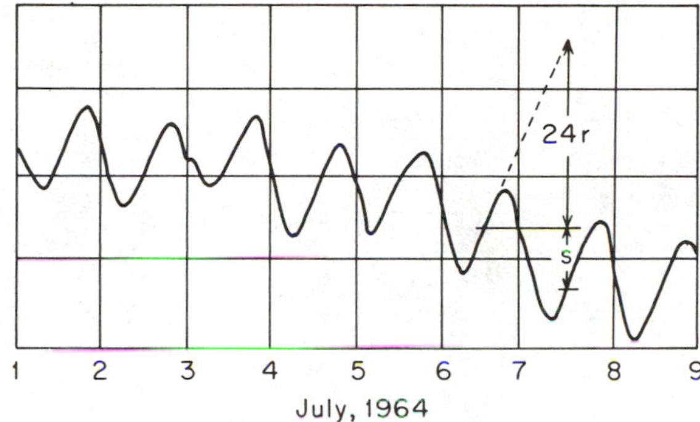

Figure 6.24 Calculation of evapotranspiration in a discharge area from water-table fluctuations induced by phreatophytic consumptive use (after Meyboom, 1967).

place during the day as a result of phreatophytic consumption (in this case by Manitoba maple); the recoveries take place during the night when the plant stomata are closed. White (1932) suggested an equation for calculating evapotranspiration on the basis of such records. The quantity of groundwater withdrawn by evapotranspiration during a 24-h period is

$$E = S_y(24r \pm s) \tag{6.16}$$

where E is the actual daily evapotranspiration ($[L]$/day), S_y the specific yield of the soil (% by volume), r the hourly rate of groundwater inflow ($[L]$/h), and s the net rise or fall of the water table during the 24-h period $[L]$. The r and s values are graphically illustrated in Figure 6.24. The value of r, which must represent the average rate of groundwater inflow for the 24-h period, should be based on the water-table rise between midnight and 4 A.M. Meyboom (1967) suggests that the S_y value in Eq. (6.16) should reflect the *readily available specific yield.* He estimates that this figure is 50% of the true specific yield as defined in Section 2.10. If laboratory drainage experiments are utilized to measure specific yield, the value used in Eq. (6.16) should be based on the drainage that occurs in the first 24 h. With regard to Figure 6.24, the total evapotranspiration for the period July 2–8 according to the White method is 1.73 ft (0.52 m).

Air Entrapment During Groundwater Recharge

Many field workers have observed an anomalously large rise in water levels in observation wells in shallow unconfined aquifers during heavy rainstorms. It is now recognized that this type of water-level fluctuation is the result of air entrapment in the unsaturated zone (Bianchi and Haskell, 1966; McWhorter, 1971). If the rainfall is sufficiently intense, an inverted zone of saturation is created at the ground surface, and the advancing wet front traps air between itself and the water table. Air pressures in this zone build up to values much greater than atmospheric.

As a schematic explanation of the phenomenon, consider Figure 6.25(a) and (b). In the first figure, the air pressure, p_A, in the soil must be in equilibrium both

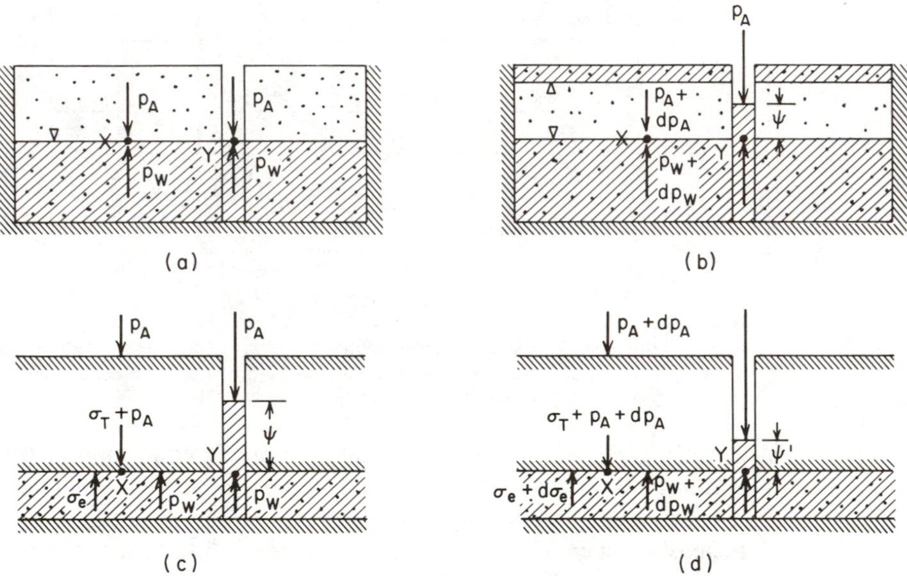

Figure 6.25 Water-level fluctuations due to (a) and (b) air entrapment during groundwater recharge in an unconfined aquifer; (c) and (d) atmospheric pressure effects in a confined aquifer.

with the atmosphere and with the fluid pressure, p_w. This will hold true at every point X on the water table within the porous medium and at the point Y in the well bore. If, as is shown in Figure 6.25(b), the advancing wet front creates an increase, dp_A, in the pressure of the entrapped air, the fluid pressure on the water table at point X must increase by an equivalent amount, dp_w. The pressure equilibrium in the well at point Y is given by

$$p_A + \gamma\psi = p_w + dp_w \tag{6.17}$$

Since $p_A = p_w$ and $dp_A = dp_w$, we have

$$\gamma\psi = dp_A \tag{6.18}$$

For $dp_A > 0$, $\psi > 0$, proving that an increase in entrapped air pressure leads to a rise in water level in an observation well open to the atmosphere.

This type of water-level rise bears no relation to groundwater recharge, but because it is associated with rainfall events, it can easily be mistaken for it. The most diagnostic feature is the magnitude of the ratio of water-level rise to rainfall depth. Meyboom (1967) reports values as high as 20:1. The anomalous rise usually dissipates within a few hours, or at most a few days, owing to the lateral escape of entrapped air to the atmosphere outside the area of surface saturation.

Atmospheric Pressure Effects

Changes in atmospheric pressure can produce large fluctuations in wells or piezometers penetrating confined aquifers. The relationship is an inverse one; increases in atmospheric pressure create declines in observed water levels.

Jacob (1940) invoked the principle of effective stress to explain the phenomenon. Consider the conditions shown in Figure 6.25(c), where the stress equilibrium at the point X is given by

$$\sigma_T + p_A = \sigma_e + p_w \tag{6.19}$$

In this equation, p_A is atmospheric pressure, σ_T the stress created by the weight of overlying material, σ_e the effective stress acting on the aquifer skeleton, and p_w the fluid pressure in the aquifer. The fluid pressure, p_w, gives rise to a pressure head, ψ, that can be measured in a piezometer tapping the aquifer. At the point Y in the well bore,

$$p_A + \gamma\psi = p_w \tag{6.20}$$

If, as is shown in Figure 6.25(d), the atmospheric pressure is increased by an amount dp_A, the change is stress equilibrium at X is given by

$$dp_A = d\sigma_e + dp_w \tag{6.21}$$

from which it is clear that $dp_A > dp_w$. In the well bore, we now have

$$p_A + dp_A + \gamma\psi' = p_w + dp_w \tag{6.22}$$

Substitution of Eq. (6.20) in Eq. (6.22) leads to

$$dp_A - dp_w = \gamma(\psi - \psi') \tag{6.23}$$

since $dp_A - dp_w > 0$, so too is $\psi - \psi' > 0$, proving that an increase in atmospheric pressure leads to a decline in water levels.

In a horizontal, confined aquifer the change in pressure head, $d\psi = \psi - \psi'$ in Eq. (6.23), is numerically equivalent to the change in hydraulic head, dh. The ratio

$$B = \frac{\gamma\, dh}{dp_A} \tag{6.24}$$

is known as the *barometric efficiency* of the aquifer. It usually falls in the range 0.20–0.75. Todd (1959) provides a derivation that relates the barometric efficiency, B, to the storage coefficient, S, of a confined aquifer.

It has also been observed that changes in atmospheric pressure can cause small fluctuations in the water table in unconfined aquifers. As the air pressure increases, water tables fall. Peck (1960) ascribes the fluctuations to the effects of the changed pressures on air bubbles entrapped in the soil-moisture zone. As the pressure increases, the entrapped air occupies less space, and it is replaced by soil water, thus inducing an upward movement of moisture from the water table. Turk (1975) measured diurnal fluctuations of up to 6 cm in a fine-grained aquifer with a shallow water table.

External Loads

It has long been observed (Jacob, 1939; Parker and Stringfield, 1950) that external loading in the form of passing railroad trains, construction blasting, and earthquakes can lead to measureable but short-lived oscillations in water levels recorded in piezometers tapping confined aquifers. These phenomena are allied in principle with the effects of atmospheric pressure. Following the notation introduced in Figure 6.25(c) and (d), note that a passing train creates transient changes in the total stress, σ_T. These changes induce changes in p_w, which are in turn reflected by changes in the piezometric levels. In a similar fashion, seismic waves set up by earthquakes create a transient interaction between σ_e and p_w in the aquifer. The Alaskan earthquake of 1964 produced water-level fluctuations all over North America (Scott and Render, 1964).

Time Lag in Piezometers

One source of error in water-level measurements that is often overlooked is that of time lag. If the volume of water that is required to register a head fluctuation in a piezometer standpipe is large relative to the rate of entry at the intake, there will be a time lag introduced into piezometer readings. This factor is especially pertinent to head measurements in low-permeability formations. To circumvent this problem, many hydrogeologists now use piezometers equipped with down-hole pressure transducers that measure head changes directly at the point of measurement without a large transfer of water. Reducer tubes that decrease the diameter of the standpipe above the intake have also been suggested (Lissey, 1967). In cases where these approaches are not feasible, the time-lag corrections suggested by Hvorslev (1951) are in order.

Suggested Readings

FREEZE, R. A. 1969. The mechanism of natural groundwater recharge and discharge: 1. One-dimensional, vertical, unsteady, unsaturated flow above a recharging or discharging groundwater flow system. *Water Resources Res.*, 5, pp. 153–171.

FREEZE, R. A. 1974. Streamflow generation. *Rev. Geophys. Space Phys.*, 12, pp. 627–647.

FREEZE, R. A., and P. A. WITHERSPOON. 1967. Theoretical analysis of regional groundwater flow: 2. Effect of water-table configuration and subsurface permeability variation. *Water Resources Res.*, 3, pp. 623–634.

HALL, F. R. 1968. Baseflow recessions—a review. *Water Resources Res.*, 4, pp. 973–983.

MEYBOOM, P. 1966. Unsteady groundwater flow near a willow ring in hummocky moraine. *J. Hydrol.*, 4, pp. 38–62.

RUBIN, J., and R. STEINHARDT. 1963. Soil water relations during rain infiltration: I. Theory. *Soil Sci. Soc. Amer. Proc.*, 27, pp. 246–251.

TÓTH, J. 1963. A theoretical analysis of groundwater flow in small drainage basins. *J. Geophys. Res.*, 68, pp. 4795–4812.

Problems

1. Consider a region of flow such as $ABCDEA$ in Figure 6.1. Set $BC = 1000$ m and make the length of CD equal to twice the length of AB. Draw flow nets for the following homogeneous, isotropic cases:
   (a) $AB = 500$ m, AD a straight line.
   (b) $AB = 500$ m, AD a parabola.
   (c) $AB = 100$ m, AD a straight line.
   (d) $AB = 200$ m, AE and ED straight lines with the slope of AE twice that of ED.
   (e) $AB = 200$ m, AE and ED straight lines with the slope of ED twice that of AE.

2. (a) Label the recharge and discharge areas for the flow nets in Problem 1 and prepare a recharge-discharge profile for each.
   (b) Calculate the volumetric rates of flow through the system (per meter of section perpendicular to the flow net) for cases in which $K = 10^{-8}$, 10^{-6}, and 10^{-4} m/s.

3. Assume a realistic range of values for P and E in Eqs. (6.2) through (6.6) and assess the reasonableness of the values calculated in Problem 2(b) as components of a hydrologic budget in a small watershed.

4. What would be the qualitative effect on the position of the hinge line, the recharge-discharge profile, and the baseflow component of runoff if the following geological adaptations were made to the system described in Problem 1(d)?
   (a) A high-permeability layer is introduced at depth.
   (b) A low-permeability layer is introduced at depth.
   (c) A high-permeability lense underlies the valley.
   (d) The region consists of a sequence of thin horizontally bedded aquifers and aquitards.

5. On the basis of the flow-net information in this chapter, how would you explain the occurrence of hot springs?

6. Label the areas on the flow nets constructed in Problem 1 where wells would produce flowing artesian conditions.

7. A research team of hydrogeologists is attempting to understand the role of a series of ponds and bogs on the regional hydrologic water balance. The long-term objective is to determine which of the surface-water bodies are permanent and which may diminish significantly in the event of long-term drought. The immediate objective is to assess which surface-water bodies are points of recharge and which are points of discharge, and to make calculations of the monthly and annual gains or losses to the groundwater system. Outline a field measurement program that would satisfy the immediate objectives at one pond.

8. On the flow net drawn in Problem 1(b), sketch in a series of water-table positions representing a water-table decline in the range 5- to 10-m/month (i.e., point A remains fixed, point D drops at this rate). For $K = 10^{-8}$, 10^{-6}, and 10^{-4} m/s, prepare a baseflow hydrograph for a stream flowing perpendicular to the diagram at point A. Assume that all groundwater discharging from the system becomes baseflow.

9. (a) Prove that a decrease in atmospheric pressure creates rising water levels in wells tapping a confined aquifer.

(b) Calculate the water-level fluctuation (in meters) that will result from a drop in atmospheric pressure of 5.0×10^3 Pa in a well tapping a confined aquifer with barometric efficiency of 0.50.

CHAPTER 7

Chemical Evolution of Natural Groundwater

7.1 Hydrochemical Sequences and Facies

Nearly all groundwater originates as rain or snowmelt that infiltrates through soil into flow systems in the underlying geologic materials. The soil zone has unique and powerful capabilities to alter the water chemistry, as infiltration occurs through this thin, biologically active zone. In recharge areas the soil zone undergoes a net loss of mineral matter to the flowing water. As groundwater moves along flowlines from recharge to discharge areas, its chemistry is altered by the effects of a variety of geochemical processes. In this section, the major changes in water chemistry that commonly occur as groundwater moves along its flow paths are described. A prerequisite to this discussion is consideration of the chemistry of rain and snow, which is the input to the subsurface hydrochemical system.

Chemistry of Precipitation

The chemical composition of water that arrives on the ground surface can be determined by inspection of chemical analyses of rain and snow. Table 7.1 lists some representative results of chemical analyses of precipitation in various parts of North America. This table indicates that the dissolved solids in rain range from several milligrams per liter in continental nonindustrial areas to several tens of milligrams per liter in coastal areas. Snowmelt that contributes water to the groundwater zone can have greater dissolved solids than rain because of dissolution of dust particles that accumulate in the snow as a result of atmospheric fallout.

Rainwater and melted snow in nonurban, nonindustrial areas have pH values normally between 5 and 6. The equilibrium pH for nonsaline water in contact with CO_2 at the earth's atmospheric value of $10^{-3.5}$ bar is 5.7. This can be demonstrated by substituting this P_{CO_2} in Eq. (3.18) to obtain the activity of H_2CO_3 and then solving for H^+ using Eq. (3.31). Because the water must be acidic, it is apparent from Figure 3.5(a) that HCO_3^- is the only ionic species of dissolved inorganic carbon present in a significant amount; therefore, $(H^+) = (HCO_3^-)$. In industrial

Table 7.1 Composition of Rain and Snow (mg/ℓ)*

Constituent	1	2	3	4	5	6	7
SiO_2	0.0	0.1	—	0.29	0.6	—	0.9
Ca	0.0	0.9	1.20	0.77	0.53	1.42	0.42
Mg	0.2	0.0	0.50	0.43	0.15	0.39	0.09
Na	0.6	0.4	2.46	2.24	0.35	2.05	0.26
K	0.6	0.2	0.37	0.35	0.14	0.35	0.13
NH_4	0.0	—	—	—	0.6	0.41	0.48
HCO_3	3	2.0	—	1.95	—	—	—
SO_4	1.6	2.0	—	1.76	0.45	2.19	3.74
Cl	0.2	0.2	4.43	3.75	0.22	3.47	0.38
NO_3	0.1	—	—	0.15	0.41	0.27	1.96
TDS	4.8	5.1	—	12.4	—	—	—
pH	5.6	—	—	5.9	5.3	5.5	4.1

*(1) Snow, Sponer Summit, U.S. Highway 50, Nevada (east of Lake Tahoe), altitude 7100 ft., Nov. 20, 1958; (2) rain, at eight sites in western North Carolina, average of 33 events, 1962–1963; (3) rain in southeastern Australia, 28 sites over 36 months, 1956–1957; (4) rain at Menlo Park, Calif., winters of 1957–1958; (5) rain, near Lake of the Woods, NW Ontario, average of 40 rain events, 1972; (6) rain and snow, northern Europe, 60 sites over 30 months, 1955–1956; (7) rain and snow at a site 20 km north of Baltimore, Maryland, average for 1970–1971.

SOURCE: Feth et al., 1964 (1); Laney, 1965 (2); Carroll, 1962 (3); Whitehead and Feth, 1964 (4); Bottomley, 1974 (5); Carroll, 1962 (6); and Cleaves et al., 1974 (7).

areas the pH of precipitation is much below 5.7, frequently as low as 3–4. In fact, *acid rains* are viewed as a major environmental problem in some regions of Europe and North America. The main cause of this increased acidity is sulfur spewed into the atmosphere from factories, mine processing plants, and coal- or petroleum-fired electrical generating stations. The occurrence of acid rain has now spread from industrial areas far out into the countryside. Emissions of sulfur into the atmosphere occur mainly as particulate S and gaseous SO_2. In the atmosphere this leads to increased concentrations of H^+ and SO_4^{2-} in rain and snow,

$$S + O_2 \longrightarrow SO_2(g) \tag{7.1}$$

$$SO_2(g) + H_2O + \tfrac{1}{2}O_2 \longrightarrow SO_4{}^{2-} + 2H^+ \tag{7.2}$$

In addition to CO_2 and SO_2, the earth's atmosphere contains other gases such as O_2, N_2, and Ar. The water becomes saturated with respect to these gases. In groundwater systems, the most important of these gases is O_2 because it imparts an appreciable oxidizing capability to the water.

In conclusion, it can be stated that rain and snowmelt are extremely dilute, slightly to moderately acidic, oxidizing solutions that can quickly cause chemical alterations in soils or in geologic materials into which they infiltrate.

Carbon Dioxide in the Soil Zone

Almost all water that infiltrates into natural groundwater flow systems passes through the soil zone. In this context the term *soil* is used as a designation of the layer at the surface of the earth that has been sufficiently weathered by physical, chemical, and biological processes to provide for the growth of rooted plants. This is a pedological definition, emphasizing that soil is a biologic as well as a geologic medium. The soil zone exerts a strong influence on the chemistry of water that infiltrates through it. The most important effects occur as a result of the processes summarized schematically in Figure 7.1. The soil has a capability to generate relatively large amounts of acid and to consume much or all of the available dissolved oxygen in the water that infiltrates it.

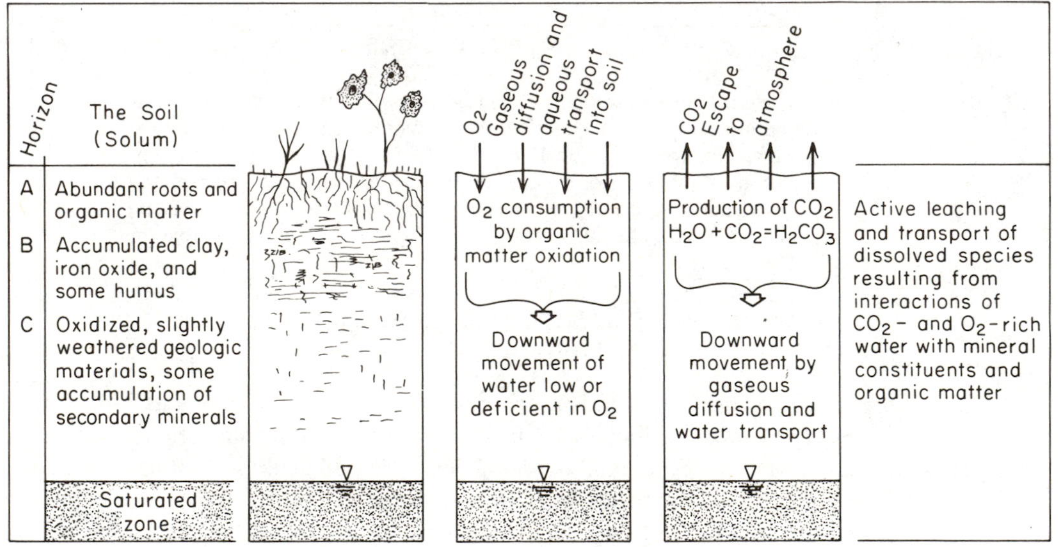

Figure 7.1 Schematic representation of major hydrochemical processes in the soil zone of recharge areas.

Geochemically, the most important acid produced in the soil zone is H_2CO_3, derived from the reaction of CO_2 and H_2O. The CO_2 is generated by the decay of organic matter and by respiration of plant roots. Organic-matter decay is the main source and can be represented by the reaction

$$O_2(g) + CH_2O = CO_2(g) + H_2O \tag{7.3}$$

where the simple carbohydrate CH_2O is used to designate organic matter. Other organic compounds can also be used in oxidation equations to represent CO_2 production. Anaerobic reactions such as the reduction of sulfate and nitrate also generate CO_2 (Table 3.11). These processes, however, make only minor contributions to the CO_2 budget of the soil atmosphere.

Measurements of the composition of gas samples from soils at locations in North America, Europe, and elsewhere have established that the CO_2 partial pressure of the soil atmosphere is normally much higher than that of the earth's atmosphere. Values in the range 10^{-3}–10^{-1} bar are typical. Because of variations of temperature, moisture conditions, microbial activity, availability of organic matter, and effects of soil structure on gas diffusion, CO_2 pressures are quite variable, both spatially and temporally. More detailed discussions of the occurrence and effect of CO_2 production in the soil zone are presented by Jakucs (1973) and Trainer and Heath (1976). When CO_2 at these partial pressures reacts with water, the pH of the water declines dramatically. For example, using the method indicated above, it can be shown that at a CO_2 partial pressure of 10^{-1} bar, water in the temperature range 0–25°C will have an equilibrium pH in the range 4.3–4.5. This is much below the pH of uncontaminated rainwater.

Carbon dioxide-charged water infiltrating through the soil zone commonly encounters minerals that are dissolvable, under the influence of H_2CO_3 which is consumed by the mineral-water reactions. Examples of some H_2CO_3 leaching reactions are given by Eqs. (3.49) and (3.54). Other examples are described later in this chapter. As H_2CO_3 is consumed in the soil zone, oxidation of organic matter and root respiration is a source of replenishment of CO_2 to the soil air. The CO_2 combines with water to produce more H_2CO_3 [Eqs. (3.14) and (3.15)]. As new water from recharge events passes through the soil, biochemical and hydrochemical processes in the soil are therefore capable of providing a continuing supply of acidity to promote mineral-water reactions. The reaction of free oxygen with reduced iron minerals such as pyrite (FeS_2) is another source of acidity. In some areas production of H^+ by this oxidation reaction plays an important role in mineral weathering in the subsoil. The soil zone can therefore be thought of as an acid pump operating in the very thin but extensive veneer of organic-rich material that covers most of the earth's surface.

In addition to the inorganic acid, there are many organic acids produced in the soil zone by biochemical processes. These substances, such as humic acids and fulvic acids, can play a major role in the development of soil profiles and in the transport of dissolved constituents downward toward the water table. It is believed by most geochemists however, that as a source of H^+ involved in mineral dissolution, these acids play a minor role compared to the effect of dissolved CO_2.

Major-Ion Evolution Sequence

As groundwater moves along its flow paths in the saturated zone, increases of total dissolved solids and most of the major ions normally occur. As would be expected from this generalization, it has been observed in groundwater investigations in many parts of the world that shallow groundwater in recharge areas is lower in dissolved solids than the water deeper in the same system and lower in dissolved solids than water in shallow zones in the discharge areas.

In a classic paper based on more than 10,000 chemical analyses of well samples from Australia, Chebotarev (1955) concluded that groundwater tends to evolve chemically toward the composition of seawater. He observed that this evolution is normally accompanied by the following regional changes in dominant anion species:

Travel along flow path ⟶

$$HCO_3^- \longrightarrow HCO_3^- + SO_4{}^{2-} \longrightarrow SO_4{}^{2-} + HCO_3^- \longrightarrow SO_4{}^{2-} + Cl^- \longrightarrow Cl^- + SO_4{}^{2-} \longrightarrow Cl^-$$

Increasing age ⟶

These changes occur as the water moves from shallow zones of active flushing through intermediate zones into zones where the flow is very sluggish and the water is old. This sequence, like many others in the geological sciences, must be viewed in terms of the scale and geology of the specific setting, with allowances for interruption and incompletion. Schoeller (1959) refers to the sequence above as the Ignatovich and Souline Sequence, in recognition of the fact that two hydrogeologists in the Soviet Union developed similar generalizations independent of the contributions by Chebotarev.

For large sedimentary basins, the Chebotarev sequence can be described in terms of three main zones, which correlate in a general way with depth (Domenico, 1972):

1. *The upper zone*—characterized by active groundwater flushing through relatively well-leached rocks. Water in this zone has HCO_3^- as the dominant anion and is low in total dissolved solids.
2. *The intermediate zone*—with less active groundwater circulation and higher total dissolved solids. Sulfate is normally the dominant anion in this zone.
3. *The lower zone*—with very sluggish groundwater flow. Highly soluble minerals are commonly present in this zone because very little groundwater flushing has occurred. High Cl^- concentration and high total dissolved solids are characteristic of this zone.

These three zones cannot be correlated specifically with distance of travel or time, other than to say that travel distance and time tend to increase from the upper zone to the lower zone. In some sedimentary basins, groundwater in the upper zone may be years or tens of years old, whereas in other basins ages of hundreds or thousands of years are common. Saline, chloride-rich water in the lower zone is usually very old, but the actual ages may vary from thousands to millions of years.

From a geochemical viewpoint the anion-evolution sequence described above can be explained in terms of two main variables, mineral availability and mineral solubility. The HCO_3^- content in groundwater is normally derived from soil zone CO_2 and from dissolution of calcite and dolomite. The partial pressures of CO_2 generated in the soil zone and the solubility of calcite and dolomite are normally the limiting constraints on the level of total dissolved solids attained. Figure 3.7

indicates that at the CO_2 partial pressures typical of the soil zone (10^{-3}–10^{-1} bar), calcite and dolomite are only moderately soluble, with equilibrium HCO_3^- concentrations in the range 100–600 mg/ℓ. Since calcite or dolomite occur in significant amounts in nearly all sedimentary basins, and because these minerals dissolve rapidly when in contact with CO_2-charged groundwater, HCO_3^- is almost invariably the dominant anion in recharge areas.

Table 3.6 indicates that there are several soluble sedimentary minerals that release SO_4^{2-} or Cl^- upon dissolution. The most common of the sulfate-bearing minerals are gypsum, $CaSO_4 \cdot 2H_2O$, and anhydrite, $CaSO_4$. These minerals dissolve readily when in contact with water. The dissolution reaction for gypsum is

$$CaSO_4 \cdot 2H_2O \longrightarrow Ca^{2+} + SO_4^{2-} + 2H_2O \tag{7.4}$$

Gypsum and anhydrite are considerably more soluble than calcite and dolomite but much less soluble than the chloride minerals such as halite (NaCl) and sylvite (KCl). If calcite (or dolomite) and gypsum dissolve in fresh water at 25°C, the water will become brackish, with total dissolved solids of about 2100 and 2400 mg/ℓ for a P_{CO_2} range of 10^{-3}–10^{-1} bar. The dominant anion will be SO_4^{2-}, so in effect we have moved into the SO_4^{2-}–HCO_3^- composition phase in the Chebotarev evolution sequence. If sufficient calcite and/or dolomite and gypsum are present to enable dissolution to proceed to equilibrium, the water will evolve quickly and directly to this phase and will not evolve beyond this phase unless it comes into contact with other soluble minerals or undergoes evaporation.

The reason that in most sedimentary terrain groundwater travels a considerable distance before SO_4^{2-} becomes a dominant anion is that gypsum or anhydrite are rarely present in more than trace amounts. In many shallow zones these minerals have never been present or have been previously removed by groundwater flushing. Therefore, although HCO_3^- and SO_4^{2-} stages can be described in terms of simple solubility constraints exerted by only two or three minerals, the process of evolution from stage to stage is controlled by the availability of these minerals along the groundwater flow paths. Given enough time, dissolution and groundwater flushing will eventually cause the readily soluble minerals such as calcite, dolomite, gypsum, and anhydrite to be completely removed from the active-flow zone in the groundwater system. Subsurface systems rarely advance to this stage, because of the rejuvenating effects of geologic processes such as continental uplift, sedimentation, and glaciation.

In deep groundwater flow systems in sedimentary basins and in some shallower systems, groundwater evolves past the stage where SO_4^{2-} is the dominant anion to a Cl^--rich brine. This occurs if the groundwater comes into contact with highly soluble chloride minerals such as halite or sylvite, which in deep sedimentary basins can occur as salt strata originally deposited during the evaporation of closed or restricted marine basins many millions of years ago. The solubilities of other chloride minerals of sedimentary origin are very high. In fact, as indicated in Table 3.6, these solubilities are orders of magnitude higher than the solubilities of calcite, dolomite, gypsum, and anhydrite. Chloride minerals of sedimentary origin dissolve rapidly in water. The general occurrence of Cl^- as a dominant anion only in deep groundwater or groundwater that has moved long distances therefore can generally be accounted for by the paucity of these minerals along the flow paths. If groundwater that has not traveled far comes into contact with abundant amounts of halite, the water will evolve directly to the Cl^- phase, regardless of the other minerals present in the system. In strata of siltstone, shale, limestone, or dolomite, where Cl^- is present in minerals occurring in only trace amounts, the rate of Cl^- acquisition by the flowing groundwater is to a large extent controlled by the process of diffusion. Cl^- moves from the small pore spaces, dead-end pores, and, in the case of fractured strata, from the matrix of the porous media to the main pores or fractures in which the bulk flow of groundwater takes place. As indicated in Section 3.4, diffusion is an extremely slow process. This, and the occurrence of sulfate- and chloride-bearing minerals in limited amounts, can account for the observation that in many groundwater systems the chemical evolution of groundwater from the HCO_3^- to SO_4^{2-} and Cl^- stages proceeds very gradually rather than by distinct steps over short distances as would be expected on the basis of solubility considerations alone.

The anion evolution sequence and the tendency for total dissolved solids to increase along the paths of groundwater flow are generalizations that, when used in the context of more rigorous geochemical reasoning, can provide considerable information on the flow history of the water. At this point we wish to emphasize, however, that in some groundwater flow systems the water does not evolve past the HCO_3^- stage or past the SO_4^{2-} stage. It is not uncommon in some sedimentary regions for water to undergo reversals in the sequence of dominant anions. Most notable in this regard is the increase in HCO_3^- and decrease in SO_4^{2-} that can occur as a result of biochemical SO_4^{2-} reduction. These processes are described in Section 7.5.

Large variations in the major cations commonly occur in groundwater flow systems. Since cation exchange commonly causes alterations or reversals in the cation sequences, generalization of cation evolution sequences in the manner used by Chebotarev for anions would be of little use because there would be so many exceptions to the rule. For major cation and anion data to provide greatest insight into the nature of groundwater flow systems, interpretations must include consideration of specific hydrochemical processes that can account for the observed concentrations. Examples of this approach are included in Sections 7.3 through 7.5.

Electrochemical Evolution Sequence

Recognition of the anion evolution sequence as a characteristic feature of many groundwater systems resulted from the compilation and interpretation of chemical data from regional flow systems. It is a generalization initially founded on observation and later supported by geochemical theory. We will now look briefly at another evolution sequence, referred to as the electrochemical evolution sequence. This sequence is founded on geochemical theory, but as yet has not been rigorously appraised on the basis of field measurements.

The electrochemical evolution sequence refers to the tendency for the redox potential of groundwater to decrease as the water moves along its flow paths. This tendency was first recognized by Germanov et al. (1958). As water from rain and snow enters the subsurface flow system, it initally has a high redox potential as a result of its exposure to atmospheric oxygen. The initial redox conditions reflect high concentrations of dissolved oxygen, with pE values close to 13, or, expressed as Eh, close to 750 mV at pH 7. In the organic-rich layers of the soil zone, the oxidation of organic matter commonly removes most of the dissolved oxygen. This process, represented by Eq. (7.3), causes the redox potential to decline. The question can be asked: How far does the redox potential decline as the water passes through the soil zone to the water table? It is reasonable to expect that the consumption of oxygen in the soil zone will vary depending on numerous factors, such as the soil structure, porosity and permeability, nature and depth distribution of organic matter, frequency of infiltration events, depth to water table, and temperature. Although dissolved oxygen is an important factor in the characterization of the hydrochemical nature of groundwater, very few studies of dissolved oxygen in groundwater have been reported in the literature. From the data that are available, however, the following generalizations can be drawn:

1. In recharge areas with sandy or gravelly soils or in cavernous limestones, shallow groundwater commonly contains detectable dissolved oxygen (i.e., greater than about 0.1 mg/).
2. In recharge areas in silty or clayey soils, shallow groundwater commonly does not contain detectable dissolved oxygen.
3. In areas with little or no soil overlying permeable fractured rock, dissolved oxygen at detectable levels commonly persists far into the flow system. In some cases the entire flow system is oxygenated.

The common occurrence of appreciable dissolved oxygen in shallow groundwater in sandy deposits is probably a result of low contents of organic matter in the soil and rapid rates of infiltration through the soil.

Even after dissolved oxygen is consumed to levels below detection by normal means, the redox potential can still be very high, as indicated in Section 3.9. The consumption of free molecular oxygen by bacterially catalyzed reactions that oxidize organic matter may continue until dissolved O_2 levels are considerably below the normal limits of detection. Eventually the point is reached where the aerobic bacteria involved in these reactions can no longer thrive. In the aerobic zone there are other reactions, such as those represented in Table 3.10 for the oxidation of ferrous iron, ammonia, manganese, and sulfide, that consume oxygen. Even though these oxidation processes may consume only a small portion of the total oxygen relative to the oxidation by organic matter, they can have a major effect on the chemical evolution of the water.

We will now consider what may happen as water moves deeper into the groundwater flow system. Stumm and Morgan (1970) state that in closed aqueous systems containing organic material and the other nutrients necessary for growth of bacteria, the oxidation of organic matter accompanied by consumption of O_2 is followed by reduction of NO_3^-. Reduction of MnO_2, if present, should occur at about the same p*E* or *Eh* as NO_3^- reduction, followed by the reduction of the ferric iron minerals, such as the various compounds represented by $Fe(OH)_3$. When sufficiently negative redox levels have been reached, the reduction of SO_4^{2-} to H_2S and HS^- and the reduction of organic matter to the dissolved gaseous species CO_2 and CH_4 may occur almost simultaneously. This electrochemical sequence of reduction processes is summarized in Table 3.11, with the initial processes of oxygen consumption represented in Table 3.10. Stumm and Morgan present the electrochemical evolution sequence as a phenomenon based on thermodynamic theory. They indicate that this sequence is consistent with observations of the chemical nature of nutrient-enriched lakes and batch digestors in sewage treatment facilities. With the possible exception of MnO_2 and $Fe(OH)_3$ reduction, the reactions described in the electrochemical evolution sequence are biologically catalyzed. The sequence of redox reactions is paralleled by an ecological succession of microorganisms, with various bacterial species adapted to the different stages of the redox sequence.

From a hydrogeologic viewpoint, the important question is whether or not the electrochemical evolution sequence occurs in the groundwater environment and, if so, where and why? The sequence, or at least parts of the sequence, are known to occur in the groundwater zone. It is known, for example, that in many areas dissolved oxygen is absent from water that recharges the groundwater zone. This is indicated by the absence of detectable dissolved oxygen in shallow wells. Presumably, the oxygen has been consumed by the processes described above. In some groundwater systems, NO_3^- occurs at shallow depth and diminishes in concentration as the water moves deeper into the flow system. Edmunds (1973) and Gillham and Cherry (1978) attributed this type of NO_3^- trend to the process of denitrification, which requires denitrifying bacteria and a moderate redox potential. It is known that in some regions groundwater has a very low redox potential. This is indicated by low SO_4^{2-} concentrations and H_2S odor from the water, characteristics that are attributed to the process of sulfate reduction in the presence of sulfate-reducing bacteria. Methane (CH_4) is a common constituent of deep groundwater in sedimentary basins and is observed at many locations, even in shallow groundwater. Its origin is attributed to bacterial fermentation of organic matter within the groundwater system. It is known that in some groundwater flow systems, the redox potential measured by the platinum-electrode method decreases along the apparent paths of regional flow. To illustrate this type of trend, redox potential data from two regional flow systems are presented in Figure 7.2.

In the Chebotarev evolution sequence, the gradual changes in anion composition and total dissolved solids were attributed to two limiting factors: mineral availability and rate of molecular diffusion. In the electrochemical evolution

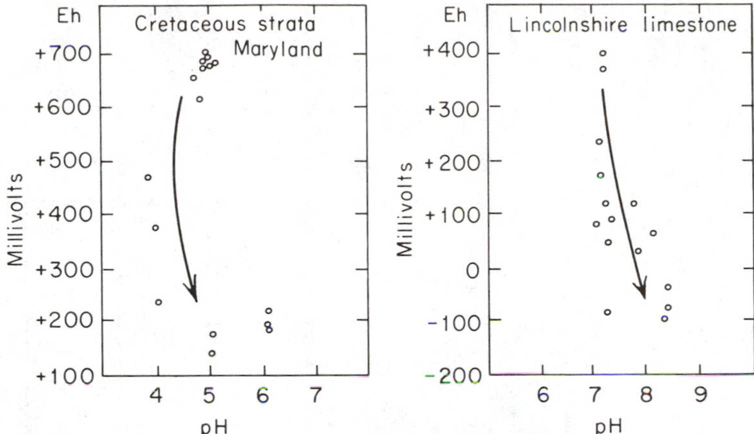

Figure 7.2 Trends in measured platinum-calomel electrode potential along regional flow paths in two aquifer systems. (a) Cretaceous strata, Maryland; (b) Lincolnshire limestone, Great Britain (adapted from R. E. Jackson, written communication, 1977, based on data from Back and Barnes, 1969, Edmunds, 1973).

sequence, other factors must control the amount and rate of decline of the redox potential along the flow paths. Because H_2S (or HS^-) and CH_4 are not present in significant amounts in many groundwater zones and because considerable $SO_4{}^{2-}$ is present in these zones, it appears that it is common for groundwater not to evolve to conditions of low redox potential even during long periods of residence time. The redox reactions that would lead progressively to low redox potential probably do not proceed in many areas because of the inability of the necessary redox bacteria to thrive. The hostility of groundwater environments to bacteria is probably caused by the lack of some of the essential nutrients for bacterial growth. It may be that even in hydrogeologic regimes in which organic carbon is abundant, the carbon may not be in a form that can be utilized by the bacteria. As the emphasis in hydrochemical investigations is broadened to include organic and biochemical topics, a much greater understanding of the redox environment of subsurface systems will be developed.

7.2 Graphical Methods and Hydrochemical Facies

An important task in groundwater investigations is the compilation and presentation of chemical data in a convenient manner for visual inspection. For this purpose several commonly used graphical methods are available. The simplest of these is the bar graph. Two examples are shown in Figure 7.3. For a single sample these two graphs represent the major-ion composition in equivalents per cubic meter (or milliequivalents per liter) and in percentage of total equivalents. The same

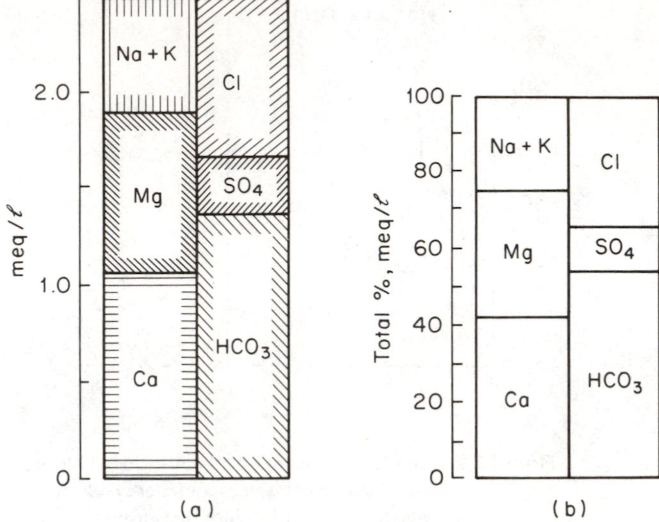

Figure 7.3 Chemical analyses of groundwater represented by bar graphs: (a) milliequivalents per liter; (b) percentage of total equivalents per liter (after Davis and De Wiest, 1966).

analysis is shown on a circular graph in Figure 7.4. In Figure 7.5(a) the analysis is shown in a manner that facilitates rapid comparison as a result of distinctive graphical shapes. This is known as a *Stiff diagram*, named after the hydrogeologist who first used it. Analysis of water with a much different composition is shown in Figure 7.5(b). The bar, circular, radial, and Stiff diagrams are all easy to construct and provide quick visual comparison of individual chemical analyses. They are not, however, convenient for graphic presentation of large numbers of analyses.

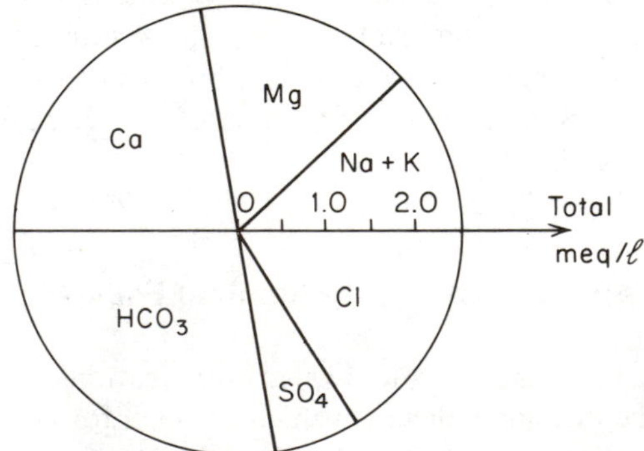

Figure 7.4 Chemical analysis of groundwater represented by a circular diagram. The radial axis is proportional to the total milliequivalents. Same chemical analysis as represented in Figure 7.3 (after Davis and De Wiest, 1966).

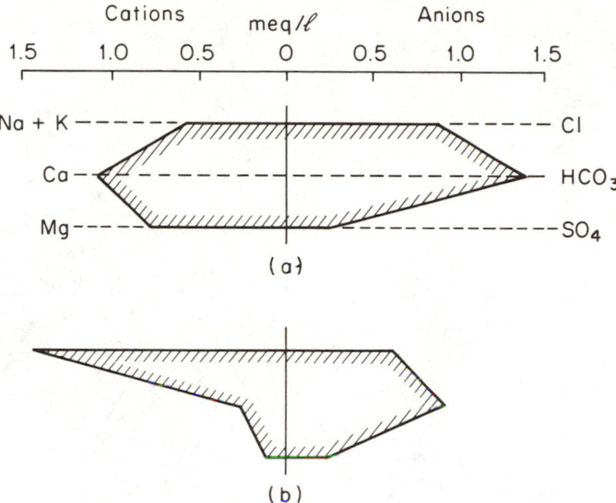

Figure 7.5 Two chemical analyses represented in the manner originated by Stiff. (a) The same analysis as in the previous three figures; (b) second analysis, illustrating contrast in shape of the graphical representation (after Davis and De Wiest, 1966).

For this purpose two other diagrams are in common use. The first one, developed by Piper (1944) from a somewhat similar design by Hill (1940), is shown in Figure 7.6; the second one, introduced into the groundwater literature by Schoeller (1955, 1962), is shown in Figure 7.7. Both of these diagrams permit the cation and anion compositions of many samples to be represented on a single graph in which major groupings or trends in the data can be discerned visually. The Schoeller semi-logarithmic diagram shows the total concentrations of the cations and anions. The trilinear diagram represents the concentrations as percentages. Because each analysis is represented by a single point, waters with very different total concentrations can have identical representations on this diagram. A single trilinear diagram has greater potential to accommodate a larger number of analyses without becoming confusing and is convenient for showing the effects of mixing two waters from different sources. The mixture of two different waters will plot on the straight line joining the two points. The semilogarithmic diagram has been used to directly determine the saturation indices of groundwaters with respect to minerals such as calcite and gypsum (Schoeller, 1962; Brown et al., 1972). This approach, however, is often not advisable, because of errors introduced by neglecting the effects of ion complexes and activity coefficients.

Some of the shortcomings of the trilinear graphs of the type developed by Hill and Piper are removed in the diagram introduced into the Soviet literature by S. A. Durov and described in the English-language literature by Zaporozec (1972). The basis of this diagram, shown in Figure 7.8, is percentage plotting of cations and anions in separate triangles, which in this respect is similar to the Piper diagram.

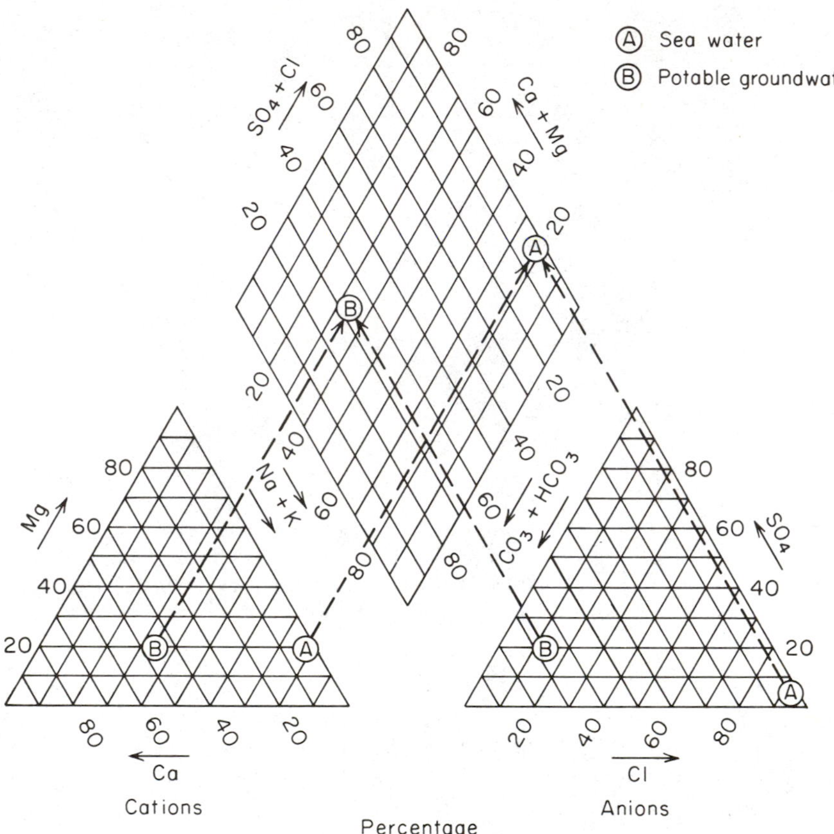

Figure 7.6 Chemical analyses of water represented as percentages of total equivalents per liter on the diagram developed by Hill (1940) and Piper (1944).

The intersection of lines extended from the two sample points on the triangles to the central rectangle gives a point that represents the major-ion composition on a percentage basis. From this point, lines extending to the two adjacent scaled rectangles provide for representation of the analysis in terms of two parameters selected from possibilities such as total major-ion concentration, total dissolved solids, ionic strength, specific conductance, hardness, total dissolved inorganic carbon, or pH. Total dissolved solids and pH are represented in Figure 7.8.

The diagrams presented above are useful for visually describing differences in major-ion chemistry in groundwater flow systems. There is also a need to be able to refer in a convenient manner to water compositions by identifiable groups or categories. For this purpose, the concept of hydrochemical facies was developed by Back (1961, 1966), Morgan and Winner (1962), and Seaber (1962). The definition of *hydrochemical facies* is a paraphrase of the definition of *facies* as used by geologists: *facies* are identifiable parts of different nature belonging to any genet-

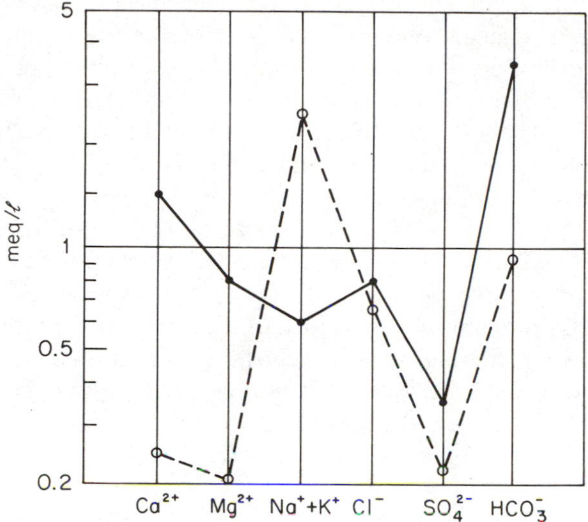

Figure 7.7 Chemical analyses of water represented on a Schoeller semi-logarithmic diagram (same analyses as in Figure 7.6).

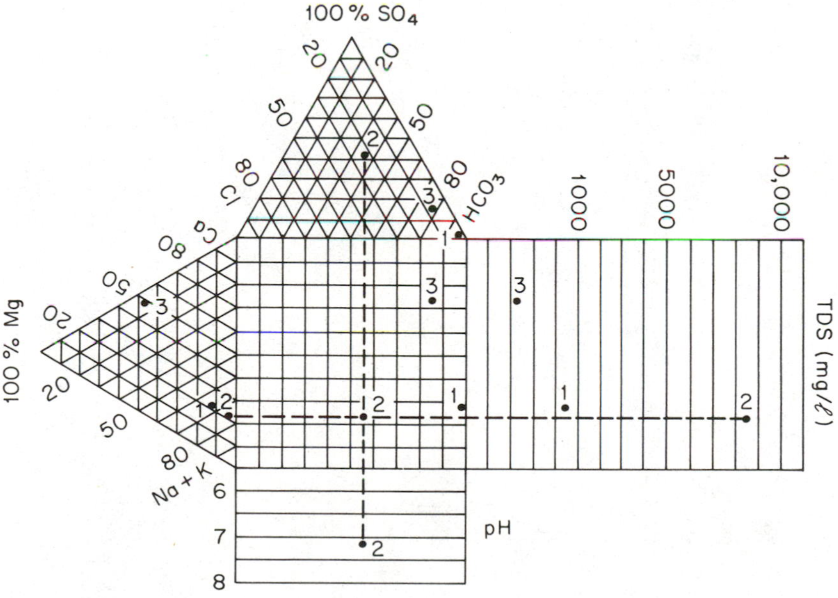

Figure 7.8 Chemical analyses represented as milliequivalents per liter on the diagram originated by Durov as described by Zaporozec (1972).

ically related body or system. Hydrochemical facies are distinct zones that have cation and anion concentrations describable within defined composition categories. The definition of a composition category is commonly based on subdivisions of the trilinear diagram in the manner suggested by Back (1961) and Back and Hanshaw (1965). These subdivisions are shown in Figure 7.9. If potassium is present in significant percentages, sodium and potassium are normally plotted as a single parameter. Definition of separate facies for the 0–10% and 90–100% domains on the diamond-shaped cation-anion graph is generally more useful than using equal 25% increments. The choice of percent categories should be made so as to best display the chemical characteristics of the water under consideration. In some situations, more subdivisions than those shown in Figure 7.9 are useful.

After arriving at a convenient classification scheme for the designation of hydrochemical facies, it is often appropriate, using maps, cross sections, or fence diagrams, to show the regional distribution of facies. An example of a fence dia-

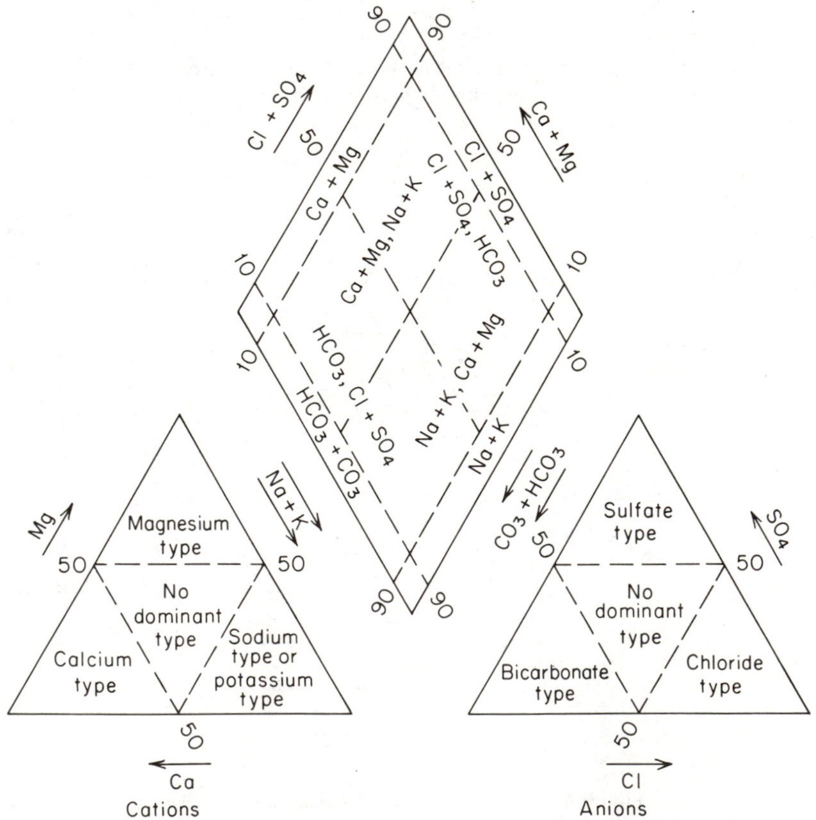

Figure 7.9 Classification diagram for anion and cation facies in terms of major-ion percentages. Water types are designated according to the domain in which they occur on the diagram segments (after Morgan and Winner, 1962; and Back, 1966).

gram showing the distribution of cation facies in the northern Atlantic Coastal Plain of the United States is shown in Figure 7.10. Also shown on this diagram is the generalized direction of regional groundwater flow.

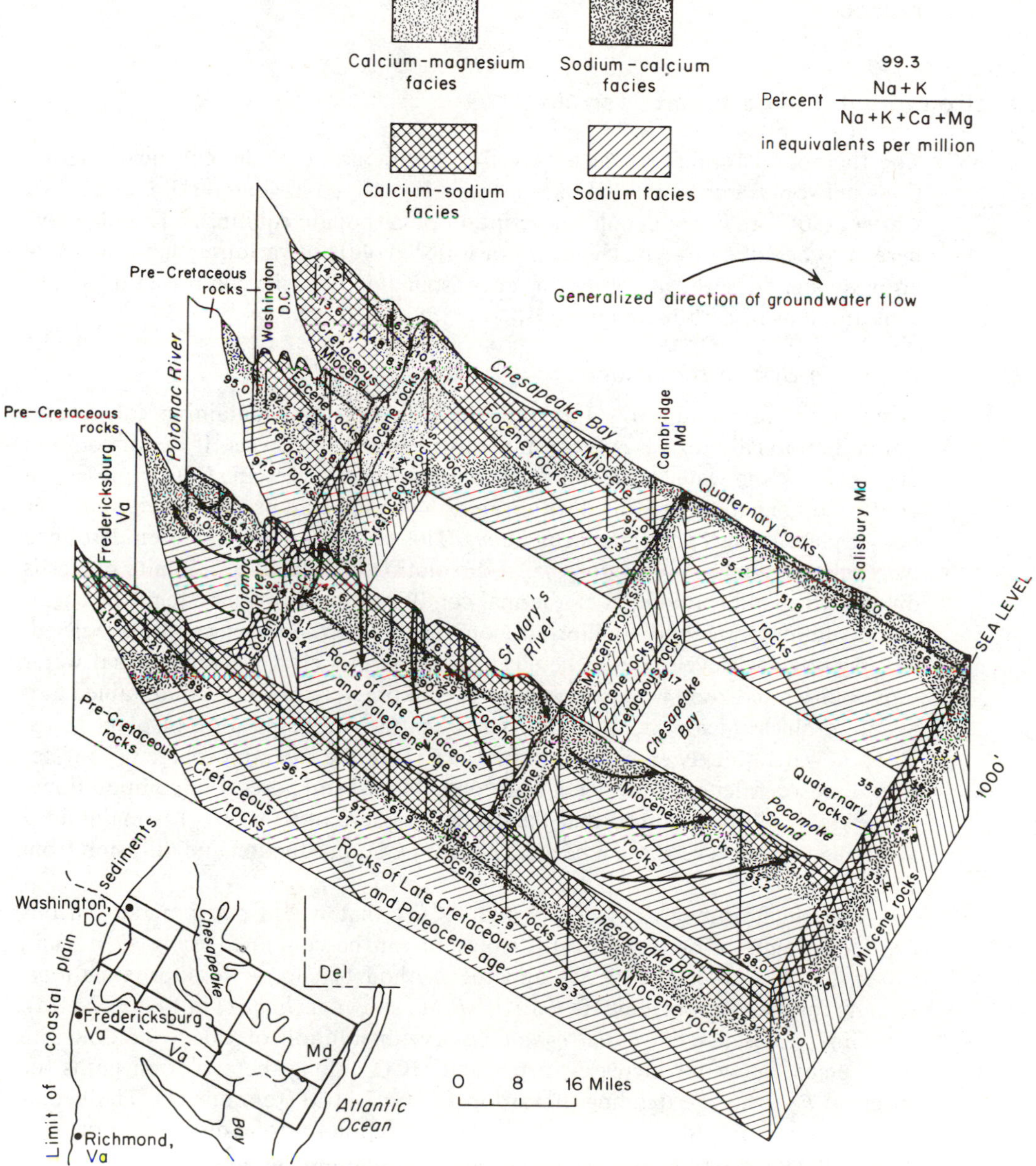

Figure 7.10 Fence diagram showing cation facies and generalized directions of groundwater flow in part of the northern Atlantic Coastal Plain (after Back, 1961).

In conclusion, it can be stated that there are many ways in which chemical analyses can be displayed graphically and there are many types of classifications that can be used for defining hydrochemical facies. The specific nature of the particular system under investigation usually governs the choice of graphical methods.

7.3 Groundwater in Carbonate Terrain

The theoretical framework necessary for consideration of the chemical interactions between water and carbonate minerals is outlined in Chapter 3. Garrels and Christ (1965) provide a detailed description of carbonate equilibria. The objective here is to describe how the chemistry of water evolves in various situations where groundwater flows through rocks or unconsolidated deposits comprised of significant amounts of carbonate minerals.

Open-System Dissolution

Water from rain and snow that infiltrates into terrain containing calcite and dolomite normally dissolves these minerals to saturation levels. If the dissolution occurs above the water table under conditions where abundant CO_2 is present in voids that are not entirely filled with water, the dissolution process is referred to as taking place under *open-system conditions*. This type of system has been described in geochemical terms in Section 3.5. If dissolution of calcite or dolomite proceeds directly to equilibrium under isothermal conditions in the open system, the chemical evolution paths and equilibrium composition of the water can be predicted. For purposes of developing a chemical evolution model, it is assumed that water moves into a soil zone where a constant partial pressure of CO_2 is maintained as a result of biochemical oxidation of organic matter and respiration of plant roots. The soil water quickly equilibrates with the CO_2 in the soil atmosphere. The water then dissolves calcite with which it is in contact in the soil pores. For computational purposes, it will be assumed that the partial pressure of CO_2 (P_{CO_2}) is maintained at a fixed value as a result of a balance between CO_2 production and diffusion from soil.

The equilibrium values of pH and HCO_3^- that would occur under various P_{CO_2} constraints prior to mineral dissolution can be computed using Eqs. (3.5), (3.18), (3.19), (3.31), and (3.32), and the method of successive approximations. Results for the P_{CO_2} range of 10^{-4}–10^{-1} bar are shown as line (1) in Figure 7.11(a). This line represents the *initial conditions*. As dissolution of calcite or dolomite takes place, the water increases in pH and HCO_3^- along the evolution paths for specified P_{CO_2} values extending upward in Figure 7.11(a) from line (1). The evolution paths are computed using a mass-balance relation for total dissolved inorganic carbon in combination with the equations indicated above. Steps along the paths are made by hypothetically dissolving small arbitrary amounts of calcite or dolomite in the water. The water composition evolves along these paths until the

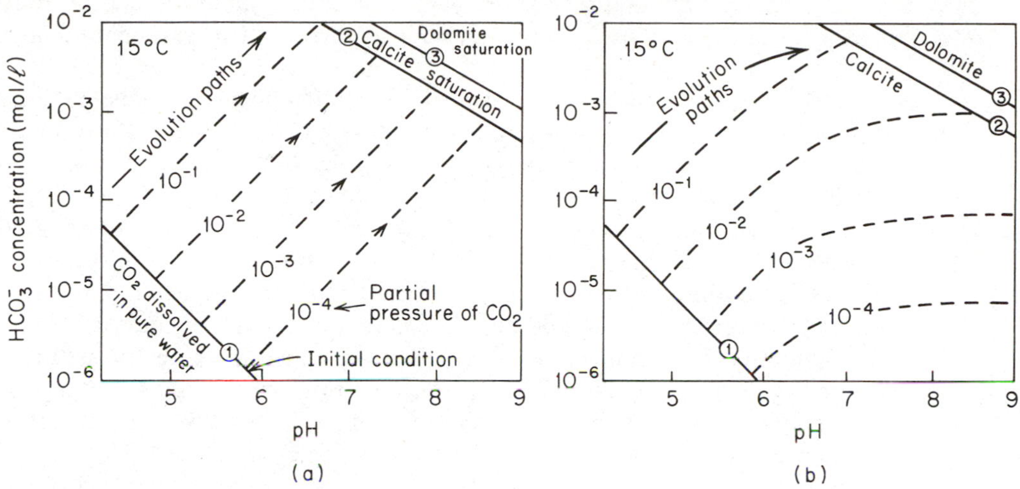

Figure 7.11 Chemical evolution paths for water dissolving calcite at 15°C. (a) Open-system dissolution; (b) closed-system dissolution. Line (1) represents the initial condition for the CO_2 charged water; line (2) represents calcite saturation; line (3) represents dolomite saturation if dolomite is dissolved under similar conditions.

water is saturated. The conditions of saturation for 15°C are represented by lines (2) and (3) for calcite and dolomite, respectively. If the temperature is higher, the saturation lines will be lower; if the temperature is lower, the lines will be higher because the solubility is greater. The positions of the evolution paths and saturation lines will be somewhat different if dissolution takes place in solutions of higher ionic strength.

Water infiltrating through the soil zone may evolve to a position on the saturation line and then evolve to disequilibrium positions off the line. For example, water under a high P_{CO_2} may become equilibrated with respect to calcite or dolomite in the upper horizons of the soil and then move deeper into the unsaturated zone, where different partial pressures of CO_2 exist in the soil air. If there are lower CO_2 partial pressures deeper in the unsaturated zone, the infiltrating water will lose CO_2 to the soil air. This is referred to as degassing or off-gassing. Its occurrence would cause a rise in pH of the soil water. Since degassing would normally occur much more rapidly than precipitation of carbonate minerals, the water would become supersaturated with respect to calcite or dolomite. The water would evolve above the saturation lines shown in Figure 7.11(a).

The partial pressure of CO_2 in the unsaturated zone below the CO_2 production zone in the soil is controlled mainly by the rate of gas diffusion downward from the organic-rich layers in the soil and the rate of escape to the atmosphere by way of short-circuiting paths such as deep desiccation cracks, frost cracks, root holes, and animal burrows. As a result, at some times of the year, low P_{CO_2} conditions can occur in the unsaturated zone below the soil. If deep infiltration occurs during these periods, conditions of supersaturation with respect to carbonate minerals will occur.

It has been indicated above that if significant amounts of carbonate minerals are present in the soil or subsoil of the unsaturated zone, dissolution to or near saturation would be expected to occur. Using a mass-balance approach, we will now determine how much mineral material must be present in order for saturation to be attained. From Figure 3.7 it is evident that at a relatively high CO_2 partial pressure of 10^{-1} bar, 6.3 mmol of Ca^{2+} will occur in solution after dissolution of calcite to equilibrium under open-system conditions. Because each mole of calcite that dissolves produces 1 mol of Ca^{2+} in solution, it can be concluded that under these conditions, 6.3 mmol (0.63 g) of dissolved calcite per liter of solution is required to produce equilibrium. To determine this amount of calcite as a percentage by weight of geologic materials, it will be assumed that the materials are granular and have a porosity of 33% and a specific gravity of 2.65. Therefore, the volumetric ratio of voids to solids is 1 : 2 and the mass of solids per liter of voids is 5300 g. If the voids are full of water and if 0.63 g of calcite from the bulk solids mass of 5300 g dissolves, the water would be saturated with respect to calcite. The 0.63 g of calcite is 0.01% by weight of the total solids. Calcite contents of this magnitude are well below detection by the methods of mineralogical analysis normally used by geologists. This example serves to illustrate that mineral constituents present in what normally are regarded as small or even insignificant amounts can exert a strong influence on the chemistry of groundwater flowing through the materials. This generalization also applies to many other mineral species that occur in the groundwater zone.

Closed-System Conditions

In situations where there are essentially no carbonate minerals in the soil zone or in the unsaturated zone below the soil zone, infiltration water charged with CO_2 can travel to the saturated zone without much CO_2 consumption. During infiltration the CO_2 will exist in solution as H_2CO_3 and $CO_2(aq)$ and will not be converted to HCO_3^-. In this discussion the minor amount of CO_2 to HCO_3^- conversion that can occur as a result of leaching of aluminosilicate materials is neglected. The effect of these minerals on the dissolved inorganic carbon in groundwater is considered in Section 7.4.

If the recharge water infiltrates to the water table without significant CO_2 consumption and then encounters carbonate minerals along its flow paths in the saturated zone, dissolution will take place in this zone under *closed-system conditions*. As H_2CO_3 is converted to HCO_3^- (see Eq. 3.54), the H_2CO_3 concentration and the CO_2 partial pressure will decline. In carbonate rocks and in most other calcareous strata there is no CO_2 replenishment below the water table. Exceptions to this generalization are discussed in Section 7.5. As in the open-system case, the water will proceed from some initial condition defined by the P_{CO_2} and pH to a condition of saturation with respect to the carbonate minerals present in the system. Figure 7.11(b) shows initial conditions, saturation conditions, and some representative evolution paths for closed-system dissolution. The evolution paths and saturation lines were calculated using procedures similar to the open-system case, except that P_{CO_2} is variable and total dissolved inorganic carbon is the sum of the initial CO_2 and the carbon from carbonate-mineral dissolution.

Comparison of Figure 7.11(a) and (b) indicates that the equilibrium pH and HCO_3^- values that result from calcite or dolomite dissolution under open- and closed-system conditions are different. Because the solubility is greater for a given initial P_{CO_2}, open-system pH values at saturation are lower and the HCO_3^- and Ca^{2+} concentrations are higher. At high initial P_{CO_2} conditions, the difference in water chemistry between the two systems is smallest. Figure 7.11(b) indicates that it is possible, within part of the soil-zone P_{CO_2} range, to evolve under closed-system conditions to pH values well above 8. Under open-system conditions, however, equilibrium pH values are below 8. In carbonate terrain the pH of natural groundwater is almost invariably between 7.0 and 8.0, which suggests that open-system conditions are common.

In the discussion above, the chemical evolution of groundwater was considered within the convenient constraints of open and closed CO_2 systems under specified time-independent boundary conditions. In nature, variations in CO_2 partial pressures, soil temperatures, and slow diffusion-controlled reaction processes can cause significant deviations from the conditions prescribed in our hydrochemical models. In some situations infiltrating groundwater may proceed partway along open-system evolution paths and then move below the water table and evolve the rest of the way to saturation under closed-system conditions. Other factors, such as adsorption, cation exchange, gas diffusion, and mechanical dispersion, may influence the chemical evolution of the water. Closed-system or partly closed-system dissolution above the water table can arise in some situations. Nevertheless, open- and closed-system models provide a useful framework within which to interpret chemical data from many hydrogeologic settings. There is a need, however, for more studies of the distribution of CO_2 partial pressures in the unsaturated zone to provide a better basis for the adoption of these models in the interpretation of field data.

Incongruent Dissolution

The concept of incongruent dissolution is introduced in Section 3.5. Specific application of this concept to the calcite–dolomite–water system is presented below. The preceding discussion was based on the premise that calcite and dolomite, if present in the hydrogeologic system, dissolve independently of each other. Although Figure 7.11 shows saturation lines for both calcite and dolomite, it should be kept in mind that these lines were calculated for dissolution of these minerals in separate systems. If these two minerals occur in the same hydrogeologic system, which is often the case, they may dissolve simultaneously or sequentially. This can lead to much different equilibrium relations than those displayed in these diagrams.

In terms of equilibrium constants and activities, the saturation conditions for calcite and dolomite can be expressed as

$$K_c = [Ca^{2+}][CO_3{}^{2-}] \tag{7.5}$$

$$K_d = [Ca^{2+}][Mg^{2+}][CO_3{}^{2-}]^2 \tag{7.6}$$

where the subscripts *c* and *d* designate calcite and dolomite, respectively. If groundwater at 10°C dissolves dolomite to saturation and then flows into a zone that contains calcite, no calcite dissolution will occur, because the water would be saturated with respect to calcite. This conclusion is based on the K_c and K_d values listed in Table 3.7. At equilibrium with respect to dolomite, the ion-activity product $[Ca^{2+}][CO_3{}^{2-}]$ would equal $K_d^{1/2}$, which at about 10°C is equal to K_c (Table 3.7). Comparison of $K_d^{1/2}$ and K_c values using data from Table 3.7, indicates that $K_d^{1/2} > K_c$ at temperatures below 10°C, and $K_d^{1/2} < K_c$ above 10°C. Therefore, if groundwater dissolves dolomite to equilibrium at the lower temperatures, the water will become supersaturated with respect to calcite, causing precipitation. The system evolves toward a condition whereby the rate of dolomite dissolution equals the rate of calcite precipitation. The coexisting processes of dolomite dissolution and calcite precipitation are implied in the expression *incongruent dissolution of dolomite*.

When groundwater dissolves dolomite to equilibrium at temperatures above 10°C and then moves into a zone where calcite exists, the water will be able to dissolve calcite because $K_d^{1/2} < K_c$. Calcite dissolution increases $[Ca^{2+}]$ and $[CO_3{}^{2-}]$ and the water will therefore become supersaturated with respect to dolomite. Because the dolomite precipitation reaction is so sluggish, supersaturation could persist for long periods of time with little or no dolomite precipitation. If significant amounts of dolomite were to form, however, the calcite dissolution process would be incongruent. Over long periods of time, the incongruency of the dolomite and calcite reactions may exert an important influence on the chemical evolution of the water and on the mineralogical evolution of the host rock.

When groundwater dissolves calcite to equilibrium first and then encounters dolomite, dolomite dissolves regardless of the temperature because the water must acquire appreciable Mg^{2+} activity before dolomite equilibrium is attained [Eq. (7.6)]. Even from the initial stages of dolomite dissolution, however, the water becomes supersaturated with respect to calcite as a result of the influx of $[Ca^{2+}]$ and $[CO_3{}^{2-}]$ from the dolomite, and calcite precipitates. The dolomite dissolution would therefore be incongruent. As calcite precipitates, the water would tend to become undersaturated with respect to dolomite. If this occurs in a zone in which dolomite exists, dolomite would continue to dissolve as calcite precipitates. As a result, there would be a decrease in the Ca^{2+}/Mg^{2+} molal ratio.

In the above discussion, the concept of incongruent dissolution of calcite and dolomite was illustrated by assuming that the groundwater encounters calcite and dolomite sequentially along its flow paths; that is, we let the water react with one mineral and then the other. In many hydrogeologic systems, such as those in glacial till or partly dolomitized limestone, calcite and dolomite exist together in the same strata and hence would be expected to dissolve simultaneously if contacted by water undersaturated with respect to both of these minerals. The incongruency relations would therefore be controlled by the differences in effective dissolution rates as well as temperature and P_{CO_2}. The differences in dissolution rates determine the Ca^{2+}/Mg^{2+} molal ratio. For example, if water infiltrates into a soil and becomes charged with CO_2, and then moves into a soil horizon that contains both calcite and dolomite, dissolution of both of these minerals occurs. If calcite dissolves much more rapidly than dolomite, calcite saturation would be attained much before dolomite saturation. After calcite saturation is reached, dolomite would continue to dissolve, but incongruently, until dolomite saturation is reached. The Ca^{2+}/Mg^{2+} molal ratio would evolve from a high initial value to a much lower value as Mg^{2+} is contributed from dolomite and as Ca^{2+} is lost through calcite precipitation. Under these circumstances, calcite supersaturation may persist a considerable distance along the groundwater flow paths.

If, on the other hand, dolomite dissolves more rapidly than calcite, the Ca^{2+}/Mg^{2+} molal ratio would be much smaller than in the case described above. This could occur if dolomite is much more abundant in the geologic materials, so the surface area of reaction would be much larger than that of calcite. If dolomite saturation is attained quickly, there would be little opportunity for calcite to dissolve. The reasoning here is similar to the case of sequential dolomite-calcite dissolution described above. The water temperature could also affect the tendency for incongruency to develop.

In conclusion, it can be stated that in groundwater systems that contain calcite and dolomite, Ca^{2+}/Mg^{2+} molal ratios can develop within a wide range, both above and below unity, depending on the influence of sequential distribution, simultaneous dissolution, incongruent dissolution, CO_2 partial pressures, temperatures, and other factors. Interpretation of water composition from carbonate systems within the strict confines of simple open- and closed-system dissolution models can in some situations be misleading. Cation exchange reactions can also produce variations in the Ca^{2+}/Mg^{2+} ratio of groundwater but, as indicated in the discussion above, they need not necessarily be invoked to explain these variations.

Other Factors

As indicated in Table 3.7, the solubilities of calcite and dolomite are quite strongly dependent on temperature. The solubility-temperature relations for these minerals are unusual in that larger solubilities occur at lower temperatures, because CO_2 is more soluble at lower temperature and because values of K_c and K_d are larger. Nearly all other mineral types have the reverse relationship; they are more soluble at higher temperatures. In the preceding section, the effect of temperature on the incongruency relations was noted. Our purpose now is to take a broader look at the effects of temperature on the chemical evolution of groundwater in carbonate terrain.

In climatic regions where snowfall accumulates on the ground during the winter, the largest groundwater recharge event commonly takes place in the springtime as snowmelt infiltrates through the cold or even partially frozen soil zone and moves downward to the water table. In vast areas of Canada and the northern United States, carbonate minerals occur in abundant amounts in the soil or subsoil at very shallow depths. During spring recharge, dissolution can proceed at very low temperatures under open- or nearly open-system conditions. At depths greater than a few meters below ground surface, temperatures are normally at least several degrees higher. Calcite and dolomite-saturated water that infiltrates downward from the cold upper zone into the underlying warmer zones will become supersaturated with respect to these minerals as a result of this temperature increase.

A situation where water is initially saturated with respect to calcite at a temperature of 0°C is taken as an example. This may represent water that infiltrates through the soil during a period of spring snowmelt. If the water moves below the frost zone to depths at which the geologic materials have temperatures closer to the average annual air temperature, the water will become supersaturated with respect to calcite. If the water moves deeper in the groundwater flow system, the temperature will continue to rise as a result of the regional geothermal gradient. The water will become progressively more supersaturated, unless the temperature effect is balanced by Ca^{2+} and $CO_3{}^{2-}$ losses as a result of calcite precipitation. At 25°C and at a CO_2 partial pressure of 10^{-2} bar, calcite is half as soluble as at 0°C. This example illustrates that the total dissolved solids in groundwater do not necessarily increase along the flow paths. If the groundwater chemistry is controlled almost entirely by interactions with carbonate minerals, it is possible for differences in temperature along the flow paths to cause decreases in total dissolved solids. In nature, however, decreases caused by carbonate-mineral precipitation can be masked by increases in dissolved solids caused by dissolution of other minerals.

Considering carbonate mineral dissolution and precipitation in terms of a regional groundwater flow system, water moves from recharge areas in which temperatures can be low, to deeper zones at higher temperatures, and then back to shallow colder zones in the discharge areas. For waters that become saturated with respect to calcite and dolomite in the recharge areas, the deeper zones would be calcite or dolomite precipitation zones. In the colder discharge areas, dissolution would once again occur if carbonate minerals are present and if other mineral-water interactions have not appreciably altered the saturation levels.

In the hydrochemical evolution processes considered above, the effects of noncarbonate salts in the water were neglected. If the water contains significant contents of noncarbonate mineral-forming ions such as Na^+, K^+, Cl^-, and $SO_4{}^{2-}$, the carbonate-mineral equilibrium is influenced by the effects of ionic strength and complex-ion formation. This can be deduced from the discussions in Sections 3.3 and 3.5. Greater salinity is reflected in higher ionic strength, which in turn causes lower values for the activity coefficients of all major ions in solution (Figure 3.3).

The solubilities of calcite and dolomite therefore increase. In the development of hydrochemical models, the effect of ionic strength in the fresh- to brackish-water salinity range can be taken into account quantitatively. The modeling of saline or brine solutions, however, involves greater uncertainties associated with the activity coefficient relations.

Groundwater that is influenced chemically by dissolution of calcite or dolomite is also commonly influenced by other minerals that exert some control on the concentrations of Ca^{2+} and Mg^{2+}. For example, dissolution of gypsum ($CaSO_4 \cdot 2H_2O$) can cause large increases in Ca^{2+} concentrations. Through the common-ion effect described in Section 3.5, this can cause supersaturation of the water with respect to calcite or dolomite or greatly limit the amount of calcite or dolomite that will dissolve when the water encounters these minerals along its flow paths. If clay minerals are present, cation exchange processes may cause large changes in the cation ratios and thereby alter the saturation levels of the water with respect to carbonate minerals. The roles of the common-ion effect and cation exchange in the chemical evolution of groundwater are described in more detail in Section 7.5.

In the development of geochemical models to describe equilibria between groundwater and carbonate minerals, the use of thermodynamic data obtained from experiments on relatively pure forms of the minerals is common practice. In natural systems, however, calcite and dolomite may deviate significantly from the ideal composition. For example, calcite can contain as much as several percent Mg in solid solution with Ca. Impurities such as Sr and Fe commonly occur in carbonate minerals. Although these impurities may be an important source of these elements in groundwater, their effect on the equilibrium constants of calcite and dolomite is generally small. In some situations, however, oxidation and hydrolysis reactions with the impurities may cause a significant production of H^+ and therefore a lower pH. It is necessary to emphasize that in our consideration of the carbonate system, the processes of dissolution and precipitation of carbonate minerals were isolated from the many other processes that in nature commonly occur concurrently within the hydrochemical system. In the interpretation of chemical data from real groundwater systems, it is usually necessary to take into account a more complex set of interacting hydrochemical processes.

Interpretation of Chemical Analyses

Calcite and dolomite exist in virtually all regions of the world in which sedimentary rocks are abundant. To describe the chemical evolution of groundwater in all these regions or even in a representative number of regions would be an insurmountable task, even if many chapters could be devoted to the cause. Instead, our approach will be to briefly summarize the hydrochemical characteristics of groundwater in a small number of carbonate-rock systems and then describe some geochemical interpretations developed for these systems. For this purpose, carbonate-rock aquifers in central Pennsylvania, central Florida, and south-central Manitoba were chosen. These three aquifers are located in very different climatic and hydrologic settings. The locations, geology, and groundwater conditions of the three aquifer systems are summarized in Table 7.2. Information on the aquifer system in Pennsylvania was obtained from Jacobson and Langmuir (1970) and Langmuir (1971). The Floridan system is described by Back and Hanshaw (1970) and Hanshaw et al. (1971). The hydrogeology of the portion of the Manitoban dolomite aquifer used in this comparison has been described by Goff (1971). Render (1970) reported on a regional study of this aquifer system.

Table 7.2 Hydrogeological Characteristics of the Carbonate-Rock Aquifer Systems for Which Water Composition Data Are Summarized in Table 7.3

	Areas of investigation		
	Pennsylvania*	Florida	Manitoba
Geography	Appalachian section of Valley and Ridge Province, Central Province	Central Florida regional limestone aquifer	Glaciated plain in interlake area of south-central Manitoba
Climate and annual precipitation	Humid continental, 990 mm	Tropical and subtropical, 1400 mm	Semihumid, continental, 500 mm
Aquifer type and age	Beds of dolomite and limestone between shale and sandstone	Tertiary limestone overlain by 0–50 m of clay, sand, and gravel	Silurian dolomite overlain by 0–30 m of glacial till
Water-table depth	10–100 m	0–30 m	0–10 m
Aquifer thickness	Very variable	100–700 m	5–50 m
Recharge areas	Sinkholes, streambed, seepage, thin soil, outcrops, and infiltration through glacial drift	Outcrop areas and areas of sand and gravel	Areas of thin glacial till and local outcrops
Depth of wells	30–150 m	50–400 m	10–50 m
Age of groundwater	Local springs: days Regional springs: months Wells: weeks to months	From months and years in recharge areas to many thousands of years elsewhere	Months to many years

*Samples in this study area were collected from local springs, regional springs, and wells. The local springs issue from carbonate rock at the base of mountain slopes; regional springs discharge at down-valley locations.

SOURCES: Back and Hanshaw, 1970; Goff, 1971; and Langmuir, 1971.

In the hydrochemical investigations, careful pH measurements were made in the field. The charge-balance errors of the chemical analyses used in the data compilation are less than 5% (acceptable limit of charge-balance error indicated in Section 3.3).

In the study in Pennsylvania, chemical analyses of 29 springs and 29 wells were conducted. Of the 29 wells, 20 are in dolomite and 9 in limestone. Twenty-two of the springs discharge from limestone and 7 from dolomite. In the hydrochemical investigation of the Floridan aquifer, samples from 53 wells were analyzed. In this discussion, data from 39 of the wells are used. The other wells were excluded to avoid the effects of mixing in saltwater zones near the ocean. In the hydrochemical investigation of the Manitoban aquifer, samples from 74 wells were analyzed.

The mean values and standard deviations of the temperature, major-ion concentrations, pH, P_{CO_2}, and saturation indices for calcite, dolomite, and gypsum for groundwater in the three study areas are listed in Table 7.3, which indicates important similarities and differences between the three areas. In each of the areas HCO_3^- is the dominant anion and SO_4^{2-} the second most abundant anion. Concentrations of Cl^- are generally very low. The average HCO_3^- content in the Manitoban aquifer is more than twice as large as the averages for the Floridan aquifer and for the spring samples from the Pennsylvania aquifer. The average HCO_3^- value for the well samples from the Pennsylvanian study area is between these two extremes. The average pH values for the Manitoban and Floridan aquifers are similar. The pH of the Pennsylvanian aquifer is slightly lower. Nearly all samples, however, lie in the relatively narrow pH range 7–8. In the Manitoban aquifer, the average

Table 7.3 Summary of Chemical Data From the Carbonate-Rock Aquifers in Central Florida, Central Pennsylvania, and Southcentral Manitoba

	Pennsylvania limestone and dolomite aquifer				Florida limestone aquifer		Manitoba dolomite aquifer	
	Springs		Wells					
Parameter	$\bar{X}$*	σ†	$\bar{X}$	σ	$\bar{X}$	σ	$\bar{X}$	σ
Temperature (°C)	10.9	1.3	18.0	1.2	24.4	1.2	5.1	0.9
pH	7.37	1.5	7.47	0.3	7.69	0.25	7.61	0.25
K^+	1.6	0.6	1.5	1.4	1.0	0.8	9	7
Na^+	3.8	1.8	3.1	3	7.9	5.3	37	36
Ca^{2+}	48	11	55	22	56	25	60	15
Mg^{2+}	14	11	28	14	12	13	60	21
HCO_3^-	183	43	265	83	160	40	417	101
Cl^-	8.2	3.5	10	9	12	9	27	26
SO_4^{2-}	22	5	20	15	53	94	96	127
P_{CO_2} (atm)	$10^{-2.2\pm0.15}$		$10^{-2.15\pm0.43}$		$10^{-2.51\pm0.35}$		$10^{-2.11\pm0.33}$	
SI_{cal}‡	−0.39	0.25	−0.16	0.12	+0.12	0.18	+0.04	0.17
SI_{dol}‡	−1.2	0.74	−0.36	0.23	−0.23	0.49	+0.27	0.35
SI_{gyp}‡	−2.0	0.14	−2.2	0.46	−2.3	0.8	−1.8	0.53

*$\bar{X}$, mean.

†σ, standard deviation.

‡Saturation index expressed in logarithmic form:

$$SI_{cal} = \log([Ca^{2+}][CO_3^{2-}]/K_{cal})$$

$$SI_{dol} = \log([Ca^{2+}][Mg^{2+}][CO_3^{2-}]^2/K_{dol})$$

$$SI_{gyp} = \log([Ca^{2+}][SO_4^{2-}]/K_{gyp})$$

SOURCE: Back, written communications; Goff, 1971; and Langmuir, 1971.

Ca^{2+}/Mg^{2+} molal ratio is less than unity, whereas in the other areas it is greater than unity.

In our interpretation of this hydrochemical information we will begin by noting that the calculated P_{CO_2} values for the groundwaters in all three areas are considerably above the P_{CO_2} of the earth's atmosphere ($10^{-3.5}$ bar). This indicates that the groundwater in these aquifers became charged with CO_2 during infiltration through soil zones. A second important observation is that there are large groundwater temperature differences between the three areas. The Floridan aquifer is warmest, with temperatures close to 25°C. In the Pennsylvanian aquifer the average groundwater temperature is close to 11°C and in the Manitoban aquifer the temperatures are near 5°C.

The pH of water in the three aquifers is significantly above 7 and below 8. Figure 3.5(a) indicates therefore that the dissolved inorganic carbon exists almost entirely as HCO_3^-. The concentrations of HCO_3^- are highest in groundwater in the Manitoba aquifer, which indicates that more calcite or dolomite has dissolved in the water in this aquifer than in the other aquifers. The amount dissolved in the Pennsylvanian aquifer is intermediate between the Manitoban and Floridan aquifers. These differences can be attributed to three main factors. The first factor is temperature. As would be expected from the solubility considerations, the coldest water has the highest content of carbonate-mineral dissolution products. This cannot account for all the differences, however. The second factor is the partial pressure of CO_2. The Manitoban water has the highest calculated partial pressure and the Floridan water has the lowest. The differences are large enough to account for much of the difference in HCO_3^- values. Trainer and Heath (1976) have attributed the relatively low CO_2 partial pressures in groundwater in the Floridan aquifer to the occurrence of permeable sands in the main recharge areas of this aquifer. The region of major recharge is shown in Figure 7.12(a). These authors suggest that relatively little CO_2 is produced in the soil zone in these areas because of the lack of abundant organic matter. They also suggest that because of the high permeability of the sand, CO_2 readily escapes from the soil to the atmosphere.

The third factor is the degree of saturation with respect to calcite and dolomite. In this regard the procedure of Langmuir (1971) is adopted; a sample is designated as being saturated if its saturation index, expressed in logarithmic form, is in the range -0.1 to $+0.1$. Sixty-two percent of the Manitoban samples were saturated with respect to both calcite and dolomite, 12% were supersaturated, and 8% were significantly undersaturated. Sixty-six percent of the Floridan samples were supersaturated with respect to calcite, 24% were saturated, and 10% were undersaturated. With respect to dolomite, 59% were supersaturated, 21% saturated, and 20% undersaturated. Results for the Pennsylvanian springs and well samples are very different: 20% were saturated and 80% were undersaturated with respect to calcite. With respect to dolomite, 4% were saturated and the rest were undersaturated. If all the undersaturated waters in the Pennsylvanian aquifer were to be brought to saturation by dissolution of calcite or dolomite, the average HCO_3^- and pH values would be much closer to the average values for the Manitoban aquifer.

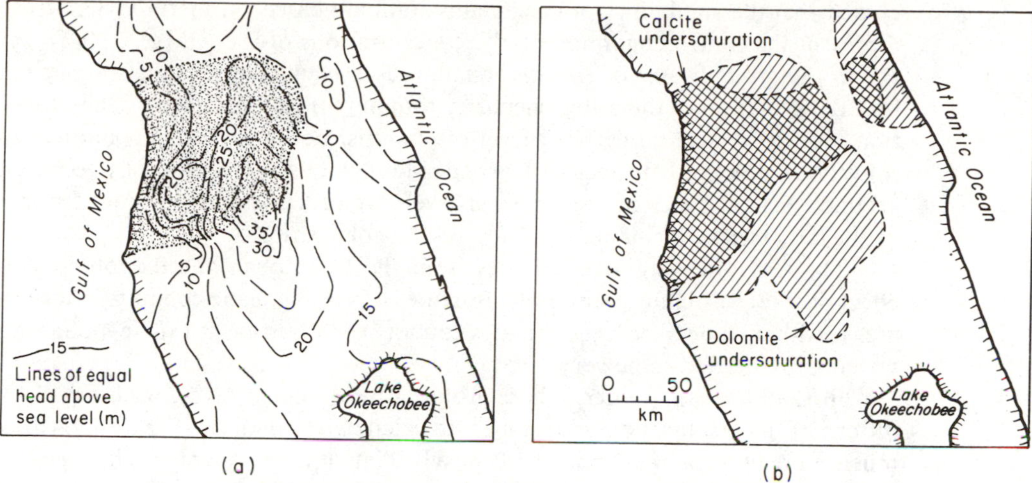

Figure 7.12 Regional limestone aquifer in south-central Florida. (a) Potentiometric surface and area of major recharge; (b) areas of undersaturation with respect to calcite and dolomite (after Hanshaw et al., 1971).

Since only 10% of the Floridan samples are undersaturated with respect to calcite, dissolution to saturation for these waters would not greatly increase the mean HCO_3^- and pH values.

It is reasonable at this point to draw the conclusion that conditions of disequilibrium (i.e., undersaturation or supersaturation) are not uncommon in carbonate aquifers. One of the most enigmatic of disequilibrium conditions in hydrogeochemical systems is the existence of undersaturation with respect to calcite and dolomite in situations where these minerals occur in abundance. Laboratory experiments on rates of calcite dissolution indicate that equilibrium should be achieved in a matter of hours or days (Howard and Howard, 1967; Rauch and White, 1977), and yet in the Pennsylvanian and Floridan carbonate-rock aquifers, much older water in contact with calcite and dolomite persists in a state of undersaturation. Dye tracer tests conducted by Jacobson and Langmuir (1970) in parts of the Pennsylvanian aquifer indicated groundwater residence times of 2–6 days over flow distances of about 7000 m. They concluded that the residence times of many of the spring waters are generally somewhat longer than 2–6 days and that the waters sampled from the wells are much older than this. Langmuir (1971) noted that the pH and HCO_3^- values of the spring waters tend to increase with their subsurface residence times. A much larger percentage of the well samples was saturated because the water had sufficient time to equilibrate with the calcite and dolomite in the aquifer. This investigation suggests that in field situations, weeks or even months of residence time can be necessary for dissolution to proceed to equilibrium with respect to calcite and dolomite. Because none of the spring or well waters was supersaturated with respect to calcite or dolomite, Langmuir concluded that the solubilities of calcite and dolomite based on thermodynamic data represent the controlling limits on the concentrations of Ca^{2+}, Mg^{2+}, HCO_3^-, and H^+ in the groundwater of this carbonate-rock system. Langmuir also concluded that the evolution of the water chemistry roughly follows the open-system dissolution model. At a more detailed level of analysis, he noted that incongruent dissolution of dolomite at times of low water-table levels and dilution by recharging groundwaters at times of higher water levels are processes that account for many of the data trends, including the Ca^{2+}/Mg^{2+} molal ratios.

The few undersaturated samples from the Manitoban aquifer represent the effect of short local flow paths along fracture or bedding plane conduits in recharge areas. Although detailed age estimates cannot be derived from existing data, these waters are expected to be very young.

In the Floridan aquifer, where groundwater is much older, widespread conditions of undersaturation with respect to calcite and dolomite [Figure 7.12(b)] are much more perplexing than in the Pennsylvanian aquifer. Average velocities in the Floridan aquifer determined by ^{14}C dates are 8 m/y (Back and Hanshaw, 1970). Within the region of undersaturation, the groundwater attains ages of hundreds to thousands of years. Back and Hanshaw suggest that perhaps in some areas a significant amount of water reaches the aquifer through sand-filled solution openings and has not been in intimate contact with the limestone. They also suggest that armoring of the limestone surface by inorganic ionic species or by organic substances may produce a state of pseudo-equilibrium between crystal surfaces and the solution. There is also the possibility that some of the well samples appear undersaturated because the well water represents a mixture of waters of different compositions that flow into the well bores from different strata or zones within the aquifer. Most of the wells in the Floridan aquifer have intake zones over large vertical intervals. The occurrence of undersaturation in waters that are a mixture of two or more saturated waters was established by Runnels (1969) and Thraikill (1968) and was demonstrated in computer simulation studies by Wigley and Plummer (1976).

In an extensive part of the Floridan aquifer, groundwater is significantly supersaturated with respect to calcite and dolomite. Back and Hanshaw (1970) and Langmuir (1971) suggest that this is caused by dissolution of trace amounts of gypsum and that the condition of supersaturation is maintained by an imbalance in rates of gypsum dissolution relative to precipitation rates of calcium carbonate (calcite or aragonite). This interpretation is consistent with the results of a kinetically based model of water chemistry evolution in this aquifer described by Palciauskas and Domenico (1976). These authors have developed a mathematical framework that indicates that the distance that groundwater must travel to attain saturation with respect to individual mineral phases increases with increasing rates of mixing and velocity and decreases with increasing rates of reaction. Their analysis shows that steady-state chemical concentrations can exist and can cause a steady level of supersaturation or undersaturation. This can occur when the rate of production of one or more dissolved species due to the dissolution of one mineral species is balanced by the rate of consumption of these species by precipitation of a second mineral species.

Much of the water in the Manitoban aquifer is supersaturated with respect to calcite and dolomite. In Figure 7.13, the water chemistry expressed in terms of pH, HCO_3^-, Ca^{2+}, and Mg^{2+} is compared to the simple open-system models for the dissolution of dolomite and calcite separately and in sequence. This comparison indicates that the data generally plot above the equilibrium lines (i.e., above the levels that would be attained if the water evolved directly to saturation under open-system dissolution). Cherry (1972) attributed this condition of disequilibrium to the combined influence of temperature change, degassing, cation exchange, and the common-ion effect caused by gypsum dissolution. Most of the recharge to the aquifer occurs in areas where the aquifer is overlain by glacial till. The till is rich in dolomite, calcite, quartz, feldspars, and clay minerals, and at shallow depth has small amounts of gypsum. A small part of the supersaturation is caused by the increase in temperature that occurs as the water moves from the colder zone in the upper meter or two of soil into the deeper parts of the flow system.

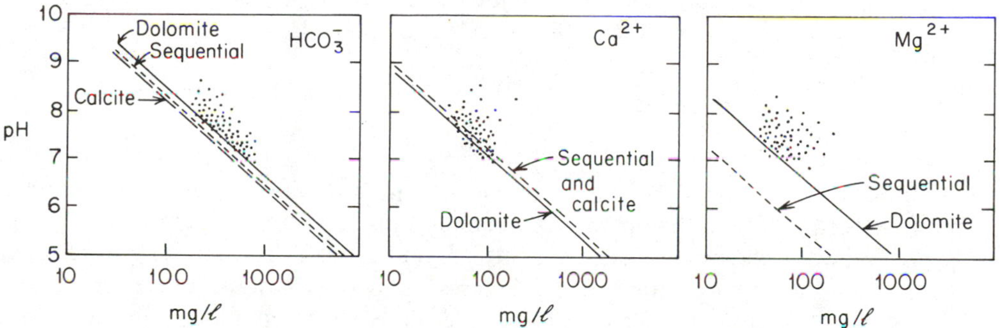

Figure 7.13 Chemical data from the Manitoban carbonate-rock aquifer plotted on diagrams showing equilibrium conditions at 5°C for open-system dissolution of calcite and dolomite and sequential dissolution of these two minerals (calcite before dolomite, which causes dolomite solubility to be depressed as a result of the common-ion effect). (After Cherry, 1972.)

In some recharge areas it is likely that during part of the year the CO_2 partial pressure in the unsaturated zone below the soil is often lower than the partial pressures in the organic-rich soil horizons in which CO_2 is generated. If CO_2 in the infiltrating recharge water degasses as it flows through this zone, the pH of the water would rise. This could account for the fact that in Figure 7.13, the water composition expressed as HCO_3^-, Ca^{2+}, and Mg^{2+} versus pH generally plots above the equilibrium lines (Cherry, 1972).

One of the most striking aspects of hydrochemical data from the three aquifer systems described above is a rather common occurrence of disequilibrium with respect to calcite-dolomite-water interactions. Because of this, one might be inclined to question the use of equilibrium concepts as an important means of interpreting hydrochemical data from carbonate rocks. However, the equilibrium concepts provided a useful framework for interpretation of the field data. Recognition of the deviations from thermodynamic equilibrium and the development of hypotheses to account for these deviations are an important part of the interpretive process. Eventually, it may be possible to interpret hydrochemical data from field settings within a quantitative framework that includes equations that describe the irreversible and kinetically controlled parts of the system, combined with the hydrodynamic processes of transport.

Hydrochemical data from aquifer systems can be used to develop interpretations of the evolution of aquifer permeability. For example, in the parts of the aquifers that are undersaturated with respect to calcite or dolomite, flowing groundwater is dissolving the aquifer rock. The porosity and permeability are therefore increasing. In terms of human history, these changes are probably imperceptible, but over periods of geologic time they are the basis for the development of permeability networks and even major landscape alterations. This aspect of groundwater processes is considered further in Chapter 11.

7.4 Groundwater in Crystalline Rocks

Crystalline rocks of igneous or metamorphic origin generally have at least one characteristic in common: they contain appreciable amounts of quartz and aluminosilicate minerals such as feldspars and micas. These minerals were originally formed at temperatures and pressures far above those occurring at or near the earth's surface. On the land surface, in the soil zone, and in the groundwater zone to depths of many hundreds or thousands of meters, these minerals are thermodynamically unstable and tend to dissolve when in contact with water. The dissolution processes cause the water to acquire dissolved constituents and the rock to become altered mineralogically.

As in the chemical evolution of groundwater in carbonate rocks, the dissolution of feldspars, micas, and other silicate minerals is strongly influenced by the chemically aggressive nature of water caused by dissolved CO_2. When CO_2 charged waters that are low in dissolved solids encounter silicate minerals high in cations, aluminum, and silica, cations and silica are leached, leaving an aluminosilicate residue with increased Al/Si ratio. This residue is usually a clay mineral such as kaolinite, illite, or montmorillonite. The cations released to the water are normally Na^+, K^+, Mg^{2+}, and Ca^{2+}. Another consequence of this process of incongruent dissolution is a rise in pH and in HCO_3^- concentration. Equations that can be used to describe these chemical changes in the water and host rock, and interpretations of hydrochemical data from igneous and metamorphic rocks, are described below.

Theoretical Considerations

Of all the minerals with which groundwater comes into contact, quartz is the most abundant, both in areal distribution and bulk composition. This discussion will therefore commence with consideration of quartz dissolution and solubility. The solubility of quartz (SiO_2) can be characterized (Stumm and Morgan, 1970) by the following equilibria (K values at 25°C):

$$SiO_2\ (\text{quartz}) + 2H_2O = Si(OH)_4 \qquad \log K = -3.7 \qquad (7.7)$$

$$Si(OH)_4 = SiO(OH)_3^- + H^+ \qquad \log K = -9.46 \qquad (7.8)$$

$$SiO(OH)_3^- = SiO_2(OH)_2^{2-} + H^+ \qquad \log K = -12.56 \qquad (7.9)$$

$$4Si(OH)_4 = Si_4O_6(OH)_6^{2-} + 2H^+ + 4H_2O \qquad \log K = -12.57 \qquad (7.10)$$

The dissolved silicon species can also be written in the form H_2SiO_4, $H_3SiO_4^-$, and so on, which portrays their acidic nature. With these equations it can be shown that in the pH range that includes nearly all groundwater (pH 6–9), the dominant dissolved silicon species is $Si(OH)_4$. At high pH values, other species are dominant in solution, and silica is more soluble. The results of analyses of Si concentrations in water are generally expressed as SiO_2. Expressed in this manner, quartz solubility is only about 6 mg/ℓ at 25°C (Morey et al., 1962). There is considerable evidence to indicate, however, that an amorphous or noncrystalline form of SiO_2, rather than quartz, controls the solubility of SiO_2 in water. The solubility of amorphous silica is approximately 115 to 140 mg/ℓ at 25°C (Krauskopf, 1956; Morey et al., 1964). The solubility increases considerably with temperature. Over long periods of time amorphous silica can evolve toward a crystalline structure and eventually become quartz.

Based on the solubility of amorphous silica and the abundance of quartz in most hydrogeologic systems, one might expect that SiO_2 would occur in major concentrations in most groundwaters. In nature, however, this is not the case. Davis (1964) compiled thousands of groundwater analyses from various areas in the United States and found that values for dissolved SiO_2 typically range from 10 to 30 mg/ℓ, with an average value of 17 mg/ℓ. Studies done elsewhere indicate that these values are reasonably representative on a global scale. Groundwater is therefore almost invariably greatly undersaturated with respect to amorphous silica. Quartz and amorphous silica generally do not exert an important influence on the level of silica in groundwater. More important in this regard are aluminosilicate minerals such as feldspars and micas.

From studies of the mineralogical and chemical nature of weathered igneous and metamorphic rocks and from thermodynamic considerations, it is known that the feldspar minerals are altered to clay minerals and other decomposition products. Table 7.4 indicates some of the common reactions that describe these

Table 7.4 Reactions for Incongruent Dissolution of Some Aluminosilicate Minerals*

Gibbsite-kaolinite	$\underline{Al_2O_3 \cdot 3H_2O} + 2Si(OH)_4 = \underline{Al_2Si_2O_5(OH)_4} + 5H_2O$
Na-montmorillonite-kaolinite	$\underline{Na_{0.33}Al_{2.33}Si_{3.67}O_{10}(OH)_2} + \frac{1}{3}H^+ + \frac{23}{6}H_2O = \frac{7}{6}\underline{Al_2Si_2O_5(OH)_4} + \frac{1}{3}Na^+ + \frac{4}{3}Si(OH)_4$
Ca-montmorillonite-kaolinite	$\underline{Ca_{0.33}Al_{4.67}Si_{7.33}O_{20}(OH)_4} + \frac{2}{3}H^+ + \frac{23}{2}H_2O = \frac{7}{3}\underline{Al_2SiO_2O_5(OH)_4} + \frac{1}{3}Ca^{2+} + \frac{8}{3}Si(OH)_4$
Illite-kaolinite	$\underline{K_{0.6}Mg_{0.25}Al_{2.30}Si_{3.5}O_{10}(OH)_2} + \frac{11}{10}H^+ + \frac{63}{60}H_2O = \frac{23}{20}\underline{Al_2Si_2O_5(OH)_4} + \frac{3}{5}K^+ + \frac{1}{4}Mg^{2+} + \frac{6}{5}Si(OH)_4$
Biotite-kaolinite	$\underline{KMg_3AlSi_3O_{10}(OH)_2} + 7H^+ + \frac{1}{2}H_2O = \frac{1}{2}\underline{Al_2Si_2O_5(OH)_4} + K^+ + 3Mg^{2+} + 2Si(OH)_4$
Albite-kaolinite	$\underline{NaAlSi_3O_8} + H^+ + \frac{9}{2}H_2O = \frac{1}{2}\underline{Al_2Si_2O_5(OH)_4} + Na^+ + 2Si(OH)_4$
Albite-Na-montmorillonite	$NaAlSi_3O_8 + \frac{6}{7}H^+ + \frac{20}{7}H_2O = \frac{3}{7}\underline{Na_{0.33}Al_{2.33}Si_{3.67}O_{10}(OH)_2} + \frac{6}{7}Na^+ + \frac{10}{7}Si(OH)_4$
Microcline-kaolinite	$\underline{KAlSi_3O_8} + H^+ + \frac{9}{2}H_2O = \frac{1}{2}\underline{Al_2Si_2O_5(OH)_4} + K^+ + 2Si(OH)_4$
Anorthite-kaolinite	$\underline{CaAl_2Si_2O_8} + 2H^+ + H_2O = \underline{Al_2Si_2O_5(OH)_4} + Ca^{2+}$
Andesine-kaolinite	$\underline{Na_{0.5}Ca_{0.5}Al_{1.5}Si_{2.5}O_8} + \frac{3}{2}H^+ + \frac{11}{4}H_2O = \frac{3}{4}\underline{Al_2Si_2O_5(OH)_4} + \frac{1}{2}Na^+ + \frac{1}{2}Ca^{2+} + Si(OH)_4$

*Solid phases are underlined.

dissolution processes. For simplicity the feldspar minerals will be considered only in terms of idealized end members; K-feldspar, Na-feldspar (albite), and Ca-feldspar (anorthite). In nature, however, feldspars contain impurities. Many feldspar minerals contain Na and Ca in various ratios as solid-solution mixtures of the two Na and Ca end members. Also included in Table 7.4 are reactions that describe the alteration of clay minerals. The incongruent dissolution reactions in Table 7.4 are written simply by introducing the appropriate dissolved species and then adjusting for mass balance in the normal manner. A major assumption inherent in this approach is the conservation of Al. That is, because the solubilities of aluminum compounds in water are extremely low, the total concentration of Al species (including complexes and polymers) removed from the solid phase is assumed to be negligible. The dissolution of feldspars is therefore assumed to produce mineral products that include all the Al removed from the feldspars. Field and laboratory studies have shown that in most circumstances this assumption is reasonable.

We will now make use of themodynamic data within an equilibrium framework to gain some insight into some of the more specific results of groundwater interactions with the feldspars and clays. Consider, for example, the albite dissolution reaction in Table 7.4. Expressed in mass-action form, it becomes

$$K_{\text{alb-kaol}} = \frac{[Na^+][Si(OH)_4]^2}{[H^+]} \tag{7.11}$$

where $K_{\text{alb-kaol}}$ is the equilibrium constant and the bracketed quantities are activities. In this development the activities of the mineral phases and water are taken as unity. This is a valid approach when considering minerals of ideal compositions in nonsaline solutions. Equation (7.11) can be expressed in logarithmic form as

$$\log K_{\text{Alb-kaol}} = \log [Na^+] + 2 \log [Si(OH)_4] - pH \tag{7.12}$$

or

$$\log K_{\text{Alb-kaol}} = \log \left(\frac{[Na^+]}{[H_+]}\right) + 2 \log [Si(OH)_4] \tag{7.13}$$

which indicates that the equilibrium condition for the albite-kaolinite reaction can be expressed in terms of pH and activities of Na^+ and $Si(OH)_4$. The kaolinite-Na montmorillonite reaction and the gibbsite-kaolinite reaction (Table 7.4) can be expressed in terms of Na^+, $Si(OH)_4$, and H^+ or pH. These equilibrium relations are the basis for construction of diagrams known as stability diagrams or as activity-activity diagrams. Examples of these diagrams are shown in Figure 7.14. The lines that separate the mineral phases of these diagrams represent equilibrium relations such as Eq. (7.11). Since minerals in real systems do not have ideal chemical compositions, the stability lines based on thermodynamic data for relatively pure mineral phases probably do not accurately represent real systems. Nevertheless, these types of diagrams have been found by many investigators to serve a useful purpose in the interpretation of chemical data from hydrogeological systems.

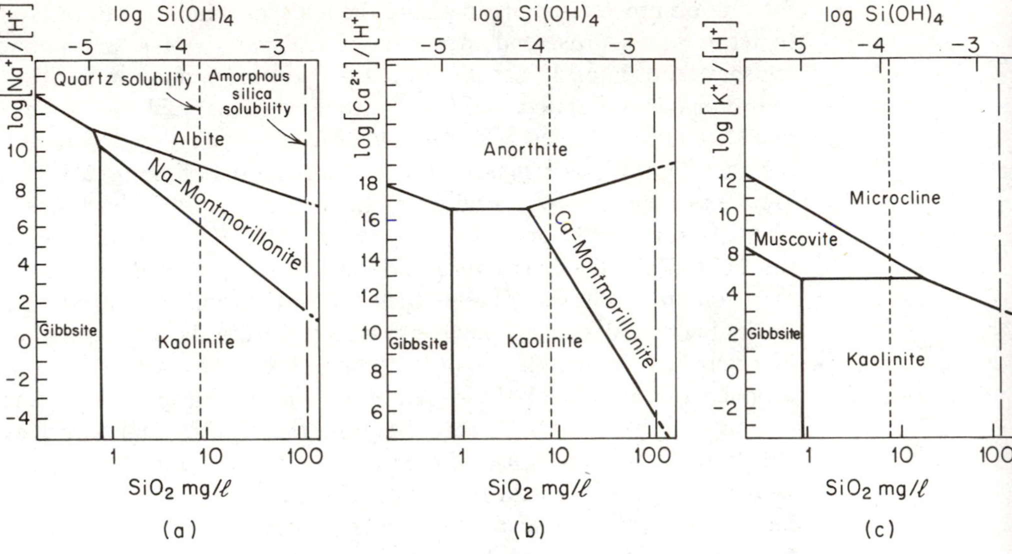

Figure 7.14 Stability relations for gibbsite, kaolinite, montmorillonite, muscovite, and feldspar at 25°C and 1 bar. (a) Gibbsite, $Al_2O_3 \cdot H_2O$; kaolinite, $Al_2Si_2O_5(OH_4)$; Na-montmorillonite, $Na_{0.33}Al_{2.33}Si_{3.67}O_{10}(CH)_2$; and albite, $NaAlSi_3O_8$. (b) Gibbsite; kaolinite; Ca-montmorillonite; and anorthite, $CaAl_2Si_2O_8$. (c) Gibbsite, kaolinite, muscovite, and microcline (after Tardy, 1971).

It is apparent from Table 7.4 that the incongruent dissolution of feldspars, micas, and clays involves consumption of H^+. Production of CO_2 in the soil zone is usually considered to be the main source of H^+. As these reactions proceed, there is a progressive increase in pH of the water. If the reactions occur in the soil zone or elsewhere in the unsaturated zone where CO_2 replenishment is significant, H_2CO_3, which is controlled by the partial pressure of CO_2 [Eq. (3.18)], remains constant while the concentrations HCO_3^- and CO_3^{2-} increase. The concentration of total dissolved inorganic carbon therefore increases. If the reactions occur below the water table, where CO_2 replenishment does not occur, H^+ consumption causes H_2CO_3 decline, P_{CO_2} decline, HCO_3^- increase, and CO_3^{2-} increase, while the concentration of total inorganic carbon remains constant. Within this theoretical framework it can be seen that if the reactions proceed far enough, groundwater in rocks comprised of feldspar and mica can acquire pH values above 7 or 8, and HCO_3^- concentrations of many tens or even hundreds of milligrams per liter.

The stoichiometries of the dissolution reactions for calcite and Ca-feldspar (anorthite) are identical; that is, for every mole of Ca^{2+} that goes into solution, 2 mol of H^+ are consumed. The charge balance of cations and anions in solution is maintained as H_2CO_3 dissociates to form HCO_3^- and CO_3^{2-}, as can be deduced from Figure 3.5(a). Although in theory Ca–HCO_3-type groundwater can evolve in rocks or unconsolidated deposits that contain Ca-feldspar, in nature this is uncommon, probably because of slow dissolution rates that develop as the feldspar acquires coatings of clay, which forms as a product at the incongruent reaction.

We will now consider the chemical evolution that may occur when fresh, slightly acidic water such as rain infiltrates through geologic materials in which feldspars are the only mineral phases that undergo significant dissolution. An initial assumption is that only Na-feldspar dissolves at a significant rate. When dissolution begins, the water contains negligible concentrations of $Si(OH)_4$ and Na. As the concentrations of these constituents increase, the water composition, expressed in terms of $Si(OH)_4$ and Na^+/H^+, will plot in the gibbsite stability field of Figure 7.14(a). This indicates that from a thermodynamic viewpoint, Na-feldspar will dissolve incongruently to produce gibbsite and dissolved products. As dissolution continues, $Si(OH)_4$ and the $[Na^+]/[H^+]$ ratio increase and the water composition moves through the gibbsite stability field into the kaolinite field. In the kaolinite field, incongruent dissolution of Na-feldspar produces kaolinite. Some of the gibbsite formed during the early stage is converted to kaolinite. As dissolution of the feldspar continues, the values of $Si(OH)_4$ and $[Na^+]/[H^+]$ increase further and the water chemistry evolves to the Na-montmorillonite stability field or more directly toward the Na-feldspar field. When the composition evolves to the boundary of the Na-feldspar field, equilibrium with respect to this feldspar is attained. Feldspar dissolution then ceases. For water to achieve equilibrium with respect to feldspar minerals, long periods of time and sluggish flow conditions are required. The water, however, is in equilibrium or near equilibrium with at least one other mineral phase. When the water composition plots in the kaolinite field, for example, equilibrium or near equilibrium exists with respect to this mineral. If the water composition plots on the boundary between kaolinite and montmorillonite, equilibrium or near equilibrium exists with respect to both these minerals.

Laboratory Experiments

The preceding discussion of silicate mineral dissolution was based on stoichiometric concentrations and on equilibrium concepts. This approach indicates nothing about the rates at which dissolution takes place or about the microscopic nature of the dissolution processes. For this type of information, laboratory experiments are useful.

Experiments on silicate mineral dissolution reported in the literature can generally be placed in two categories. Experiments in the first category involve dissolution systems where water and minerals are reacted in containers in which there is no through-flow (Garrels and Howard, 1957; Wollast, 1967; Houston, 1972). The second category includes experiments in which water is passed through mineral materials packed as porous media in cylindrical containers (Bricker, 1967; Bricker et al., 1968; Deju, 1971). Experiments in both categories have indicated that dissolution of feldspars and micas proceeds in two main stages. The first stage, which occurs in a matter of minutes after water is brought in contact with the mineral surfaces, involves the exchange of cations on the mineral surfaces for hydrogen ions in the water. This exchange is followed by a much slower and gradually decreasing rate of dissolution. This dissolution stage contributes appreciable amounts of dissolved products to the water over a period of hours or days before the dissolution rate becomes extremely small. Dissolution during the first stage of dissolution is normally congruent. During the second stage dissolution gradually becomes incongruent.

The type of results that have been obtained from silicate-mineral dissolution experiments in which there was no water through-flow (Bricker, 1967; Houston, 1972) are shown in schematic form in Figure 7.15. In these experiments the increases in the rates of concentration in the solution are controlled by the kinetics of mineral dissolution. If the mineral occurs in a porous medium, however, and if water is flowing through the medium, the concentration at a given point along a flow path will depend on dissolution kinetics *and* on the flow rate. If the flow rate is rapid compared to the rate of mineral dissolution, the concentration of dissolution products at a specified distance along the flow path may be small compared to the concentrations achieved after the same period of leaching under no-flow conditions. This topic is discussed further in Section 7.8.

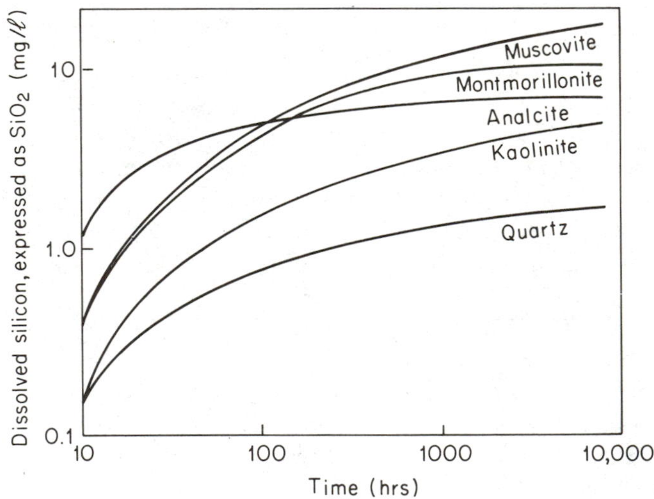

Figure 7.15 Release of silica during the dissolution of silicate minerals in distilled water at 25°C (after Bricker, 1967).

Although experiments on the dissolution of feldspars and other silicate minerals in aqueous solutions have been conducted by numerous investigators, the mechanisms that control the slow rate of dissolution of these minerals are still problematic. For feldspar, Petrovic et al. (1976) summarized the hypotheses that have been advanced to account for the slow dissolution rates. They indicate that numerous authors have suggested that the dissolution rate is controlled by the rate of diffusion of ions through layers or coatings on the mineral surfaces. For example, continuous coatings of hydrated amorphous silica-alumina precipitates have been suggested by some investigators; others favor hypotheses involving crystalline precipitates through which diffusion is presumed to occur. In another hypothesis it is presumed that diffusion occurs through a residual layer of leached feldspar consisting mainly of silica and alumina formed on feldspar surfaces. Based on detailed examination of actual surfaces of feldspar grains that had undergone appreciable dissolution in distilled water, Petrovic and coworkers concluded, however, that even when there is no significant layer or coating on feldspar surfaces, dissolution during the second stage is very slow. They did this by examining in great detail the nature of feldspar surfaces at which dissolution in laboratory vessels had occurred. As a result of this work, the concept of relatively thick continuous coatings of incongruent dissolution products on silicate minerals is now in doubt.

The discussion above indicated in a general way how silicate minerals can influence the chemical evolution of groundwater in silicate terrain. It should be kept in mind, however, that groundwater in real hydrogeologic systems reacts simultaneously with a large number of silicate minerals that have nonideal compositions. The solid products that form on the mineral surfaces as a result of incongruent dissolution are in some cases amorphous substances that require long periods of time for conversion to crystalline forms. The amorphous and clay mineral dissolution products commonly have appreciable cation exchange capacities and therefore have the capability of altering the cation ratios in the groundwater. For the water to evolve toward equilibrium with respect to the primary silicates such as feldspars, it is necessary for the concentrations of $Si(OH)_4$ and cations to progressively increase as the dissolution proceeds. If the reaction products in the pore water are continually flushed out by groundwater flow at rates that are appreciable relative to the reaction rates, equilibrium with respect to primary silicate minerals will never be attained. The nature of the clay-mineral weathering products produced in the system can therefore be dependent on the hydrodynamic and hydrochemical conditions as well as on mineralogic factors. An example of the interpretation of hydrochemical data from granitic rock using both hydrodynamic and mineral dissolution concepts is provided by Paces (1973).

In the next section, interpretations of chemical analyses of groundwater from silicate terrain are briefly appraised in light of the theoretical considerations developed above.

Interpretation of Field Data

A group of chemical analyses of samples from wells, springs, and stream baseflow from crystalline rock terrain (granites, diorites, basalts, and amphibolites) in various parts of the world are shown in Table 7.5. All these waters have very low major-ion concentrations. Without exception, HCO_3^- is the dominant anion and SiO_2 is present in major concentrations relative to the cations. The anions Cl^- and $SO_4{}^{2-}$ occur in only minor or trace concentrations. Their occurrence can normally be attributed to atmospheric sources, to the decomposition of organic matter in

Table 7.5 Mean Values of Major-Ion Composition of Groundwater and Groundwater-Derived Surface Water in Primarily Igneous Rock Areas (mg/ℓ)

Location*	Number	pH	HCO_3^-	Cl^-	SO_4^{2-}	SiO_2	Na^+	K^+	Ca^{2+}	Mg^{2+}
(1) Vosges, France	51	6.1	15.9	3.4	10.9	11.5	3.3	1.2	5.8	2.4
(2) Brittany, France	7	6.5	13.4	16.2	3.9	15.0	13.3	1.3	4.4	2.6
(3) Central Massif, France	10	7.7	12.2	2.6	3.7	15.1	4.2	1.2	4.6	1.3
(4) Alrance Spring F, France	77	5.9	6.9	<3	1.15	5.9	2.3	0.6	1.0	0.4
(5) Alrance Spring A, France	47	6.0	8.1	<3	1.1	11.5	2.6	0.6	0.7	0.3
(6) Corsica	25	6.7	40.3	22.0	8.6	13.2	16.5	1.4	8.1	4.0
(7) Senegal	7	7.1	43.9	4.2	0.8	46.2	8.4	2.2	8.3	3.7
(8) Chad	2	7.9	54.4	<3	1.4	85	15.7	3.4	8.0	2.5
(9) Ivory Coast (Korhogo, dry season)	54	5.5	6.1	<3	0.4	10.8	0.8	1.0	1.0	0.10
(10) Ivory Coast (Korhogo, wet season)	59	5.5	6.1	<3	0.5	8.0	0.2	0.6	<1	<0.1
(11) Malagasy (high plateaus)	2	5.7	6.1	1	0.7	10.6	0.95	0.62	0.40	0.12
(12) Sierra Nevada, Calif. (ephemeral springs)		6.2	2.0	0.5	1.0	16.4	3.03	1.09	3.11	0.70
(13) Sierra Nevada, Calif. (perennial springs)		6.8	54.6	1.06	2.38	24.6	5.95	1.57	10.4	1.70
(14) Kenora, NW Ontario (unconfined aquifer)	12	6.3	24.0	0.6	1.1	18.7	2.07	0.59	4.8	1.54
(15) Kenora, NW Ontario (confined aquifer)	6	6.9	59.2	0.7	0.8	22.1	3.04	1.05	11.9	4.94

*(1), A spring after thawing, 1967; (2) and (3), streams after several dry months, summer 1967; (4) and (5), two springs throughout 1966; (6), streams throughout the Island after 6 dry months, 1967; (7), streams in eastern regions, dry season 1967; (8), stream in Guera, dry season 1967; (9) and (10), streams in Korhogo area, 1965; (11), on the high plateaus and on the eastern coast, dry season 1967; (12) and (13), springs during 1961; (14) and (15), piezometers in glacial sands derived from granitic Precambrian rocks.

SOURCES: Tardy, 1971 (1) to (11); Feth et al., 1964 (12) and (13); and Bottomley, 1974 (14) and (15).

soil, and to the trace impurities in rocks and minerals. K^+ is generally the least abundant of the cations.

It should be noted that Cl^- and $SO_4{}^{2-}$ are not significant constituents in silicate rocks and there is no tendency toward development of $SO_4{}^{2-}$ or Cl^- facies as groundwater moves along flow paths in these rocks. This is the case even in regional flow systems where flow paths and water ages are very large. The Chebotarev hydrochemical evolution sequence is therefore not relevant in these systems.

The geochemical interpretation of chemical analyses of water from silicate-mineral terrain commonly involves two main approaches. The first involves plotting of data on stability diagrams to determine what may be the stable alteration products. The other approach involves calculation of reaction sequences that can account for the observed concentrations of major cations, HCO_3^-, and H^+.

Numerous investigators have observed that in igneous terrain nearly all groundwaters within several hundred meters of ground surface, and groundwater-derived surface waters such as springs and baseflow, plot in the kaolinite fields of stability diagrams such as those in Figure 7.14 (Garrels, 1967; Garrels and MacKenzie, 1967; Tardy, 1971; Bricker et al., 1968; Bottomley, 1974). A small percentage of samples plot in the montmorillonite fields and hardly any occur in the gibbsite, mica, or feldspar fields or exceed the solubility limit of amorphous silica. This suggests that alteration of feldspars and micas to kaolinite is a widespread process in groundwater flow systems in igneous materials. In a few investigations this has been substantiated by investigations of the surface coatings on the primary igneous materials. In general, however, there is little information on the weathering products that form in these subsurface systems, other than what can be inferred from water chemistry and stability diagrams. Unstable amorphous precipitates or metastable clay mineral intermediates may persist for long periods of time before clay minerals actually crystallize.

The second interpretive approach is to model the water chemistry through calculation procedures. This can be done by reacting the primary minerals to produce clay minerals and dissolved products or by reconstituting the primary minerals through combining the clay minerals with the dissolved products observed in the water. To illustrate the mineral reconstitution approach, we will use an example presented by Garrels and MacKenzie (1967) in an interpretation of the geochemical evolution of ephemeral spring water in a granite area of the Sierra Nevada of California. The calculations are summarized in Table 7.6. At the top of this table the average concentrations of dissolved constituents in the spring water are listed. Below this, the mean concentrations in snow samples are indicated. These values are assumed to be representative of water that recharges the groundwater zone feeding the springs. To obtain the concentrations derived from the rock during subsurface flow, the snow values are subtracted from the mean spring concentrations. A slight deficiency of anions results after this subtraction is made. This imbalance was corrected by assigning HCO_3^- a slightly higher value. As a first step in the reconstitution of primary granitic minerals from the water chemistry, kaolinite is converted to plagioclase in an amount that consumes all the

Table 7.6 Mean Values for Compositions of Ephemeral Springs in the Sierra Nevada, California, and Computational Steps in the Reconstitution of the Original Rock Composition From the Mean Water Composition

Reaction (coefficients × 10^{-4})	Water concentrations (mol/ℓ × 10^{-4})								Mineral products (mol/ℓ × 10^{-4})
	Na^+	Ca^{2+}	Mg^{2+}	K^+	HCO_3^-	SO_4^{2-}	Cl^-	SiO_2	
Initial concentrations in spring water	1.34	0.78	0.29	0.28	3.28	0.10	0.14	2.73	
(1) Minus concentrations in snow water	1.10	0.68	0.22	0.20	3.10	—	—	2.70	
(2) Change kaolinite back into plagioclase $1.23Al_2Si_2O_5(OH)_4 + 1.10Na^+ + 0.68Ca^{2+} + 2.44HCO_3^- + 2.20SiO_2{}^{2-} = 1.77Na_{0.62}Ca_{0.38}Al_{1.38}Si_{2.62}O_8 + 2.44CO_2 + 3.67H_2O$	0.00	0.00	0.22	0.20	0.64	0.00	0.00	0.50	$1.77Na_{0.62}Ca_{0.38}$ feldspar
(3) Change kaolinite back into biotite $0.037Al_2Si_2O_5(OH)_4 + 0.073K^+ + 0.22Mg^{2+} + 0.15SiO_2 + 0.51HCO_3^- = 0.073 KMg_3AlSi_3O_{10}(OH)_2 + 0.51CO_2 + 0.26H_2O$	0.00	0.00	0.00	0.13	0.13	0.00	0.00	0.35	0.073 biotite
(4) Change kaolinite back into K-feldspar $0.065Al_2Si_2O_5(OH)_4 + 0.13K^+ + 0.13HCO_3^- + 0.26SiO_2{}^{2-} = 0.13KAlSi_3O_8 + 0.13CO_2 + 0.195H_2O$	0.00	0.00	0.00	0.00	0.00	0.00	0.00	0.12	0.13 K-feldspar

SOURCE: Garrels and MacKenzie, 1967.

NOTE: Water concentrations for steps (2), (3), and (4) are residual dissolved concentrations after completion of reaction.

Na^+ and Ca^{2+} in the water (step 2, Table 7.6). Kaolinite is chosen as the secondary mineral because all the analyses of spring water plot in the kaolinite stability field of diagrams such as those shown in Figure 7.14. Garrels and MacKenzie (1967) note that the calculated plagioclase feldspar resulting from this step is, in fact, similar to that found in the rocks of the region. In the next step, all Mg^{2+} and enough K^+, HCO_3^-, and SiO_2 are reacted with kaolinite to form the biotite mica. The small residue of K^+, HCO_3^-, and SiO_2 is then reacted to form K-feldspar. After this step a residue of 4% of the total original silica remains. This is within the limits of error of the original values of concentration used for the mean composition of the spring water. Garrels and MacKenzie (1967) conclude that the reactions work out too well to leave much doubt that the system is indeed a closed system reacting with CO_2 and that the weathering product is kaolinite.

An alternative computational approach is to react a specified group of primary minerals with CO_2-charged water to produce the observed cation and HCO_3^- concentrations. This approach has been used by Cleaves et al. (1970) and Bottomley (1974). If reasonable assumptions can be made regarding the initial P_{CO_2}, the pH of the water can also be accounted for. The difference between the mineral reconstitution approach and the mineral dissolution approach is just a matter of bookkeeping. Adherence to stoichiometric reactions and the mass- and charge-balance principles is inherent in both methods. It is perhaps somewhat surprising that although terrain characterized by silicate minerals has many different mineral species and many variations from ideality in mineral compositions, it is often possible to account for the observed water chemistry using a relatively small number of reactions with minerals of ideal composition. More rigorous analyses of the chemical evolution of groundwater in silicate terrain are currently hindered by the lack of suitable information on reaction kinetics, on the behavior of mineral assemblages of nonideal compositions, and on the effects of dispersion and other factors. The chemical evolution of groundwater that moves to great depth in crystalline rock is influenced by increases in temperature and pressure. The necessity of incorporating these two factors renders the interpretative process much more difficult. Some groundwater systems in terrain formed primarily of silicate minerals have water compositions that are very anomalous when considered in light of the generalities presented above. For example, in some areas pH values exceed 9 or 10 and SiO_2 concentrations exceed 100 mg/ℓ. The reader is referred to Klein (1974) for an example of a study of this type of groundwater.

7.5 Groundwater in Complex Sedimentary Systems

In Sections 7.3 and 7.4, the geochemical evolution of groundwater in carbonate rocks and in relatively simple crystalline rock assemblages was described. Many sedimentary rocks or unconsolidated deposits, however, consist of mixed assemblages of minerals derived from various sedimentary, igneous, or metamorphic sources. Even individual strata commonly comprise mixed mineralogic assemblages. The assemblages can vary from bed to bed in layered sequences. These variations can cause large differences in the chemistry of groundwater from bed to bed and from region to region. We will now look at the factors that control these variations and at some approaches that can be used in the interpretation of data from these systems.

Order of Encounter

One of the most important factors in the chemical evolution of groundwater in mixed or layered assemblages is the *order of encounter*. This refers to the order in which various minerals or groups of minerals are encountered by the water as it moves through the flow system. This will be illustrated by considering chemical evolution sequences that would occur in a hypothetical hydrogeologic system that has four types of strata: sandstone, limestone, gypsum, and shale. As we proceed, further assumptions will be made.

In the first evolution sequence, the water passes through a soil developed on a limestone aquifer. The water becomes charged with CO_2 at a partial pressure of 10^{-2} bar, and then infiltrates to the water-table zone. During infiltration, saturation with respect to calcite is attained by open-system dissolution. The water passes through the limestone and enters a zone of gypsum, where dissolution to gypsum saturation occurs. From the gypsum zone the water flows into a sandstone aquifer and then into a zone of shale.

The chemical evolution of the groundwater in the various stages of this sequence was estimated based on various assumptions regarding the mineralogy and geochemical processes in the flow system. The results are tabulated in Table 7.7. The chemistry of the water in the soil zone is assumed to be controlled entirely by the carbon dioxide regime. All other chemical inputs are assumed to be insignificant. The soil-water composition was computed using the approach outlined in Section 3.5. The calculations were simplified by noting that in the charge-balance relation H^+ is balanced entirely by HCO_3^-, because the pH is below 8. Table 7.7 indicates that the pH of the soil water is 4.9.

During infiltration in the limestone, open-system dissolution of calcite to saturation causes the pH to rise to 7.3. In this zone, the water acquires a Ca–HCO_3 composition with low total dissolved solids (Table 7.7). When the water enters the gypsum bed, dissolution of gypsum to saturation causes the water to become brackish, with Ca^{2+} and $SO_4{}^{2-}$ as the dominant ions. At this stage in the calculation of the water composition, it was assumed that although gypsum dissolution causes supersaturation with respect to calcite, calcite precipitation does not proceed at a significant rate. Thus, the saturation index (SI_c) for calcite is large, as indicated in Table 7.7. Supersaturation with respect to calcite is caused by the common-ion effect.

It is assumed that as the water passes through the sandstone, gradual precipitation of calcite causes reestablishment of calcite equilibrium. The sandstone is composed of quartz and feldspar. It is assumed that these minerals have no significant effect on the water composition. The precipitation of calcite causes the pH to

Table 7.7 Estimated Groundwater Compositions Based on Mineral Dissolution and Precipitation and Cation Exchange During Flow Through a Hypothetical Sequence of Limestone, Gypsum, Sandstone, and Shale

		Groundwater chemistry* at 25°C							Saturation indices	
Zone	Geochemical processes	Na	Ca	HCO_3	SO_4	TDS	pH	P_{CO_2}	SI_c†	SI_g‡
(1) Organic soil horizon near ground surface	Water acquires CO_2 at a partial pressure of 10^{-2} bar	0	0	0.07	0	21§	4.9	10^{-2}	0	0
(2) Limestone (calcite)	Open-system dissolution of calcite by CO_2-charged water	0	66	203	0	290	7.3	10^{-2}	1	0
(3) Gypsum	Dissolution of gypsum to saturation; calcite supersaturation develops	0	670	203	1400	2330	7.3	10^{-2}	6.7	1
(4) Sandstone (quartz and plagioclase)	Precipitation of calcite caused reestablishment to calcite equilibrium	0	650	140	1400	2250	6.7	$10^{-1.6}$	1	0.95
(5) Shale (Na-montmorillonite)	Exchange of Ca^{2+} for Na^+ causes undersaturation with respect to calcite and gypsum	725	20	140	1400	2350	6.7	$10^{-1.6}$	0.06	0.2

*Concentrations expressed in mg/ℓ.

†$SI_c = [Ca^{2+}][CO_3{}^{2-}]/K$.

‡$SI_g = [Ca^{2+}][SO_4{}^{2-}]/K$.

§TDS is comprised mainly of dissolved CO_2 (i.e., H_2CO_3).

decline from 7.3 to 6.7 and the P_{CO_2} to increase from 10^{-2} to $10^{-1.6}$ bar. The Ca^{2+} and HCO_3^- concentrations decrease, and as a result the total dissolved solids also decrease, by a small percentage. Ca^{2+} and $SO_4{}^{2-}$ remain as the dominant ions.

As the water flows from the sandstone into the montmorillonitic shale, the process of cation exchange causes the Ca^{2+} concentration to decrease to a value that we arbitrarily specify as 20 mg/ℓ. This causes the Na^+ content to rise to 725 mg/ℓ. The cation exchange process is represented by Eq. (3.109). Because each mole of Ca^{2+} adsorbed is replaced by 2 mol of Na^+, cation exchange causes a slight increase in total dissolved solids but no change in pH and HCO_3^-. The loss of Ca^{2+} causes the water to become undersaturated with respect to both calcite and gypsum (Table 7.7).

In the second evolution sequence, the water passes through the surface soil and then through sandstone, shale, limestone, and finally into a gypsum zone. The P_{CO_2} in the soil is 10^{-2} bar, and the water composition is the same as in the first stage of the previous example. In the sandstone plagioclase feldspar dissolves incongruently (for the dissolution reaction, see Table 7.4). It is assumed that under closed-system conditions 0.2 mmol of H_2CO_3 is consumed by this reaction. A major portion of the dissolved CO_2 is thus converted to HCO_3^-. The water acquires small concentrations of Na^+ and Ca^{2+}, the pH rises to 6.5, and the P_{CO_2} declines to $10^{-2.4}$ bar (Table 7.8).

When the water moves from the sandstone into the limestone, dissolution of calcite to saturation causes the pH to rise to 8.9 as H^+ is consumed by the conversion of dissolved CO_2 to HCO_3^-. The P_{CO_2} decreases from $10^{-2.4}$ to $10^{-4.4}$ bar. The low Ca^{2+} and HCO_3^- concentrations of this water relative to the limestone water in the previous example illustrates the difference between closed- and open-system dissolution.

When the water enters the gypsum zone, gypsum dissolution causes it to become brackish, with Ca^{2+} and $SO_4{}^{2-}$ as the dominant ions. A high degree of supersaturation with respect to calcite develops (Table 7.8). For computational purposes it is assumed that calcite does not precipitate. In nature, however, calcite would gradually precipitate, and such high levels of supersaturation would not persist. Precipitation of calcite would cause a loss of Ca^{2+} and dissolution of more gypsum. In some situations equilibrium with respect to both gypsum and calcite can occur. Although in these examples limestone and gypsum strata control the chemical evolution of the groundwater, similar results could be attained if the water flowed through strata with only very small amounts of calcite and gypsum. It was shown in Section 7.3 that porous materials with only a fraction of a percent by weight of calcite can produce calcite-saturated water.

These two hypothetical evolution sequences indicate that the order in which groundwater encounters strata of different mineralogical composition can exert an important control on the water chemistry. As groundwater flows through strata of different mineralogical compositions, the water composition undergoes adjustments caused by imposition of new mineralogically controlled thermodynamic constraints. Although in some strata the water may attain local equilibrium with

Table 7.8 Estimated Groundwater Compositions Based on Mineral Dissolution and Precipitation and Cation Exchange; the Same Strata as in Table 7.7, Arranged in a Different Sequence

		Groundwater chemistry* at 25°C							Saturation indices	
Zone	Geochemical processes	Na	Ca	HCO_3	SO_4	TDS	pH	P_{CO_2}	SI_c†	SI_g‡
(1) Organic soil horizon near ground surface	Water acquires CO_2 at a partial pressure of 10^{-2} bar	0	0	0.07	0	21§	4.9	10^{-2}	0	0
(2) Sandstone (quartz and plagioclase)	Closed-system incongruent dissolution of plagioclase (0.2 mmol H_2CO_3 consumed)	1.6	2.8	12	0	38¶	6.5	$10^{-2.4}$	0.0005	0
(3) Shale (Na-montmorillonite)	Exchange of Ca^{2+} for Na^+	3.9	0.8	12	0	38	6.5	$10^{-2.4}$	0.0001	0
(4) Limestone (calcite)	Closed-system dissolution of calcite to saturation	3.9	8.4	31	0	55	8.9	$10^{-4.4}$	1	0
(5) Gypsum	Dissolution of gypsum to saturation	3.9	600	31	1440	2100	8.9	$10^{-4.4}$	75	1

*Concentrations expressed in mg/ℓ.

†$SI_c = [Ca^{2+}][CO_3^{2-}]/K$.

‡$SI_g = [Ca^{2+}][SO_4^{2-}]/K$.

§TDS comprised mainly of dissolved CO_2 (i.e. H_2CO_3).

¶Includes dissolved CO_2 and $Si(OH)_4$.

respect to some mineral phases, the continuous flow of the water causes disequilibrium to develop as the water moves into other strata comprised of different minerals. Considering that hydrogeologic systems contain numerous types of strata arranged in an almost limitless variety of geometric configurations, it is not unreasonable to expect that in many areas the chemistry of groundwater exhibits complex spatial patterns that are difficult to interpret, even when good stratigraphic and hydraulic head information is available.

Water Composition in Glacial Deposits

The chemistry of groundwater in glacial deposits is quite variable, because these deposits are composed of mixtures of mineralogical assemblages derived by glacial erosion of bedrock strata and of preexisting glacial sediments. Some generalizations can be made, however, with regard to the composition of groundwater in these materials. In North America there are three main composition categories into which most natural groundwaters in glacial deposits can be placed.

1. *Type I Waters*: Slightly acidic, very fresh waters (< 100 mg/ℓ TDS), in which Na^+, Ca^{2+}, and/or Mg^{2+} are the dominant cations and HCO_3^- is the abundant anion. These waters are soft or very soft. (For a definition of water hardness and softness, see Section 9.1.)
2. *Type II Waters*: Slightly alkaline, fresh waters (<1000 mg/ℓ TDS), in which Ca^{2+} and Mg^{2+} are the dominant cations and HCO_3^- is the dominant anion. These waters are hard or very hard.
3. *Type III Waters*: Slightly alkaline, brackish waters ($\approx$ 1000 to 10,000 mg/ℓ TDS), in which Na^+, Mg^{2+}, Ca^{2+}, HCO_3^-, and $SO_4{}^{2-}$ generally occur in major concentrations. Most of this water has $SO_4{}^{2-}$ as the dominant anion.

Type I water occurs in glacial deposits in parts of the Precambrian Shield in Canada and northern Minnesota, northern Wisconsin, and northern Michigan. These waters also occur in parts of Maine, Vermont, and New Hampshire, where the glacial overburden has been derived from igneous rock. Type II water is typical of the glacial materials in the midwestern region of the United States and in southern Ontario. Type III water occurs extensively in the Interior Plains region of the United States and Canada (North Dakota, Montana, Manitoba, Saskatchewan, and Alberta). (Type II water also exists in the Interior Plains region but is less common than Type III water.)

In situations where contamination from agricultural activities or sewage systems is significant, each of these water types can have appreciable concentrations of NO_3^- or Cl^-. Deposits formed as a result of glacial processes in mountainous regions also contain groundwater, but because of the more local nature and variability of these deposits, they will not be included in our discussion. The general categories above refer only to groundwaters that owe their chemical development to processes that take place in glacial deposits or in soils developed on these deposits. During the course of their flow histories, some groundwaters in glacial deposits have passed through bedrock or other nonglacial materials. The chemical composition of this water is commonly influenced by processes that occurred in these nonglacial materials.

The glacial deposits in which Type I water is common were derived from igneous or metamorphic rock. The chemical evolution of this water is controlled by interactions with aluminosilicate minerals in the manner described in Section 7.4. Because the weathering of these minerals proceeds slowly relative to the rates of groundwater movement, the groundwaters are very low in dissolved solids, with pH values that normally do not evolve above 7. Even though Ca^{2+} and Mg^{2+} are sometimes the dominant cations, the waters are soft because the total concentrations of these cations are very low. Although Type I water occurs in glacial deposits in many parts of the Precambrian shield, there are extensive areas in the Shield Region that have Type II water. This occurs because of carbonate minerals derived by glacial erosion of Paleozoic bedrock near the Shield, erosion of Paleozoic outliers on the Shield, or erosion of local zones of marble or other metamorphic rocks that contain carbonate minerals. Streams and lakes on the Shield that are fed by groundwater have acidic or alkaline waters, depending on their proximity to glacial deposits that contain carbonate minerals.

Type II waters are primarily a result of carbonate-mineral dissolution under open or partially open system CO_2 conditions. Cation exchange processes are commonly a modifying influence. Cl^- and $SO_4{}^{2-}$ concentrations rarely exceed 100 mg/ℓ because minerals such as gypsum, anhydrite, and halite are generally absent. The effects of dissolution of primary silicate minerals such as feldspars and micas are largely obliterated by the much larger concentrations of cations and HCO_3^- from the calcite and dolomite.

From a geochemical viewpoint, the most enigmatic waters in glacial deposits are those in the Type III category. This water is distinguished from Type II waters by much higher Mg^{2+}, Na^+, and $SO_4{}^{2-}$, and to a lesser extent by higher Ca^{2+} concentrations. Because of its brackish nature and high $SO_4{}^{2-}$ contents, Type III water is, from a water utilization viewpoint, an unfortunate characteristic of this region. The water is generally unsuitable for irrigation and in many cases even unsuitable for human or animal consumption.

The main chemical characteristics of Type III groundwater can be accounted for by a combination of the following processes: open or partially open system dissolution of calcite and dolomite to produce pH values in the range 7–8 and HCO_3^- values in the range 300–700 mg/ℓ, dissolution of gypsum ($CaSO_4$, $2H_2O$) and anhydrite ($CaSO_4$) to produce $SO_4{}^{2-}$ values in the range of several hundred to 2000 mg/ℓ and as much as several hundred milligrams per liter of Ca^{2+}, and alteration of the cation ratios by ion exchange. All these geochemical processes are interrelated. Cherry (1972), Grisak et al. (1976) and Davison (1976) have used stoichiometric combinations of the preceding four processes to account for Type III water in various areas in the Great Interior Plains Region of Manitoba and Saskatchewan. Since calcite and dolomite are ubiquitous in the soils and glacial deposits of the region, and since soil-zone P_{CO_2} values are generally high, dissolution of calcite and dolomite is a most reasonable way to account for the observed pH and HCO_3^- values. Type III water is characteristically saturated to moderately supersaturated with respect to calcite and dolomite. For SO_4^{2-} to occur as the dominant anion in most Type III water, only very small amounts of gypsum need to be dissolved. Type III water is generally undersaturated with respect to gypsum. As the water moves along its flow paths, the dissolution of gypsum is the major cause of increases in the total dissolved solids. Because of the common-ion effect, the additional Ca^{2+} often produces water that is supersaturated with respect to calcite and dolomite.

By means of the mineral-dissolution processes described above, the pH, TDS, HCO_3^-, and SO_4^{2-} values characteristic of Type III water can be accounted for but not the high Mg^{2+} and Na^+ concentrations and not the small but significant concentrations of Cl^-. There is also an unaccountable deficiency of Ca^{2+} in relation to SO_4^{2-}. A reasonable explanation for the major features of the Type III water involves the combined influence of carbonate-mineral dissolution by water charged with CO_2 in the soil zone, dissolution of small amounts of gypsum, and exchange of Ca^{2+} for Na^+ and Mg^{2+} on montmorillonitic clays. The availability of gypsum for dissolution is the governing factor in the evolution from freshwater to brackish-water conditions. The influx of Ca^{2+} derived from gypsum causes increased Na^+ and Mg^{2+} concentrations as the cation exchange reactions [Eqs. (3.106) and (3.109)] adjust to maintain equilibrium. Gypsum appears to be the main source of SO_4^{2-}, which is the dominant anion in most Type III waters.

The origin of the gypsum in the glacial deposits is a topic of considerable speculation. Various lines of evidence suggest that it was not contained in the glacial deposits at the time of their deposition. Cherry (1972) has suggested a hypothesis in which small but significant amounts of gypsum precipitated in these deposits as a result of penetration by brine water forced into shallow zones from deep formations during glacial loading of the regional sedimentary basin. This may have occurred in the upper zones of the various till units during the numerous episodes of glacial retreat in the Pleistocene epoch. In another hypothesis, the origin of gypsum is atributed to oxidation of small amounts of iron sulfides such as pyrite (FeS_2) in the soil and subsoil. The combined effects of infiltration, iron sulfide oxidation, calcite dissolution, and subsequent evapotranspiration cause gypsum precipitation at shallow depth. These hypotheses have not been evaluated in detail. The origin of the most important mineral in the chemical evolution of Type III water therefore remains a subject of controversy.

Groundwater in Stratified Sedimentary Rocks

Sequences of stratified sedimentary rocks of continental, deltaic, or marine origin are common in North America. These sequences normally include sandstones, siltstones, shales, limestones, and dolomites. Many of the geochemical processes that have already been considered in our discussions of other hydrogeologic environments are also important in these stratified sequences. For example, dissolution of carbonate minerals and of small amounts of gypsum, anhydrite, or halite commonly influence the major-ion composition. Alteration of feldspars, micas, and clay minerals can also be important. The purpose of this section is to describe four geochemical processes that are generally much more important in stratified sedimentary rocks than in the hydrogeologic environments considered previously. These are (1) cation exchange, (2) CO_2 generation below the soil zone, (3) biochemical reduction of sulfate, and (4) oxidation of sulfide minerals. The discussion focuses on the evolution of fresh or brackish groundwater within several hundred meters of ground surface. Deeper systems in which saline or brine waters develop are considered in Section 7.7.

A striking characteristic of many groundwaters in stratified sedimentary sequences is the occurrence of Na^+ and HCO_3^- as the dominant ions. In some situations the HCO_3^- concentrations are as high as 2500 mg/ℓ, which is more than half an order of magnitude above HCO_3^- values typical of groundwater in limestone or dolomite. Na–HCO_3-type waters occur in Tertiary and Cretaceous deposits of the Atlantic and Gulf Coastal Plains of the United States (Foster, 1950; Back, 1966), in the Tertiary and Cretaceous bedrock of western North Dakota, Montana, southern Saskatchewan, and Wyoming (Hamilton, 1970; Moran et al., 1978a; Groenewold et al., in press), and elsewhere. The occurrence of Na^+ and HCO_3^- as the dominant ions can be explained by the combined effects of cation exchange and calcite or dolomite dissolution. High Na–HCO_3 waters can be produced in sequences of strata that have significant amounts of calcite or dolomite and clay minerals with exchangeable Na^+.

The two geochemical processes are represented by the reactions

$$CaCO_3 + H_2CO_3 \longrightarrow 2HCO_3^- + Ca^{2+} \qquad (7.14)$$

$$Ca^{2+} + 2Na(ad) \rightleftharpoons 2Na^+ + Ca(ad) \qquad (7.15)$$

where (ad) denotes cations absorbed on clays. The equilibrium for Eq. (7.15) is far to the right as long as there is appreciable Na^+ on the exchange sites of the clays. Equation (7.14) proceeds to the right as long as the activity product $[Ca^{2+}][CO_3{}^{2-}]$ is less than the equilibrium constant for calcite (i.e., as long as $SI_{calcite} < 1$) and as long as calcite is available for dissolution. The removal of Ca^{2+} from solution by the exchange reaction causes the water to become or remain undersaturated with respect to calcite, thereby enabling calcite dissolution to continue. When these two processes operate below the water table, carbonate-mineral dissolution occurs under closed-system conditions. Dissolved CO_2, which is expressed as H_2CO_3, is consumed as the pH, HCO_3^-, and Na^+ values rise. The relations between pH, Ca^{2+}, and HCO_3^- concentrations for low-salinity water in which calcite has dissolved to equilibrium are shown in Figure 7.16. For the pH range that is common for groundwater, this graph indicates that when Ca^{2+} is maintained at low concentrations, equilibrium HCO_3^- concentrations are high.

In the Tertiary and Cretaceous strata in the regions mentioned above, groundwaters with less than a few tens of milligrams per liter of Ca^{2+} and Mg^{2+} and more

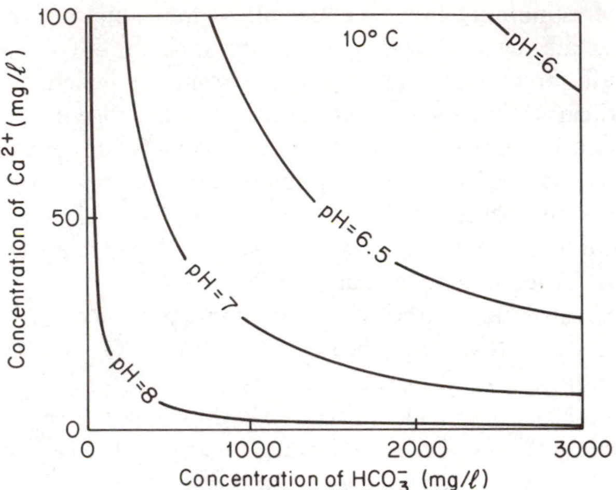

Figure 7.16 Solubility of calcite in water at 10°C expressed as a function of pH, Ca^{2+}, and HCO_3^-. Solubility lines computed from the relation $K_{eq} = (Ca^{2+})(HCO_3^-)/(H^+)$.

than 1000 mg/ℓ of HCO_3^- are common. The pH of these waters is typically in the range 7.0–8.5.

For groundwater in this pH range to evolve to such high HCO_3^- concentrations, high production of H^+ is necessary. In the Plains Regions a major source of H^+ is the oxidation of pyrite (FeS_2), which is a common constituent in the bedrock of this region (Moran et al., 1978). Hydrogen ions are released as a result of the reaction

$$FeS_2(s) + \tfrac{15}{4}O_2 + \tfrac{7}{2}H_2O \longrightarrow Fe(OH)_3(s) + 4H^+ + 2SO_4^{2-} \qquad (7.16)$$

The oxidation occurs in the soil-moisture zone as oxygen is supplied from the earth's atmosphere. Oxidation of a very small amount of pyrite relative to a given mass of the geologic material produces a large decrease in the pH of the pore water. An additional source of H^+ arises from CO_2 production in the soil zone in the usual manner.

To account for the chemical characteristics of Na–HCO_3-type groundwaters in the Atlantic and Gulf Coastal Plains, Foster (1950), Pearson and Friedman (1970), and Winograd and Farlekas (1974) have concluded that CO_2 is generated within the groundwater flow system at depths far below the water table. The CO_2 combines with H_2O to form H_2CO_3, which causes calcite dissolution. Mechanisms such as the following have been suggested for generation of CO_2 at depth:

Oxidation of organic matter by sulfate:

$$2CH_2O + SO_4^{2-} \longrightarrow HCO_3^- + HS^- + CO_2 + H_2O \qquad (7.17)$$

Coalification (diagenesis) of lignite:

$$C_nH_2O \longrightarrow \frac{n}{2} CO_2(g) + \frac{1}{2} C_nH_{2n+2} \tag{7.18}$$

In anaerobic groundwater environments, oxidation of organic matter can be accomplished through SO_4^{2-} reduction. This process, which requires the catalyzing action of anaerobic bacteria, is often identifiable in well samples by the rotten-egg smell of H_2S gas ($HS^- + H^+ = H_2S$). Groundwater in which CO_2 has been generated by SO_4^{2-} reduction typically has low SO_4^{2-} concentrations and because of this can be clearly distinguished from high-$NaHCO_3$ water that has evolved only through calcite dissolution and ion exchange processes.

Some stratified sequences of continental and deltaic deposits contain appreciable amounts of lignite. Foster (1950) and Winograd and Farlekas (1974) have suggested that lignite coalification, a process whereby temperature and pressure in an anaerobic environment progressively eliminate reduced volatile matter that subsequently oxidizes in the conversion of lignite to coal, is an important source of CO_2 in some groundwater zones in the Atlantic and Gulf Coastal Plain regions. However, this may not be an important CO_2 source in the lignite-rich deposits of the Great Plains Region.

We have illustrated how groundwater in stratified or mixed assemblages of unconsolidated sediments or rocks can follow many different geochemical evolution paths, depending on factors such as the sequence of encounter, relative rates of mineral dissolution, mineral availability and solubility, presence of organic matter and bacteria, CO_2 conditions, and temperature. Although it is often possible to explain the present-day composition of groundwater in terms of models based on the factors indicated above, the hydrochemistry of groundwater flow systems undergoes progressive and mainly irreversible changes over long periods of geologic time as groundwater continually passes through the geologic materials. The small amounts of gypsum and other soluble minerals that have strongly influenced the chemical evolution of groundwater in many hydrogeologic systems are gradually being removed from these systems. Sodium that is being exchanged for Ca^{2+} in the development of Na-rich groundwaters is being removed from the exchange sites of clay minerals. Silicate minerals such as feldspars and micas are continually being transformed to clay minerals. The characteristics of soil profiles in recharge areas that control the initial composition of groundwater as it begins to travel along its evolutionary paths are slowly changing due to flushing by repeated infiltration events. Over time periods of many thousands or millions of years, these changes affect the composition and shape of the earth's crust. The effects of groundwater geochemistry on a geological time scale are considered further in Chapter 11. On a much more recent time scale, during the last few tens or hundreds of years, man's activities have been affecting the chemical evolution of groundwater. In Chapter 9, some of the specific ways in which these activities are degrading groundwater quality are described.

7.6 Geochemical Interpretation of ^{14}C Dates

In Section 3.8, the principles of the ^{14}C method of groundwater age dating were introduced. Our purpose here is to describe some of the ways in which geochemical processes can cause adjusted or corrected groundwater ages to differ from decay ages or unadjusted ^{14}C ages. The initial step in the determination of the ^{14}C age of groundwater is to precipitate several grams or more of inorganic carbon, usually in the form of $BaCO_3$ or $SrCO_3$, from a 50- to 100-ℓ water sample. The next step is to determine by radiometric counting the percent of ^{14}C (i.e., specific activity) in the carbon extracted from the precipitate. The measured ^{14}C content is normally expressed as the ratio (R) of the ^{14}C in the sample to the ^{14}C content of modern carbon.

Based on this ratio, Eq. (3.111) indicates the unadjusted age of a ^{14}C sample. To account for the effect of dead carbon that entered the water as a result of mineral dissolution below the water table, we will use an adjustment factor denoted as Q, where

$$t = -8270 \ln R + 8270 \ln Q \tag{7.19}$$

Q is the fraction of the total dissolved inorganic carbon derived below the water table by mineral dissolution or by oxidation of organic matter. It is assumed that this added carbon contains *no* ^{14}C. Inorganic carbon that enters the water by mineral dissolution above the water table is assumed to have little influence on the ^{14}C content of the water because of rapid equilibration with the ^{14}C in the soil air, which has ^{14}C at "modern" levels. This definition of Q is consistent with that presented by Wigley (1975). This discussion will focus on the factors that affect Q and ways in which numerical estimates of Q can be obtained. Q is the ratio of inorganic carbon initially contributed to the groundwater under conditions where the carbon going into solution maintains a ^{14}C content equal to that of modern carbon *to* the total dissolved inorganic carbon in the sample.

The significance of Q will be illustrated by several examples in which it is assumed that mixing of waters from different sources does not occur. Consider a situation where water infiltrates into a soil zone and, while in the zone of CO_2 production in the soil, acquires 100 mg/ℓ of dissolved inorganic carbon from CO_2 and from calcite dissolution. The water then moves through the groundwater flow system with no further calcite dissolution, to some location where it is sampled. For this case, the Q value for the sampled water will be 1, because soil-zone dissolution of carbonate minerals produces a ^{14}C content in the water that is the same as the ^{14}C content of modern carbon, regardless of whether the calcite or dolomite that dissolves in the soil zone has significant amounts of ^{14}C. When carbonate-mineral dissolution takes place in the unsaturated zone, there is usually sufficient CO_2 generated by decay of modern organic matter to maintain equilibrium of ^{14}C contents between the water and the soil atmosphere. Whether the organic matter in the soil is tens of years old or a few hundreds of years old is of little consequence, because these time periods are short relative to the half-life of ^{14}C (5730 years).

In a second example, the water acquires 100 mg/ℓ of dissolved inorganic carbon in the soil zone under open-system conditions and then acquires another 100 mg/ℓ of inorganic carbon by dissolution of calcite and dolomite below the water table. In this case, Q has a value of 0.5. Nearly all calcite and dolomite that occurs in hydrogeologic systems is devoid of measurable ^{14}C because these minerals were originally formed hundreds of thousands or millions of years ago. When these minerals are this old, their original ^{14}C content has been lost by radioactive decay. The dissolved inorganic carbon contributed to groundwater through closed-system dissolution causes the original inorganic carbon in the water to be diluted with the nonradioactive carbon. The value of Q in this example is 0.5, because this is the ratio of carbon that initially had a modern ^{14}C value to the total carbon, which includes the original carbon plus the additional nonradioactive carbon.

In the third example the groundwater described above moves along its flow paths into a zone in which the total content of dissolved organic carbon derived from carbonate minerals is increased further. The value of Q therefore becomes smaller. For example, if the water enters a zone in which CO_2 is being generated by sulfate reduction and calcite is being dissolved as a result of the CO_2 increase and Na–Ca exchange, the additional carbon would be expected to be devoid of ^{14}C. Organic matter that occurs in geologic strata is normally very old, and therefore generally has no significant ^{14}C content. If the water acquires 100 mg/ℓ of inorganic carbon from the organic matter and from the calcite, the value of Q will be $100/300 = 0.33$.

As another example, we will consider groundwater that moves along flow paths in granitic rock that is completely devoid of carbonate minerals. The water moves through a soil zone in the recharge area, where it becomes charged with CO_2 at a partial pressure of 10^{-2} bar at 15°C. The dissolved inorganic carbon content is therefore 21 mg/ℓ, and because the pH of the water in equilibrium with this P_{CO_2} would be 5.0, the dissolved inorganic carbon is nearly all in the form of H_2CO_3. These values were obtained using the type of calculation procedure outlined in Section 3.5. The fact that most of the dissolved inorganic carbon exists as H_2CO_3 can be deduced from Figure 3.5(a). As the water moves along flow paths in the granite, the pH and HCO_3^- values will gradually rise as dissolution of silicate minerals such as feldspars and micas occurs. Eventually, the pH may rise above 7 and nearly all the dissolved inorganic carbon will exist as HCO_3^-. The value of Q, however, will remain at unity while these changes occur, because no new inorganic carbon is introduced into the groundwater from the rock mass.

These examples illustrate that as groundwater evolves chemically during its movement along regional flow paths, Q can in some situations decrease. If one wishes to obtain useful estimates of the "true" age of groundwater from ^{14}C data, it is necessary to first acquire a relatively detailed understanding of the geochemical origin of the inorganic carbon in the water. This can be done using conceptual geochemical models such as those outlined in the previous sections of this chapter. These models can be tested and improved using data on the ^{13}C content of the inorganic carbon in the groundwater and of the carbon sources in the porous media. More detailed discussions of methods for adjustment of ^{14}C ages based on interpretation of hydrochemical and ^{13}C data are provided by Pearson and Hanshaw (1970), Wigley (1975), and Reardon and Fritz (1978).

Although in the examples above, attention has been drawn to the fact that dilution of ^{14}C by geochemical processes can exert a major influence on ^{14}C dates of groundwater, it should be emphasized that ^{14}C ages can nevertheless be useful in many types of subsurface hydrologic investigations. ^{14}C data can contribute invaluable information, even if there is considerable uncertainty in the estimates of Q values. Detailed age estimates are often not necessary for a solution to a problem. For example, whether or not the water is 15,000 or 30,000 years old may not be crucial if one has a reasonable degree of confidence in information that indicates that the age is somewhere in this range, or even older. With this approach in mind, it is fortunate that large uncertainties in Q estimates for old groundwater have a relatively small influence on the calculated groundwater age.

For example, if the unadjusted age of a groundwater sample is 40,000 years and if the Q value is 0.7, the adjusted or corrected age obtained using Eq. (7.19) is 37,050 years. If the uncertainty associated with Q is large, for example ± 0.2, the corresponding range in ^{14}C age is 34,250–39,135 years. Because of the logarithmic form of the terms in Eq. (7.19), the effect of Q is small at large values of t.

In one of the hypothetical cases presented above, a Q value of 0.3 was obtained. It should be noted that in real situations this would be considered an extreme value. As a general guide, Mook (1972) has suggested that a Q value of 0.85 is a reasonable estimate for many situations. Wigley (1975) has shown that in situations where CO_2 is not generated below the water table, it is very unlikely that Q values smaller than 0.5 will develop.

We expect that in the next few decades there will be an increasing interest in the identification of zones in which old groundwater occurs. This will occur as society develops a greater desire to consume water that has been unblemished by the almost limitless variety of chemicals that are now being released into the hydrologic cycle. Some deep groundwater zones that contain saline or brine waters that can be identified as being isolated from the hydrosphere may have special value as waste disposal zones. For these and other reasons, the identification of the distribution of ^{14}C in groundwater in the upper few thousand meters of the earth's crust will be important in the years ahead.

7.7 Membrane Effects in Deep Sedimentary Basins

In this text, emphasis is placed primarily on the processes and characteristics of groundwater systems in the upper few hundred meters of the earth's crust. In these zones the temperatures are generally less than 30°C and the confining stresses are not large. However, most groundwater in the earth's crust exists at greater depths, where temperatures and pressures are much above those considered in our previous discussions. The chemical characteristics of waters at these depths are commonly very different from those at shallow depths. The effect of temperature and pressure on mineral solubilities and ion complexing and the great age of deep groundwaters are factors that produce different water compositions. Saline or brine waters are common at these depths, although in some areas brackish waters occur.

Of the various effects that have a distinctive influence on the chemical evolution of groundwater in deep flow systems, we will choose only one, known as the *membrane effect*, as a basis for further discussion. Other effects result from extensions of the chemical processes that have been described for shallow systems. The membrane effect, however, is relatively unique to deep systems in stratified sedimentary rocks. For more general discussions of the geochemistry of deep groundwaters, the reader is referred to White (1957), Graf et al. (1965), Clayton et al. (1966), van Everdingen (1968b), Billings et al. (1969), and Hitchon et al. (1971). For a review of geochemical investigations of groundwaters noted for their high temperatures, the reader is referred to Barnes and Hem (1973).

When water and solutes are driven under the influence of hydraulic head gradients across semipermeable membranes, the passage of ionic solutes through the membranes is restricted relative to the water (see Section 3.4). The concentrations of solutes on the input side of the membrane therefore increase relative to the concentrations in the output. This ion-exclusion effect is referred to as *salt filtering*, *ultrafiltration*, or *hyperfiltration*. Salt filtering can also occur in the absence of significant hydraulic gradients in situations where differential movement of ions takes place because of molecular diffusion. Salt filtration effects caused by shales were first suggested by Berry (1959) as an important process in sedimentary basins. The concept was also used by Bredehoeft et al. (1963) to explain the concentration of brine in layered sedimentary rocks. The process has been demonstrated in the laboratory by Hanshaw (1962), McKelvey and Milne (1962), Hanshaw and Coplen (1973), Kharaka and Berry (1973), and Kharaka and Smalley (1976). The membrane properties of clayey materials are believed to be caused by unbalanced surface charges on the surfaces and edges of the clay particles. As indicated in Section 3.7, the net charge on clay particles is negative. This results in the adsorption of a large number of hydrated cations onto the clay mineral surfaces. Owing to a much smaller number of positively charged sites on the edges of the clay particles and the local charge imbalance caused by the layer or layers of adsorbed cations, there is also some tendency for anions to be included in this microzone of ions and water molecules around the clay particles. The ability of compacted clays and shales to cause salt filtering develops when clay particles are squeezed so close together that the adsorbed layers of ions and associated water molecules occupy much of the remaining pore space. Since cations are the dominant charged species in the adsorbed microzones around the clay particles, the relatively immobile fluid in the compressed pores develops a net positive charge. Therefore, when an aqueous electrolyte solution moves through the pores as a result of an external gradient or molecular diffusion, cations in the solution are repelled. In order to maintain electrical neutrality across the membrane, anions are also restricted from passage through the membrane. Slight charge differences, referred to as *streaming poten-*

tials, caused by a small degree of differential migration of cations and anions, produce electrical currents across the membrane. The streaming potential also contributes to the retardation of cations in the fluid being forced through the membrane.

As a convenient way of expressing the efficiency of the clay membrane for retarding the flow of ionic species, Kharaka and Smalley (1976) have defined the *filtration ratio* as the concentration of species in the input solution divided by the concentration in the effluent solution.

Berry (1969) and van Everdingen (1968c) have described the relative factors influencing membrane filtration effects in geologic environments. Because of differences in ionic size and charge, there are relatively large differences in filtration ratios for the major cations that occur in groundwater. Divalent cations are filtered more effectively than monovalent cations. The membrane effect is stronger at lower groundwater flow rates. Distinctive differences in filtration ratios between monovalent and divalent cations do not always occur, and under some experimental conditions the trend in filtration ratios is reversed (Kharaka and Smalley, 1976). Temperature of the fluid also has a significant effect on the filtration ratios. In laboratory experiments using compacted bentonite, Kharaka and Smalley observed that the filtration ratios for alkali and alkaline earth metals decrease by as much as a factor of 2 between 25 and 80°C. They attribute this change to the effect of temperature on the nature and degree of cation hydration. Coplen (1970) observed experimentally that both hydrogen and oxygen are fractionated across montmorillonite membranes. Because of mass differences, ^{2}H and ^{18}O accumulated at the high-pressure side of the membrane. Another important conclusion derived from laboratory experiments using different types of clays is that clays with higher cation exchange capacities have greater ion-filtering efficiencies. Montmorillonitic clays, therefore, are generally much more efficient than kaolinitic clays.

If salt filtering does, in fact, exert a significant influence on the geochemical evolution of groundwater in sedimentary basins, the effects should be evident when the spatial distributions of cations and anions are investigated in deep sedimentary basins where groundwater flows or diffuses across clay or shale strata. Combinations of chemical and potentiometric data that unequivocally show the effects of salt filtering in deep zones within sedimentary basins are extremely difficult to obtain because hydrodynamic and stratigraphic conditions are generally quite complex relative to the number and distribution of groundwater monitoring wells or boreholes that are normally available for investigations of this type. Nevertheless, the salt filtering hypothesis has been found by numerous investigators to be a reasonable explanation for anomalous water compositions in various zones in deep sedimentary basins in North America and Europe. Although alternative explanations based on water-rock interactions or different hydrodynamic interpretations may be tenable as an explanation of some of these situations, there appears to be little doubt that salt filtering is in many cases an important factor. In some investigations the distributions of the stable isotopes ^{18}O and ^{2}H have been used as an aid in the interpretation of the chemical and hydrodynamic data (Graf et al. 1965; Hitchon and Friedman, 1969; Kharaka et al., 1973).

All laboratory experiments that have been conducted on the membrane properties of clays and shales have shown that large effective stresses must be applied for significant salt filtering efficiencies to be achieved. On the basis of laboratory evidence, it seems very unlikely that salt filtering will occur in most sedimentary deposits at depths below the ground surface of less than about 500–1000 m. If salt filtering were to commonly occur at shallower depths, the well-known generalization that states that total dissolved solids in groundwater tend to increase along flow paths would, of course, be invalid in many areas.

An unresolved problem, however, is whether or not clayey deposits such as clayey till that have been subjected to very high effective confining stresses at some time in the geologic past can maintain significant membrane properties long after the confining stresses have been removed. This question is of particular relevance in regions of sedimentary terrain that have been overridden by glaciers. The continental glaciers that traversed most of Canada and the northern part of the United States had thicknesses of several kilometers. In situations where the subglacial fluid pressures were able to dissipate during the period of glacial loading, the clayey deposits beneath the glaciers were subjected to large effective vertical stresses. Whether or not these deposits maintained significant membrane capabilities following deglaciation and crustal rebound remains to be established. Schwartz (1974) and Wood (1976) have invoked salt filtering in clayey glacial deposits as a hypothesis to explain some anomalous chemical trends in data from shallow wells in areas of clayey tills in southern Ontario and Michigan. Their field evidence is not unequivocal. The hypothesis has yet to be evaluated by studies in other areas or by laboratory tests on the deposits to ascertain whether they are capable of causing salt filtration.

7.8 Process Rates and Molecular Diffusion

To this point in our discussion of the chemical evolution of groundwater, emphasis has been placed on mineral dissolution and exchange reactions operating within an equilibrium framework. Equilibrium conditions are commonly observed in laboratory experiments and are conveniently amenable to description using thermodynamic concepts. In nature, however, hydrochemical processes often proceed extremely slowly, even when considered on a geological time scale. Because the rates of many reactions are slow, the bulk mass of the groundwater often remains undersaturated with respect to minerals that occur in the porous media. The rates can be slow because ions are not easily released from the crystal structures, or because the flux of water and reaction products between the bulk mass of the flowing water and the crystal surfaces is slow, or because series of reactions are involved, one of which is slow and therefore rate-determining for the system. Porous unfractured geological materials such as gravel, sand, silt, and clay are characterized by a large range of pore sizes. The bulk of the flowing water moves through the largest pores. It is this water that is obtained when wells or piezometers are sampled. The surface area over which reactions occur, however, is primarily the area that encompasses the smaller pores. The surface area that encompasses the large pores in which most of the flow occurs is usually only a small fraction of the surface area of the small pores. The processes whereby the bulk mass of the flowing groundwater acquires its chemical composition can therefore be strongly influenced by the rate of transfer of water and reaction products from the reaction surfaces in the smaller pores to the water in the larger pores. This rate can be slow. It is reasonable to expect that in many situations this rate is controlled by molecular diffusion of the reaction products through the fluid in the smaller pores into the larger pores, where they are then transported in the active, hydraulically controlled flow regime. From this line of reasoning it is apparent that the time required for equilibrium to be achieved in an experiment in which particulate mineral matter is reacted in a vessel in which the solids and liquid are stirred or agitated is normally less than the equilibration time in a situation where water is passed through a column packed with the mineral matter as a porous medium.

The effect of the rate of transfer of reaction products from the reaction surfaces to the bulk mass of flowing water is illustrated schematically in Figure 7.17. In a uniform flow field, the distance of attainment to saturation, referred to as the *saturation distance*, increases as the effective rate of transfer of reaction products between the reaction surfaces and bulk mass of flowing water decreases [Figure 7.17(a)]. In a given flow system different minerals can have different saturation distances. Figure 7.17(b) illustrates the effect of flow rate on the saturation distance for a single mineral species. In many situations the relation between the flow rate and the effective reaction rate is such that the water passes through complete groundwater flow systems without attaining saturation with respect to many of the mineral species in the host rock. This is particularly the case for aluminosilicate minerals, which have effective reaction rates that are limited by both the

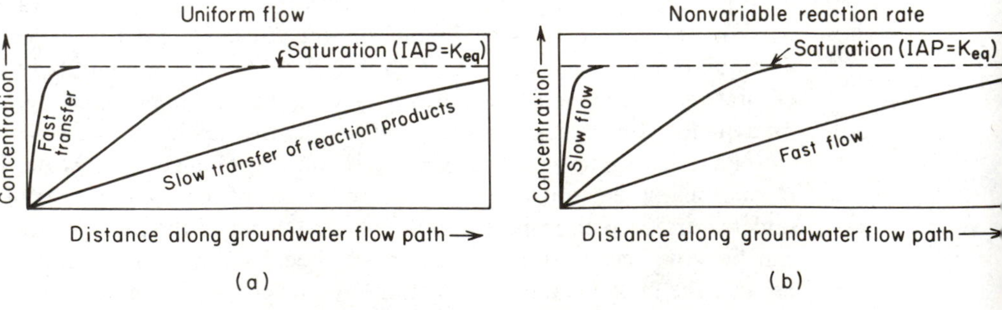

Figure 7.17 Schematic diagram showing the influence of (a) effective reaction rate and (b) flow rate on the distance along the flow path for saturation to be attained.

rate of transfer by diffusion of reaction products through the smaller to the larger pores and by slow rates of ionic release from crystal structures. For a more detailed discussion of the effect of reaction rates and diffusion on the chemical evolution of groundwater, the reader is referred to Paces (1976). Domenico (1977) has reviewed the theory of transport process and rate phenomenon in sediments. Laboratory experiments are described by Howard and Howard (1967) and Kemper et al. (1975).

In many flow systems the dominant movement of groundwater is through fractures or along bedding planes. Fractured rocks such as limestones, siltstone, shale, and basalt, and fractured nonindurated deposits such as some clays and clayey tills, have appreciable porosity in the unfractured matrix of the materials. Although the matrix has significant porosity, its permeability is commonly so low that flow in the matrix is small relative to the flow in the fracture networks. When wells or piezometers in these fractured materials are sampled, the samples represent the water flowing in the fracture system rather than the water present in the bulk mass of the porous medium. The chemistry of this water, however, can strongly reflect the results of diffusion of reaction products from the porous matrix to the fractures. Fracture openings are commonly small, even in strata with large fracture permeability. In comparison to the flux of solutes into fractures as a result of diffusion from the porous matrix, the volume of water in the fracture network is commonly small. The diffusive flux from the matrix can therefore be the controlling factor in the chemical evolution of the groundwater flowing in the fracture network. The chemical evolution of groundwater in fracture flow systems can depend on the mineralogy of the matrix material, the effective diffusion coefficients of ions in the matrix, the fracture spacing, and variations of rates of flow in the fractures. The commonly observed phenomenon of gradual increases in major ions along flow paths in regional flow systems can often be attributed to the effect of matrix diffusion.

Suggested Readings

Back, W., and B. B. Hanshaw. 1965. Chemical geohydrology. *Adv. Hydrosci.*, 1, pp. 49–109.

Garrels, R. M., and C. L. Christ. 1965. *Solutions, Minerals, and Equilibria.* Harper & Row, New York, pp. 74–91.

Hem, J. D. 1970. Study and interpretation of the chemical characteristics of natural water. *U.S. Geol. Surv. Water-Supply Paper 1973*, pp. 103–230.

Paces, T. 1976. Kinetics of natural water systems. *Proc. Symp. Interpretation of Environmental Isotopes and Hydrochemical Data in Groundwater Hydrology*, Intern. Atomic Energy Agency Spec. Publ., Vienna, pp. 85–108.

Stumm, W., and J. J. Morgan. 1970. *Aquatic Chemistry.* Wiley-Interscience, New York, pp. 383–417.

Problems

Many of the problems listed below require calculations using the law of mass action and the charge-balance and mass-balance equations. Appropriate computational methods can be deduced from material presented in Chapters 3 and 7. We suggest that students exclude the occurrence of ion pairs and complexes from their calculations. In solutions of the problems, this approach will introduce some error that could otherwise be avoided by pursuit of more tedious calculations. The errors are generally small and the instructive nature of the problems is not significantly altered. Values of equilibrium constants can be obtained from Chapter 3, or in cases where they are not included in Chapter 3, they can be calculated from free-energy data listed in well-known texts, such as Garrels and Christ (1965), Krauskopf (1967), and Berner (1971). It can be assumed that the groundwaters referred to in the problems are situated at a sufficiently shallow depth for the effect of differences in fluid pressure from the 1-bar standard value to be neglected.

1. A sample of rain has the following composition (concentrations in mg/ℓ): $K^+ = 0.3$, $Na^+ = 0.5$, $Ca^{2+} = 0.6$, $Mg^{2+} = 0.4$, $HCO_3^- = 2.5$, $Cl^- = 0.2$, $SO_4^{2-} = 15$, and $NO_3^- = 1.2$; pH 3.5; temperature 25°C. Can the pH of this water be accounted for by a hydrochemical model that is based on the assumption that the rain chemistry can be represented by equilibration of the water with atmospheric carbon dioxide as the dominant pH control? If it cannot, offer an alternative explanation for pH control.

2. Rain that infiltrates into a soil zone has pH 5.7, $K^+ = 0.3$, $Na^+ = 0.5$, $Ca^{2+} = 0.6$, $Mg^{2+} = 0.4$, $HCO_3^- = 2.5$, $Cl^- = 0.2$, and $SO_4^{2-} = 0.8$ (concentrations in mg/ℓ).

(a) In the soil zone, the water equilibrates with soil air that has a CO_2 partial pressure of 10^{-2} bar. Calculate the H_2CO_3 and HCO_3^- concentrations and pH of the water. Assume that the water does not react with solid phases in the soil.

(b) In the soil zone, the water that initially has a dissolved oxygen content in equilibrium with the above ground atmosphere (i.e., 9 mg/ℓ), has half of its dissolved oxygen consumed by oxidation of organic matter and half consumed by oxidation of iron sulfide (FeS_2). Assume that the soil is saturated with water when these processes occur, that the water reacts with no other solid phases, and that oxidation of organic matter produces CO_2 and H_2O as reaction products. Estimate the P_{CO_2}, H_2CO_3, and SO_4^{2-} contents and pH of the pore water. Which process exerts the dominant pH control, organic-matter oxidation or sulfide-mineral oxidation?

3. Water with a dissolved oxygen concentration of 4 mg/ℓ moves below the water table into geologic materials that contain 0.5% by weight pyrite (FeS_2). In this zone the dissolved oxygen is consumed by oxidation of pyrite. Estimate the pH of the water after the oxidation has occurred. The initial pH of the water is 7.9. Assume that the water reacts with no other solid phases and that the groundwater zone is at 10°C.

4. The following results were obtained from a chemical analysis of a groundwater sample (concentrations in mg/ℓ); $K^+ = 21$, $Na^+ = 12$, $Ca^{2+} = 81$, $Mg^{2+} = 49$, $HCO_3^- = 145$, $Cl^- = 17$, $SO_4{}^{2-} = 190$, and $Si = 12$; pH 7.3; temperature 15°C.
   (a) Is there any evidence suggesting that this analysis has significant analytical errors? Explain.
   (b) Represent this chemical analysis using the following diagrams: bar graph, circular graph, Stiff graph, Piper trilinear diagram, Schoeller semilogarithmic diagram, and Durov diagram.
   (c) Classify the water according to its anion and cation contents.
   (d) On which of the diagrams would the chemical analysis be indistinguishable from an analysis of water with different concentrations but similar ionic percentages?

5. Groundwater deep in a sedimentary basin has an electrical conductance of 300 millisiemens (or millimhos).
   (a) Make a rough estimate of the total dissolved solids of this water (in mg/ℓ).
   (b) What is the dominant anion in the water? Explain.

6. A sample of water from a well in a limestone aquifer has the following composition (concentrations in mg/ℓ): $K^+ = 1.2$, $Na^+ = 5.4$, $Ca^{2+} = 121$, $Mg^{2+} = 5.2$, $HCO_3^- = 371$, $Cl^- = 8.4$, and $SO_4{}^{2-} = 19$; pH 8.1; temperature 10°C.
   (a) Assuming that these data represent the true chemistry of water in the aquifer in the vicinity of the well, determine whether the water is undersaturated, saturated, or supersaturated with respect to the limestone.
   (b) The pH value listed in the chemical analysis was determined in the laboratory several weeks after the sample was collected. Comment on the reasonableness of the assumption stated in part (a).
   (c) Assuming that the concentrations and temperature indicated in part (a) are representative of *in situ* aquifer conditions, calculate the pH and P_{CO_2} that the water would have if it were in equilibrium with calcite in the aquifer.

7. In the recharge area of a groundwater flow system, soil water becomes charged with CO_2 to a partial pressure of $10^{-2.5}$ bar. The water infiltrates through quartz sand to the water table and then flows into an aquifer that contains calcite. The water dissolves calcite to equilibrium in a zone where the temperature is 15°C. Estimate the content of Ca^{2+} and HCO_3^- in the water and the pH and P_{CO_2} after this equilibrium is attained.

8. In what types of hydrogeologic conditions would you expect HCO_3^--type water to exist with little or no increase in total dissolved solids along the entire length of the regional groundwater flow paths? Explain.

9. A highly permeable carbonate-rock aquifer has natural groundwater at 5°C with the following composition (concentrations in mg/ℓ): $K^+ = 5$, $Na^+ = 52$, $Ca^{2+} = 60$, $Mg^{2+} = 55$, $HCO_3^- = 472$, $Cl^- = 16$, and $SO_4{}^{2-} = 85$; pH 7.47. The water is saturated or supersaturated with respect to calcite and dolomite. It is decided to recharge the aquifer with surface water with the following composition (concentrations in mg/ℓ): $K^+ = 2.1$, $Na^+ = 5.8$, $Ca^{2+} = 5.2$, $Mg^{2+} = 4.3$, $HCO_3^- = 48$, $Cl^- = 5$, and $SO_4{}^{2-} = 3$; pH 6.5. The recharge will take place by means of a network of injection wells that will receive the surface water from an aerated storage reservoir. Estimate the composition of the recharged water in the aquifer after it has achieved equilibrium with respect to calcite at a temperature of 20°C. Neglect the effects of mixing of the injection water and natural water in the aquifer. Explain why there is a major difference between the compositions of the two waters.

10. Water charged with CO_2 at a partial pressure of $10^{-1.5}$ bar in the soil zone infiltrates into a regional flow system in slightly fractured granitic rock. The water slowly dissolves albite incongruently until it becomes saturated with respect to this mineral. Assume that all other mineral-water interactions are unimportant. Estimate the water composition [Na^+, $Si(OH)_4$, HCO_3^-, pH, and P_{CO_2}] after albite saturation is attained.

11. Groundwater in fractured granite has the following composition (concentrations in mg/ℓ): $K^+ = 1.5$, $Na^+ = 5.8$, $Ca^{2+} = 10$, $Mg^{2+} = 6.1$, $HCO_3^- = 62$, $Cl^- = 2.1$, $SO_4{}^{2-} = 8.3$, and $Si = 12$; pH 6.8. The water has acquired this composition as a result of incongruent dissolution of plagioclase feldspar in the granite. Because of the dissolution process, clay is forming on the surfaces of the fractures.

(a) Indicate the species of clay mineral or minerals that you would expect would be forming. Assume that the solid-phase reaction product is crystallized rather than amorphous in form.

(b) Would the process of incongruent dissolution cause the permeability of the fracture to increase or decrease? Explain.

12. Estimate the water composition that results from the reaction in 1 ℓ of water with 1 mmol of H_2CO_3 with (a) calcite, (b) dolomite, (c) albite, (d) biotite, and (e) anorthite. Express your answers in millimoles per liter and milligrams per liter. For each case, indicate the direction that the pH will evolve as dissolution occurs.

13. Studies of a regional groundwater flow system in sedimentary terrain indicate that in part of the system there is a large decrease in $SO_4{}^{2-}$ and a large increase in HCO_3^- in the direction of regional flow.

(a) What geochemical processes could cause these changes in anion concentrations?

(b) Indicate other chemical characteristics of the water that should show trends that would support your explanation.

14. Groundwater in a sandstone bed within a layered sedimentary sequence comprised of shale, siltstone, lignite, and sandstone, all of continental depositional origin, has the following composition (concentrations in mg/ℓ): $K^+ = 1.2$, $Na^+ = 450$, $Ca^{2+} = 5.8$, $Mg^{2+} = 7.9$, $HCO_3^- = 1190$, $SO_4^{2-} = 20$, and $Cl^- = 12$; pH 7.5; temperature 15°C. What combination of hydrogeochemical processes could account for this type of water chemistry? Write the chemical reactions that form the framework of your answer.

15. Groundwater moves into a clayey stratum that is characterized by a selectivity coefficient of 0.7 with respect to the Mg–Ca exchange reaction described by Eqs. (3.105) and (3.107). The cation exchange capacity is 10 meq/100 g. The mole fractions of adsorbed Ca^{2+} and Mg^{2+} are both 0.5. The groundwater entering the clayey stratum has a Ca^{2+} concentration of 120 mg/ℓ and a Mg^{2+} concentration of 57 mg/ℓ. Assume that the concentration of other cations is negligible. Estimate the equilibrium concentrations of Ca^{2+} and Mg^{2+} that will occur after the water composition is altered by the Ca–Mg exchange reaction.

16. Groundwater at a temperature of 25°C moves through a limestone bed where it attains saturation with respect to calcite. It then moves into strata which contain considerable gypsum. Estimate the composition of the water after it has attained equilibrium with respect to gypsum. Assume that the rate of calcite precipitation is very slow relative to the rate of gypsum dissolution. Prior to moving into the gypsiforous strata, the water has the following composition (concentrations in mg/ℓ): $K^+ = 3.3$, $Na^+ = 8.1$, $Ca^{2+} = 101$, $Mg^{2+} = 9.2$, $HCO_3^- = 310$, $Cl^- = 12$, and $SO_4^{2-} = 36$.

17. Prepare a graph that shows the relation between the uncorrected (unadjusted) ^{14}C age, corrected (adjusted) ^{14}C age, and the parameter designated as Q in Sections 3.8 and 7.6. For a specified value of Q, are the differences between the corrected and uncorrected ages largest at young ages or old ages? Explain why the ^{14}C method is generally not useful for dating groundwater that is younger than several thousand years.

18. As water passes through the soil zone, it acquires, as a result of open-system dissolution, a HCO_3^- content of 96 mg/ℓ and a pH of 6.1. The water then moves below the water table into a dolomite aquifer. In the aquifer the HCO_3^- content rapidly increases to 210 mg/ℓ.

(a) What will be the value of Q for use in the adjustment of ^{14}C dates of water from the aquifer?

(b) The water has an uncorrected age of 43,300 years. What is the corrected age based on the Q value obtained from part (a)?

19. A horizontal sandstone aquifer occurs between two thick beds of shale. The sandstone is composed of quartz and a small percent of feldspar. Water in the sandstone is not capable of acquiring an appreciable concentration of dissolved solids by reaction with the aquifer minerals. Pore water in the shale, however, has high concentrations of dissolved solids. By considering various combinations of aquifer thickness, velocity of groundwater in the aquifer, concentration gradients in the aquitards, and diffusion coefficients, determine conditions under which the water chemistry in the aquifer would be controlled by the vertical flux by molecular diffusion of dissolved solids from the shale into the aquifer. Assume that dissolved solids that enter the aquifer are distributed uniformly over the aquifer thickness as a result of dispersion. Do you think such conditions could occur in nature?

20. Groundwater *A*, at $P_{CO_2} = 10^{-2}$, has a composition that results from the open-system dissolution of siderite ($FeCO_3$) in a stratum with no calcite or dolomite. Groundwater *B*, at the same P_{CO_2}, has a composition that results from open-system dissolution of calcite in a stratum with no siderite or dolomite. These two waters, each having been in equilibrium with their respective solid phase, are intercepted by a well in which they are mixed in equal proportions as pumping occurs. The system has a temperature of 25°C.

(a) Compute the cation and anion concentrations and pH of each of these waters.

(b) Compute the composition of the mixed water in the well.

(c) Is this mixture capable of producing calcite or siderite by precipitation?

(d) After discharge from the well into an open-air storage tank at 25°C, would calcite and/or siderite precipitate?

CHAPTER 8

Groundwater Resource Evaluation

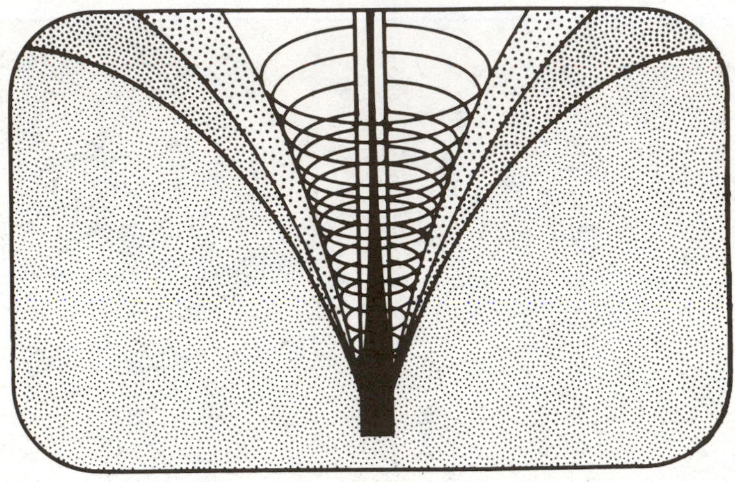

In the first seven chapters of this book we have examined the physical and chemical principles that govern groundwater flow and we have investigated the interrelationships that exist between the geological environment, the hydrologic cycle, and natural groundwater flow. In this chapter and the two that follow, we will turn to the interactions between groundwater and man. We will look at the utilization of groundwater as a resource, we will examine its role as an agent for subsurface contamination, and we will assess the part it plays in a variety of geotechnical problems.

8.1 Development of Groundwater Resources

Exploration, Evaluation, and Exploitation

The development of groundwater resources can be viewed as a sequential process with three major phases. First, there is an *exploration* stage, in which surface and subsurface geological and geophysical techniques are brought to bear on the search for suitable aquifers. Second, there is an *evaluation* stage that encompasses the measurement of hydrogeologic parameters, the design and analysis of wells, and the calculation of aquifer yields. Third, there is an *exploitation*, or *management* phase, which must include consideration of optimal development strategies and an assessment of the interactions between groundwater exploitation and the regional hydrologic system.

It is worth placing these three phases in a historical perspective. In North America and Europe, nearly all major aquifers have already been located and are being used to some extent. The era of true exploration for regional aquifers is over. We are now in a period in which detailed evaluation of known aquifers and careful management of known resources will take on greater importance. The layout of this chapter reflects this interpretation of current needs. We will treat aquifer exploration in a single section, and place heavier emphasis on the evaluation and management stages.

Let us assume that we have located an aquifer that has some apparent potential. The scope of groundwater resource evaluation and management studies might best be indicated by the following series of questions:

1. Where should the wells be located? How many wells are needed? What pumping rates can they sustain?
2. What will be the effect of the proposed pumping scheme on regional water levels?
3. What are the long-term yield capabilities of the aquifer?
4. Will the proposed development have any detrimental influence on other components of the hydrologic cycle?
5. Are there likely to be any undesirable side effects of development, such as land subsidence or seawater intrusion, that could serve to limit yields?

This chapter is designed to provide the methodology needed to answer questions of this type. The measurement and estimation of hydrogeologic parameters is treated in Sections 8.4 through 8.7. Predictions of drawdown in an aquifer under a proposed pumping scheme can be carried out for simple situations with the analytical methods presented in Section 8.3. More complex hydrogeological environments may require the application of numerical-simulation techniques, as presented in Section 8.8, or electrical-analog techniques, as presented in Section 8.9. Land subsidence is discussed in Section 8.12, and seawater intrusion in Section 8.13.

Well Yield, Aquifer Yield, and Basin Yield

The techniques of groundwater resource evaluation require an understanding of the concept of *groundwater yield*, and, perhaps surprisingly, this turns out to be a difficult and ambiguous term to address. The concept is certainly pertinent, in that one of the primary objectives of most groundwater resource studies is the determination of the maximum possible pumping rates that are compatible with the hydrogeologic environment from which the water will be taken. This need for compatibility implies that yields must be viewed in terms of a balance between the benefits of groundwater pumpage and the undesirable changes that will be induced by such pumpage. The most ubiquitous change that results from pumping is lowered water levels, so in the simplest cases groundwater yield can be defined in terms of the maximum rate of pumpage that can be allowed while ensuring that water-level declines are kept within acceptable limits.

This concept of yield can be applied on several scales. If our unit of study is a single well, we can define a well yield; if our unit of study is an aquifer, we can define an aquifer yield; and if our unit of study is a groundwater basin, we can define a basin yield. *Well yield* can be defined as the maximum pumping rate that can be supplied by a well without lowering the water level in the well below the pump intake. *Aquifer yield* can be defined as the maximum rate of withdrawal that can be sustained by an aquifer without causing an unacceptable decline in the hydraulic head in the aquifer. *Basin yield* can be defined as the maximum rate of withdrawal that can be sustained by the complete hydrogeologic system in a groundwater basin without causing unacceptable declines in hydraulic head in the system or causing unacceptable changes to any other component of the hydrologic cycle in the basin. In light of the effects of well interference that are discussed in Section 8.3, it is clear that aquifer yield is highly dependent on the number and spacing of wells tapping an aquifer. If all the wells in a highly developed aquifer pump at a rate equal to their well yield, it is likely that the aquifer yield will be exceeded. In light of the effects of aquitard leakage and aquifer interference that are also discussed in Section 8.3, it is clear that basin yield is highly dependent on the number and spacing of exploited aquifers in a basin. If all the aquifers are pumped at a rate equal to their aquifer yield, it is likely that the basin yield will be exceeded.

These simple concepts should prove useful to the reader in the early sections of this chapter. However, the concept of basin yield deserves reconsideration in greater depth, and this is presented in Section 8.10.

8.2 Exploration for Aquifers

An aquifer is a geological formation that is capable of yielding economic quantities of water to man through wells. It must be porous, permeable, and saturated. While aquifers can take many forms within the wide variety of existing hydrogeological environments, a perusal of the permeability and porosity data of Tables 2.2 and 2.4 and consideration of the discussions of Chapter 4 make it clear that certain geological deposits are of recurring interest as aquifers. Among the most common are unconsolidated sands and gravels of alluvial, glacial, lacustrine, and deltaic origin; sedimentary rocks, especially limestones and dolomites, and sandstones and conglomerates; and porous or fractured volcanic rocks. In most cases, aquifer exploration becomes a search for one or other of these types of geological deposits. The methods of exploration can be grouped under four headings: surface geological, subsurface geological, surface geophysical, and subsurface geophysical.

Surface Geological Methods

The initial steps in a groundwater exploration program are carried out in the office rather than in the field. Much can be learned from an examination of available maps, reports, and data. There are published geologic maps on some scale for almost all of North America; there are published soils maps or surficial geology maps for most areas; and there are published hydrogeological maps for some areas. Geologic maps and reports provide the hydrogeologist with an initial indication of the rock formations in an area, together with their stratigraphic and structural interrelationships. Soils maps or surficial geology maps, together with topographic maps, provide an introduction to the distribution and genesis of the surficial unconsolidated deposits and their associated landforms. Hydrogeologic maps provide a summarized interpretation of the topographic, geologic, hydrogeologic, geochemical, and water resource data available in an area.

Airphoto interpretation is also widely used in groundwater exploration. It is usually possible to prepare maps of landforms, soils, land use, vegetation, and drainage from the airphoto coverage of an area. Each of these environmental properties leads to inferences about the natural groundwater flow systems and/or the presence of potential aquifers. Way (1973) and Mollard (1973) each provide a handbook-style treatment of airphoto-interpretation methods, and both include a large number of interpreted photos, many of which illustrate significant hydrogeological features.

However, even in areas where there is a considerable amount of published information, it is usually necessary to carry out geologic mapping in the field. In view of the importance of unconsolidated sands and gravels as potential aquifers, special attention must be paid to geomorphic landforms and to the distribution of glacial and alluvial deposits. Where sand and gravel deposits are sparse, or where these deposits are shallow and unsaturated, more detailed attention must be paid to the lithology, stratigraphy, and structure of the bedrock formations.

The methods of hydrogeologic mapping outlined in Section 6.1 are useful in determining the scale and depth of natural groundwater flow systems and in mapping the extent of their recharge and discharge areas.

Subsurface Geological Methods

It is seldom sufficient to look only at the surficial manifestations of a hydrogeological environment. It is unlikely that subsurface stratigraphic relationships will be fully revealed without direct subsurface investigation. Once again, the initial step usually involves scanning the available records. Many state and provincial governments now require that geological logs of all water wells be filed in a central bank for the use of other investigators. These data, while varying widely in quality, can often provide the hydrogeologist with considerable information on past successes and failures in a given region.

In most exploration programs, especially those for large-scale industrial or municipal water supplies, it is necessary to carry out test-drilling to better delineate subsurface conditions. Test holes provide the opportunity for geological and geophysical logging and for the coring or sampling of geological materials. Test holes can also be used to obtain water samples for chemical analysis and to indicate the elevation of the water table at a site. Test-drilling programs, together with published geological maps and available well-log records, can be interpreted in terms of the local and regional lithology, stratigraphy, and structure. Their logs can be used to prepare stratigraphic cross sections, geological fence diagrams, isopach maps of overburden thickness or formation thickness, and lithofacies maps. Hydrogeological interpretations might include water-table contours and isopachs of saturated thickness of unconfined aquifers. The results of chemical analyses of groundwater samples, when graphically displayed using the methods of Chapter 7, can provide important evidence on the natural geochemical environment as well as a direct measure of water quality.

Surface Geophysical Methods

There are two regional geophysical techniques that are used to some extent in the exploration for aquifers. These are the *seismic refraction* method and the *electrical resistivity* method. The design of geophysical surveys that utilize these approaches, and the interpretation of the resulting geophysical measurements, is a specialized branch of the earth sciences. It is not expected that a groundwater hydrologist become such a specialist, and for this reason our discussion is brief. On the other hand, it *is* necessary that the hydrogeologist be aware of the power and limitations of the methods. If this brief presentation fails to meet that objective, the reader is directed to a standard geophysics textbook such as Dobrin (1960), or to one of several review papers that deal specifically with geophysical applications in groundwater exploration, such as McDonald and Wantland (1961), Hobson (1967), or Lennox and Carlson (1967).

The *seismic refraction method* is based on the fact that elastic waves travel through different earth materials at different velocities. The denser the material, the higher the wave velocity. When elastic waves cross a geologic boundary between two formations with different elastic properties, the velocity of wave propagation changes and the wave paths are refracted according to Snell's law. In seismic exploration, elastic waves are initiated by an energy source, usually a small explosion, at the ground surface. A set of receivers, called *geophones*, is set up in a line radiating outward from the energy source. Waves initiated at the surface and refracted at the critical angle by a high-velocity layer at depth will reach the more distant geophones more quickly than waves that travel directly through the low-velocity surface layer. The time between the shock and the arrival of the elastic wave at a geophone is recorded on a *seismograph*. A set of seismograph records can be used to derive a graph of arrival time versus distance from shot point to geophone, and this, in turn, with the aid of some simple theory, can be used to calculate layer depths and their seismic velocities.

In groundwater investigations the seismic refraction method has been used to determine such features as the depth to bedrock, the presence of buried bedrock channels, the thickness of surficial fracture zones in crystalline rock, and the areal extent of potential aquifers. The interpretations are most reliable in cases where there is a simple two-layer or three-layer geological configuration in which the layers exhibit a strong contrast in seismic velocity. The velocities of the layers must increase with depth; the method cannot pick up a low-velocity layer (which might well be a porous potential aquifer) that underlies a high-velocity surface layer. The depth of penetration of the seismic method depends on the strength of the energy source. For shallow investigations (say, up to 30 m) hydrogeologists have often employed *hammer seismic* methods, in which the energy source is simply a hammer blow on a steel plate set on the ground surface.

The *electrical resistivity* of a geological formation is defined as $\rho = RA/L$, where R is the resistance to electrical current for a unit block of cross-sectional area A and length L. The resistivity controls the gradient in electrical potential that will be set up in a formation under the influence of an applied current. In a saturated rock or soil, the resistivity is largely dependent on the density and porosity of the material and on the salinity of the saturating fluid. In an electrical resistivity survey an electric current is passed into the ground through a pair of *current electrodes* and the potential drop is measured across a pair of *potential electrodes*. The spacing of the electrodes controls the depth of penetration. At each setup an apparent resistivity is calculated on the basis of the measured potential drop, the applied current, and the electrode spacing. Sets of measurements are taken either in the form of *lateral profiling* or *depth profiling*. In lateral profiling the electrode spacing is kept constant as electrodes are leapfrogged down a survey line. This method provides areal coverage at a given depth of penetration. It can be used to define aquifer limits or to map areal variations in groundwater salinity. In depth profiling a series of readings is taken at different electrode spacings at a single station. Apparent resistivities are plotted against electrode spacing, and stratigraphic interpretations are made by comparing the resulting curve against published theoretical curves for simple layered geometries. Depth profiling has been widely used to determine the thickness of sand and gravel aquifers that overlie bedrock. It can also be used to locate the saltwater-freshwater interface in coastal aquifers. It is often claimed that the method can "feel" the water table, but this is questionable except in very homogeneous deposits. In urban areas the method is often hampered by the presence of pipes, rails, and wires that interfere with the current fields.

Surface geophysical methods cannot replace test drilling, although by providing data that lead to a more intelligent selection of test-hole drilling, they may lead to a reduction in the amount of drilling required. Stratigraphic interpretations based on seismic or electrical resistivity measurements must be calibrated against test-hole information.

Subsurface Geophysical Methods

There is one geophysical approach that has now become a standard tool in groundwater exploration. This approach involves the logging of wells and test holes by the methods of *borehole geophysics*. The term encompasses all techniques in which a sensing device is lowered into a hole in order to make a record that can be interpreted in terms of the characteristics of the geologic formations and their contained fluids. The techniques of borehole geophysics were originally developed in the petroleum industry and the standard textbooks on the interpretation of geophysical logs (Pirson, 1963; Wyllie, 1963) emphasize petroleum applications. Fortunately, there are several excellent review articles (Jones and Skibitzke, 1956; Patton and Bennett, 1963; Keys, 1967, 1968) that deal specifically with the application of geophysical logging techniques to groundwater problems.

A complete borehole geophysics program as it is carried out in the petroleum industry usually includes two electric logs (spontaneous potential and resistivity), three radiation logs (natural gamma, neutron, and gamma-gamma), and a caliper log that indicates variations in hole diameter. In hydrogeological applications, emphasis is usually placed on the electric logs.

The simplest electric log is the *spontaneous potential* (or self-potential) log. It is obtained with the single-point electrode arrangement shown in Figure 8.1 with the current source disconnected. It provides a measure of the naturally occurring potential differences between the surface electrode and the borehole electrode. The origin of these natural electric potentials is not well understood, but they are apparently related to electrochemical interactions that take place between the borehole fluid and the *in situ* rock-water complex.

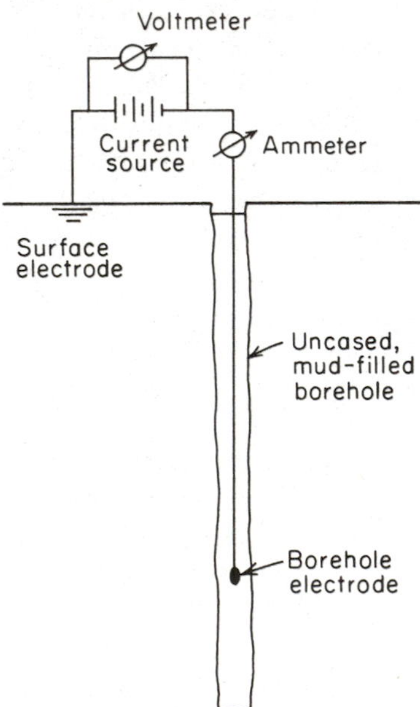

Figure 8.1 Single-point electrode arrangement for spontaneous potential and resistivity logging in a borehole.

The second electric log is a *resistivity* log. There are several electrode arrangements that can be used, but the simplest and the one most widely used in the water well industry is the single-point arrangement shown in Figure 8.1. The potential difference recorded at different depths for a given current strength leads to a log of apparent resistivity versus depth.

The two electric logs can be jointly interpreted in a qualitative sense in terms of the stratigraphic sequence in the hole. Figure 8.2 shows a pair of single-point electric logs measured in a test hole in an unconsolidated sequence of Pleistocene and Upper Cretaceous sediments in Saskatchewan. The geologic descriptions and

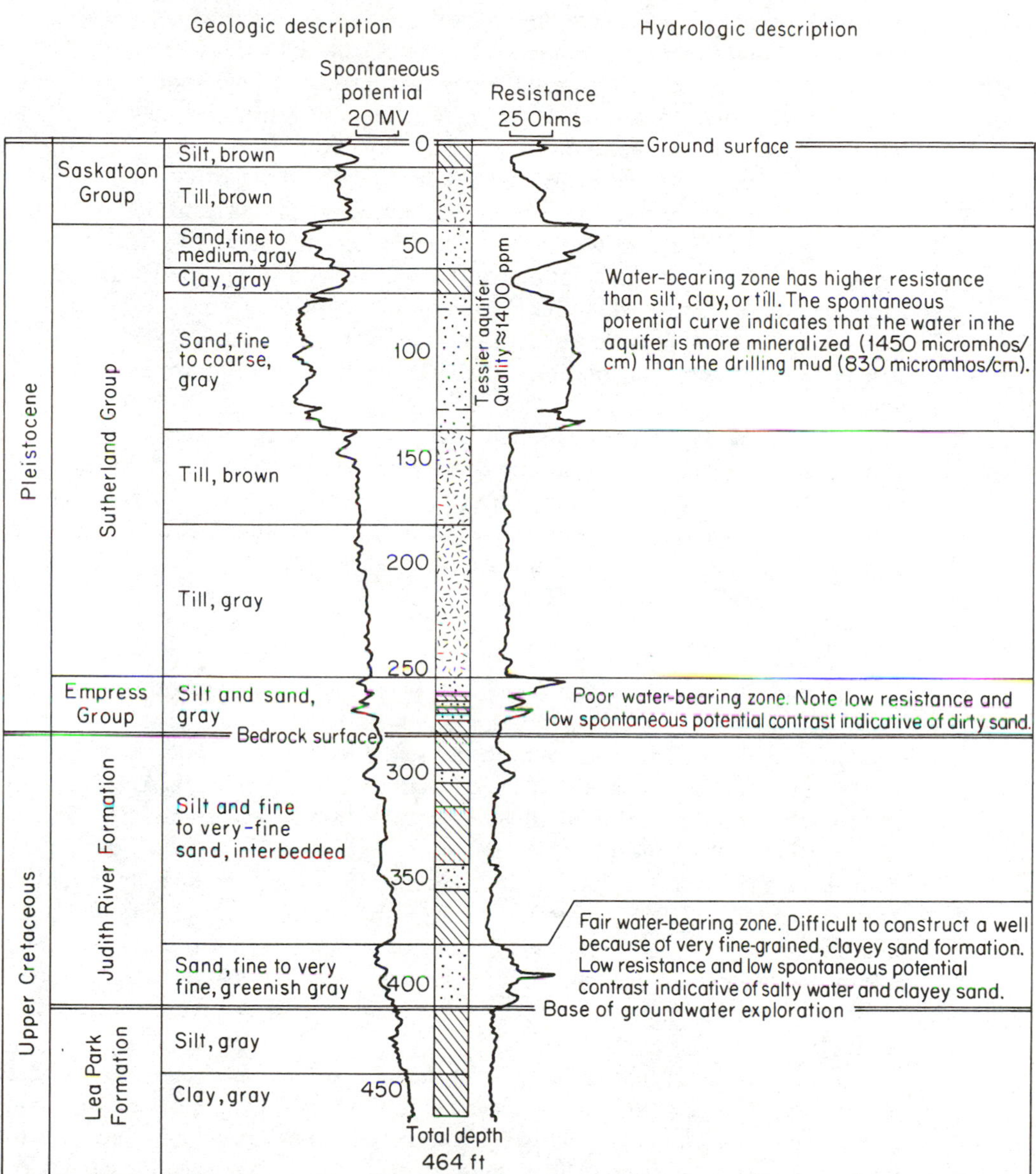

Figure 8.2 Geologic log, electric logs, geologic description, and hydrologic description of a test hole in Saskatchewan (after Christiansen et al., 1971).

the geologic log in the center are based on a core-sampling program. The hydrologic description of the potential aquifers at the site is based on a joint interpretation of the geologic and geophysical logs. In most common geological environments, the best water-yielding zones have the highest resistivities. Electric logs often provide the most accurate detail for the selection of well-screen settings.

Dyck et al. (1972) pointed out three disadvantages to single-point electric logs. They do not provide quantitative values of formation resistivity; they are affected by hole diameter and borehole fluid resistivity; and they have only a shallow radius of investigation. To emphasize the first point, the resistivity log on Figure 8.2 records simply the resistance measured between the two electrodes rather than an apparent resistivity. Multiple-point electric logs are more versatile. They can be used for quantitative calculations of the resistivity of formation rocks and their enclosed fluids. These calculations lie beyond the scope of this presentation. Campbell and Lehr (1973) provide a good summary of the techniques. Dyck et al. (1972) provide some sample calculations in the context of a groundwater exploration program.

Keys (1967, 1968) has suggested that radiation logs, especially the natural gamma log, may have applications to groundwater hydrology. A logging suite that might be considered complete for hydrogeological purposes would include a driller's log (including drilling rate), a geologic log, a spontaneous potential log, a resistivity log, a natural gamma log, and a caliper log.

Drilling and Installation of Wells and Piezometers

The drilling of piezometers and wells, and their design, construction, and maintenance, is a specialized technology that rests only in part on scientific and engineering principles. There are many books (Briggs and Fiedler, 1966; Gibson and Singer, 1971; Campbell and Lehr, 1973; U.S. Environmental Protection Agency, 1973a, 1976) that provide a comprehensive treatment of water well technology. In addition, Walton (1970) presents material on the technical aspects of groundwater hydrology, and his text includes many case histories of water well installations and evaluations. Reeve (1965), Hvorslev (1951), Campbell and Lehr (1973), and Kruseman and de Ridder (1970) discuss methods of piezometer construction and installation. In this text we will limit ourselves to a brief overview of these admittedly important practical matters. Most of what follows is drawn from Campbell and Lehr (1973).

Water wells are usually classified on the basis of their method of construction. Wells may be *dug* by hand, *driven* or *jetted* in the form of well points, *bored* by an earth auger, or *drilled* by a drilling rig. The selection of the method of construction hinges on such questions as the purpose of the well, the hydrogeological environment, the quantity of water required, the depth and diameter envisaged, and economic factors. Dug, bored, jetted and driven wells are limited to shallow depths, unconsolidated deposits, and relatively small yields. For deeper, more productive wells in unconsolidated deposits, and for all wells in rock, drilling is the only feasible approach.

There are three main types of drilling equipment: *cable tool*, *rotary*, and *reverse rotary*. The cable tool drills by lifting and dropping a string of tools suspended on a cable. The bit at the bottom of the tool string rotates a few degrees between each stroke so that the cutting face of the bit strikes a different area of the hole bottom with each stroke. Drilling is periodically interrupted to bail out the cuttings. With medium- to high-capacity rigs, 40- to 60-cm-diameter holes can be drilled to depths of several hundred meters and smaller diameter holes to greater depths. The cable-tool approach is successful over a wide range of geological materials, but it is not capable of drilling as quickly or as deeply as rotary methods. With the conventional rotary method, drilling fluid is forced down the inside of a rapidly rotating drill stem and out through openings in the bit. The drilling fluid flows back to the surface, carrying the drill cuttings with it, by way of the annulus formed between the outside of the drill pipe and the hole wall. In a reverse rotary system, the direction of circulation is reversed. Reverse rotary is particularly well suited to drilling large-diameter holes in soft, unconsolidated formations.

The conventional rotary rig is generally considered to be the fastest, most convenient, and least expensive system to operate, expecially in unconsolidated deposits. Penetration rates for rotary rigs depend on such mechanical factors as the weight, type, diameter, and condition of the bit, and its speed of rotation; the circulation rate of the drilling fluid and its properties; and the physical characteristics of the geological formation. In rock formations, *drillability* (defined as depth of penetration per revolution) is directly related to the compressive strength of the rock.

The direct rotary method is heavily dependent on its hydraulic circulation system. The most widely used drilling fluid is a suspension of bentonitic clay in water, known as *drilling mud.* During drilling, the mud coats the hole wall and in so doing contributes to the hole stability and prevents losses of the drilling fluid to permeable formations. When even heavy drilling mud cannot prevent the caving of hole walls, well casing must be emplaced as drilling proceeds. Caving, lost circulation, and conditions associated with the encounter of flowing artesian water constitute the most common drilling problems.

The design of a deep-cased well in an unconsolidated aquifer must include consideration of the surface housing, the casing, the pumping equipment, and the intake. Of these, it is the intake that is most often of primary concern to groundwater hydrologists. In the first half of this century it was quite common to provide access for the water to the well by a set of perforations or hand-sawn slots in the casing. It is now recognized that well yields can be significantly increased by the use of *well screens.* The size of the intake slots in a properly designed well screen is related to the grain-size distribution of the aquifer. Development of a screened well by pumping, surging, or backwashing draws the fines out of the aquifer, through the well screen, and up to the surface. By removing the fines from the formation in the vicinity of the well, a *natural gravel pack* is created around the screen that increases the efficiency of the intake. In some cases, an *artificial gravel pack* is emplaced to improve intake properties. Figure 8.3 shows several typical designs for wells in consolidated and unconsolidated formations.

The productivity of a well is often expressed in terms of the *specific capacity,*

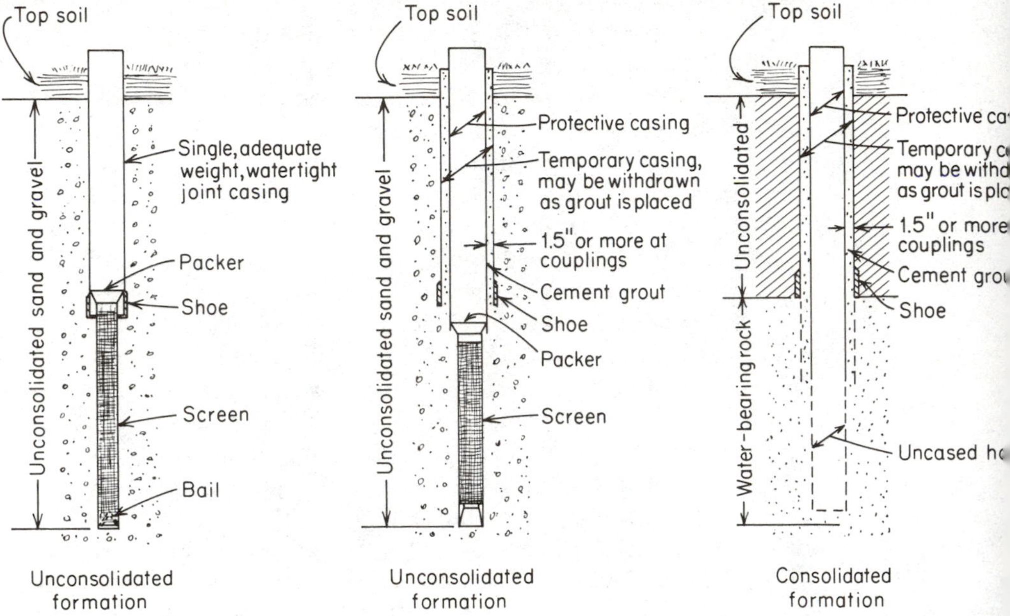

Figure 8.3 Typical well designs for consolidated and unconsolidated formations.

C_s, which is defined as $C_s = Q/\Delta h_w$, where Q is the pumping rate and Δh_w is the drawdown in the well. In this equation, $\Delta h_w = \Delta h + \Delta h_L$, where Δh is the drawdown in hydraulic head in the aquifer at the well screen boundary, and Δh_L is the *well loss* created by the turbulent flow of water through the screen and into the pump intake. Δh is calculated from the standard well-hydraulics equations developed in Section 8.3. Δh_L can be estimated by methods outlined in Walton (1970) and Campbell and Lehr (1973). In general, $\Delta h_L \ll \Delta h$.

8.3 The Response of Ideal Aquifers to Pumping

The exploitation of a groundwater basin leads to water-level declines that serve to limit yields. One of the primary goals of groundwater resource evaluation must therefore be the prediction of hydraulic-head drawdowns in aquifers under proposed pumping schemes. In this section, the theoretical response of idealized aquifers to pumping will be examined. We will investigate several types of aquifer configuration, but in each case the geometry will be sufficiently regular and the boundary conditions sufficiently simple to allow the development of an analytical solution to the boundary-value problem that represents the case at hand. These solutions, together with solutions to more complex boundary-value problems that describe less ideal conditions, constitute the foundation of the study of *well hy-*

draulics. This section provides an introduction to the topic, but the material covered is far from all-inclusive. There is a massive literature in the field and the committed reader is directed to Walton's (1970) comprehensive treatment, to Hantush's (1964) monograph, or to the excellent handbooks of Ferris et al. (1962) and Kruseman and de Ridder (1970).

Radial Flow to a Well

The theoretical analyses are based on an understanding of the physics of flow toward a well during pumping. All the necessary concepts have been introduced in Chapter 2. The distinction between *confined* and *unconfined* aquifers was explained there, as was the relation between the general concept of *hydraulic head* in a three-dimensional geologic system and the specific concept of the *potentiometric surface* on a two-dimensional, horizontal, confined aquifer. Definitions were presented for the fundamental hydrogeologic parameters: *hydraulic conductivity*, *porosity*, and *compressibility*; and for the derived aquifer parameters: *transmissivity* and *storativity*. It was explained there that pumping induces horizontal hydraulic gradients toward a well, and as a result hydraulic heads are decreased in the aquifer around a well during pumping. What is required now is that we take these fundamental concepts, put them into the form of a boundary-value problem that represents flow to a well in an aquifer, and examine the theoretical response.

At this point it is worth recalling from Section 2.10 that the definition of storativity invokes a one-dimensional concept of aquifer compressibility. The α in Eq. (2.63) is the aquifer compressibility in the vertical direction. The analyses that follow in effect assume that changes in effective stress induced by aquifer pumping are much larger in the vertical direction than in the horizontal.

The concept of aquifer storage inherent in the storativity term also implies an instantaneous release of water from any elemental volume in the system as the head drops in that element.

Let us begin our analysis with the simplest possible aquifer configuration. Consider an aquifer that is (1) horizontal, (2) confined between impermeable formations on top and bottom, (3) infinite in horizontal extent, (4) of constant thickness, and (5) homogeneous and isotropic with respect to its hydrogeological parameters.

For the purposes of our initial analysis, let us further limit our ideal system as follows: (1) there is only a single pumping well in the aquifer, (2) the pumping rate is constant with time, (3) the well diameter is infinitesimally small, (4) the well penetrates the entire aquifer, and (5) the hydraulic head in the aquifer prior to pumping is uniform throughout the aquifer.

The partial differential equation that describes saturated flow in two horizontal dimensions in a confined aquifer with transmissivity T and storativity S was developed in Section 2.11 as Eq. (2.77):

$$\frac{\partial^2 h}{\partial x^2} + \frac{\partial^2 h}{\partial y^2} = \frac{S}{T}\frac{\partial h}{\partial t}$$

Since it is clear that hydraulic-head drawdowns around a well will possess radial symmetry in our ideal system, it is advantageous to convert Eq. (2.77) into radial coordinates. The conversion is accomplished through the relation $r = \sqrt{x^2 + y^2}$ and the equation of flow becomes (Jacob, 1950)

$$\frac{\partial^2 h}{\partial r^2} + \frac{1}{r}\frac{\partial h}{\partial r} = \frac{S}{T}\frac{\partial h}{\partial t} \tag{8.1}$$

The mathematical region of flow, as illustrated in the plan view of Figure 8.4, is a horizontal one-dimensional line through the aquifer, from $r = 0$ at the well to $r = \infty$ at the infinite extremity.

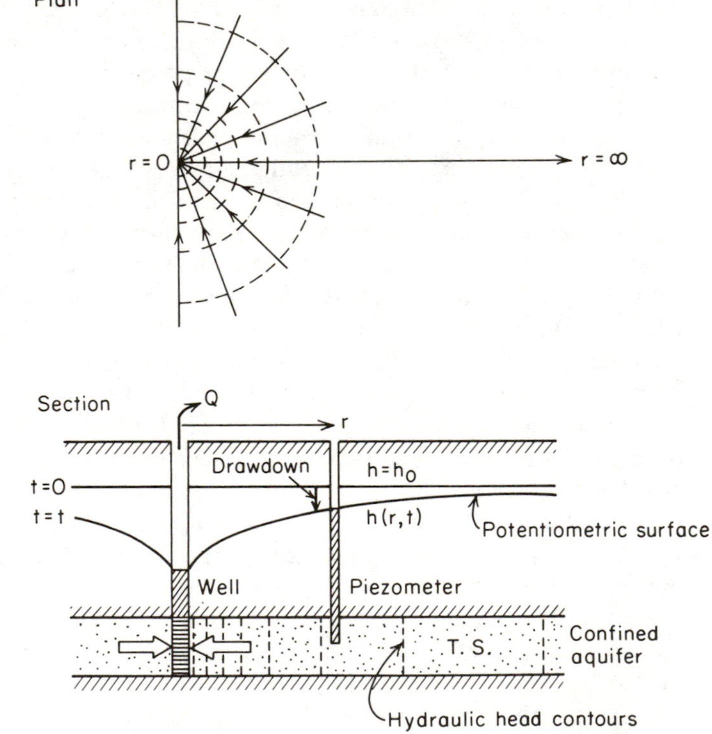

Figure 8.4 Radial flow to a well in a horizontal confined aquifer.

The initial condition is

$$h(r, 0) = h_0 \qquad \text{for all } r \tag{8.2}$$

where h_0 is the constant initial hydraulic head.

The boundary conditions assume no drawdown in hydraulic head at the infinite boundary:

$$h(\infty, t) = h_0 \qquad \text{for all } t \tag{8.3}$$

and a constant pumping rate Q $[L^3/T]$ at the well:

$$\lim_{r \to 0} \left(r \frac{\partial h}{\partial r} \right) = \frac{Q}{2\pi T} \qquad \text{for } t > 0 \tag{8.4}$$

Condition (8.4) is the result of a straightforward application of Darcy's law at the well face.

The solution $h(r, t)$ describes the hydraulic head field at any radial distance r at any time after the start of pumping. For reasons that should be clear from a perusal of Figure 8.4, solutions are often presented in terms of the drawdown in head $h_0 - h(r, t)$.

The Theis Solution

Theis (1935), in what must be considered one of the fundamental breakthroughs in the development of hydrologic methodology, utilized an analogy to heat-flow theory to arrive at an analytical solution to Eq. (8.1) subject to the initial and boundary conditions of Eqs. (8.2) through (8.4). His solution, written in terms of the drawdown, is

$$h_0 - h(r, t) = \frac{Q}{4\pi T} \int_u^\infty \frac{e^{-u}\, du}{u} \tag{8.5}$$

where

$$u = \frac{r^2 S}{4Tt} \tag{8.6}$$

The integral in Eq. (8.5) is well known in mathematics. It is called the *exponential integral* and tables of values are widely available. For the specific definition of u given by Eq. (8.6), the integral is known as the *well function*, $W(u)$. With this notation, Eq. (8.5) becomes

$$h_0 - h = \frac{Q}{4\pi T} W(u) \tag{8.7}$$

Table 8.1 provides values of $W(u)$ versus u, and Figure 8.5(a) shows the relationship $W(u)$ versus $1/u$ graphically. This curve is commonly called the *Theis curve*.

If the aquifer properties, T and S, and the pumping rate, Q, are known, it is possible to predict the drawdown in hydraulic head in a confined aquifer at any distance r from a well at any time t after the start of pumping. It is simply necessary to calculate u from Eq. (8.6), look up the value of $W(u)$ on Table 8.1, and calculate

Table 8.1 Values of $W(u)$ for Various Values of u

u	1.0	2.0	3.0	4.0	5.0	6.0	7.0	8.0	9.0
$\times 1$	0.219	0.049	0.013	0.0038	0.0011	0.00036	0.00012	0.000038	0.000012
$\times 10^{-1}$	1.82	1.22	0.91	0.70	0.56	0.45	0.37	0.31	0.26
$\times 10^{-2}$	4.04	3.35	2.96	2.68	2.47	2.30	2.15	2.03	1.92
$\times 10^{-3}$	6.33	5.64	5.23	4.95	4.73	4.54	4.39	4.26	4.14
$\times 10^{-4}$	8.63	7.94	7.53	7.25	7.02	6.84	6.69	6.55	6.44
$\times 10^{-5}$	10.94	10.24	9.84	9.55	9.33	9.14	8.99	8.86	8.74
$\times 10^{-6}$	13.24	12.55	12.14	11.85	11.63	11.45	11.29	11.16	11.04
$\times 10^{-7}$	15.54	14.85	14.44	14.15	13.93	13.75	13.60	13.46	13.34
$\times 10^{-8}$	17.84	17.15	16.74	16.46	16.23	16.05	15.90	15.76	15.65
$\times 10^{-9}$	20.15	19.45	19.05	18.76	18.54	18.35	18.20	18.07	17.95
$\times 10^{-10}$	22.45	21.76	21.35	21.06	20.84	20.66	20.50	20.37	20.25
$\times 10^{-11}$	24.75	24.06	23.65	23.36	23.14	22.96	22.81	22.67	22.55
$\times 10^{-12}$	27.05	26.36	25.96	25.67	25.44	25.26	25.11	24.97	24.86
$\times 10^{-13}$	29.36	28.66	28.26	27.97	27.75	27.56	27.41	27.28	27.16
$\times 10^{-14}$	31.66	30.97	30.56	30.27	30.05	29.87	29.71	29.58	29.46
$\times 10^{-15}$	33.96	33.27	32.86	32.58	32.35	32.17	32.02	31.88	31.76

SOURCE: Wenzel, 1942.

$h_0 - h$ from Eq. (8.7). Figure 8.5(b) shows a calculated plot of $h_0 - h$ versus t for the specific set of parameters noted on the figure. A set of field measurements of drawdown versus time measured in a piezometer that is set in an ideal confined aquifer with these properties would show this type of record.

The shape of the function $h_0 - h$ versus t, when plotted on log-log paper as in Figure 8.5(b), has the same form as the plot of $W(u)$ versus $1/u$ shown in Figure 8.5(a). This is a direct consequence of the relations embodied in Eqs. (8.6) and (8.7), where it can be seen that $h_0 - h$ and $W(u)$, and t and $1/u$, are related to one another through a constant term.

It is also possible to calculate values of $h_0 - h$ at various values of r at a given time t. Such a calculation leads to a plot of the *cone of depression* (or *drawdown cone*) in the potentiometric surface around a pumping well. Figure 8.4 provides a schematic example. The steepening of the slope of the cone near the well is reflected in the solution, Eq. (8.7). The physical explanation is clear if one carries out the simple flow-net construction shown in the plan view of Figure 8.4 and then carries the hydraulic head values down onto the section.

For a given aquifer the cone of depression increases in depth and extent with increasing time. Drawdown at any point at a given time is directly proportional to the pumping rate and inversely proportional to aquifer transmissivity and aquifer storativity. As shown in Figure 8.6, aquifers of low transmissivity develop tight, deep drawdown cones, whereas aquifers of high transmissivity develop shallow cones of wide extent. Transmissivity exerts a greater influence on drawdown than does storativity.

In that geologic configurations are seldom as ideal as that outlined above, the time-drawdown response of aquifers under pumpage often deviates from the

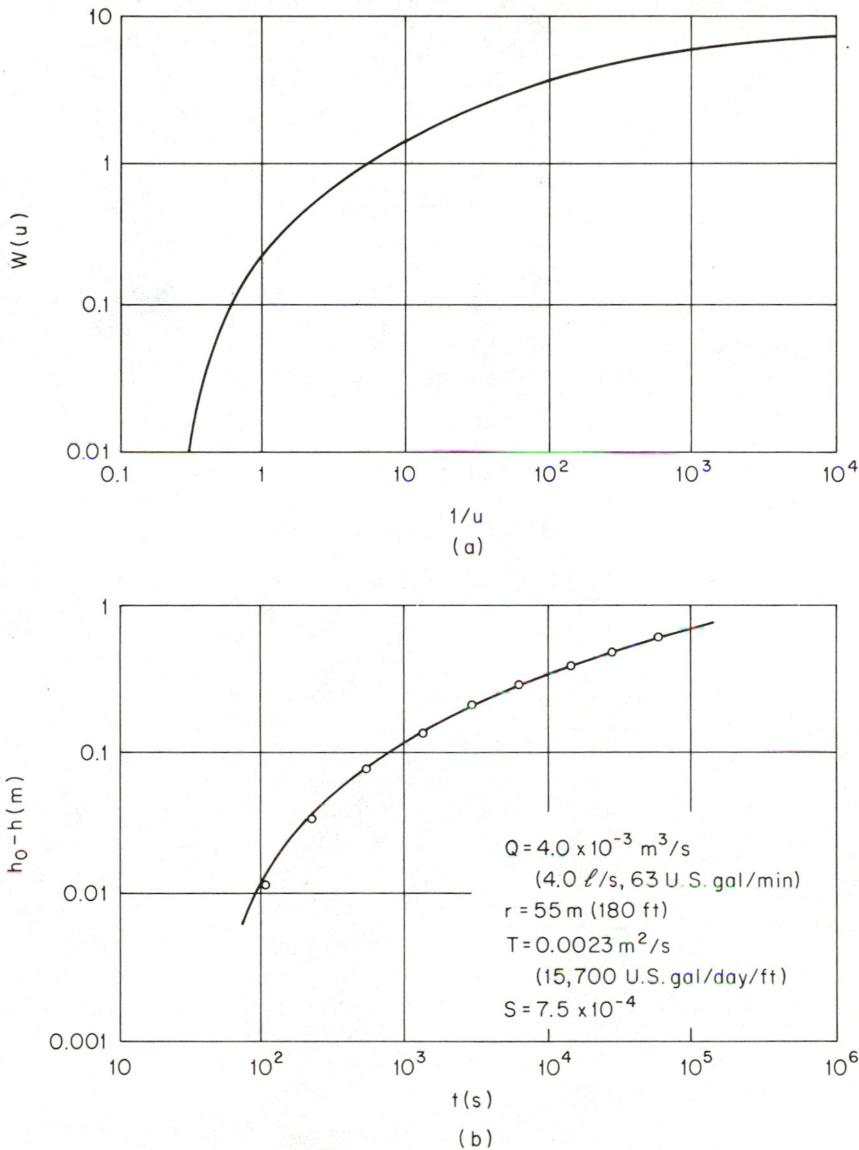

Figure 8.5 (a) Theoretical curve of $W(u)$ versus $1/u$. (b) Calculated curve of $h_0 - h$ versus t.

Theis solution shown in Figure 8.5. We will now turn to some of the theoretical response curves that arise in less ideal situations. Specifically, we will look at (1) leaky aquifers, (2) unconfined aquifers, (3) multiple-well systems, (4) stepped pumping rates, (5) bounded aquifers, and (6) partially penetrating wells.

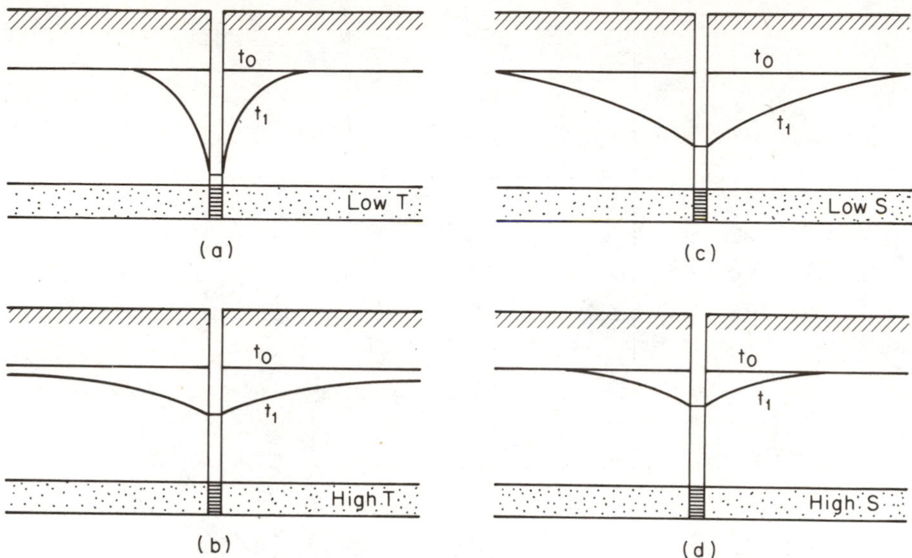

Figure 8.6 Comparison of drawdown cones at a given time for aquifers of (a) low transmissivity; (b) high transmissivity; (c) low storativity; (d) high storativity.

Leaky Aquifers

The assumption inherent in the Theis solution that geologic formations overlying and underlying a confined aquifer are completely impermeable is seldom satisfied. Even when production wells are screened only in a single aquifer, it is quite usual for the aquifer to receive a significant inflow from adjacent beds. Such an aquifer is called a *leaky aquifer*, although in reality it is the aquitard that is leaky. The aquifer is often just one part of a multiple-aquifer system in which a succession of aquifers are separated by intervening low-permeability aquitards. For the purposes of this section, however, it is sufficient for us to consider the three-layer case shown in Figure 8.7. Two aquifers of thickness b_1 and b_2 and horizontal hydraulic conductivities K_1 and K_2 are separated by an aquitard of thickness b' and vertical hydraulic conductivity K'. The specific storage values in the aquifers are S_{s_1} and S_{s_2}, while that in the aquitard is S'_s.

Since a rigorous approach to flow in multiple-aquifer systems involves boundary conditions that make the problem intractable analytically, it has been customary to simplify the mathematics by assuming that flow is essentially horizontal in the aquifers and vertical in the aquitards. Neuman and Witherspoon (1969a) report that the errors introduced by this assumption are less than 5% when the conductivities of the aquifers are more than 2 orders of magnitude greater than that of the aquitard.

The development of leaky-aquifer theory has taken place in two distinct sets of papers. The first, by Hantush and Jacob (1955) and Hantush (1956, 1960),

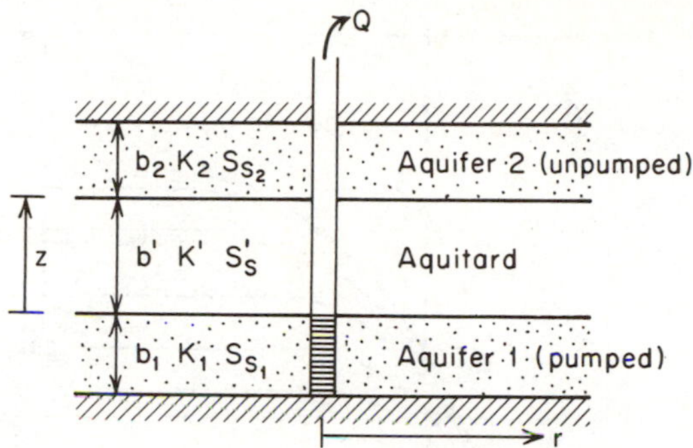

Figure 8.7 Schematic diagram of a two-aquifer "leaky" system. Recall that $T = Kb$ and $S = S_s b$.

provided the original differentiation between the Theis response and that for leaky aquifers. The second, by Neuman and Witherspoon (1969a, 1969b, 1972) evaluated the significance of the assumptions inherent in the earlier work and provided more generalized solutions.

The analytical solution of Hantush and Jacob (1955) can be couched in the same form as the Theis solution [Eq. (8.7)] but with a more complicated well function. In fact, Hantush and Jacob developed two analytical solutions, one valid only for small t and one valid only for large t, and then interpolated between the two solutions to obtain the complete response curve. Their solution is presented in terms of the dimensionless parameter, r/B, defined by the relation

$$\frac{r}{B} = r\sqrt{\frac{K'}{K_1 b_1 b'}} \tag{8.8}$$

In analogy with Eq. (8.7), we can write their solution as

$$h_0 - h = \frac{Q}{4\pi T} W(u, r/B) \tag{8.9}$$

where $W(u, r/B)$ is known as the *leaky well function*.

Hantush (1956) tabulated the values of $W(u, r/B)$. Figure 8.8 is a plot of this function against $1/u$. If the aquitard is impermeable, then $K' = 0$, and from Eq. (8.8), $r/B = 0$. In this case, as shown graphically in Figure 8.8, the Hantush-Jacob solution reduces to the Theis solution.

If T_1 ($= K_1 b_1$) and S_1 ($= S_{s_1} b_1$) are known for the aquifer and K' and b' are known for the aquitard, then the drawdown in hydraulic head in the pumped aquifer for any pumpage Q at any radial distance r at any time t can be calculated from Eq. (8.9), after first calculating u for the pumped aquifer from Eq. (8.6), r/B from Eq. (8.8), and $W(u, r/B)$ from Figure 8.9.

The original Hantush and Jacob (1955) solution was developed on the basis of two very restrictive assumptions. They assumed that the hydraulic head in the

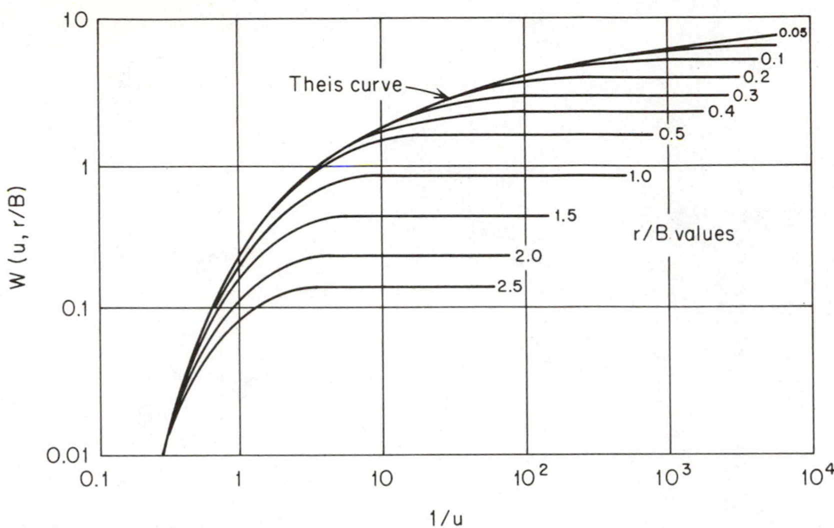

Figure 8.8 Theoretical curves of $W(u, r/B)$ versus $1/u$ for a leaky aquifer (after Walton, 1960).

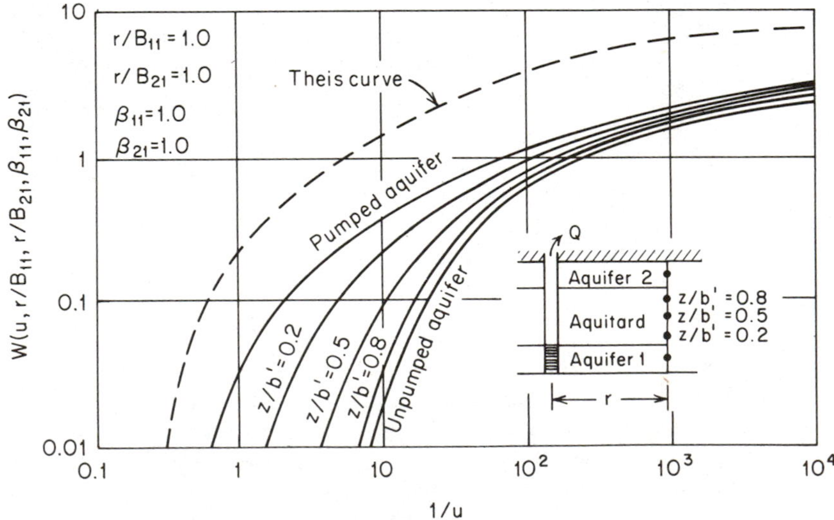

Figure 8.9 Theoretical curves of $W(u, r/B_{11}, r/B_{21}, \beta_{11}, \beta_{21})$ versus $1/u$ for a leaky two-aquifer system (after Neuman and Witherspoon, 1969a).

unpumped aquifer remains constant during the removal of water from the pumped aquifer and that the rate of leakage into the pumped aquifer is proportional to the hydraulic gradient across the leaky aquitard. The first assumption implies that the unpumped aquifer has an unlimited capacity to provide water for delivery through the aquitard to the pumped aquifer. The second assumption completely ignores the effects of the storage capacity of the aquitard on the transient solution (i.e., it is assumed that $S'_s = 0$).

In a later paper, Hantush (1960) presented a modified solution in which consideration was given to the effects of storage in the aquitard. More recently, Neuman and Witherspoon (1969a, 1969b) presented a complete solution that includes consideration of both release of water from storage in the aquitard and head drawdowns in the unpumped aquifer. Their solutions require the calculation of four dimensionless parameters, which, with reference to Figure 8.7, are defined as follows:

$$\begin{aligned} \frac{r}{B_{11}} &= r\sqrt{\frac{K'}{K_1 b_1 b'}} \\ \frac{r}{B_{21}} &= r\sqrt{\frac{K'}{K_2 b_2 b'}} \\ \beta_{11} &= \frac{r}{4b_1}\sqrt{\frac{K'S'_s}{K_1 S_{s_1}}} \\ \beta_{21} &= \frac{r}{4b_2}\sqrt{\frac{K'S'_s}{K_2 S_{s_2}}} \end{aligned} \tag{8.10}$$

Neuman and Witherspoon's solutions provide the drawdown in both aquifers as a function of radial distance from the well, and in the aquitard as a function of both radial distance and elevation above the base of the aquitard. Their solutions can be described in a schematic sense by the relation

$$h_0 - h(r, z, t) = \frac{Q}{4\pi T} W(u, r/B_{11}, r/B_{21}, \beta_{11}, \beta_{21}) \tag{8.11}$$

Tabulation of this well function would require many pages of tables, but an indication of the nature of the solutions can be seen from Figure 8.9, which presents the theoretical response curves for the pumped aquifer, the unpumped aquifer, and at three elevations in the aquitard, for a specific set of r/B and β values. The Theis solution is shown on the diagram for comparative purposes.

Because of its simplicity, and despite the inherent dangers of using a simple model for a complex system, the r/B solution embodied in Figure 8.8 is widely used for the prediction of drawdowns in leaky-aquifer systems. Figure 8.10 shows an $h_0 - h$ versus t plot for a specific case as calculated from Eq. (8.9) with the aid of Figure 8.8. The drawdown reaches a constant level after about 5×10^3 seconds. From this point on, the r/B solution indicates that steady-state conditions hold throughout the system, with the infinite storage capacity assumed to exist in the upper aquifer feeding water through the aquitard toward the well. If the overlying aquitard were impermeable rather than leaky, the response would follow the dotted line. As one would expect, drawdowns in leaky aquifers are less than those in non-

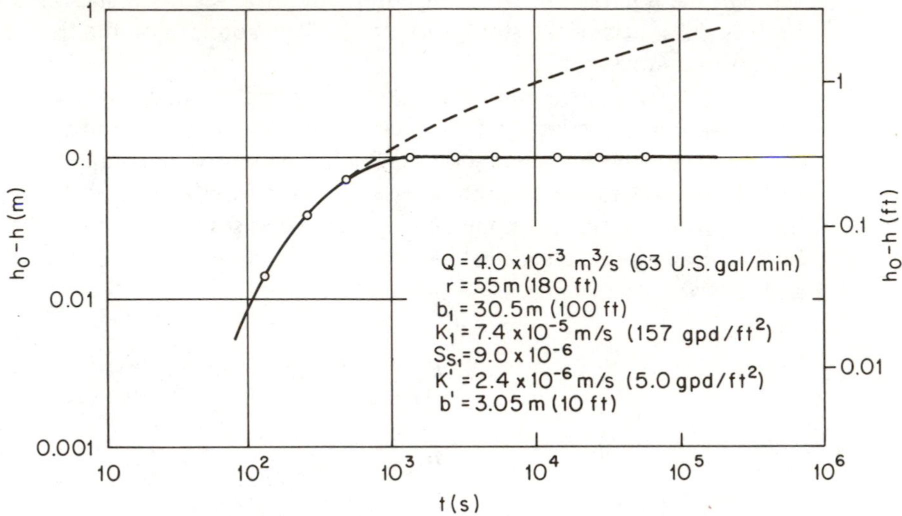

Figure 8.10 Calculated curve of $h_0 - h$ versus t for a leaky aquifer, based on Hantush-Jacob theory.

leaky aquifers, as there is now an additional source of water over and above that which can be supplied by the aquifer itself. Predictions based on the Theis equation therefore provide a conservative estimate for leaky systems; that is, they over-predict the drawdown, or, put another way, actual drawdowns are unlikely to reach the values predicted by the Theis equation for a given pumping scheme in a multiaquifer system.

Unconfined Aquifers

When water is pumped from a confined aquifer, the pumpage induces hydraulic gradients toward the well that create drawdowns in the potentiometric surface. The water produced by the well arises from two mechanisms: expansion of the water in the aquifer under reduced fluid pressures, and compaction of the aquifer under increased effective stresses (Section 2.10). There is no dewatering of the geologic system. The flow system in the aquifer during pumping involves only horizontal gradients toward the well; there are no vertical components of flow. When water is pumped from an unconfined aquifer, on the other hand, the hydraulic gradients that are induced by the pumpage create a drawdown cone in the water table itself and there are vertical components of flow (Figure 8.11). The water produced by the well arises from the two mechanisms responsible for confined delivery *plus* the actual dewatering of the unconfined aquifer.

There are essentially three approaches that can be used to predict the growth of unconfined drawdown cones in time and space. The first, which might be termed the complete analysis, recognizes that the unconfined well-hydraulics problem (Figure 8.11) involves a saturated-unsaturated flow system in which

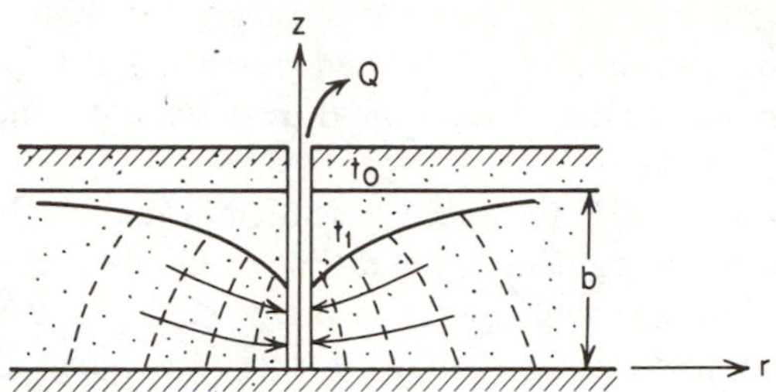

Figure 8.11 Radial flow to a well in an unconfined aquifer.

water-table drawdowns are accompanied by changes in the unsaturated moisture contents above the water table (such as those shown in Figure 2.23). The complete analysis requires the solution of a boundary-value problem that includes both the saturated and unsaturated zones. An analytical solution for this complete case was presented by Kroszynski and Dagan (1975) and several numerical mathematical models have been prepared (Taylor and Luthin, 1969; Cooley, 1971; Brutsaert et al., 1971). The general conclusion of these studies is that the position of the water table during pumpage is not substantially affected by the nature of the unsaturated flow above the water table. In other words, while it is conceptually more appealing to carry out a complete saturated-unsaturated analysis, there is little practical advantage to be gained, and since unsaturated soil properties are extremely difficult to measure *in situ*, the complete analysis is seldom used.

The second approach, which is by far the simplest, is to use the same equation as for a confined aquifer [Eq. (8.7)] but with the argument of the well function [Eq. (8.6)] defined in terms of the specific yield S_y rather than the storativity S. The transmissivity T must be defined as $T = Kb$, where b is the *initial* saturated thickness. Jacob (1950) has shown that this approach leads to predicted drawdowns that are very nearly correct as long as the drawdown is small in comparison with the saturated thickness. The method in effect relies on the Dupuit assumptions (Section 5.5) and fails when vertical gradients become significant.

The third approach, and the one most widely used in practice, is based on the concept of delayed water-table response. This approach was pioneered by Boulton (1954, 1955, 1963) and has been significantly advanced by Neuman (1972, 1973b, 1975a). It can be observed that water-level drawdowns in piezometers adjacent to pumping wells in unconfined aquifers tend to decline at a slower rate than that predicted by the Theis solution. In fact, there are three distinct segments that can be recognized in time-drawdown curves under water-table conditions. During the first segment, which covers only a short period after the start of pumping, an unconfined aquifer reacts in the same way as does a confined aquifer. Water is released instantaneously from storage by the compaction of the aquifer and by the expansion of the water. During the second segment, the effects of gravity drainage are felt. There is a decrease in the slope of the time-drawdown curve relative to the Theis curve because the water delivered to the well by the dewatering that accompanies the falling water table is greater than that which would be delivered by an equal decline in a confined potentiometric surface. In the third segment, which occurs at later times, time-drawdown data once again tend to conform to a Theis-type curve.

Boulton (1963) produced a semiempirical mathematical solution that reproduces all three segments of the time-drawdown curve in an unconfined aquifer. His solution, although useful in practice, required the definition of an empirical *delay index* that was not related clearly to any physical phenomenon. In recent years there has been a considerable amount of research (Neuman, 1972; Streltsova, 1972; Gambolati, 1976) directed at uncovering the physical processes responsible for delayed response in unconfined aquifers. It is now clear that the delay index is not an aquifer constant, as Boulton had originally assumed. It is related to the vertical components of flow that are induced in the flow system and it is apparently a function of the radius r and perhaps the time t.

The solution of Neuman (1972, 1973b, 1975a) also reproduces all three segments of the time-drawdown curve and it does not require the definition of any empirical constants. Neuman's method recognizes the existence of vertical flow components, and the general solution for the drawdown, $h_0 - h$, is a function of both r and z, as defined in Figure 8.11. His general solution can be reduced to one that is a function of r alone if an *average drawdown* is considered. His complex analytical solution can be represented in simplified form as

$$h_0 - h = \frac{Q}{4\pi T} W(u_A, u_B, \eta) \tag{8.12}$$

where $W(u_A, u_B, \eta)$ is known as the *unconfined well function* and $\eta = r^2/b^2$. Figure 8.12 is a plot of this function for various values of η. The type A curves that grow out of the left-hand Theis curve of Figure 8.12, and that are followed at early time, are given by

$$h_0 - h = \frac{Q}{4\pi T} W(u_A, \eta) \tag{8.13}$$

where

$$u_A = \frac{r^2 S}{4Tt}$$

and S is the elastic storativity responsible for the instantaneous release of water to the well. The type B curves that are asymptotic to the right-hand Theis curve of Figure 8.12, and that are followed at later time, are given by

$$h_0 - h = \frac{Q}{4\pi T} W(u_B, \eta) \tag{8.14}$$

where

$$u_B = \frac{r^2 S_y}{4Tt}$$

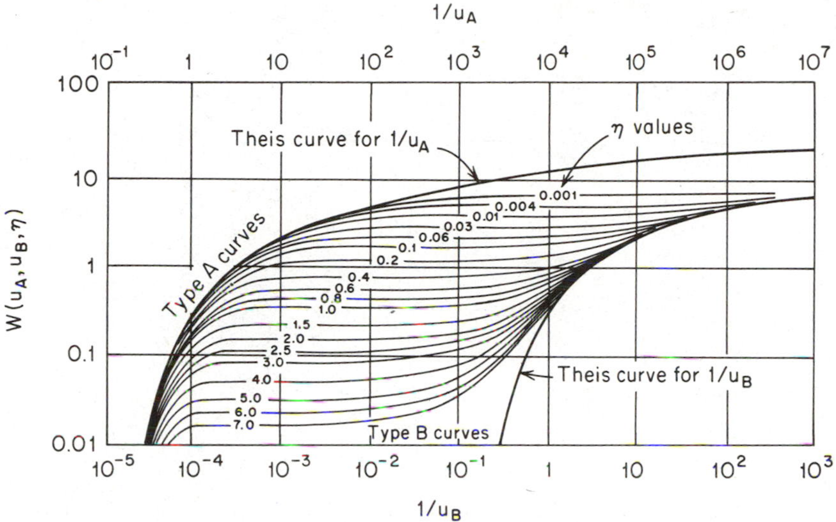

Figure 8.12 Theoretical curves of $W(u_A, u_B, \eta)$ versus $1/u_A$ and $1/u_B$ for an unconfined aquifer (after Neuman, 1975a).

and S_y is the specific yield that is responsible for the delayed release of water to the well.

For an anisotropic aquifer with horizontal hydraulic conductivity K_r and vertical hydraulic conductivity K_z, the parameter η is given by

$$\eta = \frac{r^2 K_z}{b^2 K_r} \tag{8.15}$$

If the aquifer is isotropic, $K_z = K_r$, and $\eta = r^2/b^2$. The transmissivity T is defined as $T = K_r b$. Equations (8.12) through (8.15) are only valid if $S_y \gg S$ and $h_0 - h \ll b$.

The prediction of the average drawdown at any radial distance r from a pumping well at any time t can be obtained from Eqs. (8.13) through (8.15) given Q, S, S_y, K_r, K_z, and b.

Multiple-Well Systems, Stepped Pumping Rates, Well Recovery, and Partial Penetration

The drawdown in hydraulic head at any point in a confined aquifer in which more than one well is pumping is equal to the sum of the drawdowns that would arise from each of the wells independently. Figure 8.13 schematically displays the drawdown $h_0 - h$ at a point B situated between two pumping wells with pumping rates $Q_1 = Q_2$. If $Q_1 \neq Q_2$, the symmetry of the diagram about the plane $A - A'$ would be lost but the principles remain the same.

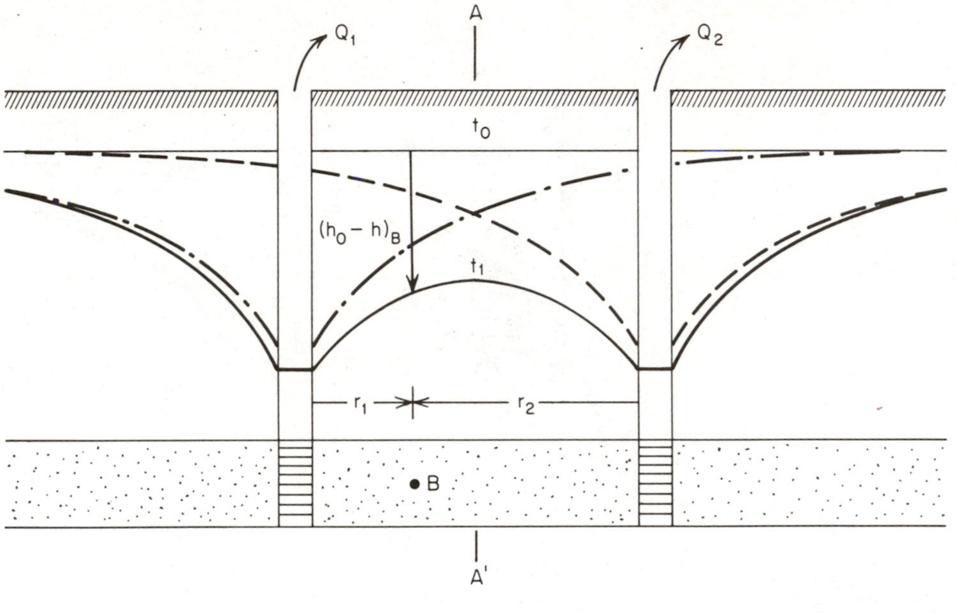

Figure 8.13 Drawdown in the potentiometric surface of a confined aquifer being pumped by two wells with $Q_1 = Q_2$.

For a system of n wells pumping at rates $Q_1, Q_2, \ldots, Q_n$, the arithmetic summation of the Theis solutions leads to the following predictive equation for the drawdown at a point whose radial distance from each well is given by $r_1, r_2, \ldots, r_n$

$$h_0 - h = \frac{Q_1}{4\pi T}W(u_1) + \frac{Q_2}{4\pi T}W(u_2) + \cdots + \frac{Q_n}{4\pi T}W(u_n) \tag{8.16}$$

where

$$u_i = \frac{r_i^2 S}{4Tt_i} \qquad i = 1, 2, \ldots, n$$

and t_i is the time since pumping started at the well whose discharge is Q_i.

The summation of component drawdowns outlined above is an application of the principle of superposition of solutions. This approach is valid because the equation of flow [Eq. (8.1)] for transient flow in a confined aquifer is linear (i.e., there are no cross terms of the form $\partial h/\partial r \cdot \partial h/\partial t$). Another application of the principle of superposition is in the case of a single well that is pumped at an initial rate Q_0 and then increased to the rates $Q_1, Q_2, \ldots, Q_m$ in a stepwise fashion by the additions $\Delta Q_1, \Delta Q_2, \ldots, \Delta Q_m$. Drawdown at a radial distance r from the pumping well is given by

$$h_0 - h = \frac{Q_0}{4\pi T}W(u_0) + \frac{\Delta Q_1}{4\pi T}W(u_1) + \cdots + \frac{\Delta Q_m}{4\pi T}W(u_m) \tag{8.17}$$

where

$$u_j = \frac{r^2 S}{4Tt_j} \qquad j = 0, 1, 2, \ldots, m$$

and t_j is the time since the start of the pumping rate Q_j.

A third application of the superposition principle is in the recovery of a well after pumping has stopped. If t is the time since the start of pumping and t' is the time since shutdown, then the drawdown at a radial distance r from the well is given by

$$h_0 - h = \frac{Q}{4\pi T}[W(u_1) - W(u_2)] \tag{8.18}$$

where

$$u_1 = \frac{r^2 S}{4Tt} \quad \text{and} \quad u_2 = \frac{r^2 S}{4Tt'}$$

Figure 8.14 schematically displays the drawdowns that occur during the pumping period and the residual drawdowns that remain during the recovery period.

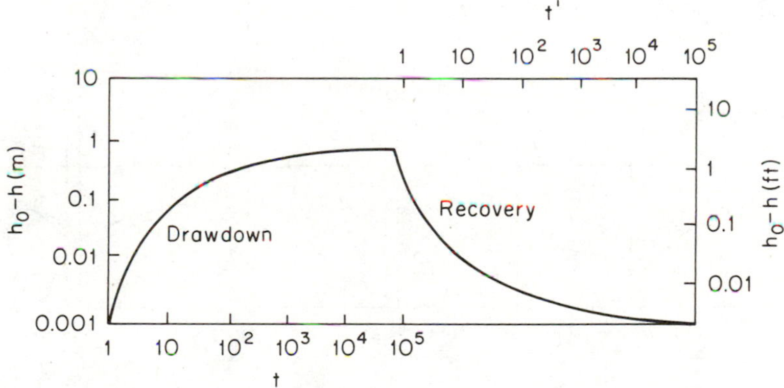

Figure 8.14 Schematic diagram of the recovery in hydraulic head in an aquifer after pumping is stopped.

It is not always possible, or necessarily desirable, to design a well that fully penetrates the aquifer under development. This is particularly true for unconfined aquifers, but may also be the case for thick confined aquifers. Even for wells that are fully penetrating, screens may be set over only a portion of the aquifer thickness.

Partial penetration creates vertical flow gradients in the vicinity of the well that render the predictive solutions developed for full penetration inaccurate. Hantush (1962) presented adaptations to the Theis solution for partially penetrating wells, and Hantush (1964) reviewed these solutions for both confined and leaky-confined aquifers. Dagan (1967), Kipp (1973), and Neuman (1974) considered the effects of partial penetration in unconfined aquifers.

Bounded Aquifers

When a confined aquifer is bounded on one side by a straight-line impermeable boundary, drawdowns due to pumping will be greater near the boundary [Figure 8.15(a)] than those that would be predicted on the basis of the Theis equation for an aquifer of infinite areal extent. In order to predict head drawdowns in such systems, the method of images, which is widely used in heat-flow theory, has been adapted for application in the groundwater milieu (Ferris et al., 1962). With this approach, the real bounded system is replaced for the purposes of analysis by an

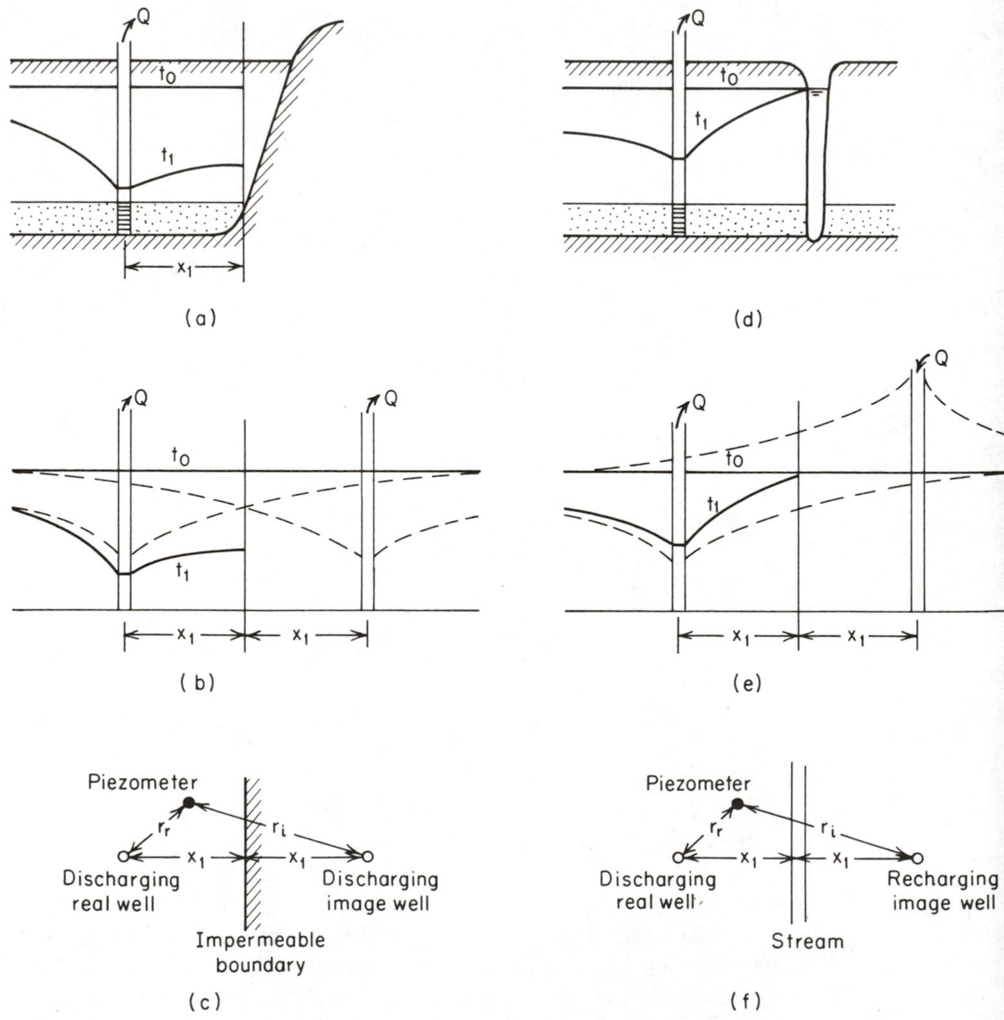

Figure 8.15 (a) Drawdown in the potentiometric surface of a confined aquifer bounded by an impermeable boundary; (b) equivalent system of infinite extent; (c) plan view.

imaginary system of infinite areal extent [Figure 8.15(b)]. In this system there are two wells pumping: the real well on the left and an image well on the right. The image well pumps at a rate, Q, equal to the real well and is located at an equal distance, x_1, from the boundary. If we sum the two component drawdowns in the infinite system (in identical fashion to the two-well case shown in Figure 8.13), it becomes clear that this pumping geometry creates an imaginary impermeable boundary (i.e., a boundary across which there is no flow) in the infinite system at the exact position of the real impermeable boundary in the bounded system. With reference to Figure 8.15(c), the drawdown in an aquifer bounded by an impermeable boundary is given by

$$h_0 - h = \frac{Q}{4\pi T}[W(u_r) + W(u_i)] \tag{8.19}$$

where

$$u_r = \frac{r_r^2 S}{4Tt} \quad \text{and} \quad u_i = \frac{r_i^2 S}{4Tt}$$

One can use the same approach to predict the decreased drawdowns that occur in a confined aquifer in the vicinity of a constant-head boundary, such as would be produced by the slightly unrealistic case of a fully penetrating stream [Figure 8.15(d)]. For this case, the imaginary infinite system [Figure 8.15(e)] includes the discharging real well and a recharging image well. The summation of the cone of depression from the pumping well and the cone of impression from the recharge well leads to an expression for the drawdown in an aquifer bounded by a constant-head boundary:

$$h_0 - h = \frac{Q}{4\pi T}[W(u_r) - W(u_i)] \tag{8.20}$$

where u_r and u_i are as defined in connection with Eq. (8.19).

It is possible to use the image well approach to provide predictions of drawdown in systems with more than one boundary. Ferris et al. (1962) discuss several geometric configurations. One of the more realistic (Figure 8.16) applies to a pumping well in a confined alluvial aquifer in a more-or-less straight river valley. For this case, the imaginary infinite system must include the real pumping well R, an image well I_1 equidistant from the left-hand impermeable boundary, and an image well I_2 equidistant from the right-hand impermeable boundary. These image wells themselves give birth to the need for further image wells. For example, I_3 reflects the effect of I_2 across the left-hand boundary, and I_4 reflects the effect of I_1 across the right-hand boundary. The result is a sequence of imaginary pumping wells stretching to infinity in each direction. The drawdown at point P in Figure 8.16 is the sum of the effects of this infinite array of wells. In practice, image wells need only be added until the most remote pair produces a negligible effect on water-level response (Bostock, 1971).

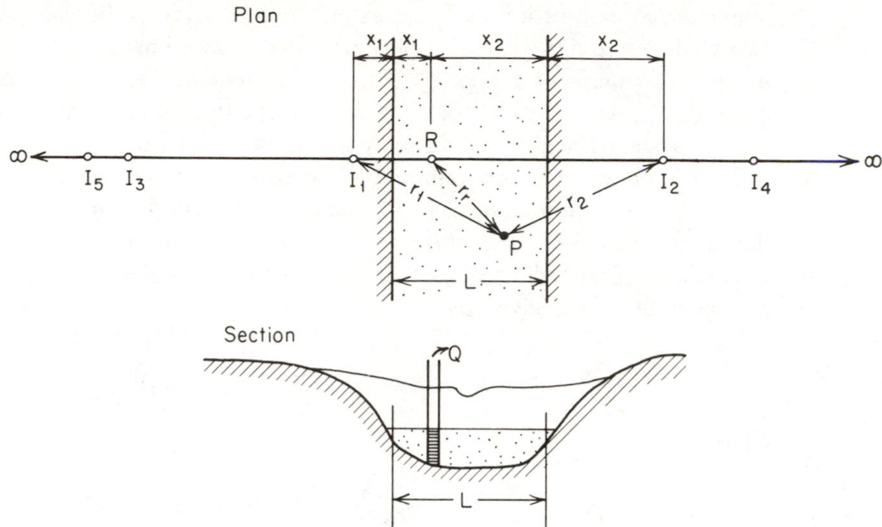

Figure 8.16 Image-well system for pumpage from a confined aquifer in a river valley bounded by impermeable boundaries.

The Response of Ideal Aquitards

The most common geological occurrence of exploitable confined aquifers is in sedimentary systems of interbedded aquifers and aquitards. In many cases the aquitards are much thicker than the aquifers and although their permeabilities are low, their storage capacities can be very high. In the very early pumping history of a production well, most of the water comes from the depressurization of the aquifer in which the well is completed. As time proceeds the leakage properties of the aquitards are brought into play and at later times the majority of the water being produced by the well is aquitard leakage. In many aquifer-aquitard systems, the *aquitards provide* the water and the *aquifers transmit* it to the wells. It is thus of considerable interest to be able to predict the response of aquitards as well as aquifers.

In the earlier discussion of leaky aquifers, two theories were introduced: the Hantush-Jacob theory, which utilizes the $W(u, r/B)$ curves of Figure 8.8, and the Neuman-Witherspoon theory, which utilizes the $W(u, r/B_{11}, r/B_{21}, \beta_{11}, \beta_{21})$ curves of Figure 8.9. In that the Hantush-Jacob theory does not include the storage properties of the aquitard, it is not suitable for the prediction of aquitard response. The Neuman-Witherspoon solution, in the form of Eq. (8.11) can be used to predict the hydraulic head $h(r, z, t)$ at any elevation z in the aquitard (Figure 8.7) at any time t, at any radial distance r, from the well. In many cases, however, it may be quite satisfactory to use a simpler approach. If the hydraulic conductivity of the aquitards is at least 2 orders of magnitude less than the hydraulic conductivity in the aquifers, it can be assumed that flow in the aquifers is horizontal and leakage in the aquitards is vertical. If one can predict, or has measurements of, $h(r, t)$ at some point in an aquifer, one can often predict the hydraulic head $h(z, t)$

at an overlying point in the aquitard by the application of a one-dimensional flow theory, developed by Karl Terzaghi, the founder of modern soil mechanics.

Consider an aquitard of thickness b' (Figure 8.17) sandwiched between two producing aquifers. If the initial condition is a constant hydraulic head $h = h_0$ in the aquitard, and if the drawdowns in hydraulic head in the adjacent aquifers can be represented as an instantaneous step function Δh, the system can be represented by the following one-dimensional boundary-value problem.

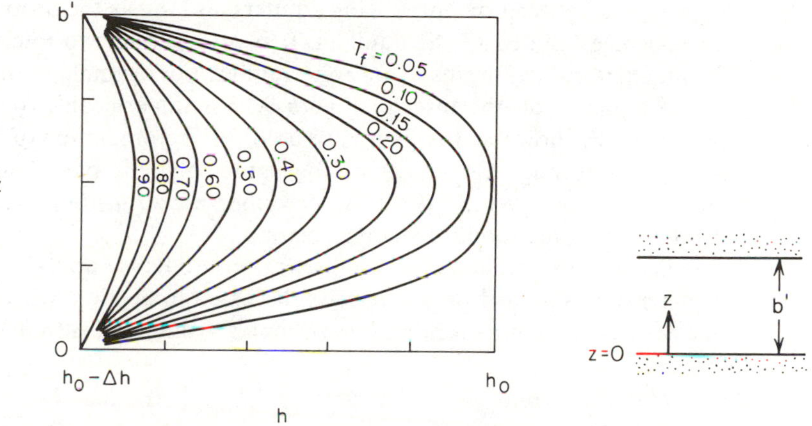

Figure 8.17 Response of an ideal aquitard to a step drawdown in head in the two adjacent aquifers.

From Eq. (2.76), the one-dimensional form of the flow equation is

$$\frac{\partial^2 h}{\partial z^2} = \frac{\rho g(\alpha' + n'\beta)}{K'} \frac{\partial h}{\partial t} \tag{8.21}$$

where the primed parameters are the aquitard properties. The initial condition is

$$h(z, 0) = h_0$$

and the boundary conditions are

$$h(0, t) = h_0 - \Delta h$$

$$h(b', t) = h_0 - \Delta h$$

Terzaghi (1925) provided an analytical solution to this boundary-value problem. He noted that for clays $n'\beta \ll \alpha'$ in Eq. (8.21). He grouped the remaining aquitard parameters into a single parameter c_v, known as the *coefficient of consolidation* and defined as

$$c_v = \frac{K'}{\rho g \alpha'} \tag{8.22}$$

He further defined the dimensionless time factor, T_f, as

$$T_f = \frac{4c_v t}{(b')^2} \tag{8.23}$$

Given the aquitard parameter c_v and the geometric parameter b', one can calculate T_f for any time t.

Figure 8.17 is a graphical presentation of Terzaghi's solution $h(z, T_f)$. It allows the prediction of the hydraulic head at any elevation z at any time t in an aquitard sandwiched between two producing aquifers, as long as the drop in hydraulic head Δh can be estimated in the aquifers. It is also possible to interpret this solution for an aquitard that drains to only one aquifer. For example, if the lower boundary of the aquitard on the inset to Figure 8.17 is impermeable, only the upper half of the curves shown in the figure are used for the prediction of $h(z, t)$. The $z = 0$ line passes through the center of the figure, and the parameters c_v and T_f are defined as above. Wolff (1970) has described a case history that utilizes the concepts of one-dimensional aquitard response.

Predictions of aquitard response, and the inverse application of this theory to determine aquitard parameters, as discussed in Section 8.6, are also important in assessing contaminant migration (Chapter 9) and land subsidence (Section 8.12).

The Real World

Each of the analytical solutions presented in this section describes the response to pumping in a very idealized representation of actual aquifer configurations. In the real world, aquifers are heterogeneous and anisotropic; they usually vary in thickness; and they certainly do not extend to infinity. Where they are bounded, it is not by straight-line boundaries that provide perfect confinement. In the real world, aquifers are created by complex geologic processes that lead to irregular stratigraphy, interfingering of strata, and pinchouts and trendouts of both aquifers and aquitards. The predictions that can be carried out with the analytical expressions presented in this section must be viewed as best estimates. They have greater worth the more closely the actual hydrogeological environment approaches the idealized configuration.

In general, well-hydraulics equations are most applicable when the unit of study is a well or well field. They are less applicable on a larger scale, where the unit of study is an entire aquifer or a complete groundwater basin. Short-term yields around wells are very dependent on aquifer properties and well-field geometry, both of which are emphasized in the well-hydraulics equations. Long-term yields on an aquifer scale are more often controlled by the nature of the boundaries. Aquifer studies on the larger scale are usually carried out with the aid of models based on numerical simulation or electric-analog techniques. These approaches are discussed in Sections 8.8 and 8.9.

The predictive formulas developed in this section and the simulation techniques described in later sections allow one to calculate the drawdowns in hydraulic head that will occur in an aquifer in response to groundwater development through wells. They require as input either the three basic hydrogeological parameters: hydraulic conductivity, K, porosity, n, and compressibility, α; or the two derived aquifer parameters: transmissivity, T, and storativity, S. There is a wide variety of techniques that can be used to measure these parameters. In the next section, we will discuss *laboratory tests*; in Section 8.5, *piezometer tests*; and in Section 8.6, *pumping tests*. In Section 8.7, we will examine some *estimation techniques*, and in Section 8.8, the determination of aquifer parameters by *inverse simulation*. The formulas presented in this section are the basis for the pumping-test approach that is described in Section 8.6.

8.4 Measurement of Parameters: Laboratory Tests

The laboratory tests described in this section can be considered as providing point values of the basic hydrogeologic parameters. They are carried out on small samples that are collected during test-drilling programs or during the mapping of surficial deposits. If the samples are undisturbed core samples, the measured values should be representative of the *in situ* point values. For sands and gravels, even disturbed samples may yield useful values. We will describe testing methods for the determination of hydraulic conductivity, porosity, and compressibility in the saturated state; and we will provide references for the determination of the characteristic curves relating moisture content, pressure head, and hydraulic conductivity in the unsaturated state. We will emphasize principles; for a more complete description of each testing apparatus and more detailed directions on laboratory procedures, the reader is directed to the soil-testing manual by Lambe (1951), the permeability handbook of the American Society of Testing Materials (1967), or the pertinent articles in the compendium of soil analysis methods edited by Black (1965). Our discussions relate more to soils than to rocks, but the principles of measurement are the same. The rock mechanics text by Jaeger (1972) discusses rock-testing procedures.

Hydraulic Conductivity

The hydraulic conductivity, K, was defined in Section 2.1, and its relationship to the permeability, k, was explored in Section 2.3. The saturated hydraulic conductivity of a soil sample can be measured with two types of laboratory apparatus. The first type, known as a *constant-head permeameter*, is shown in Figure 8.18(a); the second type, a *falling-head permeameter*, is shown in Figure 8.18(b).

In a constant-head test, a soil sample of length L and cross-sectional area A is enclosed between two porous plates in a cylindrical tube, and a constant-head differential H is set up across the sample. A simple application of Darcy's law leads to the expression

$$K = \frac{QL}{AH} \tag{8.24}$$

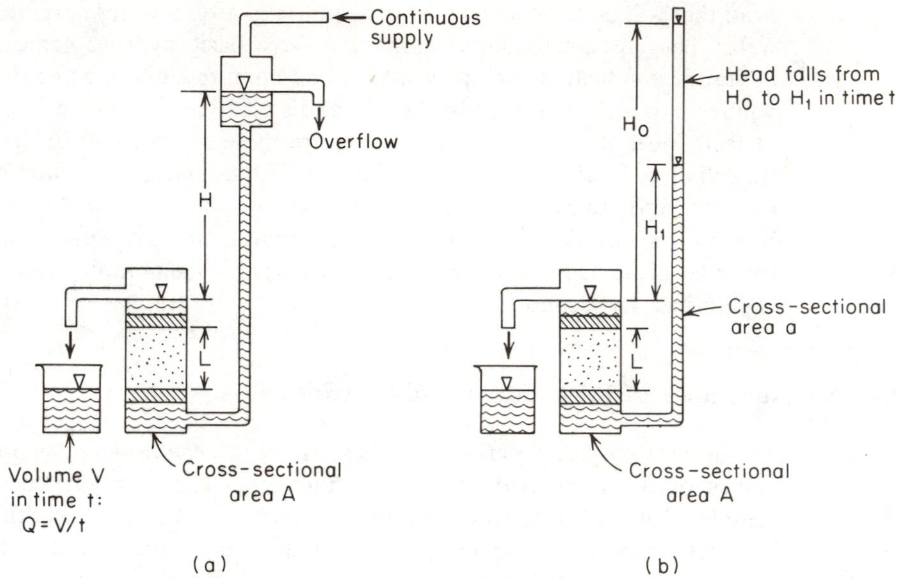

Figure 8.18 (a) Constant-head permeameter; (b) falling-head permeameter (after Todd, 1959).

where Q is the steady volumetric discharge through the system. It is important that no air become entrapped in the system, and for this reason it is wise to use deaired water. If disturbed samples are being tested in the permeameter, they should be carefully saturated from below as they are emplaced.

In a falling-head test [Figure 8.18(b)], the head, as measured in a tube of cross-sectional area a, is allowed to fall from H_0 to H_1 during time t. The hydraulic conductivity is calculated from

$$K = \frac{aL}{At} \ln \left(\frac{H_0}{H_1}\right) \tag{8.25}$$

This equation can be derived (Todd, 1959) from the simple boundary-value problem that describes one-dimensional transient flow across the soil sample. In order that the head decline be easily measurable in a finite time period, it is necessary to choose the standpipe diameter with regard to the soil being tested. Lambe (1951) suggests that for a coarse sand a standpipe whose diameter is approximately equal to that of the permeameter is usually satisfactory, whereas a fine silt may necessitate a standpipe whose diameter is one-tenth the permeameter diameter. Lambe also suggests that the point $\sqrt{H_0 H_1}$ be marked on the standpipe. If the time required for the head decline from H_0 to $\sqrt{H_0 H_1}$ is not equal to that for the decline from $\sqrt{H_0 H_1}$ to H_1, the test has not functioned correctly and a check should be made for leaks or entrapped air.

Klute (1965a) notes that the constant-head system is best suited to samples with conductivities greater than 0.01 cm/min while the falling-head system is best suited to samples with lower conductivity. He also notes that elaborate, painstaking measurements are not generally required for conductivity determinations on field samples. The variability among samples is usually large enough that precise determination of the conductivity of a given sample is not warranted.

For clayey materials the hydraulic conductivity is commonly determined from a consolidation test, which is described in the subsection on compressibility below.

Porosity

In principle, the porosity, n, as defined in Section 2.5, would be most easily measured by saturating a sample, measuring its volume, V_T, weighing it and then oven drying it to constant weight at 105°C. The weight of water removed could be converted to a volume, knowing the density of water. This volume is equivalent to the volume of the void space, V_v; and the porosity could be calculated from $n = V_v/V_T$.

In practice, it is quite difficult to exactly and completely saturate a sample of given volume. It is more usual (Vomocil, 1965) to make use of the relationship

$$n = 1 - \frac{\rho_b}{\rho_s} \tag{8.26}$$

which can be developed by simple arithmetic manipulation of the basic definition of porosity. In Eq. (8.26), ρ_b is the *bulk mass density* of the sample and ρ_s is the *particle mass density*. The bulk density is the oven-dried mass of the sample divided by its field volume. The particle density is the oven-dried mass divided by the volume of the solid particles, as determined by a water-displacement test. In cases where great accuracy is not required, $\rho_s = 2.65$ g/cm³ can be assumed for most mineral soils.

Compressibility

The compressibility of a porous medium was defined in Section 2.9 with the aid of Figure 2.19. It is a measure of the relative volumetric reduction that will take place in a soil under an increased effective stress. Compressibility is measured in a consolidation apparatus of the kind commonly used by soils engineers. In this test, a soil sample is placed in a loading cell of the type shown schematically in Figure 2.19(a). A load L is applied to the cell, creating a stress σ, where $\sigma = L/A$, A being the cross-sectional area of the sample. If the soil sample is saturated and the fluid pressure on the boundaries of the sample is atmospheric (i.e., the sample is free-draining), the effective stress, σ_e, which leads to consolidation of the sample, is equal to the applied stress, σ.

The reduction in sample thickness, b, is measured after equilibrium is achieved at each of several loading increments, and the results are converted into a graph of void ratio, e, versus effective stress, σ_e, as shown in Figure 2.19(b). The compressibility, α, is determined from the slope of such a plot by

$$\alpha = -\frac{de/(1 + e_0)}{d\sigma_e} \tag{8.27}$$

where e_0 is the initial void ratio prior to loading. As noted in Section 2.9, α is a function of the applied stress and it is dependent on the previous loading history.

Lambe (1951) describes the details of the testing procedure. The most common loading method is a lever system on which weights of known magnitude are hung. There are two types of loading cell in common use. In the *fixed-ring* container [Figure 8.19(a)], all the sample movement relative to the container is downward. In the *floating-ring* container [Figure 8.19(b)], compression occurs toward the middle from both top and bottom. In the floating-ring container, the effect of friction between the container wall and the soil specimen is smaller than in the fixed-ring container. In practice, it is difficult to determine the magnitude of the friction in any case, and because its effect is thought to be minor, it is normally neglected. Cohesionless sands are usually tested as disturbed samples. Cohesive clays must be carefully trimmed to fit the consolidometer ring.

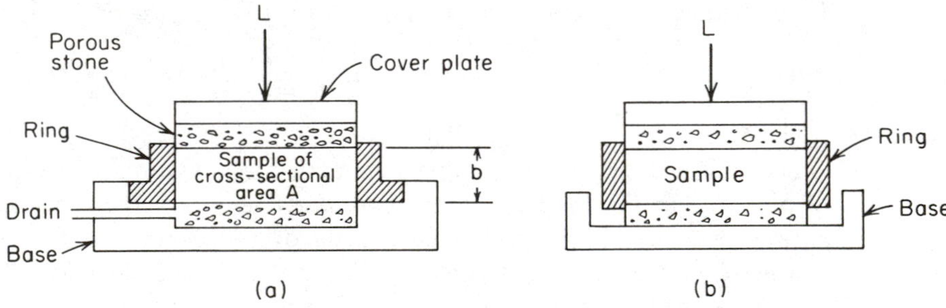

Figure 8.19 (a) Fixed-ring consolidometer; (b) floating-ring consolidometer (after Lambe, 1951).

In soil mechanics terminology, the slope of the $e - \sigma_e$ curve is called the *coefficient of compressibility*, a_v. The relationship between a_v and α is easily seen to be

$$a_v = \frac{-de}{d\sigma_e} = (1 + e_0)\alpha \tag{8.28}$$

More commonly, soils engineers plot the void ratio, e, against the logarithm of σ_e. When plotted in this manner, there is usually a significant portion of the curve that is a straight line. The slope of this line is called the *compression index*, C_c, where

$$C_c = \frac{-\,de}{d(\log \sigma_e)} \tag{8.29}$$

In most civil engineering applications the *rate* of consolidation is just as important as the *amount* of consolidation. This rate is dependent both on the compressibility, α, and the hydraulic conductivity, K. As noted in connection with Eq. (8.22), soils engineers utilize a grouped parameter known as the *coefficient of consolidation*, c_v, which is defined as

$$c_v = \frac{K}{\rho g \alpha} \tag{8.30}$$

At each loading level in a consolidation test, the sample undergoes a transient drainage process (fast for sands, slow for clays) that controls the rate of consolidation of the sample. If the rate of decline in sample thickness is recorded for each loading increment, such measurements can be used in the manner described by Lambe (1951) to determine the coefficient of consolidation, c_v, and the hydraulic conductivity, K, of the soil.

In Section 8.12, we will further examine the mechanism of one-dimensional consolidation in connection with the analysis of land subsidence.

Unsaturated Characteristic Curves

The characteristic curves, $K(\psi)$ and $\theta(\psi)$, that relate the moisture content, θ, and the hydraulic conductivity, K, to the pressure head, ψ, in unsaturated soils were described in Section 2.6. Figure 2.13 provided a visual example of the hysteretic relationships that are commonly observed. The methods used for the laboratory determination of these curves have been developed exclusively by soil scientists. It is not within the scope of this text to outline the wide variety of sophisticated laboratory instrumentation that is available. Rather, the reader is directed to the soil science literature, in particular to the review articles by L. A. Richards (1965), Klute (1965b), Klute (1965c), and Bouwer and Jackson (1974).

8.5 Measurement of Parameters: Piezometer Tests

It is possible to determine *in situ* hydraulic conductivity values by means of tests carried out in a single piezometer. We will look at two such tests, one suitable for point piezometers that are open only over a short interval at their base, and one suitable for screened or slotted piezometers that are open over the entire thickness of a confined aquifer. Both tests are initiated by causing an instantaneous change in the water level in a piezometer through a sudden introduction or removal of a known volume of water. The recovery of the water level with time is then observed. When water is removed, the tests are often called *bail tests*; when it is added, they are known as *slug tests*. It is also possible to create the same effect by suddenly introducing or removing a solid cylinder of known volume.

The method of interpreting the water level versus time data that arise from bail tests or slug tests depends on which of the two test configurations is felt to be most representative. The method of Hvorslev (1951) is for a point piezometer, while that of Cooper et al. (1967) is for a confined aquifer. We will now describe each in turn.

The simplest interpretation of piezometer-recovery data is that of Hvorslev (1951). His initial analysis assumed a homogeneous, isotropic, infinite medium in which both soil and water are incompressible. With reference to the bail test of Figure 8.20(a), Hvorslev reasoned that the rate of inflow, q, at the piezometer tip at any time t is proportional to the hydraulic conductivity, K, of the soil and to the unrecovered head difference, $H - h$, so that

$$q(t) = \pi r^2 \frac{dh}{dt} = FK(H - h) \tag{8.31}$$

where F is a factor that depends on the shape and dimensions of the piezometer intake. If $q = q_0$ at $t = 0$, it is clear that $q(t)$ will decrease asymptotically toward zero as time goes on.

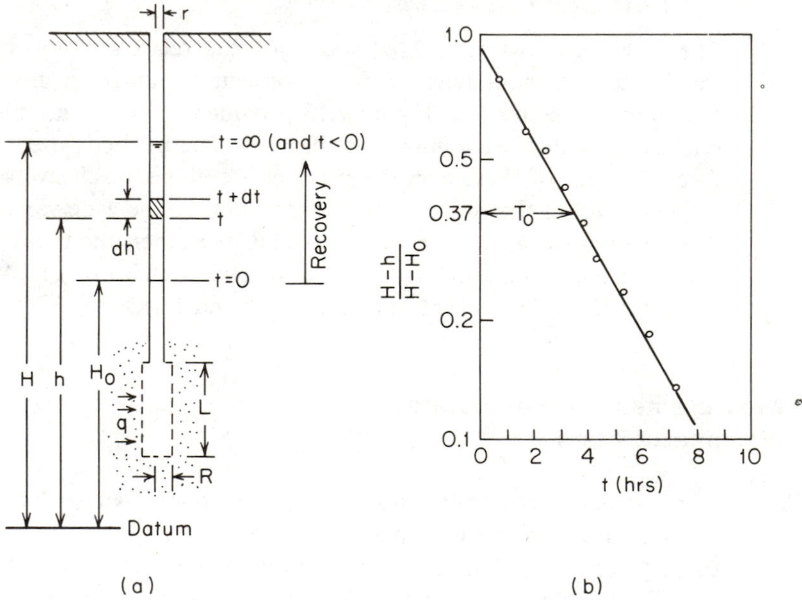

Figure 8.20 Hvorslev piezometer test. (a) Geometry; (b) method of analysis.

Hvorslev defined the *basic time lag*, T_0, as

$$T_0 = \frac{\pi r^2}{FK} \tag{8.32}$$

When this parameter is substituted in Eq. (8.31), the solution to the resulting ordinary differential equation, with the initial condition, $h = H_0$ at $t = 0$, is

$$\frac{H - h}{H - H_0} = e^{-t/T_0} \tag{8.33}$$

A plot of field recovery data, $H - h$ versus t, should therefore show an exponential decline in recovery rate with time. If, as shown on Figure 8.20(b), the recovery is normalized to $H - H_0$ and plotted on a logarithmic scale, a straight-line plot results. Note that for $H - h/H - H_0 = 0.37$, $\ln (H - h/H - H_0) = -1$, and from Eq. (8.33), $T_0 = t$. The basic time lag, T_0, can be defined by this relation; or if a more physical definition is desired, it can be seen, by multiplying both top and bottom of Eq. (8.32) by $H - H_0$, that T_0 is the time that would be required for the complete equalization of the head difference if the original rate of inflow were maintained. That is, $T_0 = V/q_0$, where V is the volume of water removed or added.

To interpret a set of field recovery data, the data are plotted in the form of Figure 8.20(b). The value of T_0 is measured graphically, and K is determined from Eq. (8.32). For a piezometer intake of length L and radius R [Figure 8.20(a)], with $L/R > 8$, Hvorslev (1951) has evaluated the shape factor, F. The resulting expression for K is

$$K = \frac{r^2 \ln (L/R)}{2LT_0} \tag{8.34}$$

Hvorslev also presents formulas for anisotropic conditions and for a wide variety of shape factors that treat such cases as a piezometer open only at its basal cross section and a piezometer that just encounters a permeable formation underlying an impermeable one. Cedergren (1967) also lists these formulas.

In the field or agricultural hydrology, several *in situ* techniques, similar in principle to the Hvorslev method but differing in detail, have been developed for the measurement of saturated hydraulic conductivity. Boersma (1965) and Bouwer and Jackson (1974) review those methods that involve auger holes and piezometers.

For bail tests of slug tests run in piezometers that are open over the entire thickness of a confined aquifer, Cooper et al. (1967) and Papadopoulos et al. (1973) have evolved a test-interpretation procedure. Their analysis is subject to the same assumptions as the Theis solution for pumpage from a confined aquifer. Contrary to the Hvorslev method of analysis, it includes consideration of both formation and water compressibilities. It utilizes a curve-matching procedure to determine the aquifer coefficients T and S. The hydraulic conductivity K can then be determined on the basis of the relation, $K = T/b$. Like the Theis solution, the method is based on the solution to a boundary-value problem that involves the transient equation of groundwater flow, Eq. (2.77). The mathematics will not be described here.

For the bail-test geometry shown in Figure 8.21(a), the method involves the preparation of a plot of recovery data in the form $H - h/H - H_0$ versus t. The plot is prepared on semilogarithmic paper with the reverse format to that of the Hvorslev test; the $H - h/H - H_0$ scale is linear, while the t scale is logarithmic. The field curve is then superimposed on the type curves shown in Figure 8.21(b). With the axes coincident, the data plot is translated horizontally into a position where the data best fit one of the type curves. A matchpoint is chosen (or rather, a vertical axis is matched) and values of t and W are read off the horizontal scales

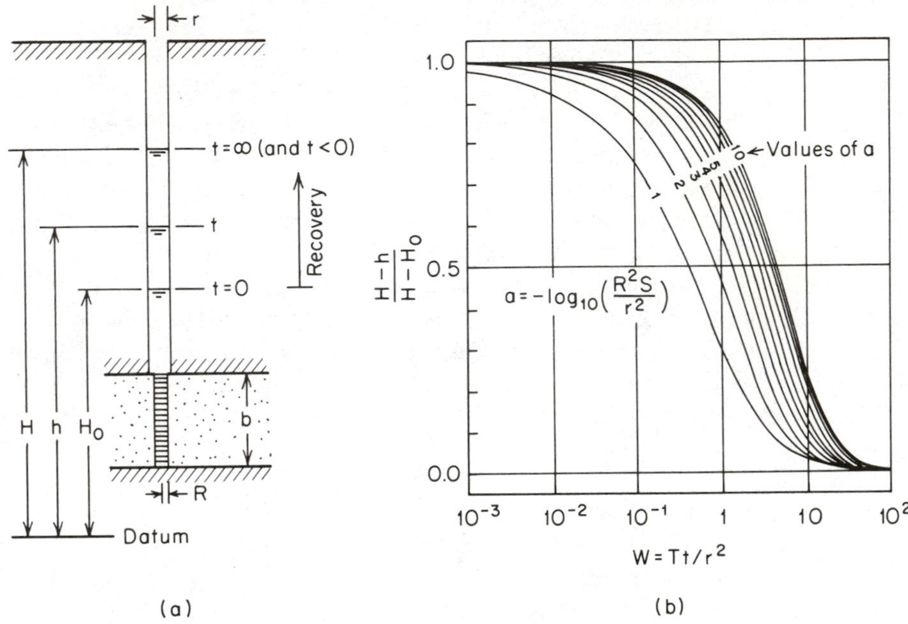

Figure 8.21 Piezometer test in a confined aquifer. (a) Geometry; (b) type curves (after Papadopoulos et al., 1973).

at the matched axis of the field plot and the type plot, respectively. For ease of calculation it is common to choose a matched axis at $W = 1.0$. The transmissivity T is then given by

$$T = \frac{Wr^2}{t} \tag{8.35}$$

where the parameters are expressed in any consistent set of units.

In principle, the storativity, S, can be determined from the a value of the matched curve and the expression shown on Figure 8.21(b). In practice, since the slopes of the various a lines are very similar, the determination of S by this method is unreliable.

The main limitation on slug tests and bail tests is that they are heavily dependent on a high-quality piezometer intake. If the wellpoint or screen is corroded or clogged, measured values may be highly inaccurate. On the other hand, if a piezometer is developed by surging or backwashing prior to testing, the measured values may reflect the increased conductivities in the artificially induced gravel pack around the intake.

It is also possible to determine hydraulic conductivity in a piezometer or single well by the introduction of a tracer into the well bore. The tracer concentration decreases with time under the influence of the natural hydraulic gradient that exists in the vicinity of the well. This approach is known as the *borehole dilution method*, and it is described more fully in Section 9.4.

8.6 Measurement of Parameters: Pumping Tests

In this section, a method of parameter measurement that is specifically suited to the determination of transmissivity and storativity in confined and unconfined aquifers will be described. Whereas laboratory tests provide point values of the hydrogeological parameters, and piezometer tests provide *in situ* values representative of a small volume of porous media in the immediate vicinity of a piezometer tip, pumping tests provide *in situ* measurements that are averaged over a large aquifer volume.

The determination of T and S from a pumping test involves a direct application of the formulas developed in Section 8.3. There, it was shown that for a given pumping rate, if T and S are known, it is possible to calculate the time rate of drawdown, $h_0 - h$ versus t, at any point in an aquifer. Since this response depends solely on the values of T and S, it should be possible to take measurements of $h_0 - h$ versus t at some observational point in an aquifer and work backward through the equations to determine the values of T and S.

The usual course of events during the initial exploitation of an aquifer involves (1) the drilling of a test well with one or more observational piezometers, (2) a short-term pumping test to determine the values of T and S, and (3) application of the predictive formulas of Section 8.3, using the T and S values determined in the pumping test, to design a production well or wells that will fulfill the pumpage requirements of the project without leading to excessive long-term drawdowns. The question of what constitutes an "excessive" drawdown and how drawdowns and well yields are related to groundwater recharge rates and the natural hydrologic cycle are discussed in Section 8.10.

Let us now examine the methodology of pumping-test interpretation in more detail. There are two methods that are in common usage for calulating aquifer coefficients from time-drawdown data. Both approaches are graphical. The first involves curve matching on a log-log plot (the *Theis method*), and the second involves interpretations with a semilog plot (the *Jacob method*).

Log-Log Type-Curve Matching

Let us first consider data taken from an aquifer in which the geometry approaches that of the idealized Theis configuration. As was explained in connection with Figure 8.5, the time-drawdown response in an observational piezometer in such an aquifer will always have the shape of the Theis curve, regardless of the values of T and S in the aquifer. However, for high T a measurable drawdown will reach the observation point faster than for low T, and the drawdown data will begin to march up the Theis curve sooner. Theis (1935) suggested the following graphical procedure to exploit this curve matching property:

1. Plot the function $W(u)$ versus $1/u$ on log-log paper. (Such a plot of dimensionless theoretical response is known as a *type curve*.)

2. Plot the measured time-drawdown values, $h_0 - h$ versus t, on log-log paper of the same size and scale as the $W(u)$ versus $1/u$ curve.
3. Superimpose the field curve on the type curve keeping the coordinate axes parallel. Adjust the curves until most of the observed data points fall on the type curve.
4. Select an arbitrary match point and read off the paired values of $W(u)$, $1/u$, $h_0 - h$, and t at the match point. Calculate u from $1/u$.
5. Using these values, together with the pumping rate Q and the radial distance r from well to piezometer, calculate T from the relationship

$$T = \frac{QW(u)}{4\pi(h_0 - h)} \tag{8.36}$$

6. Calculate S from the relationship

$$S = \frac{4uTt}{r^2} \tag{8.37}$$

Equations (8.36) and (8.37) follow directly from Eqs. (8.7) and (8.6). They are valid for any consistent system of units. Some authors prefer to present the equations in the form

$$T = \frac{AQW(u)}{h_0 - h} \tag{8.38}$$

$$S = \frac{uTt}{Br^2} \tag{8.39}$$

where the coefficients A and B are dependent on the units used for the various parameters. For SI units, with $h_0 - h$ and r measured in meters, t in seconds, Q in m^3/s, and T in m^2/s, $A = 0.08$ and $B = 0.25$. For the inconsistent set of practical units widely used in North America, with $h_0 - h$ and r measured in feet, t in days, Q in U.S. gal/min, and T in U.S. gal/day/ft, $A = 114.6$ and $B = 1.87$. For Q and T in terms of Imperial gallons, A remains unchanged and $B = 1.56$.

Figure 8.22 illustrates the curve-matching procedure and calculations for a set of field data. The alert reader will recognize these data as being identical to the calculated data originally presented in Figure 8.5(b). It would probably be intuitively clearer if the match point were taken at some point on the coincident portions of the superimposed curves. However, a few quick calculations should convince doubters that it is equally valid to take the matchpoint anywhere on the overlapping fields once they have been fixed in their correct relative positions. For ease of calculation, the matchpoint is often taken at $W(u) = 1.0$, $u = 1.0$.

The log-log curve-matching technique can also be used for leaky aquifers (Walton, 1962) and unconfined aquifers (Prickett, 1965; Neuman, 1975a). Figure 8.23 provides a comparative review of the geometry of these systems and the types of $h_0 - h$ versus t data that should be expected in an observational piezometer

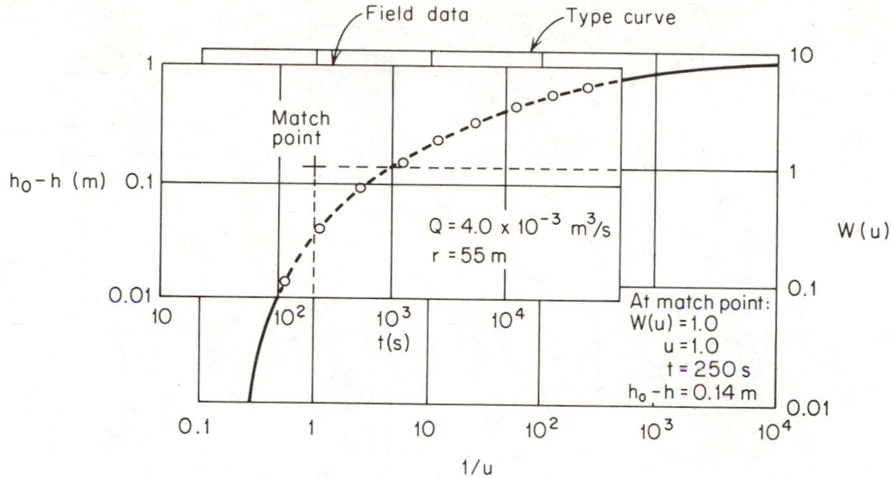

$$T = \frac{QW(u)}{4\pi(h_0 - h)} = \frac{(4.0 \times 10^{-3})(1.0)}{(4.0)(3.14)(0.14)} = 0.0023\ \text{m}^2/\text{s}\quad (15{,}700\ \text{U.S. gal/day/ft})$$

$$S = \frac{4uTt}{r^2} = \frac{(4.0)(1.0)(0.0023)(250)}{(55.0)^2} = 7.5 \times 10^{-4}$$

Figure 8.22 Determination of T and S from $h_0 - h$ versus t data using the log-log curve-matching procedure and the $W(u)$ versus $1/u$-type curve.

in each case. Sometimes time-drawdown data unexpectedly display one of these forms, thus indicating a geological configuration that has gone unrecognized during the exploration stage of aquifer evaluation.

For leaky aquifers the time-drawdown data can be matched against the leaky type curves of Figure 8.8. The r/B value of the matched curve, together with the matchpoint values of $W(u, r/B)$, u, $h_0 - h$, and t, can be substituted into Eqs. (8.6), (8.8), and (8.9) to yield the aquifer coefficients T and S. Because the development of the r/B solutions does not include consideration of aquitard storativity, an r/B curve matching approach is not suitable for the determination of the aquitard conductivity K'. As noted in the earlier subsection on aquitard response, there are many aquifer-aquitard configurations where the leakage properties of the aquitards are more important in determining long-term aquifer yields than the aquifer parameters themselves. In such cases it is necessary to design a pumping-test configuration with observational piezometers that bottom in the aquitards as well as in the aquifers. One can then use the pumping-test procedure outlined by Neuman and Witherspoon (1972), which utilizes their more general leaky-aquifer solution embodied in Eqs. (8.6), (8.10), and (8.11). They present a ratio method that obviates the necessity of matching field data to type curves as complex as

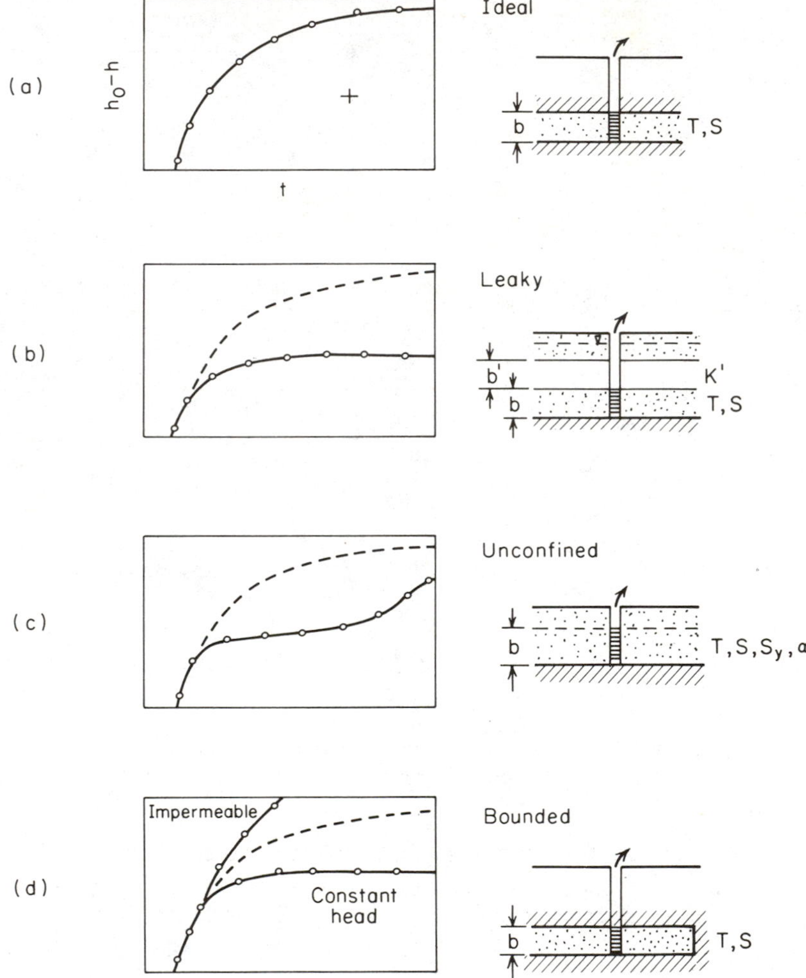

Figure 8.23 Comparison of log-log $h_0 - h$ versus t data for ideal, leaky, unconfined, and bounded systems.

those of Figure 8.9. The method only requires matching against the Theis curve, and calculations are relatively easy to carry out.

As an alternative approach (Wolff, 1970), one can simply read off a T_f value from Figure 8.17 given a hydraulic head value h measured in an aquitard piezometer at elevation z at time t. Knowing the aquitard thickness, b', one can solve Eq. (8.23) for c_v. If an α value can be estimated, Eq. (8.22) can be solved for K'.

For unconfined aquifers the time-drawdown data should be matched against the unconfined type curves of Figure 8.12. The η value of the matched curve, together with the match-point values of $W(u_A, u_B, \eta)$, u_A, u_B, $h_0 - h$, and t can be substituted into Eqs. (8.13) through (8.15) to yield the aquifer coefficients T, S, and S_y. Moench and Prickett (1972) discuss the interpretation of data at sites where lowered water levels cause a conversion from confined to unconfined conditions.

Figure 8.23(d) shows the type of log-log response that would be expected in the vicinity of an impermeable or constant-head boundary. However, bounded systems are more easily analyzed with the semilog approach that will now be described.

Semilog Plots

The semilog method of pump-test interpretation rests on the fact that the exponential integral, $W(u)$, in Eqs. (8.5) and (8.7) can be represented by an infinite series. The Theis solution then becomes

$$h_0 - h = \frac{Q}{4\pi T}\left(-0.5772 - \ln u + u - \frac{u^2}{2\cdot 2!} + \frac{u^3}{3\cdot 3!} + \cdots\right) \tag{8.40}$$

Cooper and Jacob (1946) noted that for small u the sum of the series beyond ln u becomes negligible, so that

$$h_0 - h = \frac{Q}{4\pi T}(-0.5772 - \ln u) \tag{8.41}$$

Substituting Eq. (8.6) for u, and noting that $\ln u = 2.3 \log u$, that $-\ln u = \ln 1/u$, and that $\ln 1.78 = 0.5772$, Eq. (8.41) becomes

$$h_0 - h = \frac{2.3Q}{4\pi T} \log \frac{2.25Tt}{r^2 S} \tag{8.42}$$

Since Q, r, T, and S are constants, it is clear that $h_0 - h$ versus $\log t$ should plot as a straight line.

Figure 8.24(a) shows the time-drawdown data of Figure 8.22 plotted on a semilog graph. If Δh is the drawdown for one log cycle of time and t_0 is the time intercept where the drawdown line intercepts the zero drawdown axis, it follows from further manipulation with Eq. (8.42) that the values of T and S, in consistent units, are given by

$$T = \frac{2.3Q}{4\pi\, \Delta h} \tag{8.43}$$

$$S = \frac{2.25Tt_0}{r^2} \tag{8.44}$$

As with the log-log methods, these equations can be reshaped as

$$T = \frac{CQ}{\Delta h} \tag{8.45}$$

$$S = \frac{DTt_0}{r^2} \tag{8.46}$$

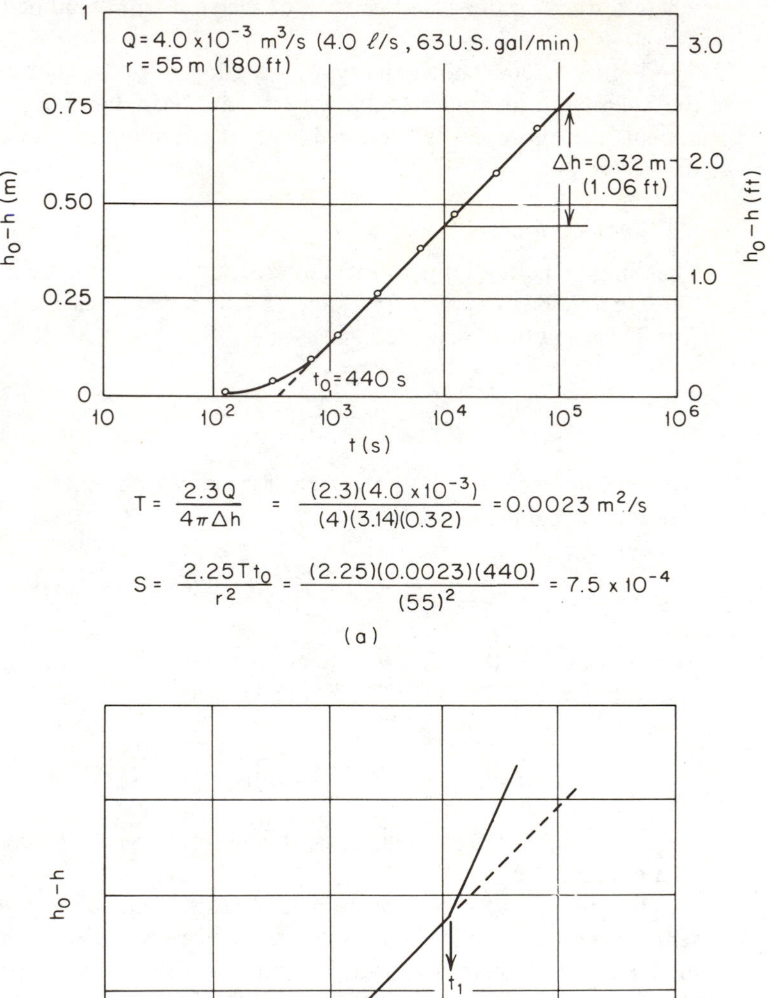

Figure 8.24 (a) Determination of T and S from $h_0 - h$ versus t data using the semilog method; (b) semilog plot in the vicinity of an impermeable boundary.

where C and D are coefficients that depend on the units used. For Δh and r in meters, t in seconds, Q in m³/s, and T in m²/s, $C = 0.18$ and $D = 2.25$. For Δh and r in feet, t in days, Q in U.S. gal/min, and T in U.S. gal/day/ft, $C = 264$ and $D = 0.3$. For Q and T in terms of Imperial gallons, $C = 264$ and $D = 0.36$.

Todd (1959) states that the semilog method is valid for $u < 0.01$. Examination of the definition of u [Eq. (8.6)] shows that this condition is most likely to be satisfied for piezometers at small r and large t.

The semilog method is very well suited to the analysis of bounded confined aquifers. As we have seen, the influence of a boundary is equivalent to that of a recharging or discharging image well. For the case of an impermeable boundary, for example, the effect of the additional imaginary pumping well is to double the slope of the $h_0 - h$ versus log t plot [Figure 8.24(b)]. The aquifer coefficients S and T should be calculated from Eqs. (8.43) and (8.44) on the earliest limb of the plot (before the influence of the boundary is felt). The time, t_1, at which the break in slope takes place can be used together with Eqs. (8.19) to calculate r_i, the distance from piezometer to image well [Figure 8.15(c)]. It takes records from three piezometers to unequivocally locate the position of the boundary if it is not known from geological evidence.

Advantages and Disadvantages of Pumping Tests

The determination of aquifer constants through pumping tests has become a standard step in the evaluation of groundwater resource potential. In practice, there is much art to successful pump testing and the interested reader is directed to Kruseman and de Ridder (1970) and Stallman (1971) for detailed advice on the design of pumping-test geometries, and to Walton's (1970) many case histories.

The advantages of the method are probably self-evident. A pumping test provides *in situ* parameter values, and these values are, in effect, averaged over a large and representative aquifer volume. One obtains information on both conductivity (through the relation $K = T/b$) and storage properties from a single test. In aquifer-aquitard systems it is possible to obtain information on the very important leakage properties of the system if observations are made in the aquitards as well as the aquifers.

There are two disadvantages, one scientific and one practical. The scientific limitation relates to the nonuniqueness of pumping-test interpretation. A perusal of Figure 8.23(b),(c), and (d) indicates the similarity in time-drawdown response that can arise from leaky, unconfined, and bounded systems. Unless there is very clear geologic evidence to direct groundwater hydrologists in their interpretation, there will be difficulties in providing a unique prediction of the effects of any proposed pumping scheme. The fact that a theoretical curve can be matched by pumping test data in no way proves that the aquifer fits the assumptions on which the curve is based.

The practical disadvantage of the method lies in its expense. The installation of test wells and observational piezometers to obtain aquifer coefficients is probably only justified in cases where exploitation of the aquifer by wells at the test site is contemplated. In most such cases, the test well can be utilized as a production well in the subsequent pumping program. In geotechnical applications, in contamination studies, in regional flow-net analysis, or in any flow-net approach that requires hydraulic conductivity data but is not involved with well development, the use of the pumping-test approach is usually inappropriate. It is our opinion that the method is widely overused. Piezometer tests are simpler and cheaper, and they can provide adequate data in many cases where pumping tests are not justified.

8.7 Estimation of Saturated Hydraulic Conductivity

It has long been recognized that hydraulic conductivity is related to the grain-size distribution of granular porous media. In the early stages of aquifer exploration or in regional studies where direct permeability data are sparse, this interrelationship can prove useful for the estimation of conductivity values. In this section, we will examine estimation techniques based on grain-size analyses and porosity determinations. These types of data are often widely available in geological reports, agricultural soil surveys, or reports of soil mechanics testing at engineering sites.

The determination of a relation between conductivity and soil texture requires the choice of a representative grain-size diameter. A simple and apparently durable empirical relation, due to Hazen in the latter part of the last century, relies on the effective grain size, d_{10}, and predicts a power-law relation with K:

$$K = Ad_{10}^2 \tag{8.47}$$

The d_{10} value can be taken directly from a grain-size gradation curve as determined by sieve analysis. It is the grain-size diameter at which 10% by weight of the soil particles are finer and 90% are coarser. For K in cm/s and d_{10} in mm, the coefficient A in Eq. (8.47) is equal to 1.0. Hazen's approximation was originally determined for uniformly graded sands, but it can provide rough but useful estimates for most soils in the fine sand to gravel range.

Textural determination of hydraulic conductivity becomes more powerful when some measure of the spread of the gradation curve is taken into account. When this is done, the *median grain size*, d_{50}, is usually taken as the representative diameter. Masch and Denny (1966) recommend plotting the gradation curve [Figure 8.25(a)] using Krumbein's ϕ units, where $\phi = -\log_2 d$, d being the grain-size diameter (in mm). As a measure of spread, they use the *inclusive standard deviation*, σ_I, where

$$\sigma_I = \frac{d_{16} - d_{84}}{4} + \frac{d_5 - d_{95}}{6.6} \tag{8.48}$$

For the example shown in Figure 8.25(a), $d_{50} = 2.0$ and $\sigma_I = 0.8$. The curves shown in Figure 8.25(b) were developed experimentally in the laboratory on prepared samples of unconsolidated sand. From them, one can determine K, knowing d_{50} and σ_I.

For a fluid of density, ρ, and viscosity, μ, we have seen in Section 2.3 [Eq. (2.26)] that the hydraulic conductivity of a porous medium consisting of uniform

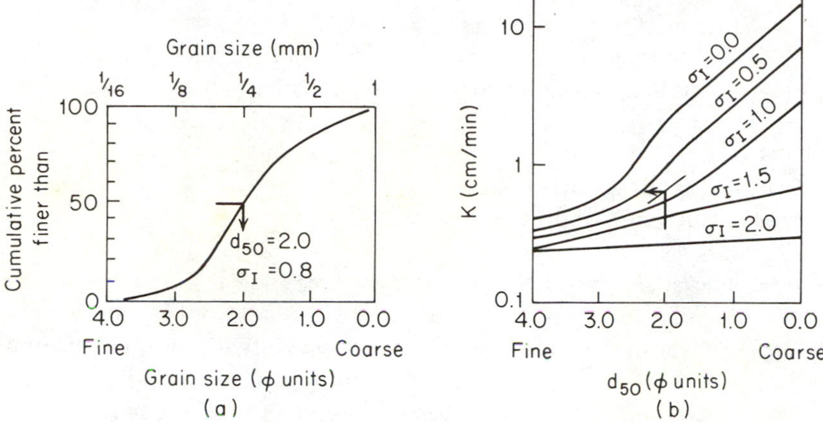

Figure 8.25 Determination of saturated hydraulic conductivity from grain-size gradation curves for unconsolidated sands (after Masch and Denny, 1966).

spherical grains of diameter, d, is given by

$$K = \left(\frac{\rho g}{\mu}\right) C d^2 \tag{8.49}$$

For a nonuniform soil, we might expect the d in Eq. (8.49) to become d_m, where d_m is some representative grain size, and we would expect the coefficient C to be dependent on the shape and packing of the soil grains. The fact that the porosity, n, represents an integrated measure of the packing arrangement has led many investigators to carry out experimental studies of the relationship between C and n. The best known of the resulting predictive equations for hydraulic conductivity is the *Kozeny-Carmen equation* (Bear, 1972), which takes the form

$$K = \left(\frac{\rho g}{\mu}\right)\left[\frac{n^3}{(1-n)^2}\right]\left(\frac{d_m^2}{180}\right) \tag{8.50}$$

In most formulas of this type, the porosity term is identical to the central element of Eq. (8.50), but the grain-size term can take many forms. For example, the *Fair-Hatch equation*, as reported by Todd (1959), take the form

$$K = \left(\frac{\rho g}{\mu}\right)\left[\frac{n^3}{(1-n)^2}\right]\left[\frac{1}{m\left(\frac{\theta}{100}\sum\frac{P}{d_m}\right)^2}\right] \tag{8.51}$$

where m is a packing factor, found experimentally to be about 5; θ is a sand shape factor, varying from 6.0 for spherical grains to 7.7 for angular grains; P is the percentage of sand held between adjacent sieves; and d_m is the geometric mean of the rated sizes of adjacent sieves.

Both Eqs. (8.50) and (8.51) are dimensionally correct. They are suitable for application with any consistent set of units.

8.8 Prediction of Aquifer Yield by Numerical Simulation

The analytical methods that were presented in Section 8.3 for the prediction of drawdown in multiple-well systems are not sophisticated enough to handle the heterogeneous aquifers of irregular shape that are often encountered in the field. The analysis and prediction of aquifer performance in such situations is normally carried out by numerical simulation on a digital computer.

There are two basic approaches: those that involve a *finite-difference* formulation, and those that involve a *finite-element* formulation. We will look at finite-difference methods in moderate detail, but our treatment of finite-element methods will be very brief.

Finite-Difference Methods

As with the steady-state finite-difference methods that were described in Section 5.3, transient simulation requires a discretization of the continuum that makes up the region of flow. Consider a two-dimensional, horizontal, confined aquifer of constant thickness, b; and let it be discretized into a finite number of blocks, each with its own hydrogeologic properties, and each having a node at the center at which the hydraulic head is defined for the entire block. As shown in Figure 8.26(a), some of these blocks may be the site of pumping wells that are removing water from the aquifer.

Let us now examine the flow regime in one of the interior nodal blocks and its four surrounding neighbors. The equation of continuity for transient, saturated flow states that the net rate of flow into any nodal block must be equal to the time rate of change of storage within the nodal block. With reference to Figure 8.26(b), and following the developments of Section 2.11, we have

$$Q_{15} + Q_{25} + Q_{35} + Q_{45} = S_{s_5}\,\Delta x\,\Delta y\,b\,\frac{\partial h_5}{\partial t} \tag{8.52}$$

where S_{s_5} is the specific storage of nodal block 5. From Darcy's law,

$$Q_{15} = K_{15}\frac{h_1 - h_5}{\Delta y}\,\Delta x\,b \tag{8.53}$$

where K_{15} is a representative hydraulic conductivity between nodes 1 and 5. Similar expressions can be written for Q_{25}, Q_{35}, and Q_{45}.

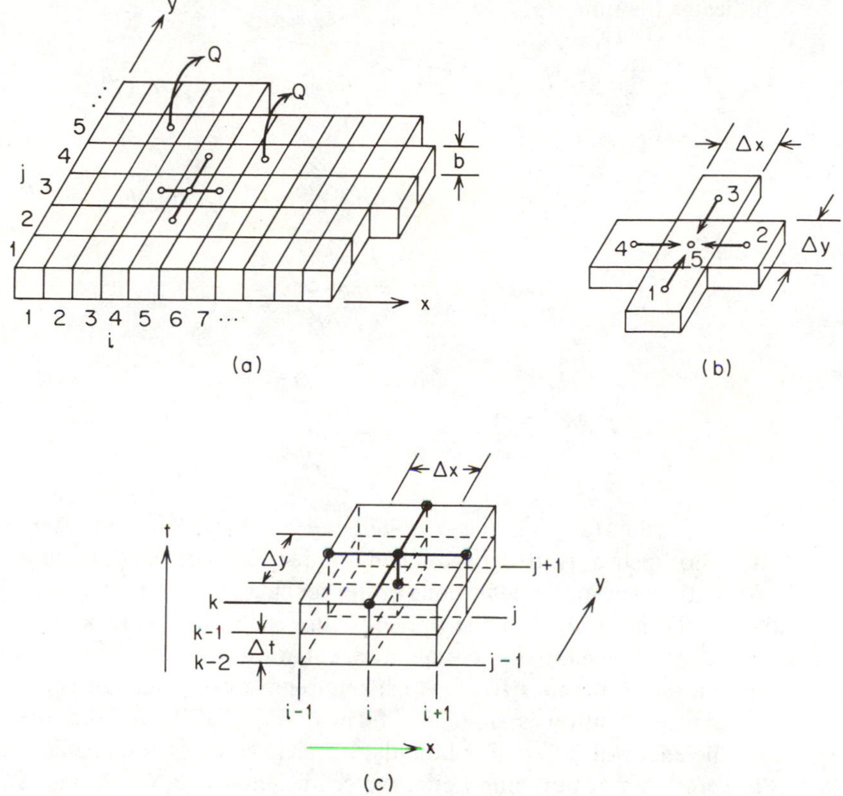

Figure 8.26 Discretization of a two-dimensional, horizontal, confined aquifer.

Let us first consider the case of a homogeneous, isotropic medium for which $K_{15} = K_{25} = K_{35} = K_{45} = K$ and $S_{s_1} = S_{s_2} = S_{s_3} = S_{s_4} = S_s$. If we arbitrarily select a square nodal grid with $\Delta x = \Delta y$, and note that $T = Kb$ and $S = S_s b$, substitution of expressions such as that of Eq. (8.53) into Eq. (8.52) leads to

$$T(h_1 + h_2 + h_3 + h_4 - 4h_5) = S\,\Delta x^2 \frac{\partial h_5}{\partial t} \tag{8.54}$$

The time derivative on the right-hand side can be approximated by

$$\frac{\partial h_5}{\partial t} = \frac{h_5(t) - h_5(t - \Delta t)}{\Delta t} \tag{8.55}$$

where Δt is the time step that is used to discretize the numerical model in a time-wise sense. If we now convert to the ijk notation indicated on Figure 8.26(c), where the subscript (i, j) refers to the nodal position and the superscript $k = 0, 1, 2, \ldots$

indicates the time step, we have

$$h^k_{i,j-1} + h^k_{i+1,j} + h^k_{i-1,j} + h^k_{i,j+1} - 4h^k_{i,j} = \frac{S\,\Delta x^2}{T\,\Delta t}(h^k_{i,j} - h^{k-1}_{i,j}) \tag{8.56}$$

In a more general form,

$$Ah^k_{i,j} = Bh^k_{i,j-1} + Ch^k_{i+1,j} + Dh^k_{i,j+1} + Eh^k_{i-1,j} + F \tag{8.57}$$

where

$$A = \frac{S\,\Delta x^2}{T\,\Delta t} + 4 \tag{8.58}$$

$$B = C = D = E = 1 \tag{8.59}$$

$$F = \frac{S\,\Delta x^2}{T\,\Delta t} \cdot h^{k-1}_{i,j} \tag{8.60}$$

Equation (8.57) is the *finite-difference* equation for an internal node (i, j) in a homogeneous, isotropic, confined aquifer. Each of the parameters S, T, Δx, and Δt that appear in the definitions of the coefficients are known, as is the value of the hydraulic head, $h_{i,j}$, at the previous time step, $k - 1$. In a similar fashion, it is possible to develop finite-difference equations for boundary nodes and corner nodes, and for nodes from which pumping takes place. In each case, the finite-difference equation is similar in form to Eq. (8.57), but the expressions for the coefficients will differ. For boundary nodes, some of the coefficients will be zero. For an internal pumping node, the coefficients A, B, C, D, and E are as given in Eqs. (8.58) and (8.59), but

$$F = \frac{\Delta x^2}{T}\left(\frac{S}{\Delta t} \cdot h^{k-1}_{i,j} + W_{i,j}\right) \tag{8.61}$$

where $W_{i,j}$ is a sink term with units $[L/T]$. W is related to the pumping rate, Q $[L^3/T]$, by

$$W_{i,j} = \frac{Q_{i,j}}{\Delta x^2} \tag{8.62}$$

Sometimes W is given a more general definition,

$$W_{i,j} = \frac{Q_{i,j}}{\Delta x^2} - R_{i,j} \tag{8.63}$$

where $R_{i,j}$ is a source term with units $[L/T]$ that represents vertical leakage into the aquifer from overlying aquitards. In this case Eq. (8.61) is used for all nodes in the system and $W_{i,j}$ is specified for every node. It will be negative for nodes accepting leakage and positive for nodes undergoing pumping.

It is possible to develop Eq. (8.57) in a more rigorous way, starting with the partial differential equation that describes transient flow in a horizontal confined aquifer. In Appendix IX, the rigorous approach is used to determine the values for the coefficients A, B, C, D, E, and F, in the general finite-difference equation for an internal node in a heterogeneous, anisotropic aquifer. In such a system each node (i, j) may be assigned its own specific values of $S_{i,j}$, $(T_x)_{i,j}$, and $(T_y)_{i,j}$, where T_x and T_y are the principal components of the transmissivity tensor in the x and y coordinate directions. The derivation of Appendix IX is carried out for a rectangular nodal grid in which $\Delta x \neq \Delta y$. A further sophistication, which is not considered there, would allow an irregular nodal grid in which the Δx and Δy values are themselves a function of nodal position. Irregular nodal spacings are often required in the vicinity of pumping wells where hydraulic gradients tend to be large. The concepts that underlie the development of these more complex finite-difference formulations is identical to that which led to Eq. (8.57). The more complex the finite-difference equations embodied in the computer program, the more versatile is that program as a numerical simulator of aquifer performance.

It is possible, then, to develop a finite-difference equation, at some degree of sophistication, for every node in the nodal grid. If there are N nodes, there are N finite-difference equations. At each time step, there are also N unknowns: namely, the N values of $h_{i,j}$ at the N nodes. At each time step, we have N linear, algebraic equations in N unknowns. This set of equations must be solved simultaneously at each time step, starting from a set of initial conditions wherein $h_{i,j}$ is known for all (i, j), and proceeding through the time steps, $k = 1, 2, \ldots$. Many methods are available for the solution of the system of equations, and numerical aquifer models are often classified on the basis of the approach that is used. For example, the method of *successive overrelaxation* that was described in Section 5.3 for the numerical simulation of steady-state flow nets is equally applicable to the system of equations that arises at each time step of a transient aquifer model. More commonly a method known as the *alternating-direction implicit procedure* is used. Remson et al. (1971) and Pinder and Gray (1977) provide a systematic and detailed presentation of these various methods as they pertain to aquifer simulation. Advanced mathematical treatment of the methods is available in the textbook by Forsythe and Wasow (1960). The original development of most numerical-simulation techniques took place in the petroleum engineering field, where the primary application is in the simulation of oil-reservoir behavior. Pinder and Bredehoeft (1968) adapted the powerful alternating-direction implicit procedure to the needs of groundwater hydrologists.

There are two aquifer-simulation programs that have been completely documented and widely applied in North America. One is the U.S. Geological Survey model, which is an outgrowth of Pinder and Bredehoeft's original work. Trescott et al. (1976) provide an updated manual for the most recent version of the computer program. The other is the Illinois State Water Survey model, which is fully documented by Prickett and Lonnquist (1971). Bredehoeft and Pinder (1970) have also shown how a sequence of two-dimensional aquifer models can be coupled together to form a quasi-three-dimensional model of an aquifer-aquitard system.

As a practical example, we will consider the analysis carried out by Pinder and Bredehoeft (1968) for an aquifer at Musquoduboit Harbour, Nova Scotia. The aquifer there is a glaciofluvial deposit of limited areal extent. Figure 8.27(a) shows the initial estimate of the areal distribution of transmissivity for the aquifer as determined from the rather sparse hydrogeological data that were available. Simulations with this transmissivity matrix failed to reproduce the drawdown patterns observed during a pumping test that was run near the center of the aquifer. The aquifer parameters were then adjusted and readjusted over several computer runs until a reasonable duplication was achieved between the measured time-drawdown data and the results of the digital model. Additional test-well logs tended to support the adjusted parameters at several points. The final transmissivity distribution is shown in Figure 8.27(b). The model was then put into prediction mode; Figure 8.27(c) is a plot of the predicted drawdown pattern 206.65 days after the start of exploitation by a proposed production well pumping at a rate of $Q = 0.963$ ft^3/s.

Render (1971, 1972) and Huntoon (1974) provide additional case histories of interest.

Finite-Element Methods

The finite-element method, first noted in Section 5.3 in connection with the simulation of steady-state flow nets, can also be used for the simulation of transient aquifer performance. As in the finite-difference approach, the finite-element approach leads to a set of N algebraic equations in N unknowns at each time step, where the N unknowns are the values of the hydraulic heads at a set of nodal points distributed through the aquifer. The fundamental difference lies in the nature of the nodal grid. The finite-element method allows the design of an irregular mesh that can be hand-tailored to any specific application. The number of nodes can often be significantly reduced from the number required for a finite-difference simulation. The finite-element approach also has some advantages in the way it treats boundary conditions and in the simulation of anisotropic media.

The development of the finite-element equations for each node requires an understanding of both partial differential equations and the calculus of variations. Remson, Hornberger, and Molz (1971) provide an introductory treatment of the method as it applies to aquifer simulation. Pinder and Gray (1977) provide an advanced treatment. Zienkiewicz (1967) and Desai and Abel (1972) are the most widely quoted general reference texts. The finite-element method was introduced into the groundwater literature by Javandel and Witherspoon (1969). Pinder and Frind (1972) were among the first to utilize the method for the prediction of regional aquifer performance. Gupta and Tanji (1976) have reported an application of a three-dimensional finite-element model for the simulation of flow in an aquifer-aquitard system in the Sutter Basin, California.

Model Calibration and the Inverse Problem

If measurements of aquifer transmissivity and storativity were available at every nodal position in an aquifer-simulation model, the prediction of drawdown patterns would be a very straightforward matter. In practice, the data base on which models

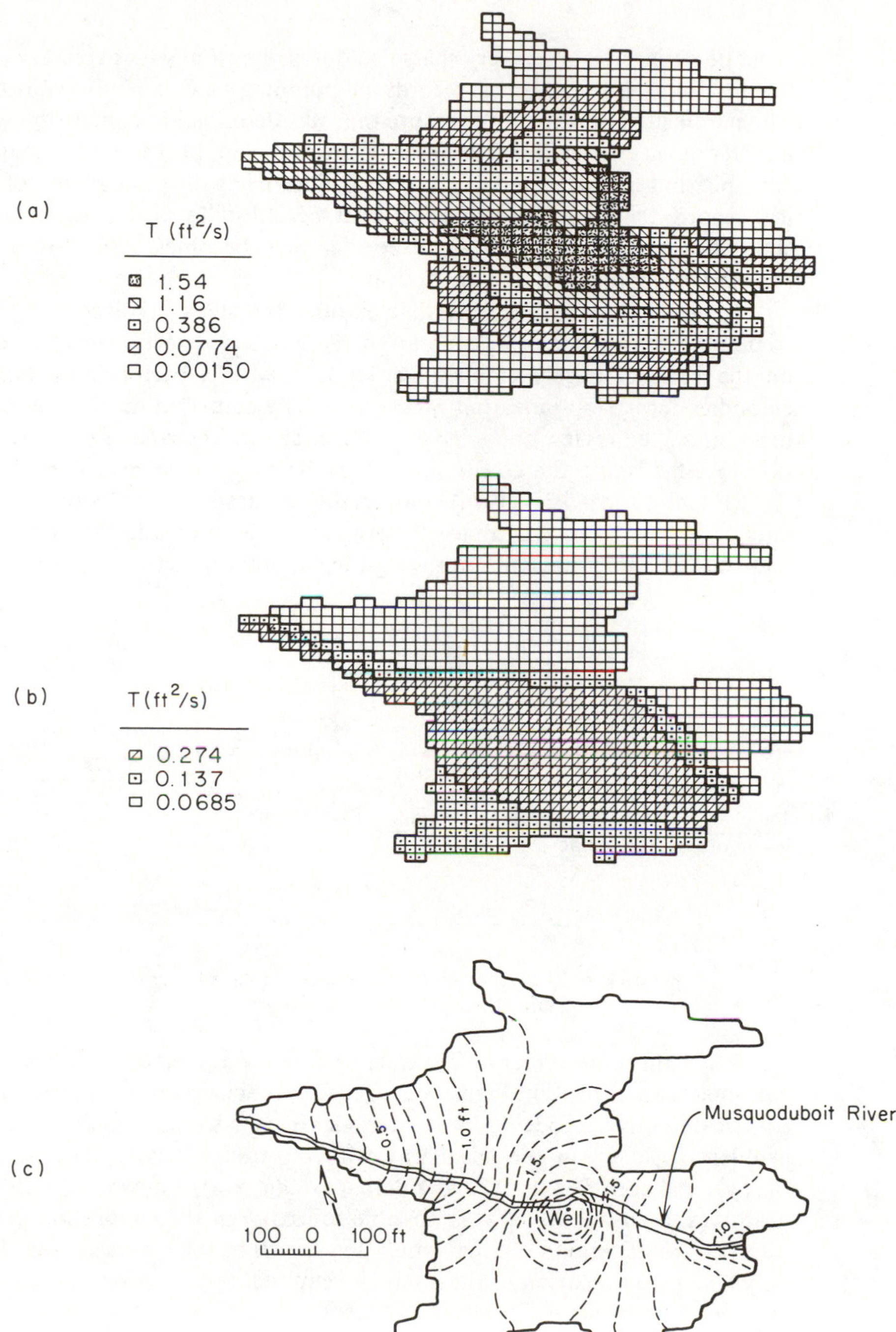

Figure 8.27 Numerical simulation of aquifer performance at Musquodoboit Harbour, Nova Scotia (after Pinder and Bredehoeft, 1968).

must be designed is often very sparse, and it is almost always necessary to calibrate the model against historical records of pumping rates and drawdown patterns. The parameter adjustment procedure that was described in connection with Figure 8.27 represents the calibration phase of the modeling procedure for that particular example. In general, a model should be calibrated against one period of the historical record, then verified against another period of record. The application of a simulation model for a particular aquifer then becomes a three-step process of *calibration*, *verification*, and *prediction*.

Figure 8.28 is a flowchart that clarifies the steps involved in the repetitive trial-and-error approach to calibration. Parameter correction may be carried out on the basis of purely empirical criteria or with a performance analyzer that embodies formal optimization procedures. The contribution by Neuman (1973a) includes a good review and a lengthy reference list. The role of subjective information in establishing the constraints for optimization was treated by Lovell et al. (1972). Gates and Kisiel (1974) considered the question of the worth of additional data. They analyzed the trade-off between the cost of additional measurements and the value they have in improving the calibration of the model.

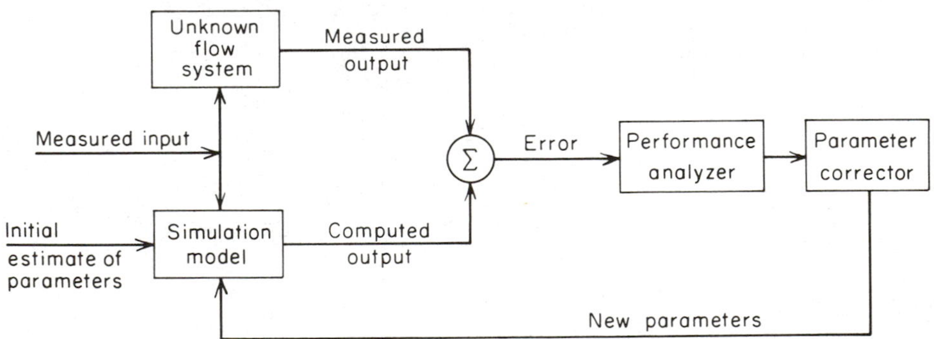

Figure 8.28 Flowchart of the trial-and-error calibration process (after Neuman, 1973a).

The term *calibration* usually refers to the trial-and-error adjustment of aquifer parameters as outlined in Figure 8.28. This approach involves the repetitive application of the aquifer model in its usual mode. In each simulation the boundary-value problem is set up in the usual way with the transmissivity, $T(x, y)$, storativity, $S(x, y)$, leakage, $R(x, y, t)$, and pumpage, $Q(x, y, t)$, known, and the hydraulic head, $h(x, y, t)$, unknown. It is possible to carry out the calibration process more directly by utilizing an aquifer simulation model in the inverse mode. In this case only a single application of the model is required, but the model must be set up as an inverse boundary-value problem where $h(x, y, t)$ and $Q(x, y, t)$ are known and $T(x, y)$, $S(x, y)$, and $R(x, y, t)$ are unknown. When posed in this fashion, the calibration process is known as the *inverse problem*.

In much of the literature, the term *parameter identification* is used to encompass all facets of the problem at hand. What we have called *calibration* is often called the *indirect* approach to the parameter identification problem, and what we have called the *inverse problem* is called the *direct* approach.

The solution of the inverse formulation is not, in general, unique. In the first place there may be too many unknowns; and in the second place, $h(x, y, t)$ and $Q(x, y, t)$ are not known for all (x, y). In practice, pumpage takes place at a finite number of points, and the historical records of head are available at only a finite number of points. Even if $R(x, y, t)$ is assumed constant or known, the problem remains ill-posed mathematically. Emsellem and de Marsily (1971) have shown, however, that the problem can be made tractable by using a "flatness criteria" that limits the allowable spatial variations in T and S. The mathematics of their approach is not simple, but their paper remains the classic discussion of the inverse problem. Neuman (1973a, 1975b) suggests using available measurements of T and S to impose constraints on the structure of $T(x, y)$ and $S(x, y)$ distributions. The contributions of Yeh (1975) and Sagar (1975) include reviews of more recent developments.

There is another approach to inverse simulation that is simpler in concept but apparently open to question as to its validity (Neuman, 1975b). It is based on the assumption of steady-state conditions in the flow system. As first recognized by Stallman (1956), the steady-state hydraulic head pattern, $h(x, y, z)$ in a three-dimensional system can be interpreted inversely in terms of the hydraulic conductivity distribution, $K(x, y, z)$. In a two-dimensional, unpumped aquifer, $h(x, y)$ can be used to determine $T(x, y)$. Nelson (1968) showed that the necessary condition for the existence and uniqueness of a solution to the steady-state inverse problem is that, in addition to the hydraulic heads, the hydraulic conductivity or transmissivity must be known along a surface crossed by all streamlines in the system. Frind and Pinder (1973) have pointed out that, since transmissivity and flux are related by Darcy's law, this criterion can be stated alternatively in terms of the flux that crosses a surface. If water is being removed from an aquifer at a steady pumping rate, the surface to which Nelson refers occurs around the circumference of the well and the well discharge alone provides a sufficient boundary condition for a unique solution. Frind and Pinder (1973) utilized a finite-element model to solve the steady-state inverse problem. Research is continuing on the question of what errors are introduced into the inverse solution when a steady-state approach is used for model calibration for an aquifer that has undergone a transient historical development.

8.9 Prediction of Aquifer Yield by Analog Simulation

Numerical simulation of aquifer performance requires a moderately large computer and relatively sophisticated programming expertise. Electric-analog simulation provides an alternative approach that circumvents these requirements at the expense of a certain degree of versatility.

Analogy Between Electrical Flow and Groundwater Flow

The principles underlying the physical and mathematical analogy between electrical flow and groundwater flow were introduced in Section 5.2. The application under discussion was the simulation of steady-state flow nets in two-dimensional vertical cross sections. One of the methods described there utilized a *resistance-network* analog that was capable of handling heterogeneous systems of irregular shape. In this section, we will pursue analog methods further, by considering the application of two-dimensional *resistance-capacitance networks* for the prediction of transient hydraulic-head declines in heterogeneous, confined aquifers of irregular shape.

Consider a horizontal confined aquifer of thickness b. If it is overlaid with a square grid of spacing, Δx_A [as in Figure 8.26(a)], any small homogeneous portion of the discretized aquifer [Figure 8.29(a)] can be modeled by a scaled-down array of electrical resistors and capacitors on a square grid of spacing, Δx_M [Figure 8.29(b)]. The analogy between electrical flow in the resistance-capacitance network and groundwater flow in the horizontal confined aquifer can be revealed by examining the finite-difference form of the equations of flow for each system. For groundwater flow, from Eq. (8.54),

$$T(h_1 + h_2 + h_3 + h_4 - 4h_5) = S\,\Delta x_A^2 \frac{\partial h_5}{\partial t_A} \tag{8.64}$$

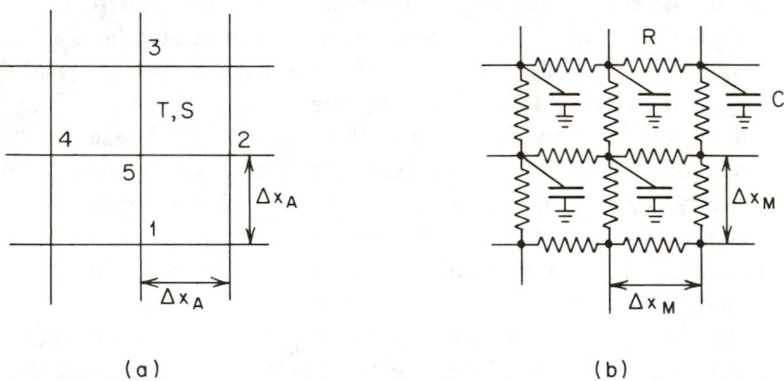

Figure 8.29 Small homogeneous portion of discretized aquifer and analogous resistor-capacitor network (after Prickett, 1975).

For the electrical circuit, from Kirchhoff's laws:

$$\frac{1}{R}(V_1 + V_2 + V_3 + V_4 - 4V_5) = C\frac{\partial V_5}{\partial t_M} \tag{8.65}$$

Comparison of Eqs. (8.64) and (8.65) leads to the analogous quantities:

1. Hydraulic head, h; and voltage, V.
2. Transmissivity, T; and the reciprocal of the resistance, R, of the resistors.

3. The product of the storativity, S, times the nodal block area, Δx_A^2; and the capacitance, C, of the capacitors.
4. Aquifer coordinates, x_A and y_A (as determined by the spacing, Δx_A); and model coordinates, x_M and y_M (as determined by the spacing, Δx_M).
5. Real time, t_A; and model time, t_M.

In addition, if pumpage is considered, there is an analogy between:

6. Pumping rate, Q, at a well; and current strength, I, at an electrical source.

Resistance-Capacitance Network

The network of resistors and capacitors that constitutes the analog model is usually mounted on a Masonite pegboard perforated with holes on approximately 1-inch centers. There are four resistors and one capacitor connected to each terminal. The resistor network is often mounted on the front of the board, and the capacitor network, with each capacitor connected to a common ground, on the back. The boundary of the network is designed in a stepwise fashion to approximate the shape of the actual boundary of the aquifer.

The design of the components of the analog requires the choice of a set of scale factors, F_1, F_2, F_3, and F_4, such that

$$F_1 = \frac{h}{V} \tag{8.66}$$

$$F_2 = \frac{\Delta x_A}{\Delta x_M} \tag{8.67}$$

$$F_3 = \frac{t_A}{t_M} \tag{8.68}$$

$$F_4 = \frac{Q}{I} \tag{8.69}$$

Heterogeneous and transversely anisotropic aquifers can be simulated by choosing resistors and capacitors that match the transmissivity and storativity at each point in the aquifer. Comparison of the hydraulic flow through an aquifer section and the electrical flow through an analogous resistor [Figure 8.30(a)] leads to the relation

$$R = \frac{F_4}{F_1 T} \tag{8.70}$$

Comparison of the storage in an aquifer section and the electrical capacitance of an analogous capacitor [Figure 8.30(b)] leads to the relation

$$C = \frac{F_1 S\, \Delta x_A^2}{F_4 F_3} \tag{8.71}$$

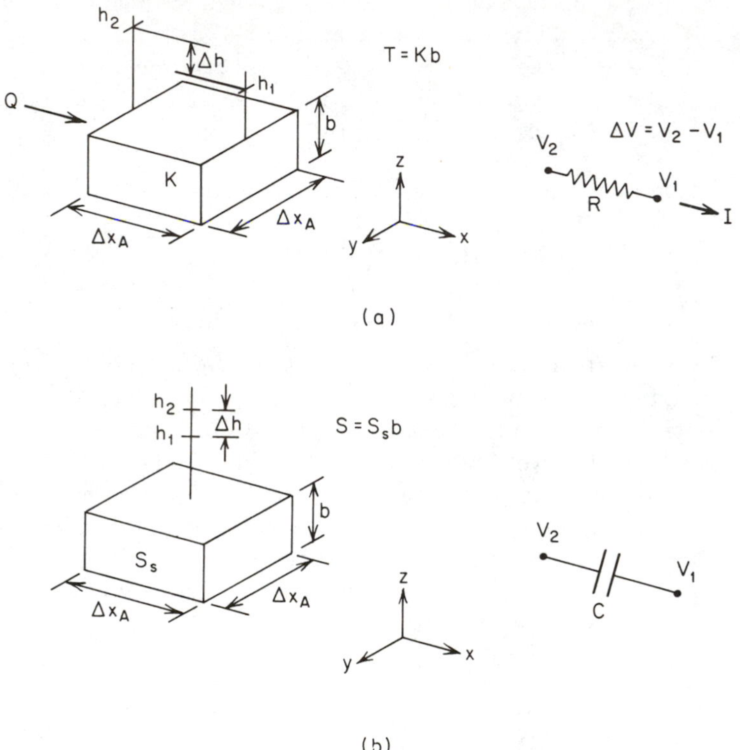

Figure 8.30 Aquifer nodal block and analogous (a) resistor and (b) capacitor (after Prickett, 1975).

The resistors and capacitors that make up the network are chosen on the basis of Eqs. (8.70) and (8.71). The scale factors, F_1, F_2, F_3, and F_4, must be selected in such a way that (1) the resistors and capacitors fall within the range of inexpensive, commercially available components; (2) the size of the model is practical; and (3) the response times of the model are within the range of available excitation-response equipment.

Figure 8.31 is a schematic diagram that shows the arrangement of excitation-response apparatus necessary for electric-analog simulation using a resistance-capacitance network. The pulse generator, in tandem with a waveform generator, produces a rectangular pulse of specific duration and amplitude. This input pulse is displayed on channel 1 of a dual-channel oscilloscope as it is fed through a resistance box to the specific terminal of the resistance-capacitance network that represents the pumped well. The second channel on the oscilloscope is used to display the time-voltage response obtained by probing various observation points in the network. The input pulse is analogous to a step-function increase in pumping rate; the time-voltage graph is analogous to a time-drawdown record at an observational piezometer. The numerical value of the head drawdown is calculated from the

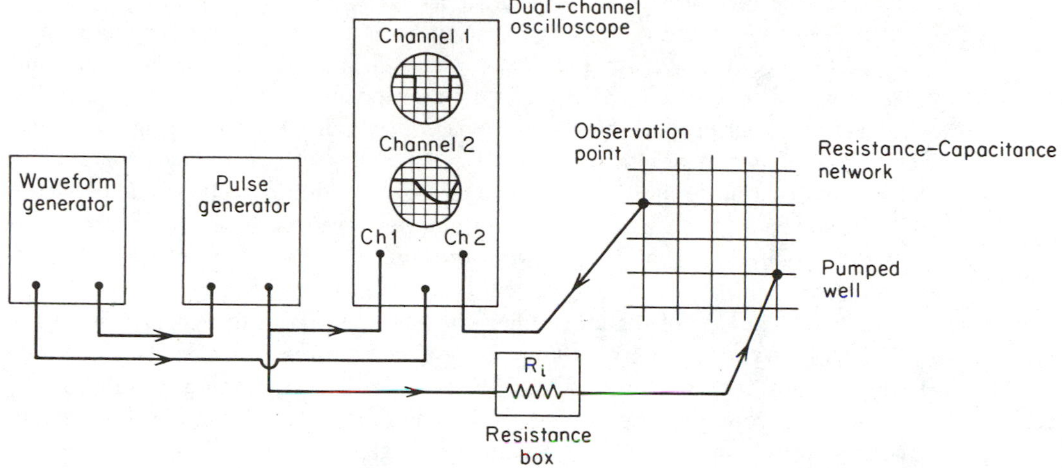

Figure 8.31 Excitation-response apparatus for electrical-analog simulation using a resistance-capacitance network.

voltage drawdown by Eq. (8.66). The time at which any specific drawdown applies is given by Eq. (8.68). Any pumping rate, Q, may be simulated by setting the current strength, I, in Eq. (8.69). This is done by controlling the resistance, R_i, of the resistance box in Figure 8.31. The current strength is given by $I = V_i/R_i$, where V_i is the voltage drop across the resistance box.

Walton (1970) and Prickett (1975) provide detailed coverage of the electric-analog approach to aquifer simulation. Most groundwater treatments owe much to the general discussion of analog simulation by Karplus (1958). Results of analog simulation are usually presented in the form of maps of predicted water-level drawdowns similar to that shown in Figure 8.27(c). Patten (1965), Moore and Wood (1967), Spieker (1968), and Render (1971) provide case histories that document the application of analog simulation to specific aquifers.

Comparison of Analog and Digital Simulation

Prickett and Lonnquist (1968) have discussed the advantages, disadvantages, and similarities between analog and digital techniques of aquifer simulation. They note that both methods use the same basic field data, and the same method of assigning hydrogeologic properties to a discretized representation of the aquifer. Analog simulation requires knowledge of specialized electronic equipment; digital simulation requires expertise in computer programming. Digital simulation is more flexible in its ability to handle irregular boundaries and pumping schemes that vary through time and space. It is also better suited to efficient data readout and display.

The physical construction involved in the preparation of a resistance-capacitance network is both the strength and the weakness of the analog method. The fact that the variables of the system under study are represented by analogous physical quantities and pieces of equipment is extremely valuable for the purposes of teaching or display, but the cost in time is large. The network, once built, describes only one specific aquifer. In digital modeling, on the other hand, once a general computer program has been prepared, data decks representing a wide variety of aquifers and aquifer conditions can be run with the same program. The effort involved in designing and keypunching a new data deck is much less than that involved in designing and building a new resistance-capacitance network. This flexibility is equally important during the calibration phase of aquifer simulation.

The advantages of digital simulation weigh heavily in its favor, and with the advent of easy accessibility to large computers, the method is rapidly becoming the standard tool for aquifer management. However, analog simulation will undoubtedly continue to play a role for some time, especially in developing countries where computer capacities are not yet large.

8.10 Basin Yield

Safe Yield and Optimal Yield of a Groundwater Basin

Groundwater yield is best viewed in the context of the full three-dimensional hydrogeologic system that constitutes a groundwater basin. On this scale of study we can turn to the well-established concept of *safe yield* or to the more rigorous concept of *optimal yield.*

Todd (1959) defines the *safe yield* of a groundwater basin as the amount of water that can be withdrawn from it annually without producing an undesired result. Any withdrawal in excess of safe yield is an *overdraft*. Domenico (1972) and Kazmann (1972) review the evolution of the term. Domenico notes that the "undesired results" mentioned in the definition are now recognized to include not only the depletion of the groundwater reserves, but also the intrusion of water of undesirable quality, the contravention of existing water rights, and the deterioration of the economic advantages of pumping. One might also include excessive depletion of streamflow by induced infiltration and land subsidence.

Although the concept of safe yield has been widely used in groundwater resource evaluation, there has always been widespread dissatisfaction with it (Thomas, 1951; Kazmann, 1956). Most suggestions for improvement have encouraged consideration of the yield concept in a socioeconomic sense within the overall framework of optimization theory. Domenico (1972) reviews the development of this approach, citing the contributions of Bear and Levin (1967), Buras (1966), Burt (1967), Domenico et al. (1968), and others. From an optimization viewpoint, groundwater has value only by virtue of its use, and the *optimal yield* must be determined by the selection of the optimal groundwater management scheme from a set of possible alternative schemes. The optimal scheme is the one that best meets a set of economic and/or social objectives associated with the uses to which the water is to be put. In some cases and at some points in time, consideration of the present and future costs and benefits may lead to optimal yields that involve mining groundwater, perhaps even to depletion. In other situations, optimal yields may reflect the need for complete conservation. Most often, the optimal groundwater development lies somewhere between these extremes.

The graphical and mathematical methods of optimization, as they relate to groundwater development, are reviewed by Domenico (1972).

Transient Hydrologic Budgets and Basin Yield

In Section 6.2 we examined the role of the average annual groundwater recharge, R, as a component in the steady-state hydrologic budget for a watershed. The value of R was determined from a quantitative interpretation of the steady-state, regional, groundwater flow net. Some authors have suggested that the safe yield of a groundwater basin be defined as the annual extraction of water that does not exceed the average annual groundwater recharge. This concept is not correct. As pointed out by Bredehoeft and Young (1970), major groundwater development may significantly change the recharge-discharge regime as a function of time. Clearly, the basin yield depends both on the manner in which the effects of withdrawal are transmitted through the aquifers *and* on the changes in rates of groundwater recharge and discharge induced by the withdrawals. In the form of a transient hydrologic budget for the saturated portion of a groundwater basin,

$$Q(t) = R(t) - D(t) + \frac{dS}{dt} \tag{8.72}$$

where $Q(t)$ = total rate of groundwater withdrawal
$R(t)$ = total rate of groundwater recharge to the basin
$D(t)$ = total rate of groundwater discharge from the basin
dS/dt = rate of change of storage in the saturated zone of the basin.

Freeze (1971a) examined the response of $R(t)$ and $D(t)$ to an increase in $Q(t)$ in a hypothetical basin in a humid climate where water tables are near the surface. The response was simulated with the aid of a three-dimensional transient analysis of a complete saturated-unsaturated system such as that of Figure 6.10 with a pumping well added. Figure 8.32 is a schematic representation of his findings. The diagrams show the time-dependent changes that might be expected in the various terms of Eq. (8.72) under increased pumpage. Let us first look at the case shown in Figure 8.32(a), in which withdrawals increase with time but do not become excessive. The initial condition at time t_0 is a steady-state flow system in which the recharge, R_0, equals the discharge, D_0. At times t_1, t_2, t_3, and t_4, new wells begin to tap the system and the pumping rate Q undergoes a set of stepped increases. Each increase is initially balanced by a change in storage, which in an unconfined aquifer takes the form of an immediate water-table decline. At the

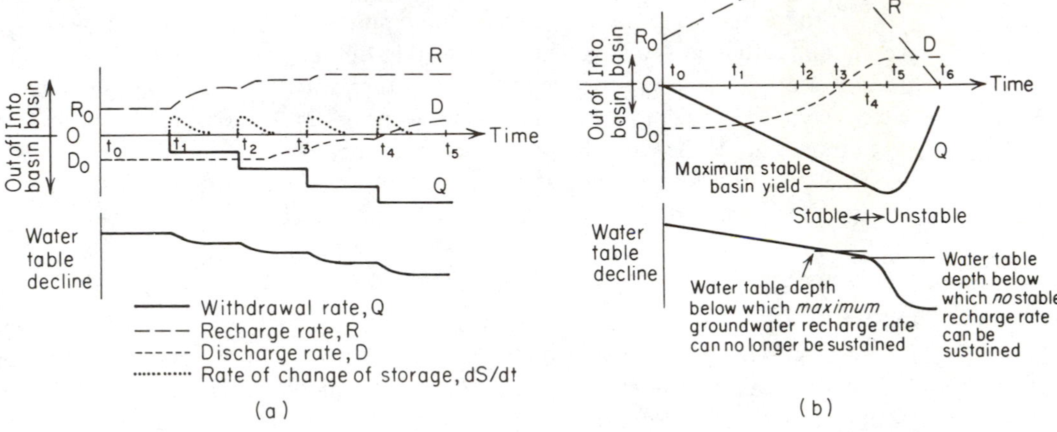

Figure 8.32 Schematic diagram of transient relationships between recharge rates, discharge rates, and withdrawal rates (after Freeze, 1971a).

same time, the basin strives to set up a new equilibrium under conditions of increased recharge, R. The unsaturated zone will now be induced to deliver greater flow rates to the water table under the influence of higher gradients in the saturated zone. Concurrently, the increased pumpage may lead to decreased discharge rates, D. In Figure 8.32(a), after time t_4, all natural discharge ceases and the discharge curve rises above the horizontal axis, implying the presence of induced recharge from a stream that had previously been accepting its baseflow component from the groundwater system. At time t_5, the withdrawal Q is being fed by the recharge, R, and the induced recharge, D; and there has been a significant decline in the water table. Note that the recharge rate attains a maximum between t_3 and t_4. At this rate, the groundwater body is accepting all the infiltration that is available from the unsaturated zone under the lowered water-table conditions.

In Figure 8.32(a), steady-state equilibrium conditions are reached prior to each new increase in withdrawal rate. Figure 8.32(b) shows the same sequence of events under conditions of continuously increasing groundwater development over several years. This diagram also shows that if pumping rates are allowed to increase indefinitely, an unstable situation may arise where the declining water table reaches a depth below which the *maximum* rate of groundwater recharge R can no longer be sustained. After this point in time the same annual precipitation rate no longer provides the same percentage of infiltration to the water table. Evapotranspiration during soil-moisture-redistribution periods now takes more of the infiltrated rainfall before it has a chance to percolate down to the groundwater zone. At t_4 in Figure 8.32(b), the water table reaches a depth below which *no* stable recharge rate can be maintained. At t_5 the maximum available rate of *induced* recharge is attained. From time t_5 on, it is impossible for the basin to supply increased rates of withdrawal. The only source lies in an increased rate of change of storage that manifests itself in rapidly declining water tables. Pumping rates can no longer be maintained at their original levels. Freeze (1971a) defines the value of Q at which instability occurs as the *maximum stable basin yield.* To develop a basin to its limit of stability would, of course, be foolhardy. One dry year might cause an irrecoverable water-table drop. Production rates must allow for a factor of safety and must therefore be somewhat less than the maximum stable basin yield.

The discussion above emphasizes once again the important interrelationships between groundwater flow and surface runoff. If a groundwater basin were developed up to its maximum yield, the potential yields of surface-water components of the hydrologic cycle in the basin would be reduced. It is now widely recognized that optimal development of the water resources of a watershed depend on the *conjunctive use* of surface water and groundwater. The subject has provided a fertile field for the application of optimization techniques (Maddock, 1974; Yu and Haimes, 1974). Young and Bredehoeft (1972) describe the application of digital computer simulations of the type described in Section 8.8 to the solution of management problems involving conjunctive groundwater and surface-water systems.

8.11 Artificial Recharge and Induced Infiltration

In recent years, particularly in the more populated areas of North America where water resource development has approached or exceeded available yield, there has been considerable effort placed on the management of water resource systems. Optimal development usually involves the conjunctive use of groundwater and surface water and the reclamation and reuse of some portion of the available water resources. In many cases, it involves the importation of surface water from areas of plenty to areas of scarcity, or the conservation of surface water in times of plenty for use in times of scarcity. These two approaches require storage facilities, and there is often advantage to storing water underground where evaporation losses are minimized. Underground storage may also serve to replenish groundwater resources in areas of overdraft.

Any process by which man fosters the transfer of surface water into the groundwater system can be classified as *artificial recharge.* The most common method involves infiltration from spreading basins into high-permeability, unconfined, alluvial aquifers. In many cases, the spreading basins are formed by the construction of dikes in natural channels. The recharge process involves the growth of a *groundwater mound* beneath the spreading basin. The areal extent of the mound and its rate of growth depend on the size and shape of the recharging basin, the duration and rate of recharge, the stratigraphic configuration of subsurface formations, and the saturated and unsaturated hydraulic properties of the geologic materials. Figure 8.33 shows two simple hydrogeological environments and the type of groundwater mound that would be produced in each case beneath a circular spreading basin. In Figure 8.33(a), recharge takes place into a horizontal unconfined aquifer bounded at the base by an impermeable formation. In Figure 8.33(b),

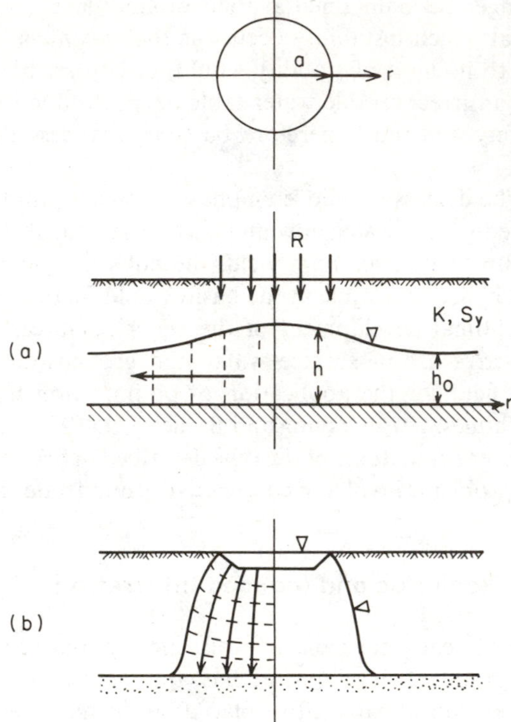

Figure 8.33 Growth of a groundwater mound beneath a circular recharge basin.

recharge takes place through a less-permeable formation toward a high-permeability layer at depth.

Both cases have been the subject of a large number of predictive analyses, not only for circular spreading basins but also for rectangular basins and for recharge from an infinitely long strip. The latter case, with boundary conditions like those shown in Figure 8.33(b), also has application to canal and river seepage. It has been studied in this context by Bouwer (1965), Jeppson (1968), and Jeppson and Nelson (1970). The case shown in Figure 8.33(a), which also has application to the development of mounds beneath waste disposal ponds and sanitary landfills, has been studied in even greater detail. Hantush (1967) provides an analytical solution for the prediction of $h(r, t)$, given the initial water-table height, h_0, the diameter of the spreading basin, a, the recharge rate, R, and the hydraulic conductivity and specific yield, K and S_y, of the unconfined aquifer. His solution is limited to homogeneous, isotropic aquifers and a recharge rate that is constant in time and space. In addition, the solution is limited to a water-table rise that is less than or equal to 50% of the initial depth of saturation, h_0. This requirement implies that $R \ll K$. Bouwer (1962) utilized an electric-analog model to analyze the same problem, and Marino (1975a, 1975b) produced a numerical simulation. All three of these analyses have two additional limitations. First, they neglect unsaturated flow by assuming that the recharge pulse traverses the unsaturated zone vertically and reaches the water table unaffected by soil moisture-conditions above the water table. Second, they utilize the Dupuit-Forchheimer theory of unconfined flow (Section 5.5) which neglects any vertical flow gradients that develop in the saturated zone in the vicinity of the mound. Numerical simulations carried out on the complete saturated-unsaturated system using the approaches of Rubin (1968), Jeppson and Nelson (1970), and Freeze (1971a) would provide a more accurate approach to the problem, but at the expense of added complexity in the calculations.

Practical research on spreading basins has shown that the niceties of predictive analysis are seldom reflected in the real world. Even if water levels in spreading ponds are kept relatively constant, the recharge rate almost invariably declines with time as a result of the buildup of silt and clay on the basin floor and the growth of microbial organisms that clog the soil pores. In addition, air entrapment between the wetting front and the water table retards recharge rates. Todd (1959) notes that alternating wet and dry periods generally furnish a greater total recharge than does continuous spreading. Drying kills the microbial growths, and tilling and scraping of the basin floor during dry periods reopens the soil pores.

There are several excellent case histories that provide an account of specific projects involving artificial recharge from spreading basins. Seaburn (1970) describes hydrologic studies carried out at two of the more than 2000 recharge basins that are used on Long Island, east of New York City, to provide artificial recharge of storm runoff from residential and industrial areas. Bianchi and Haskell (1966, 1968) describe the piezometric monitoring of a complete recharge cycle of mound growth and dissipation. They report relatively good agreement between the field data and analytical predictions based on Dupuit-Forchheimer theory. They note, however, that the anomalous water-level rises that accompany air entrapment (Section 6.8) often make it difficult to accurately monitor the growth of the groundwater mound.

While water spreading is the most ubiquitous form of artificial recharge, it is limited to locations with favorable geologic conditions at the surface. There have also been some attempts made to recharge deeper formations by means of injection wells. Todd (1959) provides several case histories involving such diverse applications as the disposal of storm-runoff water, the recirculation of air-conditioning water, and the buildup of a freshwater barrier to prevent further intrusion of seawater into a confined aquifer. Most of the more recent research on deep-well injection has centered on utilization of the method for the disposal of industrial wastewater and tertiary-treated municipal wastewater (Chapter 9) rather than for the replenishment of groundwater resources.

The oldest and most widely used method of conjunctive use of surface water and groundwater is based on the concept of *induced infiltration*. If a well produces water from alluvial sands and gravels that are in hydraulic connection with a stream, the stream will act as a constant-head line source in the manner noted in Figures 8.15(d) and 8.23(d). When a new well starts to pump in such a situation, the pumped water is initially derived from the groundwater zone, but once the cone of depression reaches the stream, the source of some of the pumped water will be streamflow that is induced into the groundwater body under the influence of the gradients set up by the well. In due course, steady-state conditions will be reached, after which time the cone of depression and the drawdowns within it remain constant. Under the steady flow system that develops at such times, the source of all the pumped groundwater is streamflow. One of the primary advantages of induced infiltration schemes over direct surface-water utilization lies in the chemical and biological purification afforded by the passage of stream water through the alluvial deposits.

8.12 Land Subsidence

In recent years it has become apparent that the extensive exploitation of groundwater resources in this century has brought with it an undesired environmental side effect. At many localities in the world, groundwater pumpage from unconsolidated aquifer-aquitard systems has been accompanied by significant land subsidence. Poland and Davis (1969) and Poland (1972) provide descriptive summaries of all the well-documented cases of major land subsidence caused by the withdrawal of fluids. They present several case histories where subsidence has been associated with oil and gas production, together with a large number of cases that involve groundwater pumpage. There are three cases–the Wilmington oil field in Long Beach, California, and the groundwater overdrafts in Mexico City, Mexico, and in the San Joaquin valley, California–that have led to rates of subsidence of the land surface of almost 1 m every 3 years over the 35-year period 1935–1970. In the San Joaquin valley, where groundwater pumpage for irrigation purposes is to blame, there are three separate areas with significant subsidence problems. Taken together, there is a total area of 11,000 km^2 that has subsided more than 0.3 m. At Long Beach, where the subsiding region is adjacent to the ocean, subsidence has resulted in repeated flooding of the harbor area. Failure of surface structures, buckling of pipe lines, and rupturing of oil-well casing have been reported. Remedial costs up to 1962 exeeded $100 million.

Mechanism of Land Subsidence

The depositional environments at the various subsidence sites are varied, but there is one feature that is common to all the groundwater-induced sites. In each case there is a thick sequence of unconsolidated or poorly consolidated sediments forming an interbedded aquifer-aquitard system. Pumpage is from sand and gravel aquifers, but a large percentage of the section consists of high-compressibility clays. In earlier chapters we learned that groundwater pumpage is accompanied by vertical leakage from the adjacent aquitards. It should come as no surprise to find that the process of aquitard drainage leads to compaction* of the aquitards just as the process of aquifer drainage leads to compaction of the aquifers. There are two fundamental differences, however: (1) since the compressibility of clay is 1–2 orders of magnitude *greater* than the compressibility of sand, the total potential compaction of an aquitard is much greater than that for an aquifer; and (2) since the hydraulic conductivity of clay may be several orders of magnitude *less* than the hydraulic conductivity of sand, the drainage process, and hence the compaction process, is much slower in aquitards than in aquifers.

Consider the vertical cross section shown in Figure 8.34. A well pumping at

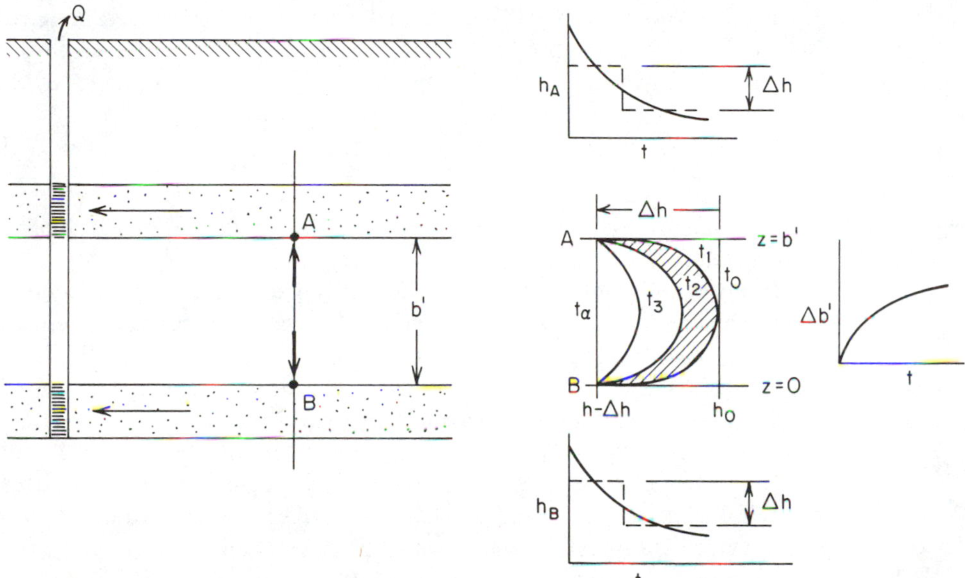

Figure 8.34 One-dimensional consolidation of an aquitard.

a rate Q is fed by two aquifers separated by an aquitard of thickness b. Let us assume that the geometry is radially symmetric and that the transmissivities in the two aquifers are identical. The time-dependent reductions in hydraulic head in the aquifers (which could be predicted from leaky-aquifer theory) will be identical at points A and B. We wish to look at the hydraulic-head reductions in the aquitard along the line AB under the influence of the head reductions in the aquifers at A and B. If $h_A(t)$ and $h_B(t)$ are approximated by step functions with a step Δh (Figure 8.34), the aquitard drainage process can be viewed as the one-dimensional, transient boundary-value problem described in Section 8.3 and presented as Eq. (8.21). The initial condition is $h = h_0$ all along AB, and the boundary conditions are

*Following Poland and Davis (1969), we are using the term "compaction" in its geological sense. In engineering jargon the term is often reserved for the increase in soil density achieved through the use of rollers, vibrators or other heavy machinery.

$h = h_0 - \Delta h$ at A and at B for all $t > 0$. A solution to this boundary-value problem was obtained by Terzaghi (1925) in the form of an analytical expression for $h(z, t)$. An accurate graphical presentation of his solution appears as Figure 8.17. The central diagram on the right-hand side of Figure 8.34 is a schematic plot of his solution; it shows the time-dependent decline in hydraulic head at times $t_0, t_1, \ldots, t_\infty$ along the line AB. To obtain quantitative results for a particular case, one must know the thickness b', the vertical hydraulic conductivity K', the vertical compressibility α', and the porosity n' of the aquitard, together with the head reduction Δh on the boundaries.

In soil mechanics the compaction process associated with the drainage of a clay layer is known as *consolidation*. Geotechnical engineers have long recognized that for most clays $\alpha \gg n\beta$, so the latter term is usually omitted from Eq. (8.21). The remaining parameters are often grouped into a single parameter c_v, defined by

$$c_v = \frac{K'}{\rho g \alpha'} \tag{8.73}$$

The hydraulic head $h(z, t)$ can be calculated from Figure 8.17 with the aid of Eq. (8.23) given c_v, Δh, and b.

In order to calculate the compaction of the aquitard given the hydraulic head declines at each point on AB as a function of time, it is necessary to recall the effective stress law: $\sigma_T = \sigma_e + p$. For $\sigma_T =$ constant, $d\sigma_e = -dp$. In the aquitard, the head reduction at any point z between the times t_1 and t_2 (Figure 8.34) is $dh = h_1(z, t_1) - h_2(z, t_2)$. This head drop creates a fluid pressure reduction: $dp = \rho g\, d\psi = \rho g\, d(h - z) = \rho g\, dh$, and the fluid pressure reduction is reflected by an increase in the effective stress $d\sigma_e = -dp$. It is the change in effective stress, acting through the aquitard compressibility α', that causes the aquitard compaction $\Delta b'$. To calculate $\Delta b'$ along AB between the times t_1 and t_2, it is necessary to divide the aquitard into m slices. Then, from Eq. (2.54),

$$\Delta b'_{t_1-t_2} = b' \sum_{i=1}^{m} \rho g \alpha'\, dh_i \tag{8.74}$$

where dh_i is the average head decline in the ith slice.

For a multiaquifer system with several pumping wells, the land subsidence as a function of time is the summation of all the aquitard and aquifer compactions. A complete treatment of consolidation theory appears in most soil mechanics texts (Terzaghi and Peck, 1967; Scott, 1963). Domenico and Mifflin (1965) were the first to apply these solutions to cases of land subsidence.

It is reasonable to ask whether land subsidence can be arrested by injecting groundwater back into the system. In principle this should increase the hydraulic heads in the aquifers, drive water back into the aquitards, and cause an expansion of both aquifer and aquitard. In practice, this approach is not particularly effective because aquitard compressibilities in expansion have only about one-tenth the value they have in compression. The most successful documented injection scheme is the one undertaken at the Wilmington oil field in Long Beach, California (Poland and Davis, 1969). Repressuring of the oil reservoir was initiated in 1958 and by 1963 there had been a modest rebound in a portion of the subsiding region and the rates of subsidence were reduced elsewhere.

Field Measurement of Land Subsidence

If there are any doubts about the aquitard-compaction theory of land subsidence, they should be laid to rest by an examination of the results of the U.S. Geological Survey subsidence research group during the last decade. They have carried out field studies in several subsiding areas in California, and their measurements provide indisputable confirmation of the interrelationships between hydraulic head declines, aquitard compaction, and land subsidence.

Figure 8.35 is a contoured map, based on geodetic measurements, of the land subsidence in the Santa Clara valley during the period 1934–1960. Subsidence is

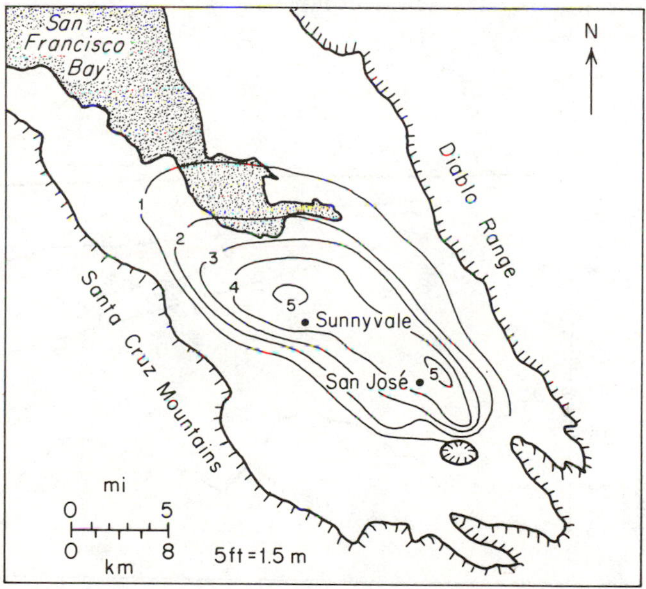

Figure 8.35 Land subsidence in feet, 1934–1960, Santa Clara valley, California (after Poland and Davis, 1969).

confined to the area underlain by unconsolidated deposits of alluvial and shallow-marine origin. The centers of subsidence coincide with the centers of major pumping, and the historical development of the subsidence coincides with the period of settlement in the valley and with the increased utilization of groundwater.

Quantitative confirmation of the theory is provided by results of the type shown in Figure 8.36. An ingeniously simple compaction-recorder installation [Figure 8.36(a)] produces a graph of the time-dependent growth of the total

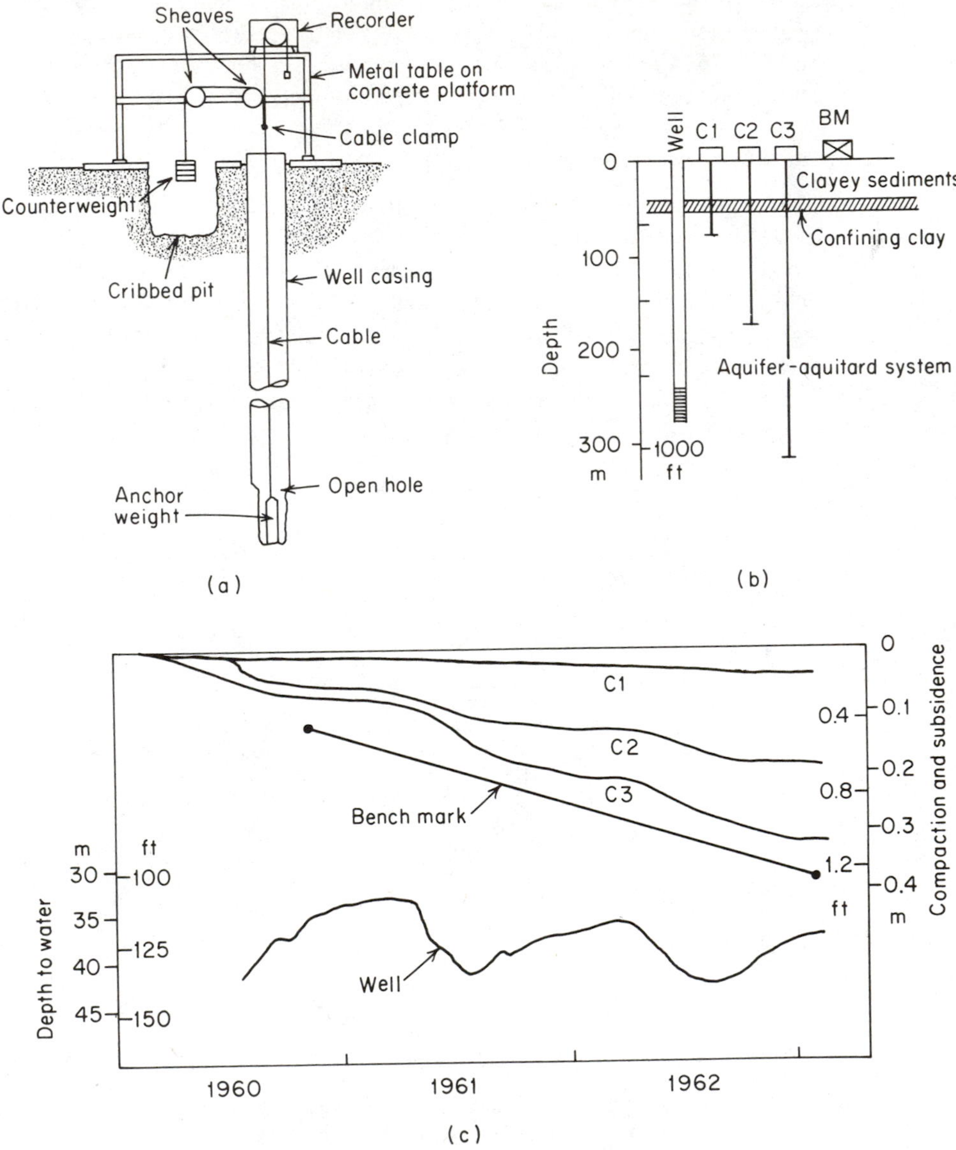

Figure 8.36 (a) Compaction-recorder installation; (b) compaction measurement site near Sunnyvale, California; (c) measured compactions, land subsidence, and hydraulic head variations at the Sunnyvale site, 1960–1962 (after Poland and Davis, 1969).

compaction of all material between the land surface and the bottom of the hole. Near Sunnyvale in the Santa Clara valley, three compaction recorders were established at different depths in the confined aquifer system that exists there [Figure

8.36(b)]. Figure 8.36(c) shows the compaction records together with the total land subsidence as measured at a nearby benchmark, and the hydraulic head for the 250- to 300-m-depth range as measured in an observation well at the measurement site. Decreasing hydraulic heads are accompanied by compaction. Increasing hydraulic heads are accompanied by reductions in the rate of compaction, but there is no evidence of rebound. At this site "the land subsidence is demonstrated to be equal to the compaction of the water-bearing deposits within the depth tapped by water wells, and the decline in artesian head is proved to be the sole cause of the subsidence" (Poland and Davis, 1969, p. 259).

Riley (1969) noted that data of the type shown on Figure 8.36(c) can be viewed as the result of a large-scale field consolidation test. If the reductions in aquitard volume reflected by the land subsidence are plotted against the changes in effective stress created by the hydraulic-head declines, it is often possible to calculate the average compressibility and the average vertical hydraulic conductivity of the aquitards. Helm (1975, 1976) has carried these concepts forward in his numerical models of land subsidence in California.

It is also possible to develop predictive simulation models that can relate possible pumping patterns in an aquifer-aquitard system to the subsidence rates that will result. Gambolati and Freeze (1973) designed a two-step mathematical model for this purpose. In the first step (the hydrologic model), the regional hydraulic-head drawdowns are calculated in an idealized two-dimensional vertical cross section in radial coordinates, using a model that is a boundary-value problem based on the equation of transient groundwater flow. Solutions are obtained with a numerical finite-element technique. In the second step of the modeling procedure (the subsidence model), the hydraulic head declines determined with the hydrologic model for the various aquifers are used as time-dependent boundary conditions in a set of one-dimensional vertical consolidation models applied to a more refined geologic representation of each aquitard. Gambolati et al. (1974a, 1974b) applied the model to subsidence predictions for Venice, Italy. Recent measurements summarized by Carbognin et al. (1976) verify the model's validity.

8.13 Seawater Intrusion

When groundwater is pumped from aquifers that are in hydraulic connection with the sea, the gradients that are set up may induce a flow of salt water from the sea toward the well. This migration of salt water into freshwater aquifers under the influence of groundwater development is known as *seawater intrusion*.

As a first step toward understanding the nature of the processes involved, it is necessary to examine the nature of the saltwater-freshwater interface in coastal aquifers under natural conditions. The earliest analyses were carried out independently by two European scientists (Ghyben, 1888; Herzberg, 1901) around the turn of the century. Their analysis assumed simple hydrostatic conditions in a homogeneous, unconfined coastal aquifer. They showed [Figure 8.37(a)] that the

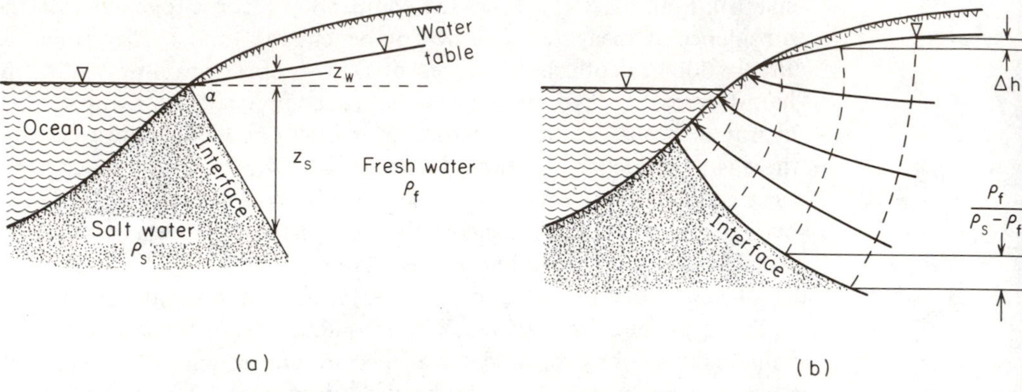

Figure 8.37 Saltwater-freshwater interface in an unconfined coastal aquifer (a) under hydrostatic conditions; (b) under conditions of steady-state seaward flow (after Hubbert, 1940).

interface separating salt water of density ρ_s and fresh water of density ρ_f must project into the aquifer at an angle $\alpha < 90°$. Under hydrostatic conditions, the weight of a unit column of fresh water extending from the water table to the interface is balanced by a unit column of salt water extending from sea level to the same depth as the point on the interface. With reference to Figure 8.37(a), we have

$$\rho_s g z_s = \rho_f g(z_s + z_w) \tag{8.75}$$

or

$$z_s = \frac{\rho_f}{\rho_s - \rho_f} z_w \tag{8.76}$$

For $\rho_f = 1.0$ and $\rho_s = 1.025$,

$$z_s = 40 z_w \tag{8.77}$$

Equation (8.77) is often called the *Ghyben-Herzberg relation*.

If we specify a change in the water-table elevation of Δz_w, then from Eq. (8.77), $\Delta z_s = 40\Delta z_w$. If the water table in an unconfined coastal aquifer is lowered 1 m, the saltwater interface will rise 40 m.

In most real situations, the Ghyben-Herzberg relation underestimates the depth to the saltwater interface. Where freshwater flow to the sea takes place, the hydrostatic assumptions of the Ghyben-Herzberg analysis are not satisfied. A more realistic picture was provided by Hubbert (1940) in the form of Figure 8.37(b) for steady-state outflow to the sea. The exact position of the interface can be determined for any given water-table configuration by graphical flow-net construction, noting the relationships shown on Figure 8.37(b) for the intersection of equipotential lines on the water table and on the interface.

The concepts outlined in Figure 8.37 do not reflect reality in yet another way. Both the hydrostatic analysis and the steady-state analysis assume that the interface separating fresh water and salt water in a coastal aquifer is a sharp boundary. In reality, there tends to be a mixing of salt water and fresh water in a zone of diffusion around the interface. The size of the zone is controlled by the dispersive characteristics of the geologic strata. Where this zone is narrow, the methods of solution for a sharp interface may provide a satisfactory prediction of the freshwater flow pattern, but an extensive zone of diffusion can alter the flow pattern and the position of the interface, and must be taken into account. Henry (1960) was the first to present a mathematical solution for the steady-state case that includes consideration of dispersion. Cooper et al. (1964) provide a summary of the various analytical solutions.

Seawater intrusion can be induced in both unconfined and confined aquifers. Figure 8.38(a) provides a schematic representation of the saltwater wedge that would exist in a confined aquifer under conditions of natural steady-state outflow. Initiation of pumping [Figure 8.38(b)] sets up a transient flow pattern that leads to declines in the potentiometric surface on the confined aquifer and inland migration of the saltwater interface. Pinder and Cooper (1970) presented a numerical mathematical method for the calculation of the transient position of the saltwater front in a confined aquifer. Their solution includes consideration of dispersion.

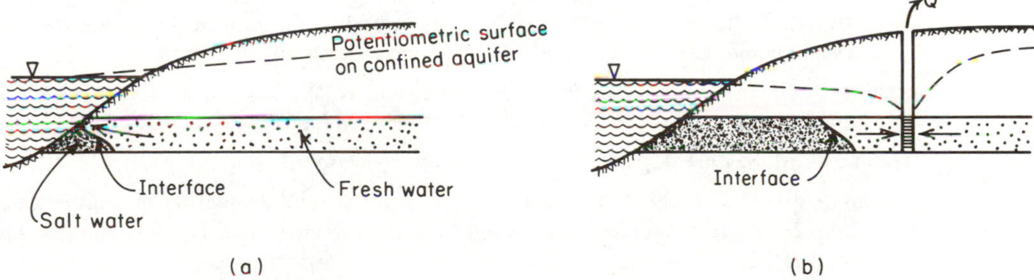

Figure 8.38 (a) Saltwater-freshwater interface in a confined coastal aquifer under conditions of steady-state seaward flow; (b) seawater intrusion due to pumping.

One of the most intensively studied coastal aquifers in North America is the Biscayne aquifer of southeastern Florida (Kohout, 1960a, 1960b). It is an unconfined aquifer of limestone and calcareous sandstone extending to an average depth of 30 m below sea level. Field data indicate that the saltwater front undergoes transient changes in position under the influence of seasonal recharge patterns and the resulting water-table fluctuations. Lee and Cheng (1974) and Segol and Pinder (1976) have simulated transient conditions in the Biscayne aquifer with finite-element numerical models. Both the field evidence and the numerical modeling confirm the necessity of considering dispersion in the steady-state and transient analyses. The nature of dispersion in groundwater flow will be considered more fully in Chapter 9 in the context of groundwater contamination.

Todd (1959) summarizes five methods that have been considered for controlling seawater intrusion: (1) reduction or rearrangement of the pattern of groundwater pumping, (2) artificial recharge of the intruded aquifer from spreading basins or recharge wells, (3) development of a pumping trough adjacent to the coast by means of a line of pumping wells parallel to the coastline, (4) development of a freshwater ridge adjacent to the coast by means of a line of recharge wells parallel to the coastline, and (5) construction of an artificial subsurface barrier. Of these five alternatives, only the first has been proven effective and economic. Both Todd (1959) and Kazmann (1972) describe the application of the freshwater-ridge concept in the Silverado aquifer, an unconsolidated, confined, sand-and-gravel aquifer in the Los Angeles coastal basin of California. Kazmann concludes that the project was technically successful, but he notes that the economics of the project remain a subject of debate.

Suggested Readings

Bouwer, H., and R. D. Jackson. 1974. Determining soil properties. *Drainage for Agriculture*, ed. J. van Schilfgaarde. American Society of Agronomy, Madison, Wis., pp. 611–672.

Cooper, H. H. Jr., F. A. Kohout, H. R. Henry, and R. E. Glover. 1964. Sea water in coastal aquifers. *U.S. Geol. Surv. Water-Supply Paper 1613C*, 84 pp.

Ferris, J. G., D. B. Knowles, R. H. Browne, and R. W. Stallman. 1962. Theory of aquifer tests. *U.S. Geol. Surv. Water-Supply Paper 1536E*.

Hantush, M. S. 1964. Hydraulics of wells. *Adv. Hydrosci.*, 1, pp. 281–432.

Kruseman, G. P., and N. A. de Ridder. 1970. Analysis and evaluation of pumping test data. *Intern. Inst. for Land Reclamation and Improvement Bull. 11*, Wageningen, The Netherlands.

Neuman, S. P., and P. A. Witherspoon. 1969. Applicability of current theories of flow in leaky aquifers. *Water Resources Res.*, 5, pp. 817–829.

Poland, J. F., and G. H. Davis. 1969. Land subsidence due to withdrawal of fluids. *Geol. Soc. Amer. Rev. Eng. Geol.*, 2, pp. 187–269.

Prickett, T. A. 1975. Modeling techniques for groundwater evaluation. *Adv. Hydrosci.*, 11, pp. 46–66, 91–116.

Remson, I., G. M. Hornberger, and F. J. Molz. 1971. *Numerical Methods in Subsurface Hydrology*. Wiley Interscience, New York, pp. 56–122.

Stallman, R. W. 1971. Aquifer-test design, observation and data analysis. *Techniques of Water Resources Investigations of the U.S. Geological Survey*, Chapter B1. Government Printing Office, Washington, D.C.

Young, R. A., and J. D. Bredehoeft. 1972. Digital computer simulation for solving management problems of conjunctive groundwater and surface-water systems. *Water Resources Res.*, 8, pp. 533–556.

Problems

1. (a) Show by dimensional analysis on Eq. (8.6) that u is dimensionless.
   (b) Show by dimensional analysis on Eq. (8.7) that $W(u)$ is dimensionless.
   (c) Show that the values of the coefficients A and B given in connection with Eqs. (8.38) and (8.39) are correct for the engineering system of units commonly used in North America in which volumes are measured in U.S. gallons.

2. A fully penetrating well pumps water from an infinite, horizontal, confined, homogeneous, isotropic aquifer at a constant rate of 25 ℓ/s. If T is 1.2×10^{-2} m²/s and S is 2.0×10^{-4}, make the following calculations.
   (a) Calculate the drawdown that would occur in an observation well 60 m from the pumping well at times of 1, 5, 10, 50, and 210 min after the start of pumping. Plot these values on a log-log graph of $h_0 - h$ versus t.
   (b) Calculate the drawdown that would occur in a set of observation wells at distances 1 m, 3 m, 15 m, 60 m, and 300 m from the pumping well at a time 210 min after the start of pumping. Plot these values on a semilog graph of $h_0 - h$ versus r.

3. A confined aquifer with $T = 7.0 \times 10^{-3}$ m²/s and $S = 5.0 \times 10^{-4}$ is pumped by two wells 35 m apart. One well is pumped at 7.6 ℓ/s and one at 15.2 ℓ/s. Plot the drawdown $h_0 - h$ as a function of position along the line joining the two wells at a time 4 h after the start of pumping.

4. (a) Why is a 10-day pumping test better than a 10-h pumping text?
   (b) Why are storativities for unconfined aquifers so much larger than those for confined aquifers?
   (c) What kind of pumping-test arrangement would be required to determine the exact location of a straight, vertical impermeable boundary?

5. (a) List the assumptions underlying the Theis solution.
   (b) Sketch two plots that show the approximate shape you would expect for the time drawdown curve from a confined aquifer if:
      (1) The aquifer pinches out to the west.
      (2) The overlying confining formations are impermeable, but the underlying formations are leaky.
      (3) The pumping well is located near a fault that is in hydraulic connection to a surface stream.
      (4) The well is on the shore of a tidal estuary.
      (5) The pump broke down halfway through the test.
      (6) The barometric pressure increased at the pump test site.

6. (a) Plot the values of u versus $W(u)$ given in Table 8.1 on a log-log graph. It is only necessary to plot those values lying in the range 10^{-}[illegible] $< u < 1$.
   (b) Plot these same values as $1/u$ versus $W(u)$ on a log-log graph.

7. The thickness of a horizontal, confined, homogeneous, isotropic aquifer of infinite areal extent is 30 m. A well fully penetrating the aquifer was continuously pumped at a constant rate of 0.1 m^3/s for a period of 1 day. The drawdowns given in the attached table were observed in a fully penetrating observation well 90 m from the pumping well. Compute the transmissivity and the storativity by using:

(a) The Theis method of log-log matching [using the type curve prepared in Problem 6(b)].

(b) The Jacob method of semilog plotting.

t (min)	$h_0 - h$ (m)	t	$h_0 - h$	t	$h_0 - h$	t	$h_0 - h$
1	0.14	7	0.39	40	0.66	100	0.81
2	0.22	8	0.40	50	0.70	200	0.90
3	0.28	9	0.42	60	0.71	400	0.99
4	0.32	10	0.44	70	0.73	800	1.07
5	0.34	21	0.55	80	0.76	1000	1.10
6	0.37	30	0.62	90	0.79		

8. A homogeneous, isotropic, confined aquifer is 30.5 m thick and infinite in horizontal extent. A fully penetrating production well is pumped at a constant rate of 38 ℓ/s. The drawdown in an observation well 30.5 m from the production well after 200 days is 2.56 m.

(a) Assume a reasonable value for the storativity and then calculate the transmissivity T for the aquifer.

(b) Calculate the hydraulic conductivity and the compressibility of the aquifer. (Assume reasonable values for any unknown parameters.)

9. (a) A well pumps at 15.7 ℓ/s from a horizontal, confined, homogeneous, isotropic aquifer. The attached table lists the drawdown observed in an observation well 30 m from the pumping well. Plot these data on a semilogarithmic graph and use the Jacob method on the early data to calculate T and S.

(b) What kind of boundary is indicated by the break in slope? Measure the slope of the two limbs and note that the second limb has twice the slope of the first limb. In this case, how many image wells must be needed to provide an equivalent aquifer of infinite extent? Draw a sketch showing a possible configuration of pumping well, image well(s), and boundary, and note whether the image wells are pumping wells or recharge wells.

t (min)	$h_0 - h$ (m)	t	$h_0 - h$	t	$h_0 - h$	t	$h_0 - h$
11	2.13	21	2.50	52	3.11	88	3.70
14	2.27	28	2.68	60	3.29	100	3.86
18	2.44	35	2.80	74	3.41	112	4.01
						130	4.14

10. The straight-line portion of a semilog plot of drawdown versus time taken from an observation well 200 ft from a pumping well ($Q = 500$ U.S. gal/min) in a confined aquifer goes through the points ($t = 4 \times 10^{-4}$ day, $h_0 - h = 1.6$ ft) and ($t = 2 \times 10^{-2}$ day, $h_0 - h = 9.4$ ft).

(a) Calculate T and S for the aquifer.

(b) Calculate the drawdown that would occur 400 ft from the pumping well 10 h after the start of pumping.

11. (a) The hydraulic conductivity of a 30-m-thick confined aquifer is known from laboratory testing to have a value of 4.7×10^{-4} m/s. If the straight-line portion of a Jacob semilogarithmic plot goes through the points ($t = 10^{-3}$ day, $h_0 - h = 0.3$ m) and ($t = 10^{-2}$ day, $h_0 - h = 0.6$ m) for an observation well 30 m from a pumping well, calculate the transmissivity and storativity of the aquifer.

(b) Over what range of time values is the Jacob method of analysis valid for this observation well in this aquifer?

12. You are asked to design a pump test for a confined aquifer in which the transmissivity is expected to be about 1.4×10^{-2} m²/s and the storativity about 1.0×10^{-4}. What pumping rate would you recommend for the test if it is desired that there be an easily measured drawdown of at least 1 m during the first 6 h of the test in an observation well 150 m from the pumping well?

13. (a) Venice, Italy, has subsided 20 cm in 35 years; San Jose, California, has subsided 20 ft in 35 years. List the hydrogeological conditions that these two cities must have in common (in that they have both undergone subsidence), and comment on how these conditions may differ (to account for the large difference in total subsidence).

(b) The following data were obtained from a laboratory consolidation test on a core sample with a cross-sectional area of 100.0 cm² taken from a confining clay bed at Venice. Calculate the compressibility of the sample in m²/N that would apply at an effective stress of 2.0×10^6 N/m².

Load (N)	0	2000	5000	10,000	15,000	20,000	30,000
Void ratio	0.98	0.83	0.75	0.68	0.63	0.59	0.56

(c) Calculate the coefficient of compressibility, a_v, and the compression index, C_c, for these data. Choose a K value representative of a clay and calculate the coefficient of consolidation, c_v.

14. It is proposed to construct an unlined, artificial pond near the brink of a cliff. The geological deposits are unconsolidated, interbedded sands and clays. The water table is known to be rather deep.

(a) What are the possible negative impacts of the proposed pond?

(b) List in order, and briefly describe, the methods of exploration you would recommend to clarify the geology and hydrogeology of the site.

(c) List four possible methods that could be used to determine hydraulic conductivities. Which methods would be the most reasonable to use? The least reasonable? Why?

15. An undisturbed cylindrical core sample of soil 10 cm high and 5 cm in diameter weights 350 gm. Calculate the porosity.

16. If the water level in a 5-cm-diameter piezometer standpipe recovers 90% of its bailed drawdown in 20 h, calculate K. The intake is 0.5 m long and the same diameter as the standpipe. Assume that the assumptions underlying the Hvorslev point test are met.

17. Assume that the grain-size curve of Figure 8.25(a) is shifted one ϕ unit to the left. Calculate the hydraulic conductivity for the soil according to both the Hazen relation and the Masch and Denny curves.

18. (a) Develop the transient finite-difference equation for an internal node in a three-dimensional, homogeneous, isotropic nodal grid where $\Delta x = \Delta y = \Delta z$.

(b) Develop the transient finite-difference equation for a node adjacent to an impermeable boundary in a two-dimensional homogeneous, isotropic system with $\Delta x = \Delta y$. Do so:

(1) Using the simple approach of Section 8.8.

(2) The more sophisticated approach of Appendix IX.

19. Assume that resistors in the range 10^4–10^5 Ω and capacitors in the range 10^{-12}–10^{-11} F are commercially available. Choose a set of scale factors for the analog simulation of an aquifer with $T \simeq 10^5$ U.S. gal/day/ft and $S \simeq 3 \times 10^{-3}$. The aquifer is approximately 10 miles square and drawdowns of 10's of feet are expected over 10's of years in response to total pumping rates up to 10^6 U.S. gal/day.

Groundwater Contamination

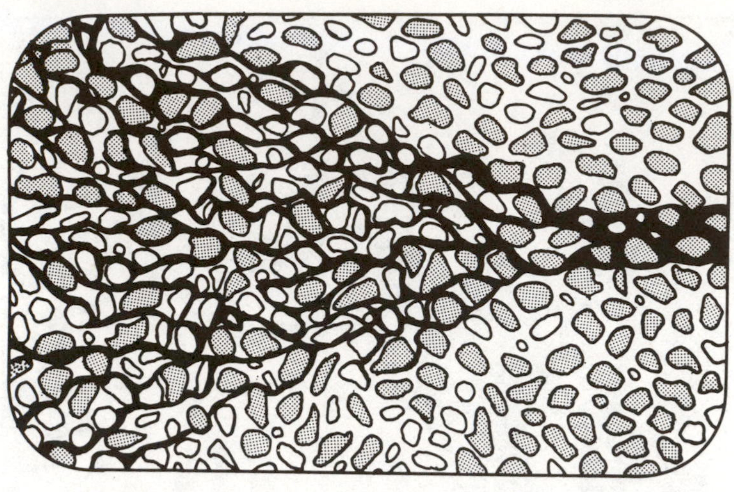

During recent years much of the emphasis in groundwater investigations in industrialized countries has shifted from problems of groundwater supply to considerations of groundwater quality. As a result of our consumptive way of life, the groundwater environment is being assaulted with an ever-increasing number of soluble chemicals. Current data indicate that in the United States there are at least 17 million waste disposal facilities emplacing more than 6.5 billion cubic meters of liquid into the ground each year (U.S. Environmental Protection Agency, 1977). As time goes on, the vast subsurface reservoir of fresh water, which a few decades ago was relatively unblemished by man's activities, is gradually becoming degraded.

The problem of water quality degradation of rivers and lakes has been evident for a long time. In general, solutions to this problem have been found in the implementation of effective legislation for discontinuing contaminant emissions. Already in some parts of the world, effective emission abatement measures have led to great improvements in surface-water quality. Unfortunately, problems of groundwater quality degradation are in many ways more difficult to overcome. Because of the heterogeneities inherent in subsurface systems, zones of degraded groundwater can be very difficult to detect. The U.S. Environmental Protection Agency (1977) has reported that almost every known instance of aquifer contamination has been discovered only after a water-supply well has been affected. Often by the time subsurface pollution is conclusively identified, it is too late to apply remedial measures that would be of much benefit. From a water quality viewpoint, degradation of groundwater often requires long periods of time before the true extent of the problem is readily detectable. Long periods of groundwater flow are often required for pollutants to be flushed from contaminated aquifers. Groundwater pollution often results in aquifers or parts of aquifers being damaged beyond repair.

Whereas the problem of achieving acceptable quality of surface waters focuses mainly on decreasing the known emissions of pollutants to these systems, the problem facing scientists and engineers involved in the protection of groundwater resources is to identify the areas and mechanisms by which pollutants can enter groundwater flow systems and to develop reliable predictions of the transport of contaminants within the flow systems. This is necessary as a basis for minimizing the impact of existing or proposed industrial, agricultural, or municipal activities on groundwater quality.

The purpose of this chapter is to provide some insight into the physical and chemical factors that influence the subsurface migration of dissolved contaminants. To this end the behavior of nonreactive solutes and of solutes that undergo reactions during subsurface migration will be considered. Following this, more specific contamination problems related to activities such as agriculture, mining, nuclear power development, and disposal of refuse, sewage, and industrial wastes will be briefly reviewed.

Throughout this chapter all solutes introduced into the hydrologic environment as a result of man's activities are referred to as *contaminants*, regardless of whether or not the concentrations reach levels that cause significant degradation of water quality. The term *pollution* is reserved for situations where contaminant concentrations attain levels that are considered to be objectionable.

The emphasis in this chapter is on the occurrence and processes that control the migration of *dissolved* contaminants in groundwater. Groundwater can also be contaminated by oily substances that exist in a liquid state in contact with water in a manner that does not lead to mixing of the oils in a dissolved form. The oily liquid is said to be *immiscible* in the water. The physical processes that control the movement of immiscible fluids in subsurface systems are described by Bear (1972) and are introduced in Section 9.5.

9.1 Water Quality Standards

Before proceeding with discussions of the principles of contaminant behavior in groundwater flow systems and of sources of groundwater contamination, we will briefly examine some of the more important water quality standards. These standards serve as a basis for appraisal of the results of chemical analyses of water in terms of suitability of the water for various intended uses. The most important of these standards are those established for drinking water (Table 9.1). The recommended limits for concentrations of inorganic constituents in drinking water have existed for many years. Limits for organic constutuents such as pesticide residues are a recent addition. There is considerable controversy with regard to the specific organic constituents that should be included in drinking water standards and the concentration limits that should be established for them.

In Table 9.1 the major constituents for which recommended permissible limits are listed are total dissolved solids (TDS), sulfate, and chloride. Consumption by humans of waters with concentrations somewhat above these limits is generally not harmful. In many regions groundwater used for drinking-water supply exceeds the limits of one or more of these parameters. Several hundred milligrams per liter of chloride must be present in order for saltiness to be detected by taste.

Table 9.1 Drinking Water Standards

Constituent	Recommended concentration limit* (mg/ℓ)
Inorganic	
Total dissolved solids	500
Chloride (Cl)	250
Sulfate (SO_4^{2-})	250
Nitrate (NO_3^-)	45†
Iron (Fe)	0.3
Manganese (Mn)	0.05
Copper (Cu)	1.0
Zinc (Zn)	5.0
Boron (B)	1.0
Hydrogen sulfide (H_2S)	0.05
	Maximum permissible concentration‡
Arsenic (As)	0.05
Barium (Ba)	1.0
Cadmium (Cd)	0.01
Chromium (Cr^{VI})	0.05
Selenium	0.01
Antimony (Sb)	0.01
Lead (Pb)	0.05
Mercury (Hg)	0.002
Silver (Ag)	0.05
Fluoride (F)	1.4–2.4§
Organic	
Cyanide	0.05
Endrine	0.0002
Lindane	0.004
Methoxychlor	0.1
Toxaphene	0.005
2,4-D	0.1
2,4,5-TP silvex	0.01
Phenols	0.001
Carbon chloroform extract	0.2
Synthetic detergents	0.5
Radionuclides and radioactivity	Maximum permissible activity (pCi/ℓ)
Radium 226	5
Strontium 90	10
Plutonium	50,000
Gross beta activity	30
Gross alpha activity	3
Bacteriological	
Total coliform bacteria	1 per 100 mℓ

SOURCES: U.S. Environmental Protection Agency, 1975 and World Health Organization, European Standards, 1970.

*Recommended concentration limits for these constituents are mainly to provide acceptable esthetic and taste characteristics.

†Limit for NO_3^- expressed as N is 10 mg/ℓ according to U.S. and Canadian standards; according to WHO European standards, it is 11.3 mg/ℓ as N and 50 mg/ℓ as NO_3^-.

Hardness of water is defined as its content of metallic ions which react with sodium soaps to produce solid soaps or scummy residue and which react with negative ions, when the water is evaporated in boilers, to produce solid boiler scale (Camp, 1963). Hardness is normally expressed as the total concentration of Ca^{2+} and Mg^{2+} as milligrams per liter equivalent $CaCO_3$. It can be determined by substituting the concentration of Ca^{2+} and Mg^{2+}, expressed in milligrams per liter, in the expression

$$\text{total hardness} = 2.5(Ca^{2+}) + 4.1(Mg^{2+}) \tag{9.1}$$

Each concentration is multiplied by the ratio of the formula weight of $CaCO_3$ to the atomic weight of the ion; hence the factors 2.5 and 4.1 are included in the hardness relation. Water with hardness values greater than 150 mg/ℓ is designated as being very hard. Soft water has values less than 60 mg/ℓ. Water softening is common practice in many communities where the water supply has a hardness greater than about 80–100 mg/ℓ. Water used for boiler feed will cause excessive scale formation (carbonate-mineral precipitation) if the hardness is above about 60–80 mg/ℓ.

Of the recommended limits specified for minor and trace inorganic constituents in drinking water, many have been established for reasons other than direct hazard to human health. For example iron and manganese are both essential to the human body. Their intake through drinking water is normally an insignificant part of the body requirement. The recommended limits placed on these metals in the Standards is for the purpose of avoiding, in household water use, problems associated with precipitates and stains that form because oxides of these metals are relatively insoluble (Camp, 1963). The recommended limit for zinc is set at 5 mg/ℓ to avoid taste produced by zinc at higher concentrations. Concentrations as high as 40 mg/ℓ can be tolerated with no apparent deteriment to general health. Zinc concentrations as low as 0.02 mg/ℓ are, however, toxic to fish. Zinc contamination can be regarded as severe pollution in ecological systems where fish are of primary interest but may be only of minor significance if human consumption is the primary use of the water.

The most common identifiable contaminant in groundwater is nitrate (NO_3^-). The recommended limit for nitrate in drinking water is 45 mg/ℓ expressed as NO_3^- or 10 mg/ℓ expressed as N. In Europe the limit recommended by the World Health Organization is 50 mg/ℓ as NO_3^- and 11.3 mg/ℓ as N. Excessive concentrations of NO_3^- have potential to harm infant human beings and livestock if consumed on a regular basis. Adults can tolerate much higher concentrations. The extent to which NO_3^- in water is viewed as a serious pollutant therefore depends on the water use.

The constituents for which *maximum permissible concentration limits* have

---

‡Maximum permissible limits are set according to health criteria.

§Limit depends on average air temperature of the region; fluoride is toxic at about 5–10 mg/ℓ if water is consumed over a long period of time.

been set in drinking water standards (Table 9.1) are all considered to have significant potential for harm to human health at concentrations above the specified limits. The specified limits are not to be exceeded in public water supplies. If the limits for one or more of the constituents are exceeded, the water is considered to be unfit for human consumption. The limits indicated in Table 9.1 are representative of the current standards in the United States and Canada. The limits are continually being appraised and modifications occur from time to time. As more is learned about the role of trace constituents in human health, the list of constituents for which maximum permissible limits exist may expand, particularly in the case of organic substances.

In many regions the most important uses of groundwater are for agriculture. In these situations it is appropriate to appraise the quality of groundwater relative to criteria or guidelines established for livestock or irrigation. Recommended concentration limits for these uses are listed in Table 9.2. The list of constituents and the concentration limits are not as stringent as for drinking water. These water quality criteria do serve to indicate, however, that concentration increases in a variety of constituents due to man's activities can cause serious degradation of groundwater quality even if the water is not used for human consumption.

Table 9.2 Recommended Concentration Limits for Water Used for Livestock and Irrigation Crop Production

	Livestock: Recommended limits (mg/ℓ)	Irrigation crops: Recommended limits (mg/ℓ)
Total dissolved solids		
Small animals	3000	700
Poultry	5000	
Other animals	7000	
Nitrate	45	—
Arsenic	0.2	0.1
Boron	5	0.75
Cadmium	0.05	0.01
Chromium	1	0.1
Fluoride	2	1
Lead	0.1	5
Mercury	0.01	—
Selenium	0.05	0.02

SOURCE: U.S. Environmental Agency, 1973b.

9.2 Transport Processes

The common starting point in the development of differential equations to describe the transport of solutes in porous materials is to consider the flux of solute into and out of a fixed elemental volume within the flow domain. A conservation of mass statement for this elemental volume is

$$\begin{bmatrix}\text{net rate of}\\ \text{change of mass}\\ \text{of solute within}\\ \text{the element}\end{bmatrix} = \begin{bmatrix}\text{flux of}\\ \text{solute out}\\ \text{of the}\\ \text{element}\end{bmatrix} - \begin{bmatrix}\text{flux of}\\ \text{solute into}\\ \text{the}\\ \text{element}\end{bmatrix} \pm \begin{bmatrix}\text{loss or gain}\\ \text{of solute mass}\\ \text{due to}\\ \text{reactions}\end{bmatrix} \tag{9.2}$$

The physical processes that control the flux into and out of the elemental volume are *advection* and *hydrodynamic dispersion.* Loss or gain of solute mass in the elemental volume can occur as a result of chemical or biochemical *reactions* or radioactive decay.

Advection is the component of solute movement attributed to transport by the flowing groundwater. The rate of transport is equal to the average linear groundwater velocity, $\bar{v}$, where $\bar{v} = v/n$, v being the specific discharge and n the porosity (Section 2.12). The advection process is sometimes called *convection,* a term that in this text is reserved for use in discussion of thermally driven groundwater flow as described in Chapter 11. The process of hydrodynamic dispersion, which is described in Section 2.13, occurs as a result of mechanical mixing and molecular diffusion.

Mathematical descriptions of dispersion are currently limited to materials that are isotropic with respect to dispersion properties of the medium. The principal differential equation that describes transport of dissolved reactive constituents in saturated isotropic porous media is derived in Appendix X. This equation is known as the *advection-dispersion equation.* Our purpose here is to examine the physical significance of the terms in this equation (advection, dispersion, and reaction). We will start with the physical processes and then turn our attention to the chemical processes.

Nonreactive Constituents in Homogeneous Media

The one-dimensional form of the advection-dispersion equation for nonreactive dissolved constituents in saturated, homogeneous, isotropic, materials under steady-state, uniform flow [Eq. (A10.11), Appendix X] is

$$D_l \frac{\partial^2 C}{\partial l^2} - \bar{v}_l \frac{\partial C}{\partial l} = \frac{\partial C}{\partial t} \tag{9.3}$$

where l is a curvilinear coordinate direction taken along the flowline, $\bar{v}$ is the average linear groundwater velocity, D_l is the coefficient of hydrodynamic dispersion in the longitudinal direction (i.e., along the flow path), and C is the solute concentration. The effects of chemical reactions, biological transformations, and radioactive decay are not included in this form of the transport equation.

The coefficient of hydrodynamic dispersion can be expressed in terms of two components,

$$D_l = \alpha_l \bar{v} + D^* \tag{9.4}$$

where α_l is a characteristic property of the porous medium known as the dynamic dispersivity, or simply as *dispersivity* $[L]$, and D^* is the *coefficient of molecular diffusion* for the solute in the porous medium $[L^2/T]$. The relation between D^* and the coefficient of diffusion for the solute species in water is described in Section 3.4. Some authors have indicated that a more accurate form of the mechanical component of the dispersion coefficient is $\alpha\bar{v}^m$, where m is an empirically determined constant between 1 and 2. Laboratory studies indicate that for practical purposes m can generally be taken as unity for granular geologic materials.

The classical experiment shown in Figure 9.1(a) is one of the most direct ways of illustrating the physical meaning of the one-dimensional form of the advection-dispersion equation. In this experiment, a nonreactive tracer at concentration C_0 is continuously introduced into a steady-state flow regime at the upstream end of a column packed with a homogeneous granular medium. For illustrative purposes it is assumed that the tracer concentration in the column prior to the introduction of the tracer is zero. It is convenient to express the tracer concentration in the

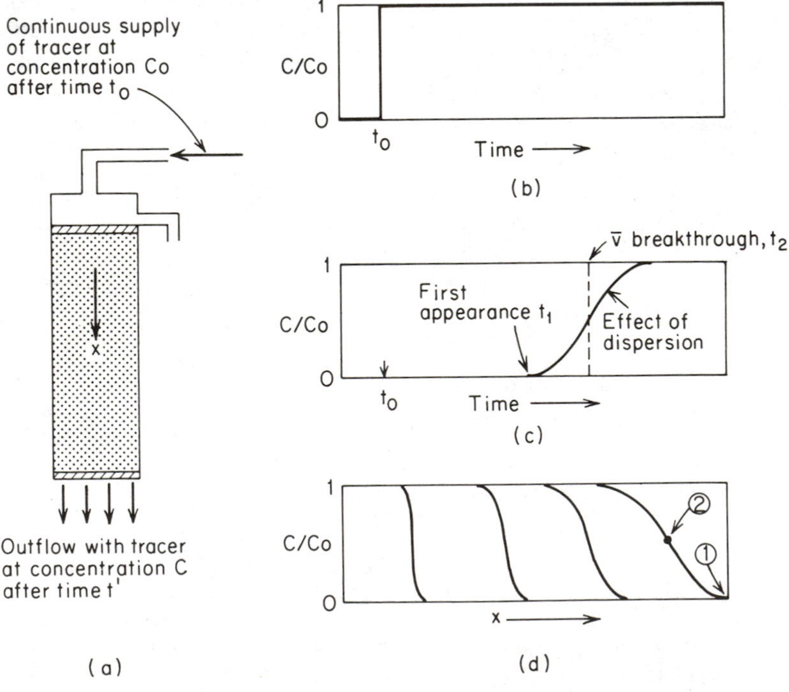

Figure 9.1 Longitudinal dispersion of a tracer passing through a column of porous medium. (a) Column with steady flow and continuous supply of tracer after time t_0; (b) step-function-type tracer input relation; (c) relative tracer concentration in outflow from column (dashed line indicates plug flow condition and solid line illustrates effect of mechanical dispersion and molecular diffusion); (d) concentration profile in the column at various times.

column as a relative concentration, defined as C/C_0, where C is the concentration in the column or in the output. The tracer input can therefore be represented as a step function, as shown in Figure 9.1(b). The concentration versus time relation of the column outflow, known as *the breakthrough curve*, is shown in Figure 9.1(c). If it is assumed that the tracer moves through the column with no mechanical dispersion or molecular diffusion, the tracer front will pass through as a plug and will exit from the column as a step function. This condition is shown as a vertical dashed line in Figure 9.1(c). In real situations, however, mechanical dispersion and molecular diffusion occur and the breakthrough curve spreads out causing the tracer to begin to appear in the outflow from the column (at time t_1) before the arrival of water traveling at the velocity of $\bar{v}$ (time t_2). This is represented in Figure 9.1(c).

Figure 9.1(d) shows instantaneous "pictures" of the dispersion interface inside the column at various times prior to breakthrough. The tracer front is spread out along the flow path. The spread of the profile increases with travel distance. The positions represented by points 1 and 2 in Figures 9.1(d) correspond to times t_1 and t_2 in Figure 9.1(c). Mechanical dispersion and molecular diffusion cause some of the tracer molecules to move faster than the average linear velocity of the water and some to move slower. The average linear velocity of the water in the column is determined by dividing the water input rate (Q) by nA, where A is the cross-sectional area of the column and n is the porosity [Eq. (2.82)].

The boundary conditions represented by the step-function input are described mathematically as

$$C(l, 0) = 0 \qquad l \geq 0$$

$$C(0, t) = C_0 \qquad t \geq 0$$

$$C(\infty, t) = 0 \qquad t \geq 0$$

For these boundary conditions the solution to Eq. (9.3) for a saturated homogeneous porous medium is (Ogata, 1970)

$$\frac{C}{C_0} = \frac{1}{2}\left[\operatorname{erfc}\left(\frac{l - \bar{v}t}{2\sqrt{D_l t}}\right) + \exp\left(\frac{\bar{v}l}{D_l}\right)\operatorname{erfc}\left(\frac{l + \bar{v}t}{2\sqrt{D_l t}}\right)\right] \tag{9.5}$$

where erfc represents the complementary error function, which is tabulated in Appendix V; l is the distance along the flow path; and $\bar{v}$ is the average linear water velocity. For conditions in which the dispersivity of the porous medium is large or when l or t is large, the second term on the right-hand side of the equation is negligible. Equation (9.5) can be used to compute the shapes of the breakthrough curves and concentration profiles illustrated in Figure 9.1(c) and (d). Analytical solutions for Eq. (9.3) with other boundary conditions are described by Rifai et al. (1956), Ebach and White (1958), Ogata and Banks (1961), Ogata (1970), and others.

The spreading out of the concentration profile and breakthrough curve of tracers or contaminants migrating through porous materials is caused by both mechanical dispersion and molecular diffusion. Figure 9.2 shows a concentration profile for the experimental conditions represented in Figure 9.1(a). In this graph

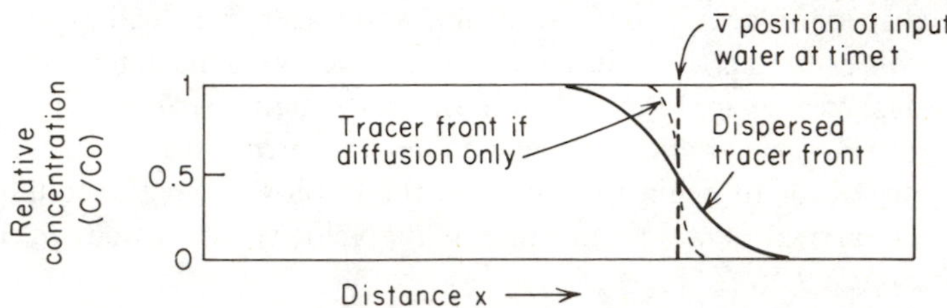

Figure 9.2 Schematic diagram showing the contribution of molecular diffusion and mechanical dispersion to the spread of a concentration front in a column with a step-function input.

the contribution of molecular diffusion to the spread of the curves is indicated schematically. At a low velocity, diffusion is the important contributor to the dispersion, and therefore the coefficient of hydrodynamic dispersion equals the diffusion coefficient ($D_l = D^*$). At a high velocity, mechanical mixing is the dominant dispersive process, in which case $D_l = \alpha_l \bar{v}$. Larger dispersivity of the medium produces greater mixing of the solute front as it advances. Laboratory experiments on tracer migration in saturated homogeneous granular materials have established relations between the influence of diffusion and mechanical dispersion, as illustrated in Figure 9.3. The dimensionless parameter $\bar{v}d/D^*$ is known as the *Peclet number*, where the average particle diameter is denoted by d. The exact shape of the relation between the Peclet number and D_l/D^* depends on the nature of the

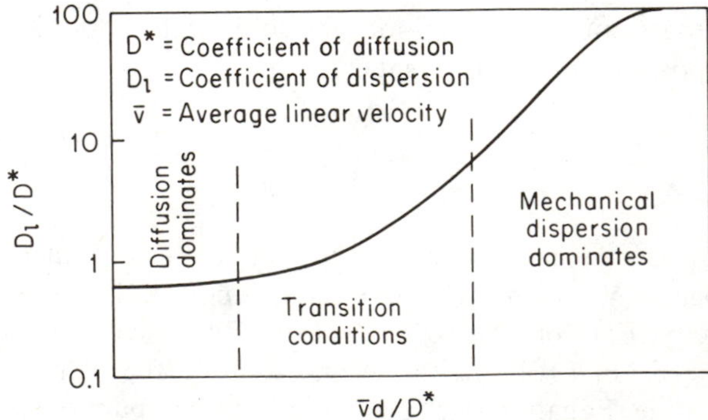

Figure 9.3 Relation between the Peclet number and the ratio of the longitudinal dispersion coefficient and the coefficient of molecular diffusion in a sand of uniform-sized grains (after Perkins and Johnston, 1963).

porous medium and on the fluid used in the experiments. The general shape illustrated in Figure 9.3 has been established by various investigators on the basis of experiments using different media (Bear, 1972).

In situations where the boundary conditions specified for Eq. (9.5) are applicable and where the groundwater velocity is so small that mechanical dispersion is negligible relative to molecular diffusion, Eq. (9.5) reduces to the one-dimensional solution to Fick's second law. This "law" is described in Section 3.4. The rate at which one-dimensional diffusion occurs is expressed graphically in Figure 9.4, which shows, for periods of diffusion of 100 and 10,000 years, diffusion distances as a function of relative concentration. The diffusion distances were obtained using Eq. (3.47) with diffusion coefficient values of 1×10^{-10} and 1×10^{-11} m^2/s.

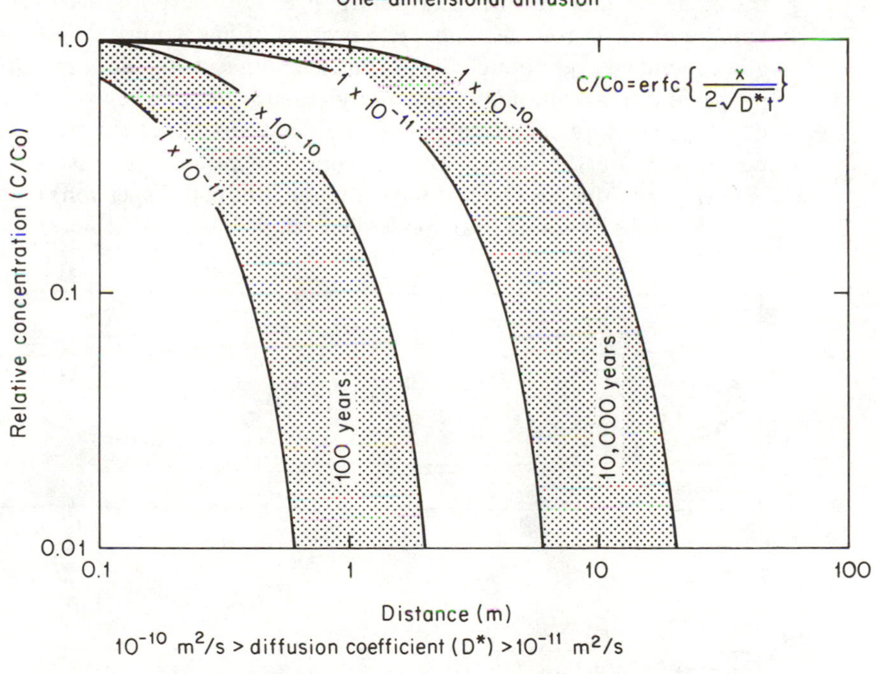

Figure 9.4 Positions of contaminant front migrating by molecular diffusion away from a source where $C = C_0$ at $t > 0$. Migration times are 100 and 10,000 years.

These values are representative of a range typical of nonreactive chemical species in clayey geologic deposits. Values for coarse-grained unconcolidated materials can be somewhat higher than 1×10^{-10} m^2/s but are less than the coefficients for the chemical species in water (i.e., $< 2 \times 10^{-9}$ m^2/s). Figure 9.4 indicates that over long periods of time, diffusion can cause contaminants to move considerable distances, even through low-permeability materials. Whether contaminant migration on this time scale is important depends on the nature of the problem. In the case of subsurface disposal of radioactive wastes or highly toxic inorganic or organic compounds, diffusion can be an important process.

One of the characteristic features of the dispersive process is that it causes spreading of the solute, if the opportunity is available, in directions transverse to the flow path as well as in the longitudinal flow direction. This is illustrated schematically for a two-dimensional horizontal flow field in Figure 9.5(a). In this experimental sand box, a nonreactive tracer is introduced as a continuous steady input to the uniform flow field. Dispersion in this two-dimensional flow domain is illustrated in a different manner by the experiment shown in Figure 9.5(b). In this case the tracer is introduced as an instantaneous point source (i.e., a slug of tracer) into the uniform flow regime. As the tracer is transported along the flow path, it spreads in all directions in the horizontal plane. The total mass of the tracer in the flow regime does not change, but the mass occupies an increasing volume of the porous medium. The process of mechanical dispersion is directionally dependent even though the porous medium is isotropic with respect to textural properties and hydraulic conductivity. Figure 9.5(b) shows that the tracer zone develops an elliptical shape as the tracer is transported through the system. This occurs because the *process* of mechanical dispersion is anisotropic. Dispersion is stronger in the direction of flow (the longitudinal dispersion) than in directions normal to the flow line (transverse dispersion).

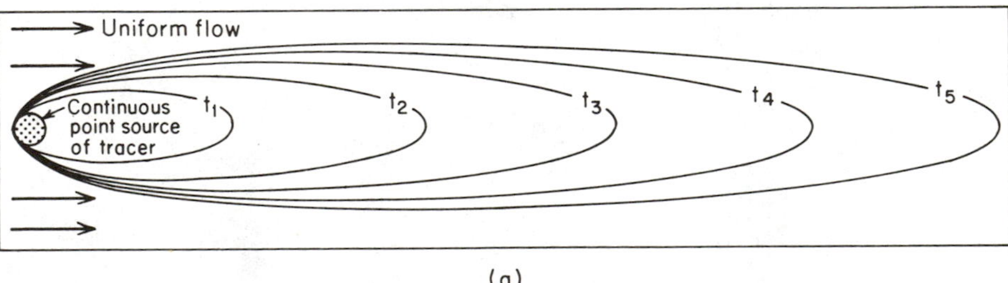

(a)

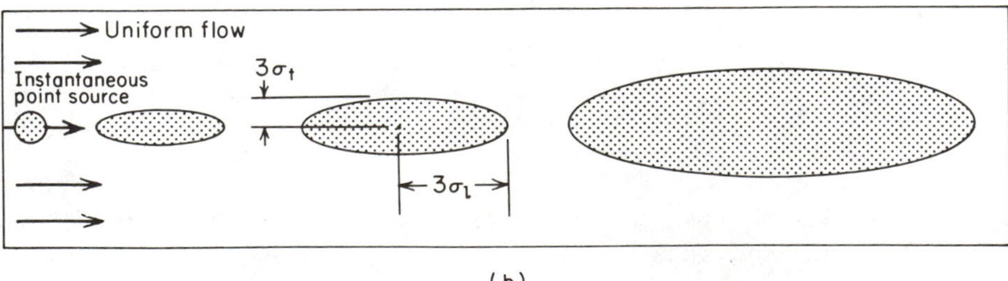

(b)

Figure 9.5 Spreading of a tracer in a two-dimensional uniform flow field in an isotropic sand. (a) Continuous tracer feed with step-function initial condition; (b) instantaneous point source.

One-dimensional expressions for the transport of dissolved constituents, such as Eq. (9.5), are useful in the interpretation of laboratory column experiments, but are of limited use in the analysis of field problems because dispersion occurs in the transverse directions as well as in the longitudinal direction. As an example of a solution to the advection-dispersion equation in three-dimensions [Eq. (A10.9), Appendix X], we will follow an approach described by Baetsle (1969). As in Figure 9.5(b), the contaminant is assumed to originate as an instantaneous slug at a point source at $x = 0$, $y = 0$, $z = 0$. The mass of contaminant is then carried away from the source by transport in a steady-state uniform flow field moving in the x-direction in a homogeneous isotropic medium. As the contaminant mass is transported through the flow system, the concentration distribution of the contaminant mass at time t is given by

$$C(x, y, z, t) = \frac{M}{8(\pi t)^{3/2}\sqrt{D_x D_y D_z}} \exp\left(-\frac{X^2}{4D_x t} - \frac{Y^2}{4D_y t} - \frac{Z^2}{4D_z t}\right) \tag{9.6}$$

where M is the mass of contaminant introduced at the point source, D_x, D_y, and D_z are the coefficients of dispersion in the x, y, z directions and X, Y, and Z are distances in the x, y, z directions from the center of gravity of the contaminant mass. The position of the center of gravity of the contaminant mass at time t will lie along the flow path in the x direction at coordinates (x_t, y_t, z_t), where $y_t = z_t = 0$ and $x_t = \bar{v}t = vt/n$, where $\bar{v}$ is the average linear velocity, v is the specific discharge, and n is the porosity. In Eq. (9.6), $X = x - \bar{v}t$, $Y = y$, and $Z = z$. It is apparent from Eq. (9.6) that the maximum concentration is located at the center of gravity of the contaminant cloud, where $X = 0$, $Y = 0$, and $Z = 0$. The mass of the contaminant introduced at the source equals C_0V_0, where C_0 is the initial concentration and V_0 is the initial volume. In the mathematical formulation of the initial conditions, the contaminant input occurs at a point and therefore has mass but no volume. In practice, however, this is expressed by the quantity C_0V_0.

From Eq. (9.6) it follows that the peak concentration that occurs at the center of gravity of the contaminant cloud is given by

$$C_{\max} = \frac{C_0V_0}{8(\pi t)^{3/2}\sqrt{D_x D_y D_z}} \tag{9.7}$$

The zone in which 99.7% of the contaminant mass occurs is described by the ellipsoid with dimensions, measured from the center of mass, of $3\sigma_x = \sqrt{2D_x t}$, $3\sigma_y = \sqrt{2D_y t}$, $3\sigma_z = \sqrt{2D_z t}$, where σ is the standard deviation of the concentration distribution. This is illustrated in the xy plane in Figure 9.5(b). At low velocities molecular diffusion is the dominant dispersive mechanism, in which case the migrating contaminant cloud is circular. Because these equations are based on idealized conditions, such as the instantaneous point source and uniform flow, they have limited use in the analysis of most field situations. In simple hydrogeologic settings, however, they can be used to obtain preliminary estimates of the migration patterns that may arise from small contaminant spills or from leaching of buried wastes (Baetsle, 1969). A variety of other analytical solutions describing the migration of contaminants in two- and three-dimensional space are described by Fried (1975) and Codell and Schreiber (in press).

Mechanical dispersion in the transverse direction is a much weaker process than dispersion in the longitudinal direction, but at low velocities where molecular diffusion is the dominant dispersive mechanism, the coefficients of longitudinal and transverse dispersion are nearly equal. This is illustrated by the experimental results shown in Figure 9.6, which indicates small dispersion coefficients over a range of low velocities. Because mechanical dispersion in the transverse direction is much weaker than in the longitudinal direction, the transverse dispersion coefficient remains diffusion-controlled until the flow velocity is quite high.

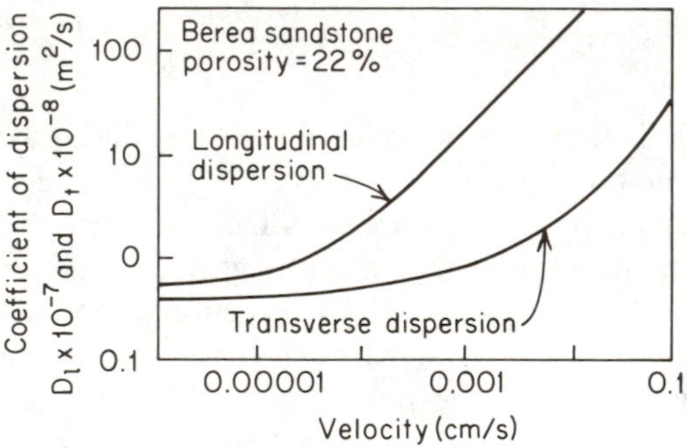

Figure 9.6 Coefficients of longitudinal and transverse dispersion for transport in a homogeneous sandstone at various flow rates (after Crane and Gardner, 1961).

The forms of the transport equation described above are based on the assumption that there is no significant density contrast between the contaminant or tracer fluid and the groundwater in the surrounding flow domain. Equations that make allowance for density contrasts are more complex. As a qualitative example of the effect of density contrasts, consider the sinking contaminant plume in an initially uniform flow field, as illustrated in Figure 9.7. If the contaminant solution

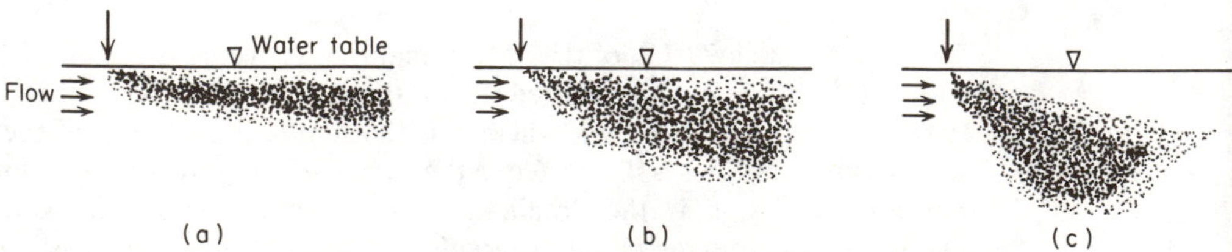

Figure 9.7 Effect of density on migration of contaminant solution in uniform flow field. (a) Slightly more dense than groundwater; (b) and (c) larger density contrasts.

entering this flow regime has the same density as the groundwater, the contaminant plume will spread in a shallow zone close to the water table. If the contaminant solution is considerably more dense than the groundwater, the plume will sink steeply downward into the groundwater flow system. Prediction of contaminant migration patterns requires accurate knowledge of the density of the contaminant solution as well as that of the groundwater.

Nonreactive Constituents in Heterogeneous Media

If it were not for the effects of heterogeneities in natural geological materials, the problem of prediction and detection of contaminant behavior in groundwater flow systems would be easily solved. Advection is the process whereby solutes are transported by the bulk mass of the flowing fluid. Advection is normally considered on the macroscopic scale in terms of the patterns of groundwater flow. These patterns are defined by the spatial and temporal distributions of the average linear velocity of the fluid. Flow patterns and flow nets have been described extensively in Chapters 5 and 6. Our purpose here is to consider in more detail the effects on flow lines and velocities exerted by various types of heterogeneities.

To illustrate the effect of simple layered heterogeneities on transport patterns, the cross-section flow-domain illustrated in Figure 9.8(a) is used. It is assumed that steady-state groundwater flow occurs through the cross section and that the

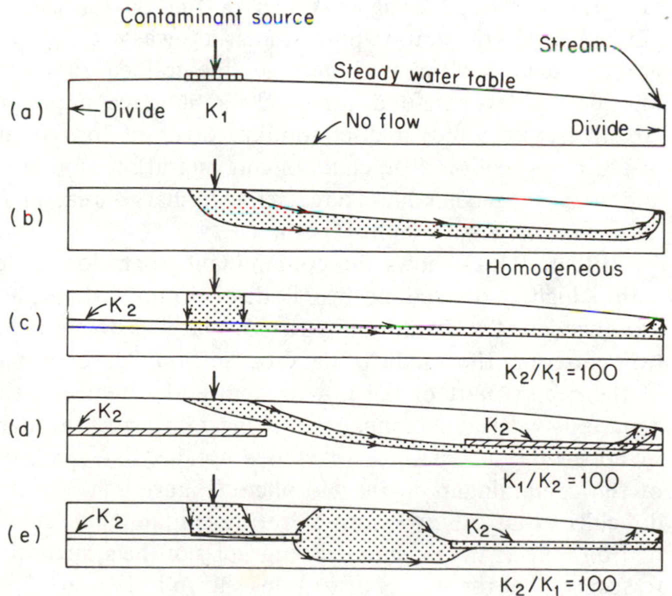

Figure 9.8 Effect of layers and lenses on flow paths in shallow steady-state groundwater flow systems. (a) Boundary conditions; (b) homogeneous case; (c) single higher-conductivity layer; (d) two lower-conductivity lenses; (e) two higher-conductivity lenses.

flow domain is isotropic with respect to hydraulic conductivity. To illustrate the effect of stratigraphic variations on the transport pattern of contaminants entering the system in a recharge area, a contaminant input zone is located on the cross section. In field situations this could represent seepage from a waste lagoon, sanitary landfill, or some other surface or near-surface source. Figure 9.8(b), (c), (d), and (e) shows the patterns of contaminant transport that would occur with various hypothetical stratigraphic configurations. The contaminant is assumed to be nonreactive and the effect of dispersion is neglected. The flow lines that depict the limits of the contaminant migration patterns were obtained by solution of the two-dimensional form of the steady-state groundwater flow equation [Eq. (2.69)], using the finite-element method in the manner described by Pickens and Lennox (1976). Figure 9.8(b) indicates that in situations where the flow domain is homogeneous, the contaminant migration pattern would be simple and relatively easy to monitor. The conditions for the flow system shown in Figure 9.8(c) are similar to the previous case, with the exception of the inclusion of a thin, higher-conductivity horizontal layer that extends across the flow domain. This would cause the contaminants to move through the flow system almost entirely in this thin layer. The total travel time would be one-fifth of the nonstratified case illustrated in Figure 9.8(b). The thin higher conductivity bed has a conductivity 100 times larger than the rest of the system and exerts a very strong influence on the migration patterns and velocity distribution. If the lower-K medium (K_1) represents a very fine-grained sand, the higher-K bed (K_2) could represent a medium- or coarse-grained sand. In stratigraphic studies of waste disposal sites, a thin medium-grained sand bed in an otherwise fine-grained sand deposit could easily be unnoticed unless careful drilling and sampling techniques are used.

In Figure 9.8(d) a discontinuous layer of low-conductivity material exists in the cross section. The contaminant migration zone moves over the first lense and under the second one. To reach the discharge area, it passes through the second lense near the end of its flow path.

Figure 9.8(e) shows the contaminant migration pattern that would exist if a thin higher-conductivity bed is discontinuous through the central part of the cross section. The discontinuity causes a large distortion in the contaminant migration pattern in the middle of the cross section. The contaminated zone spreads out in the central part of the flow system and extends to the water-table zone. In situations where contaminants can be transferred through the unsaturated zone by advection, diffusion, or vegetative uptake, this condition could lead to spread of the contaminants in the biosphere. Figure 9.8(e) also illustrates some of the difficulties that can arise in monitoring contaminated flow systems. If little information were available on the stratigraphy of the system, there would be no reason to suspect that the type of distortions shown in Figure 9.8(e) would occur. Lack of this information could result in inadequate monitoring of the system. In nature, geologic cross sections typically include many stratigraphic units with different hydraulic conductivities. Large conductivity contrasts across sharp discontinuities are common. Relative to real situations, the effects of stratification illustrated in Figure 9.8 are very simple.

In the discussion above, layered heterogeneities on the scale that could, if necessary, be identified and mapped by careful drilling, sampling, and geophysical logging were considered. Heterogeneities in another category also exist in most geologic settings. These are known as small-scale heterogeneities. They cannot be identified individually by conventional methods of field testing. Even if identification is possible using special coring techniques, these heterogeneities usually cannot be correlated from borehole to borehole. In granular aquifers, heterogeneities of this type are ubiquitous. Hydraulic conductivity contrasts as large as an order of magnitude or more can occur as a result of almost unrecognizable variations in grain-size characteristics. For example, a change of silt or clay content of only a few percent in a sandy zone can have a large effect on the hydraulic conductivity.

Figure 9.9 illustrates the effect of two types of small-scale heterogeneities on the pattern of migration of a tracer or contaminant in granular porous media. In Figure 9.9(a) the pattern of dispersion is regular and predictable using the methods described above. In Figure 9.9(b) the lense-type heterogeneities cause the

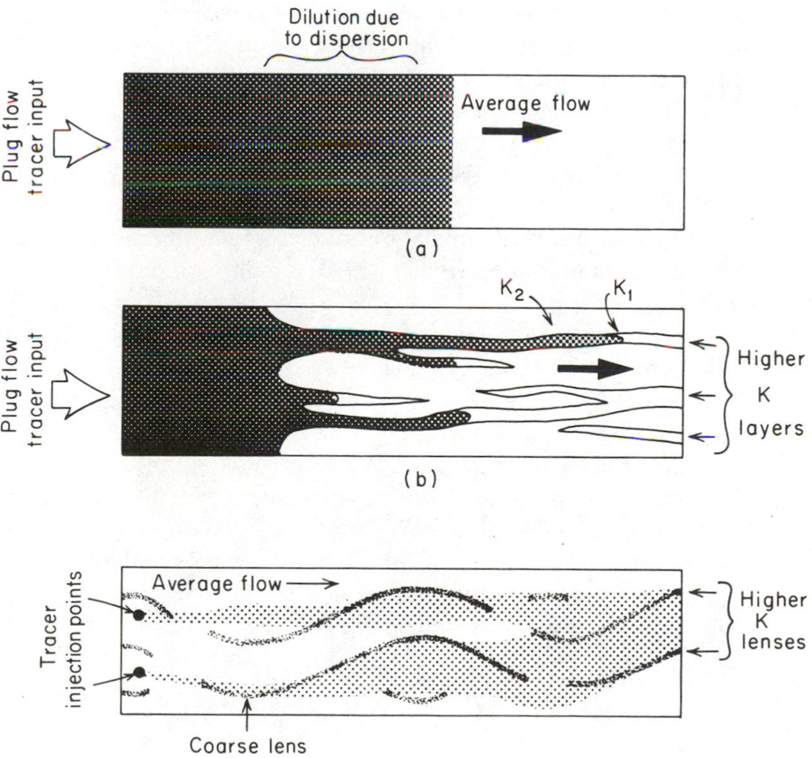

Figure 9.9 Comparison of advance of contaminant zones influenced by hydrodynamic dispersion. (a) Homogeneous granular medium; (b) fingering caused by layered beds and lenses; (c) spreading caused by irregular lenses.

tracer front to advance in a pattern commonly referred to as *fingering*. In this case the contaminant is transported more rapidly in the lenses or beds of higher hydraulic conductivity. Figure 9.9(c) illustrates results obtained by Skibitzke and Robertson (1963) using dye tracers in a box model packed with fine sand and long sinuous lenses of coarser sand. These authors observed that a large angle of refraction at the boundary between sand of contrasting permeability caused accelerated spreading of the tracer zone.

In one of the very few detailed three-dimensional studies of contaminant movement in sandy deposits, Childs et al. (1974) observed that "plumes migrate along zones . . . that, although they are texturally similar, show subtle differences in fabric that result in slight variations in permeability. Bifurcations indicate that detection of a shallow plume does not negate the existence of the other plumes of the same constituent at depth" (p. 369).

Nearly all studies of dispersion reported in the literature have involved relatively homogeneous sandy materials under controlled conditions in the laboratory. These studies have indicated that the dispersivity of these materials is small. Values of longitudinal dispersivity are typically in the range of 0.1 to 10 mm, with transverse dispersivity values normally lower by a factor of 5–20. Whether or not these values are at all indicative of dispersivities in field systems is subject to considerable controversy at the present time. Many investigators have concluded that values of longitudinal and transverse dispersivities in field systems are significantly larger than values obtained in laboratory experiments on homogeneous materials or on materials with simple heterogeneities. Values of longitudinal dispersivity as large as 100 m and lateral dispersivity values as large as 50 m have been used in mathematical simulation studies of the migration of large contaminant plumes in sandy aquifers (Pinder, 1973; Konikow and Bredehoeft, 1974; Robertson, 1974).

To illustrate the effect of large dispersivities on the migration of contaminants in a hypothetical groundwater flow system, a cross-sectional flow domain similar to that shown in Figure 9.8(a) and (b) will be used. Figure 9.10 shows the effect of dispersivity on the spreading of a contaminant plume that emanates from a source in the recharge area of the flow system. Although the cross sections shown in Figure 9.10 are homogeneous, dispersivities for the system are assumed to be large as a result of small-scale heterogeneities. With assigned values of dispersivity the patterns of contaminant distribution can be simulated using a finite-element approximation to the transport equation expressed in two-dimensional form for saturated heterogeneous isotropic media [Eq. (A10.13), Appendix X]:

$$\frac{\partial}{\partial s_l}\left(D_l \frac{\partial C}{\partial s_l}\right) + \frac{\partial}{\partial s_t}\left(D_t \frac{\partial C}{\partial s_t}\right) - \frac{\partial}{\partial s_l}(\bar{v}_l C) = \frac{\partial C}{\partial t} \tag{9.8}$$

where s_l and s_t are the directions of the groundwater flowlines and the normals to these lines, respectively. The finite-element model used to obtain the contaminant distributions shown in Figure 9.10 is described by Pickens and Lennox (1976).

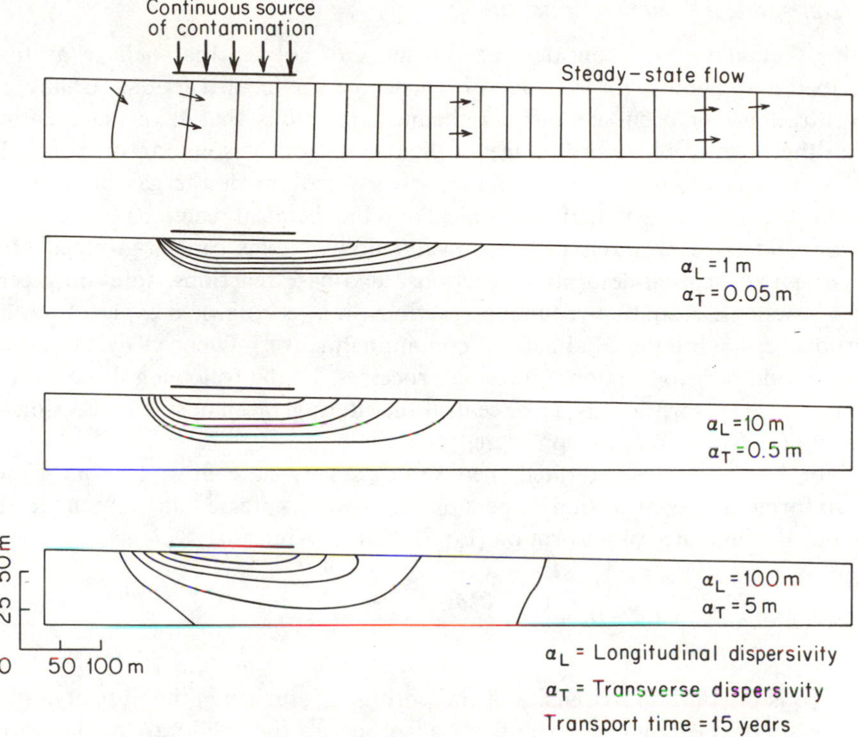

Figure 9.10 Dispersion of a contaminant during transport in a shallow groundwater flow system. Porosity 30%; hydraulic conductivity 0.5 m/day; α_l/α_t = 20; transport time 15 years; concentration contours at C/C_0 = 0.9, 0.7, 0.5, 0.3, and 0.1 (after Pickens and Lennox, 1976).

Other numerical models have been developed by Reddell and Sunada (1970), Bredehoeft and Pinder (1973), Pinder (1973), and Schwartz (1975). The simulations presented in Figure 9.10 indicate that if dispersivity is large, contaminants can spread to occupy a portion of the flow system many times larger than would be the case in the presence of advection alone. If the transverse dispersivity is very large as indicated in Figure 9.10, contaminants transported along relatively horizontal flow paths can migrate deep into the flow system. The longitudinal and transverse dispersivities represented in the simulated contaminant transport patterns shown in Figure 9.10 indicate that if dispersivity values are orders of magnitude larger than the values obtained from laboratory experiments, dispersion will exert a strong influence on contaminant transport. Whether or not dispersivities in nonfractured geologic materials under field conditions have magnitudes that are this large remains to be established by detailed field experiments. This topic is discussed further in Section 9.4.

Transport of Reactive Constituents

In this section we will consider the transport of solutes that behave as those described above, but with the added influence of chemical reactions. Changes in concentration can occur because of chemical reactions that take place entirely within the aqueous phase or because of the transfer of the solute to or from other phases such as the solid matrix of the porous medium or the gas phase in the unsaturated zone. The myriad of chemical and biochemical reactions that can alter contaminant concentrations in groundwater flow systems can be grouped in six categories: adsorption-desorption reactions, acid-base reactions, solution-precipitation reactions, oxidation-reduction reactions, ion pairing or complexation, and microbial cell synthesis. Radioactive contaminants are influenced by radioactive decay in addition to the nonradiogenic processes. In the following discussion we will focus on adsorption as a concentration-altering mechanism. In Section 9.3 other types of reactions are considered.

For homogeneous saturated media with steady-state flow, the one-dimensional form of the advection-dispersion equation expressed in a manner that includes the influence of adsorption [Eq. (A10.14), Appendix X] is

$$D_l \frac{\partial^2 C}{\partial l^2} - \bar{v}_l \frac{\partial C}{\partial l} + \frac{\rho_b}{n}\frac{\partial S}{\partial t} = \frac{\partial C}{\partial t} \tag{9.9}$$

where ρ_b is the bulk mass density of the porous medium, n is the porosity, and S is the mass of the chemical constituent adsorbed on the solid part of the porous medium per unit mass of solids. $\partial S/\partial t$ represents the rate at which the constituent is adsorbed $[M/MT]$, and $(\rho_b/n)(\partial S/\partial t)$ represents the change in concentration in the fluid caused by adsorption or desorption

$$\frac{M}{L^3}\frac{M}{MT} = \frac{M}{L^3}\frac{1}{T}$$

Adsorption reactions for contaminants in groundwater are normally viewed as being very rapid relative to the flow velocity. The amount of the contaminant that is adsorbed by the solids (i.e., the degree of adsorption) is commonly a function of the concentration in solution, $S = f(C)$. It follows that

$$-\frac{\partial S}{\partial t} = \frac{\partial S}{\partial C} \cdot \frac{\partial C}{\partial t} \tag{9.10}$$

and

$$-\frac{\rho_b}{n} \cdot \frac{\partial S}{\partial t} = \frac{\rho_b}{n} \cdot \frac{\partial S}{\partial C} \cdot \frac{\partial C}{\partial t} \tag{9.11}$$

in which the term $\partial S/\partial C$ represents the partitioning of the contaminant between the solution and the solids.

The partitioning of solutes between liquid and solid phases in a porous medium as determined by laboratory experiments is commonly expressed in two-ordinate graphical form where mass adsorbed per unit mass of dry solids is plotted against the concentration of the constituent in solution. These graphical relations of S versus C and their equivalent mathematical expressions are known as isotherms. This term derives from the fact that adsorption experiments are normally conducted at constant temperature.

Results of adsorption experiments are commonly plotted on double-logarithmic graph paper. For solute species at low or moderate concentrations, straight-line graphical relations are commonly obtained over large ranges of concentration. This condition can be expressed as

$$\log S = b \log C + \log K_d$$

or

$$S = K_d C^b \tag{9.12}$$

where S is the mass of the solute species adsorbed or precipitated on the solids per unit bulk dry mass of the porous medium, C is the solute concentration, and K_d and b are coefficients that depend on the solute species, nature of the porous medium, and other conditions of the system. Equation (9.12) is known as the *Freundlich isotherm*. The slope of the log-log adsorption relation is represented by the term b in Eq. (9.12). If $b = 1$ (i.e., if the straight-line relationship between S and C on a log-log plot has a slope of 45°), then the S versus C data will also plot as a straight line on an arithmetic plot. Such an isotherm is termed *linear*, and from Eq. (9.12) with $b = 1$,

$$\frac{dS}{dC} = K_d \tag{9.13}$$

where K_d is known as the *distribution coefficient*. This parameter is widely used in studies of groundwater contamination. K_d is a valid representation of the partitioning between liquid and solids only if the reactions that cause the partitioning are fast and reversible and only if the isotherm is linear. Fortunately, many contaminants of interest in groundwater studies meet these requirements. A comprehensive treatment of adsorption isotherms is presented by Helfferich (1962), who provides detailed information on many important types of isotherms in addition to the Freundlich isotherm.

The transfer by adsorption or other chemical processes of contaminant mass from the pore water to the solid part of the porous medium, while flow occurs, causes the advance rate of the contaminant front to be retarded. To illustrate this concept, the classical column experiment shown in Figure 9.1(a) will again be considered. It is assumed that two tracers are added to the water passing through the column. One tracer is not adsorbed and therefore moves with the water. The other tracer undergoes adsorption, and as it travels through the column part of its mass is taken up by the porous medium. The two tracers are added instantaneously to the water at the column input [step-function input as shown in Figure 9.1(b)]. As transport occurs, the two tracers are distributed in the column in the manner represented schematically in Figure 9.11. The transporting water mass represented by the nonreactive tracer moves ahead of the reactive tracer. The concentration profile for the nonadsorbed tracer spreads out as a result of dispersion. The concentration profile of the front of the reactive tracer also spreads out but travels behind the front of the nonreactive tracer. The adsorbed tracer is therefore said to be *retarded.*

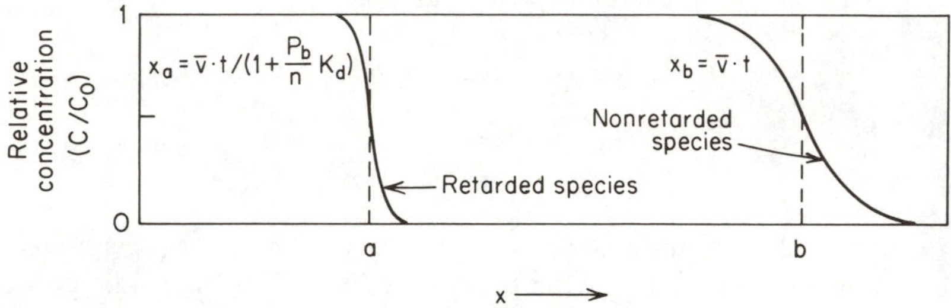

Figure 9.11 Advance of adsorbed and nonadsorbed solutes through a column of porous materials. Partitioning of adsorbed species is described by K_d. Relative velocity = $1/[1 + (\rho_b/n)K_d]$. Solute inputs are at concentration C_0 at $t > 0$.

For cases where the partitioning of the contaminant can be adequately described by the distribution coefficient (i.e., fast reversible adsorption, with linear isotherm), the retardation of the front relative to the bulk mass of water is described by the relation

$$\frac{\bar{v}}{\bar{v}_c} = 1 + \frac{\rho_b}{n} \cdot K_d \tag{9.14}$$

where $\bar{v}$ is the average linear velocity of the groundwater and $\bar{v}_c$ is the velocity of the $C/C_0 = 0.5$ point on the concentration profile of the retarded constituent. Equation (9.14) is commonly known as the *retardation equation.* The term $1 + (\rho_b/n) \cdot K_d$ is referred to as the *retardation factor.* The reciprocal of the retardation factor is known as the relative velocity ($\bar{v}_c/\bar{v}$). Equation (9.14) was originally developed on an empirical basis for use in chemical engineering by Vermeulen and Hiester (1952). It was first applied to groundwater problems by Higgins (1959) and Baetsle (1967, 1969). Baetsle indicated that it can be used to determine the retardation of the center of mass of a contaminant moving from a point source while undergoing adsorption.

To gain a more quantitative appreciation for the effects of chemical retardation on contaminant migration, some representative parameter values will be used in conjunction with Eq. (9.14). For unconsolidated granular deposits, porosity, expressed as a fraction, is commonly in the range 0.2–0.4. The average mass density of minerals that constitute unconsolidated deposits is approximately 2.65. The range of bulk mass densities, ρ_b, that correspond to the porosity range above is 1.6–2.1 g/cm^3. For these ranges of porosity and bulk mass density, ρ_b/n values range from 4 to 10 g/cm^3. An approximation to Eq. (9.14) is therefore

$$\frac{\bar{v}}{\bar{v}_c} = (1 + 4K_d) \text{ to } (1 + 10K_d) \tag{9.15}$$

The only major unknown in Eq. (9.15) is the distribution coefficient K_d. The distribution coefficient can be expressed as

$$K_d = \frac{\text{mass of solute on the solid phase per unit mass of solid phase}}{\text{concentration of solute in solution}}$$

The dimensions for this expression reduce to L^3/M. Measured K_d values are normally reported as milliliters per gram (mℓ/g).

Distribution coefficients for reactive solutes range from values near zero to 10^3 mℓ/g or greater. From Eq. (9.15) it is apparent that if $K_d = 1$ mℓ/g, the midconcentration point of the solute would be retarded relative to the bulk groundwater flow by a factor between 5 and 11. For K_d values that are orders of magnitude larger than 1, the solute is essentially immobile.

To further illustrate the effect of liquid- to solid-phase partitioning, a cross-sectional flow domain similar to the one represented in Figures 9.8 and 9.10 is used. The pattern of contamination in this cross section caused by an influx of water with contaminant species of different distribution coefficients is shown in Figure 9.12. The patterns were obtained by Pickens and Lennox (1976) using a finite-element solution to the transport equation with the reaction term described by Eq. (9.11). The case in which $K_d = 0$ shows the zone occupied by a contaminant species that is not affected by chemical reactions. Under this condition the processes of advection and dispersion cause the contaminant to gradually occupy a large part of the flow domain. The transport pattern is controlled by the contaminant input history, by the velocity distribution, and by dispersion. Contaminant species with K_d values greater than zero occupy a much smaller portion of the flow domain. If $K_d = 10$ mℓ/g, most of the contaminant mass migrates only a very short distance from the input zone during the specified migration period. This situation can be anticipated from consideration of the magnitude of this K_d value in Eq. (9.15). There is an extensive zone beyond the $C/C_0 = 0.1$ contours shown in Figure 9.12 in which the contaminant occurs at very low concentrations. If the contaminant is harmful at low concentrations, this zone can be extremely important, even though it includes only a small portion of the total contaminant mass in the flow system.

When a mixture of reactive contaminants enters the groundwater zone, each species will travel at a rate depending on its relative velocity, $\bar{v}_c/\bar{v}$. After a given time t, the original contaminant cloud will have segregated into different zones,

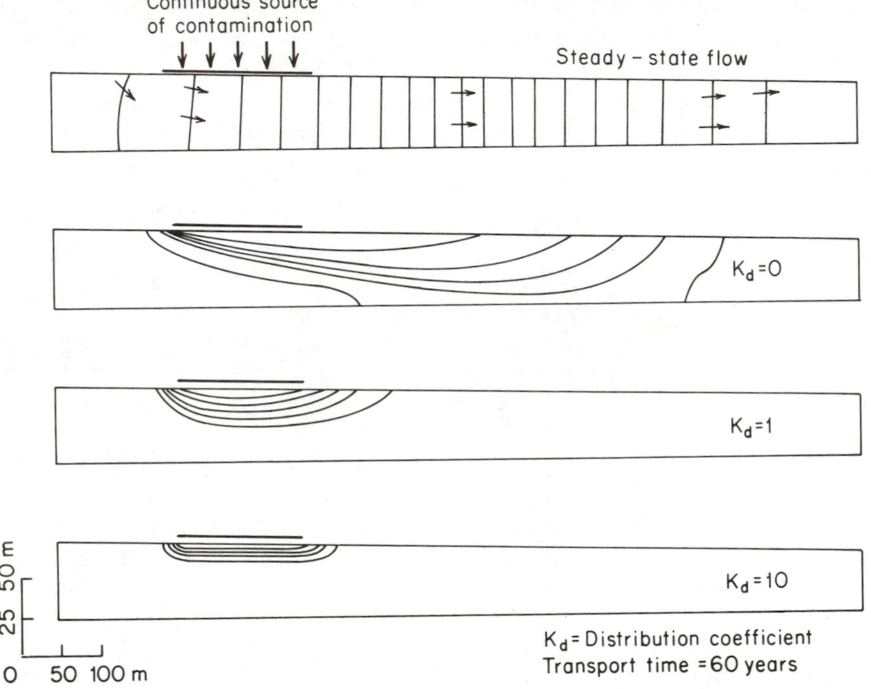

Figure 9.12 Effect of the distribution coefficient on contaminant retardation during transport in a shallow groundwater flow system. Porosity 0.3; hydraulic conductivity 0.5 m/day; α_l = 10 m; α_t = 0.5 m; transport time 60 years; concentration contours at C/C_0 = 0.9, 0.7, 0.5, 0.3, and 0.1 (after Pickens and Lennox, 1976).

each advancing in the same direction at different velocities. Considering the instantaneous point-source example described by Eqs. (9.6) and (9.7), the position of the center of mass of the migrating cloud is obtained from the relative velocity defined by the reciprocal of $\bar{v}/\bar{v}_c$ calculated from Eq. (9.14). Equation (9.6) can be used to calculate the concentration distribution of the dissolved reactive species, with substitution of τ for t, where $\tau = t(\bar{v}_c/\bar{v})$. Since the total standard deviation of a given distribution is a function of time as well as distance traveled, both parameters influence the dispersion pattern of each retarded species (Baetsle, 1969).

The distribution coefficient approach to the representation of chemical partitioning of contaminants in groundwater flow systems is based on the assumption that the reactions that partition the contaminants between the liquid and solid phases are completely reversible. As a contaminant plume advances along flow paths, the front is retarded as a result of transfer of part of the contaminant mass to the solid phase. If the input of contaminant mass to the system is discontinued, contaminants will be transferred back to the liquid phase as lower-concentration water flushes through the previously contaminated zone. In this situation the contaminant moves as a cloud or enclave through the flow system. This is illustrated

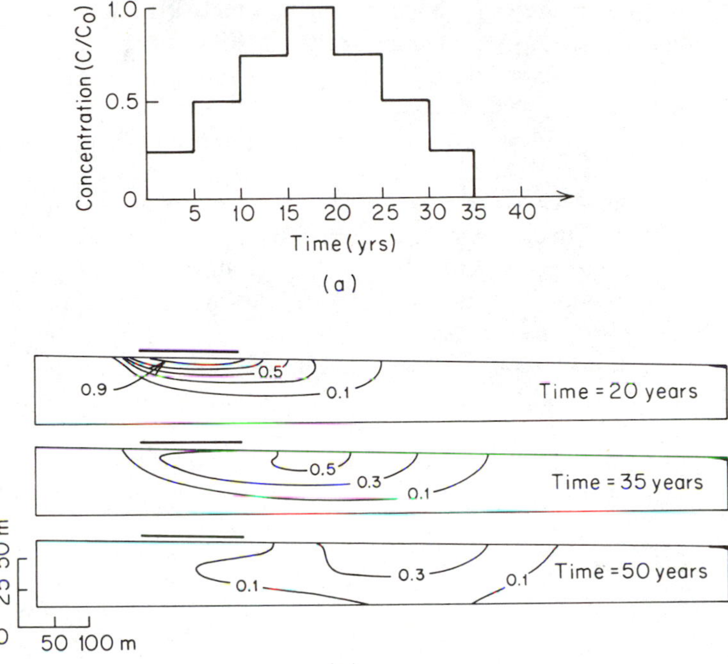

Figure 9.13 Migration of a reactive contaminant through a shallow groundwater flow system. (a) Concentration versus time relation for the contaminant source; (b) concentration distributions after 20, 35, and 50 years. Porosity 0.3; hydraulic conductivity 0.5 m/day; $\alpha_l = 10$ m; $\alpha_t = 0.5$ m; concentration contours at $C/C_0 = 0.9$, 0.7, 0.5, 0.3, and 0.1 (after Pickens and Lennox, 1976).

in Figure 9.13, which shows the migration of a contaminant enclave through the cross section illustrated in Figures 9.10 and 9.12. Initially, the contaminated zone is localized beneath the input area. After the input of contaminated water is discontinued, the contaminant mass moves along the flow paths, leaving a zone of less-contaminated water beneath the input area. As time goes on, the contaminants are flushed out of the flow system. If the partitioning reactions are completely reversible, all evidence of contamination is eventually removed from the system as complete desorption occurs. Thus, if the reactions are reversible, contaminants cannot be permanently isolated in the subsurface zone, even though retardation of the concentration front may be strong. In some situations a portion of the contaminant mass transferred to the solid part of the porous material by adsorption or precipitation is irreversibly fixed relative to the time scale of interest. This portion is not transferred back to the pore water as new water passes through the system and is therefore isolated in the subsurface environment.

When the distribution coefficient is used to determine contaminant retardation, it is assumed that the partitioning reactions are very fast relative to the rate of groundwater movement. Many substances, however, do not react sufficiently fast with the porous medium for this assumption to be valid. When contaminants of this type move through porous media, they advance more rapidly than would be the case if the reactions produce K_d type partitioning relations. This is illustrated in Figure 9.14, which shows the nonequilibrium front in a position between the front of a nonretarded tracer and the front of a retarded tracer described by the K_d relation. Analysis of the movement of contaminants that undergo partitioning in a manner that cannot be described by equilibrium relations requires information on the rates of reaction between the contaminant and the porous medium. This information is difficult to obtain. In field studies the retardation equation described above is often used because of its simplicity or because there is a lack of information on reaction rates. This can lead to serious errors in prediction of rates of contaminant migration in systems where kinetic factors are important.

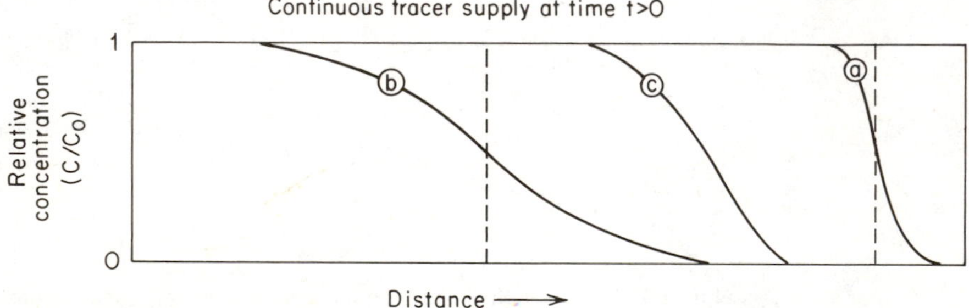

Figure 9.14 Advance of reactive and nonreactive contaminants through a column. (a) Dispersed front of nonretarded solute; (b) front of solute that undergoes equilibrium partitioning between liquid and solids; (c) front of solute that undergoes slower rate of transfer to the solids.

Transport in Fractured Media

Although contaminant transport in fractured geologic materials is governed by the same processes as in granular media—namely, advection, mechanical dispersion, molecular diffusion, and chemical reactions—the effects in fractured media can be quite different. The effective fracture porosity of fractured rocks and of consolidated cohesive materials that are fractured, such as jointed till, silt, or clay, is normally very small. Values in the order of 1–0.001%, or 10^{-2}–10^{-5} expressed as a fraction, are not unusual. Although the porosities are small, the groundwater velocities can be large. The reason for this can be deduced from the modified Darcy relation (Section 2.12)

$$\bar{v} = -\frac{K}{n_f}\frac{dh}{dl} \tag{9.16}$$

where $\bar{v}$ is the average linear velocity of water in the fractures, K the bulk hydraulic conductivity of the fractured medium, n_f the bulk fracture porosity, and dh/dl

the hydraulic gradient. This relation treats the fractured medium as an equivalent porous medium. The parameters in the equations relate to a volume segment of the medium that is sufficiently large to be described by hydraulic conductivity and porosity averaged over the bulk mass. In this approach each fracture opening is considered to be very small relative to the bulk volume of the domain over which K is measured. The number of fractures in this domain therefore must be large.

For illustrative purposes we will consider a medium that has a bulk hydraulic conductivity of 10^{-8} m/s and a fracture porosity of 10^{-4}. These values could represent conditions in a slightly fractured granite. Using a hydraulic gradient of 10^{-2}, which is within the range commonly observed in field situations, the groundwater velocity computed from Eq. (9.16) is 10 m/yr. Compared to velocities in fine-grained unfractured granular materials, this velocity is very large. For example, an unfractured granular medium, such as a silt deposit, with this hydraulic conductivity and gradient and an intergranular porosity of 0.3 would have a groundwater velocity of about 0.003 m/yr. The flux of water (volume of water per unit time passing through a specified cross-sectional area) in these two cases is the same and is extremely small. Although Eq. (9.16) can be used to compute average velocities in fractured media, it provides no indication of the velocities in individual fractures. Depending on the fracture aperture and wall roughness, the velocity of groundwater may deviate from the average by orders of magnitude.

It was indicated above that in the mathematical analysis of mechanical dispersion in granular media, the media are assumed to be isotropic with respect to dispersivity. That is, longitudinal dispersivity at a point in the medium has a single value regardless of the direction of the velocity vector. Each of the transverse dispersivities has a single value relative to the longitudinal dispersivity. The differences between longitudinal and transverse dispersivities are related to the mechanism of dispersion rather than to directional properties of the medium. Fractured geologic materials, however, are notoriously anisotropic with respect to the orientation and frequency of fractures. It can be expected that the dispersion of solutes during transport through many types of fractured rocks cannot be described by the equations developed for homogeneous granular materials. Little is known about dispersion in fractured media. A common approach in field investigations of contaminant migration in fractured rock is to treat the problem in the same manner mathematically as for granular porous media. The scale at which this approach becomes valid in the analysis of field situations is not known. As a concluding comment on this topic, the statement by Castillo et al. (1972) is appropriate:

> Although the basic theoretical aspects of . . . (dispersion) . . . have been treated at length for the case where the permeable stratum is composed of granular materials, the classical concept of flow through a porous medium is generally inadequate to describe the flow behaviour in jointed rock, and it becomes increasingly unsuitable for the analysis of dispersion. Despite these limitations, little work has been directed toward extending these ideas to handle flow through jointed rock formations (p. 778).

A modification in approach is necessary for the distribution coefficient or isotherm concept to be applicable in the analysis of the migration of reactive contaminants through fractured media. For granular materials the amount of solute adsorbed on the solid part of the porous media is expressed per unit mass of the bulk medium in a dry state. For convenience the unit mass of the porous medium is used as a reference quantity. A more mechanistic but less convenient approach would be to use a unit surface area of the porous medium as the reference quantity. This would be a reasonable approach because adsorption reactions are much more closely related to the surface area of the solid medium than to the mass of the medium. Nevertheless, for granular materials such as sands, silts, and clays, the use of mass density in the definition of the distribution coefficient normally produces acceptable results. With this approach, measurements of effective surface area are not necessary.

In the case of contaminant migration through fractured materials, it is more appropriate, as suggested by Burkholder (1976), to express the distribution coefficient K_a on a per-unit-surface-area basis.

It is therefore defined as

$$K_a = \frac{\text{mass of solute on the solid phase per unit area of solid phase}}{\text{concentration of solute in solution}}$$

The dimensions for this expression are $[M/L^2 \cdot L^3/M]$ or $[L]$. The units that are commonly used are milliliters per square centimeter.

The retardation equation therefore becomes

$$\frac{\bar{v}}{\bar{v}_c} = 1 + AK_a \tag{9.17}$$

where A is the surface area to void-space (volume) ratio $[1/L]$ for the fracture opening through which the solute is being transported. It is apparent from this relation that fractures with smaller apertures produce greater retardation of reactive solutes. The distribution coefficient in this retardation expression has the same inherent assumptions as Eq. (9.14), namely: the partitioning reactions are reversible and fast relative to the flow velocity.

Equation (9.17) is simple in conceptual terms, but it is difficult to apply to natural systems. If information can be obtained on the aperture of a fracture and if the fracture surface is assumed to be planar, $A = 2/b$, where b is the aperture width (Section 2.12). Fracture surfaces usually have small-scale irregularities and therefore can have much larger surface areas than planar surfaces. In the determination of the adsorption isotherm or distribution coefficient for the fracture, the partitioning of the contaminant between fluid in contact with the fracture and the fracture surface is measured. If the fracture surface is irregular or contains a coating of weathered material or chemical precipitates, the actual surface area with which the contaminant reacts is unknown. Without an elaborate experimental effort, it is indeterminant. A practical approach is to express the K_a relative to the area of an assumed planar fracture surface, in which case the retardation relation becomes

$$\frac{\bar{v}}{\bar{v}_c} = 1 + \frac{2K_a}{b} \tag{9.18}$$

It should be kept in mind that Eq. (9.17) is only valid for fractured materials in which the porosity of the solid mass between fractures is insignificant. When contaminants occur in fractures, there is a gradient of contaminant concentration between the fracture fluid and the fluid in the unfractured material adjacent to the fracture. If the solid matrix is porous, a portion of the contaminant mass will move by molecular diffusion from the fracture into the matrix. This mass is therefore removed, at least temporarily, from the flow regime in the open fracture.

Figure 9.15 illustrates the effect of matrix diffusion on the concentration distribution of nonreactive and reactive contaminants migrating through a fracture in a medium with a porous matrix. For illustrative purposes it is assumed that dispersion within the fracture is insignificant. Comparison of Figure 9.15(a) and (b) indicates that diffusion into the matrix causes the concentration in the fracture to diminish gradually toward the front of the advancing contaminant zone. The bulk mass of the advancing contaminant zone in the fracture appears to be retarded

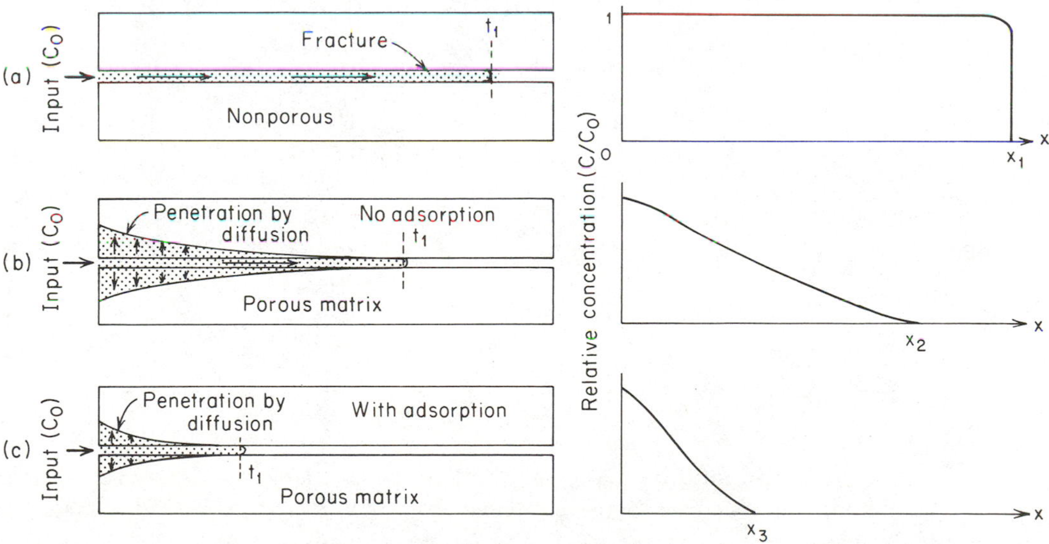

Figure 9.15 Effect of diffusion on contaminant migration in porous fractured medium. (a) Unidirectional hydraulic transport in a fracture in a nonporous medium; (b) unidirectional hydraulic transport with migration into matrix as a result of molecular diffusion; (c) unidirectional hydraulic transport with molecular diffusion and adsorption (profiles of relative concentration of reactive contaminant within fracture shown at time t_1).

because part of the contaminant mass is transferred to the matrix. The general shape of the longitudinal profile is somewhat similar to that produced by longitudinal dispersion in granular materials. If the contaminant undergoes adsorption, the effect of diffusion is to cause adsorption to occur on a much larger surface area than would be the case if the contaminant mass remained entirely within the fracture. A portion of the contaminant is adsorbed on the surface of the fracture and as diffusion occurs a portion is adsorbed in the matrix. The combined effect of adsorption on the fracture surface and adsorption in the matrix is to cause the contaminant mass in the fracture to be retarded relative to the advance that would occur in the absence of adsorption [Figure 9.15(c)].

The contaminant distribution in a porous fractured aquifer receiving waste from a surface source is illustrated schematically in Figure 9.16. As time goes on, the zone of contamination will diffuse farther into the porous matrix. If the source of contamination is discontinued, the contaminant mass in the porous matrix will eventually diffuse back to the fracture openings as fresh water flushes through the fracture network.

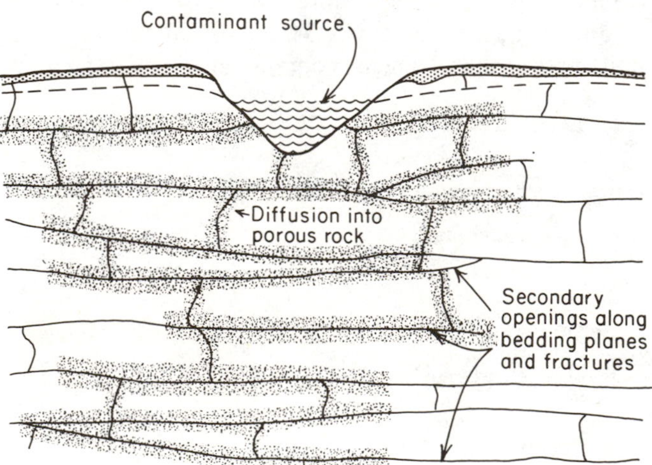

Figure 9.16 Schematic representation of contaminant migration from a surface source through fractured porous limestone.

Molecular diffusion is a process that occurs at a sufficiently rapid rate to exert a strong influence on contaminant behavior in many types of fractured materials. Even granite has appreciable primary porosity and permeability, with porosity values commonly as large as 0.05–1.0% and hydraulic conductivity in the order of 10^{-12} m/s. In the main limestone aquifer in Britain, detailed studies show that subsurface distributions of tritium and nitrate in the limestone are strongly influenced by diffusion of these constituents from the fractures, where rapid flow occurs, into the porous rock matrix (Foster, 1975). In the Plains Region of North America, deposits of glacial till and glaciolacustrine clay are commonly fractured (Section

4.4). Grisak et al. (1976) indicate that although the fractures are generally a major avenue of groundwater flow, the chemical evolution of groundwater is controlled by diffusion of dissolved reaction products from the clayey matrix into the fracture network.

9.3 Hydrochemical Behavior of Contaminants

In this section, the hydrochemical behavior of groundwater contaminants will be discussed. It is not feasible for all the hydrochemical processes that affect contaminants in groundwater to be considered in this text. Our purpose is to illustrate some of the most important processes that control the behavior of several groups of contaminants with different hydrochemical properties. The origin and causes of groundwater contamination are discussed in Section 9.4.

Nitrogen

The most common contaminant identified in groundwater is dissolved nitrogen in the form of nitrate (NO_3^-). This contaminant is becoming increasingly widespread because of agricultural activities and disposal of sewage on or beneath the land surface. Its presence in undesirable concentrations is threatening large aquifer systems in many parts of the world. Although NO_3^- is the main form in which nitrogen occurs in groundwater, dissolved nitrogen also occurs in the form of ammonium (NH_4^+), ammonia (NH_3), nitrite (NO_2^-), nitrogen (N_2), nitrous oxide (N_2O), and organic nitrogen. Organic nitrogen is nitrogen that is incorporated in organic substances.

Nitrate in groundwater generally originates from nitrate sources on the land surface, in the soil zone, or in shallow subsoil zones where nitrogen-rich wastes are buried (Figure 9.17). In some situations NO_3^- that enters the groundwater system originates as NO_3^- in wastes or fertilizers applied to the land surface. These are designated as direct nitrate sources in Figure 9.18. In other cases, NO_3^- originates by conversion of organic nitrogen or NH_4^-, which occur naturally or are introduced to the soil zone by man's activities. The processes of conversion of organic nitrogen to NH_4^- is known as *ammonification*. Through the process of *nitrification*, NH_4^+ is converted to NO_3^- by oxidation. Ammonification and nitrification are processes that normally occur above the water table, generally in the soil zone, where organic matter and oxygen are abundant. Thus, in Figure 9.18 these processes are represented as NO_3^- producers outside the boundaries of the groundwater flow system.

Concentrations of NO_3^- in the range commonly reported for groundwater are not limited by solubility constraints. Because of this and because of its anionic form, NO_3^- is very mobile in groundwater. In groundwater that is strongly oxidizing, NO_3^- is the stable form of dissolved nitrogen. It moves with the groundwater with no transformation and little or no retardation. Very shallow groundwater in highly permeable sediment or fractured rock commonly contains considerable

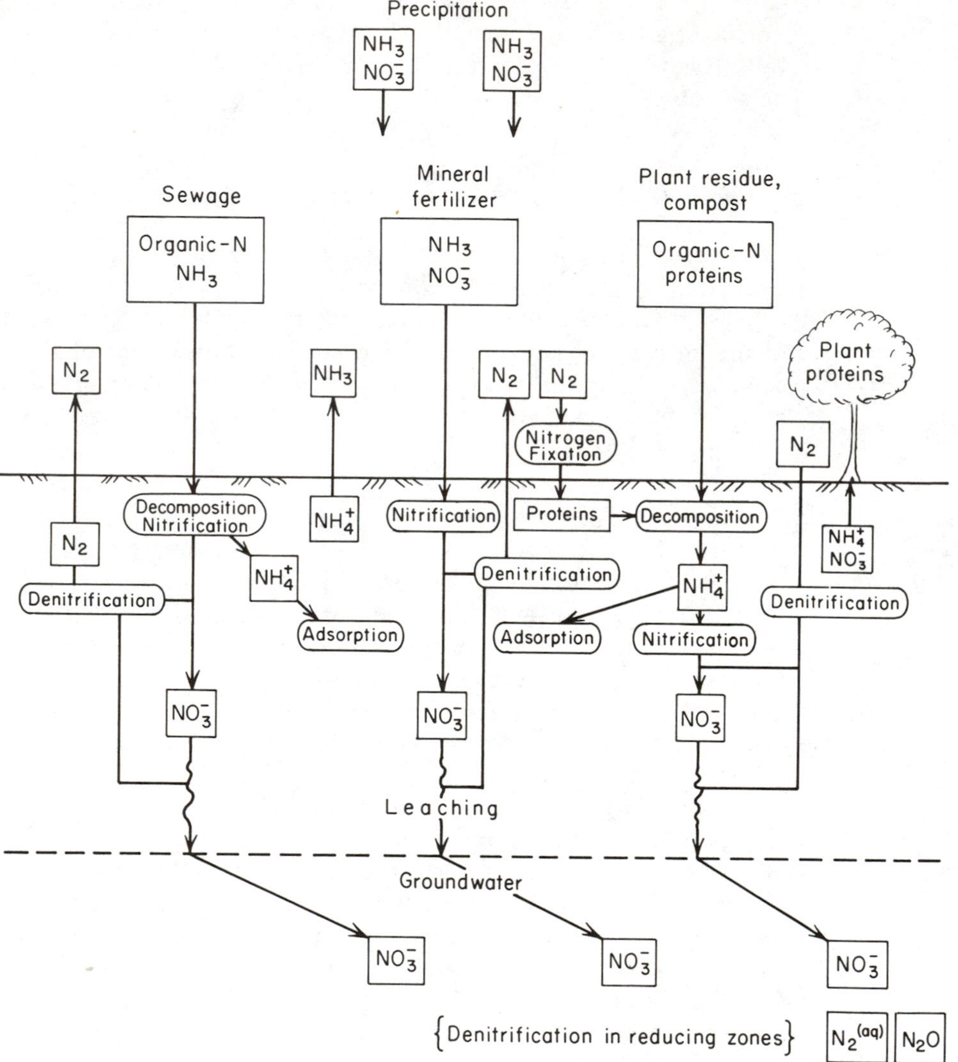

Figure 9.17 Sources and pathways of nitrogen in the subsurface environment.

dissolved O_2. It is in these hydrogeologic environments where NO_3^- commonly migrates large distances from input areas.

A decline in the redox potential of the groundwater can, in some situations, cause denitrification, a process in which NO_3^- is reduced to N_2O or N_2 (Figure 9.17). This process is represented chemically in Table 3.11. In an ideal system, which can be described by reversible thermodynamics, denitrification would occur at a redox potential of about 4.2 as pE (or +250 mv as Eh) in water at pH 7 and

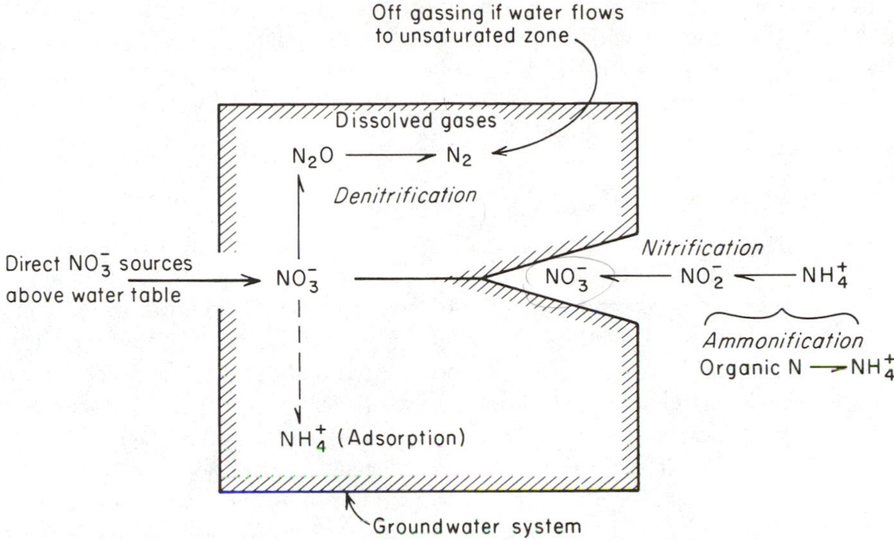

Figure 9.18 Nitrogen inputs and transformations in the groundwater system.

25°C. At this redox potential, the water would be devoid of dissolved O_2 (i.e., below the detection limit). The NO_3^- is reduced to N_2O and then, if the redox potential declines further, the N_2O is reduced to N_2. These reaction products exist as dissolved species in the groundwater. If the water moves into the unsaturated zone, a portion of the N_2O or N_2 may be lost by off-gassing to the soil air (Figure 9.18).

Figure 9.18 indicates that in addition to the denitrification pathway for the reduction of NO_3, there is a pathway that leads to NH_4^+. For biochemical reasons only a small fraction of the NO_3^- that undergoes reduction follows this reduction path. If NH_4^+ is produced in groundwater by this process, most of it would eventually be adsorbed on clay or silt-sized particles in the geologic materials.

From a water-quality viewpoint, denitrification in groundwater is a desirable process. Increased concentrations of dissolved N_2 and N_2O are not detrimental to drinking water. In contrast, NO_3^- at concentrations above 45 mg/ℓ renders water unfit for consumption by human infants. If water has more than 450 mg/ℓ of NO_3^-, it is unsuitable for consumption by livestock.

Denitrification is a process that has been observed in numerous investigations of soil systems in the laboratory and in the field. Given a source of organic matter and abundant NO_3^-, bacterial systems in soil are capable of denitrifying large amounts of NO_3^-. Denitrification in the groundwater zone, however, is a process about which little is known. It appears that a lack of suitable types or amounts of organic matter in the groundwater zone commonly inhibit the growth of denitrifying bacteria in groundwater. This limits the rate of denitrification, even if the redox system has evolved toward reducing conditions. However, since groundwater commonly flows at low velocity, a slow rate of denitrification may nevertheless be significant with respect to the nitrate budget of the subsurface environment. For discussions of field situations in which evidence indicating denitrification in aquifers is presented, the reader is referred to Edmunds (1973) and Gillham and Cherry (1978).

Trace Metals

In recent years the mobility of trace metals in groundwater has received considerable attention. Of special interest are the trace metals for which maximum permissible or recommended limits have been set in drinking water standards. These include Ag, Cd, Cr, Cu, Hg, Fe, Mn, and Zn (see Table 9.1). During the next decade this list may grow as more is learned about the role of trace metals in human health and ecology. Although these elements rarely occur in groundwater at concentrations large enough to comprise a significant percentage of the total dissolved solids, their concentrations can, depending on the source and hydrochemical environment, be above the limits specified in drinking water standards. Most of the elements listed above are in an elemental group referred to by chemists as the *transition elements*. Many of these elements are also known as *heavy metals*.

Trace metals in natural or contaminated groundwaters, with the exception of iron, almost invariably occur at concentrations well below 1 mg/ℓ. Concentrations are low because of constraints imposed by solubility of minerals or amorphous substances and adsorption on clay minerals or on hydrous oxides of iron and manganese or organic matter. Isomorphous substitution or coprecipitation with minerals or amorphous solids can also be important.

A characteristic feature of most trace metals in water is their tendency to form hydrolyzed species and to form complexed species by combining with inorganic anions such as HCO_3^-, CO_3^{2-}, SO_4^{2-}, Cl^-, F^-, and NO_3^-. In groundwater environments contaminated with dissolved organic compounds, organic complexes may also be important. Expressed in terms of the products of hydrolysis, the total concentration of a trace metal M_T that in the unhydrolyzed form exists as M^{n+} is

$$M_T = (M^{n+}) + (MOH^{(n-1)+}) + (M(OH)_2^{(n-2)+}) + \dots$$

If the total concentration, M_T, is known, the concentrations of the other species can be computed using mass-action equations with equilibrium constants derived from thermodynamic data (Leckie and James, 1974). Using zinc as an example, the hydrolyzed species *and* inorganic complexes that would form would include $ZnOH^+$, $Zn(OH)_2^\circ$, $Zn(OH)_4^{2-}$, $ZnCl^-$, $ZnSO_4^\circ$, and $ZnCO_3^\circ$. The occurrence and mobility of zinc in groundwater requires consideration of these and other dissolved species. Chemical analyses of zinc in groundwater provide direct information only on the total zinc content of water. The percent of the total concentration existing as hydrolyzed species increases with increasing pH of the water. Complexes of zinc with Cl^-, SO_4^{2-}, and HCO_3^- increase with increasing concentrations of these anions in solution. In Section 3.3, it was shown that dissolved species in groundwater resulting from the formation of complexes with major ions can be computed from analyses of total concentrations of major constituents. In much the same manner, the concentration of trace-metal complexes can be computed using concentration data from laboratory analyses. Capability to predict the mobility of trace metals in groundwater can depend on the capability for prediction of the concentrations of the most important complexes formed by the element in the water. Although information on the free and complexed forms are often required for an understanding of the mobility of trace metals, the concentration values listed in water quality standards are total concentrations.

Nearly all the trace metals of interest in groundwater problems are influenced by redox conditions, as a result either of changes in the oxidation state of the trace metal or of nonmetallic elements with which it forms complexes. The redox environment may also indirectly influence trace-metal concentrations as a result of changes in solid phases in the porous medium that cause adsorption of the trace metal. In the following discussion, mercury is used for illustration of the influence of redox conditions and complexing. Diagrams of p*E*–pH for Hg, in water that contains Cl^- and dissolved sulfur species, are shown in Figure 9.19. Figure 9.19(a) indicates

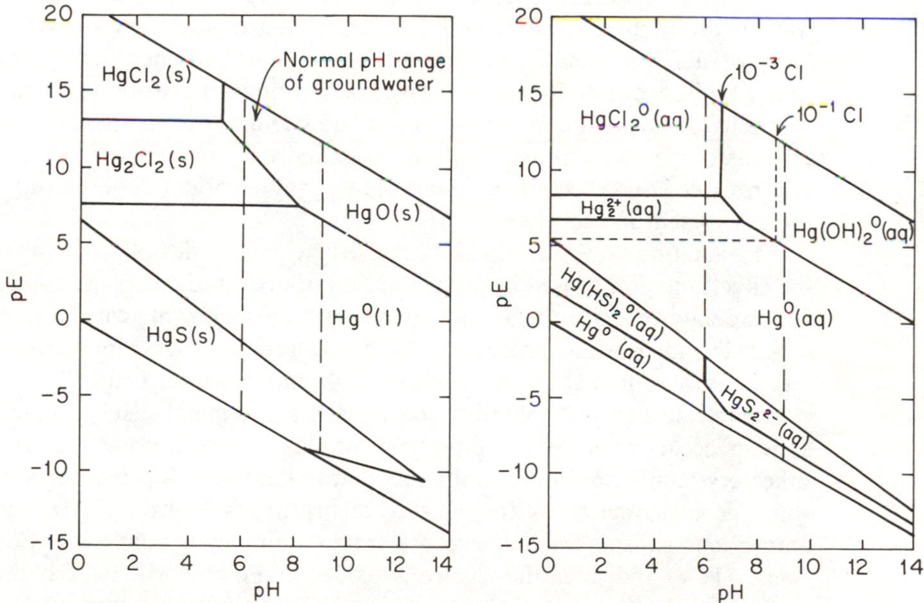

Figure 9.19 Stability fields of solid phases and aqueous species of mercury as a function of pH and p*E* at 1 bar total pressure. (a) Solid phases calculated for conditions of 10^{-3} molal Cl^- and $SO_4{}^{2-}$ in solution; (b) aqueous species calculated for conditions of 10^{-3} molal $SO_4{}^{2-}$ and 10^{-3} and 10^{-1} molal Cl^-. Fine dashed line indicates expanded boundary of $HgCl_2^\circ(aq)$ field at higher Cl^- concentration (after Leckie and James, 1974).

the main solid compounds of mercury that occur in the various pH–pE stability fields and the field in which liquid mercury occurs. The dominant aqueous Hg species in equilibrium with these solid phases containing appreciable concentrations of SO_4^{2-} and Cl^- are shown in Figure 9.19(b). In high-Cl^- water, $HgCl_2^\circ$ is the dominant dissolved species of Hg in the normal pH range of groundwater under oxidizing conditions. At low Cl^- concentrations, HgO is the equilibrium solid phase and $Hg(OH)_2^\circ$ is the dominant dissolved species at high redox potential. The main equilibrium reaction in this pH-pE environment is

$$HgO + H_2O = Hg(OH)_2^\circ \tag{9.19}$$

At 25°C, log K for this reaction is -3.7. The equilibrium $Hg(OH)_2^\circ$ concentration from this reaction is therefore 47 mg/ℓ. This concentration is 4 orders of magnitude above the maximum permissible level for drinking water. In most of the pH-pE domain below the HgO(s) stability field, solubility constraints produce equilibrium concentrations of total dissolved mercury considerably below this level. In much of the redox domain, the equilibrium concentrations are below the maximum levels permitted in drinking water.

Some of the other trace metals also have large equilibrium concentrations in waters with high redox potential. In anaerobic groundwaters, the relative insolubility of sulfide minerals can limit trace metals to extremely low concentrations. In nonacidic groundwaters with high concentrations of dissolved inorganic carbon, solubility of carbonate materials will, if equilibrium is achieved, maintain concentrations of metals such as cadmium, lead, and iron at very low levels. This is the case provided that excessive amounts of inorganic or organic complexing substances are not present in the water.

In addition to the constraints exerted by solubilities of solid substances and the effects thereon caused by the formation of dissolved complexes, the occurrence and mobility of trace metals in groundwater environments can be strongly influenced by adsorption processes. In some groundwaters, many of the trace metals are maintained by adsorption at concentrations far below those that would exist as a result only of solubility constraints. Trace-metal adsorption in subsurface systems occurs because of the presence of clay minerals, organic matter, and the other crystalline and amorphous substances that make up the porous media. In some geologic materials trace-metal adsorption is controlled by crystalline or amorphous substances that are present in only small quantities. For example, Jenne (1968) indicates that hydrous oxides of Fe and Mn furnish the principal control on the fixation of Co, Ni, Cu, and Zn in soils and freshwater sediments. In oxidizing environments, these oxides occur as coatings on grains and can enhance the adsorptive capability of the medium far out of proportion to their percent occurrence relative to the other solids. The hydrous oxide coatings can act as scavengers with respect to trace metals and other toxic constituents.

Hydrous iron and manganese oxide precipitates are usually denoted as $Fe(OH)_3$(s) and MnO_2(s). FeOOH(s) is sometimes used to designate the iron oxide precipitates. The oxides of iron and manganese may be X-ray amorphous (i.e., noncrystalline) or crystalline. In crystalline form hydrous iron oxide is known as the mineral goethite, or if the composition is Fe_2O_3, as hematite. Hydrous iron oxide precipitates are generally mixtures of different phases. Crystalline forms such as goethite and hematite form as a result of long-term aging of amorphous precipitates (Langmuir and Whittemore, 1971).

A pH–*Eh* diagram for iron in water that contains dissolved inorganic carbon and dissolved sulfur species is shown in Figure 9.20. Within the pH range typical of groundwater, $Fe(OH)_3(s)$ is thermodynamically stable at moderate to high p*E* values. In groundwaters with appreciable dissolved inorganic carbon and sulfur, $FeCO_3(s)$ (siderite) and FeS_2(pyrite or marcasite) are stable at lower p*E* values. In Figure 9.20 the boundaries of the $Fe(OH)_3(s)$ field have considerable uncertainty because of uncertainty in free-energy data for $Fe(OH)_3(s)$. Nevertheless, the pH–p*E* diagram serves to illustrate that the existence of $Fe(OH)_3(s)$ is dependent on

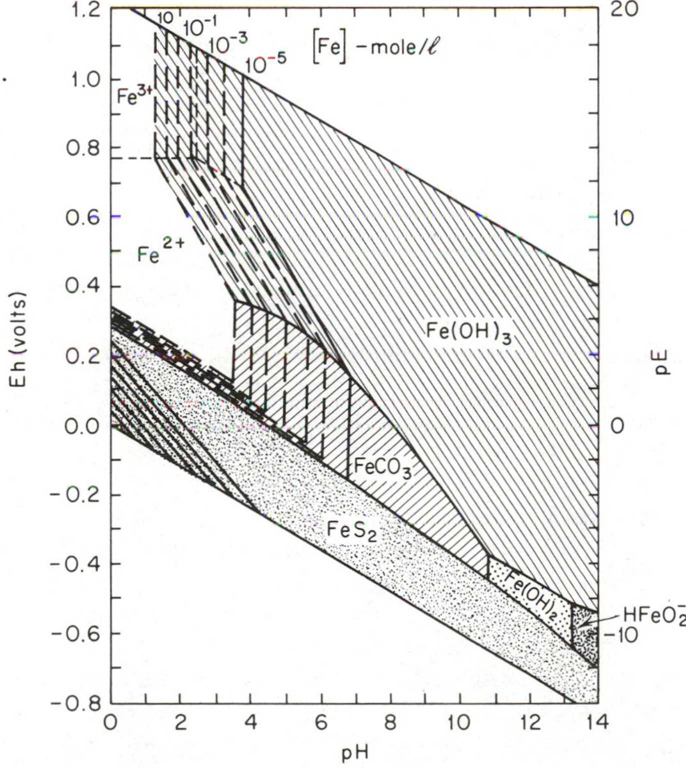

Figure 9.20 Stability fields for main solid phases and aqueous species of iron in water as a function of pH and p*E*, 25°C and 1 bar. Dashed lines represent solubility of iron. Stability fields and iron solubility calculated for conditions of total dissolved sulfur = 10^{-4} mol/ℓ and bicarbonate = 10^{-2} mol/ℓ (after Hem, 1967).

the redox conditions. From this it follows that the trace-metal adsorption capability of a groundwater system may vary greatly from one zone to another. If man's activities disturb the pH–pE regime, a zone that initially has a strong capability for trace-metal adsorption may lose this capability, or the reverse situation may occur.

In summary, it can be concluded that the environmental chemistry of trace metals is complex. It is difficult to predict their transport behavior within groundwater flow systems. In many subsurface environments adsorption and precipitation reactions cause fronts of these elements to move very slowly relative to the velocity of the groundwater. It is not surprising, therefore, that relatively few instances of trace-metal pollution of groundwater have been reported (Kaufman, 1974). In situations where trace-metal contamination does occur, however, the consequences can be serious.

More comprehensive reviews of trace-metal behavior in aqueous systems are provided by Leckie and James (1974) and Leckie and Nelson (1977). The occurrence and controls of trace metals in natural and contaminated groundwaters have been reviewed by Matthess (1974).

Trace Nonmetals

Of the many nonmetals listed in the periodic table of the elements, only a few have received much attention in groundwater investigations. These include carbon, chlorine, sulfur, nitrogen, fluorine, arsenic, selenium, phosphorus, and boron. Dissolved forms of carbon (HCO_3^-, CO_3^{2-}, CO_2, H_2CO_3), of chlorine (Cl^-), and of sulfur (SO_4^{2-}, HS^-, H_2S) occur in abundance in most natural and contaminated groundwaters. The geochemical origin and behavior of these constituents is described in Chapters 3 and 7 and need not be pursued here. Nitrogen in groundwater was discussed previously in this chapter. Our purpose here is to briefly review the hydrochemical behavior of the other important nonmetallic, inorganic constituents that occur as contaminants or as toxic natural constituents in groundwater. The following constituents will be considered: arsenic, fluoride, selenium, boron, and phosphate. These constituents are rarely present in natural or contaminated waters at concentrations above 1 mg/ℓ. Limits for the first four constituents on this list are included in the drinking water standards (Table 9.1).

Arsenic and its compounds have been widely used in pigments, as insecticides and herbicides, as an alloy in metals, and as chemical warfare agents (Ferguson and Gavis, 1972). Synthetic organic compounds have now replaced arsenic in most of these uses, but because of past usage and contributions from ore processing wastes and from natural sources, arsenic is still an element of interest in terms of environmental quality. Based on a review of arsenic data from water supply and surface water environments, Ferguson and Gavis (1972) concluded that arsenic concentrations in natural waters often approach or exceed the limits specified in drinking water standards.

The geochemistry of arsenic has been described by Onishi and Sandell (1955). Ferguson and Gavis (1972) have reviewed the arsenic cycle in natural waters. Arsenic occurs in four oxidation states, +V, +III, 0, and −III. The −III state is stable only at extremely low pE values. In the pH range typical of groundwater, the stable solid arsenic forms are As_2O_5(s) and As_2O_3(s). These solids are soluble enough for dissolved arsenic species to exist at concentrations well above the permissible concentration in drinking water. Under oxidizing conditions, the following species of dissolved arsenic are stable: $H_3AsO_4^\circ$, $H_2AsO_4^-$, $HAsO_4{}^{2-}$, and $AsO_4{}^{3-}$. Under mildly reducing conditions, $H_3AsO_3^\circ$, $H_2AsO_3^-$, and $HAsO_3{}^{2-}$ are predominant. At low pE values in waters with moderate or large concentrations of dissolved sulfur species, the sulfides As_2S_3 and AsS are stable. Under these conditions, total dissolved arsenic is limited by solubility constraints to concentrations far below the limit for drinking water. At higher pE conditions, however, dissolved arsenic species can occur at equilibrium concentrations that are much above the permissible limit for drinking water. The fact that the dominant dissolved species are either uncharged or negatively charged suggests that adsorption and ion exchange will cause little retardation as these species are transported along groundwater flow paths.

Of the various nonmetals for which maximum permissible limits are set in drinking water standards, two of these, fluoride and selenium, are of interest primarily because of contributions from natural sources rather than from man-derived sources. Although within the strict usage of the term, these constituents derived from natural sources are not *contaminants* even if they do occur at toxic levels, their occurrence will be discussed in this section.

Selenium is a nonmetallic element that has some geochemical properties similar to sulfur. Selenium can exist in the +VI, +IV, and −II oxidation states, and occurs in appreciable concentrations in such rocks as shale, in coal, in uranium ores, and in some soils (Lewis, 1976). The aqueous solubilities of selenium salts are in general greater than those of sulfate salts. In dissolved form in groundwater, selenium is present primarily as $SeO_3{}^{2-}$ and $SeO_4{}^{4-}$ ions. Experimental studies by Moran (1976) indicate that selenium concentrations in groundwater can be controlled by adsorption on coatings or colloidal particles of hydrous iron oxide. In many groundwater systems, however, there is so little selenium present in the rocks or soils that availability is the main limiting factor. There are, however, exceptions to this generalization. For example, Moran (1976) has described an area in Colorado in which waters from many wells have selenium concentrations that exceed the permissible limits for drinking water.

Fluoride, because of the beneficial effects on dental health that have been claimed for it and consequently because of its use as a municipal water-supply additive in many cities, is a constituent that has received much attention in recent decades. Fluoride is a natural constituent of groundwater in concentrations varying from less than 0.1 mg/ℓ to values as high as 10–20 mg/ℓ. Maximum permissible limits specified for drinking water range from 1.2 to 2.4 mg/ℓ (Table 9.1), depending on the temperature of the region. Concentrations recommended for optimum dental health are close to 1 mg/ℓ, but also vary slightly depending on the temperature of the region. Natural concentrations of F^- in groundwater depend on the availability of F^- in the rocks or minerals encountered by the water as it moves along its flow paths and on solubility constraints imposed by fluorite (CaF_2) or fluorapatite, $Ca_3(PO_4)_2 \cdot CaF_2$. Equilibrium dissolution-precipitation relations for these minerals in water are

$$K_{\text{flourite}} = [Ca^{2+}][F^-]^2 \qquad \log K_{25^\circ C} = -9.8 \qquad (9.20)$$

$$K_{\text{flourapatite}} = [Ca^{2+}]^5[F^-][PO_4{}^{3-}]^3 \qquad \log K_{25^\circ C} = -80 \qquad (9.21)$$

Because of the lack of $PO_4{}^{3-}$ in most groundwater environments, CaF_2 is probably the mineral phase that exerts the solubility constraint in situations where F^- is available from the host rock. However, as can be determined by substituting values in Eq. (9.20), Ca^{2+} concentrations of many hundreds of milligrams per liter are required for this solubility constraint to limit F^- concentrations to levels below drinking water standards. The fact that nearly all groundwaters are undersaturated with respect to fluorite and fluorapatite suggests that the F^- content of groundwater is generally limited by the availability of F^- in the rocks and sediments through which the groundwater moves rather than by the solubility of these minerals. Groundwater with F^- contents that exceed drinking water standards is common in the Great Plains region of North America and in parts of the southwestern United States. This suggests that F^- is more readily available from the rocks of these regions than in most other areas of North America.

Although phosphorus is not a harmful constituent in drinking water, its presence in groundwater can be of considerable environmental significance. Phosphorus additions to surface-water bodies in even small amounts can, in some circumstances, produce accelerated growth of algae and aquatic vegetation, thereby causing eutrophication of the aquatic system. Because of this, phosphorus is regarded as a pollutant when it migrates into ponds, lakes, reservoirs, and streams. The occurrence and mobility of phosphorus in groundwater is important in situations where there is a potential for groundwater to feed phosphorus into surface-water environments. Through the widespread use of fertilizers and disposal of sewage on land, the potential for phosphorus influx to surface-water systems as a result of transport through the groundwater zone is increasing.

Dissolved inorganic phosphorus in water occurs primarily as H_3PO_4, $H_2PO_4^-$, $HPO_4{}^{2-}$, and $PO_4{}^{3-}$. Since H_3PO_4 is a polyprotic acid [see discussion in Section 3.3 and Figure 3.5(b)], the relative occurrence of each of these forms of dissolved phosphorus is pH-dependent. In the normal pH range of groundwater, $H_2PO_4^-$ and $HPO_4{}^{2-}$ are the dominant species. Because these species are negatively charged, the mobility of dissolved phosphorus in groundwater below the organic-rich horizons of the soil zone is not strongly limited by adsorption. The dominant control on phosphorus in the groundwater zone is the solubility of slightly soluble phosphate minerals.

Solubility control is usually attributed to one or more of the following minerals: hydroxylapatite, $Ca_5(OH)(PO_4)_3$; strengite, $FePO_4 \cdot 2H_2O$; and varisite, $AlPO_4 \cdot 2H_2O$. From the law of mass action, equilibrium expressions for precipitation-dissolution reactions of these minerals in water can be expressed as

$$K_n = [Ca^{2+}]^5[OH^-][PO_4{}^{3-}]^3 \qquad \log K_n = -58.5 \tag{9.22}$$

$$K_s = [Fe^{3+}][H_2PO_4^-][OH^-]^2 \qquad \log K_s = -34.9 \tag{9.23}$$

$$K_v = [Al^{3+}][H_2PO_4^-][OH^-]^2 \qquad \log K_v = -30.5 \tag{9.24}$$

where K_n, K_s, and K_v are the equilibrium constants for hydroxylapatite, strengite, and varisite, respectively. The log K values are for 25°C and 1 bar. These solubility relations indicate that the concentrations of Ca^{2+}, Fe^{3+}, and Al^{3+} can control the equilibrium concentration of dissolved phosphorus in solution. Equilibrium concentrations of total dissolved phosphorus computed from the solubility relations above are shown in Figure 9.21. Since the solubilities of hydroxylapatite, strengite,

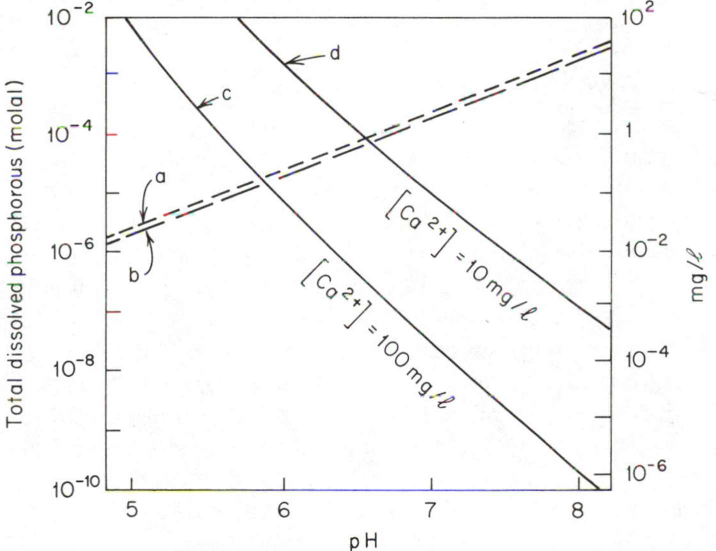

Figure 9.21 Total phosphate solubility as a function of pH. Dissolved phosphate in equilibrium with (a) variscite; (b) strengite; (c) and (d) hydroxylapatite at two calcium activities.

and varisite depend on the concentrations of Ca^{2+}, Fe^{3+}, and Al^{3+}, respectively, each solubility line is valid only for a specified concentration of these ions. Two solubility lines for hydroxylapatite (lines c and d) are shown in order to illustrate the influence of Ca^{2+} on the equilibrium phosphate concentration. The varisite solubility line (line a) is based on the assumption that the Al^{3+} concentration is governed by the solubility of gibbsite, $Al(OH)^3(s)$. For strengite solubility (line b) it is assumed that $Fe(OH)_3(s)$ limits the Fe^{3+} concentration.

From Figure 9.21 it is apparent that equilibrium concentrations of total dissolved phosphate are large in waters that have low Ca^{2+} concentrations and pH values near or below 7. In anerobic groundwater, Fe^{2+} rather than Fe^{3+} is the dominant form of dissolved iron. In this situation strengite solubility is not a limiting factor in phosphate occurrence. Groundwaters with these characteristics, namely, low Ca^{2+} concentrations and reducing redox conditions, occur in many regions that are underlain by crystalline igneous rocks or by deposits derived from these rocks. In regions such as the Precambrian Shield region of Michigan, Minnesota, and parts of Canada, migration of dissolved phosphorus from septic systems through shallow groundwater regimes into clear-water lakes poses a significant water-quality problem. Small increases in phosphorus influx to many of these lakes can cause extensive growth of algae and undesirable aquatic vegetation. Phosphorus mobility in groundwater can be a significant factor in the environmental impact of cottage and recreational developments near lakes.

For a more extensive review of the hydrochemical controls on phosphorus in aqueous systems and soils, the reader is referred to Stumm and Morgan (1970) and Beek and De Haan (1974).

Organic Substances

In Chapter 3, it was indicated that all groundwater normally contains small amounts of dissolved organic substances of natural origin. These substances, which are referred to as humic and fulvic acids, are of little concern from a water quality viewpoint. Organic substances produced by man, however, are of great concern. The number of identified man-made organic compounds now totals near 2 million and is growing at a rate of about 250,000 new formulations annually, of which 300–500 reach commercial production (Giger and Roberts, 1977).

Increasing numbers of these substances are relatively resistant to biological degradation. Many resist removal in sewage treatment plants. It is estimated that up to one-third of the total production of today's synthetic organic compounds eventually enters the biosphere (Iliff, 1973). More than 1200 individual man-made organic substances have been identified in drinking water supplies (Shackelford and Keith, 1976). This number is increasing rapidly as investigations of organic compounds in water supplies are intensified.

The question that should be addressed here is: To what extent and under what circumstances are organic compounds causing degradation of groundwater quality? Unfortunately, this question cannot be answered at present. Since there have been so few investigations of organic compounds in groundwater, it is not possible at the present time to draw any general conclusions. Our purpose here is to briefly review some of the factors that are expected to play a major role in the migration of organic compounds into groundwater systems.

Organic chemicals make their way to the land surface as a result of the use of pesticides, the use of land for sewage disposal, the use of sanitary landfills or refuse dumps for disposal of organic compounds, burial of containers with organic compounds at special burial sites, leakage from liquid waste storage ponds, and accidental spills along highways or other transportation routes. There are hundreds of thousands of locations in North America and Europe in which organic compounds may be a threat to groundwater quality.

Fortunately, there are several mechanisms that tend to prevent or retard the migration of most organic substances from the land surface or soil zone into deeper parts of the subsurface environment. These mechanisms include chemical precipitation, chemical degradation, volatilization, biological degradation, biological uptake, and adsorption.

Many organic substances have extremely low solubility in water. This generally limits the possibility for appreciable migration of large quantities in groundwater. However, because many of these substances are toxic at very low concentrations, solubility constraints are often not capable of totally preventing migration at significant concentration levels. For example, comparison of the solubilities and maximum permissible concentrations in drinking water of some of the common pesticides (Table 9.3) indicates that the solubilities generally exceed the permissible concentrations of these pesticides. A more comprehensive description of pesticide compositions and solubilities is presented by Oregon State University (1974).

Table 9.3 Comparison of Maximum Permissible Concentration Limits in Drinking Water and the Solubilities of Six Pesticides

Compound	Maximum permissible concentration (mg/ℓ)	Solubility in water (mg/ℓ)
Endrine	0.0002	0.2
Lindane	0.004	7
Methoxychlor	0.1	0.1
Toxaphene	0.005	3
2,4-D	0.1	620
2,4,5-TP silvex	0.01	—

NOTE: Solubilities from Oregon State University, 1974.

Many organic substances are lost from the soil zone as a result of volatilization (i.e., conversion to the vapor state). When the substances transform from the solid phase or from the dissolved phase to the vapor phase, they are lost by diffusion to the atmosphere. This process can greatly reduce the concentrations available for transport in subsurface water. For volatilization to occur, however, a gas phase must be present. Therefore, this process cannot be effective if the compounds migrate below the water table, where species occur only in dissolved form.

Nearly all pesticides and many other organic substances that make their way to the land surface and hence into the soil zone undergo biochemical degradation. The soil zone contains a multitude of bacteria that can convert and consume countless numbers of organic compounds. If it were not for these organisms, the biosphere would long ago have become intolerably polluted with organic compounds. In terms of environmental contamination, the major concern is focused on those organic substances that are not readily degraded by bacteria, either in the soil zone or in sewage treatment facilities. These substances are known as *refractory compounds*. Their presence in the surface environment is becoming increasingly pervasive.

The organic substances that pose the greatest threat to the quality of groundwater resources are those that are relatively soluble, nonvolatile, and refractory. The main mechanism that prevents most of these compounds from readily migrating from the land surface into aquifer systems is adsorption. Minerals and amorphous inorganic and organic substances in the soil zone and in deeper geologic materials all provide surfaces for adsorption of organic compounds. Unfortunately, adsorption isotherms are available for only a small percentage of the existing organic chemicals that are entering the biosphere. These isotherms relate to only a small number of permeable geological materials under a limited range of hydrochemical conditions. Because of this paucity of adsorption data, it is not possible to draw general conclusions on the potential magnitude of the hazard to groundwater resources posed by increasing use and dependence on organic chemicals.

For readers interested in obtaining more information on the occurrence, classification, and movement of organic substances in groundwater and surface waters, the following introduction to the literature may be useful. Giger and Roberts (1977) describe the problems associated with characterizing refractory organic compounds in contaminated waters. A classification scheme for organic compounds in water is presented by Leenheer and Huffman (1977). The chemical, ecological, and adsorptive properties of a wide variety of insecticides and herbicides are described by the Oregon State University (1974). Malcolm and Leenheer (1973) indicate the usefulness of dissolved organic carbon measurement as a contamination indicator in groundwater and surface-water investigations. Based on an extensive literature review, Shackelford and Keith (1976) summarized the reported occurrences of organic compounds in groundwater and other waters used for drinking water supplies. The behavior of petroleum substances such as oil and gasoline in water is described by McKee (1956). Adsorption isotherms for several organic compounds in selected soils are described by Kay and Elrick (1967), Hamaker and Thompson (1972), Davidson et al. (1976), and Hague et al. (1974).

9.4 Measurement of Parameters

Velocity Determination

There are three groups of methods for determination of the velocity of groundwater. The first group includes all techniques that are directly dependent on use of the Darcy equation. The second group involves the use of artificial tracers. The third group consists of groundwater age-dating methods using environmental isotopes such as tritium and carbon 14. Darcy-based techniques require information on the hydraulic conductivity, hydraulic gradient, and porosity in the portion of the flow field in which velocity estimates are desired. From these data the average linear velocity $\bar{v}$ can be computed using Eq. (2.82). Methods by which field values of hydraulic conductivity, hydraulic gradient, and porosity are determined are described in Chapters 2, 6, and 8. They need not be pursued here, other than to note that velocity estimates based on use of these parameters in Darcy-based equations have large inherent uncertainties that generally cannot be avoided. In field situations hydraulic conductivity determinations often have large uncertainties. Errors in hydraulic conductivity measurements combined with the errors associated with determination of the gradient and porosity result in considerable error being associated with the computed velocity. In some situations better accuracy can normally be achieved by use of artificial groundwater tracers, although this may involve greater expense.

The most direct method for groundwater velocity determination consists of introducing a tracer at one point in the flow field and observing its arrival at other points. After making adjustments for the effect of dispersion, the groundwater velocity can be computed from the travel time and distance data. The literature is replete with descriptions of experiments of this type. Many types of nonradioactive and radioactive tracers have been used, ranging from such simple tracers as salt (NaCl or $CaCl_2$), which can be conveniently monitored by measurements of electrical conductance, to radioisotopes such as ^3H, ^{131}I, ^{29}Br, and ^{51}Cr-EDTA (an organic complex with ^{51}Cr), which can be accurately monitored using radioactivity detectors. Radioisotopes have the disadvantage of government licensing requirements for their use and of being hazardous when used by careless workers. Fluorescent dyes (fluoroscein and rhodamine compounds) have been used by many investigators. In field tests, visual detection of the dye can sometimes yield adequate results. Dye concentrations can be measured quantitatively to very low concentrations when necessary. Recent work suggests that Freon (Cl_3CF) may be one of the best of the artificial tracers for use in groundwater velocity tests (Thompson et al., 1974). It is nonreactive with geologic materials and can be used in extremely small concentrations that are nonhazardous in public waters. For reviews of tracer techniques in groundwater investigations, the reader is referred to Knutson (1966), Brown et al. (1972), and Gaspar and Oncescu (1972).

The direct tracer method of groundwater velocity determination described above has four main disadvantages: (1) because groundwater velocities are rarely large under natural conditions, undesirably long periods of time are normally required for tracers to move significant distances through the flow system; (2) because geological materials are typically quite heterogeneous, numerous observation points (piezometers, wells, or other sampling devices) are usually required to adequately monitor the passage of the tracer through the portion of the flow field under investigation; (3) because of (1), only a small and possibly nonrepresentative sample of the flow field is tested; and (4) because of (2), the flow field may be significantly distorted by the measuring devices. As a result of these four factors, tracer experiments of this type commonly require considerable effort over extended periods of time and are rarely performed.

A tracer technique that avoids these disadvantages was developed in the USSR in the late 1940's. This technique, which has become known as the borehole dilution or point-dilution method, is now used extensively in Europe. Borehole dilution tests can be performed in relatively short periods of time in a single well or piezometer. The test provides an estimate of the horizontal average linear velocity of the groundwater in the formation near the well screen. A schematic representation of a borehole dilution test is shown in Figure 9.22(a). The test is performed in a segment of a well screen that is isolated by packers from overlying and underlying portions of the well. Into this isolated well segment a tracer is quickly introduced and is then subjected to continual mixing as lateral groundwater flow gradually removes the tracer from the well bore. The combined effect of groundwater through-flow and mixing within the isolated well segment produces a dilution versus time relation as illustrated in Figure 9.22(b). From this relation the average horizontal velocity of groundwater in the formation beyond the sand or gravel pack

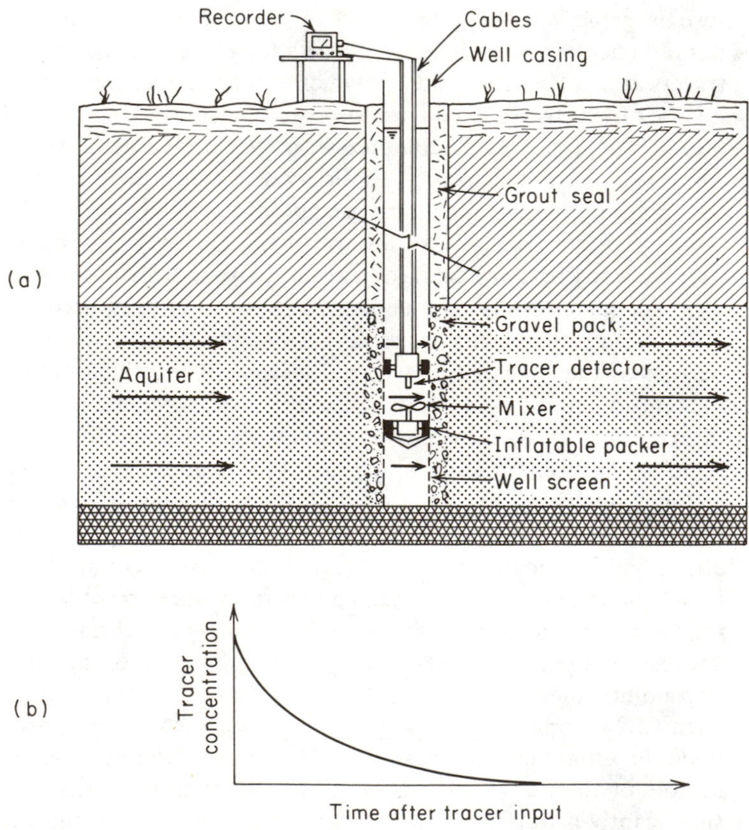

Figure 9.22 Borehole dilution test. (a) Schematic diagram of apparatus; (b) dilution of tracer with time.

but close to the well screen is computed. The theory on which the computational methods are based is described below.

Although adjustments in field technique and analytical methods can be made to take into account the effects of flow with a significant vertical component, the borehole dilution method is best suited for velocity determination in steady-state lateral flow regimes. We will proceed on this basis and with the additional stipulation that complete mixing of the tracer in the well-screen segment is maintained with no significant disturbance of the flow conditions in the formation.

The effect of the well bore and sand pack in a lateral flow regime is shown in Figure 9.23. The average linear velocity of the groundwater in the formation

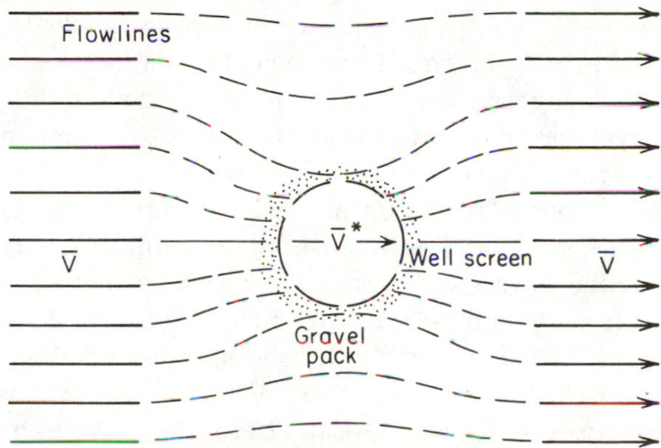

Figure 9.23 Distortion of flow pattern caused by the presence of the well screen and sand or gravel pack.

beyond the zone of disturbance is $\bar{v}$. The average bulk velocity across the center of the well bore is denoted by $\bar{v}^*$. It will be assumed that the tracer is nonreactive and that it is introduced instantaneously at concentration C_0 into the isolated segment of the well screen. The vertical cross-sectional area through the center of the isolated segment is denoted as A. The volume of this well segment is W. At time $t > 0$, the concentration C in the well decreases at a rate

$$\frac{dC}{dt} = -\frac{A \cdot \bar{v}^* \cdot C}{W} \tag{9.25}$$

which, upon rearrangement, yields

$$\frac{dC}{C} = -\frac{A \cdot \bar{v}^* \cdot dt}{W} \tag{9.26}$$

Integration and use of the initial condition, $C = C_0$ at $t = 0$, leads to

$$\bar{v}^* = -\frac{W}{A \cdot t} \ln\left(\frac{C}{C_0}\right) \tag{9.27}$$

Thus, from concentration versus time data obtained during borehole dilution tests, values of $\bar{v}^*$ can be computed. The objective of the test, however, is to obtain estimates of $\bar{v}$. This is accomplished using the relation

$$\bar{v} = \frac{\bar{v}^*}{n\bar{\alpha}} \tag{9.28}$$

where n is the porosity and α is an adjustment factor that depends on the geometry of the well screen, and on the radius and hydraulic conductivity of the sand or gravel pack around the screen. The usual range of α for tests in sand or gravel aquifers is from 0.5 to 4 (Drost et al., 1968).

Borehole dilution tests performed at various intervals within a well screen can be used to identify zones of highest groundwater velocity. These zones are often of primary interest because contaminants can move through them at velocities much higher than in other parts of the system. Identification of the high-velocity zones, which may occur in only a thin segment of an aquifer system, can provide for efficient design of monitoring networks for groundwater quality.

The borehole dilution method is described in detail by Halevy et al. (1967) and Drost et al. (1968). In most borehole dilution tests described in the literature, radioactive tracers were used. The recent advent of commercially available electrodes for use with portable pH meters for rapid down-hole measurement of Cl^- or F^- has made it feasible to conduct borehole-dilution tests with these readily available tracers in a more convenient manner than was previously the case. An example is described by Grisak et al. (1977). An even simpler approach involves the use of salt as the tracer with down-hole measurement of electrical conductance as the salt is flushed from the well screen. Borehole dilution tests, like many other types of field tests used in groundwater studies, can be accomplished using simple, inexpensive equipment or more elaborate instrumentation. The choice of method depends on factors such as the hydrogeologic setting, availability of instrumentation, and the experimental precision and reproducibility that is desired.

Dispersivity

From a measurement point of view, the most elusive of the solute transport parameters is dispersivity. Longitudinal dispersivity can be measured in the laboratory by passing a nonreactive tracer through cylindrical samples collected from boreholes or excavations. These experiments produce break-through curves as illustrated in Figure 9.1(c). The dispersivity of the sample can be computed by fitting solutions of the advection-dispersion equation to the experimentally determined breakthrough curve. If the breakthrough curve is obtained from a column test with step-function tracer input, Eq. (9.5) can be used in the analyses of the curve. The velocity is obtained by dividing the specific discharge of water through the column by the porosity. Dispersivity is then obtained as the remaining unknown in the equation. Dispersivity values obtained from column tests on disturbed or undisturbed samples of unconsolidated geological materials invariably yield values in the range 0.01–2 cm. Based on 2500 column dispersion tests of this type, Klotz and Moser (1974) observed that values of longitudinal dispersivity depend on grain size and grain-size distribution and are independent of grain shape, roughness, and angularity.

Longitudinal dispersivity values determined by column tests are generally viewed as providing little indication of the *in situ* dispersivity of the geologic materials. Dispersivity has the distinction of being a parameter for which values determined on borehole-size samples are commonly regarded as having little relevance in the analysis of problems at the field scale.

It is generally accepted that longitudinal and transverse dispersivities under field conditions are larger than those indicated by tests on borehole samples. In other words, tracer or contaminant spreading in the field as a result of dispersion is greater than is indicated by laboratory measurements. This difference is normally attributed to the effects of heterogeneities on the macroscopic flow field. Since most heterogeneities in geological materials occur at a larger scale than can be included in borehole samples, dispersivity values from tests on small samples can be viewed as representing a property of the medium but at a scale of insufficient size for general use in prediction of dispersion in the field.

Studies of contaminant migration under field conditions require dispersivity measurements in the field. Although this premise is generally accepted, there is little agreement on the types of field dispersivity tests or methods for test analysis that are most appropriate. This state of affairs may be the result of the fact that relatively few detailed field dispersivity tests have been conducted, rather than a result of excessive difficulties of the task. It has only been in recent years that dispersivity at the field scale has received much attention. In comparison to the many thousands of field hydraulic conductivity and transmissivity tests that have been conducted in the common types of geologic materials, only a few tens of field dispersivity tests are reported in the literature.

There are four main types of field dispersivity tests. These are (1) single-well withdrawal-injection tests, (2) natural-gradient tracer tests, (3) two-well recirculating withdrawal-injection tests, and (4) two-well pulse tests. In each of these tests a nonreactive tracer is introduced into the groundwater system. In the single-well test, the tracer is pumped in for a set time period followed by pumping from the well and monitoring the concentration levels. The dispersivity of the formation near the well screen is computed from the concentration response data (Percious, 1969; Fried et al., 1974). In the natural-gradient test the tracer is introduced into the system without much disturbance of the flow regime. Its migration is then monitored at one or more sampling points (Fried, 1976). In the two-well recirculating test, the tracer is injected into the flow regime at one well. It is pumped out of a second well and then recirculated through the withdrawal-injection system. The concentration versus time response at the withdrawal well serves as a basis for computation of the dispersivity using analytical or numerical models (Grove and Beetem, 1971; Pickens et al., in press). In the two-well pulse test a tracer is introduced into a well situated within the drawdown cone caused by pumping of a second well. Concentration data from the pumping well are used for calculation of a dispersivity value for the segment of the formation between the two wells (Zuber, 1974).

Fried (1975) presents an outline of the test methods and the mathematical basis for analysis of data from the first three types of tests indicated above. In each case dispersivity values are obtained by fitting an analytical or numerical model to the experimental data. Zuber (1974) emphasizes that the dispersivity value obtained from a given field experiment depends, sometimes to a high degree, on the mathematical model used in the analysis, and on the scale of the experiment. Aquifers are commonly stratified and tracers travel at different rates through the different layers. Even though the differences in hydraulic conductivity between layers may be almost imperceptible, the design of wells used for sampling can exert a dominant influence on the dispersivity values computed from concentration response data. Tests in which monitoring wells with large screened intervals are used can yield large apparent dispersivities because of mixing in the well screen. Pickens et al. (in press) describe a multilevel point-sampling device that is well suited for use in dispersion tests in sandy aquifers. Castillo et al. (1972) show that the dispersive nature of fractured rocks can exhibit great complexities in comparison with that expected of granular materials.

Chemical Partitioning

Reactive contaminants transported by groundwater are distributed between the solution phase and other phases. Reactions between the dissolved species and the geological materials may cause a portion of the dissolved species to be transferred to the solids as a result of adsorption or ion exchange. Reactions primarily among the contaminant, other dissolved constituents, and the geological materials may cause a portion of the contaminant concentration to be incorporated into a solid form as a result of chemical precipitation. Above the capillary fringe, where a continuous gas phase normally exists in part of the void space, reactions may cause some of the contaminant mass to be transferred from the solution phase to the gas phase, such as occurs during denitrification in the unsaturated zone. In each of these processes the contaminant is partitioned between the solution and other phases. The ultimate fate of the contaminant in the subsurface zone can depend on the degree of irreversibility of the reactions. Prediction of the rate and concentrations at which a contaminant will be transported in groundwater requires knowledge of the rates and extent to which this partitioning will occur.

In this brief discussion of this broad topic, we will focus on the partitioning of contaminants between the liquid and solid phases. There are four main approaches to the determination of this type of partitioning. These include (1) use of computational models based primarily on thermodynamically derived constants or coefficients for equilibrium systems, (2) laboratory experiments in which the contaminant in solution is allowed to react under controlled conditions with samples of the geologic materials of interest, (3) field experiments in which the degree of partitioning is determined during passage of contaminant solutions through a small segment of the groundwater system and (4) studies of existing sites at which contamination has already occurred.

Insight into the computational approach based on equilibrium thermodynamics can be acquired from Chapters 3 and 7. If it is expected that the concentration of the contaminant in solution is controlled by precipitation-dissolution reactions and if the necessary thermodynamic data on the aqueous and solid components of the system are available, the equilibrium concentration of the contaminant in solution under specified conditions can be computed. Although the necessary computational techniques are well developed, this method has limited application for many types of contaminants because of uncertainties with regard to chemical composition and the free energies of the controlling solid phases or because of sluggish rates of the dominant reactions. In many cases the contaminant species of interest are transported in solutions that are chemically very complex. The presence of organic compounds can cause contaminant mobility to be enhanced considerably beyond that predicted based on inorganic considerations alone.

In the laboratory, the degree of contaminant partitioning is determined in *column experiments* and in what are known as *batch experiments*. In column experiments (Figure 9.1), prepared solutions or natural waters to which the contaminant is added are passed through cylindrical samples of the geologic materials of interest. If the flow rate and input water chemistry is regulated to approximate the field conditions and if disturbance of the sample prior to emplacement in the column has not caused the material to acquire properties that deviate significantly from field conditions, the degree of partitioning and retardation obtained from this type of experiment provides an indication of what will take place in the field. Column experiments, however, are rarely conducted with adherence to all of these requirements. There is considerable uncertainty, therefore, in application of the results to field situations. Column experiments are described by Rovers and Farquhar (1974) and Griffin et al. (1976) using sanitary landfill leachate, by Routson and Serne (1972) using trace concentrations of radionuclides, by Kay and Elrick (1967) and Huggenberger et al. (1972) using lindane (a pesticide), by Doner and McLaren (1976) using urea, and by many other investigators using various chemical constituents.

In batch experiments the contaminated solution and the geologic material in a disaggregated state are brought into contact in a reaction vessel. After a period of time that normally ranges from hours to days, the degree of partitioning of the contaminant between the solution and the geologic materials is determined. For partitioning data from these experiments to be applied with confidence in the analysis of field situations, comparisons with results of column or field tests are necessary. Batch tests have the advantage of being relatively quick and inexpensive to conduct. For some contaminants, the batch test is a standard method for establishing adsorption isotherms or selectivity coefficients in ion exchange reactions. Sample disturbance and the lack of representation of field flow conditions can detract from the validity of the results in the analysis of field situations. Samples used in batch tests are usually exposed to oxidizing conditions (i.e., to oxygen in the air) during sample preparation and during the tests. Since the adsorptive capabilities of oxidized materials can be much different than reduced materials, the test results can be invalid for analysis of contaminant behavior in field systems.

The most direct but rarely the most convenient method for determining the partitioning and retardation of the contaminant is to conduct field tests. Injection of a solution of appropriate composition into a small segment of the groundwater system followed by monitoring of its behavior can provide, in favorable circumstances, a basis for prediction of contaminant behavior elsewhere in the system. Field tests of this type can be time-consuming and expensive. In order to obtain adequate information, numerous tests may be required. In some situations the need to obtain reliable information on contaminant behavior is great enough to justify this effort.

Another approach for obtaining information on the partitioning and retardation of contaminants during transport in groundwater is to conduct investigations at existing sites where groundwater pollution has already occurred. For results of these investigations to have more than site-specific significance, not only must the distributions of the contaminants in the water and on the porous media be determined, but the factors that influence these distributions must also be investigated. During recent years an appreciable number of detailed studies of sites with subsurface contamination have been reported in the literature. Some of the more notable examples are those by McKee et al. (1972), Childs et al. (1974), Suarez (1974), Ku et al. (1978), Goodall and Quigley (1977), and Gillham and Cherry (1978).

9.5 Sources of Contamination

Land Disposal of Solid Wastes

In North America approximately 3 kg of refuse per capita is produced daily. More than 20,000 landfills across the continent accommodate more than 90% of the solid waste that is produced by municipal and industrial activities. According to Yen and Scanlon (1975), a city of 1 million people generates refuse with an annual volume equivalent to 80 ha covered 5 m deep. Although materials recovery and incineration may eventually decrease the amount of waste that is disposed of by landfilling, landfills will continue to be the primary method of disposal of these wastes during at least the next few decades.

The design, construction, and operational aspects of land disposal of refuse are described by Mantell (1975). For purposes of this discussion this information is not required, other than to recognize that much of the solid waste (refuse) that is now disposed of on land is emplaced in engineered disposal systems known as *sanitary landfills*. In sanitary landfills, solid waste is reduced in volume by compaction and then is covered with earth. Ideally, the earth cover is placed over the refuse at the conclusion of each day's operation, but in practice less frequent cover application is common. The landfill, consisting of successive layers of compacted waste and earth, may be constructed on the ground surface or in excavations. In North America a large number of the older sites that receive municipal wastes are open dumps or poorly operated landfills. Newer sites are generally better situated and better operated. It is estimated that 90% of the industrial wastes that are considered to be hazardous are landfilled, primarily because it is the least expensive waste management option.

Our purpose here is to consider some of the effects that refuse disposal can have on the groundwater environment. With the exception of arid areas, buried refuse in sanitary landfills and dumps is subject to leaching by percolating water derived from rain or snowmelt. The liquid that is derived from this process is known as *leachate*. Table 9.4 indicates that leachate contains large numbers of inorganic contaminants and that the total dissolved solids can be very high. Leachate also contains many organic contaminants. For example, Robertson et al. (1974) identified more than 40 organic compounds in leachate-contaminated groundwater in a sandy aquifer in Oklahoma. These authors concluded that many of these compounds were produced by leaching of plastics and other discarded manufactured items within the refuse. Not only do the leachates emanating from

Table 9.4 Representative Ranges for Various Inorganic Constituents in Leachate From Sanitary Landfills

Parameter	Representative range (mg/ℓ)
K^+	200–1000
Na^+	200–1200
Ca^{2+}	100–3000
Mg^+	100–1500
Cl^-	300–3000
SO_4^{2-}	10–1000
Alkalinity	500–10,000
Fe (total)	1–1000
Mn	0.01–100
Cu	<10
Ni	0.01–1
Zn	0.1–100
Pb	<5
Hg	<0.2
NO_3^-	0.1–10
NH_4^+	10–1000
P as PO_4	1–100
Organic nitrogen	10–1000
Total dissolved organic carbon	200–30,000
COD (chemical oxidation demand)	1000–90,000
Total dissolved solids	5000–40,000
pH	4–8

SOURCES: Griffin et al., 1976; Leckie et al., 1975.

landfills contain contaminants derived from solids, but many leachates contain toxic constituents from liquid industrial wastes placed in the landfill.

Concern has developed in recent years with regard to the effect of landfills on the quality of groundwater resources. Garland and Mosher (1975) cite several examples where groundwater pollution has been caused by landfills. A case where leachate migration caused serious pollution of a large aquifer used as a city's water supply is described by Apgar and Satherthwaite (1975). It is expected that the cost of rectifying this situation will eventually total many millions of dollars.

Numerous investigations in North America and Europe have shown that in nonarid regions, infiltration of water through refuse causes water table mounding within or below the landfill. The mounding process is similar to that described in Section 8.11. Water-table mounding causes leachate to flow downward and outward from the landfill as illustrated in Figure 9.24. Downward flow of leachate may threaten groundwater resources. Outward flow normally causes leachate springs at the periphery of the landfill or seepage into streams or other surface-water bodies. If the paths of leachate migration do not lead to aquifers containing potable water, downward movement of leachate will not pose a threat to groundwater resources.

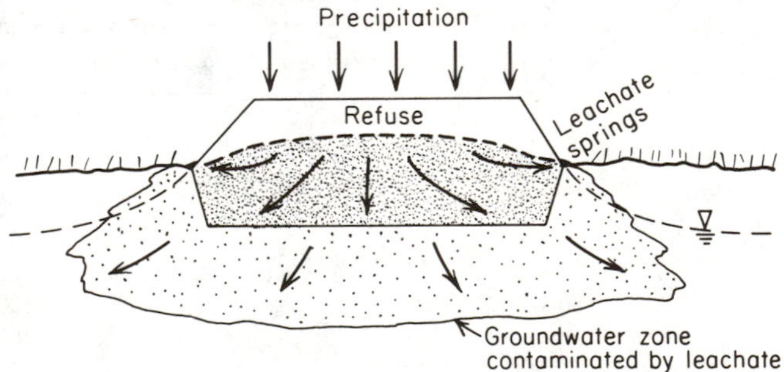

Figure 9.24 Water-table mound beneath a landfill, causing leachate springs and migration of contaminants deeper into the groundwater zone.

In situations where landfills are located in relatively permeable materials such as sand, gravel, or fractured rock, leachate migration may cause contamination over areas many times larger than the areas occupied by the landfills. An example of such a case is shown in Figure 9.25. At this landfill site on moderately permeable glaciodeltaic sand, a large plume of leachate-contaminated water, represented in Figure 9.25 by the Cl^- distribution, has penetrated deep into the aquifer and has moved laterally several hundreds of meters along the paths of groundwater flow. This contamination developed over a period of 35 years. Infiltration of water through the landfill will continue to produce leachate for many decades. Transport by groundwater flow in the sand will cause the zone of contamination to greatly expand. In this particular case, however, the aquifer is not

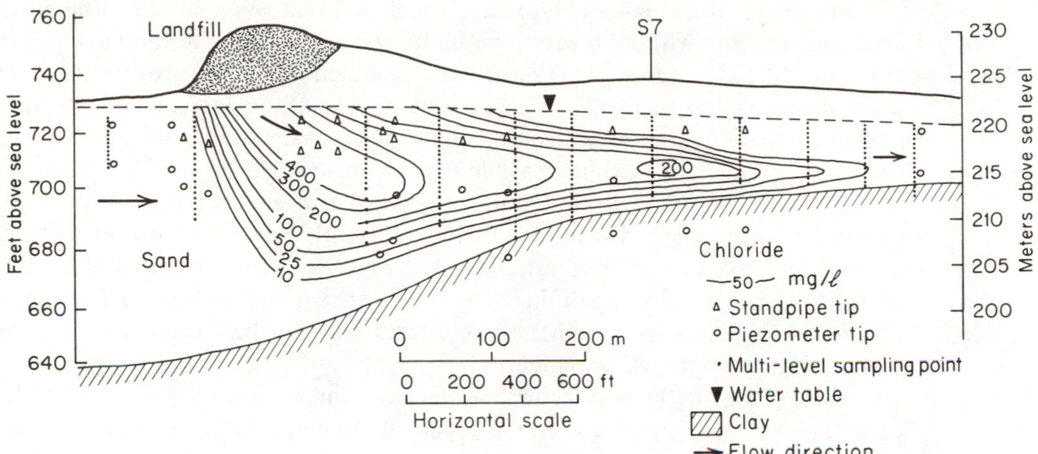

Figure 9.25 Plume of leachate migrating from a sanitary landfill on a sandy aquifer; contaminated zone is represented by contours of Cl^- concentration in groundwater.

used for water supply. The spreading contaminant plume is therefore not regarded as a significant problem. At a landfill on sand and gravel on Long Island, N.Y., Kimmel and Braids (1974) delineated a leachate plume that is more than 3000 m long and greater than 50 m in depth. These two examples and others described in the literature indicate that if leachate has access to active groundwater flow regimes, pollution can spread over very large subsurface zones. Physical and chemical processes are sometimes incapable of causing appreciable attenuation of many of the toxic substances contained within the leachate plume.

If landfills are situated in appropriate hydrogeologic settings, both groundwater and surface-water pollution can be avoided. It is commonly not possible, however, to choose sites with ideal hydrogeologic characteristics. In many regions land of this type is not available within acceptable transportation distances, or it may not be situated in an area that is publicly acceptable for land filling. For these and other reasons most landfills are located on terrain that has at least some unfavorable hydrogeologic features.

Although it is well established that landfills in nonarid regions produce leachate during at least the first few decades of their existence, little is known about the capabilities for leachate production over much longer periods of time. In some cases leachate production may continue for many decades or even hundreds of years. It has been observed, for example, that some landfills from the days of the Roman Empire are still producing leachate. Many investigators have concluded that at the present time there have been very few occurrences of leachate contamination of aquifers that are used for water supply. Whether or not it will be possible to draw similar conclusions many years from now remains to be established.

Farvolden and Hughes (1976) have concluded that solid waste can be buried at almost any site without creating an undue groundwater pollution hazard, provided that the site is properly designed and operated. A testing program to define the hydrogeological environment is essential. These authors indicate that if uncontrolled leachate migration is unacceptable, the leachate should be collected and treated as a liquid waste. One feasible way to ensure that no leachate leaves the site is to establish a hydraulic gradient toward the site, perhaps by pumping. Liners for emplacement beneath landfills are currently being evaluated as a control method but have not yet been established in practice. Some examples of controls on leachate migration using drains or wells are shown in Figure 9.26. These types of control measures require that the collected leachate be treated or otherwise managed in an appropriate manner.

In addition to the production of leachate, infiltration of water into refuse causes gases to be generated as biochemical decomposition of organic matter

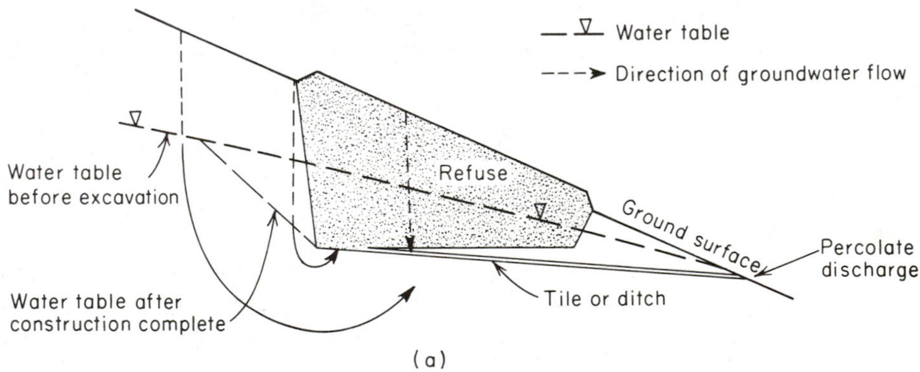

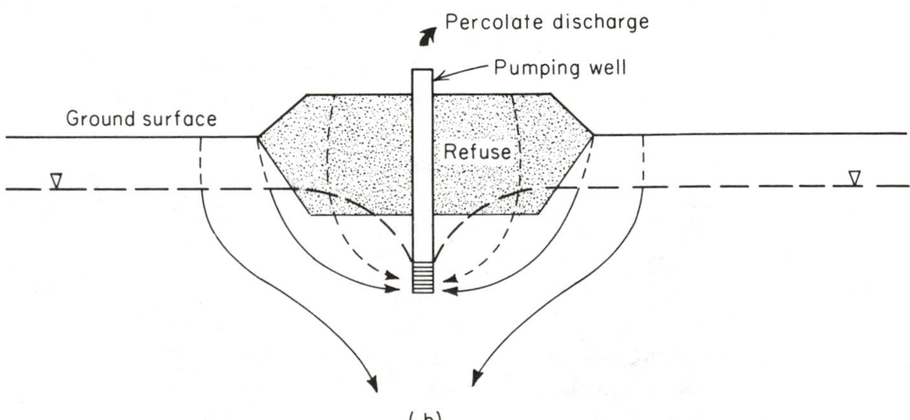

Figure 9.26 Control of leachate in a sanitary landfill by (a) tile drain or ditch and (b) pumped well (after Hughes et al., 1971).

occurs. Gases such as CO_2, CH_4, H_2S, H_2, and N_2 are commonly observed. CO_2 and CH_4 are almost invariably the most abundant of these gases. CH_4 (methane) has a low solubility in water, is odorless, and generally is of little influence on groundwater quality. In the environmental impact of landfills, however, it can be of great importance because of its occurrence in gaseous form in the zone above the water table. It is not uncommon for CH_4 to attain explosive levels in the refuse air. In some situations CH_4 at dangerous levels can move by gaseous diffusion from the landfill through the unsaturated zone in adjacent terrain. Migration of CH_4 at combustible levels from landfills through soils into residences has occurred in urban areas. In recent years, installation of gas vents in landfills to prevent buildup of methane in the zone above the water table has become a common practice.

In addition to hazards caused by the potential for methane explosion, gaseous migration from landfills can result in extensive damage to vegetation and odor problems. Case histories of gas migration from landfills have been described by Flower (1976). Mohsen (1975) has presented a theoretical analysis of subsurface gas migration from landfill sources. The interactions of the various factors that influence gas production in landfills have been described by Farquhar and Rovers (1973).

Sewage Disposal on Land

Sewage is placed on or below the land surface in a variety of ways. Widespread use of septic tanks and drains in rural, recreational, and suburban areas contributes filtered sewage effluent directly to the ground. Septic tanks and cesspools are the largest of all contributors of wastewater to the ground and are the most frequently reported sources of groundwater contamination in the United States (U.S. Environmental Protection Agency, 1977). Twenty-nine percent of the U.S. population disposes of its domestic waste through residential disposal systems. An increasing percentage of the municipal sewage in industrialized countries is being processed in primary and secondary sewage treatment plants. Although this decreases surface-water pollution, it produces large volumes of solid residual materials known as *sewage sludge*. In many areas this sludge, which contains a large number of potential contaminants, is spread on agricultural or forested lands. In some regions liquid sewage that has not been treated or that has undergone partial treatment is sprayed on the land surface. Application of liquid sewage and sewage sludge to the land provides nutrients such as nitrogen, phosphorus, and heavy metals to the soil. This can stimulate growth of grasses, trees, and agricultural crops. Land that is infertile can be made fertile by this practice. One of the potential negative impacts of this type of sewage disposal is degradation of groundwater quality.

Primary- and secondary-treated sewage is being spread on forested land and crop land in an increasing number of areas in Europe and North America. For example, in Muskegon County, Michigan, more than 130 million liters per day of sewage effluent is sprayed on the land surface (Bauer, 1974). For many decades cities such as Berlin, Paris, Milan, Melbourne, Fresno, and many others have been using sewage for irrigation of crops. Not only are the nutrients in sewage effluent valuable, but the water itself is a valuable resource in many regions. In some situations intensely treated sewage effluent may be used as a source of artificial recharge for aquifers that serve for municipal water supply. Injection of treated sewage into coastal aquifers may serve as a means of controlling the intrusion of salt water.

Considering the many ways in which liquid and solid constituents from sewage reach the land surface and subsurface zones, it is reasonable to expect that over the long term the quality of groundwater resources in many areas will reflect the extent to which hydrogeologic factors are considered in the overall planning and operation of sewage management systems. In a textbook of this type it is not feasible to look specifically at the hydrogeologic and geochemical factors that are important in each of the land-application or disposal-of-sewage options that are in use. Before proceeding to other topics, however, we will provide a brief guide to some of the more important studies that have been conducted. For a detailed guide to the literature in this area, the reader is referred to the U.S. Environmental Protection Agency (1974a).

During the 1950's and early 1960's it was observed that one of the most serious consequences of land disposal of sewage by way of septic systems was contamination of groundwater by alkyl benzene sulfonate (ABS), which was a major component of household detergents. ABS is relatively nonbiodegradable and exists in water in anionic form. In the 1960's, numerous cases of shallow contamination of sand and gravel aquifers were reported. The problem was most acute in areas where septic systems were draining into unconfined aquifers in which there were numerous shallow water supply wells. Case histories of this type of problem in Long Island and in Southern California are described by Perlmutter et al. (1964) and Klein (1964).

In the mid-1960's the detergent industry replaced ABS with linear alykl sulfonate (LAS), a compound that is readily biodegradable in aerobic environments. Cases of LAS and ABS contamination of wells have been a rare occurrence since LAS gained widespread use, a somewhat surprising situation considering that many septic systems drain into anaerobic groundwater environments where the effects of biodegradation are probably minimal. LAS may undergo considerable retardation as a result of adsorption.

Effluent from septic systems includes many other types of contaminants. One of the most frequently reported of these contaminants in groundwater is nitrate. As indicated in Section 9.3, nitrate commonly does not undergo complete biochemical reduction to N_2 even if the groundwater system is anaerobic. Nitrate emanating from septic systems into groundwater is transported along the groundwater flow paths. A detailed case history of the migration of nitrate and other contaminants in groundwater as a result of discharge from septic systems was presented by Childs et al. (1974).

In some areas the primary concern with regard to contaminant migration from septic systems is surface-water quality rather than groundwater quality.

This is particularly the case in areas of recreational lakes where cottages and tourist facilities use septic systems located near lakes. Transport of nitrogen and phosphorus through the groundwater zone into lakes can cause lake eutrophication manifested by accelerated growth of algae and decrease in water clarity. Some examples of hydrogeologic investigations in recreational lake environments are described by Dudley and Stephenson (1973) and Lee (1976).

Another concern associated with the disposal of treated or untreated sewage on or below the land surface revolves around the question of how far and how fast pathogenic bacteria and viruses can move in subsurface flow systems. This problem is also crucial in the development of municipal water supplies by extraction of water from wells located adjacent to polluted rivers. The literature is replete with investigations of movement of bacteria through soils or granular geological materials. As bacteria are transported by water flowing through porous media, they are removed by straining (filtering), die-off, and adsorption. The migration of the bacterial front is greatly retarded relative to the velocity of the flowing water. Although bacteria can live in an adsorbed state or in clusters that clog parts of the porous medium, their lives are generally short compared to groundwater flow velocities. In medium-grained sand or finer materials, pathogenic and coliform organisms generally do not penetrate more than several meters (Krone et al., 1958). Field studies have shown, however, that in heterogeneous aquifers of sand or gravel, sewage-derived bacteria can be transported tens or hundreds of meters along the groundwater flow paths (Krone et al., 1957; Wesner and Baier, 1970).

Viruses are very small organic particles (0.07–0.7 μm in diameter) that have surface charge. There is considerable evidence from laboratory investigations indicating that viruses are relatively immobile in granular geological materials (Drewry and Eliassen, 1968; Robeck, 1969; Gerba et al., 1975; Lance et al., 1977). Adsorption is a more important retardation mechanism than filtering in highly permeable granular deposits. Problems associated with sampling and identification of viruses in groundwater systems have restricted the understanding of virus behavior under field conditions. Advances in sampling technology (Wallis et al., 1972; Sweet and Ellender, 1972) may lead to a greatly improved understanding of virus behavior in aquifers recharged with sewage effluent.

Although there is considerable evidence indicating that bacteria and viruses from sewage have small penetration distances when transported by groundwater through granular geologic materials, similar generalizations cannot be made for transport in fractured rock. It is known that these microorganisms can live for many days or even months below the water table. In fractured rocks, where groundwater velocities can be high, this is sufficient time to produce transport distances of many kilometers.

As man relies more heavily on land application as a means of disposal for municipal sewage effluent and sludge, perhaps the greatest concern with regard to groundwater contamination will be the mobility of dissolved organic matter. Sewage effluent contains many hundreds of dissolved organic compounds, of which very little is known about their toxicity and mobility. Some of these compounds may eventually be shown to be more significant in terms of degradation of groundwater quality than nitrate, trace metals, bacteria, or viruses.

Agricultural Activities

Of all the activities of man that influence the quality of groundwater, agriculture is probably the most important. Among the main agricultural activities that can cause degradation of groundwater quality are the usage of fertilizers and pesticides and the storage or disposal of livestock or fowl wastes on land. The most widespread effects result from the use of fertilizer. In industrialized countries most fertilizer is manufactured chemically. This type of fertilizer is known as inorganic fertilizer. In less developed countries, animal or human wastes are widely used as organic fertilizer.

Fertilizers are categorized with respect to their content of nitrogen (N), phosphorus (P), and potassium (K). These are the three main nutrients required by crops. The annual application rates of fertilizers vary greatly from region to region and from crop to crop. Nitrogen applications, (expressed as N), generally vary from about 100 to 500 kg/ha·yr. Because fertilizer is used year after year, it is to be expected that in many areas some of the N, P, or K is carried by infiltrating water downward to the water table, where it can migrate in the groundwater flow regime. For reasons explained in Section 9.3, nitrogen in the form of NO_3^- is generally much more mobile in subsurface flow systems than dissolved species of phosphorus. Cation exchange causes K^+ to have low mobility in most nonfractured geologic materials.

Of the three main nutrients in fertilizer, N in the form of NO_3^- is the one that most commonly causes contamination of groundwater beneath agricultural lands. High NO_3^- concentrations have been delineated in extensive areas in many parts of the world, including Israel (Saliternik, 1972), England (Foster and Crease, 1972), Germany (Groba and Hahn, 1972), California (Calif. Bureau Sanitary Eng., 1963; Nightingale, 1970; Ayers and Branson, 1973), Nebraska (Spalding et al., 1978), southern Ontario, and southern Alberta. Many wells in these areas have NO_3^- concentrations that exceed the recommended limit for drinking water. In areas where NO_3^- contamination is areally extensive, fertilizer rather than animal wastes from feedlots or lagoons or septic field seepage is usually identified as the primary nitrogen source. Nitrate is the principal dissolved nitrogen component, with ammonium and organic nitrogen present in much lower concentrations. Although in many aquifers that are contaminated by NO_3^-, the concentrations are below the limits recommended for drinking water, it is disturbing to note that gradual increases in NO_3^- have been observed. The widespread use of inorganic fertilizers began after World War II. The major impact on groundwater quality resulting from this change in agricultural practice is probably not yet fully developed. Nitrate contamination is rarely reported at depths of more than about 10–100 m below the water table. As time goes on, however, NO_3^- contamination may extend to greater depth in areas where there are significant downward flow components. For example, NO_3^- in deep wells in California, ranging in depth from 240 to 400 m below ground surface, increased from approximately 1 mg/ℓ in 1950 to 10–17 mg/ℓ in 1962 (Broadbent, 1971). The extent to which denitrification occurs as water moves along regional flow paths is a major uncertainty inherent in predictions of long-term NO_3^- increases in aquifers.

In England, NO_3^- contamination of a large regional carbonate-rock aquifer is widespread. Analysis of the occurrence and movement of NO_3^- in this aquifer is complicated by the fact that NO_3^- is carried in groundwater flowing in a network of joints and solution channels, while some of the NO_3^- is lost from the active flow regime as a result of diffusion into the porous matrix of the limestone (Young et al., 1977). If at some time in the future the NO_3^- concentration in the flow network declines, NO_3^- will diffuse from the matrix back into the flow regime.

Although extensive NO_3^- contamination of shallow groundwater can often be attributed to leaching of fertilizer, NO_3^- in shallow groundwater in large areas in southern Alberta (Grisak, 1975), southern Saskatchewan, Montana (Custer, 1976), and Texas (Kreitler and Jones, 1975) is not caused by fertilizer use. In these areas it appears that most of the NO_3^- is derived by oxidation and leaching of natural organic nitrogen in the soil. The greater abundance and deeper penetration of oxygen into the soil has occurred as a result of cultivation. In some areas the initial turning of the sod as settlers moved on the land was probably a major factor. In other areas continual deep cultivation during the modern era of farming has been a major influence.

In many agricultural areas shallow groundwater has become contaminated locally as a result of leaching of NO_3^- from livestock and fowl wastes. The conversion of organic nitrogen in these wastes to NO_3^- takes place through biochemical processes. Relatively small source areas such as farm manure piles, fowl-waste lagoons, and feedlots contribute NO_3^- to groundwater, but if these contaminant sources are not directly underlain by aquifers, the contamination is rarely very significant. Specific cases of groundwater contamination from animal wastes are reported by Hedlin (1972) and by Gillham and Webber (1969). In agricultural areas contamination of shallow wells by NO_3^- and other consituents commonly occurs because of faulty well construction. If wells are not properly sealed by grout or clay along the well bore above the screen, contaminated runoff can easily make its way to the aquifer zone near the well screen.

Concurrent with the widespread increase in the use of chemical fertilizers since World War II has been the rapid development and use of a multitude of organic pesticides and herbicides. In a report on groundwater pollution in the southwestern United States, Fuhriman and Barton (1971) concluded that pollution by pesticides must be listed as an important potential hazard. However, they obtained no direct evidence indicating significant pesticide contamination of groundwater. Kaufman (1974), in a review of the status of groundwater contamination in the United States, indicates that this conclusion appears to characterize today's situation—that of a potential but as yet-unrealized problem. Based on a literature review and field studies in Kent, England, Croll (1972) arrived at a similar conclusion. It is well known from laboratory experiments that many pesticides and herbicides with appreciable solubility in water have significant mobility in some types of geologic materials, particularly clean sands and gravels (Burns and Mclaren, 1975; Adams, 1973). It is not unreasonable to expect that the use of these chemicals in agriculture will eventually cause parts of some aquifers to become contaminated. Davidson et al. (1976) have pointed out that because of the immense size to which the pesticide industry has grown, the problems associated with the disposal of surplus and waste pesticide materials and empty or partially empty pesticide containers has become acute. High concentrations of pesticides in groundwater can result in greater mobility than at lower concentrations. At higher concentrations, the exchange sites are more readily saturated with the pesticide, or the biodegradation capabilities of the medium may be exceeded.

Petroleum Leakage and Spills

In industrialized countries hundreds of thousands of steel gasoline storage tanks lie buried at filling stations. Many thousands of kilometers of underground pipelines carry petroleum products across continents. Tanker trucks with oil and gasoline are continually on the move. It is not surprising, therefore, that leakages and spills from these sources are an increasing threat to groundwater quality. Most of the buried storage tanks at filling stations were placed in the ground since World War II. Because stringent requirements for tank testing and replacement are only gradually being implemented in most countries, leakage problems caused by older tanks are common, particularly in regions of high water tables and frequent infiltration.

Contamination of groundwater by petroleum products from leaky tanks, from pipelines, or from spills is a much different type of problem than those described elsewhere in this chapter. The major difference lies in the fact that oils and gasoline are less dense than water and are *immiscible* in water. As a consequence of this, oil or gasoline from leakages or spills migrate almost exclusively in the unsaturated zone. The processes of petroleum movement in the unsaturated zone have been described in detail by Schwille (1967), van Dam (1967), and Dietz (1971). The following discussion is based primarily on these references.

Figure 9.27 illustrates the main subsurface migration stages that occur when oil seeps in the ground. In this case, the hydrogeologic conditions are simple. There is appreciable depth of unsaturated zone beneath the level of oil entry into the system. The term "oil" is used here to refer to both crude oil proper and its liquid derivatives, such as gasoline.

In the first migration stage, the oil movement is primarily downward under the influence of gravitational forces. During this seepage stage, capillary forces produce some lateral migration. This causes a zone, referred to as the *oil wetting zone*, around the core of the infiltration body. It is comparable in origin to the natural capillary fringe on the water table. In the oil wetting zone, the degree of oil saturation decreases outwardly and capillary forces (surface tension) are dominant. In the main seepage zone, only gravitational forces exist.

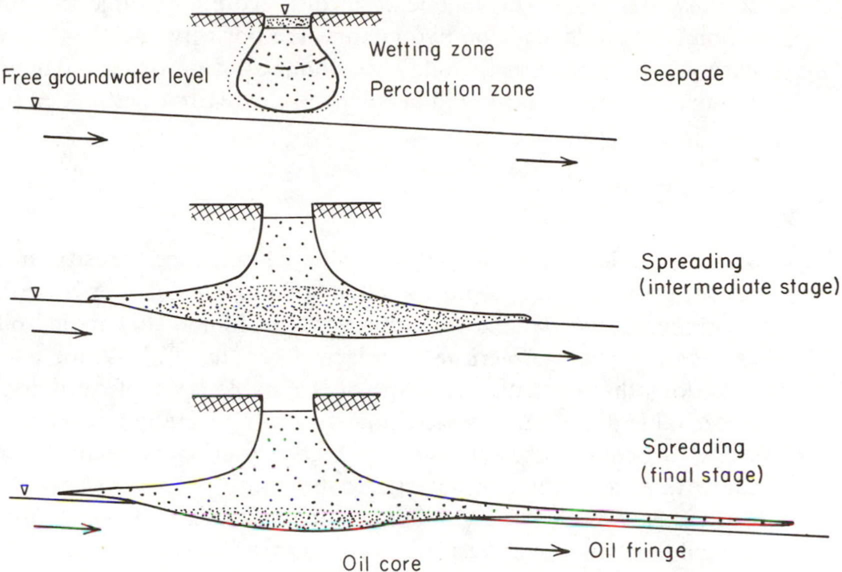

Figure 9.27 Stages of migration of oil seeping from a surface source (after Schwille, 1967).

Downward seepage of oil ceases when the seepage front reaches the water table. Although it might be expected that the oil would spread laterally on top of the capillary fringe rather than along the water table, experimental and field evidence indicates that considerable migration occurs within the capillary fringe at or very near the water table. Since oil is immiscible in water and is less dense than water, it may slightly depress the water table. Except for small amounts of hydrocarbons that go into solution, the oil does not penetrate below the water table. As oil accumulates on the water table, the oil zone spreads laterally, initially under the influence of gradients caused by gravity and later in response mainly to capillary forces. Capillary spreading becomes very slow and eventually a relatively stable condition is attained. In theory, stability occurs when a condition known as *residual oil saturation* or *immobile saturation* is reached. The experience of oil production engineers is that below a certain degree of saturation, oil is held in a relatively immobile state in the pore spaces. If the percent oil saturation is reduced further, isolated islands or globules of oil become the dominant mode of oil occurrence. Over the range of pressure gradients that can occur, these islands are stable. As the mass of oil spreads laterally due to capillary forces, the residual oil saturation condition must eventually be attained, provided that the influx of oil from the source ceases. This is referred to as the *stable* stage.

When the oil spill or leakage volume is small relative to the surface available for contact as the oil moves through the zone above the water table, the oil migration zone may attain residual saturation and become immobile before penetrating to the water table. The volume of porous medium required to immobilize a given amount of oil depends on two factors: the porosity and the nature of the hydrocarbons that comprise the oil. The volume B of porous geologic materials that is required to immobilize a spill or leakage volume can be estimated from the relation

$$B = \frac{B_0}{nS_0} \tag{9.29}$$

where B_0 is the volume of oil entering the system, n the porosity, and S_0 the residual oil saturation. If the depth to water table and values for n and S_0 are known, this relation can be used to estimate the likelihood that spilled oil will penetrate to the water table (American Petroleum Institute, 1972). Van Dam (1967) presents equations that describe the shape of the stable layer of oil if penetration to the water table occurs. In practice, however, it is generally not possible in field situations to obtain sufficient data on the distribution of relative permeabilities for more than a qualitative analysis to be made (Dietz, 1971). Laboratory model experiments by Schwille (1967) have demonstrated that minor differences in permeabilities laterally or vertically can cause strong distortions in the shape of the oil migration zone.

Because oil leakages or spills usually do not involve large fluid volumes of oil, and because migration is limited by residual oil saturation, one might expect that oil is not a significant threat to groundwater quality. This is unfortunately not the case. Crude oil and its derivatives contain hydrocarbon components that have significant solubility in water. In general, the lighter the petroleum derivative, the greater is the solubility. Commercial gasoline, for example, has solubility of 20–80 mg/ℓ. It can be detected by taste and odor at concentrations of less than 0.005 mg/ℓ (Ineson and Packham, 1967). Because the solubility of the lighter hydrocarbons greatly exceeds the concentration levels at which water is considered to be seriously polluted, it is not difficult to envision situations where the effect of hydrocarbon dissolution is of much greater concern in terms of groundwater quality than the localized immobile zone of immiscible hydrocarbons on and above the water table. For example, in the situation represented in Figure 9.28, the lateral flow of groundwater beneath the zone of immobilized oil could cause soluble hydrocarbons to be transported large distances along the groundwater flow paths.

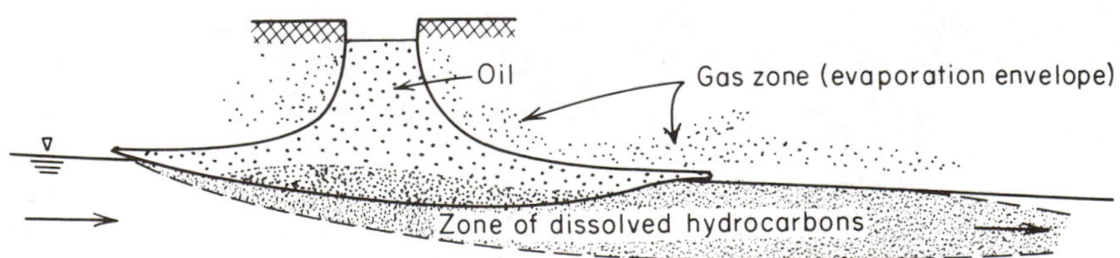

Figure 9.28 Migration of dissolved and gaseous hydrocarbons from a zone of oil above the water table (after Schwille, 1967).

In situations where oil penetrates to the water-table zone and then spreads out and becomes relatively immobile, the effect of water table fluctuations can be important. If the water table falls, much of the oil remains in the newly created zone as a thin coating on the surface of the porous medium. This film is not removed by water flushing or air ventilation. The problem can be ameliorated by the effect of bacteria. There exist species of aerobic and anaerobic bacteria that grow rapidly in the presence of oil or gasoline if the other necessary nutrients are also available. In favorable circumstances, bacterial oxidation can consume much of the oil or gasoline that accumulates above the water table as a result of leakages or spills. A fluctuating groundwater table is believed to promote the processes of biodegradation.

Of the many examples of petroleum contamination of groundwater that have been reported, the case history described by McKee et al. (1972) is particularly illustrative of the problems and processes involved and remedial measures that can be used to minimize the damage to subsurface water quality. Control and remedial procedures are described by the American Petroleum Institute (1972).

Disposal of Radioactive Waste

It has been several decades since nuclear engineers and scientists at an Idaho research station watched four household light bulbs flicker to life as a result of man's first generation of electricity from atomic energy. From this modest beginning, nuclear-power-generating facilities have grown to the point where they now produce more than 15% of the electricity in the United States and Canada and larger percentages in some European countries. By the turn of this century it is expected that in North America and Europe, the percentages will be much larger. Feared by many as a threat to mankind's future and hailed by others as the answer to the world's energy problems, nuclear generation of electricity has sparked controversy around the globe. At the present time there are several uncertainties inherent in activities associated with the generation of nuclear power. One of them is man's capability to safely isolate radioactive wastes from the biosphere for long periods of time. Because of its hydrogeologic nature, this nuclear-power-related topic is worthy of discussion in this text.

The hydrogeologic aspects of the nuclear power industry will be considered within the framework of the *nuclear fuel cycle*. This expression refers to all the stages in the nuclear power industry in which nuclear fuel is developed and used and in which radioactive wastes are generated. This includes uranium mining, milling, refining, uranium enrichment, fuel fabrication, fuel consumption in reactors, fuel reprocessing, waste solidification, and burial of solidified waste or unreprocessed spent fuel in deep geological repositories. The "front end" of the nuclear fuel cycle involves mining and milling of uranium ore. An undesirable by-product of these activities is the production of large volumes (hundreds of millions of cubic meters per year in North America) of waste rock from mining and tailings from milling. Waste rock and tailings are usually placed in piles on the land surface or as fill material in topographic depressions confined by small earth embankments or dams. Because they contain isotopes of uranium, thorium, and radium, waste rock and tailings are a form of solid low-level radioactive waste. Radium 226 (^{226}Ra), with a half-life of 1620 years, poses the greatest environmental hazard. Table 9.1 indicates that the maximum permissible concentration of ^{226}Ra in drinking water is 3 pCi/ℓ, which is equivalent to 10^{-9} mg/ℓ. This concentration is so small that it is orders of magnitude below the maximum permissible concentrations for trace metals such as lead, mercury, or cadmium (Table 9.1). Extremely small amounts of ^{226}Ra leached from waste rock or tailings into groundwater can therefore cause the water to be unfit for human consumption. Uranium mining in North America generally occurs in areas remote from population centers and from industrial or agricultural developments. In these areas, groundwater quality has, until recently, not been a subject of significant concern. The extent to which ^{226}Ra and other hazardous constituents from waste rock or tailings are entering groundwater and their fate within groundwater flow systems is not known. We can expect, however, that in the next decade hydrogeological factors will play a much greater role in the design and evaluation of disposal sites for uranium mining and milling wastes than has previously been the case.

The next stage in the nuclear fuel cycle is uranium refining, a process in which the mill product is upgraded in preparation for uranium enrichment into nuclear fuel (the U.S. and European approach) or unenriched fuel fabrication (the Canadian approach). In the refining process, small quantities of solid or semisolid, low-level radioactive wastes are generated. The chemical nature of these wastes varies from refinery to refinery, but the wastes generally contain ^{226}Ra, ^{230}Th, and ^{238}U in what are normally small but significant concentrations. As in the case of mining and milling wastes, ^{226}Ra is the isotope of main concern. The refinery wastes are assigned to near-surface burial grounds that are located near the refineries. After more than 20 years of use, the burial ground at the principal uranium refinery in Canada (Port Hope, Ontario) was found in the mid-1970's to be emitting leachate with ^{226}Ra, as well as other nonradioactive contaminants. Although no aquifers were in jeopardy, remedial measures include excavation of the wastes with subsequent reburial at a site with hydrogeologic capability for longer-term isolation of the wastes from the biosphere.

The next major waste-generation stage in the nuclear fuel cycle is the operation of nuclear reactors for power production, weapons production, or research. In this stage, low-level solid radioactive wastes in the form of discarded equipment, assorted slightly radioactive refuse, and ion-exchange materials from decontamination facilities are produced. These wastes are known as *reactor wastes*. The term "low-level" is used here in a qualitative sense to distinguish these wastes from the very highly radioactive materials in used nuclear fuel or material derived directly from the used fuel.

Since the start of nuclear power production on a commercial basis in North America, the total volume of accumulated reactor wastes has amounted to about 40,000 m^3 (as of 1975). These wastes have been emplaced in shallow burial sites at 11 major locations in the United States and four major locations in Canada.

In the United States, the volume is expected to rise to 50,000 m^3 by 1980 and then to more than 300,000 m^3 by the year 2000 (*Nuclear News*, 1976). By use of existing and economically viable technology, the projected volumes can be reduced by a factor of 2 or 3. It is hoped, of course, that new waste processing methods will be developed to provide further volume reduction.

In spite of the improvements that may develop, the volumes of reactor waste are expected to be enormous in comparison with the volumes that have been handled in the past. Since the standard method of managing reactor wastes is to assign them to shallow burial sites, these mounting waste volumes can be viewed as a potential source of contamination to groundwater and other environments. The past history of shallow low-level waste burial in the United States is less than satisfactory. Of the 11 existing sites at which radioactive wastes resulting from commercial power production have been buried, three are leaking radioactive constituents to the environment (*Ground Water Newsletter*, 5, no. 3, 1976). Although at present this leakage to subsurface flow systems does not present a hazard to potable water supplies, it is striking evidence that undesirable consequences of inadequate hydrogeologic studies of waste management sites can become evident many years or decades after site usage begins. There is now little doubt that at the time most of these sites were established many years ago, more attention was given to the economics of handling the wastes, the ready availability of land for burial use, and proximity to transportation routes than was given to the ultimate fate of the wastes. With these lessons in hand, the problem facing hydrogeologists now is to ensure, through use of proper site search and evaluation methodologies, that future sites for shallow burial of solid low-level wastes have adequate radionuclide containment capabilities and that proper subsurface monitoring facilities are installed and operated.

Reactor wastes contain a variety of radionuclide species, with half-lives ranging from seconds to many decades or longer. Of these nuclides, ^{137}Cs, ^{90}Sr, and ^{60}Co, with half-lives of 28, 33, and 6 years, respectively, are usually regarded as posing the most significant environmental hazard. Wastes with these radionuclides need several hundred years to decay to very low radioactivity levels.

Figure 9.29 illustrates several types of waste-burial options. Although other situations exist, these will serve as a basis for discussion of some of the main concepts used in the development of burial facilities. In Figure 9.29(a), the wastes are placed in strong engineered containers constructed of materials such as concrete and steel situated on the ground surface. In these containers they can remain in storage in areas from which the public is excluded. Deterioration of the containers can be readily monitored. If problems arise, the containers can be repaired or the wastes can be placed in new containers, provided, of course, that a responsible organization remains in charge of the facilities. The facilities shown in Figure 9.29(b) are similar, except that the storage containers are covered with earth materials. If these materials are properly designed, they will protect the containers from weathering and thereby extend their life expectancy. The earth materials can be thought of as an engineered hydrogeological environment. In Figure 9.29(c) and

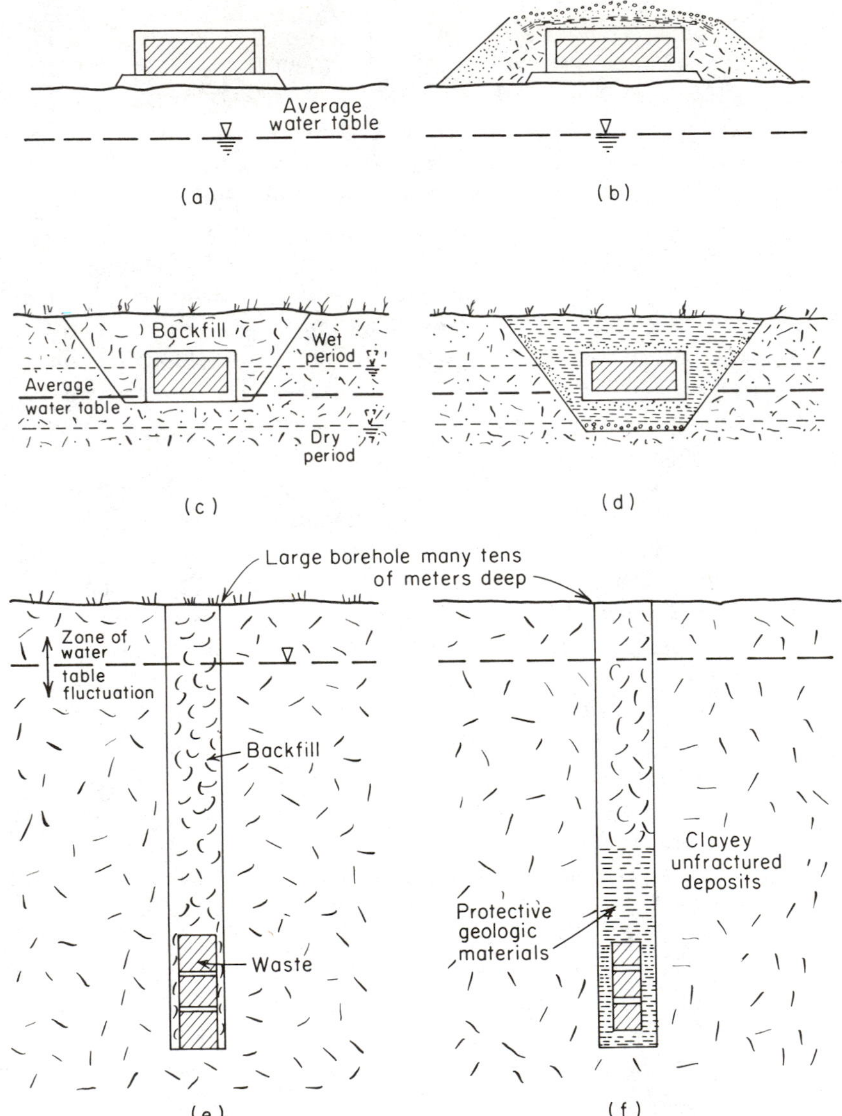

Figure 9.29 Schematic diagrams illustrating methods of low-level radioactive waste storage or disposal in shallow-water-table areas. (a) Above ground container storage; (b) above ground container storage with protection by geologic materials; (c) shallow burial in trench with backfill; (d) shallow burial in trench with additional containment provided by engineered zone of special geological materials; (e) deeper burial with backfill; (f) deeper burial in large-diameter borehole with protection by high-retardation geologic material.

(d), the wastes are stored in containers situated a few meters below ground surface, either above or below the water table. In the case shown in Figure 9.29(c), earth material from the excavation is used as backfill around the containers. In Figure 9.29(d), part of the fill in the excavation is designed to provide enhanced containment capabilities for the system. If the containers are located above the water table and if there is good reason to believe that during periods of water-table fluctuation, the water table will not rise into the burial zone, there will be little possibility of radionuclides escaping into the environment. If the water table fluctuates within the zone of burial, the containers are subjected to variable hydraulic and geochemical conditions. Their life expectancy and the fate of radionuclides in the event of container failure is much less certain than in the case of burial entirely above the water table.

In Figure 9.29(e) and (f), the containers are buried in large holes about 10 to 20 m deeper than in the previous examples. In nonarid regions the permanent water table in these situations would normally be above the containers. In the case depicted in Figure 9.29(e), the hole is simply backfilled with material originally removed from the site. In Figure 9.29(f) the excavated zone around the containers is packed with geological material such as bentonitic clay chosen to improve the long-term containment capabilities of the burial facility.

Nearly all of the burial sites for reactor wastes in the United States and Canada are in the category represented by Figure 9.29(c), with the water table within or just slightly below the burial zone. Some of the wastes have been placed directly in the ground without the protection of watertight containers. Most of the sites are located in poor hydrogeologic settings. It is not surprising, therefore, that subsurface migration of radionuclides from the burial zones is a common circumstance.

To avoid problems of subsurface radionuclide migration at future sites, numerous investigators have proposed that future sites be located in hydrogeologic environments that are shown to have long-term containment capability. To achieve this capability, the site should have the following characteristics: geomorphic and structural stability, isolation from fractured bedrock or other subsurface flow regimes that are too complex for development of reliable pathway analyses (i.e., the site should have a simple hydrogeologic framework), absence of subsurface flowlines that lead directly to the biosphere or to subsurface zones of potable water, and low predicted radionuclide velocities resulting from favorable combinations of groundwater velocity and chemical retardation. In addition to hydrogeologic criteria such as these, various investigators have indicated that the water table should be deep enough to permit waste burial to occur entirely in the unsaturated zone (Cherry et al., 1974). The predicted upper level of the range of water-table fluctuation over many centuries should be below the bottom of the burial zone. The criteria outlined above, if adhered to, would lead to the development of waste-burial facilities that would provide considerably better long-term containment than is the case for existing sites. Unfortunately, in most humid and semihumid regions of North America, this desired ideal combination of hydrogeologic properties is rare, or even nonexistent, within areas that also meet social and economic criteria. Water-table depth is generally shallow, which prevents establishment of burial zones at appreciable depth below ground surface. Wastes that are buried within a few meters, or even within 5–10 m of ground surface, are generally not considered to be in isolation from future generations during the hundreds of years or more that will be necessary for decay to reduce radioactivity in the waste to low levels.

As an alternative approach to the siting of burial grounds for low-level solid radioactive waste, Cherry et al. (in press) proposed that in humid and semihumid regions, burial zones be located in unfractured clayey aquitards in the manner represented in Figure 9.29(f). According to this scheme, burial would occur at the bottom of large (2–5 m in diameter) auger holes more than 15 m below ground surface. The burial zone would be at considerable depth below the water table and below the zone of active weathering. The wastes would therefore be isolated from the biosphere in a hydrogeologic environment in which groundwater velocity is extremely low and chemical retardation is great. Accidental unearthment of the waste by future generations would be much less probable than is the case for shallow, above-water-table waste burial. For this approach to be evaluated in detail it will be necessary to direct research toward the hydrogeologic properties of clayey aquitards such as clayey till, glaciolacustrine clay, and soft shale.

For additional information on the hydrogeologic aspects of the management of low-level solid radioactive waste (primarily reactor waste), the reader is referred to Peckham and Belter (1962), Richardson (1962a, 1962b), Mawson and Russell (1971), Cherry et al. (1973), and Pferd et al. (1977).

In the nuclear reactors that are currently used for power production, fuel rods composed of solid uranium oxide undergo fission reactions that release heat and decay particles. After a period of time in the reactor, the fuel rods are replaced. The *spent fuel* contains a wide variety of toxic radioactive isotopes produced from the uranium and from other elements. The ultimate fate of these man-made radionuclides is the core of what has become known as the *high-level radioactive waste disposal problem*. Numerous proposed solutions to this problem have been suggested, each with the objective of isolating the radionuclides from the biosphere for the life span of their radioactivity. This is the final stage of the nuclear fuel cycle.

Disposal options such as burial within the Antarctic ice cap, emplacement in the ocean floor at locations where natural burial beneath migrating continental plates will occur, and rocket transport beyond the earth's gravity field have been excluded because of impracticalities during the next few decades. It is now generally hoped that a satisfactory solution can be obtained by emplacing the radioactive material in an engineered repository in geologic strata in which it will be isolated from zones of active groundwater flow. This approach is commonly referred to as *terminal storage*. This implies that for a generation or two the repository environment will be monitored, and that if all goes well the wastes will then be regarded as having been permanently disposed.

In the United States, four main hydrogeologic possibilities are being investigated for suitability for respository development. These are (1) deep salt beds, (2) deep crystalline igneous rocks, (3) deep shale strata, and (4) thick unsaturated zones in arid regions. Because of differences in climate, option (4) is not available in Canada or in European countries. The most critical question in the evaluation of these options is whether or not the wastes will be isolated from the biosphere for periods of time that are considered to be acceptable.

The waste will contain numerous radionuclide species, but by the year 2020, 99% of the projected accumulation of radioactivity will be due to the presence of ^{90}Sr and ^{137}Cs (Gera and Jacobs, 1972) which have half-lives of 28 and 33 years, respectively, and which will decay to very low levels within about 1000 years. Much longer periods of time, however, are required for decay to low levels of long-lived transuranic nuclides in the waste, namely, ^{238}Pu, ^{239}Pu, ^{240}Pu, ^{241}Am, and ^{243}Am, with half-lives ranging from 89 years for ^{238}Pu to 24,000 years for ^{239}Pu. Radioactive decay of these elements in the waste produces other radionuclides, known as daughter products (^{237}Np, ^{226}Ra, ^{129}I, ^{99}Tc, and others). If these are taken into account, the material will remain hazardous for millions of years, although at much lower radioactivity levels than will occur during the first thousand years.

The radionuclides can be placed in the repository in their original form as spent fuel or they may be incorporated into other materials after the spent fuel has been *reprocessed*. Reprocessing is a chemical treatment in which spent fuel is dissolved in acid and plutonium is separated from the other radionuclides. Plutonium is viewed by the nuclear power industry as a valuable commodity because it can be used to produce power in fast-breeder reactors. After the extraction of plutonium, a hot, highly radioactive waste solution containing the remaining radionuclide species and some plutonium residue remains as waste. It is now generally agreed by the nuclear regulatory agencies in the various countries working on the problem that these wastes must be solidified and incorporated into solid relatively inert materials such as ceramics or glass. This must be done before proceeding with commitment to any scheme for long-term subsurface storage or disposal. From the chemical reprocessing plant the waste will proceed, after a period of interim storage, through a solidification plant. After being solidified and encapsulated, the waste will then be ready for emplacement in a subsurface geological repository. Although reprocessing removes most of the plutonium from the spent fuel, the waste that remains after reprocessing is still highly radioactive. From a hydrogeological viewpoint, isolation of spent fuel or isolation of solidified waste from reprocessing are fundamentally the same problem.

For a subsurface repository to be viewed as satisfactory within the abovementioned time framework, it must be capable of protecting the wastes from the effects of landscape erosion caused by wind, water, and even glaciers. It must be located in an area that does not have a significant seismic hazard or potential for volcanic activity. The protective materials within which the wastes are placed in the repository and the hydrogeologic environment outside the repository must be capable, within an exceptionally high degree of predictive confidence, of preventing migration of radionuclides in groundwater from the repository into the biosphere. It is this latter criterion that is the most difficult to establish at the level of confidence that is required. Never before in the history of mankind have engineers and scientists been asked to provide safety analyses relevant to such a long period of time. The feasibility of achieving long-term waste isolation in each of the four hydrogeologic repository options listed above is currently under evaluation. The deep-salt option is discussed by Bradshaw and McClain (1971) and by Blomeke et al. (1973). The potential for repository development in shale is described by Ferro et al. (1973). Winograd (1974) reviewed the hydrogeologic aspects of the arid region/unsaturated zone option. For a broader view of the high-level radioactive-waste-disposal problem, the reader is referred to Kubo and Rose (1973) and Cohen (1977).

Deep-Well Disposal of Liquid Wastes

Injection of liquid wastes, mainly of industrial origin, has been widely adopted as a waste disposal practice in North America. The purpose of this procedure is to isolate hazardous substances from the biosphere. As the discharge of pollutants to rivers and lakes has become increasingly objectionable, and as legislation for protection of surface water resources has become more stringent, the use of deep permeable zones for liquid waste disposal has become an increasingly attractive waste management option for many industries. Inventories of industrial liquid-waste injection wells in the United States were conducted in 1964, 1967, 1968, 1972, and 1973. During this period the number of waste injection wells increased from 30 in 1964 to at least 280 in 24 states in 1973 (Warner and Orcutt, 1973). In Canada in 1976 at least 80 injection wells were in use. Injection wells used for return of brines extracted during oil or gas field pumping are normally not categorized as waste injection wells. There are more than 100,000 of these wells in North America. Chemical, petrochemical, and pharmaceutical companies are the largest users of waste injection wells. Other important users are petroleum refineries, gas plants, steel mills, potash mines, uranium mills, and processing plants. In Florida, Hawaii, Louisiana, and Texas, injection of sewage effluent into saline-water aquifers occurs on a minor scale. Nearly all the waste injection wells are in the depth range of 200–4000 m. Most are between 300 and 2000 m deep. The injection zones are generally located in sandstones, carbonate rocks, and basalt.

Most injection wells are operated at injection pressures less than 7×10^3 kN/m^2. The trend in recent years is toward lower injection pressures and injection rates in the range of 500–1400 ℓ/min. The effect of an injection well on the hydrodynamic conditions in a hypothetical horizontal aquifer in which there is regional flow, are shown in Figure 9.30. The injection well causes a mound in the potentiometric surface. The mound extends unsymmetrically in the direction of regional flow in the aquifer. As injection continues the areal extent of the mound spreads to occupy an ever-increasing area. This process can be viewed as the inverse of the effect of a pumping well in a confined aquifer, and in fact is described mathemat-

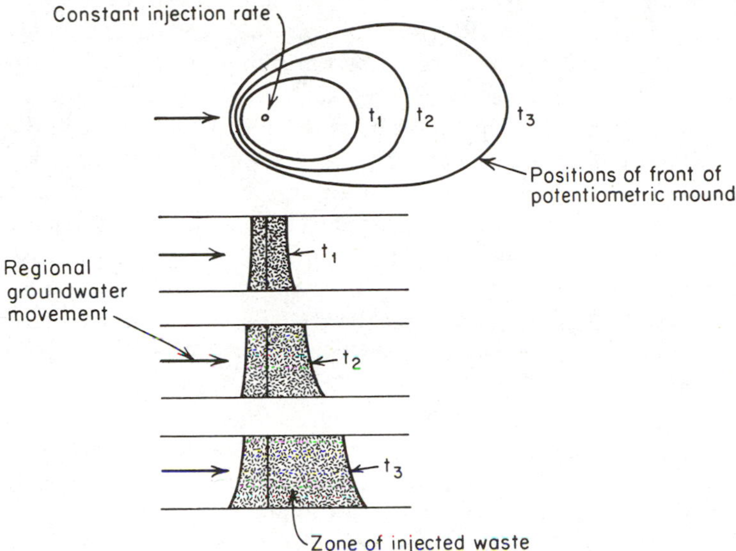

Figure 9.30 Potentiometric mound caused by waste disposal well and expansion of zone occupied by waste. Positions at times t_1, t_2, and t_3 (after Kazmann, 1974).

ically by the same equations, modified for the effect of injection rather than pumping (Warner, 1965). If injection wells are located close together, the potentiometric mounds coalesce in a manner analogous to the drawdown interface in fields of pumping wells. The spread of the front of the potentiometric mound is very rapid in comparison to the spread of the zone of injected waste. The front of the potentiometric mound spreads by pressure translation. The front of the waste zone spreads as volume displacement occurs. The waste zone spreads in direct proportion to the cumulative volume of waste that is forced into the aquifer. The interface between the formation water and the waste will be gradational as a result of dispersion.

Based on reviews of status and impact of waste injection practices in the United States and Canada, Warner and Orcutt (1973) and Simpson (1976) concluded that documented cases of even minor disposal system failure and related contamination of surface and near-surface waters are rare. This may not continue to be the case, however, as waste injection becomes an increasingly common waste disposal practice and as the length of time that strata have received wastes increases. One of the few reported instances in which waste injection has caused surface contamination occurred in southern Ontario near Sarnia, where most of the Canadian petrochemical industry is located. Contamination was caused by five injection wells, the first of which was drilled in 1958 and the others in 1960. Refinery caustic and phenolic waters were injected down the wells at rates less than 400 ℓ/min into carbonate rock strata at depths between 200 and 260 m below surface. In the late 1960's and early 1970's, phenol occurrences in fluids reaching the surface in the Sarnia district were observed (Simpson, 1976). It is believed that the contamination occurred because of upward leakage of the wastes through abandoned unplugged wells. There may be as many as 30,000 unplugged wells in southwestern Ontario in the vicinity of Sarnia (van Everdingen and Freeze, 1971). Many of these were drilled decades ago before plugging of abandoned wells was required. The hazard represented by unplugged wells in areas of waste injection is a particularly insidious one because the location of many of the wells is unknown. Some no longer even exist at ground surface but provide vertical connections below ground surface. There may be more than 1 million unplugged, unlocated wells in North America. In regard to the long-term effect of deep-well injection of noxious wastes, van Everdingen and Freeze (1971) have suggested that vertical connections provided by unplugged wells may well be the most important hazard.

Another major hazard associated with the practice of waste injection is the inducement of earthquakes as a result of increasing pore-water pressures along faults. This topic is discussed in Chapter 11.

As a concluding statement on deep-well disposal, the comments by A. M. Piper (1970) of the United States Geological Survey seem appropriate:

> In its predilection for grossly oversimplifying a problem and seeking to resolve all variants by a single massive attack, the United States appears to verge on accepting deep injection of wastes as a certain cure for all the ills of water pollution (p. 2).
>
> Injection does not constitute permanent disposal. Rather, it detains in storage and commits to such storage—for all time in the case of the most intractable wastes—under-ground space of which little is attainable in some areas, and which definitely is exhaustible in most areas (p. 6).
>
> Admittedly, injecting liquid wastes deep beneath the land surface is a potential means for alleviating pollution of rivers and lakes. But, by no stretch of the imagination is injection a panacea that can encompass all wastes and resolve all pollution even if economic limitations should be waived. Limitations on the potentials for practical injection are stringent indeed—physical, chemical, geologic, hydrologic, economic and institutional (including legal) limitations (p. 5).

Other Sources

There are many other sources that contribute contaminants to the groundwater zone. In the northern United States and in Canada large quantities of salts are applied to roads to combat adverse ice conditions during the winter months. Contamination of shallow aquifers along roads that receive salt is not uncommon in these regions. Since salts such as $NaCl$ and $CaCl_2$ are highly soluble and relatively mobile in groundwater, there is little that can be done to prevent this situation, other than to decrease the amount of salt usage.

Activities of the mining industry are another potential cause of groundwater contamination. The effects range from changes in groundwater chemistry caused by mining to infiltration of leachate from tailings and other wastes. The extensive occurrence of acid water drainage from abandoned coal mines in the Appalachian region of the United States is the most visible example of adverse effects of mining on groundwater and surface water.

Seepage from industrial waste lagoons is another cause of groundwater contamination. Across North America there are thousands of artificial ponds and lagoons that contain countless types of liquid wastes. In many cases the lagoons are not lined with impermeable barriers, thereby providing opportunity for seepage of wastes downward into the subsurface environment. In situations where potable-water aquifers are located nearby, this can cause serious problems. Many years may pass, however, before the extent of the problem becomes evident.

In some regions urbanization is spreading into the recharge areas of major aquifers. Even if centralized sewage treatment facilities rather than septic systems are used, urban activities produce numerous sources of contamination to the ground. Prediction of the long-term effect of urbanization on groundwater quality is a difficult task but a necessary one if we are to develop methods of land use planning that will minimize adverse impacts on the groundwater environment.

As an introduction to the literature on groundwater contamination and related topics, the reader is referred to Hall (1972), Summers and Spiegal (1974), Todd and McNulty (1976), and Wilson et al. (1976). An indication of the nature and regional extent of groundwater contamination in the United States is presented in the reviews by Fuhriman and Barton (1971), Scalf et al. (1973), Miller et al. (1974), and Scalf (1977). A summary of waste disposal practices and their effects on groundwater in the United States has been described by U.S. Environmental Protection Agency (1977). A review of recent research activities related to chemical problems in hydrogeology has been presented by Back and Cherry (1976).

Suggested Readings

BAETSLÉ, L. H. 1969. Migration of radionuclides in porous media, *Progress in Nuclear Energy, Series XII, Health Physics*, ed. A. M. F. Duhamel. Pergamon Press, Elmsford, N.Y., pp. 707–730.

CHERRY, J. A., R. W. GILLHAM, and J. F. PICKENS. 1975. Contaminant hydrogeology: Part 1: Physical processes, *Geosci. Can.*, 2, pp. 76–84.

FRIED, J. J. 1976. *Ground Water Pollution*. Elsevier, Amsterdam, pp. 1–47.

OGATA, AKIO. 1970. Theory of dispersion in a granular medium. *U.S. Geol. Surv. Prof. Paper 411-I*, p. 134.

U.S. Environmental Protection Agency. 1977. Waste disposal practices and their effects on ground water. *The Report to Congress*, pp. 81–107.

Problems

1. A chemical analysis of groundwater yields the following results (mg/ℓ): $K^+ = 3$, $Na^+ = 110$, $Ca^{2+} = 80$, $Mg^{2+} = 55$, $HCO_3^- = 420$, $Cl^- = 220$, $SO_4{}^{2-} = 35$, $NO_3^- = 15$, Fe(total) = 0.8, Mn(total) = 0.2, $F^- = 0.6$, As = 0.03, Pb = 0.08, B = 0.9. Comment on the suitability of this water for the following uses:
   (a) Municipal water supply.
   (b) Irrigation of vegetable crops.
   (c) Livestock.
   (d) Brewing of beer.

2. Would water with the composition indicated in Problem 1 normally be softened for household use? How would the process of water softening be expected to alter the composition?

3. Using a cylindrical column (10 cm in diameter and 30 cm long) of relatively homogeneous sand, an experiment with a step-function input of a nonreactive tracer is conducted (see Figure 9.1). The porosity of the sand is 35%, the steady-state flow rate is 1 ℓ/h, and the hydraulic gradient is 0.1. The $C/C_0 = 0.5$ point on the breakthrough curve arrived 0.8 h after the tracer initially entered the column. The $C/C_0 = 0.25$ point arrived at 0.7 h and the $C/C_0 = 0.75$ point at 0.9 h. Estimate the dispersivity of the sand.

4. A contaminant zone is migrating through an aquifer composed of medium-grained sand. The average hydraulic gradient is 0.01. A representative value of the hydraulic conductivity of the sand is 1×10^{-5} m/s. Is the movement of nonreactive contaminants influenced primarily by advection and mechanical dispersion or by molecular diffusion? Explain.

5. A sanitary landfill is located on a deposit of dense clay that is 5 m thick overlying an aquifer that provides drinking water to a small town. A zone of leachate-contaminated groundwater has accumulated at the base of the landfill on the clay surface. Observations in the aquifer indicate a steady piezometric level of 250.5 m above mean sea level. The surface of the water table in the landfill is at about 251.3 m. The hydraulic conductivity of the clay is approximately 2×10^{-11} m/s, and the porosity is 19%. Estimate how long it will take for nonreactive contaminants to move through the clay into the aquifer. Express your answer as a range of values that you consider to be reasonable in light of the available data.

6. As a result of the rupture of a storage tank, 10 m³ of liquid waste containing 100 kg of dissolved arsenic rapidly infiltrated into a shallow, unconfined, sandy aquifer in which the flow is horizontal. The average groundwater velocity in the aquifer is 0.5 m/day, the dispersivity is 0.1 m, and the coefficient of molecular diffusion is 2×10^{-10} m²/s. As the contaminated zone moves through the

aquifer, the arsenic does not undergo significant adsorption or precipitation. Estimate the maximum arsenic concentration after the contaminant cloud has moved a distance of 500 m. What will be the approximate dimensions of the cloud? Assume that the leakage from the storage tank can be approximated as a point source and that the aquifer can be treated as a homogeneous medium with uniform flow.

7. High-level radioactive waste is buried in a cavern in unfractured shale at a depth of 1000 m below ground surface. The burial zone is separated from the nearest overlying aquifer by a vertical thickness of 100 m of shale. The shale has a hydraulic conductivity of the order of 10^{-12} m/s and vertical hydraulic gradient of about 10^{-2} directed upward. In the shale, nonreactive radionuclides have effective diffusion coefficients in the order of 10^{-10} m^2/s. It is expected that the wastes will become wet at some time during the next 1000 years and will then move slowly out into the shale. Is it reasonable to expect that radionuclides will remain entirely within the shale during the next 100,000 years? Ignore the potential effects of faulting, glaciation, and so on, as a cause of radionuclide transfer through the shale. Consider only the influence of flow, mechanical dispersion, and molecular diffusion.

8. Field observations in a granitic area indicate bulk hydraulic conductivities on the order of 10^{-6} cm/s. The granite has a cubic array of joints with a representative spacing of 10 cm between joint planes. Estimate the average groundwater velocity for a zone in which the flow is horizontal and the hydraulic gradient is 10^{-2}.

9. In laboratory experiments using a pesticide and samples from a sandy aquifer, it is observed that when water with the pesticide is equilibrated at various concentrations with the sand samples, the partitioning of the pesticide between the liquid and solid phases is as follows: test 1, 100 μg/g adsorbed at 10 mg/mℓ in solution; test 2, 300 μg/g adsorbed at 220 mg/mℓ in solution; test 3, 600 μg/g adsorbed at 560 μg/mℓ in solution; test 4, 1000 μg/g adsorbed at 1000 mg/mℓ in solution. What distribution coefficient is indicated by these data? Express your answer in milliliters per gram. In sand (porosity = 35%) below the water table, estimate the relative velocity at which the pesticide would migrate in an advection-controlled system.

10. Studies of the behavior of a toxic chemical compound in a sandstone aquifer indicate the following parameter values: porosity 10%, average linear velocity 0.1 cm/day, and distribution coefficient 75 mℓ/g. Estimate the velocity at which the center of mass of a zone contaminated with the compound would travel.

11. Hydrogeological studies of a site for a proposed lagoon for storage of toxic liquid wastes indicate that the hydraulic gradient at the site is downward. The water table is located at a depth of 4 m below the ground surface. Samples from piezometers at depths of 5, 10, 15, 20, 25, 30, 40, and 50 m below ground surface have tritium concentrations of 75, 81, 79, 250, 510, 301, 50, and 10 tritium units. The site is located in the interior of North America. The piezometers are situated in a thick deposit of shale. Provide an interpretation of the tritium data. What is the nature of the permeability of the shale?

12. Groundwater in a sandstone aquifer at a temperature of 25°C has the following composition (mg/ℓ): $K^+ = 12$, $Na^+ = 230$, $Ca^{2+} = 350$, $Mg^{2+} = 45$, $HCO_3^- = 320$, $Cl^- = 390$, and $SO_4^{2-} = 782$; pH 7.6. If F^- is supplied to the water from minerals in the aquifer and if the F^- concentration is not limited by availability, will solubility constraints be expected to maintain the F^- concentration at a level below the limit specified for drinking water? Explain.

13. Effluent from a septic (sewage disposal) system infiltrates into an unconfined gravel aquifer. Upon mixing with the groundwater, the contaminated part of the aquifer has the following content of inorganic constituents (mg/ℓ): $K^+ = 3.1$, $Na^+ = 106$, $Ca^{2+} = 4.2$, $Mg^{2+} = 31$, $HCO_3^- = 81$, $Cl^- = 146$, and $SO_4^{2-} = 48$; pH 6.3, $Eh = -0.1$ V, DO = 0, temperature 23°C. Assuming that equilibrium occurs and that mineral precipitation-dissolution reactions control the concentration of dissolved inorganic phosphorus, indicate (a) the mineral that would provide the solubility constraint on the phosphorus concentration; (b) the dominant dissolved species of inorganic phosphorus; (c) the equilibrium concentration of dissolved phosphorus.

14. A borehole-dilution test is conducted in a well with an inside diameter of 10 cm. The packer-isolated interval in which the tracer is introduced is 100 cm long. After 2 h the tracer concentration declines to one-half of its initial value. The flow in the formation is horizontal, the well has no sand or gravel pack, and the tracer is nonradioactive and nonreactive. Estimate the average linear velocity in the formation.

15. A disposal well for liquid industrial waste commences operation in a horizontal isotropic confined limestone aquifer that has the following characteristics: thickness = 10 m, secondary porosity = 0.1%, bulk hydraulic conductivity = 5×10^{-5} m/s, specific storage = 10^{-6} cm^{-1}. The injection rate is 100 ℓ/min.

(a) To what distance from the injection well will the front of the potentiometric mound have extended after 1 month?

(b) To what distance from the injection well will the front of contamination have moved after 1 month? Neglect the effects of regional groundwater flow and dispersion. Assume that the aquifer is homogeneous and that the primary permeability of the limestone matrix is negligible.

16. Studies of an unconfined aquifer indicate a shallow zone that contains dissolved oxygen in the range 2–6 mg/ℓ and NO_3^- in the range 30–50 mg/ℓ. The source of the NO_3^- is fertilizer. Below this zone there is no detectable dissolved oxygen and no detectable NO_3^-. Hydraulic-head data indicate that groundwater flows from the upper zone to the lower zone. All the water in the aquifer is very young. Suggest a hydrochemical hypothesis to account for the large decrease in NO_3^- as the water moves downward in the aquifer. What additional data would be desirable as a basis for testing your hypothesis?

17. Salt (as NaCl only) applied to a highway during the winter for prevention of icing problems has caused contamination of a shallow unconfined aquifer near the highway. It has been observed that the Cl^- content of water from many domestic wells, which was formerly soft, has become hard as the Cl^- has risen. The large increase in hardness can be attributed to the effect of road-salt contamination. Outline a geochemical hypothesis to explain the hardness increase.

18. Natural water in a deep sandstone aquifer composed of quartz, feldspar, and a small amount of clay has the following composition (mg/ℓ): $K^+ = 18$, $Na^+ = 850$, $Ca^{2+} = 41$, $Mg^{2+} = 120$, $HCO_3^- = 820$, $Cl^- = 470$, and $SO_4^{2-} = 1150$; pH 8.1. Wastewater that contains abundant dissolved organic matter is put into the aquifer through a disposal well. After injection commences, observation wells in the aquifer near the disposal well yield water in which abundant H_2S and CH_4 are detected. Prior to waste injection, the observation wells showed no detectable H_2S and CH_4. The wastewater injected into the aquifer did not contain these gases. Outline a geochemical hypothesis to account for the occurrence of H_2S and CH_4 in the observation wells. Would you expect the pH of the water to increase or decrease? Explain, with the aid of appropriate chemical equations.

19. An unlined lagoon is used intermittently for recharge of water from a secondary sewage treatment plant. The water has moderate concentrations of major ions (K^+, Na^+, Ca^{2+}, Mg^{2+}, Cl^-, SO_4^{2-}, and HCO_3^-) and an appreciable content of NH_4^+, bacteria, and organic matter. The treated sewage infiltrates from the lagoon downward into a sandy aquifer. Observation wells used to monitor the change in groundwater chemistry caused by the artificial discharge system indicate that the aquifer water in the zone of influence has total hardness, nitrate, and dissolved inorganic carbon concentrations that are considerably above the natural levels in the aquifer. No bacteria are detected in the aquifer. The NH_4^+ concentrations are very low. Account for these chemical characteristics of the water. Include appropriate chemical equations as part of your explanation.

20. In an area of strip mining for coal, the noncoal geologic materials removed during mining (referred to as cast overburden) are returned to the stripped areas as part of a land reclamation program. The average porosity of the cast overburden is 30%. The average degree of saturation of the material is 40%. Assuming that the water and entrapped air in the voids do not migrate, and assuming that all the oxygen in the air and water in the voids is consumed by oxidation of pyrite, estimate the following:

(a) The SO_4^{2-} content of the pore water.

(b) The pH of the pore water (assume that no carbonate-mineral buffering occurs.

(c) The amount (weight percent) of calcite that would be necessary in the cast overburden to neutralize the acid product by pyrite oxidation.

(d) The amount of pyrite necessary for consumption of all the oxygen by the oxidation reaction.

(e) Would the amounts of calcite and pyrite obtained in parts (c) and (d) be noticeable by normal means of examination of the geologic materials?

Groundwater and Geotechnical Problems

10.1 Pore Pressures, Landslides, and Slope Stability

Landslides have always been viewed with a mixture of fascination and respect. Together with earthquakes and volcanoes, they represent one of the few natural geologic events with the speed and power to affect the course of man. In this section, we shall learn that groundwater plays an important role in the generation of landslides; in Section 11.1, we shall learn that its role in the generation of earthquakes is conceptually similar.

Landslides are of great interest to both geomorphologists and geotechnical engineers. The geomorphologist's interest centers on the role of landslides as a process in landform evolution. For a geotechnical engineer, a large landslide is simply the extreme event in the spectrum of slope stability hazards that he must consider in his engineering design. More often he is concerned with the analysis of much smaller man-made slopes in such projects as highway cuts, earth dams, or open pit mines.

The concepts and failure mechanisms that underlie slope stability analysis hold on both natural slopes and man-made slopes. They are equally valid for large potentially catastrophic landslides and for simple embankment slipouts. The influence of groundwater conditions, which is the central focus of this section, is the same in all cases. There *are* some significant differences between the analysis of *soil* slopes and the analysis of slopes in *rock*, and following a review of the basic limit equilibrium techniques, the role of groundwater is examined under separate headings for each of these two geotechnical environments.

This presentation attempts to distill the essence of a very large literature. Many of the concepts were originated or clarified in Terzaghi's (1950) classic analysis of the mechanism of landslides. A text by Zaruba and Mencl (1969) places its emphasis on the engineering geology aspects of large landslides, and one by Carson and

Kirkby (1972) reviews the geomorphological implications. Eckel (1958), Coates (1977), and Schuster and Krizek (in press) provide a comprehensive review of slope-stability engineering, and a recent text by Hoek and Bray (1974) emphasizes rock slope engineering. Standard soil mechanics texts such as Terzaghi and Peck (1967) treat the subject in some detail. Throughout the literature there is generous recognition of the importance of fluid pressures, but this recognition is not always coupled with an up-to-date understanding of the probable patterns of steady-state and transient subsurface flow in slopes.

We will begin with an examination of the mechanisms of subsurface movement on planar surfaces.

Mohr-Coulomb Failure Theory

Let us first consider the failure criteria on a well-defined plane of weakness at depth. Consider such a plane [Figure 10.1(a)] in a regional stress field with *maximum principal stress* σ_1 in the vertical direction and *minimum principal stress* σ_3 in the horizontal direction. If we wish to calculate the shear stress τ and the normal stress σ acting on the plane *in the absence of water*, we can put Figure 10.1(a) in the form of the free-body diagram of Figure 10.1(b). The shear stresses on planes

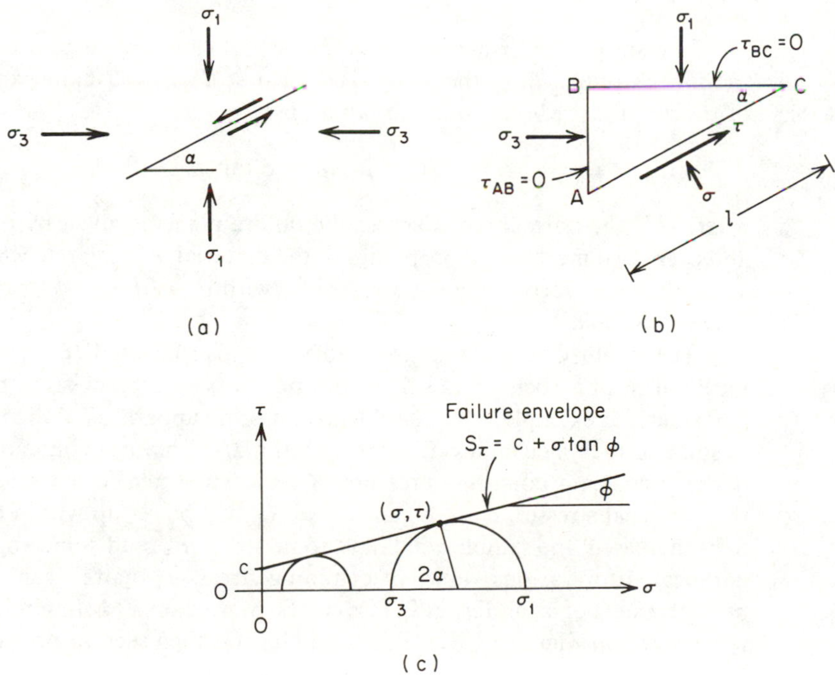

Figure 10.1 Stress equilibrium on a plane of weakness in a regional stress system, and the Mohr circle representation.

parallel to the principal stresses are zero, and the conditions of force equilibrium in the horizontal and vertical directions yield:

$$\sigma_3 l \sin\alpha - \sigma l \sin\alpha + \tau l \cos\alpha = 0 \tag{10.1}$$

$$\sigma_1 l \cos\alpha - \sigma l \cos\alpha - \tau l \sin\alpha = 0 \tag{10.2}$$

Solving Eqs. (10.1) and (10.2) for σ and τ,

$$\sigma = \frac{\sigma_3 \sin^2\alpha + \sigma_1 \cos^2\alpha}{\cos^2\alpha + \sin^2\alpha} \tag{10.3}$$

$$\tau = (\sigma_1 - \sigma_3)\sin\alpha\cos\alpha \tag{10.4}$$

Trignometric ingenuity can be applied to Eqs. (10.3) and (10.4) to produce the usual Mohr circle formulations [Figure 10.1(c)]:

$$\sigma = \frac{\sigma_1 + \sigma_3}{2} + \frac{\sigma_1 - \sigma_3}{2}\cos 2\alpha \tag{10.5}$$

$$\tau = \frac{\sigma_1 - \sigma_3}{2}\sin 2\alpha \tag{10.6}$$

The shear stress τ acting on the plane will cause movement only if it exceeds the shear strength S_τ of the plane. The shear strength is usually expressed in terms of the empirical Mohr-Coulomb failure law:

$$S_\tau = c + \sigma \tan\phi \tag{10.7}$$

where σ is the normal stress across the failure plane as given by Eq. (10.5), and c and ϕ are two mechanical properties of the material, c being the *cohesion* (the shear strength under zero confining stress, i.e., with $\sigma = 0$) and ϕ being the *angle of internal friction.*

The Mohr-Coulomb failure theory can also be used to describe the failure mechanism in a rock or soil that does not possess a preexisting plane of failure. Consider, for example, a standard triaxial testing apparatus of the type widely used in soil and rock mechanics (Figure 10.2). If a dry, homogeneous soil or rock is kept under a constant confining pressure, S_3, and subjected to a vertical pressure, S_1, then internal stresses, $\sigma_3 = S_3$ and $\sigma_1 = S_1$, will be set up within the specimen. If S_1 is increased, the sample will fail at some stress, σ_1, and some angle, α. If the test is repeated for various values of confining stress, σ_3, paired values of the experimental results of σ_3 and σ_1 at failure can be plotted on a Mohr circle diagram of the type shown in Figure 10.1(c). Equation (10.7) is then seen to be the equation for a failure envelope that can be obtained experimentally for any given soil or rock. The relation between the angle of failure, α, and the angle of internal friction, ϕ, can be determined graphically from Figure 10.1(c) as $\alpha = 45° - \phi/2$. For the analysis of slope stability at a field site, the values of c and ϕ for the soils or rocks that

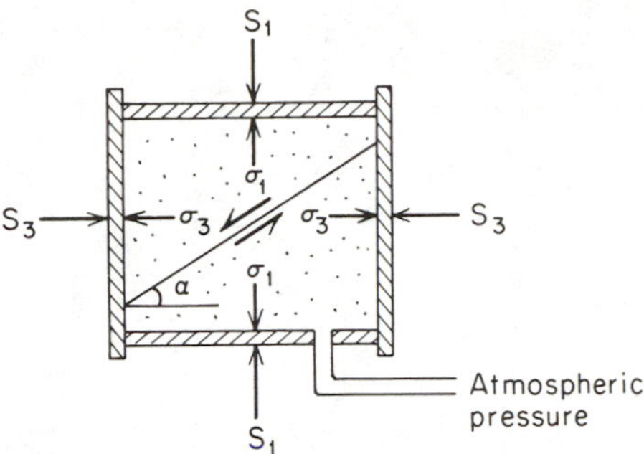

Figure 10.2 Schematic representation of laboratory setup for drained triaxial test.

make up the slope must be measured in the laboratory in a triaxial apparatus of the type just described. If failure is anticipated on a specific type of surface, such as a joint plane in fractured rock, the c and ϕ values should refer to the rock-rock interface, and must be measured on a sample that includes such a feature. If the soil or rock is free of incipient failure planes, the c and ϕ values must be measured on a homogeneous sample. As should be clear from Figure 10.1(c), higher c and higher ϕ both lead to greater shear strength and less likelihood of failure. For sands and fractured rocks, $c \longrightarrow 0$ and the material strength arises almost totally out of the ϕ term. For clays, $\phi \longrightarrow 0$ and the strength is due almost totally to cohesion.

The foregoing paragraphs describe the failure mechanisms in dry soils and rocks. Our primary interest lies in materials that are saturated with groundwater. If the preexisting or incipient plane of failure is water-bearing, and if water exists there under a fluid pressure, p, the principle of effective stress must be invoked. The total normal stress σ in Figure 10.1(b) must be replaced by the effective stress $\sigma_e = \sigma - p$. The failure law becomes

$$S_\tau = c' + (\sigma - p) \tan \phi' \tag{10.8}$$

where the primes on c and ϕ indicate that these mechanical properties must now be determined for saturated conditions using a "drained" triaxial test. In a drained test, water that is expelled from the sample under the influence of the increased vertical pressures is allowed to drain to the atmosphere as in Figure 10.2. If drainage is not provided, the fluid pressure p must be monitored in the cell, and $\sigma_1 = S_1 - p$ and $\sigma_3 = S_3 - p$. Equation (10.8) makes it clear that increases in fluid pressure tend to decrease shear strength on failure planes.

Limit Equilibrium Methods of Slope Stability Analysis

Consider the stress conditions in a homogeneous soil with no preexisting failure planes. Near the surface in flat-lying terrain [Figure 10.3(a)], the direction of maximum principal stress σ_1 (due to the weight of overlying material) is vertical, and the direction of minimum principal stress σ_3 is horizontal. In the vicinity of a slope, on

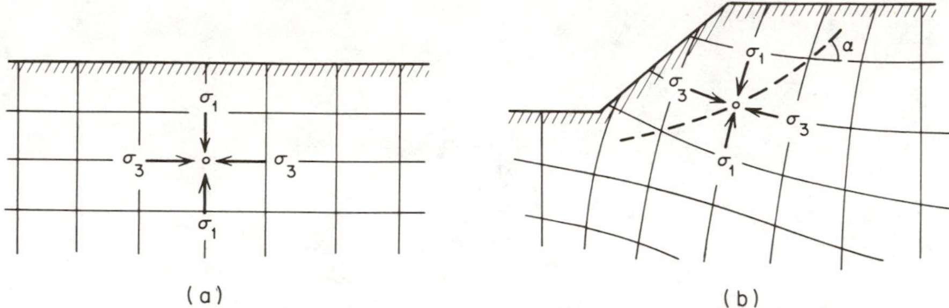

Figure 10.3 Orientation of principal stresses. (a) Beneath flat terrain; (b) adjacent to slopes.

the other hand, the stress distribution becomes skewed, in the manner shown in Figure 10.3(b). As shown there, one consequence of this stress pattern is that the planes of incipient failure, oriented at $\alpha = 45^\circ - \phi/2$ from the σ_3 direction, are curved. In soil mechanics these possible planes of failure are called *slip circles* or *slip surfaces*. The limit-equilibrium approach to slope stability analysis involves the arbitrary selection of a set of several possible slip surfaces for a given slope. For each slip surface an equilibrium analysis is carried out using the Mohr-Coulomb failure criteria, and a factor of safety, F_S, defined as the ratio of shearing *strength* on the slip surface to shearing *stress* on the slip surface, is calculated. If $F_S > 1$, the slope is considered to be stable with respect to that slip surface. The slip surface with the lowest value of F_S is regarded as the incipient failure plane. If $F_S \leq 1$ on the critical surface, failure is imminent.

Consider a homogeneous, isotropic clay soil for which the angle of internal friction approaches zero. In such cases, the shear strength of the soil is derived solely from its cohesion c, and the Mohr-Coulomb failure law [Eq. (10.7)] becomes simply $S_\tau = c$. For such soils, the slip surface can be closely approximated by a circle [Figure 10.4(a)]. The factor of safety will be given by the ratio of the resisting moment to the disturbing moment about the point O. The disturbing force is simply the weight W of the potential slide, and the resisting force is that of the cohesive strength c acting along the length l between points A and B. For this simple case,

$$F_S = \frac{Wd}{clr} \tag{10.9}$$

For more complex situations a more sophisticated method of analysis is needed and this is provided by the *conventional method of slices*. It can be applied to slip surfaces of irregular geometry and to cases where c and ϕ (or c' and ϕ') vary along the slip surface. This method also invokes the effective stress principle by considering the reduction in soil strength along the slip surface due to the fluid pressures (or *pore pressures*, as they are commonly called in the slope stability literature) that exist there on saturated slopes. For the conventional method, the

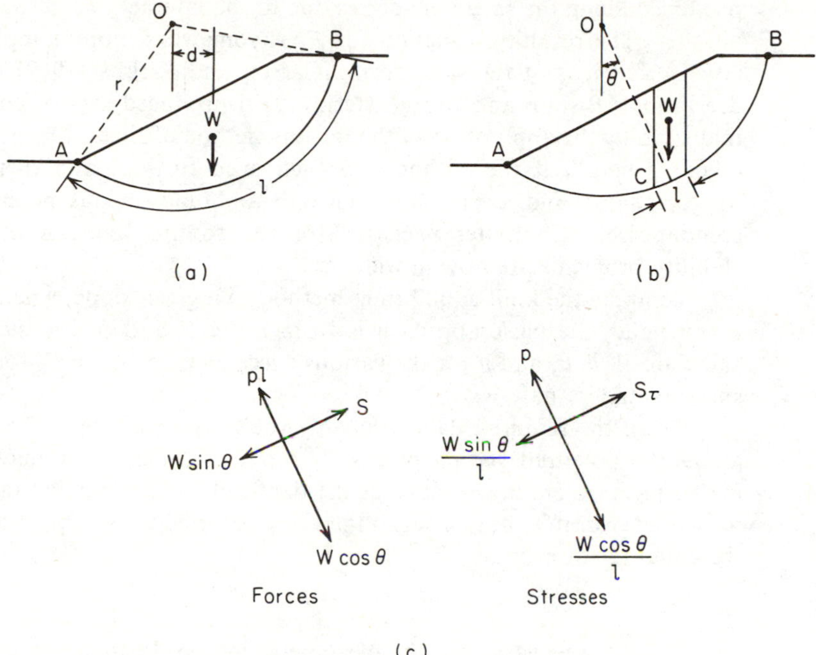

Figure 10.4 Slope stability analysis by (a) circular arc and (b) conventional method of slices; (c) forces acting at point *C*.

slide is divided into a series of vertical slices. Figure 10.4(b) shows the geometry of an individual slice, and Figure 10.4(c) indicates the conditions of force equilibrium and stress equilibrium that exist at point *C* on the slip surface at the base of the slice. At *C*, the shearing stress is $(W \sin \theta)/l$ and the shearing strength S_τ is given, as before, by

$$S_\tau = c' + (\sigma - p) \tan \phi' \tag{10.10}$$

For $\sigma = (W \cos \theta)/l$, Eq. (10.10) becomes

$$S_\tau = \frac{c'l + (W \cos \theta - pl) \tan \phi'}{l} \tag{10.11}$$

and the factor of safety is given by

$$F_S = \frac{\sum_A^B [c'l + (W \cos \theta - pl) \tan \phi']}{\sum_A^B W \sin \theta} \tag{10.12}$$

The conventional method of slices was improved by Bishop (1955), who recognized the need to take into account the horizontal and vertical stresses produced along the slice boundaries due to the interactions between one slice and another. The resulting equation for F_S is somewhat more complicated than Eq. (10.12), but it is of the same form. [Carson and Kirkby (1972) present a simple derivation.] Bishop and Morgenstern (1960) produced sets of charts and graphs that simplify the application of the Bishop method of slices. Morgenstern and Price (1965) generalized the Bishop approach even further, and their technique for irregular slopes and general slip surfaces in nonhomogeneous media has been widely computerized. Computer packages for the routine analysis of complex slope stability problems are now in wide use.

To apply the limit equilibrium method to a given slope, whether by computer or by hand, the basic approach is to measure c' and ϕ' for the slope material, calculate W, l, θ, and p for the various slices, and calculate F_S for the various slip surfaces under analysis.

Of all the required data, probably the most sensitive is the pore pressure p along the potential sliding planes. If economics permit, it may be possible to install piezometers in the slope at the depth of the anticipated failure plane. The measured hydraulic heads, h, can then be converted to pore pressures by means of the usual relationship:

$$p = \rho g(h - z) \tag{10.13}$$

where z is the elevation of the piezometer intake. In many cases, however, field instrumentation is not feasible, and it behooves us to reexamine the hydrogeology of slopes in light of the needs of slope stability analysis.

Effect of Groundwater Conditions on Slope Stability in Soils

The hydraulic heads (and pore pressures) on a slope reflect the steady-state or transient groundwater flow system that exists there. From the considerations of Chapter 6, it should be clear that if reasonable estimates can be made of the water-table configuration and of the distribution of soil types, it should be possible to predict the pore pressure distributions along potential slip surfaces by means of flow-net construction or with the aid of analytical, numerical, or analog simulation.

Patton and Hendron (1974) have pointed out that geotechnical slope stability analyses often invoke incorrect pore pressure distributions such as that which arises from the static case [Figure 10.5(a)], or from an "earth dam" type of flow net [Figure 10.5(b)], that seldom occurs on natural slopes. A more realistic steady-state flow pattern for homogeneous isotropic material would be that of Figure 10.5(c). For complex slopes and water-table configurations, or for more complex soil configurations, the various techniques for steady-state flow-net construction discussed in Chapter 5, including those that consider the seepage face, are at the disposal of the geotechnical engineer. For a slope with a factor of safety that approaches 1, the differences between the pore pressure distributions that would arise from the choice of the various hillslope flow systems of Figure 10.5 could well control whether the analysis predicts stability or failure.

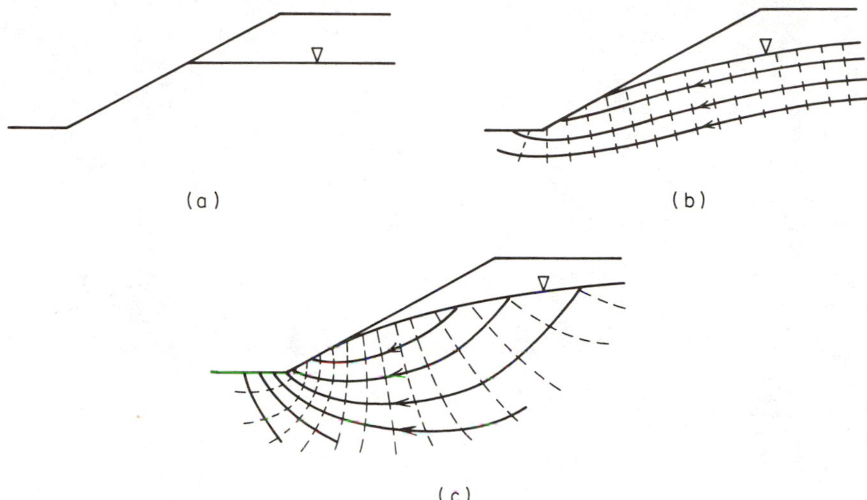

Figure 10.5 Groundwater flow systems on slopes. (a) Static; (b) commonly assumed, but incorrect, flow system; (c) typical flow system in slopes (after Patton and Hendron, 1974).

If a steady-state flow net is to be used to predict pore pressure distributions for slope-stability analysis, it should be constructed for the most critical case, that is, for a case with the water table at its highest possible position. For slopes where little is known of the water-table configuration, the most conservative course is to assume a water table coincident with the ground surface.

There are some cases where a local flow net of the type shown in Figure 10.5(c), even with the water table at the ground surface, will underestimate the pore pressures in the slope. If the design slope lies at the base of a much longer slope in a deep valley, for example, the whole slope may be part of a regional discharge area, and anomalously high pore pressures may exist there. Considerations of regional groundwater flow systems become pertinent to the analysis of slope stability in such cases.

In Section 6.5, we learned that subsurface hydrologic systems in hillslopes are seldom simple, and that they seldom exist in the steady state. The hydrologic response of a hillslope to rainfall, for example, involves a complex, transient, saturated-unsaturated interaction that usually leads to a water-table rise, albeit one that may be very difficult to predict. The amount of rise, the duration of the rise, and the time lag between the rainfall event and the resulting rise may vary widely depending on the hillslope configuration, the rainfall duration and intensity, the initial moisture conditions, and the saturated and unsaturated hydrogeologic properties of the hillslope materials.

Consider the hillslope in Figure 10.6(a). Rising water tables over the period $t_0, t_1, t_2, \ldots$ due to the rainfall R will lead to increasing pore pressures p_c as a function of time at point C on the potential slip surface [Figure 10.6(b)]. If pore

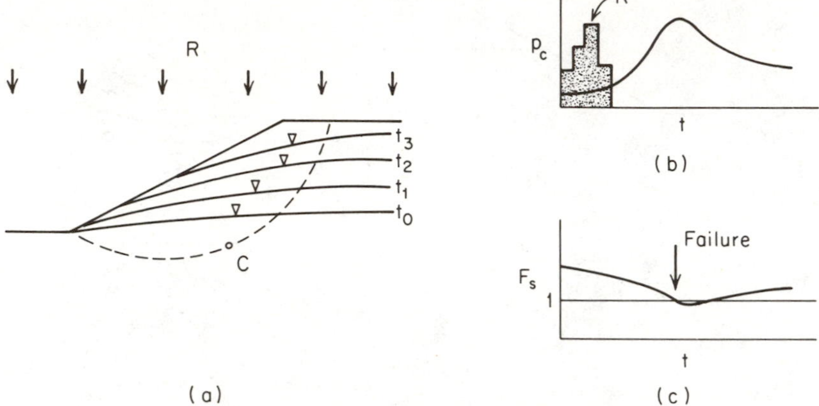

Figure 10.6 (a) Transient position of the water table during rainfall on a slope; (b) pore pressure at point *C* during and following a rainfall period; (c) factor of safety as a function of time during and following a rainfall period.

pressures rise along the entire length of the slip surface, as they would on a small slope during a widespread rainfall, the factor of safety F_S will decrease with time [Figure 10.6(c)]. At the point in time when F_S becomes less than unity, failure will occur. It is a very common observation (Terzaghi, 1950) that slope failures often occur during the wet season, or following major rainfall or snowmelt events. The triggering mechanism of such failures is the increase in pore pressures along the potential failure planes.

Effect of Groundwater Conditions on Slope Stability in Rock

For the purpose of slope stability analysis, one might divide rock slopes into three categories: (1) those consisting of solid rock, (2) those with a small finite number of intersecting joint surfaces, and (3) those that constitute heavily jointed rock masses. The first of these—solid rock—is rare, and in any case most rock mechanics texts show that extremely high vertical rock faces are possible in the absence of joints (see, for example, Jaeger, 1972). Clearly, fluid pressures can play no important role in the stability of solid rock slopes.

The analysis of slope stability for the third case, that of a heavily jointed rock mass, is not significantly different from that for soil. It is possible to define potential circular slip surfaces in this type of slope, and the usual flow-net approach to the prediction of pore pressures on the slip surface is valid.

We are left with rock slopes that possess a small number of preferential failure surfaces due to a well-developed but relatively widely spaced jointing pattern. The analysis of slope stability in this type of geologic environment has been the subject of much recent research in the field of rock mechanics (Jaeger, 1971; John, 1968;

Londe et al., 1969; Patton and Hendron, 1974; Hoek and Bray, 1974). As an example, consider the simple potential wedge failure of Figure 10.7. The rock block under analysis is bounded by a basal joint with shear strength dependent on the c' and ϕ' values of the planar surface and a vertical tension crack that has no shear strength. If a steady-state "flow system" exists in this simple fracture system, it is one in which the height of water in the tension crack remains at the level shown, and the joint surface remains saturated (presumably in the presence of a small spring discharging at some rate Q at the point where the joint intersects the slope). The pore pressure distributions will be as shown, and the resultant pore pressure forces acting against the stability of the rock wedge are those shown as U and V. Hoek and Bray (1974) calculate the factor of safety for cases of this type from the relation

$$F_S = \frac{c'l + (W\cos\theta - U - V\sin\theta)\tan\phi'}{W\sin\theta + V\cos\theta} \tag{10.14}$$

Our primary interest lies in the nature of the pore pressure distributions in rock slopes of this type and in the way in which groundwater flow systems differ in such slopes from those in soil slopes of comparable geometry. Patton and Deere (1971) make two important points in this regard. First, they suggest that one might expect very irregular pore pressure distributions in jointed slopes [Figure 10.8(a)] under the influence of the individual structural features. Second, they note that porosities of jointed rock are extremely small (0–10%) in comparison with those for soils (20–50%). This leads to large, rapid, water-table fluctuations in jointed slopes [Figure 10.8(b)] in response to rainfall or snowmelt events. Pore pressure increases are therefore higher in rock slopes than in soil slopes for a given rainfall, and the potential capacity of rainfall events as a triggering mechanism for slope failures is correspondingly higher in rock slopes. Lumb (1975) and Bjerrum and Jorstad (1964) present statistical results that show a high correlation between infiltration events and slope failures on weathered and unweathered rock slopes.

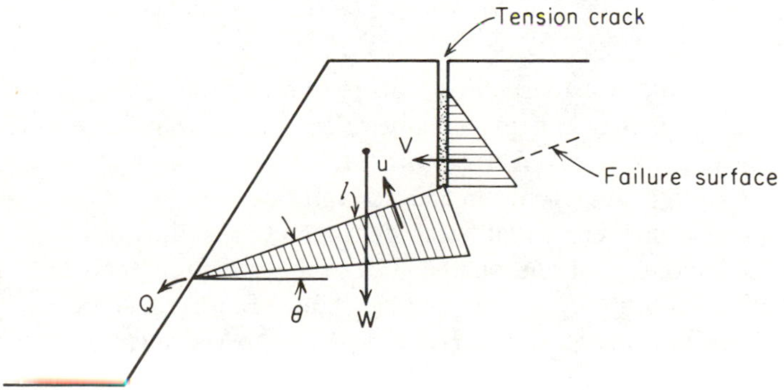

Figure 10.7 Slope stability analysis of a rock wedge (after Hoek and Bray, 1974).

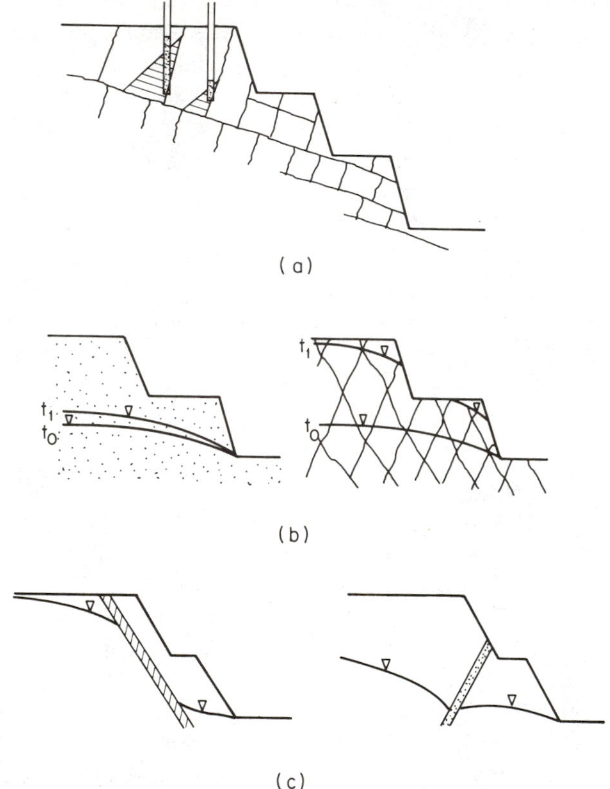

Figure 10.8 Some aspects of groundwater flow in rock slopes. (a) Possible large differences in fluid pressures in adjacent rock joints; (b) comparison of transient water-table fluctuations in porous soil slopes and low-porosity rock slopes; (c) fault as a low-permeability groundwater barrier and as a high-permeability subsurface drain (after Patton and Deere, 1971).

Faults are a structural feature that may be present on rock slopes, and hydrologically, they can play many roles. Faults that have developed thick zones of sheared and broken rock with little fault gouge may be highly permeable, while those that possess a thin (but continuous) layer of gouge may form almost impermeable barriers. Figure 10.8(c) schematically illustrates the effect of two possible fault configurations on the water-table position (and thus on the pore pressure distribution on potential sliding planes) in a rock slope.

Sharp, Maini, and Harper (1972) have carried out numerical simulations of groundwater flow in heavily jointed, homogeneous rock slopes that possess anisotropy in their hydraulic conductivity values. Horizontally bedded slopes in which the principal direction of anisotropy is horizontal do not develop pore pressures as great as slopes where the bedding and the principal direction of anisotropy dip parallel to the slope face. The very large divergence in the hydraulic head distribution that they show between the two cases illustrates the importance of a detailed understanding of the hydrogeological regime on a slope for the purposes of stability analysis.

Hodge and Freeze (1977) have presented several finite-element simulations of hydraulic-head distributions in regional geologic environments that are prone to major slope-stability failures. Figure 10.9 shows the hydraulic-head pattern in a hypothetical thrusted sedimentary sequence of the type common in the Western Cordillera (Deere and Patton, 1967). The high pore pressures indicated by the potentiometric line on the base of unit *A* lead to a low factor of safety for the overlying dip slope.

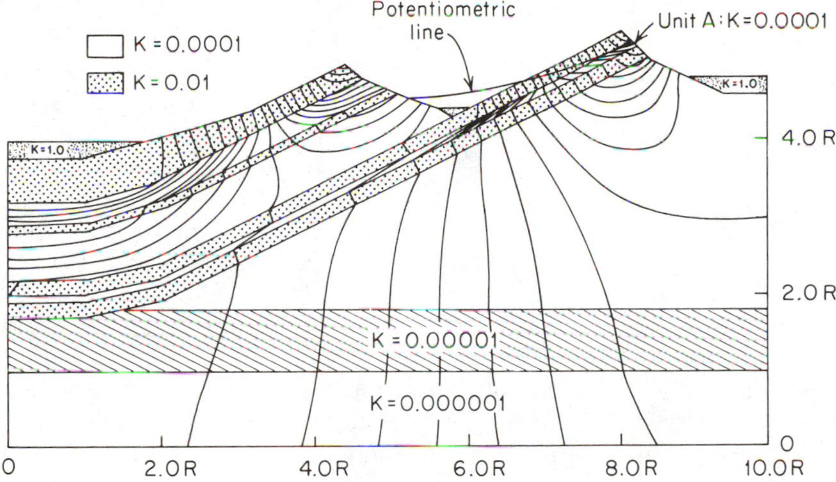

Figure 10.9 Hydrogeologic regime in thrusted sedimentary environment. Potentiometric line indicates hydraulic-head values on base of unit *A* (after Hodge and Freeze, 1977).

10.2 Groundwater and Dams

It is probably safe to say that few engineering projects have greater ability to stir men's minds than the design and construction of a large dam. Within the engineering profession, there is the excitement that a massive, integrated engineering endeavor creates. The usual tasks of ensuring the technical accuracy of the engineering calculations and assessing the economic ramifications of the engineering decisions take on a special importance when they involve the taming of a great river. Outside the profession, in society at large, equal passions are often aroused. They may be those of support—for an improved water supply system, or cheaper power, or the security of a new flood control scheme—or they may be those of concern—for the potential impact of man's added encroachment on the natural environment.

The engineering concerns are usually centered on the damsite, and in the first part of this section we will look at the role of groundwater in the engineering aspects of dam design. The environmental concerns are more often associated with the reservoir, and in the later part of the section we will examine the environmental interactions that take place between an artificially impounded reservoir and the regional hydrogeologic regime.

Types of Dams and Dam Failures

No two dams are exactly alike. Individual dams differ in their dimensions, design, and purpose. They differ in the nature of the site they occupy and in the size of the reservoir they impound. One obvious initial classification would separate large multipurpose dams, few in number but great in impact, from the much larger number of smaller structures such as tailings dams, coffer dams, floodwalls, and overflow weirs. In this presentation the role of groundwater is examined in the context of the larger dams, but the principles are equally applicable to the smaller structures.

Krynine and Judd (1957) classify large dams into four categories: gravity dams, slab and buttress dams, arch dams, and earth and rockfill dams. The first three represent impermeable concrete structures that do not permit the percolation of water through them or the buildup of pore pressures within them. These three are differentiated on the basis of their geometry and by the mechanisms through which they transfer the water loads to their foundations. A gravity dam has an axis that extends across a valley from one abutment to the other in a straight line, or very nearly so. Its structural cross section is massive, usually trapezoidal, but approaching a triangle in some cases. A slab-and-buttress dam has a cross section that is considerably thinner than a full gravity dam, but one that is buttressed by a set of vertical walls aligned at right angles to the dam axis. An arch dam has a curved axis, its convex face upstream. In the most spectacular cases, its section may be little more than a reinforced concrete wall, often less than 20 ft thick. In a gravity dam, the water load is transmitted to the foundations through the dam itself; in a buttress dam the load is transmitted through the buttresses; and in an arch dam the load is transmitted to the rock abutments by the thrusting action of the arch. All three types of concrete dam must be founded on rock, and the role of subsurface water is thus limited to the groundwater flow and pore pressure development that can occur in the abutment rocks and in the rock foundations.

In the first half of this century, most of the earth's large dams were contructed as concrete dams on rock foundations. However, in recent years, as the better damsites have become exhausted and the economic trade-off between the costs of concrete construction and the cost of earthmoving has changed, there has been a sweeping move toward earth and rockfill dams. These dams derive their stability from a massive cross section and as a result they can be built at almost any site, on either rock or soil foundations. From a hydrologic point of view, the primary property that differentiates earth dams from concrete dams is that they are permeable to some degree. They allow a limited flow of water through their cross section and they permit the development of pore pressures within their mass.

There are essentially five events that can lead to a catastrophic dam failure: (1) overtopping of the dam by a flood wave due to insufficient spillway capacity, (2) movement within the rock foundations or abutments on planes of geological weakness, (3) the development of large uplift pressures on the base of the dam, (4) piping at the dam toe, and (5) slope failures on the upstream or downstream face of the dam. The first three of these failure mechanisms can occur in both concrete and earth dams; the last two are limited to earth and rockfill dams.

There is also a sixth mode of failure—excessive leakage from the reservoir—that is seldom catastrophic but which represents just as serious a breakdown in design as any of the first five. Of course, leakage always takes place to some degree through the foundation rocks beneath a dam, and in earth dams, there is always some loss through the dam itself. These losses are usually carefully analyzed during the design of a dam. Losses that are unexpected more often take place through the abutment rocks or from the reservoir at some point distant from the dam. There are even a few case histories (Krynine and Judd, 1957) that report leakage so excessive that the dams were effectively unable to hold water. Engineers need methods of prediction for the rates of leakage, great or small, because leakage values are an important part of the hydrologic balance on which cost-benefit analyses of dams are based.

The presence of groundwater is a typical feature at damsites. Of the six modes of failure we have listed, groundwater plays an important role in five of them—geological failure, uplift failure, piping failure, slope failure, and leakage failure. Under the next few headings we will examine the various failure mechanisms in more detail, and we will look at some of the design features that are incorporated in dams to provide safeguards against failure. For the first section, which deals with concrete dams on rock foundations, Jaeger (1972), Krynine and Judd (1957), and Legget (1962) provide useful reference texts. For the discussions that deal with earth and rockfill dams, the specialty texts of Sherard et al. (1963) and Cedergren (1967) are invaluable. A recent book by Wahlstrom (1974) on damsite exploration techniques contains much material of hydrogeologic interest.

Seepage Under Concrete Dams

In order to examine the failure mechanisms on geological planes of weakness, consider the cross section of Figure 10.10(a) through an impermeable, concrete, gravity dam and its underlying rock foundation. If the elevation of the reservoir level on the upstream side of the dam is h_1 and the elevation of the tailwater pond on the downstream side of the dam is h_2, a steady-state flow net can be constructed in the infinite half-space on the basis of the boundary conditions $h = h_1$ on AB, $h = h_2$ on CD, and BC impermeable. The flow net shown in Figure 10.10(a) is for a homogeneous, isotropic medium. The hydraulic heads, pressure heads, and fluid pressures (or pore pressures) that exist at any point in the system are independent

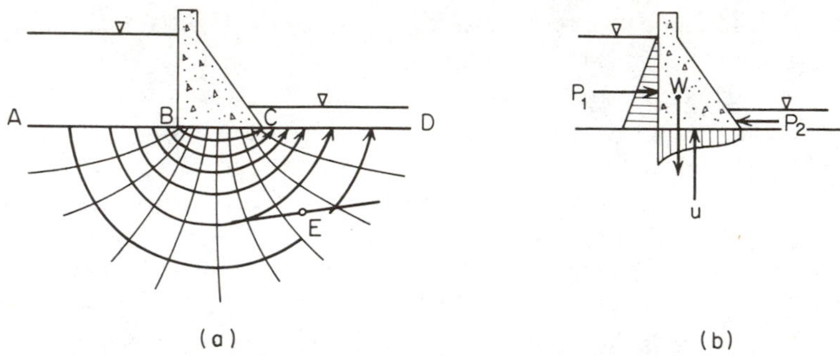

Figure 10.10 (a) Flow net; (b) force equilibrium diagram for concrete dam on a permeable foundation.

of the hydraulic conductivity of the medium, although the flow velocities and quantities of leakage would of course depend on this parameter. It should be clear that the hydraulic heads and pore pressures at any point E will be higher after the impounding of the reservoir than they were prior to dam construction, and that they will be higher when the reservoir is at full supply level than when the reservoir is at lower levels. If there is a preexisting plane of geological weakness (a fault, shear zone, or major joint plane) passing through point E, the discussions of Section 10.1 apply. Higher pore pressures at E will create reduced shear strengths on the plane and reduced resistance to potential displacements. The failure of the Malpasset thin-arch dam near Frejus, France, in December 1959 (Jaeger, 1972) is the classic example of a dam failure triggered by small movements on a plane of weakness in the foundation rocks beneath the dam. In the Malpasset disaster, over 400 people were killed and much of the town of Frejus was destroyed.

The flow-net approach can also be utilized to examine uplift pressures on the base of a dam. Given the hydraulic-head values from the flow net, one can calculate the fluid pressures along the line BC in Figure 10.10(a) from the usual relation $p = \rho g(h - z)$. In that the elevations of all points on BC are identical, the gradient of the fluid pressure along BC will be the same as that for the hydraulic head. In Figure 10.10(b), the nature of the uplift pressures on the base of the dam and the resulting uplift force U are schematically illustrated.

To analyze the stability of the dam against sliding, it would also be necessary to consider the water loads P_1 and P_2 and the weight W of the dam. The Mohr-Coulomb failure law, or any other criterion that describes the frictional resistance to movement along the basal contact between the dam and its foundation, could then be used to calculate a factor of safety. Sliding failures of concrete dams are seldom the result of an incorrect assessment of P_1, P_2, or W. They are usually the result of unexpected uplift pressures in the foundation rocks. The failure of the St. Francis dam near Saugus, California in March 1928 led to 236 deaths and several million dollars of property damage (Krynine and Judd, 1957). The primary cause of the failure was the softening and disintegration of a conglomerate formation that formed the foundation rock on one abutment, but percolating water was the primary erosive agent, and uplift pressures may also have contributed to the failure.

Grouting and Drainage of Dam Foundations

The grouting of rock formations is probably more an art than a science. It is, however, an art that is based on an understanding of the hydrogeological properties of rock and the nature of groundwater flow at damsites. The term *grouting* refers to the injection of a sealing agent into the permeable features of a rock foundation. Usually, the grout is a mixture of neat cement and water, with a cement/water ratio in the range 1: 7 to 1: 10. Some grout mixtures contain lime, clay, or asphalt, and in recent years, chemical grouts have come into use. In most cases, the permeable feature that is being grouted is the joint system that exists in the foundation rock at the damsite. In other cases, a grouting program may be specifically aimed at fault zones, solution cavities, or high-permeability horizons in sedimentary or volcanic rocks.

Grouting is carried out for three reasons: (1) to reduce leakage under the dam, (2) to reduce the uplift pressures, and (3) to strengthen jointed rock foundations. To these ends, there are two types of grouting that are carried out at most damsites: *consolidation grouting* and *curtain grouting*. The purpose of consolidation grouting is to strengthen the foundation rock. It is carried out with low injection pressures in shallow holes, with the aim of sealing major crevices and openings. At Norris dam in Tennessee, which is founded on limestone and dolomite (Krynine and Judd, 1957), consolidation grouting was carried out in holes 7–15 m deep arranged in a grid on 1–3 m centers beneath the entire structure.

Curtain grouting is designed to reduce both leakage and uplift pressures. Grouting is carried out at higher injection pressures in holes up to 100 m deep. The curtain is usually created with a single or double line of holes located near the heel of the dam and aligned parallel to the dam axis. A split-spacing approach is often used, whereby the initial grouting is carried out from holes on, say, 8-m centers; and then later holes are inserted on 4-m, 2-m, and even 1-m centers. Piezometer tests are carried out on secondary and tertiary holes prior to grouting to test the efficiency of the grout already in place. Grouting specifications usually specify both the minimum allowable grout take and the maximum allowable injection pressures. In some cases, grouting is carried out until the grout take is zero ("grout to refusal"). Injection pressures must be limited to avoid rock uplift, blowouts, and weakening of the foundation rocks.

There is no question that an effective grout curtain reduces leakage under a dam, but there is considerable controversy over the role of a grout curtain in the reduction of uplift pressures. Figure 10.11(a) shows the flow net beneath a concrete dam in a homogeneous, isotropic foundation rock bounded by an impermeable formation at the base. The uplift pressures along the line AB at the base of the dam are shown schematically on the inset to the right. If a vertical grout curtain is established [Figure 10.11(b)], the flow net is altered considerably, and, in theory, the uplift-pressure profile along AB is significantly reduced. However, Casagrande

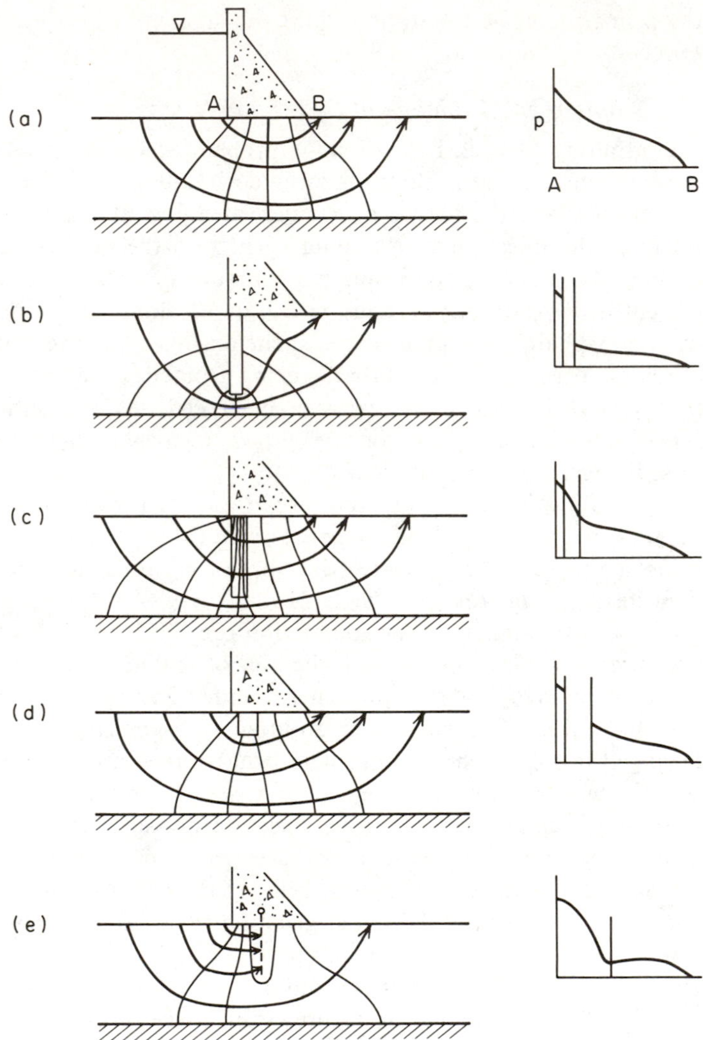

Figure 10.11 Uplift pressure profiles and flow nets for various grout curtain and drainage configurations.

(1961) pointed out that the theoretical efficiency suggested by Figure 10.11(b) is seldom realized. In the first place, it is not possible to develop a grout curtain that reduces the hydraulic conductivity to zero along its length; and in the second place, the grout curtain geometry shown in Figure 10.11(b) is somewhat misleading with regard to scale. Figure 10.11(c) shows the flow net that would exist for a grout curtain with a hydraulic conductivity one-tenth that of the ungrouted rock; Figure 10.11(d) shows the flow net that would exist for a grouted zone more in keeping with usual dam dimensions. The reductions in uplift pressure in both these cases is significantly less than that shown in Figure 10.11(b). Casagrande notes that uplift pressures are actually more effectively reduced by drainage [Figure 10.11(e)]. However, the presence of a drain induces even greater leakage from the reservoir than would occur under natural conditions. It is common practice now to use an integrated design with a grout curtain to reduce leakage and drainage behind the curtain to reduce uplift pressures.

The Grand Rapids hydroelectric project in Manitoba provides a grouting case history second to none (Grice, 1968; Rettie and Patterson, 1963). The project involved 25 km of earth dikes, enclosing a reservoir greater than 5000 km² in area, in a region underlain by highly fractured dolomites. A grout curtain up to 70 m in depth was emplaced from holes on less than 2-m centers over the entire length of the dikes. Grice (1968) notes that the grout curtain reduced leakage through the grouted formation by 83%, but it induced greater flows through the underlying ungrouted rock. He estimates that the grouting program reduced net leakage from the reservoir by 63%.

Steady-State Seepage Through Earth Dams

Failures of earth or rockfill dams can result from excessive leakage, from piping at the toe, or from slope failures on the dam face. All three can be analyzed with the aid of steady-state flow nets. For those rare situations where an earth dam is constructed on an impervious formation [Figure 10.12(a)], the flow net can be limited to the dam itself. Where the foundation materials are also permeable [Figure 10.12(b)], the flow net must include the entire dam-foundation system.

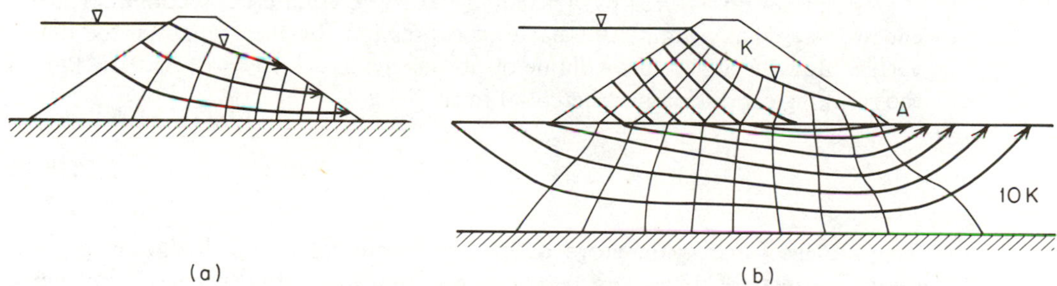

Figure 10.12 Flow nets for a homogeneous, isotropic earth dam on (a) impermeable foundation and (b) permeable foundation.

While it is recognized that a dam cross section constitutes a saturated-unsaturated flow regime, it is not common in engineering analysis to consider the unsaturated portions of the system. The free-surface approach outlined in Section 5.5 and Figure 5.14 is almost universally used. In Figure 10.13, flow is assumed to be concentrated in the saturated portion *ABEFA*. The water table *BE* is assumed to be a flow line. The specified heads are $h = h_1$ on *AB* and $h = z$ on the seepage face *EF*. The position of the exit point must be determined by trial and error. The flow nets of Figure 10.12 exemplify the type of flow nets that result. Engineering texts

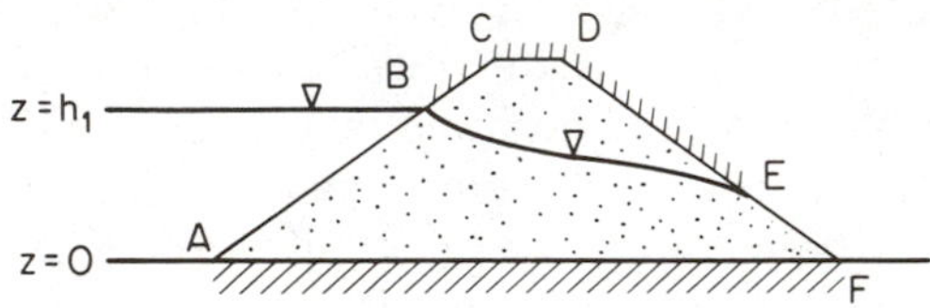

Figure 10.13 Boundary-value problem for saturated-unsaturated flow system in earth dam.

on groundwater seepage such as Harr (1962) or Cedergren (1967) provide many examples of flow nets for earth dams.

Let us now consider the question of piping. The mechanism of piping can be explained in terms of the forces that exist on an individual soil grain in a porous medium during flow. The flow of water past the soil grain occurs in response to an energy gradient. (Recall from Section 2.2 that the hydraulic potential was defined in terms of the energy per unit mass of flowing fluid.) A measure of this gradient is provided by the difference in hydraulic head Δh between the front and back faces of the grain. The force that acts on the grain due to the differential head is known as the *seepage force*. It is exerted in the direction of flow and can be calculated (Cedergren, 1967) from the expression

$$F = \rho g \, \Delta h \, A \tag{10.15}$$

where A is the cross-sectional area of the grain and ρ is the mass density of water. If we multiply Eq. (10.15) by $\Delta z/\Delta z$ and let A refer to a cross-sectional area that encompasses many grains, we have an expression for the seepage force during vertical flow through a unit volume of porous media with $V = A \, \Delta z = 1$. Putting the resulting expression in differential form yields

$$F = \rho g \frac{\partial h}{\partial z} \tag{10.16}$$

The seepage force is therefore directly proportional to the hydraulic gradient $\partial h/\partial z$. In areas of downward-percolating groundwater, the seepage forces act in the same direction as the gravity forces, but in areas of upward-flowing water, they oppose the gravity forces. If the upward-directed seepage force at any discharge point in a flow system [say, at point A in Figure 10.12(b)] exceeds the downward-directed gravity force, piping will occur. Soil grains will be carried away by the discharging seepage and the dam will be undermined.

The downward-directed gravity force is due to the buoyant weight of the saturated porous medium. A soil with a dry density $\rho_S = 2.0$ g/cm^3 has a buoyant density ($\rho_b = \rho_S - \rho$) that is almost exactly equal to the density of water, $\rho = 1.0$ g/cm^3. For this very representative ρ_S value, the seepage force will exceed the gravity force for all hydraulic gradients greater than 1.0. One simple test for piping is therefore to examine the flow net for a proposed dam design and calculate the hydraulic gradients at all discharge points. If there are exit gradients that approach 1.0, an improved design is required.

The ultimate failure mode in cases of piping is usually a slope failure on the downstream face. Slope failures can also occur there if the pore pressures created near the face by the internal flow system are too great. The limit equilibrium methods of slope stability analysis, introduced in the previous section, are just as applicable to earth dams as they are to natural slopes.

To avoid the hydraulic conditions that lead to piping or slope failures in earth dams, dam designers can incorporate many different design features. Figure 10.14(a) and (b) shows how an internal drainage system or a rock toe can serve to reduce hydraulic heads on the downstream slope of an earth dam. Figure 10.14(c) illustrates a zoned dam with a downstream shell five times more permeable than a central core. One consequence of such a design is a lowered exit point on the downstream face. If the contrast between core and shell is even greater, and drainage is added, [Figure 10.14(d)], the internal flow analysis is reduced to consideration of flow through the core itself. The shell and drain act as if they are infinitely permeable. Figure 10.14(e) shows the influence of a partial cutoff, or downward extension of the central core, on flow through a permeable foundation material. An extension of the core all the way to the basal boundary of the permeable layer would be even more effective.

Transient Seepage Through Earth Dams

Slope failures on the upstream face of a dam are usually the result of rapid drawdowns in the reservoir level. At full supply levels the effects of high pore pressures in the face are offset by the weight of overlying reservoir water. Following rapid drawdown, the high pore pressures remain, but the support has been removed. Unless the transient dissipation of these pore pressures is rapid, that is, unless the transient drainage of the dam face is rapid, instabilities may develop on the critical slip surface and slope failures can occur. Figure 10.15(a) is a schematic illustration of the transient response to rapid drawdown in an unzoned earth dam. Figure 10.15(b) shows the nature of the insurance offered against this type of failure by the presence of a high-permeability shell.

Freeze (1971a) pointed out that transient flow in earth dams is a saturated-unsaturated process; and, especially in the case of clay cores, the flow regime may be highly dependent on the unsaturated hydraulic properties of the earthfill material. However, it is not common in engineering practice to investigate the unsaturated properties of fill materials, so the free-surface approaches of De Wiest (1962) and Dicker (1969), which consider only the saturated flow, are of great practical importance in the analysis of transient seepage through earth dams.

There is another failure mechanism in earth dams that has transient overtones, and that is the triggering of slope failures by liquefaction during earthquake shocks. Cedergren (1967) notes that maximum security against liquefaction is

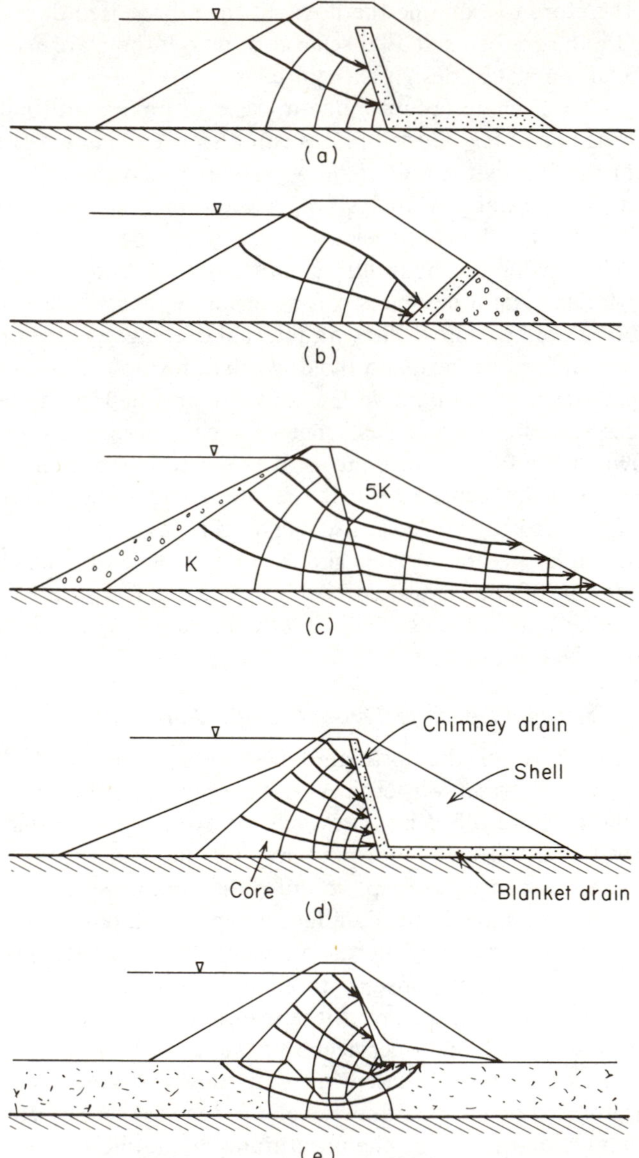

Figure 10.14 Design features for earth and rockfill dams (after Cedergren, 1967).

provided by dams with the smallest zones of saturation in their downstream shells. He concludes that every dam should be well drained, if for no other reason than to improve stability during earthquakes.

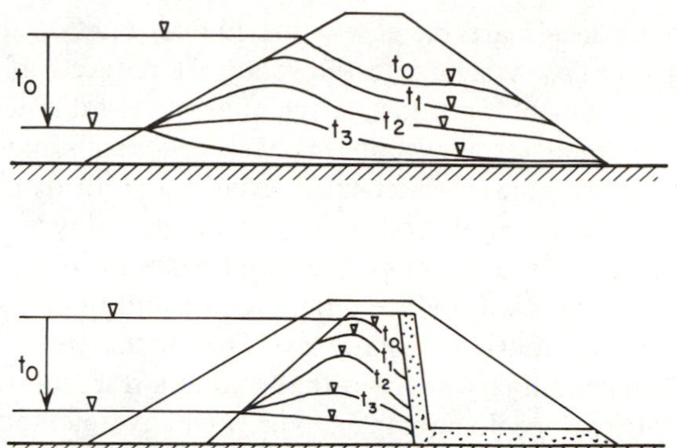

Figure 10.15 Transient response of the water table in an earth dam to rapid drawdown of the reservoir. (a) Homogeneous dam; (b) zoned dam with permeable shell.

Hydrogeologic Impact of Reservoirs

The impoundment of a reservoir behind a dam can have a significant impact on the various environmental systems that exist in a watershed. The hydrologic regime is affected in the most direct way: runoff patterns are influenced both above the reservoir and below it, and streamflow rates are altered in both time and space. A new reservoir also generates a massive readjustment in the erosion-sedimentation regime in a river basin. The upstream sediment load is trapped by the reservoir and downstream erosion is enhanced. These environmental consequences have been recognized since the earliest days of dam building, but it is only recently that ecologists have been able to document an impact of perhaps greater importance. It is now clear that reservoirs often cause alarming disruptions in a wide variety of ecosystems, including fish and wildlife regimes and vegetation patterns. In many cases, the nature of the ecological readjustment is controlled by the availability of water, and this in turn is dependent to some degree on the changes that have been induced in the hydrogeological regime.

The introduction of a reservoir into a valley that is acting as a regional discharge area produces both transient readjustment and long-term permanent change in the hydrogeologic system adjacent to the reservoir. During the initial rise of the reservoir level, a transient flow system is induced in the reservoir banks. As the hydraulic heads are raised at the reservoir boundary, there is a reversal of flow directions and an influx of water from the reservoir into the groundwater system. The mechanism is identical to that of bank storage in stream banks during flood stages (Section 6.6). For reservoirs that may be tens or even hundreds of kilometers long and water-level increases that may be 30 m or more, the quantitative significance of these transient flow processes can be considerable.

The end result of the initial transient readjustment is a set of long-term, permanent changes in the regional hydrogeologic regime. Water tables are higher, hydraulic heads are increased in aquifers, and the rates of discharge from the subsurface flow system into the valley are reduced. If water-table elevations prior to impoundment were low, a regional water-table rise can be beneficial in that improved moisture conditions in near-surface soils may aid agricultural production. On the other hand, if water-table levels were already close to the surface, the influence may be harmful. Soils may become waterlogged, and there is the possibility of soil salinization through increased evaporation. In deeper aquifers, increased hydraulic heads will reduce pumping lifts and in rare situations can cause wells to flow that previously had static levels below the ground surface.

The preliminary analyses that lead to a reservoir design should include predictions of hydrogeologic impact. The predictive methods of simulation currently in use have largely been adapted from the methods of analysis of bank storage. The initial transient response of the water table can be modeled with a saturated model based on the Dupuit-Forchheimer assumptions (Hornberger et al., 1970) or with a more complex saturated-unsaturated analysis (Verma and Brutsaert, 1970). Transient increases in hydraulic head in a hydraulically connected confined aquifer can be predicted with the subsurface portion of Pinder and Sauer's (1971) coupled model of streamflow-aquifer interaction. All these methods require knowledge of the time rate of fluctuation of reservoir stage and the saturated and/or unsaturated hydrogeologic properties of the geologic formations in the vicinity of the reservoir. Similar methods can be used to predict hydrogeologic response to operating fluctuations in reservoir level. This application has much in common with the assessment of transient flow through earth dams, as discussed earlier in this section.

Once a reservoir has attained its operating level, seasonal and operational fluctuations in water level are usually relatively small in comparison with the initial rise, and transient effects become less important. Prediction of long-term, permanent changes in the hydrogeologic regime can be carried out with a steady-state model, in which the head on the reservoir boundary is taken as the full supply level of the reservoir. Simulations can be performed on vertical two-dimensional cross sections aligned at right angles to the reservoir axis, or in two-dimensional horizontal cross sections through specific aquifers. Solutions are usually obtained numerically with the aid of a computer (Remson et al., 1965) or with analog models of the type described in Section 5.2 (van Everdingen, 1968a).

If the presence of a reservoir influences the hydrogeologic environment, so does the hydrogeologic environment influence a reservoir. In the eyes of a dam designer, the question of interaction is framed in the latter sense. In addition to the primary question of hydrologic supply and the secondary question of sedimentation in the reservoir, dam designers must consider three potential geotechnical problems in connection with reservoir design: (1) leakage from the reservoir at points other than the dam itself, (2) slope stability of the reservoir banks, and (3) earthquake generation. Each of these phenomena is influenced by groundwater conditions, either directly or through pore pressure effects.

Leakage from reservoirs at points distant from the dam is not uncommon.

It was a recurring problem in several of the dams constructed in limestone terrain by the Tennessee Valley Authority during the first half of this century.

The slope stability of reservoir banks, particularly under conditions of fluctuating water levels, is an important aspect of dam design. This has been especially true since the spectacular failure of the Vaiont reservoir in Italy in 1963. At that site a massive slide into the reservoir involving 200–300 million m^3 of material created a wave 250 m high that overtopped the dam and delivered 300 million m^3 of water into the downstream valley. Jaeger (1972) reports that almost 2500 lives were lost in the disaster.

The impoundment of a reservoir also changes the stress conditions at depth. The water load of the reservoir increases the total stresses, and this effect, together with the increase in fluid pressures brought about by hydrogeologic readjustment, influences effective stresses at depth. Carder (1970) documents a large number of case histories where reservoir impoundment has led to seismic activity.

10.3 Groundwater Inflows Into Tunnels

There is probably no engineering project that requires a more compatible marriage between geology and engineering than the construction of a tunnel. Consideration of the local and regional lithology, stratigraphy, and geologic structure influence not only the choice of routes but also the methods of excavation and support. A recent text by Wahlstrom (1973) outlines the history and development of tunneling and emphasizes the role of geology in tunnel planning. Krynine and Judd (1957) and Legget (1962) provide informative but less-detailed discussions of tunneling within the context of an overall treatment of engineering geology.

The tunneling literature contains references to many case histories in a wide variety of geologic environments, but while lithology, stratigraphy, and structure vary from case to case, there is one feature that is remarkably common. In case after case, the primary geotechnical problem encountered during tunnel construction involved the inflow of groundwater. Some of the most disastrous experiences in tunneling have been the result of interception of large flows of water from highly fractured, water-saturated rocks. Tunnelers the world over know that in planning for a tunnel it is essential to make every attempt to identify the nature of the groundwater conditions that are likely to be encountered.

If groundwater inflows are predicted in advance, it is usually possible to design suitable drainage systems. Where tunnels can be driven upgrade, the tunnel itself provides a primary drainage facility. Where tunnels must be driven downgrade or from internal headings serviced by a shaft, more complex drainage facilities involving pumps and piping systems are required. In either case, the requirements of design make it an important task to correctly predict the amounts and rates of water inflow that are likely to appear in the tunnel. In some cases it has proven possible to reduce groundwater inflows after the fact by grouting, but this approach is seldom of value when very large, unexpected inflows occur.

In this section we will first examine the role that a tunnel plays within the regional hydrogeologic system. In later subsections we will describe two famous case histories, and we will review some methods of predictive analysis.

A Tunnel as a Steady-State or Transient Drain

In the simplest light, a tunnel acts as a drain. Consider, for illustrative purposes, an infinitely long tunnel in a homogeneous, isotropic porous medium. If the pressure heads on the tunnel walls are taken as atmospheric and the water table is maintained at constant elevation, a steady-state flow net of the type shown in Figure 10.16(a) can be constructed. If the hydraulic conductivity of the medium is known,

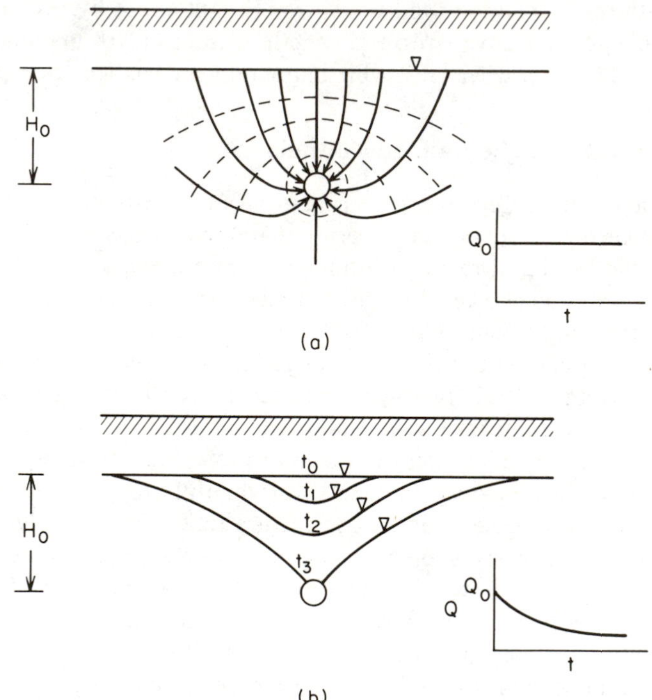

Figure 10.16 Tunnel as (a) a steady-state and (b) a transient drain.

the rate of groundwater inflow Q_0 per unit length of tunnel can be calculated from a quantitative analysis of the flow net. In fact, even if the geologic formations are heterogeneous and anisotropic, a flow-net analysis, albeit a more complicated one, could provide the steady-state rates of inflow, provided that the necessary hydraulic conductivity values can be determined for the various formations.

The steady-state approach is valid as long as the water table is not drawn down by the existence of the tunnel. However, for rock formations with low porosity and low specific storage, it is unlikely that steady-state conditions could be maintained for long in the presence of a tunnel. It is more likely [Figure 10.16(b)] that a transient flow system will develop with declining water tables above the tunnel. In that case the initial steady-state inflow rate Q_0 per unit length of tunnel will decrease as a function of time.

If geologic conditions were always simple and an infinitely long tunnel could be instantaneously driven, the calculation of tunnel inflows would be a simple matter. Unfortunately, the geology along a tunnel line is seldom as homogeneous as the use of the two-dimensional cross sections of Figure 10.16 would imply. There is usually an alternating sequence of more-permeable and less-permeable formations, and the inflows along a tunnel line are seldom constant through space, let alone time. Very often it is extreme inflows from one small, unexpected, high-permeability zone that leads to the greatest difficulties. Unconsolidated sand and gravel deposits and permeable sedimentary strata such as sandstone or limestone can lead to water problems. More often, it is very localized secondary features such as solution cavities, and fracture zones associated with faults or other structural features, that lead to the largest inflows at the face.

In short, then, tunnelers must be ready to cope with two main types of groundwater inflow: (1) regional inflows along the tunnel line, and (2) catastrophic inflows at the face. The first type can usually be analyzed with a steady-state flow-net analysis. Flows are relatively small and decrease slowly with time. It is usually possible to design for them with tunnel drainage systems. Flows of the second type are very difficult to predict. They may be very large but decrease rapidly with time. It is difficult to design economic drainage systems for them, and they create an especially dangerous hazard if the tunnel is being driven downhill or from a closed heading. Inflows greater than 1000 ℓ/s have been recorded at tunnel headings in several tunnels during their construction (Goodman et al., 1965).

Hydrogeologic Hazards of Tunneling

Large groundwater inflows into tunnels are sometimes associated with high temperatures and noxious gases. The first usually occurs in deep tunnels under the influence of the thermal gradient, or in areas of recent volcanic or seismic activity. Explosive gases such as methane are known to occur in coals and shales, and the coal-mining industry long ago learned to respect their power. In tunneling, however, it is usually difficult to anticipate their occurrence.

At the Tecolate Tunnel (Trefzger, 1966), all the main hazards of tunneling occurred together to create what has become the classic case history in the field. The Tecolate Tunnel was driven through the Santa Ynez mountains 19 km northwest of Santa Barbara, California, during the period 1950–1955. It is 10.3 km long and 2.1 m in diameter. It is an aqueduct that carries water from a supply reservoir to the Santa Barbara metropolitan district. The tunnel penetrates a sequence of poorly consolidated shales, siltstones, sandstones, and conglomerates, and crosses one major fault and several minor ones. Major groundwater inflows were encountered with the temperature of the inflows ranging between 11 and 41°C at the face. The largest inflows at the face reached 580 ℓ/s at temperatures up to 40°C. One inflow at 180 ℓ/s held up construction for 16 months and resisted all grouting attempts. All flows came from intensely fractured siltstones and sandstones. The temperatures are thought to have been caused by residual heat from geologically recent faulting. To cope with the almost unbearable conditions in the tunnel, workmen traveled to and from the heading in mine cars immersed up to their necks in cold water.

The San Jacinto tunnel near Banning, California (Thompson, 1966), is one component of the Colorado River Aqueduct, which delivers water from the Colorado watershed to the Los Angeles area. Preconstruction geologic studies led to the conclusion that the predominant rock type would be massive granite. Although some suggestion of faulting was noted, no one visualized the tremendous volumes of water that were later found to be associated with these structural features.

The tunnel was driven from two headings serviced by a central shaft. With the tunnel advanced only about 50 m from the shaft, a heavy flow of water, estimated at 480 ℓ/s, surged into the heading accompanied by over 760 m^3 of rock debris. The limited sections of tunnel east and west of the shaft were soon flooded and water ultimately filled the 250 m shaft to within 45 m of the surface.

The source of the water was a fractured fault zone with a particularly malicious configuration. The fault zone was bounded on the footwall side by a thin layer of impermeable clay gouge. The water-bearing zone occurred in the heavily fractured hanging wall. The initial tunnel heading intercepted the fault from the footwall side with the resultant catastrophic inflow. Subsequent mapping located 21 faults along the tunnel line, each with the same internal "stratigraphy" as the original fault. Subsequent tunneling experience showed that the headings that approached the fault zones from the hanging-wall side still encountered large inflows of groundwater, but by intercepting smaller flows sooner, and by spreading the total flow over a larger area and over a longer time, catastrophic inflows were avoided.

Predictive Analysis of Groundwater Inflows Into Tunnels

If tunnelers are to be expected to cope safely and efficiently with large groundwater inflows, hydrogeologists and geotechnical engineers are going to have to develop more reliable methods of predictive analysis. The only theoretical analyses that we could find in the literature for the prediction of groundwater inflows into tunnels are those of Goodman et al. (1965). They represent an excellent initial attack on the problem but are far from the final work. They show that for the case of a tunnel of radius r acting as a steady-state drain [Figure 10.16(a)] in a homogeneous, isotropic media with hydraulic conductivity K, the rate of groundwater inflow Q_0 per unit length of tunnel is given by

$$Q_0 = \frac{2\pi K H_0}{2.3 \log (2H_0/r)} \tag{10.17}$$

Their analysis for the transient case [Figure 10.16(b)] shows the cumulative rate of inflow $Q(t)$ per unit length of tunnel at any time t after the breakdown of steady flow to be given by

$$Q(t) = \left(\frac{8C}{3} KH_0^3 S_y t\right)^{1/2} \tag{10.18}$$

where K is the hydraulic conductivity of the medium, S_y is the specific yield, and C is an arbitrary constant. The development of Eq. (10.18), however, is based on a very restrictive set of assumptions. It assumes that the water table is parabolic in shape and that the Dupuit-Forchheimer horizontal flow assumptions hold. In addition, Eq. (10.18) is only valid for flow conditions that arise after the water-table decline has reached the tunnel, that is, after t_3 in Figure 10.16(b). On the basis of Dupuit-Forchheimer theory, the constant C in Eq. (10.18) should be 0.5, but Goodman et al. (1965), on the basis of laboratory modeling studies, found that a more suitable value approached 0.75. Equation (10.18) may be suitable for order-of-magnitude design-inflow estimates, but it should be used with a healthy dose of skepticism.

For more complex hydrogeologic environments that cannot be represented by the idealized configurations of Figure 10.16, numerical mathematical models can be prepared for each specific case. Goodman et al. (1965) provide a transient analysis for the prediction of inflows at the face from a vertical water-bearing zone. Wittke et al. (1972) describe the application of a finite-element model to a tunnel line in jointed rock. Their analysis is based on the discontinuous approach to flow in fractured rock (Section 2.12) rather than the continuous approach followed by Goodman et al. (1965).

We have, in this section, considered only those groundwater problems that arise during *construction* of a tunnel. If the tunnel is to carry water, and if that water is to be under pressure, there are design considerations that are influenced by the interactions between tunnel flow and groundwater flow during *operation*. If the tunnel is to be unlined, an analysis must be made of the water losses that will occur to the regional flow system under the influence of the high hydraulic heads that will be induced in the rocks at the tunnel boundary. If the tunnel is to be lined, its design must take into account the pressures that will be exerted on the outside of the lining by the groundwater system when the tunnel is empty.

For these purposes, steady-state and transient flow nets can once again be used to advantage. For a more detailed treatment of the design aspects, the reader is referred to texts on engineering geology or rock mechanics, such as those by Krynine and Judd (1957) or Jaeger (1972).

10.4 Groundwater Inflows Into Excavations

Any engineering excavation that must be taken below the water table will encounter groundwater inflow. The rates of inflow will depend on the size and depth of the excavation and on the hydrogeological properties of the soils or rocks being excavated. At sites where the soil or rock formations have low hydraulic conductivities, only small inflows will occur, and these can usually be handled easily by pumpage from a sump or collector trench. In such cases a sophisticated hydrogeological analysis is seldom required. In other cases, particularly in silts and sands, dewatering of excavations can become a significant aspect of engineering construction and design.

Drainage systems also serve other purposes, apart from the lowering of water tables and the interception of seepage. They reduce uplift pressures and uplift gradients on the bottom of an excavation, thus providing protection against bottom heave and piping. A dewatered excavation also leads to reduced pore pressures on its slopes so that slope stability is improved. In the design of open pit mines, this is a factor of some importance; if decreased pore pressures can lead to an increase in the design pit-slope of even 1°, the savings created by reduced excavation can be many millions of dollars.

Drainage and Dewatering of Excavations

The control of groundwater inflow to excavations can be accomplished in several ways. Sharp (in press) lists the following methods as being in current wide use: (1) horizontal drainholes drilled into the slope face; (2) vertical wells drilled behind the slope crest or from benches on the slope face; (3) drainage galleries behind the slope, with or without radial drainholes drilled from the gallery; and (4) drainage trenches constructed down or along the slope face. Figure 10.17 schematically illustrates how the first three of these techniques can be effective in lowering the water table around an excavation.

Horizontal drains are the cheapest, quickest, and most flexible method of drainage. Piteau and Peckover (in press) provide many practical suggestions for their design and emplacement in rock slopes. Galleries or wells are more expensive, but they have the advantage that they do not inferfere with the workings on the slope face. Dewatering can be carried out with these methods prior to breaking ground so that excavation can take place "in the dry." The design of a dewatering system based on a pattern of pumped wells or wellpoints must be based on the principles presented in Section 8.3 for multiple well systems. The cone of drawdown in the water table at the excavation is created by mutual interference between the individual drawdown cones of each well or wellpoint. Transmissivities and storativities are usually determined on the earliest installations and the design of the remainder of the system is based on these values. Briggs and Fiedler (1966) and Cedergren (1975) provide detailed discussions of the practical aspects of dewatering systems. The maximum drawdown that can be achieved with one stage of wellpoints has been found in practice to be about 5 m. Some deep excavations have been dewatered with as many as eight wellpoint stages.

Vogwill (1976) provides an excellent practical case history of dewatering problems at an open pit mine. At Pine Point in the North West Territories of Canada, lead-zinc ore is mined from a series of pits in a Devonian dolomite reef complex. Transmissivities are in the range 0.005–0.01 m^2/s (30,000–70,000 U.S. gal/day/ft) and dewatering, carried out through pumped wells, must remove

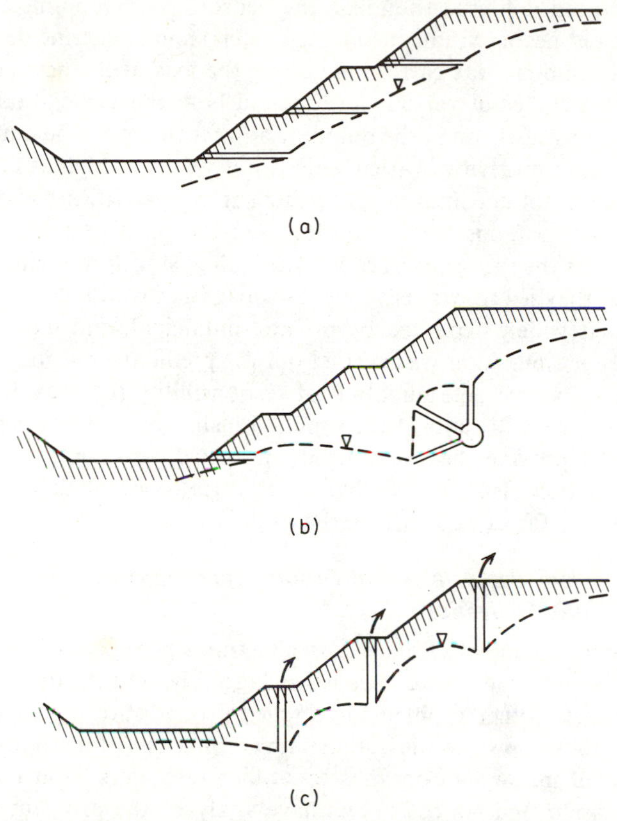

Figure 10.17 Dewatering excavations by (a) horizontal drains; (b) drainage gallery with radial drain holes; (c) three-stage wellpoint system.

between 60 and 950 ℓ/s (1,000–15,000 U.S. gal/min) from the various pits. Vogwill concludes that increasing dewatering requirements and costs could well lead to a situation in the future where mining schedules and forecasts will be determined entirely by open pit dewatering requirements.

The realignment of the Welland Canal in southern Ontario provides a dewatering case history of a different type. The Welland Canal traverses the Niagara Penninsula between Lake Erie and Lake Ontario. It is a key navigational link in the Great Lakes shipping route. A realignment of a portion of the canal in 1968 involved the excavation of about 13 km of new canal. The design required permanent depressuring of a regional aquifer at two sites to reduce the hazards of uplift and slope failure, and the temporary dewatering of some portions of the channel during excavation.

Farvolden and Nunan (1970) and Frind (1970) discuss the hydrogeological aspects of the dewatering program. The main aquifer in the area is a thin zone of fractured dolomite found on the bedrock surface immediately below 20–30 m of low-permeability, unconsolidated glacial and lacustrine deposits. Extensive drilling and sampling was carried out along the axis of the new channel, and piezometers were installed at various locations in both the surficial deposits and the bedrock. Pumping tests run in the dolomite aquifer to determine aquifer coefficients, showed that transmissivities varied widely, but values as high as 0.015 m^2/s (90,000 IGPD/ft) were not uncommon. These high transmissivities had both positive and negative implications for the project. On the positive side, they made it possible to dewater the entire construction site from just four pumping centers. On the negative side, they led to extensive areal propagation of the drawdown cones in an aquifer that is widely exploited by private, municipal, and industrial wells. A numerical aquifer simulation was carried out for predictive purposes, and one of its primary goals was the determination of responsibility for drawdowns in areas of mutual interference. Results of the simulation showed that pumping rates of about 100 ℓ/s would provide the necessary 10 m drawdown along the realignment route. The simulation also showed that an elliptical cone of drawdown would affect water levels as far as 12 km from the canal.

Predictive Analysis of Groundwater Inflows Into Excavations

The development of quantitative methods of analysis for the prediction of groundwater inflow into excavations has lagged behind the development of such methods for many other problems in applied groundwater hydrology. The only analytical methods known to the authors are adaptations of methods designed for the prediction of inflow hydrographs to surface reservoirs from large unconfined aquifers. Brutsaert and his coworkers have analyzed the problem first presented in Figure 5.14 using each of the approaches schematically illustrated there. Verma and Brutsaert (1970) solved the complete saturated-unsaturated system with a numerical method. Verma and Brutsaert (1971) solved the problem numerically as a two-dimensional, saturated, free-surface problem; and analytically as a one-dimensional, saturated problem simplified by use of the Dupuit assumptions. The predictive methodology of Figure 10.18 is based on an earlier study (Ibrahim and Brutsaert, 1965) carried out with a laboratory model. The results were later confirmed by the mathematical models of Verma and Brutsaert (1970, 1971).

Figure 10.18(a) shows the geometry of the two-dimensional vertical cross section under analysis. It has relevance to the prediction of groundwater inflows into excavations only if the following assumptions and limitations are noted: (1) the excavated face is vertical; (2) the excavation is emplaced instantaneously; (3) the boundary conditions and initial conditions on the hydrogeological system are as shown on Figure 10.18(a); (4) the geological stratum is homogeneous and isotropic; and (5) the excavation is long and lineal in shape, rather than circular, so that the two-dimensional cartesian symmetry is applicable. While these assumptions may appear restrictive, results may nevertheless be of use in estimating the probable transient response of more complex systems.

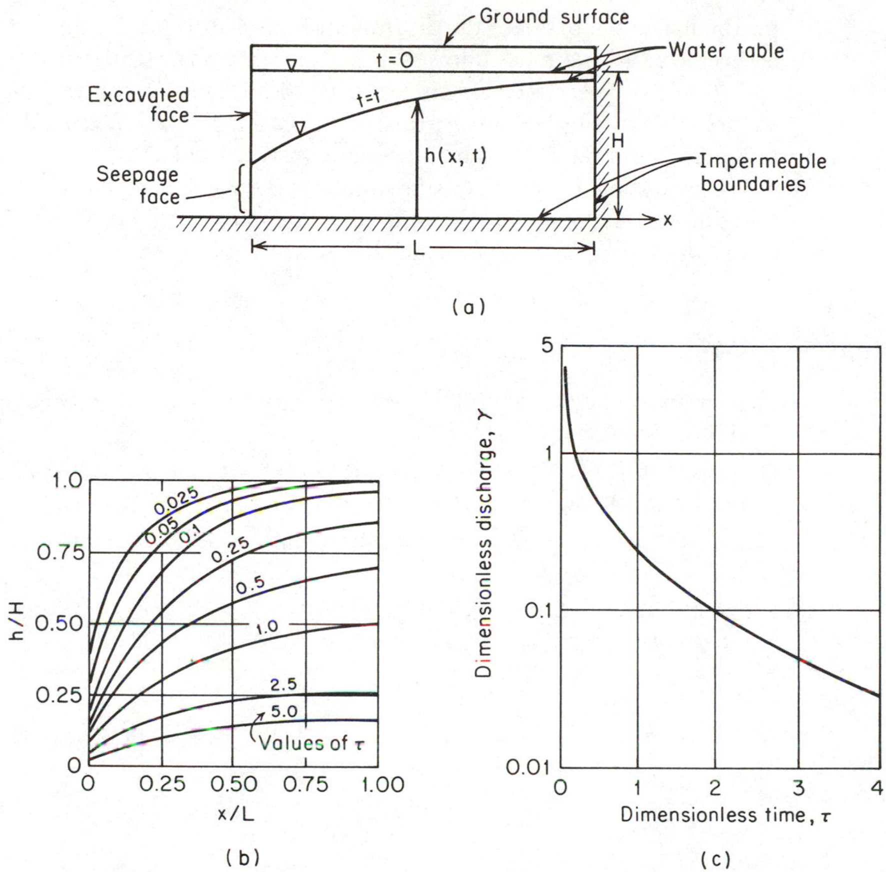

Figure 10.18 Prediction of groundwater inflows into an excavation (after Ibrahim and Brutsaert, 1965).

Figure 10.18(b) shows the transient response of the water table, plotted as dimensionless drawdown, h/H, versus dimensionless distance, x/L. The parameter τ is a dimensionless time given by

$$\tau = \frac{KH}{S_y L^2} t \tag{10.19}$$

where H and L are defined by Figure 10.18(a), K and S_y are the hydraulic conductivity and specific yield of the aquifer, and t is time. In Figure 10.18(c), the dimensionless discharge γ, defined by

$$\gamma = \frac{S_y L}{KH^2} q \tag{10.20}$$

is plotted against τ. The outflow $q = q(t)$ is the rate of flow (with dimensions L^3/T) into the excavation from the seepage face, per unit length of excavated face perpendicular to the plane of the diagram in Figure 10.18(a). To apply the method to a specific case, one must know K, S_y, H, and L. τ is calculated from Eq. (10.19) and $h(x, t)$ is determined from Figure 10.18(b). The $\gamma(\tau)$ values determined from Figure 10.18(c) can be converted to $q(t)$ values through Eq. (10.20). The formulas and graphs can be used with any set of consistent units.

It is possible to carry out similar analyses for circular pits, and for cases where the external boundary is a constant-head boundary, with $h(L, t) = H$ for all $t > 0$, rather than an impermeable boundary.

Suggested Readings

Casagrande, A. 1961. Control of seepage through foundations and abutments of dams. *Géotechnique*, 11, pp. 161–181.

Goodman, R. E., D. G. Moye, A. van Schalkwyk, and I. Javandel. 1965. Ground water inflows during tunnel driving. *Eng. Geol.*, pp. 39–56.

Jaeger, J. C. 1971. Friction of rocks and stability of rock slopes. *Géotechnique*, 21, pp. 97–134.

Terzaghi, K. 1950. Mechanism of landslides. *Berkey Volume: Application of Geology to Engineering Practice*. Geological Society America, New York, pp. 83–123.

CHAPTER 11

Groundwater and Geologic Processes

Groundwater plays an important role in many geologic processes. For example, the fluid pressures that build up on faults are now recognized to have a controlling influence on fault movement and the generation of earthquakes. On another front, subsurface flow systems are responsible for the transfer of heat and chemical constituents through geologic systems, and as a result, groundwater is important in such processes as the development of geothermal systems, the thermodynamics of pluton emplacement, and the genesis of economic mineral deposits. At depth, groundwater flow systems control the migration and accumulation of petroleum. Nearer the surface, they play a role in such geomorphologic processes as karst formation, natural slope development, and stream bed erosion.

In this chapter, we will discuss the role of groundwater in these and other geologic processes. The treatment is brief and the list of topics and references is far from exhaustive. Many of the developments we report are recent. The consequences of groundwater flow have not as yet been extensively evaluated in research on geologic processes.

11.1 Groundwater and Structural Geology

One of the most exciting recent developments in geologic thought concerns the influence of groundwater pressures on fault movement and the possible implications this has for the prediction and control of earthquakes. The concepts were first put forward by Hubbert and Rubey (1959) in their classic paper on the role of fluid pressure in the mechanics of overthrust faulting.

Hubbert-Rubey Theory of Overthrust Faulting

Hubbert and Rubey were addressing a geological mystery of long standing. It had been recognized since the early 1800's on the basis of field evidence that movements of immense thrust blocks over considerable distances had taken place along overthrust faults with extremely low dip angles. Many thrust faults had been mapped that involved stratigraphic thicknesses of thousands of meters and travel distances of tens of kilometers. What was not understood was the mechanism of movement. Many calculations had been carried out in which horizontal tectonic forces or gravitational sliding had been invoked as the mechanism of propulsion, but all had foundered on the need for unrealistically low frictional resistances on the fault plane. When more realistic coefficients of friction were used, the analyses showed that the horizontal forces necessary to cause thrusting would create stresses that greatly exceed the strength of any known rocks.

Hubbert and Rubey solved this mechanical paradox by invoking the Mohr-Coulomb failure theory, as developed in Section 10.1, in its effective stress formulation. Their analysis was the first to take into account the existence of fluid pressures on faults at depth. They utilized the relation presented in Eq. (10.8), but as seems reasonable for a smooth fault plane, they assumed the cohesive strength to be negligible and set $c' = 0$. The failure criterion then becomes

$$S_\tau = (\sigma - p) \tan \phi' \tag{11.1}$$

where S_τ is the shear strength that must be overcome to allow movement, σ the normal stress across the fault plane, p the fluid pressure, and ϕ' the angle of internal friction for the rock-rock interface. They reasoned that large values of p in Eq. (11.1) would serve to reduce the normal component of effective stress on the fault plane and hence reduce the critical value of the shear stress required to produce sliding. They showed that the horizontal forces of propulsion needed to produce these reduced shear stresses do not exceed the strength of rock. They referred to oil field measurements to support their contention that high fluid pressures are a common occurrence at depth. The more recent developments in our understanding of regional flow systems (as reported in Chapter 6) make it clear that these high fluid pressures are a natural outgrowth of the subsurface systems of fluid movement that exist in the heterogeneous geological environment in the upper few thousand meters of the earth's crust.

Figure 11.1 reproduces Hubbert and Rubey's free-body diagram for a thrust block of dimensions x_1 by z_1 being pushed from the rear down an inclined plane

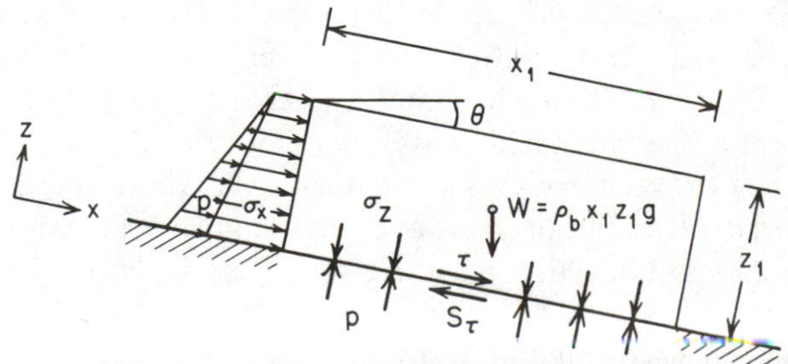

Figure 11.1 Stress equilibrium on a thrust block at incipient movement down an inclined fault plane (after Hubbert and Rubey, 1959).

of slope θ. The block is propelled jointly by the total stress, $\sigma_x + p$, applied to its rearward edge and the component of its weight parallel to the slope. A shear stress is created at the base of the block, and at the point of incipient slippage, $\tau = S_\tau$, where S_τ is the shear strength of the fault plane as given by Eq. (11.1). The equilibrium of forces acting on a section of unit thickness perpendicular to the diagram is given by

$$\int_0^{z_1} (\sigma_x + p)\, dz + \rho_b g z_1 x_1 \sin\theta - \int_0^{x_1} (\sigma_z - p) \tan\phi'\, dx = 0 \qquad (11.2)$$

where ρ_b is the bulk density of the rock. Hubbert and Rubey solved Eq. (11.2) for x_1, the maximum length of block that can be moved by this mechanism. To make such a calculation, it is necessary to know the geometrical parameters, θ and z_1, the mechanical properties, ϕ' and ρ_b, and the value of the fluid pressure, p, on the fault plane. Hubbert and Rubey expressed this last parameter in terms of the ratio $\lambda = p/\sigma_z$. They provide a table of calculated x_1 values for a rock slab 6000 m thick resting on a fault plane with representative ϕ' and ρ_b values. For θ values in the range 0–10° and λ values in the range 0–0.95, the maximum length of block that can be moved varies from 21 to 320 km. These lengths are in keeping with the observed travel distances of overthrust fault blocks. Hubbert and Rubey therefore concluded that consideration of the fluid pressures in the groundwater in the vicinity of fault planes removes the paradox surrounding the mechanism of overthrust faulting.

Earthquake Prediction and Control

Earthquakes are the physical manifestation of the movement of fault blocks. The Hubbert–Rubey theory is therefore pertinent to earthquake genesis. Dramatic confirmation of the influence of elevated fluid pressures on earthquake production came to light in the late 1960's in a somewhat unexpected way in connection with the now-famous Rocky Mountain Arsenal disposal well near Denver, Colorado.

During the period April 1962 to September 1965 there were 710 small earthquakes recorded in the Denver area. This was a seismological mystery because prior to this time the only recorded earthquake had occurred in 1882. The solution to the mystery was provided by Evans (1966), who noted that the first earthquake occurred just 1 month after the first injection of liquid waste at a disposal well at the U.S. Army's Rocky Mountain Arsenal. The injection well was designed for the disposal of contaminated wastewater from the chemical plant at the Arsenal. The well was drilled through sedimentary rocks, bottoming at a depth of 3671 m in fractured Precambrian schist and granite gneiss. Injection was carried out at rates of 12–25 ℓ/s at injection pressures of 3–7 $\times$ 10^6 N/m^2. Evans noted that the earthquake frequency in the period 1962–1965 was closely correlated with the volume of waste injected (Figure 11.2). Further investigations showed that the epicenters of almost all the shocks were located within a circular area 16 km in diameter centered at the Rocky Mountain Arsenal.

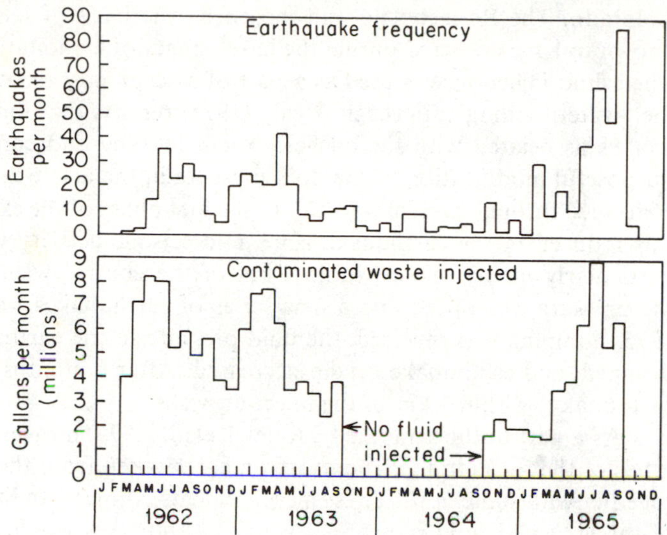

Figure 11.2 Earthquake frequency in the Denver area, 1962–1965, as correlated with the injection of contaminated waste water at the Rocky Mountain Arsenal disposal well (after Evans, 1966).

Each of the earthquakes that occurred in the Denver cluster presumably reflected movement on a preexisting fault at depth in the vicinity of the Arsenal well. Apparently, the increases in fluid pressures engendered by the injection had the effect of triggering the small fault movements. Evans' observations thus provided convincing confirmation of the validity of the theoretical calculations of Hubbert and Rubey (1959). Healy et al. (1968) after an extensive review of the evidence conclude that the Hubbert-Rubey mechanism provides a complete and satisfactory explanation for the triggering of the Denver earthquakes.

If increased fluid pressures encourage fault movement, then decreased fluid pressures should retard fault movement, and the possibility of earthquake control is raised. Seismologists and groundwater hydrologists are now working together to examine the possibility of man-made intervention in the faulting process. The ultimate scheme would be to take a fault such as the San Andreas in California, tighten it along most of its length by dewatering the fault zone, and then encourage controlled movement in a small portion by injecting water into the fault zone at that point. In this way it might be possible to move sequentially along the fault, relieving the tectonic stresses that build up along its length with a series of small controlled fault movements rather than awaiting a large catastrophic earthquake.

The social and ethical conundrums that would result from the serious consideration of such a scheme, together with the momentous implications of a technical failure, will certainly delay, and may well prevent, implementation of earthquake control in heavily populated areas. However, large-scale field experiments have already been carried out in a less populated area at an oil field near Rangely,

Colorado. The Rangely site was chosen on the basis of seismic activity that was known to have occurred during the latter stages of exploitation of the oil reservoir when fluid injection was used as a part of a secondary recovery program utilizing the "waterflooding" approach. Healy (1975) reports that monitoring of the earthquakes associated with the oil field began in 1969 and continued through 1974. Purposeful modification of the fluid pressure in the active zone began in 1970 and continued through December 1973. In the first phase of the experiment, the pressure was reduced in the earthquake zone, and seismic activity was greatly decreased, particularly in the region within 1 km of the control wells. In November 1972, the pressure was raised and a new series of earthquakes was initiated. In March 1973, pumping was reversed, the fluid pressure in the earthquake-producing zone dropped, and earthquake acivity decreased. After 6 months there were no further earthquakes within 1 km of the injection wells.

As a part of the same study, Raleigh et al. (1972) measured the frictional properties of the rocks in the laboratory on cores taken from the oil field. These data, together with some *in situ* stress measurements, allowed an independent calculation of the values of fluid pressure at which earthquakes would be expected to occur. The predicted critical level was $p = 2.57 \times 10^7$ N/m^2. The values in the seismically active part of the reservoir at a time of frequent earthquakes were measured at 2.75×10^7 N/m^2. Healy (1975) concludes that the Rangely experiments establish beyond doubt the importance of fluid pressure as a critical parameter in the earthquake mechanism.

It has also been suggested that very detailed pressure measurements on faults may provide precursory evidence of impending earthquakes. Scholz et al. (1973) review the *dilatancy model* of earthquake prediction and describe the role played by the interaction between the stress field and the fluid-pressure field just prior to the actual triggering of movement on a fault.

11.2 Groundwater and Petroleum

It is now widely accepted (Weeks, 1961; Hedberg, 1964; Levorsen, 1967) that petroleum originates as organic matter that is incorporated into fine-textured sediments at the time of their deposition. However, while organic-rich clays and shales are found throughout the sedimentary basins of the world in great areal and volumetric abundance, present-day accumulations of petroleum are found only in localized concentrations of relatively small volume. Further, they do not occur in the clays and shales themselves, but rather in coarse-textured sandstones and in porous or fractured carbonate rocks. It is apparent that petroleum must undergo significant migration from its highly dispersed points of origin to its present positions of concentration and entrapment. During this migration petroleum is an immiscible and presumably minor constituent of the subsurface water-saturated environment. It is therefore reasonable to examine the processes of migration and accumulation of petroleum in light of our understanding of regional groundwater flow systems. Such an examination has ramifications in the field of petroleum exploration.

Migration and Accumulation of Petroleum

Petroleum migration is often viewed as a two-step process. The term *primary migration* refers to the processes whereby water and entrained petroleum are expelled from the fine-grained source sediments into the more permeable aquifers of a sedimentary system. The term *secondary migration* is reserved for the movement of petroleum and water through the aquifer systems to the structural and stratigraphic traps, where oil and gas pools are formed.

Primary migration can be viewed as one result of the process of consolidation that takes place in newly deposited fine-grained sediments. Bredehoeft and Hanshaw (1968) have shown that the influence of the added load provided by the additional sediments that are continuously being emplaced at the top of a sedimentary sequence in a depositional environment is sufficient to produce significant consolidation. The mechanism is identical to that described in connection with land subsidence in Section 8.12. Once again, our understanding of the process hangs on the effective stress equation

$$\sigma_T = \sigma_e + p \tag{11.3}$$

In this case, it is the direct natural change in the total stress, σ_T, that drives the consolidation process, rather than an artificial change induced in the fluid pressure, p, as was the case where land subsidence is caused by overpumping. In either case, the result is an increase in effective stress, σ_e, and a compaction of the highly compressible fine-grained sediments. During the consolidation process, water is driven out of the fine-grained sediments into any aquifers that may be present in the system. If the temperature and pressure environments in the consolidating sediments have been conducive to the maturation processes that transform organic matter into mobile petroleum, this entrained petroleum is driven into the aquifers with the water.

It has been recognized since early in the century (see Rich, 1921) that secondary migration of petroleum is brought about by the movement of groundwater in the reservoir rocks. It is the water that provides the transporting medium for the immiscible petroleum droplets that ultimately accumulate to form oil pools. Toth (1970) has noted that the accumulation of petroleum requires the favorable interaction of at least three processes: (1) continuous import of hydrocarbons, (2) separation and preferential retention of the dispersed hydrocarbons from the transporting water, and (3) a continuous removal of water discharged of its hydrocarbon content. The first and third processes require a suitable flow system. On the second point, it is usually assumed that the separation of petroleum from water takes place under the influence of pressure changes, temperature changes, or salinity changes. Any of these can lead to a flocculation of the entrained petroleum droplets into larger discrete oil accumulations until phase continuity is achieved and buoyancy effects can come into play. Since oil and gas both have densities less than water, they become concentrated in the upper parts of the flowing aquifers. Oil pools arise where anticlinal structures or stratigraphic complexities create a trap for the low-density petroleum. Levorsen (1967) reviews the various geological conditions that give rise to traps. Hubbert (1954) discusses the capillary mechanism in a two-phase oil-water system that explains the efficiency of a low-permeability interface as a barrier to petroleum migration. In the following subsection, Hubbert's ideas will be traced further with respect to the interaction between petroleum entrapment and the subsurface hydraulic-potential field.

Hydrodynamic Entrapment of Petroleum

Movement of oil, gas, and water through a porous medium is an example of immiscible multiphase flow. As noted near the close of Section 2.6, the analysis of such systems is extremely complex. It is necessary to consider separate Darcy equations for each of the fluids flowing simultaneously through the system. It is also necessary to determine the effective permeabilities of the porous medium to each of the phases. Because the permeability of the medium is different with respect to each fluid, the magnitudes of the Darcy velocities for each phase will be different from one another.

Hubbert (1954) has shown that not only are the magnitudes of the three velocity vectors different, but so are the directions. In explanation of this point, consider first the diagram shown in Figure 11.3(a) for a single-phase fluid. The direction of movement of a unit mass of fluid at the point P is perpendicular to the

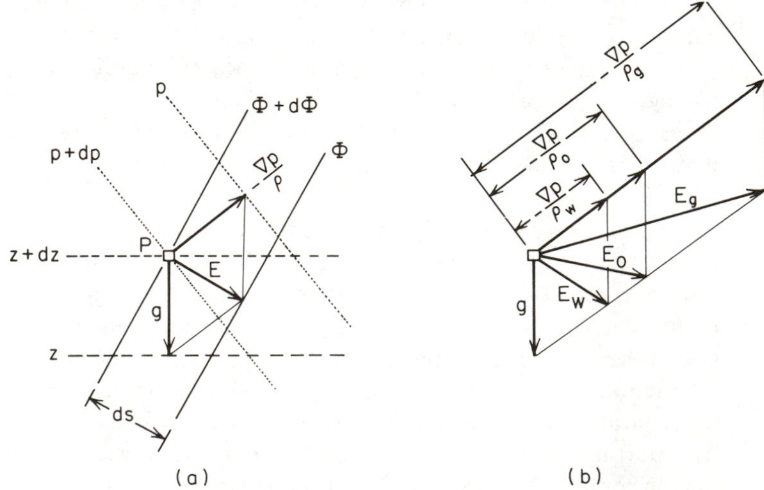

Figure 11.3 (a) Components of the impelling force E acting on a unit mass of water at a point P in a steady-state groundwater flow system; (b) impelling forces on water, oil, and gas in a three-phase steady-state flow system (after Hubbert, 1954).

lines of equal hydraulic potential. The force acting on the unit mass in the direction of movements is denoted by E. Recall from Eq. (2.15) that the hydraulic potential Φ is defined as

$$\Phi = gz + \frac{p}{\rho} \tag{11.4}$$

where p is the fluid pressure and ρ is the fluid density. In that the potential is defined in terms of energy per unit mass, the work required to move the unit mass from the potential $\Phi + d\Phi$ to the potential Φ is simply $-d\Phi$. With reference to Figure 11.3(a), it is also clear that the work is equal to $E\,ds$. Therefore, we have

$$E = -\frac{d\Phi}{ds} \tag{11.5}$$

or, invoking Eq. (11.4),

$$E = g - \frac{\nabla p}{\rho} \tag{11.6}$$

where g is a vector with components $(0, 0, -g)$ and ∇p is a vector with components $(\partial p/\partial x, \partial p/\partial y, \partial p/\partial z)$. The vector g acts vertically downward; the vector ∇p may act in any direction, and in general it will not be coincident with g. Figure 11.3(a) is a graphical presentation of Eq. (11.6).

In a three-phase system, the fluid densities are not equal. We have $\rho_w > \rho_o > \rho_g$, where the subscripts refer to water, oil, and gas, respectively. This fact leads to the vector diagram shown in Figure 11.3(b). This diagram provides a graphical explanation for the lack of coincidence of the directions and magnitudes of the impelling forces E_w, E_o, and E_g. The hydraulic gradients for each of the phases will be in the direction of their respective impelling forces.

The practical manifestation of this phenomenon is the hydrodynamic entrapment of petroleum as proposed by Hubbert (1954). In Figure 11.4 the oil and water equipotentials are shown superimposed on one another for a case where there is a buoyant upward movement of oil in an aquifer in which the flow of groundwater is from left to right. Hubbert (1954) shows that the slope of the tilted oil-water interface, dZ/dl, is given by

$$\frac{dZ}{dl} = \frac{\rho_w}{\rho_w - \rho_o}\frac{dh}{dl} \tag{11.7}$$

The interface will be horizontal only if there is no hydraulic gradient. In order for a structure or monocline to hold oil, the dip of the permeability boundary in the direction of fluid movement must be greater than the tilt of the oil-water interface. Otherwise, the oil will be free to migrate downdip along with the water. In areas where high hydraulic gradients lead to relatively fast groundwater flow, traps with steeper closing dips are required to hold oil than in areas with low hydraulic gradients and slow groundwater flow. Conversely, in areas where dips are relatively

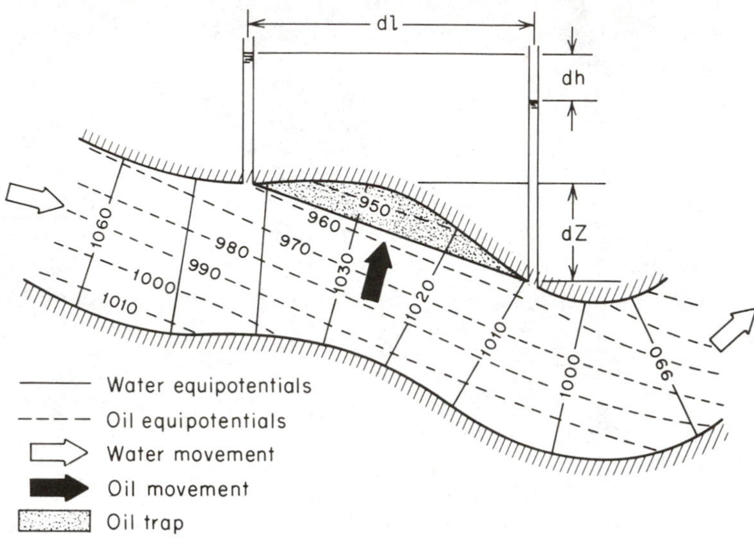

Figure 11.4 Vertical section through a plunging nose showing a hydrodynamically controlled oil trap (after Hubbert, 1954).

uniform, those parts of the basin with low hydraulic gradients produce more locations for oil entrapment than hydrodynamically more active zones.

Regional Flow Systems and Petroleum Accumulations

It should be clear from the previous two subsections that there are two sets of conditions that lead to the entrapment of petroleum. The first is the set of geologic conditions that control the occurrence of structural and stratigraphic traps, and the second is the set of groundwater flow conditions that control the hydrodynamic aspects of entrapment. In considering these latter properties, Tóth (1970) noted that petroleum accumulation should be enhanced by (1) long flow systems that encompass a sufficient volume of possible source rock; (2) static or quasi-static hydraulic zones, where hydrodynamic entrapment can be expected to be most efficient; and (3) ascending groundwater movement, whereby a continuous removal of water from the traps is ensured. In the western Canada sedimentary basin, where many petroleum fields are known, Hitchon (1969a, 1969b) and van Everdingen (1968b) showed that, on the basis of both hydraulic and geochemical evidence, large flow systems extending from the Rockies to the Canadian shield exist in the deeper formations of the basin. Tóth (1970) found statistical confirmation of his own hypotheses in several areas of Alberta. In these areas, his results indicate that the relative probability of hydrocarbons being associated with each of the three conditions are as follows: ascending limbs, 78%; quasi-static zones, 72%; and large regional systems, 72%.

Hitchon and Hays (1971) applied a similar approach in the Surat basin of Australia. They found that hydrocarbon occurrences are concentrated in one of the discharge areas of the basin. However, they are not limited to this area, and there are large portions of the same discharge area that have not yet yielded petroleum. The deposits occur at depth in an area of upward-rising groundwater but not at points with particularly low gradients.

A basic fact that must be kept in mind (van Everdingen, 1968b) when dealing with the influence of major circulation systems on petroleum accumulation is the certainty that present-day hydrodynamic potential distributions are of recent geologic origin. The current topography in western Canada, for example, probably emerged in the late Tertiary. During pre-Tertiary times, potential distributions must have been different if for no other reason than the absence of the high relief recharge areas provided by the Rocky Mountains. It may be necessary to unravel paleohydrogeologic regimes to fully understand the interactions between groundwater flow and petroleum accumulations.

Implications for Petroleum Exploration

The results of Tóth (1970) and Hitchon and Hays (1971) are probably representative of the success thus far in relating present-day groundwater flow systems and petroleum accumulations. The relationships are discernible but far from universal. What should be clear from this discussion, however, is that in the search for oil an understanding of the existing three-dimensional subsurface flow system and its genesis is of an importance comparable to the knowledge of the stratigraphy and structure of a sedimentary basin. Hubbert (1954) notes that if hydrodynamic conditions prevail, as they almost always do, it is important that their nature be determined in detail, formation by formation over the entire basin in order that the positions of traps can be better determined and otherwise obscure accumulations of petroleum are not overlooked.

Hitchon (1971) makes the additional case that interpretations of geochemical exploration for petroleum should take the regional groundwater flow systems into account. He also notes that although surface prospecting for oil has been somewhat equivocal, the lack of success may not rest on any fundamental breakdown in the logical sequence of events between the occurrence of the indicator in the oil field and its appearance at the surface but rather on the usual absence of a careful examination of the possible subsurface flow routes by which hydrocarbons may be carried to the surface.

11.3 Groundwater and Thermal Processes

On a global scale the earth's thermal regime involves the flow of heat from the deeper layers of the planet toward its surface. The *geothermal gradient* that gives evidence of this heat-flow regime has been widely measured by geophysicists involved in terrestrial heat-flow studies. On average, the temperature increases approximately 1°C for each 40 m of depth. However, this gradient is far from uniform. In the upper 10 m or so, diurnal and seasonal variations in air temperature create a zone that is thermally transient. Beneath this zone the effects of air temperature are quickly damped out, but anomalous geothermal gradients can arise in at least three additional ways: (1) as a result of variations in thermal conductivity between geological formations, (2) as a response to geologically recent volcanic or intrusive sources of heat production at depth, and (3) due to the spatial redistribution of heat by flowing groundwater. In this section we will examine this third mechanism as it pertains to natural groundwater flow, geothermal systems, and the thermal regimes that accompany pluton emplacement.

Before examining these specific cases, some general comments are in order. The simultaneous flow of heat and groundwater is a coupled process of the type introduced in Section 2.2. The flow of water is controlled by the pattern of hydraulic gradients, but there may also be additional flow induced by the presence of a thermal gradient [as indicated by Eq. (2.22)]. Heat is transported through the system both by *conduction* and *convection*. Conductive transport occurs even in static groundwater. It is controlled by the thermal conductivity of the geological formations and the contained pore water. Convective transport occurs only in moving groundwater. It is the heat that is carried along with the flowing groundwater. In most systems convective transport exceeds conductive transport.

It is common to distinguish between two limiting types of convective heat transfer. Under *forced convection*, fluid inflows and outflows are present and fluid motion is due to the hydraulic forces acting on the boundaries of the system. Under *free convection*, fluid cannot enter or leave the system. The motion of the fluid is due to density variations caused by the temperature gradients. In the analysis of forced convection, density gradients are ignored and buoyancy effects are considered negligible; in free convection the fluid motion is controlled by buoyancy effects. The transport of heat by natural groundwater flow systems is an example of forced convection. Geothermal systems in which water-steam phase transitions occur are usually analyzed as free convection. Many geothermal systems include a combination of both phenomena. One refers to such conditions as *mixed convection.*

Thermal Regimes in Natural Groundwater Flow Systems

Consider a vertical two-dimensional cross section through a geological system that is thermally and hydraulically homogeneous and isotropic. Let us first examine a case, such as that shown in Figure 11.5(a), in which groundwater conditions are static. Hydraulic heads throughout such a system will be equal to z_0, the elevation of the horizontal water table that is the upper boundary of the system. Figure 11.5(b) displays the boundary-value problem that would represent the steady-state heat-flow regime for this case. The temperature, T_s, on the upper surface is the mean annual air temperature. The vertical boundaries are insulated against horizontal heat flow. The vertical temperature gradient dT/dz, on the base of the

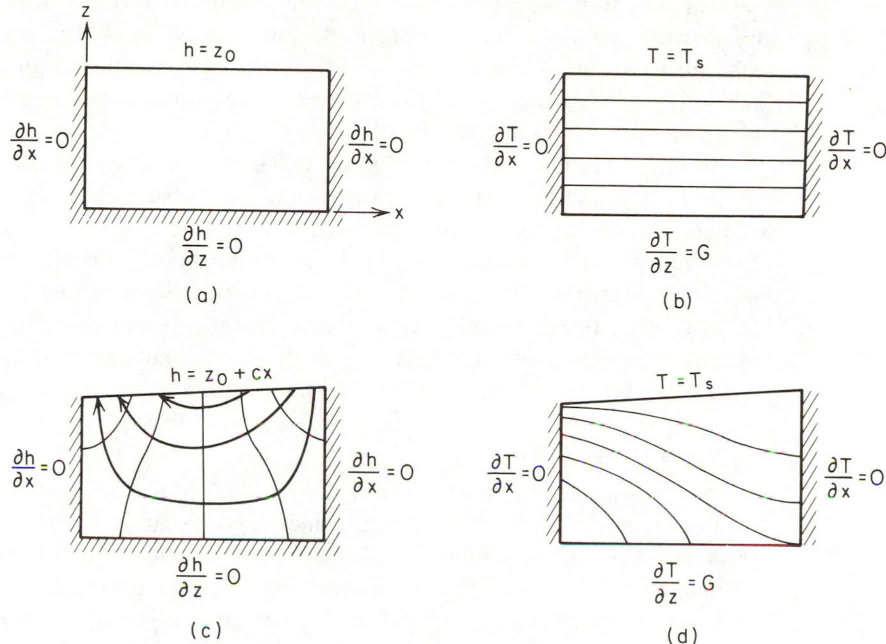

Figure 11.5 Influence of a simple regional groundwater flow system on spatial temperature distributions in a vertical section (after Domenico and Palciauskas, 1973).

system is equal to the geothermal gradient, G. The resulting isotherms are horizontal. Groundwater temperatures in the upper 100 m of the regime would be expected to be 1–2°C greater than the mean annual air temperature, in accordance with the uniform geothermal gradient.

Figure 11.5(c) and (d) is generalized from the results of Parsons (1970) and Domenico and Palciauskas (1973), each of whom studied the influence of regional groundwater flow systems on temperature distributions. Domenico and Palciauskas utilized analytical solutions to the coupled boundary-value problem; Parsons used numerical solutions and provided field evidence to support his findings. Figure 11.5(c) is the simple regional flow system first introduced in Chapter 6. Figure 11.5(d) shows how the thermal regime is altered by the convective heat transport. The geothermal gradient is greater near the surface in discharge areas than it is in recharge areas. It increases with increasing depth in recharge areas, and decreases with increasing depth in discharge areas. Domenico and Palciauskas (1973) show that the effects are more pronounced in flow regions in which the depth of the basin is of the same order of magnitude as the lateral extent, and less pronounced in shallow flow systems of large lateral extent. Parsons (1970) shows that the effects are greater in high-permeability deposits, where groundwater flow velocities are larger, than in low-permeability deposits, where velocities are small.

Cartwright (1968, 1974) has described methods whereby soil temperatures and shallow groundwater temperatures can be used in the field to distinguish recharge areas and discharge areas and to prospect for shallow aquifers. Schneider (1962) showed that local subsurface temperature anomalies can be used to detect infiltration from surface-water sources.

Stallman (1963), in presenting the equations of flow for the simultaneous flow of heat and groundwater, suggested that the measurement of vertical groundwater-temperature profiles might provide a useful method for estimating groundwater velocities. Bredehoeft and Papadopoulos (1965) provide a solution of Stallman's equations for one-dimensional, vertical, steady flow of groundwater and heat. They provide a set of type curves whereby groundwater velocities can be calculated from temperature data. If head measurements are also available, their method can be used to calculate vertical hydraulic conductivities.

Geothermal Systems

In recent years there has been considerable interest in the development of geothermal power, and this interest has led to increased research into the nature of geothermal systems. Elder (1965) and White (1973) provide excellent reviews of the characteristics of geothermal areas and the physical processes associated with them. Witherspoon et al. (1975) review the various mathematical models that have been proposed for simulating geothermal systems.

Geothermal energy is captured by removing the heat from hot waters that are pumped to the surface through wells. Geothermal reservoirs of practical interest must have temperatures in excess of 180°C, an adequate reservoir volume, and a sufficient reservoir permeability to ensure sustained delivery of fluids to wells at adequate rates. The shallower the geothermal reservoir, the more economically feasible is its exploitation. For this reason much interest has centered on an understanding of the mechanisms that can lead to high-temperature fluids at shallow depths. It is now clear that this situation is usually brought about by hydrothermal convection systems in which most of the heat is transported by circulating fluids. Two mechanisms can be envisaged. The first is the forced-convection system suggested by White (1973) and shown in Figure 11.6(a), whereby a local flow system is recharged and discharged vertically through high-permeability fracture zones and heated at depth during residence in more permeable strata. This configuration can give rise to geysers and hot springs at the surface in the discharge zone. Donaldson (1970) has described a simple quantitative model for the simulation of systems of this type.

The second mechanism is one of free convection in a confined aquifer at depth. As shown in Figure 11.6(b), a system in which the upper and lower boundaries of an aquifer are impermeable to fluid flow but conductive to heat flow will lead to the establishment of convective cells of fluid flow that distort the uniform geothermal gradient in the aquifer and create alternating hot spots and cool spots along the upper boundary. This type of convective flow has been known in pure

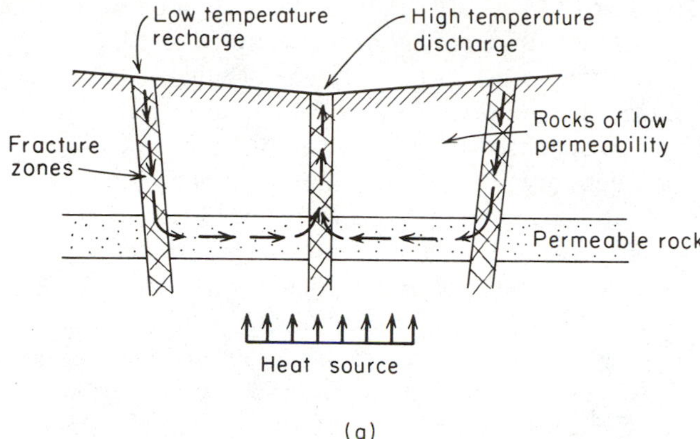

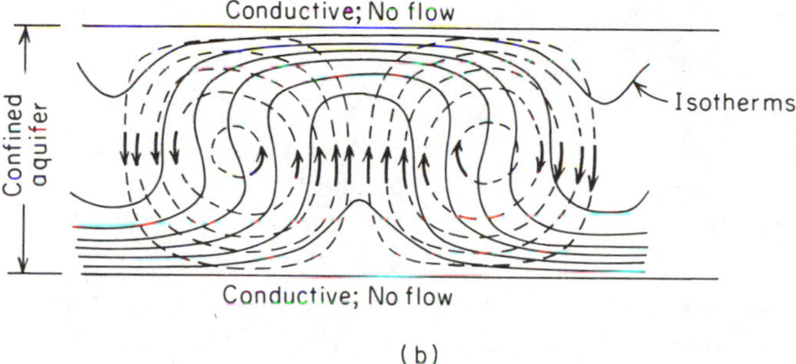

Figure 11.6 Upward migration of hot water into shallow geothermal reservoirs due to (a) forced convection through high-permeability conduits (after White, 1973) and (b) free convection in a confined aquifer (after Donaldson, 1962).

fluid mechanics since early in the century. Its importance to geothermal processes was brought to the attention of geophysicists by Donaldson (1962).

Regardless of the mechanism that brings hot fluid to shallow depths, geothermal systems can be further classified (White, 1973) into *hot-water systems* and *vapor-dominated systems*. In the hot-water type, water is the continuous phase throughout the system and thus provides the pressure control. In the vapor-dominated type, steam is the continuous pressure-controlling phase, although there is general agreement that liquid water is usually present as well. Because a few geothermal systems produce superheated steam with no associated liquid, vapor-dominated systems are sometimes called *dry-steam systems*. The thermodynamics of steam-water geothermal systems under free and forced convection is an advanced topic of current interest to hydrogeological researchers.

Because the configuration of characteristics necessary to create an exploitable geothermal field occurs only rarely, the resource does not appear to offer any kind of panacea to man's energy problems. White (1973) summarizes the geothermal-power-generating capacity of the world as of 1972.

In those areas where geothermal resources are economically significant, there is much ongoing research into the application of simulation models of heat flow/fluid flow systems. Mercer et al. (1975), for example, have developed a single-phase, two-dimensional, horizontal, finite-element model for the hot-water aquifer in the Wairakei geothermal system in New Zealand. The future hope is that models of this type will be able to increase the efficiency of exploitation of geothermal heat by aiding in the optimal design of well spacings and pumping rates in a similar manner to the conventional aquifer models discussed in Chapter 8. However, it is not yet clear whether the great expense and technical difficulty of obtaining the necessary data from great depth in hot systems can be overcome. Until its applicability is confirmed in the real world, geothermal simulation remains a potentially powerful but as yet unproven tool.

Pluton Emplacement

Norton and Knight (1977) have studied a heat flow/fluid flow system of considerable geological importance. They utilized a numerical mathematical model to simulate the thermal regime following pluton emplacement at depth. Figure 11.7 shows the boundary-value problem that they considered. The system is insulated against heat flow at the base and conductive on the other three sides. The system is one of free convection, in that the region is surrounded on all four sides by boundaries that are impermeable to fluid flow. Norton and Knight carried out transient simulations that show the time-dependent growth and decay of the anomalous thermal regime. The right-hand side of Figure 11.7 shows the temperature field 50,000 years after emplacement of a pluton at 920°C into a host rock with

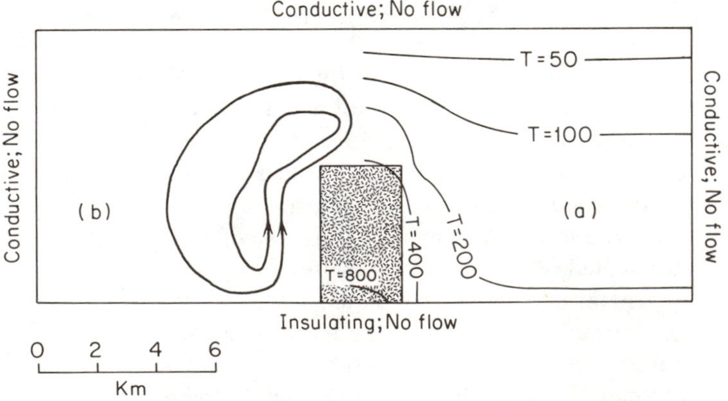

Figure 11.7 (a) Temperature field and (b) fluid circulation 50,000 years after emplacement of a pluton at 920°C into a host rock with initial geothermal gradient of 20°C/km. Permeability of the host rock is 10^{-11} cm^2 (after Norton and Knight, 1977).

an initial geothermal gradient of 20°C/km. The field is symmetric about the center line. The left-hand side of the diagram shows one of the two symmetric convective fluid-circulation cells at the same point in time. In the original paper the authors also showed some example pathlines (Section 2.8) that indicate the paths followed by individual particles of water during the transient event. They conclude that waters in natural pluton systems move away from their points of origin to positions several kilometers away in a few hundred thousand years. Such large-scale fluid circulation is of great importance in understanding the genesis of hydrothermal mineral deposits that are often associated with plutonic environments.

11.4 Groundwater and Geomorphology

Karst and Caves

A landscape that exhibits irregularities in surface form caused by rock dissolution is known as a *karst* landscape, after the characteristic Karst Region of Yugoslavia. Karst landscapes are usually formed on limestone and to a lesser extent on dolomite, but they can also develop in areas of gypsum or rock salt. Processes in carbonate rock will be the focus of this discussion.

The irregularities of the land surface in karst areas are caused by surface and subsurface removal of rock mass by dissolution of calcite or dolomite. Karst areas normally have caves developed as a result of dissolution along joints, bedding planes, or other openings. In major karst regions thousands of kilometers of caves exist, extending in places to depths of more than 1 km. In some parts of the world, networks of caves exist in areas where the original karst nature of the landscape has been obliterated by more recent geomorphic processes, such as glaciation or alluviation.

Thraikill (1968) states that investigations of limestone caves by various geologists have led to three generalizations with regard to cave origin: (1) most limestone caves are the result of solution by cold meteoric waters, (2) many of these solution caves were excavated when the rock was completely filled with water, and (3) some of these sub-water-table caves exhibit horizontal surfaces or horizontal distribution passages that are unrelated to bedding or other structures of the enclosing rocks.

It is evident that limestone at the start of the cave-forming process must have some open joints or bedding planes, or possibly well-connected pores. Of the innumerable open joints and bedding planes in karst areas, only a few are ultimately enlarged to form cavern passages. A combination of factors causes larger distances of penetration of calcite-undersaturated water in a small number of openings. This ultimately leads to preferential enlargement of these openings. This, in turn, causes more of the flow to be captured by the enlarged channels, and through the amalgamation of these channels the process of cavern development proceeds.

Figure 11.8 shows an example of a horizontal cave that cuts across joints and bedding planes. Such caves are believed to have formed in shallow sub-water-table

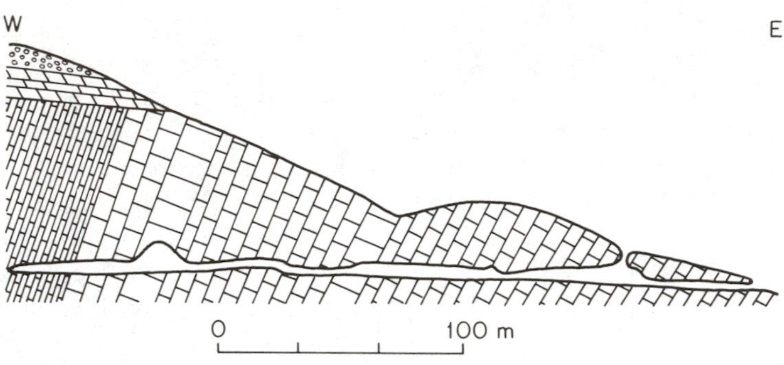

Figure 11.8 Schematic profile on the Lehman Caves, Nevada (after Moore and Nicholas, 1964).

zones. This situation is intuitively reasonable when one considers the fact that channel or cavern enlargement must be accomplished by flowing groundwater that is undersaturated with respect to calcite. As the water flows in the rock, it will approach saturation and thus have less capability for enlarging the flow passage.

The most difficult problem in understanding the origin of caves is how to account for the occurrence of undersaturated water at considerable distances from ground surface. As indicated in Chapter 7, it is well known from laboratory experiments that water in contact with limestone attains saturation quickly relative to natural flow rates in karst limestone. The laboratory experiments by Howard and Howard (1967) are particularly illustrative of this process. Thraikill (1968) has concluded that the uptake of CO_2 in the soil bears little direct relation to cave excavation in the sub-water-table zone. Observations above the water table of the chemical character of water moving downward through secondary openings indicate that this water is typically saturated or supersaturated with respect to calcite, often as a result of the combined effects of calcite dissolution and off-gassing of CO_2. If this type of subsurface water is not aggressive with respect to the rock, we are faced with a dilemma with regard to channel enlargement in the sub-water-table zone. To produce the necessary undersaturated water in shallow sub-water-table zones, the following mechanisms have been suggested: (1) changes in groundwater temperature, (2) mixing of dissimilar waters, (3) floods in surface streams or rapid snowmelt causing large rapid recharge of undersaturated water, and (4) production of acid along the paths of flow.

It can be shown with the aid of geochemical reasoning that when some types of calcite-saturated waters mix, the mixed water is slightly undersaturated, providing that the original solutions have different CO_2 partial pressure (Wigley and Plummer, 1976) or temperatures (Thraikill, 1968). Since water in the shallow sub-water-table zone is commonly a mixture of waters from various inflow areas or fracture zones, and because only very slight undersaturation is necessary to excavate a cave over geologic time, this mechanism is often cited in discussions of cave genesis. It has proved difficult, however, to obtain corroborative data in the field.

Thraikill (1968) indicates that many of the processes thought to be important in cave excavation will operate most effectively during floods. He indicates that the shapes of some caverns suggest that the most active enlargement was localized between a low and a high water table.

Moore and Nicholas (1964) suggest that in some cases oxidation of small amounts of sulfide minerals, especially pyrite, may cause a decline in groundwater pH and, consequently, create cavern enlargement by calcite dissolution. Dissolved oxygen would be the most active oxidizing agent. If this process occurs, one would expect it to be limited to shallow zones, where dissolved oxygen in the groundwater is most abundant.

In summary, karst and caves are perhaps the most dramatic evidence of the ability of flowing groundwater to alter the form of the earth's surface and subsurface. It does not require special knowledge to recognize that limestone is sculptured and excavated by chemically aggressive water. On closer inspection, however, it is clear that a fuller understanding of cave genesis offers ample room for the application of hydrologic and geochemical concepts that involve complex interactions in time and space. Holland et al. (1964), Howard (1964a, 1964b), Thraikill (1968), and Ford and Cullingford (1976) provide more comprehensive discussions of the processes of fracture enlargement and cave genesis.

Natural Slope Development

The processes that lead to natural slope development have been described both qualitatively and quantitatively in great detail by Carson and Kirkby (1972). They note that any slope morphology can be viewed as the outgrowth of a two-step process whereby material must first be loosened from the bedrock by *weathering* before it can be moved downslope by a wide variety of possible *transport* processes. The saturated-unsaturated subsurface flow regime on the hillslope is an important element in both steps of the process.

Weathering of bedrock at the base of a soil is largely a chemical process. The principles and conceptual models outlined in Chapters 3 and 7 provide a suitable basis for understanding the mineral dissolution processes that lead to soil formation. Carson and Kirkby (1972) note further that in humid regions the chemical dissolution of material by groundwater and its downslope transport in solution can be a major form of hillslope erosion in its own right, in some cases of the same order of magnitude as all forms of mechanical erosion combined. The high dissolved loads of many rivers reflect the effectiveness of chemical removal as a transport agent. Carson and Kirkby (1972) provide a synthesis of data available from the United States that relates dissolved load concentrations in streams to mean rates of surface lowering by solution. For a watershed in the southern United States with mean annual runoff of 20 cm, an average solute concentration of 200 ppm in the gaged streams represents a rate of denudation of 0.003 cm/yr.

The downslope transport of material by mechanical means occurs both as discrete mass movements in the form of landslides, slumps, and earthflows, and as sediment transport in surface runoff. The influence of pore pressure distributions created by hillslope flow systems on the occurrence of slope instabilities was treated in Section 10.1. The concepts and failure mechanisms described there, in a geotechnical context, are equally valid when examining the role of landslides as a process in landform evolution. We will not repeat that treatment here; rather, following Kirkby and Chorley (1967), we will examine the implications of the various mechanisms of streamflow generation, as outlined in Section 6.5, on the processes of surface-water erosion.

The classic analysis of hillslope erosion is a direct outgrowth of Horton's (1933) concepts of streamflow generation. The Horton model presumes the widespread occurrence of overland flow. In that, the depth and velocity of overland flow on a hillslope will increase downslope, there should be some critical point at which the flow becomes sufficient to entrain soil particles from the slope. Below this boundary, stream channels will develop as a consequence of this erosion.

Kirkby and Chorley (1967) note that the Horton model is most appropriate on bare slopes in arid regions. However, on vegetated slopes in humid regions, the transfer of rainfall to runoff by means of subsurface stormflow or by the mechanisms proposed by Dunne and Black (1970a, 1970b), whereby overland flow is restricted to near-channel wetlands, are more likely to be encountered. Under these circumstances, surface erosion due to overland flow will be restricted to lowland areas adjacent to stream courses. Headward erosion of tributary streams will occur by piping (Section 10.2) at the exit points of subsurface seepage paths. The locations of these points of seepage are controlled in large part by the subsurface distribution of hydraulic conductivity. In this indirect way, subsurface stratigraphy exerts a strong influence on the density and pattern of the drainage network that develops in such a watershed. In summary, the relative positions of the saturated wetlands, variable source areas, and subsurface seepages that control the nature of the erosive processes on a hillslope in humid climates are a direct reflection of the subsurface saturated-unsaturated hydrogeologic regime.

Fluvial Processes

The classic approach to the analysis of bedload transport in streams completely neglects the effect of seepage forces in the bed. It is well established that river beds are either losing or gaining water in terms of subsurface flow, but it is not clear whether or not the seepage forces created by these flows play a significant role in streambed processes and the evolution of river morphology. This question has been addressed in a paper by Harrison and Clayton (1970), but their results are somewhat equivocal.

The inspiration for their study was a set of observations on an Alaskan stream in which the authors noticed striking contrasts between those portions of the stream accepting influent groundwater seepage and those portions losing water by effluent seepage. The gaining reach of the stream was transporting pebbles and cobbles as large as a few inches, whereas the losing reach was transporting sediment no larger than silt or very fine sand. The *competence* of the gaining reach, defined as the maximum size of particle that will undergo incipient motion at a given stream velocity, was 500 times greater than that of the losing reach. In that this variation in competence could not be explained by differences in stream velocity, channel slope, or bank sediment, Harrison and Clayton concluded that differences in the seepage gradients in the streambed were responsible for the great increase in competence of the gaining reach. This conclusion seemed logical, in that upward seepage in the gaining reach should buoy up the grains in the streambed, reducing their effective density and allowing them to be transported at velocities much lower than normal.

To test this hypothesis, a laboratory study was initiated. Results of the experiments, contrary to expectation, showed that seepage gradients had little influence on competence. The only effect confirmed in the laboratory experiments concerned downward seepage in channels with a large suspended sediment load. Under these conditions, a mud seal tended to form on the streambed. This mud seal discouraged the entrainment of bed material in the sealed area. In retrospect, the authors concluded that the Alaskan field observations might well be explained by this mechanism.

Vaux (1968) carried out a study of the interactions between streamflow and groundwater flow in alluvial streambed deposits in a completely different context. His interest centered on the rate of interchange between stream water and groundwater as it affects the oxygen supply in salmon spawning grounds. He utilized an analog groundwater model to assess the controlling features of the system.

Glacial Processes

An understanding of glacial landforms is best achieved through an examination of the mechanisms of erosion and sedimentation that accompany the advance and retreat of glaciers and continental ice sheets. It is now widely recognized (Weertman, 1972; Boulton, 1975) that the occurrence of pore water in the soils and rocks that underlie an ice sheet exerts an important influence on its rate of movement and on its erosive power. The existence of water at the base of a glacier is a consequence of the thermal regime that exists there. Heat, sufficient to melt the basal ice, is produced by the upward geothermal gradient and by the frictional generation of heat due to sliding.

Let us consider the flow of glacier ice across saturated, permeable rock. The movement of glacier ice involves yet another application of Terzaghi's concept of effective stress as presented in Section 2.9. High pore pressures lead to reduced effective stress at the ice-rock boundary and rapid rates of advance. Low pore pressures lead to increased effective stress and slow rates of advance. Similar mechanisms have been considered in the application of Mohr-Coulomb failure theory to the analysis of landslides (Section 10.1) and in the Hubbert-Rubey theory of overthrust faulting (Section 11.1).

Glacial erosion occurs both by *abrasion* and by *quarrying*. Abrasion of surfical bedrock by sliding ice is caused by the grinding action of glacial debris that becomes embedded in the sole of the glacier. Its presence there is evidence of the quarrying abilities of flowing ice to pluck material from jointed rock and unconsolidated sediments at other points along its travel path. In areas where permeable subglacial units exist, the fluid pressures in these layers can exert considerable influence on both of these erosive mechanisms. Boulton (1974, 1975) provides a quantitative analysis of the role of subglacial water in both abrasion and quarrying.

Clayton and Moran (1974) have presented a glacial-process model that places the glacial-erosion regime of a continental ice sheet in the context of the relationships between glacial flow, heat flow, and groundwater flow. Consider an ice sheet moving across a permeable geologic unit (Figure 11.9). Well back from the margins, where the flow of ice converges toward the glacier base, free water rather than permafrost should be present at the base, and pore-water pressures may be high. Because in this zone the glacier is not frozen to its base, sliding occurs and abrasion is the only mode of erosion. Near the margins of the glacier, on the other hand, ice flow diverges from the base, pore-water pressures are lower, the ice sheet is more likely to be frozen to its bed, and quarrying is the principal mode of erosion.

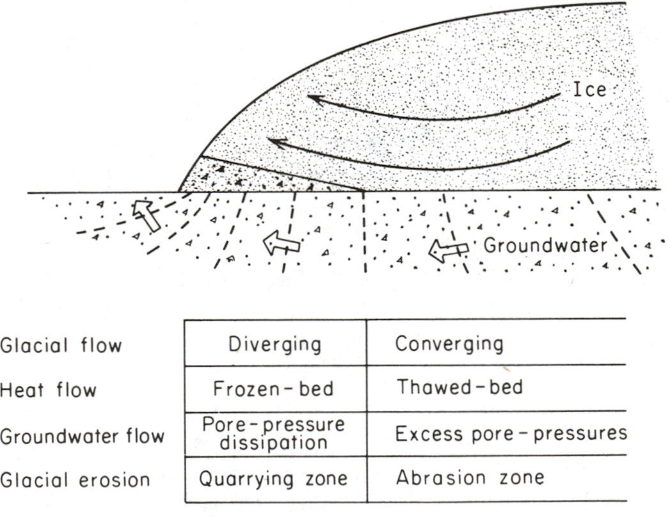

Glacial flow	Diverging	Converging
Heat flow	Frozen-bed	Thawed-bed
Groundwater flow	Pore-pressure dissipation	Excess pore-pressures
Glacial erosion	Quarrying zone	Abrasion zone

Figure 11.9 Interrelationships among glacial flow, heat flow, groundwater flow, and glacial erosion at the margin of a continental ice sheet (after Clayton and Moran, 1974).

Moran (1971) and Christiansen and Whitaker (1976) provide a detailed description of the various *glaciotectonic* structures that can develop in glacial deposits due to large-scale block inclusion and thrust faulting at the glacier margins. Among the mechanisms suggested for the generation of the high porewater pressures that are a necessary condition for the development of these features are (1) the advance of the ice sheet over the pinchout of a buried aquifer, (2) the advance of the ice sheet over debris containing buried ice blocks remaining from an earlier advance, (3) the consolidation of compressible sediments under the influence of the ice load, and (4) the rapid formation of a permafrost layer at the time of glaciation. These two latter concepts were first discussed by Mathews and MacKay (1960).

11.5 Groundwater and Economic Mineralization

The modern theories of groundwater hydrology have not as yet found widespread application in the field of mineral exploration. There is, however, great potential for their application on at least two fronts. In the first place, the genesis of many economic mineral deposits is closely bound to the physical and chemical processes that take place in the subsurface hydrologic environment. Much of the speculation into the modes of origin of various orebodies could benefit from hydrogeological analyses that utilize the flow-system approach of Chapter 6 and the hydrogeochemical concepts of Chapter 7. On a second front, it is clear that many anomalies uncovered during geochemical exploration could receive a more complete interpretation if groundwater flow theory were invoked in the search for the source. In the two subsections that follow, each of these questions will be briefly examined. There is a massive literature in the mineral exploration field and, apart from a few standard texts, our reference list is limited almost exclusively to those papers that invoke hydrogeological mechanisms or methodology.

Genesis of Economic Mineral Deposits

White (1968), Skinner and Barton (1973), and Park and MacDiarmid (1975) provide an excellent suite of recent references on economic mineral deposits and their genesis. Perusal of the ore-deposit classifications that they present makes it clear that there are very few types of deposits that do not in some way involve subsurface fluids. The direct influence of shallow groundwater is responsible for *supergene enrichment* in recharge areas, and for the deposition of *caliches* and *evaporites* in discharge areas. Residual weathering processes that lead to *laterites* also involve hydrologic processes.

By far the most important genetic mechanisms that involve subsurface flow are the ones that lead to *hydrothermal* deposits. White (1968) summarized the four-step process that leads to the generation of an ore deposit involving a hydrous fluid. First, there must be a source of ore constituents, usually dispersed in a magma or in sedimentary rocks; second, there must be solution of the ore minerals in the hydrous phase; third, a migration of the metal-bearing fluid; and fourth, selective precipitation of the ore constituents. White (1968) notes that very saline Na-Ca-Cl brines are potent solvents for such metals as copper and zinc. Proof that such brines exist lies in the fact that they are commonly encountered in deep petroleum exploration. There are three possible sources for these brines: magmatic, connate, and meteoric. *Connate* waters are those waters trapped in sediments at the time of their deposition. *Meteoric* waters are groundwaters that originated at the ground surface. Deeply circulating meteoric waters can attain sufficient salinity only through such secondary processes as the solution of evaporites or membrane concentration (Section 7.7). The precipitation of ore minerals is brought about by thermodynamic changes induced in the carrier brine under the influence of cooling, pressure reduction, or chemical reactions with host rocks or host fluids. The processes are best understood by the use of mass-transfer calculations of the type pioneered by Helgeson (1970).

With these basic introductory concepts in hand, we will now limit our further discussion to consideration of a specific type of ore deposit that has been widely ascribed to mechanisms involving groundwater flow: lead-zinc-fluorite-barite deposits of the Mississippi Valley type.

The Mississippi Valley lead-zinc deposits (White, 1968; Park and MacDiarmid, 1975) are strata-bound in nearly horizontal carbonate rocks lacking congruent tectonic structures that might control their localization. They occur at shallow depths in areas remote from igneous intrusives. The mineralogy is usually simple and nondiagnostic, with sphalerite, galena, fluorite, and barite as the principal ore minerals. A wide variety of origins has been proposed for this type of deposit, but White (1968) concludes that deposition from deeply circulating, heated, connate brine is the mechanism most compatible with the available temperature, salinity, and isotopic data.

Noble (1963) suggested that the circulation of connate water may have been controlled by diagenetic compaction of the source beds. Brines expelled from the sediments in this way would then be transported through transmissive zones (Figure 11.10), which became the loci of major ore concentrations. The brines may have contained metals in solution prior to burial as well as metals acquired during the diagenesis of the enclosing sediments. Noble's theory is attractive in that it provides an integrated mechanism for the leaching of metals from a dispersed

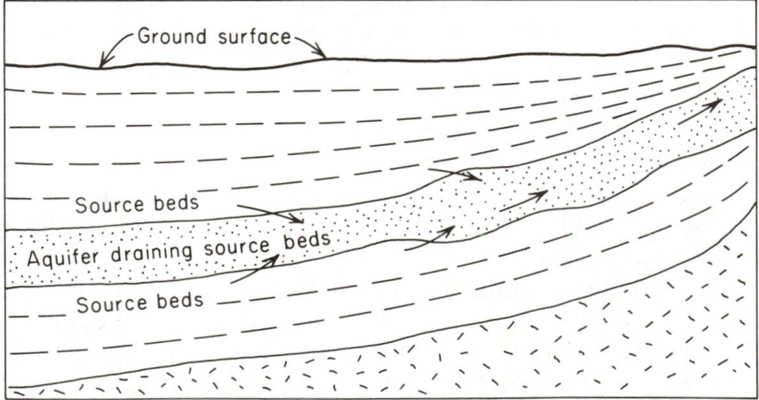

Figure 11.10 Idealized section showing aquifer transmitting mineralized brines of compaction from source beds (after Noble, 1963).

source, their migration through the geological system, and their concentration in high-permeability carbonate rocks.

McGinnis (1968) suggested a twist on Noble's theory whereby compaction of the source beds is accomplished by the loading provided by continental ice sheets. Under these circumstances, sedimentary brines would be forced to discharge near the margins of continental ice sheets in a manner similar to that described in the previous section in connection with Figure 11.9. The inspiration for McGinnis' explanation is the apparent clustering of Mississippi Valley type deposits along the southernmost extremity of continental glaciation and in the driftless area of Wisconsin.

Hitchon (1971, 1977) noted that oil pools and ore deposits in sedimentary rocks have several features in common. Both are aggregates of widely dispersed matter concentrated at specific sites where physical and chemical charges in aqueous carrier fluids caused unloading. He believes that the petroleum in the Zama-Rainbow oil field in northern Alberta and the Mississippi Valley type lead-zinc deposits of the nearby Pine Point ore body may have been sequentially unloaded from the same formation fluid. Both are located in the Middle Devonian Keg River Formation, and Pine Point is downstream from Zama-Rainbow in terms of the hydraulic head patterns that currently exist in the Keg River Formation. As an independent piece of evidence, Hitchon notes that petroleum is a common minor constituent in fluid inclusions from Mississippi Valley type lead-zinc deposits.

In closing this subsection it is worth noting, as has Hitchon (1976), that water is the fundamental fluid genetically relating all mineral deposits. It is the vehicle for the transportation of materials in solution and it takes part in the reactions that result in the original dissolution of the metals and their ultimate precipitation as ore. If the movement of subsurface water were to cease, chemical and physical equilibrium between the water and the rocks would eventually occur and there would be no further opportunities for the generation of mineral deposits. In this sense, the existence of subsurface flow is essential to the genesis of mineral deposits.

Implications for Geochemical Exploration

Hawkes and Webb (1962) define geochemical prospecting as any method of mineral exploration based on the systematic measurement of one or more chemical properties of any naturally occurring material. The material may be rock, soil, stream sediment, water, or vegetation. The objective of such a measurement program is the detection of abnormal chemical patterns, or *geochemical anomalies*, that might indicate the existence of an ore body.

Anomalous chemical patterns in groundwater or surface water are sometimes called hydrogeochemical anomalies. The most *mobile* metal elements, that is, the elements most easily dissolved and transported in water and therefore the most likely to produce hydrogeochemical anomalies, are copper, zinc, nickel, cobalt, and molybdenum (Bradshaw, 1975). Lead, silver, and tungsten are less mobile; gold and tin are virtually immobile. Because of the expense of drilling, groundwater is seldom sampled directly, but springs and seepage areas are widely used in geochemical exploration. Figure 11.11 shows the various types of geochemical anomalies that might be expected to develop in the vicinity of an ore body. Groundwater plays an important role in delivering metal ions to the hydrogeochemical concentration zones in seepage areas and in lake and stream sediments.

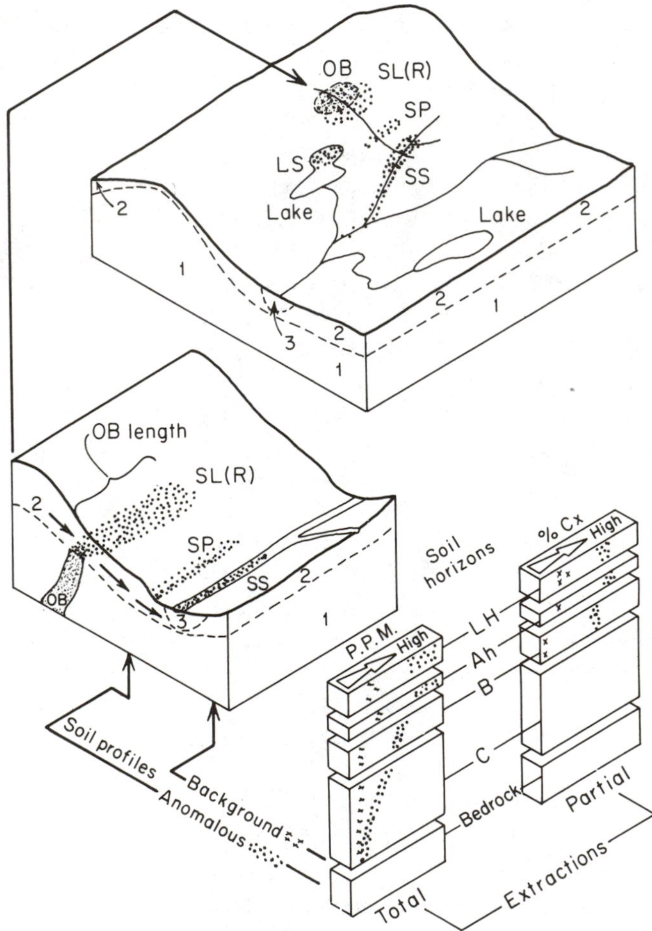

Figure 11.11 Schematic diagram showing the development of geochemical anomalies in an area where bedrock is overlain by a residual soil (after Bradshaw, 1975).
Anomaly types: SL (R), residual soil anomaly; SP, seepage anomaly; SS, stream sediment anomaly; LS, lake sediment anomaly. The density of dots indicates anomaly strength.
Geology: 1, bedrock; 2, residual soil; 3, recent alluvium.
Other: OB, ore body; PPM, parts per million; % Cx, cold extractable concentration; →, groundwater flow direction.

One of the most successful applications of spring sampling techniques is that described by de Geoffroy et al. (1967) in the upper Mississippi Valley lead-zinc district. They sampled 3766 springs over an area of 1066 km^2. An interpretation of the measurements indicated 56 zinc anomalies. Of these, 26 coincided with known zinc deposits, and drill testing of a small number of the remaining anomalies confirmed the presence of zinc ore in their vicinity. In the carbonate terrain of this area, surface-water sampling is ineffective because the heavy metals are quickly precipitated from groundwater within a short distance of its emergence into the surface environment. De Geoffroy et al. (1967) conclude that spring sampling is the most satisfactory geochemical method in the search for ore bodies of moderate size in carbonate rocks.

There have been other instances of successful groundwater-oriented geochemical exploration programs. Among the most interesting conclusions are those of Graham et al. (1975), who found that fluorine in groundwater can act as a guide to Pb-Zn-Ba-F mineralization, and Clarke and Kugler (1973), who advocate dissolved helium in groundwater as an indicator for uranium ore. On a negative note, Gosling et al. (1971) report that hydrogeochemical prospecting for gold in the Colorado Front Range is unpromising.

Hoag and Webber (1976) suggest that sulfate concentrations in groundwater, because they are indicative of the oxidation environment of the sulfides that produce them, can be used to estimate the depth of mineralization of possible ore bodies. They note that this information might help determine what types of further exploration would be most helpful in locating possible sulfide deposits.

In all this, the recent developments in physical and chemical hydrogeology that have been reviewed in this book, are extremely pertinent. The rates at which metals are taken into solution from ore bodies by passing groundwater are controlled by the principles introduced in Chapter 3 and discussed in Chapter 7. The diffusion, dispersion, and retardation processes that accompany their transport by the groundwater system are identical to those described in Chapter 9 in connection with groundwater contamination. Perhaps the most direct suggestion for the application of groundwater flow theory in geochemical exploration has come from R. E. Williams (1970). He suggests that initial hydrochemical sampling be confined to discharge areas of regional flow systems. Once a geochemical anomaly has been located, the groundwater flow paths that lead to it would be determined by hydrogeological field mapping and the mathematical modeling methods introduced in Chapter 6.

Suggested Readings

Clayton, L., and S. R. Moran. 1974. A glacial process-form model. *Glacial Geomorphology*, ed. D. R. Coates. State University of New York, Binghamton, N.Y., pp. 89–119.

DONALDSON, I. G. 1962. Temperature gradients in the upper layers of the earth's crust due to convective water flows. *J. Geophys. Res.*, 67, pp. 3449–3459.

HUBBERT, M. K. 1954. Entrapment of petroleum under hyrodynamic conditions. *Bull. Amer. Assoc. Petrol. Geol.*, 37, pp. 1954–2026.

HUBBERT, M. K., and W. W. RUBEY. 1959. Role of fluid pressures in mechanics of overthrust faulting: I. Mechanics of fluid-filled porous solids and its application to overthrust faulting. *Bull. Geol. Soc. Amer.*, 70, pp. 115–166.

THRAIKILL, J. 1968. Chemical and hydrologic factors in the excavation of limestone caves. *Bull. Geol. Soc. Amer.*, 79, pp. 19–46.

WHITE, D. E. 1968. Environments of generation of some base-metal ore deposits. *Econ. Geol.*, 63, pp. 301–335.

Appendices

Appendix I Elements of Fluid Mechanics

The analysis of groundwater flow requires an understanding of the elements of fluid mechanics. Albertson and Simons (1964) provide a useful short review; Streeter (1962) and Vennard (1961) are standard texts. Our purpose here is simply to introduce the basic fluid properties: mass density, weight density, compressibility, and viscosity; and to examine the concepts of fluid pressure and pressure head.

An examination of the principles of fluid mechanics must begin with a review of the mechanics of materials in general. Table Al.1 provides a list of the basic mechanical properties of matter, together with their dimensions and units in the SI metric system. The SI system has as its basic dimensions: mass, length, and time; with basic SI units: kilogram (kg), meter (m), and second (s). All other properties are measured in units that are derived from this basic set. Some of these properties are so widely encountered, and their basic dimensions so complex, that special SI names have been coined for their derived units. As noted on Table Al.1, force and weight are measured in newtons (N), pressure and stress in N/m^2 or pascals (Pa), and work and energy in joules (J).

Table A1.1 Definitions, Dimensions, and SI Units for Basic Mechanical Properties

Property	Symbol	Definition	SI unit	SI symbol	Dimension of unit	
					Derived	Basic
Mass	M		kilogram	kg		kg
Length	l		meter	m		m
Time	t		second	s		s
Area	A	$A = l^2$				m^2
Volume	V	$V = l^3$				m^3
Velocity	v	$v = l/t$				m/s
Acceleration	a	$a = l/t^2$				m/s^2
Force	F	$F = Ma$	newton	N	N	$kg{\cdot}m/s^2$
Weight	w	$w = Mg$	newton	N	N	$kg{\cdot}m/s^2$
Pressure	p	$p = F/A$	pascal	Pa	N/m^2	$kg/m{\cdot}s^2$
Work	W	$W = Fl$	joule	J	N·m	$kg{\cdot}m^2/s^2$
Energy		Work done	joule	J	N·m	$kg{\cdot}m^2/s^2$
Mass density	ρ	$\rho = M/V$				kg/m^3
Weight density	γ	$\gamma = w/V$			N/m^3	$kg/m^2{\cdot}s^2$
Stress	σ, τ	Internal response to external p	pascal	Pa	N/m^2	$kg/m{\cdot}s^2$
Strain	ϵ	$\epsilon = \Delta V/V$				Dimensionless
Young's modulus	E	Hooke's law			N/m^2	$kg/m{\cdot}s^2$

Table A1.2 provides an SI analysis of certain fluid properties and groundwater terms that occur in this text. Each is more fully described in Chapter 2.

Much of the technology associated with groundwater resource development in North America is still married to the FPS (foot-pound-second) system of units.

Table A1.2 Definitions, Dimensions, and SI Units for Fluid Properties and Groundwater Terms

Property	Symbol	Definition	SI unit	SI symbol	Dimensions of unit	
					Derived	Basic
Volume	V	$V = l^3$	liter (= $m^3 \times 10^{-3}$)	ℓ	ℓ	m^3
Discharge	Q	$Q = l^3/t$			ℓ/s	m^3/s
Fluid pressure	p	$p = F/A$	pascal	Pa	N/m^2	$kg/m \cdot s^2$
Head	h					m
Mass density	ρ	$\rho = M/V$				kg/m^3
Dynamic viscosity	μ	Newton's law	centipoise (= $N \cdot s/m^2 \times 10^{-3}$)	cP	cP, $N \cdot s/m^2$	$kg/m \cdot s$
Kinematic viscosity	ν	$\nu = \mu/\rho$	centistoke (= $m^2/s \times 10^{-6}$)	cSt	cSt	m^2/s
Compressibility	α, β	$\alpha = 1/E$			m^2/N	$m \cdot s^2/kg$
Hydraulic conductivity	K	Darcy's law			cm/s	m/s
Permeability	k	$k = K\mu/\rho g$			cm^2	m^2
Porosity	n					Dimensionless
Specific storage	S_s	$S_s = \rho g(\alpha + n\beta)$				1/m
Storativity	S	$S = S_s b$*				Dimensionless
Transmissivity	T	$T = Kb$*				m^2/s

*b, thickness of confined aquifer (see Section 2.10).

Table A1.3 Conversion Factors FPS (foot-pound-second) System of Units to SI Units

	Multiply	By	To obtain
Length	ft	3.048×10^{-1}	m
	ft	3.048×10	cm
	ft	3.048×10^{-4}	km
	mile	1.609×10^{3}	m
	mile	1.609	km
Area	ft^2	9.290×10^{-2}	m^2
	mi^2	2.590	km^2
	acre	4.047×10^{3}	m^2
	acre	4.047×10^{-3}	km^2
Volume	ft^3	2.832×10^{-2}	m^3
	U.S. gal	3.785×10^{-3}	m^3
	U.K. gal	4.546×10^{-3}	m^3
	ft^3	2.832×10	ℓ
	U.S. gal	3.785	ℓ
	U.K. gal	4.546	ℓ
Velocity	ft/s	3.048×10^{-1}	m/s
	ft/s	3.048×10	cm/s
	mi/h	4.470×10^{-1}	m/s
	mi/h	1.609	km/h
Acceleration	ft/s^2	3.048×10^{-1}	m/s^2
Mass	lb_m*	4.536×10^{-1}	kg
	slug*	1.459×10	kg
	ton	1.016×10^{3}	kg
Force and weight	lb_f*	4.448	N
	poundal	1.383×10^{-1}	N
Pressure and stress	psi	6.895×10^{3}	Pa or N/m^2
	lb_f/ft^2	4.788×10	Pa
	$poundal/ft^2$	1.488	Pa
	atm	1.013×10^{5}	Pa
	in Hg	3.386×10^{3}	Pa
	mb	1.000×10^{2}	Pa
Work and energy	ft-lb_f	1.356	J
	ft-poundal	4.214×10^{-2}	J
	Btu	1.055×10^{-3}	J
	calorie	4.187	J
Mass density	lb/ft^3	1.602×10	kg/m^3
	$slug/ft^3$	5.154×10^{2}	kg/m^3
Weight density	lb_f/ft^3	1.571×10^{2}	N/m^3
Discharge	ft^3/s	2.832×10^{-2}	m^3/s
	ft^3/s	2.832×10	ℓ/s
	U.S. gal/min	6.309×10^{-5}	m^3/s
	U.K. gal/min	7.576×10^{-5}	m^3/s
	U.S. gal/min	6.309×10^{-2}	ℓ/s
	U.K. gal/min	7.576×10^{-2}	ℓ/s
Hydraulic conductivity	ft/s	3.048×10^{-1}	m/s
(see also Table 2.3.)	U.S. $gal/day/ft^2$	4.720×10^{-7}	m/s
Transmissivity	ft^2/s	9.290×10^{-2}	m^2/s
	U.S. gal/day/ft	1.438×10^{-7}	m^2/s

*A body whose mass is 1 lb mass (lb_m) has a weight of 1 lb force (lb_f). 1 lb_f is the force required to accelerate a body of 1 lb_m to an acceleration of $g = 32.2$ ft/s^2. A slug is the unit of mass which, when acted upon by a force of 1 lb_f, acquires an acceleration of 1 ft/s^2.

Table A1.3 provides a set of conversion factors for converting FPS units to SI units.

The *mass density* (or simply, *density*) ρ of a fluid is defined as its mass per unit volume (Table A1.1). The *weight density* (or *specific weight*, or *unit weight*) γ of a fluid is defined as its weight per unit volume. The two parameters are related by

$$\gamma = \rho g \tag{A1.1}$$

For water, $\rho = 1.0\ \text{g/cm}^3 = 1000\ \text{kg/m}^3$; $\gamma = 9.8 \times 10^3\ \text{N/m}^3$. In the FPS system, $\gamma = 62.4\ \text{lb}_f/\text{ft}^3$.

The *specific gravity G* of any material is the ratio of its density (or specific weight) to that of water. For water, $G = 1.0$; for most soils and rocks, $G \approx 2.65$.

The *viscosity* of a fluid is the property that allows fluids to resist relative motion and shear deformation during flow. The more viscous the fluid, the greater the shear stress at any given velocity gradient. According to Newton's law of viscosity,

$$\tau = \mu \frac{dv}{dy} \tag{A1.2}$$

where τ is the shear stress, dv/dy the velocity gradient, and μ the viscosity, or *dynamic viscosity*. The *kinematic viscosity* ν is given by

$$\nu = \frac{\mu}{\rho} \tag{A1.3}$$

where ρ is the fluid density.

The *compressibility* of a fluid reflects its stress-strain properties. *Stress* is the internal response of a material to an external pressure. For fluids, stress is imparted through the fluid pressure. *Strain* is a measure of the linear or volumetric deformation of a stressed material. For fluids, it takes the form of reduced volume (and increased density) under increasing fluid pressures. The compressibility of water β is fully discussed in Section 2.9. It is defined by Eq. (2.44).

The density, viscosity, and compressibility of water are functions of temperature and pressure (Dorsey, 1940; Weast, 1972). In general, their variation is not great, and for the range of pressures and temperatures that occur in most groundwater applications, it is common to consider them as constants. At 15.5°C, $\rho = 1.0\ \text{g/cm}^3$, $\mu = 1.124$ cP, and $\beta = 4.4 \times 10^{-10}\ \text{m}^2/\text{N}$.

The *fluid pressure p* at any point in a standing body of water is the force per unit area which acts at that point. Under hydrostatic conditions, the fluid pressure at a point reflects the weight of the column of water overlying a unit cross-sectional area around the point. It is possible to express pressure relative to absolute zero pressure, but more commonly it is expressed relative to atmospheric pressure. In the latter case it is called *gage pressure*, as this is the pressure reading that is obtained on gages zeroed to the atmosphere.

The *pressure head* ψ at a point in a fluid is the height that a column of water would attain in a manometer placed at that point. In a standing body of water, ψ is equal to the depth of the point of measurement below the surface. If p is expressed as a gage pressure, ψ is defined by the relationship

$$p = \rho g \psi = \gamma \psi \tag{A1.4}$$

In effect, the pressure head ψ is a measurement of the fluid pressure p.

Fluid pressures are also developed in groundwater as it flows through porous geological formations and soils. In Section 2.2, the elements of fluid mechanics presented in this appendix are applied in the development of groundwater flow theory.

Appendix II Equation of Flow for Transient Flow Through Deforming Saturated Media

A rigorous development of the equation of flow for transient flow in saturated media must recognize the fact that transient changes in fluid pressure lead to deformations in the granular skeleton of a porous medium, and these deformations imply that the medium, as well as the water, is in motion. This realization creates the need for two refinements to the classical derivation presented by Jacob (1940) and followed in Section 2.11 of this text. First, as recognized by Biot (1955), it is necessary to cast Darcy's law in terms of the relative velocity of fluid to grains. And second, as recognized by Cooper (1966), it is necessary to consider the conservation of mass for the medium as well as for the fluid in the elemental control volume. One can develop the continuity relationships in one of three ways: (1) by considering a deforming elemental volume in deforming coordinates, (2) by considering a deforming elemental volume in fixed coordinates, or (3) by considering a fixed elemental volume in fixed coordinates. Following Gambolati and Freeze (1973), we will use a fixed elemental volume in fixed coordinates. The approach requires the use of vector notation and the material derivative (total derivative, substantial derivative). If these concepts are not familiar, Aris (1962) and Wills (1958) provide introductory treatments. The development will be presented here for a homogeneous, isotropic medium with hydraulic conductivity K, porosity n, and vertical compressibility α. The same approach is easily adapted to heterogeneous and anisotropic media.

In vector notation, the three-dimensional form of Darcy's law [Eq. (2.34)] is

$$\vec{v} = -K\nabla h \tag{A2.1}$$

where $\vec{v} = (v_x, v_y, v_z)$ is the relative velocity of fluid to grains, and $\nabla h = (\partial h/\partial x, \partial h/\partial y, \partial h/\partial z)$ is the hydraulic gradient. We can expand the vector $\vec{v}$ as

$$\vec{v} = n(\vec{v}_w - \vec{v}_s) \tag{A2.2}$$

where $\vec{v}_w$ is the fluid velocity and $\vec{v}_s$ the velocity of the deforming medium.

The equation of state for the water [Eq. (2.47)] is

$$\rho = \rho_0 e^{\beta p} \tag{A2.3}$$

and that for the soil grains, which are incompressible, is

$$\rho_s = \text{constant} \tag{A2.4}$$

The equation of continuity for the water is

$$-\nabla \cdot [n\rho\vec{v}_w] = \frac{\partial}{\partial t}[n\rho] \tag{A2.5}$$

and that for the soil is

$$-\nabla \cdot [(1-n)\rho_s \vec{v}_s] = \frac{\partial}{\partial t}[(1-n)\rho_s] \tag{A2.6}$$

In these equations $\nabla \cdot$ is the divergence operator:

$$\nabla \cdot = \frac{\partial}{\partial x} + \frac{\partial}{\partial y} + \frac{\partial}{\partial z}$$

Expanding Eq. (A2.5), we arrive at

$$-\rho \nabla \cdot (n\vec{v}_w) - n\vec{v}_w \cdot \nabla \rho = n\frac{\partial \rho}{\partial t} + \rho\frac{\partial n}{\partial t} \tag{A2.7}$$

If we cancel ρ_s from Eq. (A2.6) and rearrange that equation, we obtain an expression for $\partial n/\partial t$. This can be substituted in Eq. (A2.7) together with an expression for $n\vec{v}_w$ obtained from Eq. (A2.2). After dividing through by ρ and rearranging, Eq. (A2.7) becomes

$$-\nabla \cdot \vec{v} - \left(\frac{\vec{v}}{\rho}\right) \cdot \nabla \rho = \left(\frac{n}{\rho}\right)\left(\frac{\partial \rho}{\partial t} + \vec{v}_s \cdot \nabla \rho\right) + \nabla \cdot \vec{v}_s \tag{A2.8}$$

If we use the material derivative, $D/Dt = \partial/\partial t + \vec{v}_s \cdot \nabla$, Eq. (A2.8) can be written as

$$-\nabla \cdot \vec{v} - \left(\frac{\vec{v}}{\rho}\right) \cdot \nabla \rho = \frac{n}{\rho}\frac{D\rho}{Dt} + \nabla \cdot \vec{v}_s \tag{A2.9}$$

The first term on the right-hand side of Eq. (A2.9) can be related to the compressibility of water β by the relation

$$\frac{D\rho}{Dt} = \rho\beta\frac{Dp}{Dt} \tag{A2.10}$$

The material derivative on the right-hand side of Eq. (A2.10) can be replaced by a partial derivative only if the following inequality is satisfied:

$$\vec{v}_s \cdot \nabla p \ll \frac{\partial p}{\partial t} \tag{A2.11}$$

Let us further assume, on the left-hand side of Eq. (A2.9), that

$$\left(\frac{\vec{v}}{\rho}\right) \cdot \nabla \rho \ll \nabla \cdot \vec{v} \tag{A2.12}$$

Then, substituting Eqs. (A2.1) and (A2.10) in Eq. (A2.9) yields

$$\nabla \cdot (K\nabla h) = n\beta\frac{\partial p}{\partial t} + \nabla \cdot \vec{v}_s \tag{A2.13}$$

In a three-dimensional stress field, the grain velocity vector $\vec{v}_s = (v_{sx}, v_{sy}, v_{sz})$ is related to the deformation (or soil displacement) vector $\vec{u} = (u_x, u_y, u_z)$ by

$$\vec{v}_s = \frac{D\vec{u}}{Dt} \tag{A2.14}$$

In a one-dimensional vertical stress field,

$$v_{sx} = v_{sy} = u_x = u_y = 0 \tag{A2.15}$$

If the conditions of Eq. (A2.15) are satisfied, the final term of Eq. (A2.13) can be expanded (Cooper, 1966; Gambolati and Freeze, 1973; Gambolati, 1973a) as

$$\nabla \cdot \vec{v}_s = \frac{\partial v_{sz}}{\partial z} = \frac{\partial}{\partial z}\left(\frac{Du_z}{Dt}\right) = \frac{D}{Dt}\left(\frac{\partial u_z}{\partial z}\right) = \alpha \frac{Dp}{Dt} \tag{A2.16}$$

where α is the vertical compressibility of the porous medium. The change of derivative around the central equality is valid for a position vector but not in general. The material derivative in the right-hand expression of Eq. (A2.16) can be replaced by the partial derivative if Eq. (A2.11) is satisfied. In that case, Eq. (A2.13) becomes

$$\nabla \cdot (K\, \nabla h) = n\beta \frac{\partial p}{\partial t} + \alpha \frac{\partial p}{\partial t} \tag{A2.17}$$

Since $p = \rho g(h - z)$ and K is a constant, Eq. (A2.17) simplifies to

$$\nabla^2 h = \frac{\rho g(\alpha + n\beta)}{K} \frac{\partial h}{\partial t} \tag{A2.18}$$

Or, recalling that $S_s = \rho g(\alpha + n\beta)$ and expanding the vector notation,

$$\frac{\partial^2 h}{\partial x^2} + \frac{\partial^2 h}{\partial y^2} + \frac{\partial^2 h}{\partial z^2} = \frac{S_s}{K} \frac{\partial h}{\partial t} \tag{A2.19}$$

Equation (A2.19) is identical to Eq. (2.75) developed by Jacob (1940). The more rigorous development makes it clear that the validity of the classical equation of flow rests on the satisfaction of the inequalities of Eqs. (A2.11) and (A2.12) and the stress condition of Eq. (A2.15). It is unlikely that these conditions are always satisfied. Gambolati (1973b) shows that where the rate of consolidation $\vec{v}_s$ exceeds the rate of percolating fluid $\vec{v}_w$, as it can in thick clay layers of low permeability and high compressibility, the inequalities may not be satisfied. With regard to the stress condition, the $\nabla \cdot \vec{v}_s$ term at the end of Eq. (A2.13) is really the protruding tip of an iceberg that relates the three-dimensional flow field to the three-dimensional stress field. Biot (1941, 1955) first exposed the interrelationships and Verruijt (1969) provides a clear derivation. Schiffman et al. (1969) provides a comparison of the classical and Biot approaches, and Gambolati (1974) analyses the range of validity of the classical equation of flow.

Appendix III Example of an Analytical Solution to a Boundary-Value Problem

Consider the simple groundwater flow problem shown in Figure 2.25(a). The equation of flow for steady-state saturated flow in the xy plane is

$$\frac{\partial^2 h}{\partial x^2} + \frac{\partial^2 h}{\partial y^2} = 0 \tag{A3.1}$$

The mathematical statement of the boundary conditions is

$$\frac{\partial h}{\partial y} = 0 \quad \text{on} \quad y = 0 \text{ and } y = y_L \tag{A3.2}$$

$$h = h_0 \quad \text{on} \quad x = 0 \tag{A3.3}$$

$$h = h_1 \quad \text{on} \quad x = x_L \tag{A3.4}$$

We will solve for $h(x, y)$ using the separation-of-variables technique.

Assume that the solution is a product solution of the form

$$h(x, y) = X(x) \cdot Y(y) \tag{A3.5}$$

Equation (A3.1) then becomes

$$Y\frac{\partial^2 X}{\partial x^2} + X\frac{\partial^2 Y}{\partial y^2} = 0 \tag{A3.6}$$

Dividing through by XY yields

$$\frac{1}{X}\frac{\partial^2 X}{\partial x^2} = -\frac{1}{Y}\frac{\partial^2 Y}{\partial y^2} \tag{A3.7}$$

The left-hand side is independent of y. Therefore the right-hand side, despite its appearance, must also be independent of y, since it is identically equal to the left-hand side. Similarly, the right-hand side is independent of x, and so too is the left-hand side. If both sides are independent of x and y, each side must equal a constant. Therefore,

$$\frac{1}{X}\frac{\partial^2 X}{\partial x^2} = G \quad \text{and} \quad \frac{1}{Y}\frac{\partial^2 Y}{\partial y^2} = G \tag{A3.8}$$

The constant G may be positive, negative, or zero. All three cases lead to product solutions, but only the case $G = 0$ leads to a solution that is physically meaningful for this problem. Equations (A3.8) become

$$\frac{1}{X}\frac{\partial^2 X}{\partial x^2} = 0 \quad \text{and} \quad \frac{1}{Y}\frac{\partial^2 Y}{\partial y^2} = 0 \tag{A3.9}$$

These are ordinary differential equations whose solutions are well known:

$$X = Ax + B \quad \text{and} \quad Y = Cy + D \tag{A3.10}$$

The product solution Eq. (A3.5) becomes

$$h(x, y) = (Ax + B)(Cy + D) \tag{A3.11}$$

We can evaluate the coefficients A, B, C, and D by invoking the boundary conditions. Differentiating Eq. (A3.11) with respect to y yields

$$\frac{\partial h}{\partial y} = (Ax + B)C \tag{A3.12}$$

and invocation of Eq. (A3.2) implies that $C = 0$. From Eq. (A3.11), we are left with

$$h(x, y) = (Ax + B)D = Ex + F \tag{A3.13}$$

Invoking the boundary conditions of Eqs. (A3.3) and (A3.4) yields $F = h_0$ and $E = -(h_0 - h_1)/x_L$. The solution is, therefore,

$$h(x, y) = h_0 - (h_0 - h_1)\frac{x}{x_L} \tag{A3.14}$$

This equation is identical to Eq. (2.81), presented in Section 2.11 without proof.

It is immediately clear that Eq. (A3.14) satisfies the boundary conditions of Eqs. (A3.3) and (A3.4). Differentiation with respect to y yields zero in satisfaction of Eq. (A3.2). Differentiation with respect to x twice also yields zero, so the solution Eq. (A3.14) satisfies the equation of flow [Eq. (A3.1)].

Appendix IV Debye-Hückel Equation and Kielland Table for Ion-Activity Coefficients

Debye-Hückel expression for individual ion activities:

$$\log \gamma = \frac{-Az^2\sqrt{I}}{1 + \mathring{a}B\sqrt{I}}$$

Values of the Ion-Size Parameter $\mathring{a}$ for Common Ions Encountered in Natural Water:

$\mathring{a} \times 10^8$	Ion
2.5	NH_4^+
3.0	K^+, Cl^-, NO_3^-
3.5	OH^-, HS^-, MnO_4^-, F^-
4.0	SO_4^{2-}, PO_4^{3-}, HPO_4^{2-}
4.0–4.5	Na^+, HCO_3^-, $H_2PO_4^-$, HSO_3^-
4.5	CO_3^{2-}, SO_3^{2-}
5	Sr^{2+}, Ba^{2+}, S^{2-}
6	Ca^{2+}, Fe^{2+}, Mn^{2+}
8	Mg^{2+}
9	H^+, Al^{3+}, Fe^{3+}

Parameters A and B at 1 Bar

Temperature (°C)	A	B ($\times 10^{-8}$)
0	0.4883	0.3241
5	0.4921	0.3249
10	0.4960	0.3258
15	0.5000	0.3262
20	0.5042	0.3273
25	0.5085	0.3281
30	0.5130	0.3290
35	0.5175	0.3297
40	0.5221	0.3305
50	0.5319	0.3321
60	0.5425	0.3338

Kielland Table of Ion-Activity Coefficients at 25°C Arranged by the Size of Ions (Based on Debye-Hückel Equation)

Charge	Size,* å	Ions	$I =$ 0.0005	0.001	0.0025	0.005	0.01	0.025	0.05	0.1
1	2.5	Rb^+, Cs^+, Ag^+, NH_4^+, Tl^+	0.975	0.964	0.945	0.924	0.898	0.85	0.80	0.75
	3	K^+, Cl^-, Br^-, I^-, CN^-, NO_3^-, NO_2^-, OH^-, F^-, ClO_4^-	0.975	0.964	0.945	0.925	0.899	0.85	0.805	0.755
	4	Na^+, IO_3^-, HCO_3^-, HSO_3^-, $H_2PO_4^-$, ClO_2^-, $C_2H_3O_2^-$	0.975	0.964	0.947	0.928	0.902	0.86	0.82	0.775
	6	Li^+, $C_6H_5COO^-$	0.975	0.965	0.948	0.929	0.907	0.87	0.835	0.80
	9	H^+	0.975	0.967	0.950	0.933	0.914	0.88	0.86	0.83
2	4.5	Pb^{2+}, Hg_2^{2+}, SO_4^{2-}, CrO_4^{2-}, CO_3^{2-}, SO_3^{2-}, $C_2O_4^{2-}$, $S_2O_3^{2-}$, H citrate^{2-}	0.903	0.867	0.805	0.742	0.665	0.55	0.455	0.37
	5	Sr^{2+}, Ba^{2+}, Cd^{2+}, Hg^{2+}, S^{2-}, WO_4^{2-}	0.903	0.868	0.805	0.744	0.67	0.555	0.465	0.38
	6	Ca^{2+}, Cu^{2+}, Zn^{2+}, Sn^{2+}, Mn^{2+}, Fe^{2+}, Ni^{2+}, Co^{2+}, phthalate^{2-}	0.905	0.870	0.809	0.749	0.675	0.57	0.485	0.405
	8	Mg^{2+}, Be^{2+}	0.906	0.872	0.813	0.755	0.69	0.595	0.52	0.45
3	4	PO_4^{3-}, $Fe(CN)_6^{3-}$, $Cr(NH_3)_6^{3+}$	0.796	0.725	0.612	0.505	0.395	0.25	0.16	0.095
	9	Al^{3+}, Fe^{3+}, Cr^{3+}, Sc^{3+}, In^{3+}, and rare earths	0.802	0.738	0.632	0.54	0.445	0.325	0.245	0.18

*Note that these sizes are rounded values for the *effective* size in water solution and are not the size of the simple ions, unhydrated. For a more detailed discussion, see the original paper from which these values are taken [J. Kielland, *J. Amer. Chem. Soc.*, 59, 1675 (1937)].

References

Berner, R. A. 1971. *Principles of Chemical Sedimentology*. McGraw-Hill, New York, 240 pp.

Klotz, I. M. 1950. *Chemical Thermodynamics*. Prentice-Hall, Englewood Cliffs, N. J., 369 pp.

Manov, G. G., R. G. Bates, W. J. Hamer, and S. F. Acree. 1943. Values of the constants in the Debye-Hückel equation for activity coefficients. *J. Amer. Chem. Soc.*, 65, pp. 1765–1767.

Appendix V Complementary Error Function (erfc)

$$\text{erf}\,(\beta) = \frac{2}{\sqrt{\mu}} \int_0^{\beta} e^{-\epsilon^2}\, d\epsilon$$

$$\text{erf}\,(-\beta) = -\text{erf}\,\beta$$

$$\text{erfc}\,(\beta) = 1 - \text{erf}\,(\beta)$$

β	erf (β)	erfc (β)
0	0	1.0
0.05	0.056372	0.943628
0.1	0.112463	0.887537
0.15	0.167996	0.832004
0.2	0.222703	0.777297
0.25	0.276326	0.723674
0.3	0.328627	0.671373
0.35	0.379382	0.620618
0.4	0.428392	0.571608
0.45	0.475482	0.524518
0.5	0.520500	0.479500
0.55	0.563323	0.436677
0.6	0.603856	0.396144
0.65	0.642029	0.357971
0.7	0.677801	0.322199
0.75	0.711156	0.288844
0.8	0.742101	0.257899
0.85	0.770668	0.229332
0.9	0.796908	0.203092
0.95	0.820891	0.179109
1.0	0.842701	0.157299
1.1	0.880205	0.119795
1.2	0.910314	0.089686
1.3	0.934008	0.065992
1.4	0.952285	0.047715
1.5	0.966105	0.033895
1.6	0.976348	0.023652
1.7	0.983790	0.016210
1.8	0.989091	0.010909
1.9	0.992790	0.007210
2.0	0.995322	0.004678
2.1	0.997021	0.002979
2.2	0.998137	0.001863
2.3	0.998857	0.001143
2.4	0.999311	0.000689
2.5	0.999593	0.000407
2.6	0.999764	0.000236
2.7	0.999866	0.000134
2.8	0.999925	0.000075
2.9	0.999959	0.000041
3.0	0.999978	0.000022

Appendix VI Development of Finite-Difference Equation for Steady-State Flow in a Homogeneous, Isotropic Medium

The partial differential equation that describes steady-state flow in a two-dimensional, homogeneous, isotropic region of flow (Section 2.11) is Laplace's equation:

$$\frac{\partial^2 h}{\partial x^2} + \frac{\partial^2 h}{\partial y^2} = 0 \tag{A6.1}$$

To find the finite-difference equation for an interior node in the nodal grid used to discretize the region of flow, we must replace the second-order partial derivatives in Eq. (A6.1) by differences. Let us consider the first term of the equation first. Recall that the definition of the partial derivative with respect to x of a function of two variables $h(x, z)$ is

$$\frac{\partial h}{\partial x} = \lim_{\Delta x \to 0} \frac{h(x + \Delta x, z) - h(x, z)}{\Delta x} \tag{A6.2}$$

On a digital computer it is impossible to take the limit as $\Delta x \longrightarrow 0$, but it is possible to approximate the limit by assigning to Δx some arbitrarily small value; in fact, we can do so by designing a nodal network with a mesh spacing of Δx.

For any value of z, say z_0, we can expand $h(x, z_0)$ in a Taylor's expansion about the point (x_0, z_0) as follows:

$$h(x, z_0) = h(x_0, z_0) + (x - x_0)\frac{\partial h}{\partial x}(x_0, z_0) + \frac{(x - x_0)^2}{2}\frac{\partial^2 h}{\partial x^2}(x_0, z_0) + \dots \tag{A6.3}$$

If we let $x = x_0 + \Delta x$ (this is known as a *forward difference*) and abandon all the terms of order greater than unity, we can approximate $\partial h/\partial x$ by

$$\frac{\partial h}{\partial x}(x_0, z_0) = \frac{h(x_0 + \Delta x, z_0) - h(x_0, z_0)}{\Delta x} \tag{A6.4}$$

The abandoned terms of the Taylor expansion represent the truncation error in the finite-difference approximation.

We can obtain a similar expression to Eq. (A6.4) by substituting the *backward difference*, $x = x_0 - \Delta x$, into Eq. (A6.3). This yields

$$\frac{\partial h}{\partial x}(x_0, z_0) = \frac{h(x_0, z_0) - h(x_0 - \Delta x, z_0)}{\Delta x} \tag{A6.5}$$

To obtain the approximation for $\partial^2 h/\partial x^2$, we write the difference equation in terms of $\partial h/\partial x$ using a forward-difference expression:

$$\frac{\partial^2 h}{\partial x^2}(x_0, z_0) = \frac{\frac{\partial h}{\partial x}(x_0 + \Delta x, z_0) - \frac{\partial h}{\partial x}(x_0, z_0)}{\Delta x} \tag{A6.6}$$

and substitute the backward-difference expression of Eq. (A6.5) in Eq. (A6.6) to get

$$\frac{\partial^2 h}{\partial x^2}(x_0, z_0) = \frac{h(x_0 + \Delta x, z_0) - 2h(x_0, z_0) + h(x_0 - \Delta x, z_0)}{(\Delta x)^2} \tag{A6.7}$$

In a similar manner, we can develop a difference expression for $\partial^2 h/\partial z^2$ as

$$\frac{\partial^2 h}{\partial z^2}(x_0, z_0) = \frac{h(x_0, z_0 + \Delta z) - 2h(x_0, z_0) + h(x_0, z_0 - \Delta z)}{(\Delta z)^2} \tag{A6.8}$$

For a square grid, $\Delta x = \Delta z$; adding Eqs. (A6.7) and (A6.8) to form Laplace's equation yields

$$\frac{1}{(\Delta x)^2}[h(x_0 + \Delta x, z_0) + h(x_0 - \Delta x, z_0) + h(x_0, z_0 + \Delta z) + h(x_0, z_0 - \Delta z) - 4h(x_0, z_0)] = 0 \tag{A6.9}$$

If we let (x_0, z_0) be the nodal point (i, j), Eq. (A6.9) can be rearranged to yield

$$h_{ij} = \frac{1}{4}(h_{i+1,j} + h_{i-1,j} + h_{i,j+1} + h_{i,j-1}) \tag{A6.10}$$

which is identical to Eq. (5.24).

Appendix VII Tóth's Analytical Solution for Regional Groundwater Flow

Tóth (1962, 1963) has presented two analytical solutions for the boundary-value problem representing steady-state flow in a vertical, two-dimensional, saturated, homogeneous, isotropic flow field bounded on top by a water table and on the other three sides by impermeable boundaries (the shaded cell of Figure 6.1, reproduced in an xz coordinate system in Figure A7.1).

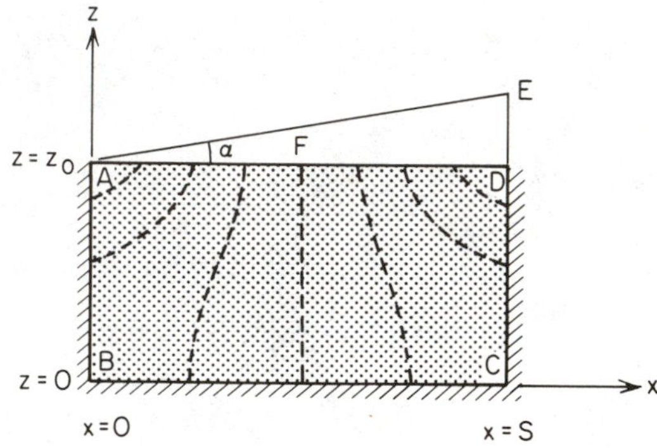

Figure A7.1 Region of flow for Tóth's analytical solution.

He first considered the case where the water-table configuration is a straight line of constant slope. For this case, the region of flow in Figure A7.1 is the region $ABCEA$. Since it is not possible to obtain an analytical solution in a trapezoidal region, Tóth approximated the actual region of flow by the shaded region $ABCDA$. He projected the hydraulic-head values that exist along the actual water table AE onto the upper boundary AD of the region of solution. The approximation is satisfactory for small α.

The equation of flow is Laplace's equation:

$$\frac{\partial^2 h}{\partial x^2} + \frac{\partial^2 h}{\partial z^2} = 0 \tag{A7.1}$$

For a region bounded by $x = s$ and $z = z_0$, the boundary conditions are

$$\begin{aligned} &\frac{\partial h}{\partial x}(0, z) = \frac{\partial h}{\partial x}(s, z) = 0 && \text{on } AB \text{ and } CD \\ &\frac{\partial h}{\partial z}(x, 0) = 0 && \text{on } BC \\ &h(x, z_0) = z_0 + cx && \text{on } AD \end{aligned} \tag{A7.2}$$

where $c = \tan \alpha$.

The analytical solution, obtained by separation of variables, is

$$h(x, z) = z_0 + \frac{cs}{2} - \frac{4cs}{\pi^2} \sum_{m=0}^{\infty} \frac{\cos [(2m+1)\pi x/s] \cosh [(2m+1)\pi z/s]}{(2m+1)^2 \cosh [(2m+1)\pi z_0/s]} \quad \text{(A7.3)}$$

This equation satisfies both the equation of flow (A7.1) and the boundary conditions (A7.2). When plotted and contoured, it leads to the equipotential net shown in Figure A7.1. Flow is from the recharge boundary DF through the field to the discharge boundary AF.

Tóth also considered the case where the water-table configuration is specified as a sine curve superimposed on the line AE (to represent hummocky topography). The final boundary condition then becomes

$$h(x, z_0) = z_0 + cx + a \sin bx \qquad \text{on } AD \quad \text{(A7.4)}$$

where $c = \tan \alpha$, $a = a'/\cos \alpha$, and $b = b'/\cos \alpha$, a' being the amplitude of the sine curve and b' being the frequency.

The analytical solution for this case takes the form

$$\begin{aligned} h(x, z) = {} & z_0 + \frac{cs}{2} + \frac{a}{sb}(1 - \cos bs) \\ & + 2 \sum_{m=1}^{\infty} \left[\frac{ab(1 - \cos bs \cos m\pi)}{b^2 - m^2\pi^2/s^2} + \frac{cs^2}{m^2\pi^2}(\cos m\pi - 1) \right] \quad \text{(A7.5)} \\ & \times \frac{\cos (m\pi x/s) \cosh (m\pi z/s)}{s \cdot \cosh (m\pi z_0/s)} \end{aligned}$$

The equipotential net described by this function is similar to that of Figure 6.2(b).

Appendix VIII Numerical Solution of the Boundary-Value Problem Representing One-Dimensional Infiltration Above a Recharging Groundwater Flow System

Consider a one-dimensional, vertical, homogeneous flow field with its upper boundary at the ground surface and its lower boundary below the water table. As outlined in Section 6.4, the equation of flow for such a system is

$$\frac{\partial}{\partial z}\left[K(\psi)\left(\frac{\partial \psi}{\partial z} + 1\right)\right] = C(\psi)\frac{\partial \psi}{\partial t} \tag{A8.1}$$

and the boundary conditions are

$$\frac{\partial \psi}{\partial z} = \frac{R}{K(\psi)} - 1 \tag{A8.2}$$

at the top, and

$$\frac{\partial \psi}{\partial z} = \frac{Q}{K_0} - 1 \tag{A8.3}$$

at the base. The solution is in terms of the pressure head, $\psi(z, t)$. The position of the water table at any time is given by the value of z at which $\psi = 0$. The required input includes the initial conditions, $\psi(z, 0)$, the rainfall rate R, the groundwater recharge rate Q, and the unsaturated characteristic curves $K(\psi)$ and $C(\psi)$. For $\psi \geq \psi_a$, $K = K_0$ and $C = 0$, ψ_a being the air entry pressure head.

The numerical finite-difference scheme used by Rubin and Steinhardt (1963), Liakopoulos (1965b), and Freeze (1969b) is the implicit scheme of Richtmyer (1957). With this method the (z, t) plane is represented by a rectangular grid of points $j = 1, 2, \ldots, L$ along the z axis, and $n = 1, 2, \ldots$ along the t axis. The distance between the vertical nodes is Δz, and between the time steps, Δt. The solution at any given grid point (j, n) is ψ_j^n. Under this notation the finite-difference form for Eq. (A8.1) for an interior node ($j = 2$ to $j = L - 1$), after suitable rearrangement, is

$$\begin{aligned} C(\psi_{\mathrm{III}})\left(\frac{\psi_j^n - \psi_j^{n-1}}{\Delta t}\right) = \frac{1}{\Delta z}\Bigg\{&\left[K(\psi_{\mathrm{I}})\left(1 + \frac{1}{2\Delta z}[\psi_{j+1}^{n-1} + \psi_{j+1}^n - \psi_j^{n-1} - \psi_j^n]\right)\right] \\ &- \left[K(\psi_{\mathrm{II}})\left(1 + \frac{1}{2\Delta z}[\psi_j^{n-1} + \psi_j^n - \psi_{j-1}^{n-1} - \psi_{j-1}^n]\right)\right]\Bigg\} \end{aligned} \tag{A8.4}$$

where the values of ψ_{I}, ψ_{II}, and ψ_{III}, which determine the K and C values to be applied for a particular node at a particular time step, are determined by extrapolation from previous time steps as described by Rubin and Steinhardt (1963).

Similar finite-difference equations, which incorporate the boundary conditions (A8.2) and (A8.3), can be written for the nodes at the upper and lower boundary ($j = 1, j = L$).

For each time step, the finite-difference approximations constitute a system of L linear algebraic equations in L unknowns. The general form is

$$-A_j\psi_{j+1}^n + B_j\psi_j^n - C_j\psi_{j-1}^n = D_j \tag{A8.5}$$

where the coefficients A, B, C, and D vary with the node ($j = 1, j = 2$ to $L - 1$, $j = L$), with the time step ($n = 1, n = 2, n > 2$), and with the saturation ($\psi \geq \psi_a$, $\psi < \psi_a$). The variables on which A, B, C, and D depend are the ψ values from the previous time step, the boundary values R and Q, and the functional relationships $K(\psi)$ and $C(\psi)$.

The values of ψ_j^n are calculated from the recurrence relation

$$\psi_j^n = E_j\psi_{j+1}^n + F_j \tag{A8.6}$$

where

$$E_j = \frac{A_j}{B_j - C_jE_{j-1}}$$

$$F_j = \frac{D_j + C_jF_{j-1}}{B_j - C_jE_{j-1}}$$

The E and F coefficients are calculated from $j = 1$ to $j = L$, and the ψ's are back-calculated from $j = L$ to $j = 1$ using Eq. (A8.6). For $j = 1$, $E_1 = A_1/B_1$ and $F_1 = D_1/B_1$. Complications arise if the upper node ($j = L$) is saturated.

If $K(\psi)$ and $C(\psi)$ are hysteretic, they are usually built into the computer program in the form of a table of values representing the main wetting, main drying, and principal scanning curves. The program must locate the correct curve on the basis of whether the node in question is wetting or drying, and by scanning the past history of the node to determine the ψ value at which a change from wetting to drying or vice versa occurred.

Appendix IX Development of Finite-Difference Equation for Transient Flow in a Heterogeneous, Anisotropic, Horizontal, Confined Aquifer

The partial differential equation that describes transient flow through a saturated anisotropic medium (Section 2.11) is

$$\frac{\partial}{\partial x}\left(K_x \frac{\partial h}{\partial x}\right) + \frac{\partial}{\partial y}\left(K_y \frac{\partial h}{\partial y}\right) + \frac{\partial}{\partial z}\left(K_z \frac{\partial h}{\partial z}\right) = S_s \frac{\partial h}{\partial t} \tag{A9.1}$$

For a horizontal, confined aquifer of thickness, b, the two-dimensional form of Eq. (A9.1) reduces to

$$\frac{\partial}{\partial x}\left(T_x \frac{\partial h}{\partial x}\right) + \frac{\partial}{\partial y}\left(T_y \frac{\partial h}{\partial y}\right) = S \frac{\partial h}{\partial t} \tag{A9.2}$$

where T_x and T_y are the principal components of the transmissivity tensor defined by $T = Kb$, and S is the storativity defined by $S = S_s b$. To find the finite-difference equation for an interior node in the nodal grid used to discretize the region of flow, we must replace the partial derivatives in Eq. (A9.2) by differences. Using the definitions developed in Appendix VI and the ijk notation of Section 8.8 and Figure 8.26(c), we can write the finite-difference expression:

$$\frac{\partial}{\partial x}\left(T_x \frac{\partial h}{\partial x}\right) \simeq \frac{1}{\Delta x}\left[\left(T_x \frac{\partial h}{\partial x}\right)^k_{i+1/2,j} - \left(T_x \frac{\partial h}{\partial x}\right)^k_{i-1/2,j}\right] \tag{A9.3}$$

where the subscript $(i + \frac{1}{2}, j)$ means that the bracketed quantity is evaluated at the midpoint between nodes (i, j) and $(i + 1, j)$, and the superscript k means that the bracketed quantity is evaluated at time step k. We can further approximate terms on the right-hand side of Eq. (A9.3) by

$$\left(T_x \frac{\partial h}{\partial x}\right)^k_{i+1/2,j} = (T_x)_{i+1/2,j}(h^k_{i+1,j} - h^k_{i,j})\frac{1}{\Delta x} \tag{A9.4a}$$

$$\left(T_x \frac{\partial h}{\partial x}\right)^k_{i-1/2,j} = (T_x)_{i-1/2,j}(h^k_{i,j} - h^k_{i-1,j})\frac{1}{\Delta x} \tag{A9.4b}$$

If we evaluate $(T_x)_{i+1/2,j}$ and $(T_x)_{i-1/2,j}$ by simple averages of the form

$$(T_x)_{i+1/2,j} \simeq \frac{(T_x)_{i+1,j} + (T_x)_{i,j}}{2} \tag{A9.5}$$

then these expressions can be substituted in Eqs. (A9.4), and Eqs. (A9.4) can in turn be substituted in Eq. (A9.3) to give

$$\frac{\partial}{\partial x}\left(T_x \frac{\partial h}{\partial x}\right) \simeq \frac{1}{2\Delta x^2}\{[(T_x)_{i+1,j} + (T_x)_{i,j}]h^k_{i+1,j}$$

$$- [(T_x)_{i+1,j} + 2(T_x)_{i,j} + (T_x)_{i-1,j}]h^k_{i,j} \tag{A9.6}$$

$$+ [(T_x)_{i,j} + (T_x)_{i-1,j}]h^k_{i-1,j}\}$$

Similarly:

$$\frac{\partial}{\partial y}\left(T_y \frac{\partial h}{\partial y}\right) \simeq \frac{1}{2\Delta y^2}\{[(T_y)_{i,j+1} + (T_y)_{i,j}]h^k_{i,j+1}$$

$$- [(T_y)_{i,j+1} + 2(T_y)_{i,j} + (T_y)_{i,j-1}]h^k_{i,j} \tag{A9.7}$$

$$+ [(T_y)_{i,j} + (T_y)_{i,j-1}]h^k_{i,j-1}\}$$

Finally, we can approximate the right-hand side of Eq. (A9.2) by

$$S\frac{\partial h}{\partial t} \simeq S_{ij}\left(\frac{h^k_{i,j} - h^{k-1}_{i,j}}{\Delta t}\right) \tag{A9.8}$$

Substituting Eqs. (A9.6), (A9.7), and (A9.8) for the three terms in Eq. (A9.2) and gathering terms leads to the general finite-difference equation for an internal node in a heterogeneous, anisotropic aquifer:

$$Ah^k_{ij} = Bh^k_{i,j-1} + Ch^k_{i+1,j} + Dh^k_{i,j+1} + Eh^k_{i-1,j} + F \tag{A9.9}$$

where

$$A = \frac{1}{2\Delta x^2}[(T_x)_{i+1,j} + 2(T_x)_{i,j} + (T_x)_{i-1,j}]$$

$$+ \frac{1}{2\Delta y^2}[(T_y)_{i,j+1} + 2(T_y)_{i,j} + (T_y)_{i,j-1}]$$

$$+ \frac{S_{i,j}}{\Delta t}$$

$$B = \frac{1}{2\Delta y^2}[(T_y)_{i,j} + (T_y)_{i,j-1}]$$

$$C = \frac{1}{2\Delta x^2}[(T_x)_{i+1,j} + (T_x)_{i,j}]$$

$$D = \frac{1}{2\Delta y^2}[(T_y)_{i,j+1} + (T_y)_{i,j}]$$

$$E = \frac{1}{2\Delta x^2}[(T_x)_{i,j} + (T_x)_{i-1,j}]$$

$$F = \frac{S_{i,j}}{\Delta t} \cdot h^{k-1}_{i,j}$$

If the aquifer is homogeneous and isotropic, then $T_x = T_y = T$ for all (i, j) and $S_{i,j} = S$ for all (i, j). Under these conditions, and for a square nodal grid with $\Delta x = \Delta y$, the coefficients of Eq. (A9.9) become

$$A = \frac{4T}{\Delta x^2} + \frac{S_{i,j}}{\Delta t}$$

$$B = C = D = E = \frac{T}{\Delta x^2}$$

$$F = \frac{S}{\Delta t} \cdot h_{i,j}^{k-1}$$

If we divide through by $T/\Delta x^2$, these coefficients are seen to be the same as those developed in a less rigorous way in Section 8.8 and presented in connection with Eq. (8.57).

Trescott et al. (1976) have suggested that there is some advantage to utilizing the harmonic mean rather than the arithmetic mean in Eq. (A9.5). This approach changes the coefficients in the finite-difference equation but does not alter the concepts underlying the development.

Equation (A9.9) is written in terms of the hydraulic head values at five nodes at time step k and one node at time step $k - 1$. It is known as a backward-difference approximation. Remson et al. (1971) note that there are some computational advantages to using a central-difference approximation, known as the *Crank-Nicholson scheme*, that utilizes head values at five nodes at time step k and five nodes at time step $k - 1$. The alternating-direction implicit procedure (ADIP) utilized by Pinder and Bredehoeft (1968) involves two finite-difference equations, one in the xt plane and one in the yt plane. Each uses head values at three nodes at time step k and three nodes at time step $k - 1$.

Appendix X Derivation of the Advection-Dispersion Equation for Solute Transport in Saturated Porous Media

The equation derived here is a statement of the law of conservation of mass. The derivation is based on those of Ogata (1970) and Bear (1972). It will be assumed that the porous medium is homogeneous and isotropic, that the medium is saturated, that the flow is steady-state, and that Darcy's law applies. Under the Darcy assumption, the flow is described by the average linear velocity, which carries the dissolved substance by advection. If this were the only transport mechanism operative, nonreactive solutes being transported by the fluid would move as a plug. In reality, there is an additional mixing process, hydrodynamic dispersion (Section 2.13), which is caused by variations in the microscopic velocity within each pore channel and from one channel to another. If we wish to describe the transport process on a macroscopic scale using macroscopic parameters, yet take into account the effect of microscopic mixing, it is necessary to introduce a second mechanism of transport, in addition to advection, to account for hydrodynamic dispersion.

To establish the mathematical statement of the conservations of mass, the solute flux into and out of a small elemental volume in the porous medium will be considered (Figure A10.1). In Cartesian coordinates the specific discharge v has components (v_x, v_y, v_z) and the average linear velocity $\bar{v} = v/n$ has components $(\bar{v}_x, \bar{v}_y, \bar{v}_z)$. The rate of advective transport is equal to $\bar{v}$. The concentration of the solute C is defined as the mass of solute per unit volume of solution. The mass of solute per unit volume of porous media is therefore nC. For a homogeneous

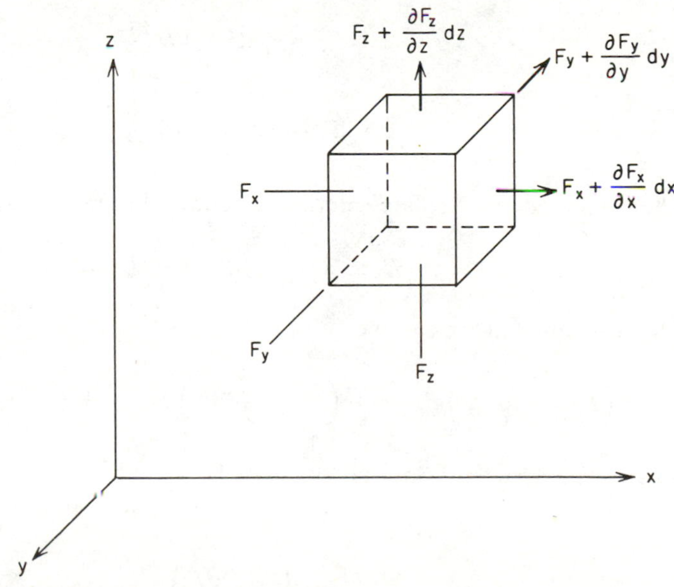

Figure A10.1 Mass balance in a cubic element in space.

medium, the porosity n is a constant, and $\partial(nC)/\partial x = n\,\partial C/\partial x$. The mass of solute transported in the x direction by the two mechanisms of solute transport can be represented as

$$\text{transport by advection} = \bar{v}_x nC\,dA \tag{A10.1}$$

$$\text{transport by dispersion} = nD_x \frac{\partial C}{\partial x}\,dA \tag{A10.2}$$

where D_x is the dispersion coefficient in the x direction and dA is the elemental cross-sectional area of the cubic element. The dispersion coefficient D_x is related to the dispersivity α_x and the diffusion coefficient D^* by Eq. (9.4):

$$D_x = \alpha_x \bar{v}_x + D^* \tag{A10.3}$$

The form of the dispersive component embodied in Eq. (A10.2) is analogous to Fick's first law.

If F_x represents the total mass of solute per unit cross-sectional area transported in the x direction per unit time, then

$$F_x = \bar{v}_x nC - nD_x \frac{\partial C}{\partial x} \tag{A10.4}$$

The negative sign before the dispersive term indicates that the contaminant moves toward the zone of lower concentration. Similarly, expressions in the other two directions are written

$$F_y = \bar{v}_y nC - nD_y \frac{\partial C}{\partial y} \tag{A10.5}$$

$$F_z = \bar{v}_z nC - nD_z \frac{\partial C}{\partial z} \tag{A10.6}$$

The total amount of solute entering the cubic element (Figure A10.1) is

$$F_x\,dz\,dy + F_y\,dz\,dx + F_z\,dx\,dy$$

The total amount leaving the cubic element is

$$\left(F_x + \frac{\partial F_x}{\partial x}dx\right)dz\,dy + \left(F_y + \frac{\partial F_y}{\partial y}dy\right)dz\,dx + \left(F_z + \frac{\partial F_z}{\partial z}dz\right)dx\,dy$$

where the partial terms indicate the spatial change of the solute mass in the specified direction. The difference in the amount entering and leaving the element is, therefore,

$$\left(\frac{\partial F_x}{\partial x} + \frac{\partial F_y}{\partial y} + \frac{\partial F_z}{\partial z}\right)dx\,dy\,dz$$

Because the dissolved substance is assumed to be nonreactive, the difference between the flux into the element and the flux out of the element equals the amount of dissolved substance accumulated in the element. The rate of mass change in the element is

$$-n\frac{\partial C}{\partial t}\,dx\,dy\,dz$$

The complete conservation of mass expression, therefore, becomes

$$\frac{\partial F_x}{\partial x}+\frac{\partial F_y}{\partial y}+\frac{\partial F_z}{\partial z}=-n\frac{\partial C}{\partial t} \qquad \text{(A10.7)}$$

Substitution of expressions (A10.4), (A10.5), and (A10.6) in (A10.7) and cancellation of n from both sides yields:

$$\left[\frac{\partial}{\partial x}\left(D_x\frac{\partial C}{\partial x}\right)+\frac{\partial}{\partial y}\left(D_y\frac{\partial C}{\partial y}\right)+\frac{\partial}{\partial z}\left(D_z\frac{\partial C}{\partial z}\right)\right]$$

$$-\left[\frac{\partial}{\partial x}(\bar{v}_xC)+\frac{\partial}{\partial y}(\bar{v}_yC)+\frac{\partial}{\partial z}(\bar{v}_zC)\right]=\frac{\partial C}{\partial t} \qquad \text{(A10.8)}$$

In a homogeneous medium in which the velocity $\bar{v}$ is steady and uniform (i.e., if it does not vary through time or space), dispersion coefficients D_x, D_y, and D_z do not vary through space, and Eq. (A10.8) becomes

$$\left[D_x\frac{\partial^2 C}{\partial x^2}+D_y\frac{\partial^2 C}{\partial y^2}+D_z\frac{\partial^2 C}{\partial z^2}\right]-\left[\bar{v}_x\frac{\partial C}{\partial x}+\bar{v}_y\frac{\partial C}{\partial y}+\bar{v}_z\frac{\partial C}{\partial z}\right]=\frac{\partial C}{\partial t} \qquad \text{(A10.9)}$$

In one dimension,

$$D_x\frac{\partial^2 C}{\partial x^2}-\bar{v}_x\frac{\partial C}{\partial x}=\frac{\partial C}{\partial t} \qquad \text{(A10.10)}$$

In some applications, the one-dimensional direction is taken as a curvilinear coordinate in the direction of flow along a flowline. The transport equation then becomes

$$D_l\frac{\partial^2 C}{\partial l^2}-\bar{v}_l\frac{\partial C}{\partial l}=\frac{\partial C}{\partial t} \qquad \text{(A10.11)}$$

where l is the coordinate direction along the flowline, D_l the longitudinal coefficient of dispersion, and $\bar{v}_l$ the average linear velocity along the flowline.

For a two-dimensional problem, it is possible to define two curvilinear coordinate directions, S_l and S_t, where S_l is directed along the flowline and S_t is orthogonal to it. The transport equation then becomes

$$D_l \frac{\partial^2 C}{\partial S_l^2} + D_t \frac{\partial^2 C}{\partial S_t^2} - \bar{v}_l \frac{\partial C}{\partial S_l} = \frac{\partial C}{\partial t} \tag{A10.12}$$

where D_l and D_t are the coefficients of dispersion in the longitudinal and transverse directions, respectively.

If v_l varies along the flowline and D_l and D_t vary through space, Eq. (A10.13) becomes

$$\frac{\partial}{\partial S_l}\left(D_l \frac{\partial C}{\partial S_l}\right) + \frac{\partial}{\partial S_t}\left(D_t \frac{\partial C}{\partial S_t}\right) - \frac{\partial}{\partial S_l}(\bar{v}_l C) = \frac{\partial C}{\partial t} \tag{A10.13}$$

Equations (A10.8) through (A10.13) represent six forms of the advection-dispersion equation for solute transport in saturated porous media. Equation (A10.11) is identical to Eq. (9.3) in Section 9.2. The solution to any one of these equations will provide the solute concentration C as a function of space and time. It will take the form $C(x, y, z, t)$ for Eqs. (A10.8) and (A10.9); $C(l, t)$ for Eq. (A10.11); and $C(S_l, S_t, t)$ for Eqs. (A10.12) and (A10.13). There are many well known analytical solutions to the simpler forms of the transport equation. Equation (9.5) in Section 9.2 is an analytical solution to Eq. (A10.11), and Eq. (9.6) is an analytical solution to Eq. (A10.9). In most field situations, two- or three-dimensional analyses are required. In addition, the velocities are seldom uniform and the dispersivities are usually variable in space. For these conditions, numerical methods adapted for digital computers must be used to obtain solutions.

The coefficient of dispersion in a three-dimensional, homogeneous, isotropic porous medium as expressed in Eq. (A10.12) is a second-order symmetrical tensor with nine components. D_l and D_t are the diagonal terms of the two-dimensional form. The directional properties of the dispersion coefficient are caused by the nature of the dispersive process in the longitudinal and transverse directions. If the medium is itself anisotropic, mathematical description of the dispersive process takes on much greater complexity. No analytical or numerical solutions are available for anisotropic systems. Only for isotropic media has the coefficient of dispersion been represented successfully by experimental methods.

It is possible to extend the transport equation to include the effects of retardation of solute transport through adsorption, chemical reaction, biological transformations, or radioactive decay. In this case the mass balance carried out on the elemental volume must include a source-sink term. For retardation due to adsorption, the transport equation in a homogeneous medium in a one-dimensional system along the direction of flow takes the form

$$D_l \frac{\partial^2 C}{\partial l^2} - \bar{v}_l \frac{\partial C}{\partial l} + \frac{\rho_b}{n} \frac{\partial S}{\partial t} = \frac{\partial C}{\partial t} \tag{A10.14}$$

where ρ_b is the bulk mass density of the porous medium, n the porosity, and S the mass of chemical constituent adsorbed on a unit mass of the solid part of the porous medium. The first term of Eq. (A10.14) is the dispersion term, the second is the advection term, and the third is the reaction term. This form of the reaction term is considered in Section 9.2 in connection with Eqs. (9.9) through (9.13). Analytical solutions to Eq. (A10.14) are presented by Codell and Schreiber (in press). Figure 9.12 presents some sample solutions carried out by Pickens and Lennox (1976) using numerical methods.

References

Adams, R. S., Jr. 1973. Factors influencing soil adsorption and bioactivity of pesticides. *J. Residue Rev.*, 47, pp. 1–54.

Albertson, M. L., and D. B. Simons. 1964. Fluid mechanics. *Handbook of Applied Hydrology*, ed. V. T. Chow. McGraw-Hill, New York, pp. 7.1–7.49.

Allen, J. R. C. 1970. *Physical Processes of Sedimentation*. George Allen & Unwin, London, 248 pp.

American Petroleum Institute. 1972. The migration of petroleum products in soil and ground water. *Amer. Petrol. Inst. Publ. 4149*, Washington, D.C.

American Society of Testing Materials. 1967. *Permeability and Capillarity of Soils*. ASTM, Philadelphia, 210 pp.

Amyx, J. W., D. M. Bass, Jr., and R. L. Whiting. 1960. *Petroleum Reservoir Engineering Physical Properties*. McGraw-Hill, New York.

Anderson, D. M., and N. R. Morgenstern. 1973. Physics, chemistry and mechanics of frozen ground. *Permafrost*, 2nd Intern. Conf. U.S. Nat. Acad. Sci.—U.S. Nat. Res. Council, Washington, D.C., pp. 257–288.

Apgar, M. A., and W. B. Satherthwaite, Jr. 1975. Ground water contamination associated with the Llangollen landfill, New Castle County, Delaware. *Proc. Res. Symp., Gas and Leachate from Landfills, New Brunswick, N.J.* U.S. Environmental Protection Agency, National Environmental Research Center, Cincinnati, Ohio.

Argersinger, W. J., A. W. Davidson, and O. D. Bonner. 1950. Thermodynamics of ion exchange phenomena. *Kansas Acad. Sci. Trans.*, 53, pp. 404–410.

Aris, R. 1962. *Vectors, Tensors and the Basic Equations of Fluid Mechanics*. Prentice-Hall, Englewood Cliffs, N.J.

Atlantic-Richfield Hanford Company. 1976. *Preliminary Feasibility Study of Storage of Radioactive Wastes in Columbia River Basalts*. Rept. ARH-ST-137, Vol. 1, 168 pp.

Atwater, G. I. 1966. The effect of decrease in porosity with depth on oil and gas reserves in sandstone reservoirs. Unpublished paper presented to School of Earth Sciences, Stanford University, Stanford, Calif., Jan. 10.

Ayers, R. S., and R. C. Branson, eds. 1973. Nitrates in the upper Santa Anna River Basin in relation to ground water pollution. *Calif. Agr. Exp. Sta. Bull. 861*, p. 60.

Babcock, K. L. 1963. Theory of the chemical properties of soil colloidal systems at equilibrium. *Hilgardia*, 34, pp. 417–542.

BABCOCK, K. L., and R. K. SCHULZ. 1963. Effect of anions on the sodium-calcium exchange in soils. *Proc. Soil Sci. Soc. Amer.*, 27, pp. 630–631.

BACK, W. 1961. Techniques for mapping of hydrochemical facies. *U.S. Geol. Surv. Prof. Paper 424–D*, pp. 380–382.

BACK, W. 1966. Hydrochemical facies and ground-water flow patterns in northern part of Atlantic Coastal Plain. *U.S. Geol. Surv. Prof. Paper 498–A*, 42 pp.

BACK, W., and I. BARNES. 1969. Relation of electrochemical potentials and iron content to groundwater flow patterns. *U.S. Geol. Survey Prof. Paper 498–C*, 16 pp.

BACK, W., and J. A. CHERRY. 1976. Chemical aspects of present and future hydrogeologic problems. *Proc. Symp. Adv. Groundwater Hydrol. Amer. Water Resources Assoc.*

BACK, W., and B. B. HANSHAW. 1965. Chemical geohydrology. *Adv. Hydrosci.*, 2, pp. 49–109.

BACK, W., and B. B. HANSHAW. 1970. Comparison of chemical hydrogeology of the carbonate peninsulas of Florida and Yucatan. *J. Hydrol.*, 10, pp. 330–368.

BAETSLÉ, L. H. 1967. Computational methods for the prediction of underground movement of radionuclides. *J. Nuclear Safety*, 8, no. 6, pp. 576–588.

BAETSLÉ, L. H. 1969. Migration of radionuclides in porous media. *Progress in Nuclear Energy, Series XII, Health Physics*, ed. A. M. F. Duhamel. Pergamon Press, Elmsford, N.Y., pp. 707–730.

BANIN, A., and D. M. ANDERSON. 1974. Effects of salt concentration changes during freezing on the unfrozen water content of porous materials. *Water Resources Res.*, 10, pp. 124–128.

BARNES, I., and J. D. HEM. 1973. Chemistry of subsurface waters. *Ann. Rev. Earth Planetary Sci.*, 1, pp. 157–301.

BAUER, W. J. 1974. Land treatment designs, present and future. *Proc. Intern. Conf. Land for Waste Management*, ed. J. Thomlinson. National Research Council, Ottawa, Canada, pp. 343–346.

BAVER, L. D., W. H. GARDNER, and W. R. GARDNER. 1972. *Soil Physics*, 4th ed. John Wiley & Sons, New York.

BEAR, J. 1972. *Dynamics of Fluids in Porous Media*. American Elsevier, New York.

BEAR, J., and G. DAGAN. 1965. The relationship between solutions of flow problems in isotropic and anisotropic soils. *J. Hydrol.*, 3, pp. 88–96.

BEAR, J., and O. LEVIN. 1967. The optimal yield of an aquifer. *Intern. Assoc. Sci. Hydrol.*, Proc. Haifa Symp. Publ. 72, pp. 401–412.

BEAR, J., D. ZASLAVSKY, and S. IRMAY. 1968. *Physical Principles of Water Percolation and Seepage*. UNESCO, Paris, 465 pp.

BEEK, J., and F. A. M. DE HAAN. 1974. Phosphate removal by soil in relation to waste disposal. *Proc. Intern. Conf. Land for Waste Management*, ed. J. Thomlinson. National Research Council, Ottawa, Canada, pp. 77–86.

BENNION, D. W., and J. C. GRIFFITHS. 1966. A stochastic model for predicting variations in reservoir rock properties. *Trans. Amer. Inst. Mining Met. Engrs.*, 237, no. 2, pp. 9–16.

BERNER, R. A. 1971. *Principles of Chemical Sedimentology*. McGraw-Hill, New York, 240 pp.

BERRY, F. A. F. 1959. *Hydrodynamics and Geochemistry of the Jurassic and Cretaceous Systems in the San Juan Basin, Northwestern New Mexico and Southwestern Colorado*. Unpublished Ph.D. thesis, Stanford University.

BERRY, F. A. F. 1969. Relative factors influencing membrane filtration effects in geologic environments. *Chem. Geol.*, 4, pp. 295–301.

BETSON, R. P. 1964. What is watershed runoff? *J. Geophys. Res.*, 69, pp. 1541–1551.

BIANCHI, W. C., and E. E. HASKELL, JR. 1966. Air in the vadose zone as it affects water movements beneath a recharge basin. *Water Resources Res.*, 2, pp. 315–322.

BIANCHI, W. C., and E. E. HASKELL, JR. 1968. Field observations compared with Dupuit-Forchheimer theory for mound heights under a recharge basin. *Water Resources Res.*, 4, pp. 1049–1057.

Billings, G. K., B. Hitchon, and D. R. Shaw. 1969. Geochemistry and origin of formation waters in western Canada sedimentary basin. *Chem. Geol.*, 4, pp. 211–223.

Biot, M. A. 1941. General theory of three-dimensional consolidation. *J. Appl. Phys.*, 12, pp. 155–164.

Biot, M. A. 1955. Theory of elasticity and consolidation for a porous anisotropic solid. *J. Appl. Phys.*, 26, pp. 182–185.

Bishop, A. W. 1955. The use of the slip circle in the stability analysis of slopes. *Geotechnique*, 5, pp. 7–17.

Bishop, A. W., and G. E. Blight. 1963. Some aspects of effective stress in saturated and partly saturated soils. *Géotechnique*, 13, pp. 177–197.

Bishop, A. W., and N. R. Morgenstern. 1960. Stability coefficients for earth slopes. *Géotechnique*, 10, pp. 29–150.

Biswas, A. 1970. *History of Hydrology*. North-Holland, Amsterdam, 336 pp.

Bjerrum, L., and F. Jorstad. 1964. *Rockfalls in Norway*. Norwegian Geotechnical Institute, Oslo.

Black, C. A., ed. 1965. *Methods in Soil Analysis, Part 1*. American Society of Agronomy, Madison, Wis., 770 pp.

Blaney, H. F., and W. D. Criddle. 1950. Determining water requirements in irrigated areas from climatological and irrigation data. *U.S. Dept. Agr. Soil Conserv. Serv., Rept. T.P. 96*.

Blatt, H., G. Middleton, and R. Murray. 1972. *Origin of Sedimentary Rocks*. Prentice-Hall, Englewood Cliffs, N.J., 634 pp.

Blomeke, J. O., J. P. Nichols, and W. C. McClain. 1973. Managing radioactive wastes. *Phys. Today*, 26, pp. 36–42.

Bodman, G. B., and E. A. Colman. 1943. Moisture and energy conditions during downward entry of water into soils. *Soil Sci. Soc. Amer. Proc.*, 8, pp. 116–122.

Boersma, L. 1965. Field measurement of hydraulic conductivity below a water table. *Methods of Soil Analysis, Part 1*, ed. C. A. Black. American Society of Agronomy, Madison, Wis., pp. 222–233.

Bolt, G. H., and P. H. Groenevelt. 1969. Coupling phenomena as a possible cause for non-Darcian behaviour of water in soil. *Bull. Intern. Assoc. Sci. Hydrol.*, 14, no. 2, pp. 17–26.

Bostock, C. A. 1971. Estimating truncation error in image well theory. *Water Resources Res.*, 7, pp. 1658–1660.

Bottomley, D. 1974. *Influence of Hydrology and Weathering on the Water Chemistry of a Small Precambrian Shield Watershed.* Unpublished M.Sc. thesis, University of Waterloo.

Boulton, G. S. 1974. Processes and patterns of glacial erosion. *Glacial Geomorphology*, ed. D. R. Coates. State University of New York, Binghamton, N.Y., pp. 41–87.

Boulton, G. S. 1975. Processes and patterns of subglacial sedimentation: a theoretical approach. *Ice Ages: Ancient and Modern*, ed. A. E. Wright and F. Moseley. Geol. J. Spec. Issue No. 6, pp. 7–42.

Boulton, N. S. 1954. The drawdown of the water table under non-steady conditions near a pumped well in an unconfined formation. *Proc. Inst. Civil Engrs.*, 3, pp. 564–579.

Boulton, N. S. 1955. Unsteady radial flow to a pumped well allowing for delayed yield from storage. *Proc. Gen. Assembly Rome, Intern. Ass. Sci. Hydrol. Publ. 37*, pp. 472–477.

Boulton, N. S. 1963. Analysis of data from nonequilibrium pumping tests allowing for delayed yield from storage. *Proc. Inst. Civil Engrs.*, 26, pp. 469–482.

Boussinesq, J. 1904. Recherches théoretiques sur l'écoulement des nappes d'eau infiltrées dans le sol et sur le débit des sources. *J. Math. Pure Appl.*, 10, pp. 5–78.

Bouwer, H. 1962. Analyzing ground-water mounds by resistance network. *J. Irrig. Drain. Div., Proc. Amer. Soc. Civ. Engrs.*, 88 (IR3), pp. 15–36.

Bouwer, H. 1965. Theoretical aspects of seepage from open channels. *J. Hydraul. Div., Proc. Amer. Soc. Civil Engrs.*, 91 (HY3), pp. 37–59.

BOUWER, H., and R. D. JACKSON. 1974. Determining soil properties. *Drainage for Agriculture*, ed. J. van Schilfgaarde. American Society of Agronomy, Madison, Wis., pp. 611–672.

BOUWER, H., and W. C. LITTLE. 1959. A unifying numerical solution for two-dimensional steady flow problems in porous media with an electrical resistance network *Soil Sci. Soc. Amer. Proc.*, 23, pp. 91–96.

BRADSHAW, P. M. D., ed. 1975. Conceptual models in exploration geochemistry: the Canadian Cordillera and Canadian Shield. *J. Geochem. Exploration*, 4, pp. 1–231.

BRADSHAW, R. L., and W. C. MCCLAIN. 1971. Project salt vault: a demonstration of the disposal of high-activity solidified wastes in underground salt mines. *Rept. ORNL-4555, Oak Ridge Nat. Lab.*, Oak Ridge, Tenn.

BREDEHOEFT, J. D., and B. B. HANSHAW. 1968. On the maintenance of anomalous fluid pressures: I. Thick sedimentary sequences. *Geol. Soc. Amer. Bull.*, 79, pp. 1097–1106.

BREDEHOEFT, J. D., and I. S. PAPADOPOULOS. 1965. Rates of vertical groundwater movement estimated from the earth's thermal profile. *Water Resources Res.*, 1, pp. 325–328.

BREDEHOEFT, J. D., and G. F. PINDER. 1970. Digital analysis of areal flow in multiaquifer groundwater systems: a quasi three-dimensional model. *Water Resources Res.*, 6, pp. 883–888.

BREDEHOEFT, J. D., and G. F. PINDER. 1973. Mass transport in flowing groundwater. *Water Resources Res.*, 9, pp. 194–209.

BREDEHOEFT, J. D., and R. A. YOUNG. 1970. The temporal allocation of groundwater: a simulation approach. *Water Resources Res.*, 6, pp. 3–21.

BREDEHOEFT, J. D., C. R. BLYTH, W. A. WHITE, and G. B. MAXEY. 1963. Possible mechanism for concentration of brines in subsurface formations. *Bull. Amer. Assoc. Petrol. Geol.*, 47, pp. 257–269.

BRICKER, O. P. 1967. Cations and silica in natural waters: control by silicate minerals. *Intern. Assoc. Sci. Hydrol. Publ. 78*, pp. 110–119.

BRICKER, O. P., A. E. GODFREY, and E. T. CLEAVES. 1968. Mineral-water interaction during the chemical weathering of silicates. *Advances in Chemistry Series No. 73*, American Chemical Society, Washington, D.C.

BRIGGS, G. F., and A. G. FIEDLER, eds. 1966. *Ground Water and Wells*. Edward E. Johnson, Inc., St. Paul, Minn.

BROADBENT, F. E. 1971. Nitrogen in soil and water. *Symp. Nitrogen Soil Water, University of Guelph, Guelph, Ontario*, p. 56.

BROCK, T. D. 1966. *Principles of Microbial Ecology*. Prentice-Hall, Englewood Cliffs, N.J., 306 pp.

BROWN, I. C., ed. 1967. Groundwater in Canada. *Geol. Surv. Can. Econ. Geol. Rept. No. 24*.

BROWN, R. H., A. A. KONOPLYANTSEV, J. INESON, and U. S. KOVALEVSKY, eds. 1972. *Ground Water Studies: An International Guide for Research and Practice, Studies and Reports in Hydrology*, UNESCO, 7, no. 10, pp. 1–18.

BRUTSAERT, W. F., E. A. BREITENBACH, and D. K. SUNADA. 1971. Computer analysis of free-surface well flow. *J. Irrig. Drain. Div., Proc. Amer. Soc. Civil Engrs.*, 97, pp. 405–420.

BURAS, N. 1966. Dynamic programming in water resource development. *Adv. Hydrosci.*, 3, pp. 367–412.

BURKHOLDER, H. C. 1976. Methods and data for predicting nuclide migration in geologic media. *Intern. Symp. Management of Wastes from the LWR Fuel Cycle, Denver, Colo.*

BURNS, E. A., and D. A. MCLAREN. 1975. Factors affecting pesticide loss from soil. *Soil Biochemistry*, Vol. 4. Marcel Dekker, New York.

BURT, O. 1967. Temporal allocation of groundwater. *Water Resources Res.*, 3, pp. 45–56.

BURT, T. P., and P. J. WILLIAMS. 1976. Hydraulic conductivity in frozen soils. *Earth Surface Processes*, 1, pp. 249–360.

CALIF. BUREAU OF SANITARY ENGINEERING. 1963. Occurrence of nitrate in ground water supplies in southern California. Bureau of Sanitary Engineering, State Department of Public Health.

Camp, T. R. 1963. *Water and Its Impurities*. Reinhold, New York, 355 pp.

Campbell, M. D., and J. H. Lehr. 1973. *Water Well Technology*. McGraw-Hill, New York.

Carbognin, L., P. Gatto, G. Mozzi, G. Gambolati, and G. Ricceri. 1976. New trend in the subsidence of Venice. *2nd Intern. Symp. Land Subsidence*, Anaheim, Calif.

Carder, D. S. 1970. Reservoir loading and local earthquakes. *Geol. Soc. Amer. Eng. Geol. Case Histories*, 8, pp. 51–61.

Carroll, D. 1962. Rainwater as a chemical of geologic processes—a review. *U.S. Geol. Surv. Water-Supply Paper 1535-G*.

Carslaw, H. S., and J. C. Jaeger. 1959. *Conduction of Heat in Solids*. Oxford University Press, London.

Carson, M. A., and M. J. Kirkby. 1972. *Hillslope Form and Process*. Cambridge University Press, Cambridge, England, 475 pp.

Cartwright, K. 1968. Thermal prospecting for groundwater. *Water Resources Res.*, 4, pp. 395–401.

Cartwright, K. 1974. Tracing shallow groundwater systems by soil temperatures. *Water Resources Res.*, 10, pp. 847–855.

Casagrande, A. 1961. Control of seepage through foundations and abutments of dams. *Géotechnique*, 11, pp. 161–181.

Casagrande, L. 1952. Electro-osmotic stabilization of soils, Boston Society of Civil Engineers. *Contrib. Soil Mech., 1941–1953*, 285 pp.

Castillo, E. R., J. Krizek, and G. M. Karadi. 1972. Comparison of dispersion characteristics in fissured rock. *Proc. 2nd Intern. Symp. Fundamentals of Transport Phenomena in Porous Media, Guelph, Ontario*, 2, pp. 778–797.

Cedergren, H. R. 1967. *Seepage, Drainage, and Flow Nets*. John Wiley & Sons, New York.

Cedergren, H. R. 1975. Drainage and dewatering. *Foundation Engineering Handbook*, ed. H. F. Winterkorn and H. Y. Fang. Van Nostrand Reinhold, New York, pp. 221–243.

Chamberlain, T. C. 1885. The requisite and qualifying conditions of artesian wells. *U.S. Geol. Surv. Ann. Rept.*, pp. 125–173.

Chebotarev, I. I. 1955. Metamorphism of natural waters in the crust of weathering. *Geochim. Cosmochim. Acta*, 8, pp. 22–48, 137–170, 198–212.

Cherry, J. A. 1972. Geochemical processes in shallow groundwater flow systems in five areas in southern Manitoba, Canada. *Proc. 24th Intern. Geol. Congr., Montreal*, Sec. 11, pp. 208–221.

Cherry, J. A., G. E. Grisak, and W. E. Clister. 1973. Hydrogeologic studies at a subsurface radioactive waste management site in west-central Canada. *Intern. Symp. Underground Waste Management Artificial Recharge, AAPG-USGS, New Orleans*, ed. J. Braunstein, pp. 436–467.

Cherry, J. A., G. E. Grisak, and R. E. Jackson. 1974. Hydrogeological factors in shallow subsurface radioactive waste management in Canada. *Proc. Intern. Conf. Land for Waste Management*, ed. J. Thomlinson. National Research Council, Ottawa, Canada, pp. 131–146.

Cherry, J. A., R. W. Gillham, and J. F. Pickens. 1975. Contaminant hydrogeology: Part 1: physical processes. *J. Geosci. Can.*, 2, pp. 76–84.

Cherry, J. A., R. W. Gillham, G. E. Grisak, and D. L. Lush. In press. Hydrogeological factors in shallow subsurface radioactive waste management. *Proc. Symp. Management Low-Level Radioactive Waste*, ed. Melvin Carter. U.S. Environmental Protection Agency.

Childs, E. C. 1943. The water table, equipotentials, and streamlines in drained land. *Soil Sci.*, 56, pp. 317–330.

Childs, E. C. 1969. *The Physical Basis of Soil Water Phenomena*. Wiley-Interscience, New York.

Childs, K. E., S. B. Upchurch, and B. Ellis. 1974. Sampling of variable waste migration patterns in groundwater. *Groundwater*, 12, pp. 369–376.

CHILINGAR, G. V. 1963. Relationship between porosity, permeability, and grain-size distribution of sands and sandstones. *Proc. Intern. Sedimentol. Congr., Amsterdam, Antwerp.*

CHOPPIN, G. 1965. *Chemistry*, 38, p. 11.

CHOW, V. T., ed. 1964a. *Handbook of Applied Hydrology*. McGraw-Hill, New York.

CHOW, V. T., ed. 1964b. Hydrology and its development. *Handbook of Applied Hydrology*, McGraw-Hill, New York, pp. 1.1–1.22.

CHRISTIANSEN, E. A., and S. H. WHITAKER. 1976. Glacial thrusting of drift and bedrock. *Glacial Till: An Interdisciplinary Study*, ed. R. F. Leggett. Roy. Soc. Can. Spec. Publ. 12, pp. 121–130.

CHRISTIANSEN, E. A., S. H. WHITAKER, and W. A. MENELEY. ca. 1971. Groundwater Program. *Saskatchewan Research Council.*

CLARKE, W. B., and G. KUGLER. 1973. Dissolved helium in groundwater: a possible method for uranium and thorium prospecting. *Econ. Geol.*, 68, pp. 243–251.

CLAYTON, L., and S. R. MORAN. 1974. A glacial process-form model. *Glacial Geomorphology*, ed. D. R. Coates. State University of New York, Binghamton, N.Y., pp. 89–119.

CLAYTON, R. N., I. FRIEDMAN, D. L. GRAF, T. K. MAYEDA, W. F. MEENTS, and N. F. SHIMP. 1966. The origin of saline formation waters. *J. Geophys. Res.*, 71, pp. 3869–3882.

CLEAVES, E. T., A. E. GODFREY, and O. P. BRICKER. 1970. Geochemical balance of a small watershed and its geomorphic implications. *Geol. Soc. Amer. Bull.*, 81, pp. 3015–3032.

CLEAVES, E. T., D. W. FISHER, and O. P. BRICKER. 1974. Chemical weathering of serpentinite in the eastern Piedmont of Maryland. *Geol. Soc. Amer. Bull.*, 85.

CLOKE, P. L. 1966. The geochemical application of *Eh–pH* diagrams. *J. Geol. Educ.*, 4, pp. 140–148.

COATES, D. R., ed. 1977. Landslides. *Geol. Soc. Am. Rev. Eng. Geol.*, 3.

CODELL, R. B., and D. L. SCHREIBER. In press. NRC models for evaluating the transport of radionuclides in groundwater. *Proc. Symp. Low-level Radioactive Waste Management*. U.S. Geological Survey and National Regulatory Commission, Atlanta, Ga.

COHEN, B. L. June 1977. The disposal of radioactive wastes from fission reactors. *Sci. Amer.*, 236, no. 6.

COLLINS, R. E. 1961. *Flow of Fluids Through Porous Materials*. Reinhold, New York.

COOLEY, R. L. 1971. A finite-difference method for unsteady flow in variably saturated porous media: application to a single pumping well. *Water Resources Res.*, 7, pp. 1607–1625.

COOPER, H. H. 1966. The equation of groundwater flow in fixed and deforming coordinates. *J. Geophys. Res.*, 71, pp. 4785–4790.

COOPER, H. H., Jr., and C. E. JACOB. 1946. A generalized graphical method for evaluating formation constants and summarizing well field history. *Trans. Amer. Geophys. Union*, 27, pp. 526–534.

COOPER, H. H., Jr., and M. I. RORABAUGH. 1963. Groundwater movements and bank storage due to flood stages in surface streams. *U.S. Geol. Surv. Water-Supply Paper 1536-J.*

COOPER, H. H., Jr., F. A. KOHOUT, H. R. HENRY, and R. E. GLOVER. 1964. Sea water in coastal aquifers. *U.S. Geol. Surv. Water-Supply Paper 1613-C*, 84 pp.

COOPER, H. H., Jr., J. D. BREDEHOEFT, and I. S. PAPADOPOULOS. 1967. Response of a finite-diameter well to an instantaneous charge of water. *Water Resources Res.*, 3, pp. 263–269.

COPLEN, T. B. 1970. Isotopic fractionation of water by ultra-filtration. Unpublished Ph.D. thesis, University of Chicago.

CRANE, F. E., and G. H. F. GARDNER. 1961. Measurements of transverse dispersion in granular media. *Chem. Eng. Data, 6*, pp. 283–287.

CRANK, J. 1956. *The Mathematics of Diffusion*. Oxford University Press, Oxford, 347 pp.

CROLL, B. T. 1972. The impact of organic pesticides and herbicides upon groundwater pollution. *Ground Water Pollution in Europe*, ed. John A. Cole. Water Information Center, Inc., Port Washington, N.Y., pp. 350–364.

Custer, G. S. 1976. Shallow ground-water salinization in dryland farm areas of Montana. *Montana Universities Joint Water Resources Res. Center Rept. No. 79*, Bozeman, Mont.

Custodio, E., and M. R. Llamas. 1974. *Hidrologia Subterranea*. Ediciones Omega, Barcelona, Spain, 2 vols.

Dagan, G. 1967. A method of determining the permeability and effective porosity of unconfined anisotropic aquifers. *Water Resources Res.*, 3, pp. 1059–1071.

Damsgaard, W. 1964. Stable isotopes in precipitation. *Tellus*, 16, pp. 436–468.

Darcy, H. 1856. *Les fontaines publiques de la ville de Dijon*. Victor Dalmont, Paris.

Davidson, J. M., Li-Tse Ou, and P. S. C. Rao. 1976. Behavior of high pesticide concentrations in soil water systems. *Proc. Res. Symp. Residual Management by Land Disposal*, ed. W. H. Fuller, University of Arizona, Tucson, Ariz.

Davis, S. N. 1964. Silica in streams and ground water. *Amer. J. Sci.*, 262, pp. 870–891.

Davis, S. N. 1969. Porosity and permeability of natural materials. *Flow Through Porous Media*, ed. R. J. M. De Wiest. Academic Press, New York, pp. 54–89.

Davis, S. N., and R. J. M. De Wiest. 1966. *Hydrogeology*. John Wiley & Sons, New York, 463 pp.

Davison, C. A. 1976. A hydrogeochemical investigation of a brine disposal lagoon—aquifer system, Esterhazy, Saskatchewan. Unpublished Master's thesis, University of Waterloo.

Deere, D. U., and F. D. Patton. 1967. Effect of pore pressure on the stability of slopes. *Geol. Soc. Amer.–Soc. Civil Engrs. Symp. New Orleans, La.*

de Geoffroy, J., S. M. Wu, and R. W. Heins. 1967. Geochemical coverage by spring sampling method in the Southwest Wisconsin zinc area. *Econ. Geol.*, 62, 679–697.

Deju, R. A. 1971. A model of chemical weathering of silicate minerals. *Geol. Soc. Amer. Bull.*, 82.

Delwiche, C. C. 1967. Energy relationships in soil biochemistry. *Soil Biochemistry*, ed. A. D. McLaren and G. H. Peterson. Marcel Dekker, New York, pp. 173–193.

Denbigh, K. B. 1966. *The Principles of Chemical Equilibrum: With Applications in Chemistry and Chemical Engineering*, 2nd ed. Cambridge University Press, Cambridge, England.

Desai, C. S., and J. F. Abel. 1972. *Introduction to the Finite Element Method.* Van Nostrand Reinhold, New York.

De Wiest, R. J. M. 1962. Free-surface flow in homogeneous porous mediums. *Trans. Amer. Soc. Civil Engrs.*, 127, pp. 1045–1084.

De Wiest, R. J. M. 1966. On the storage coefficient and the equations of groundwater flow. *J. Geophys. Res.*, 71, pp. 1117–1122.

Dewsnut, R. L., D. W. Jensen, and R. W. Swenson. 1973. *A Summary Digest of State Water Laws.* National Water Commission.

Dicker, D. 1969. Transient free-surface flow in porous media. *Flow Through Porous Media*, ed. R. J. M. De Wiest. Academic Press, New York, pp. 293–330.

Dietz, D. N. 1971. Pollution of permeable strata by oil components. *Water Pollution by Oil*, ed. Peter Hepple. Elsevier, Amsterdam, pp. 128–142.

Dobrin, M. B. 1960. *Introduction to Geophysical Prospecting*. McGraw-Hill, New York, 446 pp.

Domenico, P. A. 1972. *Concepts and Models in Groundwater Hydrology*. McGraw-Hill, New York.

Domenico, P. A. 1977. Transport phenomena in chemical rate processes in sediments. *Ann. Rev. Earth Planet Sci.*, 5, pp. 287–317.

Domenico, P. A., and M. D. Mifflin. 1965. Water from low-permeability sediments and land subsidence. *Water Resources Res.*, 1, pp. 563–576.

Domenico, P. A., and V. V. Palciauskas. 1973. Theoretical analysis of forced convective heat transfer in regional groundwater flow. *Geol. Soc. Amer. Bull.*, 84, pp. 3803–3814.

Domenico, P. A., D. Anderson, and C. Case. 1968. Optimal groundwater mining. *Water Resources Res.*, 4, pp. 247–255.

DONALDSON, I. G. 1962. Temperature gradients in the upper layers of the earth's crust due to convective water flows. *J. Geophys. Res.*, 67, pp. 3449–3459.

DONALDSON, I. G. 1970. The simulation of geothermal systems with a simple convective model. *Geothermics*, Spec. Issue 2, Vol. 2, pt. 1, pp. 649–654.

DONER, H. E., and A. D. MCLAREN. 1976. *Soil Nitrogen Transformations: A Modeling Study, Vol 1: Environmental Biogeochemistry*, ed. J. O. Nriagu. Ann Arbor Science Publishers, Ann Arbor, Mich.

DORSEY, N. E. 1940. Properties of ordinary water-substance in all its phases; water-vapor, water and all the ices. *Amer. Chem. Soc. Monogr. No. 81*, Reinhold, New York.

DREWRY, W. A., and R. E. ELIASSEN. 1968. Virus movement in groundwater. *J. Water Pollution Control Fed.*, 40, pp. 257–271.

DROST, W., D. KLOTZ, A. KOCH, H. MOSER, F. NEUMAIER, and W. RAUERT. 1968. Point dilution methods of investigating groundwater flow by means of radioisotopes. *Water Resources Res.*, 4, pp. 125–146.

DUDLEY, J. G., and D. A. STEPHENSON. 1973. Nutrient enrichment of ground water from septic tank disposal systems. Upper Great Lakes Regional Commission.

DUNNE, T. 1978. Field studies of hillslope flow processes. *Hillslope Hydrology*, ed. M. J. Kirkby. Wiley-Interscience, New York, pp. 227–293.

DUNNE, T., and R. D. BLACK. 1970a. An experimental investigation of runoff production in permeable soils. *Water Resources Res.*, 6, pp. 478–490.

DUNNE, T., and R. D. BLACK. 1970b. Partial area contributions to storm runoff in a small New England watershed. *Water Resources Res.*, 6, pp. 1296–1311.

DUPUIT, J. 1863. *Études théoriques et pratiques sur le mouvement des eaux dans les canaux découverts et à travers les terrains perméables*. Dunod, Paris.

DYCK, J. H., W. S. KEYS, and W. A. MENELEY. 1972. Application of geophysical logging to groundwater studies in southern Saskatchewan. *Can. J. Earth Sci.*, 9, pp. 78–94.

EAGLESON, P. S. 1970. *Dynamic Hydrology*. McGraw-Hill, New York, 462 pp.

EBACK, E. A., and R. R. WHITE. 1958. Mixing of fluids flowing through beds of packed solids. *Amer. Inst. Chem. Eng. J.*, 4, no. 2.

ECKEL, E. B., ed. 1958. *Landslides and Engineering Practice*. Highway Res. Board, Spec. Rept. No. 29.

EDMUNDS, W. M. 1973. Trace element variations across an oxidation-reduction barrier in a limestone aquifer. *Proc. Symp. Hydrogeochem. Biochem.*, Tokyo 1970, The Clarke Company, Washington, D.C., pp. 500–526.

ELDER, J. W. 1965. Physical processes in geothermal areas. *Terrestrial Heat Flow*, ed. W. H. K. Lee, *Amer. Geophys. Union Monogr. 8*, pp. 211–239.

ELLIS, R. R., L. DEVI, and W. A. WEIBENGA. 1968. The investigation of water flow through porous mediums by means of radiotracers, *Water Resources Res.*, 4, pp. 413–416.

EL-PRINCE, A. M., and K. L. BABCOCK. 1975. Prediction of ion exchange equilibria in aqueous systems with more than two counter ions. *Soil Sci.*, 120, pp. 332–338.

EMSELLEM, Y., and G. de MARSILY. 1971. An automatic solution for the inverse problem. *Water Resources Res.*, 7, pp. 1264–1283.

EVANS, D. M. 1966. The Denver area earthquakes and the Rocky Mountain Arsenal disposal well. *Mountain Geol.*, 3, pp. 23–26.

FARQUAR, G. F., and F. A. ROVERS. 1973. Gas production during refuse decomposition. *Water, Air Soil Pollution*, 2.

FARVOLDEN, R. N. 1963. Geologic controls on groundwater storage and base flow. *J. Hydrol.*, 1, pp. 219–249.

FARVOLDEN, R. N., and G. M. HUGHES. 1976. Hydrogeological implications in solid waste disposal. *Bull Intern. Assoc. Sci. Hydrol.*, pp. 146–158.

FARVOLDEN, R. N., and J. P. NUNAN. 1970. Hydrogeologic aspects of dewatering at Welland. *Can. Geotech. J.*, 7, pp. 194–204.

FAYERS, F. J., and J. W. SHELDON. 1962. The use of a high-speed digital computer in the study of the hydrodynamics of geologic basins. *J. Geophys. Res.*, 67, pp. 2421–2431.

FERGUSON, J. F., and J. GAVIS. 1972. A review of the arsenic cycle in natural waters. *Water Research*. Pergamon Press, Oxford, pp. 1259–1274.

FERRIS, J. G., D. B. KNOWLES, R. H. BROWNE, and R. W. STALLMAN. 1962. Theory of aquifer tests. *U.S. Geol. Surv. Water-Supply Paper 1536-E.*

FERRO, C., G. P. GIANNOTTI, M. MITTEMBERGHER, D. MUSY, G. SIDOTI, E. STAMPHONE, and C. VALLONE. 1973. Utilization of clay formations for storage of solid high-level radioactive wastes. *Management of Radioactive Wastes from Fuel Reprocessing*. Organization for Economic Cooperation and Development and International Atomic Energy Agency, Paris, pp. 887–916.

FETH, J. H., C. C. ROBERTSON, and W. L. POLZER. 1964. Sources of mineral constituents in water from granitic rocks, Sierra Nevada, California and Nevada. *U.S. Geol. Surv. Water-Supply Paper 1535-I*, 170 pp.

FLOWER, F. B. 1976. Case history of landfill gas movement through soils. *Gas and Leachate from Landfills*, ed. E. J. Genetelli and J. Cirello. U.S. Environmental Protection Agency, Cincinnati, Ohio, 600/9-76-004, pp. 177–189.

FORD, T. D., and C. H. D. CULLINGFORD, eds. 1976. *The Science of Speleology*. Academic Press, New York, 594 pp.

FORCHHEIMER, P. 1930. *Hydraulik*. Teubner Verlagsgesellschaft, Stuttgart.

FORSYTHE, G. E., and W. R. WASOW. 1960. *Finite-Difference Methods for Partial Differential Equations.* John Wiley & Sons, New York.

FOSTER, M. D. 1950. The origin of high sodium bicarbonate waters in the Atlantic and Gulf Coastal Plains. *Geochim. Cosmochim. Acta*, 1, pp. 33–48.

FOSTER, S. S. D. 1975. The chalk groundwater tritium anomaly—a possible explanation. *J. Hydrol.*, 25, pp. 159–165.

FOSTER, S. S. D., and R. I. CREASE. 1972. Nitrate pollution of chalk groundwater in East Yorkshire: a hydrogeological appraisal. *Groundwater Pollution in Europe*, ed. J. A. Cole. Water Information Center, Port Washington, N.Y., pp. 269–271.

FREEZE, R. A. 1967. Quantitative interpretation of regional groundwater flow patterns as an aid to water balance studies. *Intern. Assoc. Sci. Hydrol.*, Gen. Assembly of Berne, Publ. 78, pp. 154–173.

FREEZE, R. A. 1969a. Regional groundwater flow—Old Wives Lake Drainage Basin, Saskatchewan. *Can. Inland Waters Branch, Sci. Series No. 5*, 245 pp.

FREEZE, R. A. 1969b. The mechanism of natural groundwater recharge and discharge: 1. One-dimensional, vertical, unsteady, unsaturated flow above a recharging or discharging groundwater flow system. *Water Resources Res.*, 5, pp. 153–171.

FREEZE, R. A. 1971a. Three-dimensional, transient, saturated-unsaturated flow in a groundwater basin. *Water Resources Res.*, 7, pp. 347–366.

FREEZE, R. A. 1971b. Influence of the unsaturated flow domain on seepage through earth dams. *Water Resources Res.*, 7, pp. 929–941.

FREEZE, R. A. 1972a. Role of subsurface flow in generating surface runoff: 1. Baseflow contributions to channel flow. *Water Resources Res.*, 8, pp. 609–623.

FREEZE, R. A. 1972b. Role of subsurface flow in generating surface runoff: 2. Upstream source areas. *Water Resources Res.*, 8, pp. 1272–1283.

FREEZE, R. A. 1974. Streamflow generation. *Rev. Geophys. Space Phys.*, 12, pp. 627–647.

FREEZE, R. A. 1975. A stochastic conceptual analysis of one-dimensional groundwater flow in non-uniform homogeneous media. *Water Resources Res.*, 11, pp. 725–741.

FREEZE, R. A., and J. BANNER. 1970. The mechanism of natural groundwater recharge and discharge: 2. Laboratory column experiments and field measurements. *Water Resources Res.*, 6, pp. 138–155.

FREEZE, R. A., and P. A. WITHERSPOON. 1966. Theoretical analysis of regional groundwater flow: 1. Analytical and numerical solutions to the mathematical model. *Water Resources Res.*, 2, pp. 641–656.

FREEZE, R. A., and P. A. WITHERSPOON. 1967. Theoretical analysis of regional groundwater flow: 2. Effect of water-table configuration and subsurface permeability variation. *Water Resources Res.*, 3, pp. 623–634.

FREEZE, R. A., and P. A. WITHERSPOON. 1968. Theoretical analysis of regional groundwater flow: 3. Quantitative interpretations. *Water Resources Res.*, 4, pp. 581–590.

FRIED, J. J. 1975. *Groundwater Pollution*, Elsevier, Amsterdam, 330 pp.

FRIED, J. J., P. C. LEVEQUE, D. POITRINAL, and J. SEVERAC. 1974. Local studies of miscible pollutions of groundwater: the single well pulse technique. *Groundwater Pollution in Europe*, ed. J. A. Cole. Water Information Center, Port Washington, N.Y., pp. 388–406.

FRIND, E. O. 1970. Theoretical analysis of aquifer response due to dewatering at Welland. *Can. Geotech. J.*, 7, pp. 205–216.

FRIND, E. O., and G. F. PINDER. 1973. Galerkin solution of the inverse problem for aquifer transmissivity. *Water Resources Res.*, 9, pp. 1397–1410.

FRITZ, P., R. J. DRIMMIE, and F. W. RENDER. 1974. Stable isotope contents of a major prairie aquifer in central Manitoba, Canada. *Isotope Techniques in Groundwater Hydrology*. Intern. Atomic Energy Agency, Vienna, pp. 379–397.

FRITZ, P., J. A. CHERRY, K. U. WEYER, and M. SKLASH. 1976. Storm runoff analyses using environmental isotopes and major ions. *Interpretation of Environmental Isotope and Hydrochemical Data in Groundwater Hydrology*. Intern. Atomic Energy Agency, Vienna, pp. 111–130.

FUHRIMAN, D. K., and J. R. BARTON. 1971. Ground water pollution in Arizona, California, Nevada and Utah. *Environ. Protect. Agency. Rept. 16060 ERU.*

GAMBOLATI, G. 1973a. Equation for one-dimensional vertical flow of groundwater: 1. The rigourous theory. *Water Resources Res.*, 9, pp. 1022–1028.

GAMBOLATI, G. 1973b. Equation for one-dimensional vertical flow of groundwater: 2. Validity range of the diffusion equation. *Water Resources Res.*, 9, pp. 1385–1395.

GAMBOLATI, G. 1974. Second-order theory of flow in three-dimensional deforming media. *Water Resources Res.*, 10, pp. 1217–1228.

GAMBOLATI, G. 1976. Transient free surface flow to a well: An analysis of theoretical solutions. *Water Resources Res.*, 12, pp. 27–39.

GAMBOLATI, G., and R. A. FREEZE. 1973. Mathematical simulation of the subsidence of Venice: 1. Theory. *Water Resources Res.*, 9, pp. 721–733.

GAMBOLATI, G., P. GATTO, and R. A. FREEZE. 1974a. Mathematical simulation of the subsidence of Venice: 2. Results. *Water Resources Res.*, 10, pp. 563–577.

GAMBOLATI, G., P. GATTO, and R. A. FREEZE. 1974b. Predictive simulation of the subsidence of Venice. *Science*, 183, pp. 849–851.

GARLAND, G. A., and D. C. MOSHER. 1975. Leachate effects from improper land disposal. *J. Wasteage*, 6, pp. 42–48.

GARRELS, R. M. 1967. Genesis of some ground waters from igneous rocks. *Researches in Geochemistry*, Vol. 2, ed. P. H. Abelson. John Wiley & Sons, New York, 663 pp.

GARRELS, R. M., and C. L. CHRIST. 1965. *Solutions, Minerals and Equilibria*. Harper & Row, New York, 450 pp.

GARRELS, R. M., and P. HOWARD. 1957. Reactions of feldspars and mica with water at low temperature and pressure. *Proc. 6th Nat. Conf. Clays Clay Minerals.*

GARRELS, R. M., and F. T. MACKENZIE. 1967. Origin of the chemical compositions of some springs and lakes. *Equilibrium Concepts in Natural Water Systems*, ed. R. F. Gould. American Chemical Society Publications, Washington, D.C.

GASPAR, E., and M. ONCESCU. 1972. *Radioactive Tracers in Hydrology*. American Elsevier, New York, 90 pp.

GATES, J. S., and C. C. KISIEL. 1974. Worth of additional data to a digital computer model of a groundwater basin. *Water Resources Res.*, 10, pp. 1031–1038.

GERA, F., and D. G. JACOBS. 1972. Considerations in the long-term management of high-level radioactive wastes. *Rept. ORNL-4762*, Oak Ridge Nat. Lab., Oak Ridge, Tenn., 151 pp.

GERBA, C. P., C. WALLIS, and J. L. MELNICK. 1975. The fate of wastewater bacteria and viruses in soil. *J. Irri. Drainage Div., Proc. Amer. Soc. Civil Engrs.*, 101, pp. 157–174.

GERMANOV, A. I., G. A. VOLKOV, A. K. LISITSIN, and V. S. SEREBENNIKOV. 1958. Investigation of the oxidation reduction potential of ground waters. *Acad. Sci. USSR, Moscow, Inst. Geol. Ore Deposits, Petrogr., Mineral, Geochem.*, pp. 322–329.

GHYBEN, W. B. 1888. Nota in verband met de voorgenomen putboring nabij Amsterdam. *Tijdschrift van Let Koninklijk Inst. van Ing.*

GIBBS, H. J., and W. Y. HOLLAND. 1960. Petrographic and engineering properties of loess. *U.S. Bur. Reclamation, Eng. Monogr. No. 28.*

GIBSON, U. P., and R. D. SINGER. 1971. *Water Well Manual.* Premier Press, Berkeley, California 156 pp.

GIGER, W., and P. V. ROBERTS. 1977. Characterization of refractory organic carbon. *Water Pollution Microbiology*, Vol. 2, ed. Ralph Mitchell. Wiley-Interscience, New York.

GILLHAM, R. W., and J. A. CHERRY. 1978. Field evidence of denitrification in shallow groundwater flow systems. *Proc. 13th Can. Symp. Water Pollution Res.* McMaster University, Hamilton, Ontario.

GILLHAM, R. W., and L. R. WEBBER. 1969. Nitrogen contamination of groundwater by barnyard leachates. *J. Water Pollution Control Fed.*, 41, pp. 1752–1762.

GOFF, K. J. 1971. Hydrology and chemistry of the Shoal Lakes Basin, Interlake Area, Manitoba. Unpublished Master's thesis, University of Manitoba.

GOODALL, D. C., and R. M. QUIGLEY. 1977. Pollutant migration from two sanitary landfill sites near Sarnia, Ontario. *Can. Geotech. J.*, 14, pp. 223–236.

GOODMAN, R. E., D. G. MOYE, A. VAN SCHALKWYK, and I. JAVANDEL. 1965. Ground water inflows during tunnel driving. *Eng. Geol.*, 2, pp. 39–56.

GOSLING, A. W., E. A. JENNE, and T. T. CHAO. 1971. Gold content of natural waters in Colorado. *Econ. Geol.* 66, pp. 309–313.

GRAF, D. L., I. FRIEDMAN, and W. F. MEENTS. 1965. The origin of saline formation waters: II. Isotropic fractionation by shale micropore systems. *Illinois State Geol. Surv. Circ. 393*, 32 pp.

GRAHAM, G. S., S. E. KESLER, and J. C. van LOON. 1975. Fluorine in groundwater as a guide to Pb–Zn–Ba–F mineralization. *Econ. Geol.*, 70, pp. 396–398.

GRAY, D. M., et al. 1970. *Handbook on the Principles of Hydrology.* National Research Council, Canada.

GRAY, D. M., G. A. MCKAY, and J. M. WIGHAM. 1970. Energy, evaporation and evapotranspiration. *Handbook on the Principles of Hydrology*, ed. D. M. Gray. National Research Council, Canada.

GREEN, D. W., H. DABIRI, C. F. WEINAUG, and R. PRILL. 1970. Numerical modeling of unsaturated groundwater flow and comparison of the model to a field experiment. *Water Resources Res.*, 6, pp. 862–874.

GREENKORN, R. A., and D. P. KESSLER. 1969. Dispersion in heterogenous nonuniform anisotropic porous media. *Ind. Eng. Chem.*, 61, pp. 14–32.

GRICE, R. H. 1968. Hydrogeology of the jointed dolomites, Grand Rapids Hydroelectric Power Station, Manitoba, Canada. *Geol. Soc. Amer. Eng. Geol. Case Histories*, 6, pp. 33–48.

GRIFFIN, R. A., K. CARTWRIGHT, N. F. SHIMP, J. D. STEELE, R. R. BUCH, W. A. WHITE, G. M. HUGES, and R. H. GILKESON. 1976. Alteration of pollutants in municipal landfill leachate by clay minerals: Part I. Column leaching and field verification. *Illinois State Geol. Surv. Bull. 78.*

GRIM, R. E. 1968. *Clay mineralogy*, 2nd ed. McGraw-Hill, New York, 596 pp.

GRISAK, G. E. 1975. Nitrates in shallow groundwater near Lethbridge. *Tech. Rept. Environ. Protection Services*, Edmonton, Alberta.

GRISAK, G. E., and J. A. CHERRY. 1975. Hydrogeologic characteristics and response of fractured till and clay confining a shallow aquifer. *Can. Geotech. J.*, 12, pp. 23–43.

GRISAK, G. E., J. A. CHERRY, J. A. VONHOF, and J. P. BLEUMLE. 1976. Hydrogeologic and hydrochemical properties of fractured till in the interior plains region. *Glacial Till*, ed. R. F. Legget. Spec. Publ. Roy. Soc. Can., pp. 304–335.

GRISAK, G. E., W. F. MERRITT, and D. W. WILLIAMS. 1977. A fluoride borehole dilution apparatus for groundwater velocity measurements. *Can. Geotech. J.*, 14, pp. 554–561.

GROBA, F., and J. HAHN. 1972. Variations of groundwater chemistry by anthropogenic factors in northwest Germany. *Proc. 24th Intern. Geol. Congr.* Montreal, Sec. 11, *Hydrogeol.*, pp. 270–281.

GROENEWOLD, G., and others. In press. Geology and hydrogeology of the Knife River Basin, West-Central North Dakota. *North Dakota Geol. Surv. Rept. Investigation 65.*

GROVE, D. B., and W. A. BEETEM. 1971. Porosity and dispersion constant calculations for a fractured carbonate aquifer using the two well tracer method. *Water Resources Res.*, 7, pp. 128–134.

GUENTHER, W. B. 1968. *Quantitative Chemistry.* Addison-Wesley, Reading, Mass.

GUENTHER, W. B. 1975. *Chemical Equilibrium: A Practical Introduction for the Physical and Life Sciences.* Plenum Press, New York, 248 pp.

GUGGENHEIM, E. A. 1949. *Thermodynamics: An Advanced Treatment for Chemists and Physicists*, 4th ed. North-Holland, Amsterdam.

GUPTA, S. K., and K. K. TANGI. 1976. A three-dimensional Galerkin finite-element solution of flow through multiaquifers in Sutter Basin, California. *Water Resources Res.*, 12, pp. 155–162.

GURR, C. G., T. J. MARSHALL, and J. T. HUTTON. 1952. Water movement in soil due to a temperature gradient. *Soil Sci.*, 24, pp. 335–344.

GYMER, R. G. 1973. *Chemistry: An Ecological Approach.* Harper & Row, New York, 801 pp.

HAGMAIER, J. L. 1971. Groundwater flow, hydrochemistry, and uranium deposition in the Power River Basin, Wyoming. Ph.D. thesis, University of North Dakota, 166 pp.

HAGUE, R., D. W. SCHMEDDING, and V. H. FREED. 1974. Aqueous solubility, adsorption, and vapour behaviour of polychlorinated bipheryl arochor 1254. *J. Environ. Sci. Technol.*, 8, no. 2, pp. 139–142.

HALEVY, E., H. MOSER, O. ZELLHOFER, and A. ZUBER. 1967. Borehole dilution techniques: a critical review. *Isotopes in Hydrology.* IAEA, Vienna, pp. 531–564.

HALL, E. S. 1972. Some chemical principles of groundwater pollution. *Groundwater Pollution in Europe*, ed. J. A. Cole. Water Information Center, Port Washington, N.Y.

HALL, F. R. 1968. Baseflow recessions—a review. *Water Resources Res.*, 4, pp. 973–983.

HAMAKER, J. W., and J. M. THOMPSON. 1972. Physicochemical relationships of organic chemicals in soils—adsorption. *Organic Chemicals in the Soil Environment*, ed. C. A. Goring and J. W. Hamahu. Marcel Dekker, New York, 49 pp.

HAMBLIN, W. K. 1976. *The Earth's Dynamic Systems: A Textbook in Physical Geology.* Burgess, Minneapolis, Minn.

HAMILTON, T. M. 1970. Groundwater flow in part of the Little Missouri River Basin, North Dakota. Unpublished Ph.D. thesis, University of North Dakota.

HANSHAW, B. B. 1962. Membrane properties of compacted clays. Ph.D. thesis, Harvard University.

HANSHAW, B. B., and T. B. COPLEN. 1973. Ultrafiltration by a compacted clay membrane: II. Sodium ion exclusion at various ionic strengths. *Geochim. Cosmochim. Acta*, 37, pp. 2311–2327.

HANSHAW, B. B., W. BACK, and R. G. DEIKE. 1971. A geochemical hypothesis for dolomitization by ground water. *Econ. Geol*, pp. 710–724.

HANTUSH, M. S. 1956. Analysis of data from pumping tests in leaky aquifers. *Trans. Amer. Geophys. Union*, 37, pp. 702–714.

HANTUSH, M. S. 1960. Modification of the theory of leaky aquifers. *J. Geophys. Res.*, 65, pp. 3713–3725.

HANTUSH, M. S. 1962. Drawdown around a partially-penetrating well. *Amer. Soc. Civil Engrs. Trans.*, 127, pp. 268–283.

HANTUSH, M. S. 1964. Hydraulics of wells. *Adv. Hydrosci.*, 1, pp. 281–432.

HANTUSH, M. S. 1967. Growth and decay of groundwater mounds in response to uniform percolation. *Water Resources Res.*, 3, pp. 227–234.

HANTUSH, M. S., and C. E. JACOB. 1955. Nonsteady radial flow in an infinite leaky aquifer. *Trans. Amer. Geophys. Union*, 36, pp. 95–100.

HARLAN, R. L. 1973. Analysis of coupled heat-fluid transport in partially frozen soil. *Water Resources Res.*, 9, pp. 1314–1323.

HARR, M. E. 1962. *Groundwater and Seepage*. McGraw-Hill, New York.

HARRISON, S. S., and L. CLAYTON. 1970. Effects of groundwater seepage on fluvial processes. *Bull. Geol. Soc. Amer.*, 81, pp. 1217–1226.

HAWKES, H. E., and J. S. WEBB. 1962. *Geochemistry in Mineral Exploration*. Harper & Row, New York, 415 pp.

HEALY, J. H. 1975. Recent highlights and future trends in research on earthquake prediction and control. *Rev. Geophys. Space Phys.*, 13, pp. 361–364.

HEALY, J. H., W. W. RUBEY, D. T. GRIGGS, and C. B. RALEIGH. 1968. The Denver earthquakes. *Science*, 161, pp. 1301–1310.

HEDBERG, H. D. 1964. Geologic aspects of origin of petroleum. *Bull. Amer. Assoc. Petrol. Geol.*, 48, pp. 1755–1803.

HEDLIN, R. A. 1972. Nitrate contamination of ground water in the Neepawa-Langruth area of Manitoba. *Can. J. Soil Sci.*, pp. 75–84.

HELFFERICH, F. 1962. *Ion Exchange*. McGraw-Hill, New York.

HELGESON, H. C. 1970. A chemical and thermodynamic model of ore deposition in hydrothermal systems. *Min. Soc. Amer. Spec. Paper No. 3*, pp. 155–186.

HELM, D. C. 1975. One-dimensional simulation of aquifer system compaction near Pixley, California: 1. Constant Parameters. *Water Resources Res.*, 11, pp. 465–478.

HELM, D. C. 1976. One-dimensional simulation of aquifer system compaction near Pixley, California: 2. Stress-dependent parameters. *Water Resources Res.*, 12, pp. 375–391.

HEM, J. D. 1967. Equilibrium chemistry of iron in ground water. *Principles and Applications of Water Chemistry*, ed. S. D. Faust and J. V. Hunter. John Wiley & Sons, New York.

HEM, J. D. 1970. Study and interpretation of the chemical characteristics of natural water, *U.S. Geol. Surv. Water-Supply Paper 1473*, 363 pp.

HENRY, H. R. 1960. Salt water intrusion into coastal aquifers. *Intern. Assoc. Sci. Hydrol, Publ. 52*, pp. 478–487.

HERZBERG, A. 1901. Die Wasserversorgung einiger Nordseebader. *J. Gasbeleucht. Wasserversorg.*, 44, pp. 815–819.

HEWLETT, J. D., and A. R. HIBBERT. 1963. Moisture and energy conditions within a sloping soil mass during drainage. *J. Geophys. Res.*, 68, pp. 1081–1087.

HEWLETT, J. D., and A. R. HIBBERT. 1967. Factors affecting the response of small watersheds to precipitation in humid areas. *Forest Hydrology*, ed. W. E. Sopper and H. W. Lull. Pergamon Press, Oxford, pp. 275–290.

HEWLETT, J. D., and W. L. NUTTER. 1970. The varying source area of streamflow from upland basins. *Proc. Amer. Soc. Civil Engrs. Symp. Interdisciplinary Aspects of Watershed Management*, Montana State University, Bozeman, pp. 65–93.

HIGGINS, G. H. 1959. Evaluation of the groundwater contamination hazard from underground nuclear explosives. *J. Geophys. Res.*, 64., pp. 1509–1519.

HILL, R. A. 1940. Geochemical patterns in Coachella Valley, Calif. *Trans. Amer. Geophys. Union*, 21.

HITCHON, B. 1969a. Fluid flow in the Western Canada Sedimentary Basin: 1. Effect of topography. *Water Resources Res.*, 5, pp. 186–195.

HITCHON, B. 1969b. Fluid flow in the Western Canada Sedimentary Basin: 2. Effect of geology. *Water Resources Res.*, 5, pp. 460–469.

HITCHON, B. 1971. Origin of oil: geological and geochemical constraints. *Origin and Refinery of Petroleum*, Advances in Chemistry Series, No. 103, American Chemical Society, Washington, D.C., pp. 30–66.

HITCHON, B. 1976. Hydrogeochemical aspects of mineral deposits in sedimentary rocks. *Handbook of Stratabound and Stratiform Ore Deposits*, ed. K. H. Wolf. Elsevier, New York, pp. 53–66.

HITCHON, B. 1977. Geochemical links between oil fields and ore deposits in sedimentary rocks. *Proc. Forum on Oil and Ore in Sediments*, Imperial College, London, 34 pp.

HITCHON, B., and I. FRIEDMAN. 1969. Geochemistry and origin of formation waters in the Western Canada Sedimentary Basin: I. Stable isotopes of hydrogen and oxygen. *Geochim. Cosmochim. Acta*, 35, pp. 1321–1349.

HITCHON, B., and J. HAYS. 1971. Hydrodynamics and hydrocarbon occurrences, Surat Basin, Queensland, Australia. *Water Resources Res.*, 7, pp. 658–676.

HITCHON, B., G. K. BILLINGS, and J. E. KLOVAN. 1971. Geochemistry and origin of formation waters in the Western Canada Sedimentary Basin: III. Factors controlling chemical composition. *Geochim. Cosmochim. Acta*, 35, pp. 567–598.

HOAG, R. B., and G. R. WEBBER. 1976. Significance for mineral exploration of sulphate concentrations in groundwaters. *CIM Bull.*, 69, no. 776, pp. 86–91.

HOBSON, G. 1967. Seismic methods in mining and groundwater exploration. Proc. Can. Centennial Conf. Mining Groundwater Geophys. *Geol. Surv. Can., Econ. Geol. Rept. 26*, pp. 148–176.

HODGE, R. A. L., and R. A. FREEZE. 1977. Groundwater flow systems and slope stability. *Can. Geotech J.*, 14, pp. 466–476.

HOEK, E., and J. BRAY. 1974. *Rock Slope Engineering*. Institute of Mining and Metallurgy, London.

HOEKSTRA, P. 1966. Moisture movement in soils under temperature gradients with the cold-side temperature below freezing. *Water Resources Res.*, 2, pp. 241–250.

HOLLAND, H. D., T. V. KIRSIPU, J. S. HUEFNER, and U. M. OXBURGH. 1964. On some aspects of the chemical evolution of cave waters. *J. Geol.*, 72, no. 1, pp. 36–67.

HOLMES, R. M., and G. W. ROBERTSON. 1959. A modulated soil moisture budget. *Monthly Weather Rev.*, 87, no. 3, pp. 1–7.

HORNBERGER, G. M., I. REMSON, and A. A. FUNGAROLI. 1969. Numeric studies of a composite soil moisture groundwater system. *Water Resources Res.*, 5, pp. 797–802.

HORNBERGER, G. M., J. EBERT, and I. REMSON. 1970. Numerical solution of the Boussinesq equation for aquifer-stream interaction. *Water Resources Res.*, 6, pp. 601–608.

HORTON, R. E. 1933. The role of infiltration in the hydrologic cycle. *Trans. Amer. Geophys. Union*, 14, pp. 446–460.

HOUSTON, W. N. 1972. The surface chemistry and geochemistry of feldspar weathering. Unpublished Master's thesis, McMaster University.

HOWARD, A. D. 1964a. Processes of limestone cavern development. *Intern. J. Speleol.*, 1, pp. 47–60.

HOWARD, A. D. 1964b. Model for cavern development under artesian groundwater flow, with special reference to the Black Hills. *Nat. Speleol. Soc. Bull.*, 26, pp. 7–16.

HOWARD, A. D., and B. Y. HOWARD. 1967. Solution of limestone under laminar flow between parallel boundaries. *Caves and Karst*, 9, pp. 25–40.

HUBBERT, M. K. 1940. The theory of groundwater motion. *J. Geol.*, 48, pp. 785–944.

HUBBERT, M. K. 1954. Entrapment of petroleum under hydrodynamic conditions. *Bull. Amer. Assoc. Petrol. Geol.*, 37, pp. 1954–2026.

HUBBERT, M. K. 1956. Darcy's law and the field equations of the flow of underground fluids. *Trans. Amer. Inst. Min. Met. Eng.*, 207, pp. 222–239.

HUBBERT, M. K., and W. W. RUBEY. 1959. Role of fluid pressures in mechanics of overthrust faulting: I. Mechanics of fluid-filled porous solids and its application to overthrust faulting. *Bull. Geol. Soc. Amer.*, 70, pp. 115–166.

HUGGENBERGER, F., J. LETEY, and W. J. FARMER. 1972. Observed and calculated distribution of lindane in soil columns as influenced by water movement. *J. Soil Sci. Soc. Amer. Proc.*, 36, pp. 544–548.

HUGHES, G. M., R. A. LANDON, and R. N. FARVOLDEN. 1971. Hydrogeology of solid waste disposal sites in northeastern Illinois. *U.S. Environ. Protection Agency Rept. SW-12D*, 154 pp.

HUNTOON, P. W. 1974. Predicting water-level declines for alternative groundwater developments in the Upper Big Blue River Basin, Nebraska. *Univ. Nebraska, Inst. Agr. Natural Resources, Resource Rept.*, 106 pp.

HVORSLEV, M. J. 1951. Time lag and soil permeability in groundwater observations. *U.S. Army Corps Engrs. Waterways Exp. Sta. Bull. 36*, Vicksburg, Miss.

IBRAHIM, H. A., and W. BRUTSAERT. 1965. Inflow hydrographs from large unconfined aquifers. *J. Irr. Drain. Div., Proc. Amer. Soc. Civil Engrs.*, 91 (IR2), pp. 21–38.

ILIFF, N. A. 1973. Organic chemicals in global aspects of chemistry. *Toxicology and Technology as Applied to the Environment*, Vol. 2, ed. F. Coulson and F. Korte. Academic Press, New York, 64 pp.

INESON, J., and R. A. DOWNING. 1964. The groundwater component of river discharge and its relationship to hydrogeology. *J. Inst. Water Eng.*, 18, pp. 519–541.

INESON J., and R. F. PACKHAM. 1967. Contamination of water by petroleum products. *The Joint Problems of the Oil and Water Industry*, ed. Peter Hepple. Proc. Symp. Brighton, England, Inst. Petrol., pp. 97–116.

IRMAY, S. 1958. On the theoretical derivation of Darcy and Forchheimer formulas. *Trans. Amer. Geophys. Union*, 39, pp. 702–707.

JACOB, C. E. 1939. Fluctuations in artesian pressure produced by passing railroad trains as shown in a well on Long Island, New York. *Trans. Amer. Geophys. Union*, 20, pp. 666–674.

JACOB, C. E. 1940. On the flow of water in an elastic artesian aquifer. *Trans. Amer. Geophys. Union*, 2, pp. 574–586.

JACOB, C. E. 1950. Flow of groundwater. *Engineering Hydraulics*, ed. H. Rouse. John Wiley & Sons, New York, pp. 321–386.

JACOBSON, R. L., and D. LANGMUIR. 1970. The chemical history of some spring waters in carbonate rocks. *Ground Water*, 8, pp. 5–9.

JAEGER, J. C. 1971. Friction of rocks and stability of rock slopes. *Géotechnique*, 21, pp. 97–134.

JAEGER, C. 1972. *Rock Mechanics and Engineering*. Cambridge University Press, London.

JAKUCS, L. 1973. The karstic corrosion of naturally occurring limestones in the geomorphology of our age, *Symp. Karst-Morphogenesis Intern. Geogr. Union*, 52 pp.

JAVANDEL, I., and P. A. WITHERSPOON. 1969. A method of analyzing transient fluid flow in multilayered aquifers. *Water Resources Res.*, 5, pp. 856–869.

JENNE, E. A. 1968. Control of Mn, Fe, Ni, Cu, and Zn concentration in soils and waters: significant role of hydrous Mn and Fe oxides. *Trace Inorganics in Water*, American Chemical Society Publication 73, pp. 337–387.

JENSEN, H. E., and K. L. BABCOCK. 1973. Cation-exchange equilibria on a Yolo loam. *Hilgardia*, 41, pp. 475–488.

JEPPSON, R. W. 1968. Axisymmetric seepage through homogeneous and nonhomogeneous porous mediums. *Water Resources Res.*, 4, pp. 1277–1288.

JEPPSON, R. W., and R. W. NELSON. 1970. Inverse formulation and finite-difference solution to partially-saturated seepage from canals. *Soil Sci. Soc. Amer. Proc.*, 34, pp. 9–14.

JOHN, K. W. 1968. Graphical stability analysis of slopes in jointed rock. *J. Soil Mech. Found. Div., Proc. Amer. Soc. Civil Engrs.*, 49 (SM2), pp. 497–526.

JOHNSON, A. I., and D. A. MORRIS. 1962. Physical and hydrologic properties of water bearing deposits from core holes in the Las Banos-Kettleman City Area, California. U.S. Geol. Surv. Open-File Dept., Denver, Colo.

JOHNSON, A. I., R. P. MOSTON, and D. A. MORRIS. 1968. Physical and hydrologic properties of water-bearing deposits in subsiding areas in central California. *U.S. Geol. Surv. Prof. Paper 497A*.

JOHNSTON, P. M. 1962. Geology and ground-water resources of the Fairfax Quadrangle, Virginia. *U.S. Geol. Surv. Water-Supply Paper 1539-L*.

JONES, P. H., and H. E. SKIBITZKE. 1956. Subsurface geophysical methods in groundwater hydrology. *Adv. Geophys.*, 3, pp. 241–300.

KARPLUS, W. J. 1958. *Analog Simulation*. McGraw-Hill, New York.

KAUFMAN, W. J. 1974. Chemical pollution of ground water. *Water Technol./Quality*, pp. 152–158.

KAY, B. D., and D. L. ELRICK. 1967. Adsorption and movement of lindane in soils. *J. Soil Sci.*, *104*, pp. 314–322.

KAZMANN, R. G. 1956. Safe yield in groundwater development, reality or illusion? *J. Irr. Drain. Div., Proc. Amer. Soc. Civ. Engrs.*, 82 (IR3), 12 pp.

KAZMANN, R. G. 1972. *Modern Hydrology*, 2nd ed. Harper & Row, New York.

KAZMANN, R. G. 1974. Waste surveillance in subsurface disposal projects. *Ground Water Proc. 2nd NWWA-EPA National Ground Water Quality Symp.*, 1, pp. 418–426.

KEMPER, W. D., J. OLSEN, and C. J. DEMOODY. 1975. Dissolution rate of gypsum in flowing groundwater. *Soil Sci. Soc. Amer. Proc.*, 39, pp. 458–463.

KERN, D. B. 1960. The hydration of carbon dioxide. *J. Chem. Educ.*, 37, pp. 14–23.

KEYS, W. S. 1967. Borehole geophysics as applied to groundwater. Proc. Canadian Centennial Conf. Mining Groundwater Geophys. *Geol. Surv. Can., Econ. Geol. Rept. 26*, pp. 598–612.

KEYS, W. S. 1968. Logging in groundwater hydrology. *Groundwater*, 6, pp. 10–18.

KHARAKA, Y. K., and I. BARNES. 1973. SOLMNEQ, solution—mineral equilibrium computations. *U.S. Geol. Surv. Computer Contr. NTIS Rept. PB2-15899*.

KHARAKA, Y. K., and F. A. F. BERRY. 1973. Simultaneous flow of water and solutes through geological membranes: I. Experimental investigation. *Geochim. Cosmochim. Acta*, 37, pp. 2577–2603.

KHARAKA, Y. K., and W. C. SMALLEY. 1976. Flow of water and solutes through compacted clays. *Amer. Assoc. Petrol. Geol. Bull.*, 60, pp. 973–980.

KHARAKA, Y. K., F. A. F. BERRY, and I. FRIEDMAN. 1973. Isotopic composition of oil-field brines from Kettleman North Dome, California, and their geologic implications. *Geochim. Cosmochim. Acta*, 37, pp. 1899–1908.

KIMMEL, G. E., and O. C. BRAIDS. 1974. Leachate plumes in a highly permeable aquifer. *Ground Water*, 12, pp. 388–393.

KIPP, K. L. 1973. Unsteady flow to a partially penetrating finite radius well in an unconfined aquifer. *Water Resources Res.*, 9, pp. 448–462.

KIRKBY, M. J., and R. J. CHORLEY. 1967. Throughflow, overland flow, and erosion. *Bull. Intern. Assoc. Sci. Hydrol.*, 12, no. 3, pp. 5–21.

KIRKHAM, D. 1964. Soil Physics. *Handbook of Applied Hydrology*, ed. V. T. Chow. McGraw-Hill, New York, pp. 5.1–5.26.

KIRKHAM, D. 1967. Explanation of paradoxes in Dupuit-Forchheimer seepage theory. *Water Resources Res.*, 3, pp. 609–622.

KIRKHAM, D., and W. L. POWERS. 1972. *Advanced Soil Physics*. Wiley-Interscience, New York.

KLEIN, J. M. 1974. Geochemical behaviour of silica in the artesian ground water of the closed basin area, San Luis Valley, Colorado. Unpublished M.Sc. thesis, School of Mines, Boulder, Col.

KLEIN, S. A. 1964. The fate of detergents in septic tank systems and oxidation ponds. *Sanitary Eng. Res. Lab. SERL Rpt. No. 64-1*, University of California, Berkeley.

KLOTZ, D., and H. MOSER. 1974. Hydrodynamic dispersion as aquifer characteristic: model experiments with radioactive tracers. *Isotope Techniques in Groundwater Hydrology*, Vol. 2. Intern. Atomic Energy Agency, Vienna, pp. 341–354.

KLUTE, A. 1965a. Laboratory measurement of hydraulic conductivity of saturated soil. *Methods of Soil Analysis, Part 1*, ed. C. A. Black. American Society of Agronomy, Madison, Wis., pp. 210–221.

KLUTE, A. 1965b. Laboratory measurement of hydraulic conductivity of unsaturated soil. *Methods of Soil Analysis, Part 1*, ed. C. A. Black. American Society of Agronomy, Madison, Wis., pp. 253–261.

KLUTE, A. 1965c. Water diffusivity. *Methods of Soil Analysis, Part 1*, ed. C. A. Black. American Society of Agronomy, Madison, Wis., pp. 262–272.

KNUTSON, G. 1966. Tracers for ground water investigations. *Groundwater Problems* (*Proc. Intern. Symp. Stockholm, Sweden*). Pergamon Press, Oxford.

KOHOUT, F. A. 1960a. Cyclic flow of salt water in the Biscayne aquifer of southeastern Florida. *J. Geophys. Res.*, 65, pp. 2133–2141.

KOHOUT, F. A. 1960b. Flow pattern of fresh water and salt water in the Biscayne aquifer of the Miami area, Florida. *Intern. Assoc. Sci. Hydrol. Publ. 52*, pp. 440–448.

KONIKOW, L. F., and J. D. BREDEHOEFT. 1974. Modeling flow and chemical quality changes in an irrigated steam-aquifer system. *Water Resources Res.*, 10, no. 3, pp. 546–562.

KRAUSKOPF, K. B. 1956. Dissolution and precipitation of silica at low temperatures. *Geochim. Cosmochim. Acta*, 10, pp. 1–26.

KRAUSKOPF, K. B. 1967. *Introduction to Geochemistry*. McGraw-Hill, New York.

KREITLER, C. W., and D. C. JONES. 1975. Natural soil nitrate: the cause of the nitrate contamination of ground water in Runnels county, Texas. *Ground Water*, 13, no. 1, pp. 53–61.

KRONE, R. B., P. H. MCGAUHEY, and H. B. GOTAAS. 1957. Direct discharge of groundwater with sewage effluents. *Amer. Soc. Civil Engrs., J. Sanit. Eng. Div.*, 83 (SA4), pp. 1–25.

KRONE, R. B., G. T. ORLAB, and C. HODGKINSON. 1958. Movement of coliform bacteria through porous media. *Sewage Industrial Wastes*, 30, pp. 1–13.

KROSZYNSKI, U. I., and G. DAGAN. 1975. Well pumping in unconfined aquifers: the influence of the unsaturated zone. *Water Resources Res.*, 11, pp. 479–490.

KRUSEMAN, G. P., and N. A. de RIDDER. 1970. Analysis and evaluation of pumping test data. *Intern. Inst. Land Reclamation and Improvement Bull. 11*, Wageningen, The Netherlands.

KRYNINE, D. P., and W. R. JUDD. 1957. *Principles of Engineering Geology and Geotechnics*. McGraw-Hill, New York.

KU, H. F. H., B. G. KATZ, D. J. S. SULAM, and R. K. KRULIKAS. 1978. Scavenging of chromium and cadmium by aquifer material, South Farmingdale—Massapequa area, Long Island, New York. *Ground Water*, 16, pp. 112–118.

Kubo, A. S., and D. J. Rose. 1973. Disposal of nuclear wastes. *Science*, 182, pp. 1205–1211.

Lal, D., and B. Peters. 1962. Cosmic ray produced isotopes and their applications in problems of geophysics. *Progr. Elementary Particle Cosmic Ray Phys.*, 6, pp. 1–74.

Lal, D., and H. E. Suess. 1968. The radioactivity of the atmosphere and hydrosphere. *Ann. Rev. Nuclear Sci.*, 18, pp. 407–434.

Lambe, T. W. 1951. *Soil Testing for Engineers*. John Wiley & Sons, New York, 165 pp.

Lance, J. C., C. P. Gerba, and J. L. Melnick. 1977. Virus movement in soil columns flooded with secondary sewage effluent. *Appl. J. Environ. Microbiol.*

Laney, R. L. 1965. A comparison of the chemical composition of rainwater and ground water in western North Carolina. *U.S. Geol. Surv. Prof. Paper 525-C*, pp. C187–189.

Langmuir, D. 1970. *Eh-pH* determination. *Methods in Sedimentary Petrology*, ed. R. E. Carver. Wiley-Interscience, New York, pp. 597–635.

Langmuir, D. 1971. The geochemistry of some carbonate ground waters in central Pennsylvania. *Geochim. Cosmochim. Acta*, 35, pp. 1023–1045.

Langmuir, D., and D. O. Whittemore. 1971. Variation in the stability of precipitated ferric oxyhydroxides. *Proc. Symp. Nonequilibrium Systems in Natural Water Chem.* ed. J. D. Hem. Advances in Chemistry Series No. 106, American Chemical Society, Washington, D. C., pp. 209–234.

Lattman, L. A., and R. R. Parizek. 1964. Relationship between fracture traces and the occurrence of ground water in carbonate rocks. *J. Hydrol.*, 2, pp. 73–91.

Leckie, J. O., and R. O. James. 1974. Control mechanisms for trace metals in natural waters. *Aqueous-Environmental Chemistry of Metals*, ed. A. J. Rubin, Ann Arbor Science Publishers, Ann Arbor, Mich., pp. 1–127.

Leckie, J. O., and M. B. Nelson. In press. Role of natural heterogeneous sulfide systems in controlling the concentration and distribution of heavy metals. *Amer. J. Sci.*

Leckie, J. O., J. G. Pace, and C. Halvadakis. 1975. Accelerated refuse stabilization through controlled moisture application. Unpublished report, Dept. Environmental Engineering, Stanford University, Stanford, Calif.

Lee, C. H., and T. S. Cheng. 1974. On seawater encroachment in coastal aquifers. *Water Resources Res.*, 10, pp. 1039–1043.

Lee, D. R. 1976. The role of groundwater in eutrophication of a lake in glacial outwash terrain. *Intern. J. Speleol.*, 8, pp. 117–126.

Lee, D. R. 1977. A device for measuring seepage flux in lakes and estuaries. *Limnol. Oceanogr.*, 22, pp. 140–147.

Lee, D. R., and J. A. Cherry. In press. A field exercise on groundwater flow using seepage meters and minipiezometers. *J. Geol. Educ.*

Leenheer, J. A., and E. W. D. Huffman, Jr. 1977. Classification of organic solutes in water using macroreticular resins. *J. Res., U.S. Geol. Surv.*

Legget, R. F. 1962. *Geology and Engineering*, 2nd ed. McGraw-Hill, New York.

LeGrand, H. E. 1949. Sheet structure, a major factor in the occurrence of ground water in the granites of Georgia. *Econ. Geol.*, 44, pp. 110–118.

LeGrand, H. E. 1954. Geology and ground water in the Statesville area. *North Carolina: North Carolina Dept. Conservation Development, Div. Mineral Resources Bull. 68.*

Lennox, D. H., and V. Carlson. 1967. Integration of geophysical methods for groundwater exploration in the prairie provinces, Canada. Proc. Can. Centennial Conf. Mining Groundwater Geophys., *Geol. Surv. Can., Econ. Geol. Rept. 26*, pp. 517–533.

Levorsen, A. I. 1967. *Geology of Petroleum*, 2nd ed. W. H. Freeman, San Francisco, 724 pp.

LEWIS, B.-A. G. 1976. Selenium in biological systems, and pathways for its volatilization in higher plants. *Environmental Biogeochemistry*, ed. J. O. Nriagir. Ann Arbor Science Publishers, Ann Arbor, Mich., p. 1.

LIAKOPOULOS, A. C. 1965a. Theoretical solution of the unsteady unsaturated flow problems in soils. *Bull. Intern. Assoc. Sci. Hydrol.*, 10, pp. 5–39.

LIAKOPOULOS, A. C. 1965b. Variation of the permeability tensor ellipsoid in homogeneous anisotropic soils. *Water Resources Res.*, 1, pp. 135–141.

LINSLEY, R. K., M. A. KOHLER, and J. L. H. PAULHUS. 1975. *Hydrology for Engineers*. McGraw-Hill, New York.

LISSEY, A. 1967. The use of reducers to increase the sensitivity of piezometers. *J. Hydrol.*, 5, pp. 197–205.

LISSEY, A. 1968. Surficial mapping of groundwater flow systems with application to the Oak River Basin, Manitoba. Ph.D. thesis, University of Saskatchewan, 141 pp.

LOHMAN, S. W., ed. 1972. Definitions of selected groundwater terms—revisions and conceptual refinements. *U.S. Geol. Surv. Water-Supply Paper 1988*, 21 pp.

LONDE, P., G. VIGIER, and R. VORMERINGER. 1969. Stability of rock slopes—a three dimensional study. *J. Soil Mech. Found. Div., Proc. Amer. Soc. Civil Engrs.*, 95 (SM1), pp. 235–262.

LOUIS, C. 1969. A study of groundwater flow in jointed rock and its influence on the stability of rock masses. *Imperial College, Rock Mech. Res. Rept. 10.*

LOVELL, R. E., L. DUCKSTEIN, and C. C. KISIEL. 1972. Use of subjective information in estimation of aquifer parameters. *Water Resources Res.*, 8, pp. 680–690.

LUMB, P. 1975. Slope failures in Hong Kong. *Quart. J. Eng. Geol.*, 8, pp. 31–65.

LUTHIN, J. N. 1953. An electrical resistance network for solving drainage problems. *Soil Sci.*, 75, pp. 259–274.

LUTHIN, J. N., and P. R. DAY. 1955. Lateral flow above a sloping water table. *Soil. Sci. Soc. Amer. Proc.*, 18, pp. 406–410.

LUTHIN, J. N., and R. E. GASKELL. 1950. Numerical solutions for tile drainage of layered soils. *Trans. Amer. Geophys. Union*, 31, pp. 595–602.

LVOVITCH, M. I. 1970. World water balance: general report. *Proc. Symp. World Water Balance, Intern. Assoc. Sci. Hydrol.*, 2, pp. 401–415.

MAASLAND, M. 1957. Soil anisotropy and land drainage. *Drainage of Agricultural Lands*, ed. J. N. Luthin. American Society of Agronomy, Madison, Wis., pp. 216–285.

MADDOCK, T. 1974. The operation of a stream-aquifer system under stochastic demands. *Water Resources Res.*, 10, pp. 1–10.

MALCOLM, R. L., and J. A. LEENHEER. 1973. The usefulness of organic carbon parameters in water quality investigations. *Inst. Environ. Sci. Proc., No. 19*, Anaheim, Calif., pp. 336–340.

MANTELL, C. L., ed. 1975. *Solid Wastes: Origin, Collection, Processing, and Disposal.* Wiley-Interscience, New York, 1127 pp.

MARCUS, H., and D. E. EVENSON. 1961. Directional permeability in anisotropic porous media. *Water Resources Center Contrib. No. 31*, University of California, Berkeley.

MARINO, M. A. 1975a. Artificial groundwater recharge: I. Circular recharging area. *J. Hydrol.*, 25, pp. 201–208.

MARINO, M. A. 1975b. Artificial groundwater recharge: II. Rectangular recharging area. *J. Hydrol.*, 26, pp. 29–37.

MASCH, F. D., and K. J. DENNY. 1966. Grain-size distribution and its effect on the permeability of unconsolidated sands. *Water Resources Res.*, 2, pp. 665–677.

MATTHESS, G. 1974. Heavy metals as trace constituents in natural and polluted groundwaters. *Geol. Mijnbouw*, 53, pp. 149–155.

MATHEWS, W. H., and J. R. MACKAY. 1960. Deformation of soils by glacier ice and the influence of pore pressure and permafrost. *Roy. Soc. Can. Trans.*, 54, ser. 3, sec. 4, pp. 27–36.

MAWSON, C. A., and A. E. RUSSELL. 1971. Canadian experience with a national waste-management facility. *Management of Low- and Intermediate-Level Radioactive Wastes, IAEC, Vienna*, pp. 183–194.

MAXEY, G. B. 1964. Hydrogeology. *Handbook of Applied Hydrology*, ed. V. T. Chow. McGraw-Hill, New York, pp. 4.1–4.38.

MAXWELL, J. C. 1964. Influence of depth, temperature, and geologic age on porosity of quartzose sandstone. *Bull. Amer. Assoc. Petrol. Geol.*, 48, pp. 697–709.

MAZOR, E. 1972. Paleotemperatures and other hydrological parameters deduced from noble gases dissolved in groundwaters: Jordan Rift Valley, Israel. *Geochim. Cosmochim. Acta*, 36, pp. 1321–1336.

MCBRIDE, M. S., and H. O. PFANNKUCH. 1975. The distribution of seepage within lakebeds. *U.S. Geol. Surv. J. Res.*, 3, no. 5, pp. 505–512.

MCCARTY, P. L. 1965. Thermodynamics of biological synthesis and growth. *Advances in Water Pollution Research*, ed. J. K. Baars, Pergamon Press, New York, pp. 169–187.

MCCRACKEN, D. D., and W. S. DORN. 1964. *Numerical Methods and Fortran Programming with Applications in Engineering and Science*. John Wiley & Sons, New York.

MCDONALD, H. R., and D. WANTLAND. 1961. Geophysical procedures in ground water study. *Trans. Amer. Soc. Civil Engrs.*, 126, pp. 122–135.

MCGARY, L. M., and T. W. LAMBERT. 1962. Reconnaissance of ground-water resources of the Jackson Purchase region, Kentucky. *U.S. Geol. Surv. Hydrol. Atlas 13*.

MCGINNIS, L. D. 1968. Glaciation as a possible cause of mineral deposition. *Econ. Geol.*, 63, pp. 390–400.

MCGUINNESS, C. L. 1963. The role of groundwater in the National Water Situation. *U.S. Geol. Surv. Water-Supply Paper 1800*.

MCKEE, J. E. 1956. Oily substances and their effects on the beneficial uses of water. *State Water Pollution Control Board Publ. 16*, Sacramento, Calif.

MCKEE, J. E., F. B. LAVERLY, and R. M. HERTEL. 1972. Gasoline in groundwater. *J. Water Pollution Control Fed.*, 44, pp. 293–302.

MCKELVEY, J. G., and I. H. MILNE. 1962. The flow of salt through compacted clay. *Clay and Clay Minerals*, 9, pp. 248–259.

MCWHORTER, D. B. 1971. Infiltration affected by flow of air. *Colorado State Univ. Hydrol. Paper 49*, Fort Collins, Colo.

MEINZER, O. E. 1923. The occurrence of groundwater in the United States, with a discussion of principles. *U.S. Geol. Surv. Water-Supply Paper 489*.

MEINZER, O. E. 1927. Plants as indicators of groundwater. *U.S. Geol. Surv. Water-Supply Paper 577*, 91 pp.

MERCER, J. W., G. F. PINDER, and I. G. DONALDSON. 1975. A Galerkin finite-element analysis of the hydrothermal system at Wairakei, New Zealand. *J. Geophys. Res.* 80, pp. 2608–2621.

MEYBOOM, P. 1961. Estimating groundwater recharge from stream hydrographs. *J. Geophys. Res.*, 66, pp. 1203–1214.

MEYBOOM, P. 1964. Three observations on stream-flow depletion by phreatophytes. *J. Hydrol.*, 2, pp. 248–261.

MEYBOOM, P. 1966a. Groundwater studies in the Assiniboine River Drainage Basin: I. The evaluation of a flow system in south-central Saskatchewan. *Geol. Surv. Can. Bull. 139*, 65 pp.

MEYBOOM, P. 1966b. Unsteady groundwater flow near a willow ring in hummocky moraine. *J. Hydrol.*, 4, pp. 38–62.

MEYBOOM, P. 1967. Groundwater studies in the Assiniboine River Drainage Basin: II. Hydrologic characteristics of phreatophytic vegetation in south-central Saskatchewan. *Geol. Surv. Can. Bull 139*, 64 pp.

MEYBOOM, P. 1968. Hydrogeology: A decennial appraisal and forecast. *The Earth Sciences in Canada: A Centennial Appraisal and Forecast*, ed. E. R. W. Neale. Roy. Soc. Can. Spec. Publ. 11, pp. 203–221.

MILLER, D. W., F. A. DELUCA, and T. L. TESSIER. 1974. Groundwater contamination in the northeast states. *U.S. Environ. Protect. Agency Rept. EPA 660/2-74-056.*

MOENCH, A. F., and T. A. PRICKETT. 1972. Radial flow in an infinite aquifer undergoing conversion from artesian to water table conditions. *Water Resources Res.*, 8, pp. 494–499.

MOHSEN, M. F. N. 1975. Gas migration from sanitary landfills and associated problems. Unpublished Ph.D. thesis, University of Waterloo.

MOLLARD, J. D. 1973. *Landforms and Surface Materials of Canada: A Stereoscopic Airphoto Atlas and Glossary*. J. D. Mollard and Associates Ltd. Regina, Sask., Canada.

MOOK, W. G. 1972. Application of natural isotopes in ground water hydrology. *Geol. Mijnbouw*, 51.

MOORE, G. W., and G. NICHOLAS. 1964. *Speleology*. Heath, Boston.

MOORE, J. E., and L. A. WOOD. 1967. Data requirements and preliminary results of an analog-model evaluation—Arkansas River Valley in Eastern Colorado. *Ground Water*, 5, no. 1, pp. 20–23.

MORAN, R. E. 1976. Geochemistry of selenium in ground water near Golden, Jefferson County, Colorado. *Geol. Soc. Amer. Abst., Ann. Meeting*, 8, no. 6, 1018 pp.

MORAN, S. R. 1971. Glaciotectonic structures in drift. *Till: A Symposium*, ed. R. P. Goldthwait. Ohio State University Press, Columbus, Ohio, pp. 127–248.

MORAN, S. R., J. A. CHERRY, and J. H. ULMER. 1976. An environmental assessment of a 250 MMSCFD dry ash Surgi coal gasification facility in Dunn Country, North Dakota. *Univ. North Dakota, Eng. Exp. Sta. Bull. 76-12-EES-01.*

MORAN, S. R., J. A. CHERRY, J. H. ULMER, and W. PETERSON. 1978a. Geology and subsurface hydrology of the Dunn Center Area, North Dakota. *North Dakota Geol. Surv. Rept. Invest. 61.*

MORAN, S. R., G. GROENEWOLD, and J. A. CHERRY. 1978b. Hydrogeologic and geochemical concepts and methods in overburden investigations for reclamation of mined land. *North Dakota Geol. Surv. Rept. Invest. 62*, 150 pp.

MOREY, G. W., R. O. FOURNIER, and J. J. ROWE. 1962. The solubility of quartz in water in the temperature interval from 25°C to 300°C. *Geochim. Cosmochim. Acta*, 26, pp. 1029–1043.

MOREY, G. W., R. O. FOURNIER, and J. J. ROWE. 1964. The solubility of amorphous silica at 25°C. *J. Geophys. Res.*, 69, no. 10, pp. 1995–2002.

MORGAN, C. O. and M. D. WINNER, JR. 1962. Hydrochemical facies in the 400 foot and 600 foot sands of the Baton Rouge Area, Louisiana. *U.S. Geol. Surv. Prof. Paper 450-B*, pp. B120–121.

MORGENSTERN, N. R., and V. E. PRICE. 1965. The analysis of the stability of general slip surfaces. *Géotechnique*, 15, pp. 79–93.

MUNNICH, K. O. 1957. Messung des ^{14}C-Gehaltes von hartem Groundwasser. *Naturwissenshaften*, 44, p. 32.

MURRAY, C. R. 1973. Water use, consumption, and outlook in the U.S. in 1970. *J. Amer. Water Works Assoc.*, 65, pp. 302–308.

MURRAY, C. R., and E. B. REEVES. 1972. Estimated use of water in the United States in 1970. *U.S. Geol. Surv. Circ. 676.*

NACE, R. L., ed. 1971. Scientific framework of world water balance. *UNESCO Tech. Papers Hydrol.*, 7, 27 pp.

NARASIMHAN, T. N. 1975. A unified numerical model for saturated-unsaturated groundwater flow. Ph.D. dissertation, University of California, Berkeley.

NELSON, R. W. 1968. In place determination of permeability distribution for heterogeneous porous media through analysis of energy dissipation. *Soc. Petrol. Engrs. J.*, 8, pp. 33–42.

NEUMAN, S. P. 1972. Theory of flow in unconfined aquifers considering delayed response of the water table. *Water Resources Res.*, 8, pp. 1031–1045.

NEUMAN, S. P. 1973a. Calibration of distributed parameter groundwater flow models viewed as a multiple-objective decision process under uncertainty. *Water Resources Res.*, 9, pp. 1006–1021.

NEUMAN, S. P. 1973b. Supplementary comments on theory of flow in unconfined aquifers considering delayed response of the water table. *Water Resources Res.*, 9, pp. 1102–1103.

NEUMAN, S. P. 1974. Effect of partial penetration on flow in unconfined aquifers considering delayed gravity response. *Water Resources Res.*, 10, pp. 303–312.

NEUMAN, S. P. 1975a. Analysis of pumping test data from anisotropic unconfined aquifers considering delayed gravity response. *Water Resources Res.*, 11, pp. 329–342.

NEUMAN, S. P. 1975b. Role of subjective value judgement in parameter identification. *Modeling and Simulation of Water Resources Systems*, ed. G. C. Vansteenkiste. North-Holland, Amsterdam, pp. 59–82.

NEUMAN, S. P., and P. A. WITHERSPOON. 1969a. Theory of flow in a confined two-aquifer system. *Water Resources Res.*, 5, pp. 803–816.

NEUMAN, S. P., and P. A. WITHERSPOON. 1969b. Applicability of current theories of flow in leaky aquifers. *Water Resources Res.*, 5, pp. 817–829.

NEUMAN, S. P., and P. A. WITHERSPOON. 1972. Field determination of the hydraulic properties of leaky multiple-aquifer systems. *Water Resources Res.*, 8, pp. 1284–1298.

NEWBURY, R. W., J. A. CHERRY, and R. A. COX. 1969. Groundwater-streamflow systems in Wilson Creek Experimental Watershed. *Can. J. Earth Sci.*, 6, pp. 613–623.

NIGHTINGALE, H. I. 1970. Statistical evaluation of salinity and nitrate content and trends beneath urban and agricultural areas. *Ground Water*, 8, no. 1, pp. 22–28.

NOBLE, E. A. 1963. Formation of ore deposits by water of compaction. *Econ. Geol.* 58, pp. 1145–1156.

NORTON, D., and J. KNIGHT. 1977. Transport phenomena in hydrothermal systems: cooling plutons. *Amer. J. Sci.*, 277, pp. 937–981.

OGATA, A. 1970. Theory of dispersion in a granular medium. *U.S. Geol. Surv. Prof. Paper 411-I.*

OGATA, A., and R. B. BANKS. 1961. A solution of the differential equation of longitudinal dispersion in porous media. *U.S. Geol. Surv. Prof. Paper 411-A.*

OLSEN, H. W. 1969. Simultaneous fluxes of liquid and charge in saturated kaolinite. *Soil Sci. Soc. Amer. Proc.*, 33, pp. 334–338.

ONISHI, H., and E. B. SANDELL. 1955. Geochemistry of arsenic. *Geochim. Cosmochim. Acta*, 7, pp. 1–33.

OPPENHEIMER, C. H. 1963. Editor's Preface. *Introduction to Geological Microbiology*, by S. V. Kuznetsov, M. V. Ivanov, and N. N. Lyalikova. McGraw-Hill, New York, 251 pp.

OREGON STATE UNIVERSITY. 1974. Disposal of environmentally hazardous wastes. *Task Force Rept. Environ. Health Sci. Center Oregon State Univ.*

PACES, T. 1973. Steady-state kinetics and equilibrium between ground water and granitic rocks. *Geochim. Cosmochim. Acta*, 37, pp. 2641–2663.

PACES, T. 1976. Kinetics of natural water systems. *Proc. Symp. Interpretation of Environmental Isotopes and Hydrochemical Data in Groundwater Hydrology, Intern. Atomic Energy Agency Spec. Publ.*, Vienna.

PALCIAUSKAS, V. V., and P. A. DOMENICO. 1976. Solution chemistry, mass transfer, and the approach to chemical equilibrium in porous carbonate rocks and sediments. *Geol. Soc. Amer. Bull.*, 87, pp. 207–214.

PAPADOPOULOS, I. S., J. D. BREDEHOEFT, and H. H. COOPER. 1973. On the analysis of slug test data. *Water Resources Res.*, 9, pp. 1087–1089.

PARIZEK, R. R. 1969. Glacial ice-contact rings and ridges. *Geol. Soc. Amer. Spec. Paper 123*, INQUA volume.

PARIZEK, R. R., and L. J. DREW. 1966. Random drilling for water in carbonate rocks. *Proc. Symp. Short Course Computers Operations Res. Mineral Ind. Exp. Sta.*, Vol. 3. Pennsylvania State University. University Park, Pa., pp. 1–22.

PARK, C. F., Jr., and R. A. MACDIARMID. 1975. *Ore Deposits*, 3rd ed. W. H. Freeman, San Francisco, 529 pp.

PARKER, G. G., and V. T. STRINGFIELD. 1950. Effects of earthquakes, trains, tides, winds, and atmospheric pressure changes on water in the geologic formations of southern Florida. *Econ. Geol.*, 45, pp. 441–460.

PARKS, G. A. 1967. Aqueous surface chemistry of oxides and complex oxide minerals. *Equilibrium Concepts in Natural Water Systems*. American Chemical Society, Washington, D.C., pp. 121–160.

PARSONS, M. L. 1970. Groundwater thermal regime in a glacial complex. *Water Resources Res.*, 6, pp. 1701–1720.

PATTEN, E. P. 1965. Design, construction and use of electric analog model. *Analog Model Study of Groundwater in Houston District*, by L. A. Wood and R. K. Gabrysch. *Texas Water Comm. Bull. 6508*.

PATTEN, E. P., and G. D. BENNETT. 1963. Application of electrical and radioactive well logging to ground-water hydrology. *U.S. Geol. Surv. Water-Supply Paper 1544-D*.

PATTON, F. D., and D. U. DEERE. 1971. Geologic factors controlling slope stability in open pit mines. *Stability in Open Pit Mining*, ed. C. O. Brawner and V. Milligan. American Institute of Mining Engineers, New York, pp. 23–48.

PATTON, F. D., and A. J. HENDRON, Jr. 1974. General report on mass movements. *Proc. 2nd Intern. Congr., Intern. Assoc. Eng. Geol.*, Sao Paulo, Brazil, 2, pp. V-GR.1–V-GR.57.

PAULING, L., and P. PAULING. 1975. *Chemistry*. W. H. Freeman, San Francisco, 767 pp.

PAYNE, B. R. 1972. Isotope hydrology. *Adv. Hydrosci.*, 8, pp. 95–138.

PEARSON, F. J. Jr., and I. FRIEDMAN. 1970. Sources of dissolved carbonate in an aquifer free of carbonate minerals. *Water Resources Res.*, 6, pp. 1775–1781.

PEARSON, F. J. Jr., and B. B. HANSHAW. 1970. Sources of dissolved carbonate species in groundwater and their effects on carbon-14 dating. *Isotope Hydrology*. International Atomic Energy Agency, Vienna, pp. 271–286.

PEARSON, F. J., Jr., and D. E. WHITE. 1967. Carbon-14 ages and flow rates of water in Carrizo sand, Atascosa County, Texas. *Water Resources Res.*, 3, pp. 251–261.

PECK, A. J. 1960. The water table as affected by atmospheric pressure. *J. Geophys. Res.*, 65, pp. 2383–2388.

PECKHAM, A. E., and W. G. BELTER. 1962. Considerations for selection and operation of radioactive burial sites. *2nd Ground Disposal Radioactive Wastes Conf., USAEC Rept. TED-7668, Book 2, Chalk River, Canada*, pp. 428–436.

PELTON, W. L., K. M. KING, and C. B. TANNER. 1960. An evaluation of the Thornthwaite and mean temperature methods for determining potential evapotranspiration. *Agronomy J.*, 52, pp. 387–395.

PENMAN, H. L. 1948. Natural evaporation from open water, bare soil and grass. *Proc. Roy. Soc. Lond.*, A193, pp. 120–145.

PERCIOUS, D. J. 1969. Aquifer dispersivity by recharge-discharge of a fluorescent dye tracer through a single well. Unpublished M.Sc. thesis, University of Arizona.

PERKINS, T. K., and O. C. JOHNSTON. 1963. A review of diffusion and dispersion in porous media. *J. Soc. Petrol. Eng.*, 3, pp. 70–83.

PERLMUTTER, N. M., M. LIEBER, and H. L. FRAUENTHAL. 1964. Contamination of ground water by detergents in a suburban environment—south Farmingdale area, Long Island, New York. *U.S. Geol. Surv. Prof. Paper 501-C*, pp. 170–175.

PETERSON, R. 1954. Studies of the Bearspaw Shale at a damsite. *Proc. Amer. Soc. Civil Engrs., Soil Mech. Found. Div.*, Vol. 80, No. 476.

PETROVIC, R., R. A. BERNER, and M. B. GOLDHABER. 1976. Rate control in dissolution of alkali feldspars: 1. Study of residual feldspar grains by x-ray photoelectron microscopy. *Geochim. Cosmochim. Acta*, 40, pp. 537–548.

PFERD, J. W., R. H. FAKUNDINY, and J. F. DAVIS. 1977. Geology and integrity of low-level waste burial techniques at West Valley, New York. *Symp. Management Low-Level Radioactive Waste, Atlanta, Ga.*

PHILIP, J. R. 1957a. The theory of infiltration: 1. The infiltration equation and its solution. *Soil Sci.*, 83, pp. 345–357.

PHILIP, J. R. 1957b. The theory of infiltration: 2. The profile at infinity. *Soil Sci.*, 83, pp. 435–448.

PHILIP, J. R. 1957c. The theory of infiltration: 3. Moisture profiles and relation to experiment. *Soil Sci.*, 84, pp. 163–178.

PHILIP, J. R. 1957d. The theory of infiltration: 4. Sorptivity and algebraic infiltration equations. *Soil Sci.*, 84, pp. 257–264.

PHILIP, J. R. 1957e. The theory of infiltration: 5. The influence of the initial moisture content. *Soil Sci.*, 84, pp. 329–339.

PHILIP, J. R. 1957f. Evaporation, and moisture and heat fields in the soil. *J. Meteor.*, 14, pp. 354–366.

PHILIP, J. R. 1958a. The theory of infiltration: 6. Effect of water depth over soil. *Soil Sci.*, 85, pp. 278–286.

PHILIP, J. R. 1958b. The theory of infiltration: 7. *Soil Sci.*, 85, pp. 333–337.

PHILIP, J. R., and D. A. de VRIES. 1957. Moisture movement in porous materials under temperature gradients. *Trans. Amer. Geophys. Union*, 38, pp. 222–232.

PICKENS, J. F., and W. C. LENNOX. 1976. Numerical simulation of waste movement in steady groundwater flow systems. *Water Resources Res.*, 12, no. 2, pp. 171–180.

PICKENS, J. F., J. A. CHERRY, G. E. GRISAK, W. F. MERRITT, and B. RISTO. In press. A multilevel device for water sampling and piezometric monitoring in cohesionless deposits. *Ground Water*.

PIERSOL, R. J., L. E. WORKMAN, and M. C. WATSON. 1940. Porosity, total liquid saturation, and permeability of Illinois oil sands. *Illinois Geol. Surv. Rept. Invest.* 67.

PINDER, G. F. 1973. A Galerkin-finite element simulation of groundwater contamination on Long Island, N.Y. *Water Resources Res.*, 9, no. 6, pp. 1657–1669.

PINDER, G. F., and J. D. BREDEHOEFT. 1968. Application of the digital computer for aquifer evaluation. *Water Resources Res.*, 4, pp. 1069–1093.

PINDER, G. F., and H. H. COOPER, Jr. 1970. A numerical technique for calculating the transient position of the saltwater front. *Water Resources Res.*, 6, pp. 875–882.

PINDER, G. F., and E. O. FRIND. 1972. Application of Galerkins procedure to aquifer analysis. *Water Resources Res.*, 8, pp. 108–120.

PINDER, G. F., and W. G. GRAY. 1977. *Finite Element Simulation in Surface and Subsurface Hydrology*. Academic Press, New York, 295 pp.

PINDER, G. F., and J. F. JONES. 1969. Determination of ground-water component of peak discharge from chemistry of total runoff. *Water Resources Res.*, 5, pp. 438–445.

PINDER, G. F., and S. P. SAUER. 1971. Numerical simulation of flood-wave modification due to bank storage effects. *Water Resources Res.*, 7, pp. 63–70.

PIPER, A. M. 1960. Interpretation and current status of ground-water rights. *U.S. Geol. Surv. Circ. 432*.

PIPER, A. M. 1970. Disposal of liquid wastes by injection underground—neither myth nor millennium. *U.S. Geol. Surv. Circ. 631*.

PIPER, A. M. 1944. A graphic procedure in the geochemical interpretation of water analyses. *Trans. Amer. Geophys. Union*, 25, pp. 914–923.

PIRSON, S. J. 1958. *Oil Reservoir Engineering*. McGraw-Hill, New York.

PIRSON, S. J. 1963. *Handbook of Well Log Analysis for Oil and Gas Formation Evaluation*. Prentice-Hall, Englewood Cliffs, N.J.

PITEAU, D. R., and F. L. PECKOVER. In press. Rock slope engineering. *Landslides: Analysis and Control*, ed. R. Schuster and R. Krizek. Transportation Research Board, U.S. Department of Transportation.

POLAND, J. F. 1972. Subsidence and its control. *Underground Waste Management and Environmental Implications*. Amer. Assoc. Petrol. Geol. Mem. 18, pp. 50–71.

POLAND, J. F., and G. H. DAVIS. 1969. Land subsidence due to withdrawal of fluids. *Geol. Soc. Amer. Rev. Eng. Geol.*, 2, pp. 187–269.

POLUBARINOVA-KOCHINA, P. Ya. 1962. *Theory of Groundwater Movement*. Princeton University Press, Princeton, N.J., 613 pp.

POURBAIX, M. J. N., J. VAN MUYLDER, and N. DE ZHOUBOV. 1963. *Atlas d'Equilibres Electrochémiques à 25°C*. Gauthier-Villars, Paris. (J. A. Franklin, trans., Pergamon Press, London, 1966.)

PRICKETT, T. A. 1965. Type curve solution to aquifer tests under water table conditions. *Ground Water*, 3, no. 3, pp. 5–14.

PRICKETT, T. A. 1975. Modeling techniques for groundwater evaluation. *Adv. Hydrosci.*, 10, pp. 1–143.

PRICKETT, T. A., and C. G. LONNOQUIST. 1968. Comparison between analog and digital simulation techniques for aquifer evaluation. *Proc. Symp. Use Analog Digital Computers Hydrol., Intern. Assoc. Sci. Hydrol., Publ. 81*, pp. 625–634.

PRICKETT, T. A. and C. G. LONNQUIST. 1971. Selected digital computer techniques for groundwater resource evaluation. *Illinois State Water Surv. Bull. 55*, 62 pp.

RAGAN, R. M. 1968. An experimental investigation of partial-area contributions. *Intern. Assoc. Sci. Hydrol. Gen. Assembly, Berne, Publ. 76*, pp. 241–249.

RAINWATER, F. H., and L. O. THATCHER. 1960. Methods for collection and analysis of water samples. *U.S. Geol. Surv. Water-Supply Paper 1454*, 301 pp.

RALEIGH, C. B., J. H. HEALY, and J. D. BREDEHOEFT. 1972. Faulting and crustal stress at Rangely Colorado. *Flow and Fracture of Rocks Monogr. 16*, American Geophysical Union, Washington, D.C. pp. 275–284.

RASMUSSEN, W. C., and G. E. ANDREASEN. 1959. Hydrologic budget of the Beaverdam Creek basin, Maryland. *U.S. Geol. Surv., Water-Supply Paper 1472*.

RAUCH, H. W., and W. B. WHITE. 1977. Dissolution kinetics of carbonate rocks: 1. Effects of lithology on dissolution rate. *Water Resources Res.*, 13, no. 2, pp. 381–394.

REARDON, E. J., and P. FRITZ. 1978. Computer modelling of groundwater ^{13}C and ^{14}C, isotope compositions. *J. Hydrol.*, 36, pp. 201–224.

REDDELL, D. L., and D. K. SUNADA. 1970. Numerical simulation of dispersion in groundwater aquifers. *Hydrol. Paper No. 41*. Colorado State University, Fort Collins, 79 pp.

REEVE, R. C. 1965. Hydraulic head. *Methods in Soil Analysis, Part 1*, ed. C. A. Black. American Society of Agronomy, Madison, Wis., pp. 180–196.

REISENAUER, A. E. 1963. Methods for solving problems of multidimensional, partially-saturated steady flow in soils. *J. Geophys. Res.*, 68, pp. 5725–5733.

REMSON, I., C. A. APPEL, and R. A. WEBSTER. 1965. Ground-water models solved by digital computer. *J. Hydraul. Div., Proc. Amer. Soc. Civil Engrs.*, 91 (HY3), pp. 133–147.

REMSON, I., G. M. HORNBERGER, and F. J. MOLZ. 1971. *Numerical Methods in Subsurface Hydrology*. Wiley-Interscience, New York.

RENDER, F. W. 1970. Geohydrology of the metropolitan Winnipeg area as related to groundwater supply and construction. *Canadian Geotech. J.*, 7, pp. 243–274.

RENDER, F. W. 1971. Electric analog and digital modeling of the upper carbonate aquifer in the metropolitan Winnipeg area. *Geol. Assoc. Can. Spec. Paper 9*, pp. 311–320.

RENDER, F. W. 1972. Estimating the yield of the upper carbonate aquifer in the metropolitan Winnipeg area by means of a digital model. *Proc. 24th Intern. Geol. Congr.*, Sec. 11, pp. 36–45.

RETTIE, J. R., and F. W. PATTERSON. 1963. Some foundation considerations at the Grand Rapids Hydroelectric Project. *Eng. J. Eng. Inst. Can.*, 46, no. 12, pp. 32–38.

RICH, J. L. 1921. Moving underground water as a primary cause of the migration and accumulation of oil and gas. *Econ. Geol.*, 16, pp. 347–371.

RICHARDS, L. A. 1931. Capillary conduction of liquids through porous mediums. *Physics*, 1, pp. 318–333.

RICHARDS, L. A. 1965. Physical condition of water in soil. *Methods of Soil Analysis, Part 1*, ed. C. A. Black. American Society of Agronomy, Madison, Wis., pp. 128–152.

RICHARDS, S. J. 1965. Soil suction measurements with tensiometers. *Methods in Soil Analysis, Part 1*, ed. C. A. Black, American Society of Agronomy, Madison, Wis., pp. 153–163.

RICHARDSON, R. M. 1962a. Northeastern burial ground studies. *2nd Ground Disposal Radioactive Waste Conf. USAEC Rept. TID-7628 Book 2, Chalk River, Can.*, pp. 460–461.

RICHARDSON, R. M. 1962b. Significance of climate in relation to the disposal of radioactive waste at shallow depth below ground. *Proc. on Retention and Migration of Radioactive Ions Through the Soil.* Commissariat à l'Energie Atomique, Institut National des Sciences et Techniques Nucléaires, Saclay, France, pp. 207–211.

RICHTMYER, R. D. 1957. *Difference Methods for Initial Value Problems.* Wiley-Interscience, New York.

RIFAI, M. N. E., W. J. KAUFMAN, and D. K. TODD. 1956. Dispersion phenomena in laminar flow through porous media. *Report No. 3, I.E.R. Series 90, Sanitary Eng. Res. Lab., Univ. Calif., Berkeley.*

RILEY, F. S. 1969. Analysis of borehole extensometer data from central California. *Proc. Tokyo Symp. Land Subsidence, Intern. Assoc. Sci. Hydrol.*, 2, pp. 423–431.

ROBECK, G. G. 1969. Microbial problems in groundwater. *Ground Water*, 7, pp. 33–35.

ROBERTSON, J. B. 1974. Digital modeling of radioactive and chemical waste transport in the Snake River Plain aquifer at the National Reactor Testing Station, Idaho. *U.S. Geol. Surv., Water Resources Div., IDO-22054*, Idaho Falls, Idaho.

ROBERTSON, J. M., C. R. TOUSSAINT, and M. A. JORQUE. 1974. Organic compounds entering ground water from a landfill. *Environ. Protect. Technol. Ser. EPA 660/2-74-077.*

ROBINSON, R. A., and R. H. STOKES. 1965. *Electrolyte Solutions*, 2nd ed. Butterworth, London.

ROBINSON, T. W. 1958. Phreatophytes. *U.S. Geol. Surv. Water-Supply Paper 1423*, 84 pp.

ROBINSON, T. W. 1964. Phreatophyte research in the western States, March 1959 to July 1964. *U.S. Geol. Surv. Circ. 495.*

ROSE, H. E. 1945. An investigation into the laws of flow of fluids through beds of granular material. *Proc. Inst. Mech. Engrs.*, 153, pp. 141–148.

ROSSINI, F. D., D. D. WAGMAN, W. H. EVANS, S. LEVINE, and I. JAFFE. 1952. *Selected values of chemical thermodynamics properties.* U.S. Nat. Bur. Standards Circ. 500, 1268 pp.

ROUTSON, R. C., and R. J. SERNE. 1972. Experimental support studies for the percol and transport models. *Battelle, Pacific Northwest Laboratories BNWL-1719*, Richland, Wash.

ROVERS, F. A., and G. T. FARQUHAR. 1974. Evaluating contaminant attenuation in the soil to improve landfill selection and design. *Proc. Intern. Conf. Land for Waste Management, Ottawa, Can.*

ROWE, P. W. 1972. The relevance of soil fabric to site investigation practice. *Géotechnique*, 22, pp. 195–300.

RUBIN, J. 1968. Theoretical analysis of two-dimensional transient flow of water in unsaturated and partly unsaturated soils. *Soil Sci. Soc. Amer. Proc.*, 32, pp. 607–615.

RUBIN, J., and R. STEINHARDT. 1963. Soil water relations during rain infiltration: I. Theory. *Soil Sci. Soc. Amer. Proc.*, 27, pp. 246–251.

RUBIN, J., R. STEINHARDT, and P. REINIGER. 1964. Soil water relations during rain infiltration: II. Moisture content profiles during rains of low intensities. *Soil Sci. Soc. Amer. Proc.*, 28, pp. 1–5.

RUNNELS, D. D. 1969. Diagenesis, chemical sediments and the mixing of waters. *J. Sedimentary Petrol.*, 39, pp. 1188–1201.

SAGAR, B., S. YAKOWITZ, and L. DUCKSTEIN. 1975. A direct method for the identification of the parameters of dynamic nonhomogeneous aquifers. *Water Resources Res.*, 11, pp. 563–570.

SALITERNIK, C. 1972. Groundwater pollution by nitrogen compounds. *6th Intern. Conf. Water Pollution Res., Jerusalem.*

SCALF, M. R. 1977. Groundwater pollution problems in the southwestern United States. *U.S. Environ. Protect. Agency, Rept. 600/3-77-012.*

SCALF, M. R., J. W. KEELEY, and C. J. LAFEVERS. 1973. Groundwater pollution in the south central States. *U.S. Environ. Protect. Agency Rept. EPA-122-73-268*, June.

SCHEIDEGGER, A. 1960. *Physics of Flow Through Porous Media.* University of Toronto Press, Toronto, Canada.

SCHICHT, R. J., and W. C. WALTON. 1961. Hydrologic budgets for three small watersheds in Illinois. *Illinois State Water Surv. Rept. Invest. 40.*

SCHIFFMAN, R. L., A. T. F. CHEN, and J. C. JORDAN. 1969. An analysis of consolidation theories. *J. Soil Mech. Found. Div., Amer. Soc. Civil Engrs.*, 95 (SM1), pp. 285–312.

SCHNEIDER, R. 1962. An application of thermometry to the study of groundwater. *U.S. Geol. Surv. Water-Supply Paper 1544-B*, 16 pp.

SCHOELLER, H. 1955. Géochimie des eaux souterraines. *Rev. Inst. Franc. Pétrole, Paris*, 10, no. 3, pp. 181–213, and 10, no. 4, pp. 219–246.

SCHOELLER, H. 1959. Arid zone hydrology recent developments. *UNESCO Rev., Reicardi*, 12.

SCHOELLER, H. 1962. *Les Eaux souterraines.* Mason et Cie, Paris.

SCHOLZ, C. H., L. R. SYKES, and Y. P. AGGARWAL. 1973. Earthquake prediction: a physical basis. *Science*, 181, pp. 803–810.

SCHUSTER, R., and R. KRIZEK, eds. In press. *Landslides: Analysis and Control.* Transportation Research Board, U.S. Department of Transportation.

SCHWARTZ, F. W. 1974. The origin of chemical variations in groundwaters from a small watershed in southwestern Ontario. *Can. J. Earth Sci.*, 11, no. 7, pp. 893–904.

SCHWARTZ, F. W. 1975. On radioactive waste management: an analysis of the parameters controlling subsurface contaminant transfer. *J. Hydrol.*, 27, pp. 51–71.

SCHWILLE, F. 1967. Petroleum contamination of the subsoil—a hydrological problem. *The Joint Problems of the Oil and Water Industries.* ed. Peter Hepple. Elsevier, Amsterdam, pp. 23–53.

SCOTT, J. S., and F. W. RENDER. 1964. Effect of an Alaskan earthquake on water levels in wells at Winnipeg and Ottawa, Canada. *J. Hydrol.*, 2, pp. 262–268.

SCOTT, R. F. 1963. *Principles of Soil Mechanics.* Addison-Wesley, Reading, Mass.

SEABER, P. R. 1962. Cation hydrochemical facies of groundwater in the Englishtown Formation, New Jersey. *U.S. Geol. Surv. Prof. Paper 450-B*, pp. B124–B126.

SEABURN, G. E. 1970. Preliminary results of hydrologic studies at two recharge basins on Long Island, New York. *U.S. Geol. Surv. Prof. Paper 627-C*, 17 pp.

SEIDELL, A. 1958. *Solubilities*, 1, 4th ed. American Chemical Society, D. van Nostrand Co., Princeton, N.J.

SEGOL, G., and G. F. PINDER. 1976. Transient simulation of saltwater intrusion in southeastern Florida. *Water Resources Res.*, 12, pp. 65–70.

SHACKELFORD, W. M., and L. H. KEITH. 1976. Frequency of organic compounds identified in water. Analytical Chemistry Branch, Environmental Research Laboratory, Athens, Ga.

SHARP, J. C. In press. Groundwater. *Pit Slope Design Manual.* Canada Department of Energy Mines and Resources.

SHARP, J. C., and Y. N. T. MAINI. 1972. Fundamental considerations on the hydraulic characteristics of joints in rock. *Percolation Through Fissured Rock.* edited by W. Wittke. International Society of Rock Mechanics, Stuttgart.

SHARP, J. C., Y. N. T. MAINI, and T. R. HARPER. 1972. Influence of groundwater on the stability of rock masses: 1. Hydraulics within rock masses. *Trans. Inst. Min. and Met. Lond.*, 81, pp. A13–A20.

SHAW, F. S., and R. V. SOUTHWELL. 1941. Relaxation methods applied to engineering problems: VII. Problems relating to the percolation of fluids through porous materials. *Proc. Roy. Soc. Lond.*, A178, pp. 1–17.

SHERARD, J. L., R. J. WOODWARD, S. F. GIZIENSKI, and W. A. CLEVENGER. 1963. *Earth and Earth-Rock Dams.* John Wiley & Sons, New York.

SHTERNINA, E. B. 1960. Solubility of gypsum in aqueous solutions of salts. *Intern. Geol. Rev.*, 1, pp. 605–616.

SHTERNINA, E. B., and E. V. FROLOVA. 1945. Solubility of the system $CaCO_3$–$CaSO_4$–NaCl–CO_2–H_2O at 25°C. *Compt. Rend. (Doklady) Acad. Sci. URSS*, 47, no. 1, pp. 33–35.

SILLEN, L. G., and A. E. MARTELL. 1964. *Stability Constants of Metal Ion Complexes.* Chem. Soc. Lond. Spec. Publ. 17.

SILLEN, L. G. and A. E. MARTELL. 1971. *Stability constants of metal ion complexes (Supplement).* Chem. Soc. Lond. Spec. Publ. 25.

SIMPSON, F. 1976. Deep well injection of fluid industrial wastes in Canada. *Proc. 15th Ann. Conf., Ontario Petrol. Inst.*

SINGH, K. P. 1969. Theoretical baseflow curves. *J. Hydraul. Div., Proc. Amer. Soc. Civil Engrs.*, 95 (HY6), pp. 2029–2048.

SKEMPTON, A. W. 1961. Effective stress in soils, concrete and rocks. *Conference on Pore Pressures and Suction in Soils.* Butterworth, London, pp. 4–16.

SKIBITZKE, H. E., and G. M. ROBERTSON. 1963. Dispersion in groundwater flowing through heterogeneous materials. *U.S. Geol. Surv. Prof. Paper 386-B.*

SKINNER, B. J., and P. B. BARTON, Jr. 1973. Genesis of mineral deposits. *Ann. Rev. Earth Planetary Sci.*, 1, pp. 183–211.

SKLASH, M. G. 1978. Isotope studies of runoff from small headwater basins. Ph.D. thesis, University of Waterloo, Ontario.

SKLASH, M. G., R. N. FARVOLDEN, and P. FRITZ. 1976. A conceptual model of watershed response to rainfall developed through the use of oxygen-18 as a natural tracer. *Can. J. Earth Sci.*, 13, pp. 271–283.

SMITH, R. E., and D. A. WOOLHISER. 1971. Overland flow on an infiltrating surface. *Water Resources Res.*, 7, pp. 899–913.

SNOW, D. T. 1968. Rock fracture spacings, openings, and porosities. *J. Soil Mech. Found. Div., Proc. Amer. Soc. Civil Engrs.*, 94, pp. 73–91.

SNOW, D. T. 1969. Anisotropic permeability of fractured media. *Water Resources Res.*, 5, pp. 1273–1289.

SPALDING, R. F., J. R. GORMLY, B. H. CURTIS, and M. E. EXNER. 1978. Nonpoint nitrate contamination of groundwater in Merrick County, Nebraska. *Ground Water*, 16, pp. 86–95.

SPIEKER, A. M. 1968. Effect of increased pumping of groundwater in the Fairfield-New Baltimore area, Ohio—a prediction by analog-model study. *U.S. Geol. Surv. Prof. Paper 605-C.*

STALLMAN, R. W. 1956. Use of numerical methods for analyzing data on groundwater levels. *Intern. Assoc. Sci. Hydrol. Publ. 41*, pp. 227–231.

STALLMAN, R. W. 1963. Computation of groundwater velocity from temperature data. *Methods of Collecting and Interpreting Groundwater Data*, ed. R. Bentall, *U.S. Geol. Surv. Water-Supply Paper 1544-H*, pp. H36–H46.

STALLMAN, R. W. 1964. Multiphase fluids in porous media—a review of theories pertinent to hydrologic studies. *U.S. Geol. Surv. Prof. Paper 411E.*

STALLMAN, R. W. 1971. Aquifer-test design, observation and data analysis. *Techniques of Water Resources Investigations of the U.S. Geological Survey*. Chap. B1. Government Printing Office, Washington, D.C.

STEPHENSON, D. A. 1971. Groundwater flow system analysis in lake environments, with management and planning implications. *Water Resources Bull.*, 7, pp. 1038–1047.

STEPHENSON, G. R., and R. A. FREEZE. 1974. Mathematical simulation of subsurface flow contributions to snowmelt runoff, Reynolds Creek Watershed, Idaho. *Water Resources Res.*, 10, pp. 284–294.

STREETER, V. L. 1962. *Fluid Mechanics*, 3rd ed. McGraw-Hill, New York.

STRELTSOVA, T. D. 1972. Unsteady radial flow in an unconfined aquifer. *Water Resources Res.*, 8, pp. 1059–1066.

STUART, W. T., E. A. BROWN, and E. C. RHODEHAMEL. 1954. Groundwater investigations of the Marquette iron-mining district. *Michigan. Geol. Surv. Div. Tech. Rept. 3.*

STUMM, W., and J. J. MORGAN. 1970. *Aquatic Chemistry*. John Wiley & Sons, New York, 583 pp.

SUAREZ, D. L. 1974. Heavy metals in waters and soil associated with several Pennsylvania landfills. Unpublished Ph.D. thesis, The Pennsylvania State University, 222 pp.

SUGISAKI, R. 1959. Measurement of effective flow velocity of groundwater by means of dissolved gases. *Amer. J. Sci.*, 259, pp. 144–153.

SUGISAKI, R. 1961. Geochemical study of ground water. *Nagoya Univ. J. Earth Sci.*, 10, no. 1, pp. 1–33.

SUMMERS, W. K. 1972. Specific capacities of wells in crystalline rocks. *Ground Water*, 10, no. 6, pp. 37–47.

SUMMERS, W. K., and Z. SPIEGAL. 1974. *Groundwater Pollution: A Bibliography*. Ann Arbor Science Publishers, Ann Arbor, Mich.

SUTCLIFFE, R. C. 1970. World water balance: a geophysical problem. *Proc. Symp. World Water Balance, Intern. Assoc. Sci. Hydrol.*, 1, pp. 19–24.

SWARTZENDRUBER, D. 1962. Non-Darcy flow behaviour in liquid-saturated porous media. *J. Geophys. Res.*, 67, pp. 5205–5213.

SWEET, B. H., and R. D. ELLENDER. 1972. Electroosmosis: a new technique for concentrating viruses from water. *Water Res.*, 6, pp. 775–779.

TARDY, Y. 1971. Characterization of the principal weathering types by the geochemistry of waters from some European and African crystalline massifs. *Chem. Geol.*, 7, pp. 253–271.

TAYLOR, G. S., and J. N. LUTHIN. 1969. Computer methods for transient analysis of water table aquifers. *Water Resources Res.*, 5, pp. 144–152.

TAYLOR, S. A., and J. W. CARY. 1964. Linear equations for the simultaneous flow of matter and energy in a continuous soil system. *Soil Sci. Soc. Amer. Proc.*, 28, pp. 167–172.

TERZAGHI, K. 1925. *Erdbaumechanic auf Bodenphysikalischer Grundlage*. Franz Deuticke, Vienna.

TERZAGHI, K. 1950. Mechanism of landslides. *Berkey Volume: Application of Geology to Engineering Practice*. Geological Society of America, New York, pp. 83–123.

TERZAGHI, K., and R. B. PECK. 1967. *Soil Mechanics in Engineering Practice*, 2nd ed. John Wiley & Sons, New York.

THEIS, C. V. 1935. The relation between the lowering of the piezometric surface and the rate and duration of discharge of a well using groundwater storage. *Trans. Amer. Geophys. Union*, 2, pp. 519–524.

THOMAS, H. E. 1951. *The Conservation of Groundwater*. McGraw-Hill, New York.

THOMAS, H. E. 1958. Hydrology vs. water allocation in the eastern United States. *The Law of Water Allocation in the Eastern United States*. Ronald Press, New York.

THOMPSON, G. M., J. M. HAYES, and S. N. DAVIS. 1974. Fluorocarbon tracers in hydrology. *Geophys. Res. Letters*, 1, pp. 377–380.

THOMPSON, T. F. 1966. San Jacinto Tunnel. *Engineering Geology in Southern California*, Assoc. Eng. Geol. Spec. Publ., pp. 104–107.

THORNTHWAITE, C. W. 1948. An approach toward a rational classification of climate. *Geog. Rev.*, 38, pp. 55–94.

THRAIKILL, J. 1968. Chemical and hydrologic factors in the excavation of limestone caves. *Geol. Soc. Amer. Bull.*, 79, pp. 19–46.

TODD, D. K. 1955. Ground-water in relation to a flooding stream. *Proc. Amer. Soc. Civil Engrs.*, 81, pp. 1–20, separate 628.

TODD, D. K. 1959. *Ground Water Hydrology*. John Wiley & Sons, New York.

TODD, D. K., and D. E. O. MCNULTY. 1976. *Polluted Groundwater*. Water Information Center, Port Washington, N.Y.

TOLMAN, C. F. 1937. *Groundwater*. McGraw-Hill, New York.

TÓTH, J. 1962. A theory of groundwater motion in small drainage basins in central Alberta. *J. Geophys. Res.*, 67, pp. 4375–4387.

TÓTH, J. 1963. A theoretical analysis of groundwater flow in small drainage basins. *J. Geophys. Res.*, 68, pp. 4795–4812.

TÓTH, J. 1966. Mapping and interpretation of field phenomena for groundwater reconnaissance in a prairie environment, Alberta Canada. *Bull. Intern. Assoc. Sci. Hydrol.*, 11, no. 2, pp. 1–49.

TÓTH, J. 1968. A hydrogeological study of the Three Hills area, Alberta. *Research Council of Alberta, Geol. Div., Bull. 24*.

TÓTH, J. 1970. Relation between electric analogue patterns of groundwater flow and accumulation of hydrocarbons. *Can. J. Earth Sci.*, 7, pp. 988–1007.

TRAINER, F. W., and R. C. HEATH. 1976. Bicarbonate content of groundwater in carbonate rock in eastern North America. *J. Hydrol.*, 31, pp. 37–55.

TREFZGER, R. E. 1966. Tecolote tunnel. *Engineering Geology in Southern California*. Assoc. Eng. Geol. Spec. Publ., pp. 108–113.

TRESCOTT, P. C., G. F. PINDER, and S. P. LARSON. 1976. Finite-difference model for aquifer simulation in two-dimensions with results of numerical experiments. *Techniques of Water Resources Investigations of the U.S. Geol. Surv.*, Book 7, Chap. Cl, 116 pp.

TRUESDELL, A. H., and B. F. JONES. 1974. WATEQ, a computer program for calculating chemical equilibria of natural waters. *U.S. Geol. Surv. J. Res.*, 2, pp. 233–248.

TURK, L. J. 1975. Diurnal fluctuations of water tables induced by atmospheric pressure changes. *J. Hydrol.*, 26, pp. 1–16.

U. S. ENVIRONMENTAL PROTECTION AGENCY. 1973a. *Manual of Individual Water Supply Systems*. Report EPA-430/9-74-007, 155 pp.

U. S. ENVIRONMENTAL PROTECTION AGENCY. 1973b. *Water Quality Criteria 1972*, EPA R3 73033. Government Printing Office, Washington, D.C.

U. S. ENVIRONMENTAL PROTECTION AGENCY. 1974a. *Land Application of Sewage Effluents and Sludges: Selected Abstracts*. Government Printing Office, Washington, D.C.

U. S. ENVIRONMENTAL PROTECTION AGENCY. 1974b. *Methods for Chemical Analysis of Water and Wastes*. Office of Technology Transfer, Washington, D.C., pp. 105–106.

U. S. ENVIRONMENTAL PROTECTION AGENCY. 1975. Water programs: national interim primary drinking water regulations. *Federal Register*, 40, no. 248.

U. S. ENVIRONMENTAL PROTECTION AGENCY. 1976. *Manual of Water Well Construction Practices*. Report EPA-570/9-75-001, 156 pp.

U. S. ENVIRONMENTAL PROTECTION AGENCY. 1977. The report to Congress: waste disposal practices and their effects on ground water. USEPA Office of Water Supply, Office of Solid Waste Management Programs.

VAN BAVEL, C. H. M. 1966. Potential evapotranspiration: the combination concept and its experimental verification. *Water Resources Res.*, 2, pp. 455–467.

VAN DAM, J. 1967. The migration of hydrocarbons in a water bearing stratum. *The Joint Problems of the Oil and Water Industries*, ed. Peter Hepple. Institute of Petroleum, London.

VAN EVERDINGEN, R. O. 1968a. The influence of the South Saskatchewan Reservoir on the local groundwater regime—a prognosis. *Geol. Surv. Can. Paper 65-39*.

VAN EVERDINGEN, R. O. 1968b. Studies of formation waters in Western Canada: geochemistry and hydrodynamics. *Can. J. Earth Sci.*, 5, pp. 523–543.

VAN EVERDINGEN, R. O. 1968c. Mobility of main ion species in reverse osmosis and the modification of subsurface brines. *Canadian J. Earth Sci.*, 5, pp. 1253–1260.

VAN EVERDINGEN, R. O. 1976. Geocryological terminology. *Can. J. Earth Sci.*, 13, no. 6, pp. 862–867.

VAN EVERDINGEN, R. O., and R. A. FREEZE. 1971. Subsurface disposal of waste in Canada. *Inland Waters Branch Tech. Rept.* Department of Environment, Canada.

VAN GROSSE, A. V., W. M. JOHNSTON, R. L. WOLFGANG, and W. F. LIBBY. 1951. Tritium in nature. *Science*, 113, pp. 1–2.

VAN OLPHEN, H. 1963. *An Introduction to Colloid Chemistry*. Wiley-Interscience, New York. 301 pp.

VANSELOW, A. P. 1932. Equilibria of the base-exchange reactions of bentonites, permutites, soil-colloids and zeolites. *Soil Sci.*, 33, pp. 95–113.

VAN VOAST, W. A., and R. B. HEDGES. 1975. Hydrogeologic aspects of existing and proposed strip coal mines near Decker, Southeastern Montana. *Montana Bur. Mines Geol. Bull.*, 97.

VAUX, W. G. 1968. Intragravel flow and interchange of water in a streambed. *Fishery Bull.*, 66, pp. 479–489.

VENNARD, J. K. 1961. *Elementary Fluid Mechanics*, 4th ed. John Wiley & Sons, New York.

VERMA, R. D., and W. BRUTSAERT. 1970. Unconfined aquifer seepage by capillary flow theory. *J. Hydraul. Div., Proc. Amer. Soc. Civil Engrs.*, 96 (HY6), pp. 1331–1344.

VERMA, R. D., and W. BRUTSAERT. 1971. Unsteady free-surface groundwater seepage. *J. Hydraul. Div., Proc. Amer. Soc. Civil Engrs.*, 97 (HY8), pp. 1213–1229.

VERMEULEN, T., and N. K. HIESTER. 1952. Ion exchange chromatography of trace elements. *Ind. Eng. Chem.*, 44, pp. 636–651.

VERRUIJT, A. 1969. Elastic storage of aquifers. *Flow Through Porous Media*, ed. R. J. M. DeWiest. Academic Press, New York, pp. 331–376.

VOGWILL, R. I. J. 1976. Some practical aspects of open-pit dewatering at Pine Point. *Bull. Can. Inst. Min. Met.*, 69, 768, pp. 76–88.

VOMOCIL, J. A. 1965. Porosity. *Methods of Soil Analysis, Part 1*, ed. C. A. Black. American Society of Agronomy, Madison, Wis., pp. 299–314.

WAHLSTROM, E. E. 1973. *Tunneling in Rock*. American Elsevier, New York.

WAHLSTROM, E. E. 1974. *Dams, Dam Foundations and Reservoir Sites*. American Elsevier, New York.

WALKER, E. H. 1956. Ground-water resources of the Hopkinsville Quadrangle, Kentucky. *U.S. Geol. Surv. Water-Supply Paper 1328*.

WALLIS, C., A. HOMMA, and J. L. MELNICK. 1972. A portable virus concentrator for testing water in the field. *Water Res*, 6, pp. 1249–1256.

WALTON, W. C. 1960. Leaky artesian aquifer conditions in Illinois. *Illinois State Water Surv. Rept. Invest.* 39, 27 pp.

WALTON, W. C. 1962. Selected analytical methods for well and aquifer evaluation. *Illinois State Water Surv. Bull. 49*, 81 pp.

WALTON, W. C. 1970. *Groundwater Resource Evaluation*. McGraw-Hill, New York, 664 pp.

WARNER, D. L. 1965. Deep well injection of liquid waste. *Environmental Health Series, Water Supply and Pollution Control, No. 999-WP-21*. U.S. Department of Health, Education and Welfare, Cincinnati, Ohio.

WARNER, D. L., and D. H. ORCUTT. 1973. Industrial wastewater injection wells in United States—status of use and regulation. *Underground Waste Management and Artificial Recharge*, ed. J. Braunstein. Amer. Assoc. Petrol. Geol., U.S. Geol. Surv., Intern. Assoc. Hydrol. Sci., 2.

WARREN, J. E., and H. S. PRICE. 1961. Flow in heterogeneous porous media. *Soc. Petrol. Eng. J.*, 1, pp. 153–169.

WAY, D. S. 1973. *Terrain Analysis: A Guide to Site Selection Using Aerial Photographic Interpretation.* Dowden, Hutchison & Ross, Stroudsburg, Pa., 392 pp.

WAYMAN, C. H. 1967. Adsorption on clay mineral surfaces. *Principles and Applications of Water Chemistry*, ed. S. D. Faust and J. V. Hunter. John Wiley & Sons, New York, pp. 127–167.

WEAST, R. C., ed. 1972. *Handbook of Chemistry and Physics*, 53rd ed. CRC Press, Cleveland, Ohio.

WEEKS, L. G. 1961. Origin, migration and occurrence of petroleum. *Petroleum Exploration Handbook*, ed. G. B. Moody. McGraw-Hill, New York, pp. 5-1–5-50.

WEERTMAN, J. 1972. General theory of water flow at the base of a glacier or ice sheet. *Rev. Geophys. Space Phys.*, 10, pp. 287–333.

WENZEL, L. K. 1942. Methods of determining permeability of water-bearing materials. *U.S. Geol. Surv. Water-Supply Paper 887*, 192 pp.

WESNER, G. M., and D. C. BAIER. 1970. Injection of reclaimed wastewater into confined aquifers. *J. Amer. Water Works Assoc.*, 62, pp. 203–210.

WHIPKEY, R. Z. 1965. Subsurface stormflow from forested slopes. *Bull Intern. Assoc. Sci. Hydrol.*, 10, no. 2, pp. 74–85.

WHITE, D. E. 1957. Magmatic, connate, and metamorphic waters. *Geol. Soc. Amer. Bull.*, 68, pp. 1659–1682.

WHITE, D. E. 1968. Environments of generation of some base-metal ore deposits. *Econ. Geol.*, 63, pp. 301–335.

WHITE, D. E. 1973. Characteristics of geothermal resources. *Geothermal Energy*, ed. P. Kruger and C. Otte. Stanford University Press, Stanford, Calif., pp. 69–94.

WHITE, N. F., H. R. DUKE, and D. K. SUNADA. 1970. Physics of desaturation in porous materials. *J. Irr. Drain. Div., Amer. Soc. Civil Engrs.*, 96, pp. 165–191.

WHITE, W. N. 1932. A method of estimating groundwater supplies based on discharge by plants and evaporation from soil. *U.S. Geol. Surv. Water-Supply Paper 659-A.*

WHITEHEAD, H. C., and J. H. FETH. 1964. Chemical composition of rain, dry fallout, and bulk precipitation at Menlo Park, California. *J. Geophys. Res.*, 69, no. 16, pp. 1957–1959.

WHITFEILD, M. 1974. Thermodynamic limitations of the use of the platinum electrode in *Eh* measurements. *Limnol. Oceanogr.*, 19, pp. 857–865.

WIGLEY, T. M. L. 1975. Carbon 14 dating of groundwater from closed and open systems. *Water Resources Res.*, 11, pp. 324–328.

WIGLEY, T. M. L., and L. N. PLUMMER. 1976. Mixing of carbonate waters. *Geochim. Cosmochim. Acta*, 40, pp. 989–995.

WIKLANDER, L. 1964. Cation and anion exchange phenomena. *Chemistry of the Soil*, 2nd ed., ed. F. E. Bear. American Chemical Society; Reinhold, New York.

WILLARDSON, L. S., and R. L. HURST. 1965. Sample size estimates in permeability studies. *J. Irr. Drain. Div., Amer. Soc. Civil Engrs.*, 91, pp. 1–9.

WILLIAMS, J. R. 1970. Ground water in the permafrost regions of Alaska. *U.S. Geol. Surv. Prof. Paper 696*.

WILLIAMS, R. E. 1970. Applicability of mathematical models of groundwater flow systems to hydrogeochemical exploration. *Idaho Bur. Mines Geol. Pamphlet 144*, 13 pp.

WILLIAMS, R. E., and R. N. FARVOLDEN. 1969. The influence of joints on the movement of groundwater through glacial till. *J. Hydrol.*, 5, pp. 163–170.

WILLS, A. P. 1958. *Vector Analysis with an Introduction to Tensor Analysis.* Dover, New York.

WILSON, J. L., R. L. LENTON, and J. PORTER, eds. 1976. Ground-water pollution: technology economics and management. *Dept. Civil Eng., M.I.T., Rept. TR208.*

WILSON, L. G., and J. N. LUTHIN. 1963. Effect of air flow ahead of the wetting front on infiltration. *Soil Sci.*, 96, pp. 136–143.

WINOGRAD, I. J. 1974. Radioactive waste storage in the arid zone. *EOS, Trans. Amer. Geophys. Union*, 55, no. 10, pp. 884–894.

WINOGRAD, I. J., and G. M. FARLEKAS. 1974. Problems in ^{14}C dating of water from aquifers of deltaic origin. *Isotope Hydrology*, International Atomic Energy Agency, Vienna, pp. 69–93.

WINTER, T. C. 1976. Numerical simulation analysis of the interaction of lakes and groundwater. *U.S. Geol. Surv. Prof. Paper 1001*, 45 pp.

WISLER, C. O., and E. F. BRATER. 1959. *Hydrology*, 2nd ed. John Wiley & Sons, New York, 408 pp.

WITHERSPOON, P. A., S. P. NEUMAN, M. L. SOREY, and M. J. LIPPMAN. 1975. Modeling geothermal systems. *Lawrence Berkeley Lab. Rep. LBL-3263*, 68 pp.

WITTKE, W. 1973. General report on the symposium "Percolation Through Fissured Rock." *Bull. Intern. Assoc. Eng. Geol.*, pp. 3–28.

WITTKE, W., P. RISSLER, and S. SEMPRICH. 1972. Three-dimensional laminar and turbulent flow through fissured rock according to discontinuous and continuous models (German). *Proc. Symp. Percolation Through Fissured Rock, Intern. Soc. Rock Mech., Stuttgart.*

WOLFF, R. G. 1970. Field and laboratory determination of the hydraulic diffusivity of a confining bed. *Water Resources Res.*, 6, pp. 194–203.

WOLLAST, R. 1967. Kinetics of the alteration of K-feldspar in buffered solutions at low temperature. *Geochim. Cosmochim. Acta*, 31, pp. 635–648.

WOOD, W. W. 1976. A hypothesis of ion filtration in a potable water aquifer system. *Ground Water*, 14, pp. 233–244.

WYLLIE, M. R. J. 1963. *The Fundamentals of Well Log Interpretation.* Academic Press, New York, 238 pp.

YEH, W. W. 1975. Aquifer parameter identification. *J. Hydraul. Div., Proc. Amer. Soc. Civ. Engrs.*, 101 (HY9), pp. 1197–1209.

YEN, B. C., and B. SCANLON. 1975. Sanitary landfill settlement rates. *J. Geotech. Div., Proc. Amer. Soc. Civ. Engrs.* 101 (GT5), pp. 475–487.

YOUNG, A., P. F. LAU, and A. S. MCLATCHIE. 1964. Permeability studies of argillaceous rocks. *J. Geophys. Res.*, 69, pp. 4237–4245.

YOUNG, C. P., D. B. OAKES, and W. B. WILKINSON. 1977. Prediction of future nitrate concentrations in groundwater. *Proc. Third Nat. Groundwater Quality Symp.*, U.S. Environ. Protect. Agency and Nat. Water Well Assoc., Las Vegas, Nev., U.S. Env. Protect. Agency Rept. 600/9-77-014.

YOUNG, R. A., and J. D. BREDEHOEFT. 1972. Digital computer simulation for solving management problems of conjunctive groundwater and surface water systems. *Water Resources Res.*, 8, pp. 533–556.

YOUNGS, E. G., and A. J. PECK. 1964. Moisture profile development and air compression during water uptake by bounded porous bodies: 1. Theoretical introduction. *Soil Sci.*, 98, pp. 280–294.

YU, W., and Y. Y. HAIMES. 1974. Multilevel optimization for conjunctive use of groundwater and surface water. *Water Resources Res.*, 10, pp. 625–636.

ZAPOROZEC, A. 1972. Graphical interpretation of water quality data. *Ground Water*, 10, pp. 32–43.

ZARUBA, Q., and V. MENCL. 1969. *Landslides and Their Control.* American Elsevier, New York.

ZIENKIEWICZ, O. C. 1967. *The Finite-Element Method in Structural and Continuum Mechanics.* McGraw-Hill, New York.

ZUBER, A. 1974. Theoretical possibilities of the two-well pulse method. *Intern. Atomic Energy Agency, Rept. SM-182/45.*

INDEX

A

B

C

I

J

K

L

M

P

R

S

W

Y

Z